Micro Economic Analysis in Agriculture

Micro Economic Analysis in Agriculture

— Volume 2 —

K. Nirmal Ravi Kumar
Associate Professor & Head
Department of Agricultural Economics
Acharya N.G. Ranga Agricultural University
Agricultural College, Mahanandi
Nandyal, Kurnool District – 518 502

2012
DAYA PUBLISHING HOUSE®
New Delhi – 110 002

ISBN 978-81-7035-795-7 (Set)
ISBN 978-81-7035-796-4 (Volume 1)
ISBN 978-81-7035-797-1 (Volume 2)

Published by : **Daya Publishing House®**
4760-61/23, Ansari Road, Darya Ganj,
New Delhi - 110 002
Phone: 23245578, 23244987
Fax: (011) 23260116
e-mail : dayabooks@vsnl.com
website : www.dayabooks.com

Laser Typesetting : **Classic Computer Services**
Delhi - 110 035

Printed at : **Chawla Offset Printers**
Delhi - 110 052

PRINTED IN INDIA

– Dedicated to –

Late. K. RAMA CHANDRA RAO

ACHARYA N.G. RANGA AGRICULTURAL UNIVERSITY

ADMINISTRATIVE OFFICE, RAJENDRANAGAR, HYDERABAD-500030

V. NAGI REDDY, I.A.S
Vice-Chancellor

Phone : +91-40-24015035(O)
+91-40-24015031(F)
Email: angrau_vc@yahoo.com
GRAMS: "AGRIVARSITY"

Foreword

The subject 'Economics' has become a choice of study for many of the students in different Universities in India. The analytical aspects of this subject arouse interest among the student community and this motivates and guides them to pursue their post-graduate and doctoral programmes in the field of Economics. This text book entitled 'Micro-Economic analysis in Agriculture' is of no exception in analyzing the economic aspects in agriculture at the micro-level. The meritorious aspect of this text book is that, it presented all the topics in a more analytical manner that too considering simpler examples that we witness in our vicinity. Further, the author has quoted several examples from the field of agricultural sector and in this context, it is certainly a contributing one for the students specializing in Agricultural Economics. I complement the diligent, sincere and painstaking efforts of Dr. K. Nirmal Ravi Kumar in bringing out this important and timely publication, as it motivates and guides the students to pursue their career in the field of Economics. Further, it contains an up-dated and standard material on the various aspects of Micro-Economics duly covering the latest syllabi as prescribed by the several universities in India, thereby, it helps the Academicians to understand and explain the concepts in a more analytical manner. I hope this volume will also be very useful to the students, who desire to be competent enough in various competitive examinations.

V. Nagi Reddy

ACHARYA N.G. RANGA AGRICULTURAL UNIVERSITY

ADMINISTRATIVE OFFICE, RAJENDRANAGAR, HYDERABAD-500030

Dr. Y. ESWARA PRASAD
Ph.D
Professor & University Head &
Hon. Director
Department of Agricultural Economics
& CCS
College of Agriculture,
Rajendranagar – 500 030, Hyderabad
and
Member, Board of Management,
ANGRAU

Phone : +91-40-24015042
+91-40-24011561
Extn: 374
Mobile: 9441609517
Res: 040-23341406

Preface

In the modern era, economic aspects of agri-business are assuming paramount significance. In this context, understanding the basic concepts of micro-economic analysis with special reference to agricultural sector is essential. Moreover, in the recent period, there have been many new developments in the field of micro economic theory. This text book, *'Micro Economic Analysis in Agriculture'* covers both basic and advanced concepts of Micro Economics. Specifically, it dealt with the concepts of Micro Economics considering examples from the agricultural sector to the extent possible and the style of presentation of the concepts is very simple, analytical and easy to understand with attractive graphical representations. I complement the sincere efforts of Dr. K. Nirmal Ravi Kumar in bringing out this text book duly considering the latest syllabi, as prescribed by the IV Deans' Committee recommendations appointed by the Indian Council of Agricultural Research. This text book will be very useful and instructive for the graduates, post-graduates and doctoral students, who wish to pursue their career in the field of Economics in general and in Agricultural Economics in particular. It will also be of immense use for the Academicians, as it harbours standard and updated material pertaining to the concepts of Micro Economics. I hope this volume will even serve the students, who are aspiring high in the field of Economics and to be competent enough in various competitive examinations.

(Y. ESWARA PRASAD)

ACHARYA N.G. RANGA AGRICULTURAL UNIVERSITY
AGRICULTURAL COLLEGE, MAHANANDI
M.C. FARM – 518 502, MAHANANDI, KURNOOL DISTRICT

Dr. K. NIRMAL RAVI KUMAR
Ph.D
Associate Professor & Head
Dept. of Agricultural Economics

Phone : +91-08179696983
Email: drknrk@gmail.com

Author's Note

I consider this as a bold attempt to present a comprehensive text material on the fundamental aspects of Micro-Economics. With a limited experience in the field of teaching (5 years) in the Acharya N.G. Ranga Agricultural University, I will not dare enough to make a tall claim regarding the originality of the material incorporated in this text book entitled, '*Micro Economic Analysis in Agriculture*', but I certainly admit that, it is presented in a more analytical form to enable the students and Academicians to learn and conceptualize the concepts. As a teacher in the field of Agricultural Economics, I quoted examples from the agricultural sector to the extent possible, so as to benefit the students to understand the concepts more clearly. The present volume can serve as a standardized text book for the graduates, post-graduates and doctoral students, who wish to pursue their career in the field of Economics in different Indian Universities.

This task of writing and publishing the text book on the advanced versions of 'Micro-Economic Analysis in Agriculture' has been one of the most challenging and satisfying experiences of my life. I am especially thankful for the friendship and unfailing support of Mr. V. Sita Ram Babu, who practically demonstrated the true meaning of friendship by standing with me during the pursuit of this work. Finally, last but by no means least, a grateful kiss goes to my wife, Meena, my son, Sri Ram and my mother Janaki for their unwavering love and support. Their devotion, patience and understanding were indispensable. I gratefully dedicate this piece of work to my late father, Sri K. Rama Chandra Rao, who instilled in me at an early age the importance of education and the value of hard work. I only wish you could have been here to see me finish the journey I started with your efforts and initiatives. I know you would have been so proud.

K. Nirmal Ravi Kumar

Contents

Part V - Factor Pricing

Part IV

Product Pricing

15

Theory of Costs: Costs and their Behaviour Patterns

A farmer employs several resources to produce the output and he tries to minimize the cost in the production programme. He takes into consideration the resources relationships in allocating them in the enterprise or among the enterprise combinations. This concept was discussed earlier through analyzing the LCC or Optimum factor combination. The other important way that guides the farmer to conduct the agri-business is to analyze the cost structure both in the short run and long run production programmes. We know, production is the result of resources and resource services employed in the business. The farmer has to make payments for these resources and resource services employed in the production programme. From the point of view of the factor supplier, it is called as 'Factor income', while from the farmer or firm point of view, it is 'Factor payment' or 'Cost of inputs'. So, 'cost' refers to the money expenditure incurred by the farmer in his production programme. The determinants of cost of production include, the size of agri-business, the level of production, the nature of technology used, the quantity of resources and resource services used, managerial and labour efficiency etc. Thus, the cost of production of a commodity is the aggregate of prices paid for the factors of production used in producing a commodity.

Generally, the term 'cost of production' refers to the money expenses incurred in the production of a commodity. In the words of Gulhrie and Wallace, '*in Economics, cost of production has a special meaning. It is all of the payments or expenditures necessary to obtain the factors of production of land, labor, capital and management required to produce a commodity. It represents money costs which we want to incur in order to acquire the factors of production*'. In the words of Campbell, '*production costs are those, which must be received by resource owners in order to assume that, they will continue to supply them in a particular time of production*'. But money expenses are not the only expenses incurred in the production of a commodity and there are number of factors and services such as, entrepreneurship, land, capital etc., which are offered by the farmer himself without charging any price or receiving any payment for them. So, while computing the total cost of production, allowance should be made for such expenses. This allocation and use of resources in the farm will influence the supply of output. The supply of output along with demand in the market determines the price of the commodity. So, we can say, cost of the resources indirectly influences the price of the commodity. It is, therefore, essential to have clear understanding about different types of costs that influence the output or production of the farm.

15.1. Different Types of Costs

There are different types of costs widely used in the production programmes and they are discussed here under.

1. Explicit Costs

In the production programme, the farmer employs various resources and resource services such as seed, manures, fertilizers, pesticides, hired labour, bullock labour, machinery labour etc. To employ these resources, the farmer has to purchase them or their services through paying cash outlay from his pocket. Such costs incurred by the farmer are called explicit costs or out of pocket costs.

2. Implicit Costs

Implicit costs refer to costs of self-owned and self-employed resources. They are also called as imputed costs. Examples include, wages of farmer's own labour, wages of farm family labour, interest on owned fixed capital, rental value of owned land etc. These implicit costs are also important to compute the cost of production of a commodity because, had the farmer not employed these owned resources, he has to depend upon hired resources to use them in the production programme. The implicit costs consist of the opportunity costs of using the own resources without receiving any explicit compensation for those resources. For example, the farmer himself works along with hired labour, but does not receive any wages, thereby, he forgoes the opportunity to earn a wage working for someone else. These implicit costs are considered as costs in an economic sense, as they are a part of the firm's costs of doing business.

3. Accounting Costs

It refers to the money costs or explicit costs. That means, these are the costs incurred by the firm, which involve actual cash payment. Hence, these costs are also called as explicit costs or out of pocket costs or paid out costs. These costs are called as Accounting costs because, they can be maintained in an account form for analyzing the cost of production of a product and thereby, the performance of the business. These costs points out, how much expenditure has already been incurred on a particular process or on production as such. Since these costs relate to the past, these are generally Sunk costs. Since these costs are recorded, they are useful for meeting taxation needs as well as to calculate profit or loss of the business firm. To calculate the accounting profit of the firm, only such costs are deducted from the revenue which has been fully paid out. So, accounting profits are the profits calculated as the difference between TR and total explicit costs of a firm, given by, Accounting Profits = TR - Explicit Costs (Figure 15.1). Examples include, wages and salaries of hired labour, repairs and maintenance charges of capital equipment, rents on buildings, interest on capital borrowed, fuel charges, power charges, costs of materials etc.

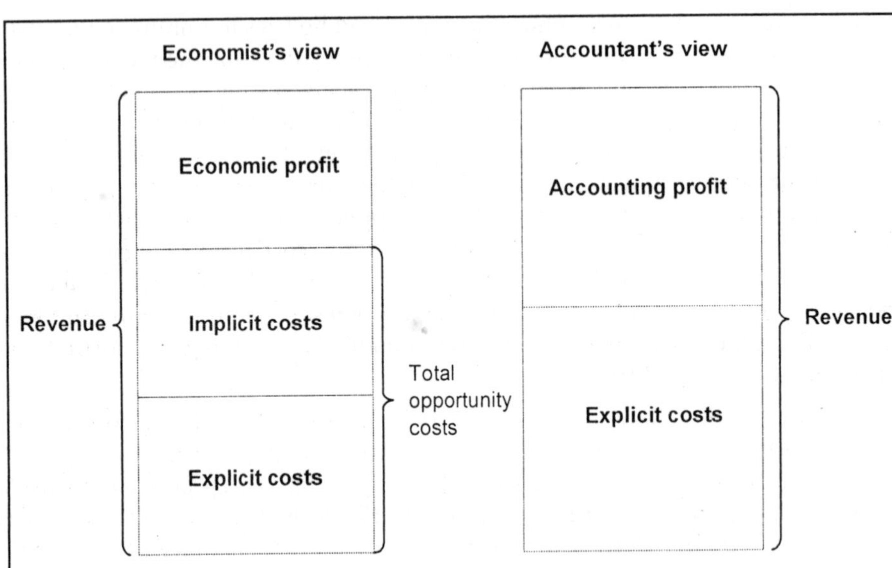

Figure 15.1: Economists vs Accountants view of costs of the firm.

4. Economic Costs

Economic costs include both explicit and implicit costs. Explicit costs refer to out of pocket cost or paid out costs or accounting costs, which involve actual cash out lay from the pocket of farmer. Implicit costs refer to costs of self-owned and self-employed resources. So, Economic cost = Explicit cost (Accounting cost) + Implicit cost. A firm will earn economic profits only, if it is making revenue in excess of economic cost. So, Economic profit = (TR – Economic Costs) or (TR – (Explicit + Implicit Costs)). So, economic profits are, therefore, smaller in value than accounting profits. Among these explicit and implicit costs, it is important for the farmer to cover explicit costs from gross returns, so as to continue the business, because explicit costs are paid out costs. Thus, Economic costs include both Accounting costs and implicit costs (Figure 15.1). If the entrepreneur wants to earn normal profits, he has to cover these economic costs from the gross returns of the business.So, both Economists and Accountants look at costs, but each takes a slightly different perspective. Accountants are retrospective. They look at explicit costs only *i.e.*, costs involving direct, out-of-pocket payments for wages, salaries, materials, hiring machinery costs etc. The implicit costs are not regarded as costs in an accounting sense, but they are a part of the firm's costs of doing business, nonetheless. But, Economists look at both explicit and implicit costs and hence, they are forward looking or related to future. So, compared to Accounting costs, Economic costs weigh more importance regarding the decision-making in the business.

5. Money Cost

In general, when we talk about the costs incurred in the firm, it implies money cost. Money cost or nominal cost is the total money expenses incurred by the farmer in producing a commodity. It includes, cost of raw materials, wages and salaries of labour, expenditure on machinery and equipment, depreciation on machinery, buildings and such other capital goods, interest on capital etc. In the words of Hanson, '*the money costs of producing a certain output of a commodity is the sum of all the payments to the factors of production engaged in the production of that commodity*'. So, money costs enter the record books of firm.

6. Real Costs

Real cost is a subjective concept. It expresses the pains and sacrifices involved in producing a commodity. Money cost refers to the money paid for the factors of productions involved in producing the commodity, whereas, the efforts and sacrifices made by the farmers to save and invest the capital and contributing their labour and managerial services through foregoing their leisure constitute real costs. So, money cost is only a time bound or piece bound wage rate. But, real cost refers to physical quantities of various factors of production employed in producing the commodity. For example, the real cost of constructing a sugar factory is composed of six months of labour working days, iron metal used, cement bags, sand, bricks, depreciation of machinery employed, bottles of paints etc. So, real cost signifies the aggregate of real productive resources employed in the production of the commodity. Classical Economists understood by real costs the pains and sacrifices of labour. Alfred Marshall calls real cost as '*social cost*' and describes it as, '*the production of a commodity generally requires many different kinds of labour and the use of capital in many forms. The exertions of all the different kinds of labour that are directly or indirectly involved in making it, together with the abstinences or rather the waitings required for saving the capital used in making it, all these efforts and sacrifices together will be called the real cost of production of a commodity*'. So, according to Marshall, concept of real cost involves more a philosophical significance. But, this concept is really difficult to measure.

7. Opportunity Cost

It is other wise called as Alternative cost. Since, it is difficult to measure the real cost, as it deals with troubles and sacrifices of different factors employed in the production programme, the concept of Opportunity cost was introduced. This concept of opportunity cost was based on scarcity and versatility (alternative applicabilities) features of the resources. We know, nearly, all decisions taken in the agri-business involve trade-offs. When a farmer chooses something, he has to give up some other. So, every action and choice made by the farmer is associated with advantages and disadvantages, costs and gains. For example, if the farmer wants to apply fertilizer on one crop (paddy), he has to forego its application on other crop (maize). So, some amount of net revenue that could be generated in the next best use of fertilizer must be given up (maize) or sacrificed due to its employment on the most profitable use (paddy). The net revenue which we forego, when making a choice or decision is called the Opportunity cost of the decision. This concept of opportunity cost was proposed by John Stuart Mill. It is otherwise called as Economic opportunity loss. According to Leftwitch, '*opportunity cost of a particular product is the value of the foregone alternative product that resources used in its production could have produced*'. In the words of Ferguson, '*the alternative or opportunity cost of producing one unit of the commodity X is the amount of commodity Y that must be sacrificed in order to use resources to produce X rather than Y*'. In the words of Lipsay, '*opportunity cost of using any factor is what is currently foregone by using it*'. So, it is the cost equivalent to the returns from the next best alternative foregone. That means, it refers to the returns from the next best missed alternative. This also implies that, the cost of using a resource in a particular venture is the benefit foregone by not using it in the next best alternative use. This is because, in general, the resources are very scarce and hence, they cannot be used simultaneously to produce several products. If we use a resource on one enterprise, its service is missed on other enterprise. The foregone opportunity is also taken as cost and is termed as opportunity cost. So, opportunity cost of this resource is the returns foregone from the next best alternative use. Hence, this concept of opportunity cost arises only when there are alternatives for the use of resources *i.e.*, it is a relative concept. This opportunity cost includes both explicit costs and implicit costs. The concept of opportunity cost differs from implicit costs in the sense that, it is more comprehensive, as it relates to all the resources (both borrowed and owned) whereas, the concept of imputed cost applies only to the self-use of the self-owned resources. Similarly, the concept of opportunity cost differs from outlay costs in the sense that, outlay costs are the explicit costs, which are entered into the record books of an account, but opportunity costs relate to sacrificed alternatives and they are not entered into the record books of an account. The following are the examples to illustrate the concept of opportunity cost

- For example, a farmer who is producing paddy can produce maize with the same factors. Therefore, the opportunity cost of a quintal of paddy is the amount of output of maize given up.
- Suppose, a person is having Rs.500/- and he can invest on shirt, pant, dhoti and short. If the price of pant is Rs.500, shirt is Rs.400, dhoti is Rs.390 and short is Rs.275, then the opportunity cost of purchasing a pant is not Rs.400 of shirt, but the satisfaction he misses from purchasing and wearing the shirt. This is because, opportunity cost is not equal to the cost of next best alternative fore gone, but equal to the returns (satisfaction in this case) missed from the next best alternative. Further, the opportunity cost is the satisfaction he misses from the shirt alone, as opportunity cost is the cost equivalent to returns from the next best alternative only and not all the commodities.
- If a sugar factory owns the land, there is no cost of using the land (*i.e.*, the rent) to the entrepreneur. But, the factory has an opportunity cost of using this land, which is equivalent to the rent forgone by not letting the land out on rent.

Limitations of Opportunity Cost

The following are the important limitations of the concept of opportunity cost:

- The concept of opportunity cost will arise only when there are alternatives for resource usage. If alternatives are absent, there is no opportunity cost for that resource.
- In case, if the resource is not willing to move to other occupations, the concept of opportunity cost will not arise. For example, a skilled labour working in a sugar factory may not show willingness to move to other occupation, thereby, the opportunity cost of that labour resource will not exist.
- Sometimes, it is very difficult to measure the opportunity cost. For example, if a smoky factory is located in an area and it offers significant employment to the people living there, but causing pollution hazards to the people living in that area. Taking the health hazards of the people into consideration, it is difficult to compute the opportunity cost of the establishment of that factory.
- In case of highly specialized capital equipment such as rice mill, mango processing unit, groundnut decorticator etc, the concept of opportunity cost will not arise because, these capital equipments have no alternative uses.
- The concept of opportunity cost assumes perfect competitive market situation, which implies that, the nature of commodity is homogenous. But in realistic situation, imperfect competition is more common.

So, the concept of opportunity cost is subjective. Only the chooser can determine the most attractive alternative for itself from its special point of view. Every farmer will have a different list of things valued in decline from top to bottom due to different priorities. Even for the same activity or enterprise, different farmers take it with different opportunity cost.

8. Private Costs

Private cost is the cost incurred by a firm for production. It includes both explicit costs and implicit costs. The various costs such as purchase of capital equipment, hiring of labour, purchase of resources etc., fall under the category of Private costs. This is from the producers' side. The private cost also arises from the consumers' side or customers' side. For example, if a person is owning a car, he will incur various costs like fuel cost, oil cost, lubricant cost, repairs and maintenance charges, wages paid to driver etc.

9. Social Cost

Economic costs can be calculated at two levels *viz.*, micro-level and macro-level. The micro-level economic costs relate to functioning of a firm as a production unit, while the macro-level economic costs are the ones that are generated by the decisions of the firm, but are paid by the society and not the by firm. For example, if the decision of a pesticide firm is to expand its output, it leads to increase in its costs and these costs are called Private costs. But, these private costs also leads to certain costs to the society, may be in the nature of greater pollution, greater congestion etc. These costs which are external to the firm are called external costs or externalities from society's point of view. Thus, private costs are those which are actually incurred by a firm in the business activity to produce the product and these private costs do not always equate with the total cost to society. Social costs, on the other hand, are the total costs to the society on account of production of a product. That means, they are not borne by the producing firm, but are incurred by others in the society. In manufacturing the pesticides say, the firm discharges its wastes into the nearby river and this leads to the following negative implications or externalities to the society:

- The people in that locality have to purchase water purifiers to have safe drinking water.

- ☆ The recreation use of this river gets affected.
- ☆ Farmers practicing aquaculture are affected.
- ☆ Pollution results in tremendous health hazards, which involve costs to the society as a whole.

These problems are mainly due to firm's or producer's negligence. Even consumers' actions can further aggravate the social cost like using the same river water for irrigation purposes, livestock maintenance etc. All these costs are called External costs, arising out of a firm and directly associated with a production of a product or a service. But, the producer will not bear these external costs, but are experienced by the society as a whole. So, social cost is the sum of private costs and external costs and the difference between social costs and private costs is called an External cost. The sum of both private costs and social costs constitute Economic costs. So, we can write,

Social costs = Private costs + External costs

External costs = Social costs - Private costs

Economic costs = Private costs - Social costs

So, if external costs > 0, then social costs > private costs. Then society tends to:

- ☆ Price the good or service too low, and
- ☆ Consume too much of the good or service.

The Social cost can also be referred to as the Total cost. It should be noted that, these costs are not exclusively monetary, as externalities such as pollution negatively affect the air quality of everyone around the polluted area. It is important that, the externalities may ne negative or positive.

Negative Externalities

When social costs of production are greater than that of the private cost of production, it is a case of negative externality of production. These negative externalities will arise when the firm externalizing their costs onto the society, in order to reduce its own total cost. As a result of negative externalities, we see that, the society will be negatively influenced by the actions of the firm. The above cited example of a pesticide firm is the case of a negative externality.

Positive Externalities

In some cases, the externalities may be positive. When social costs of production are less than that of the private cost of production, it is a case of positive externality of production. A good example of a positive externality is say, an establishment of educational institution or a yoga centre in a locality. In the beginning, few people learn healthy habits in a yoga centre. These people, in turn, disseminate these healthy habits among the people within locality and outside the locality. So, this positive externality when added to the private benefit will lead to higher social benefits.

10. Incremental Costs and Sunk Costs

The incremental costs are additions made to the existing costs due to changes made in the nature and level of business activity. For example, a sugar factory may incur incremental costs by adding or replacing the machinery, by changing the distribution network for sale of the product etc. That means, these costs can be avoided by not bringing about any change in the activity and hence, these costs are also called as Avoidable costs or Escapable costs. Moreover, these costs are additional costs resulting from a contemplated change and hence, they are also called as Differential costs.

Sunk costs are the costs that do not change by varying the nature or the level of the business activity. For example, all the past establishment costs in a sugar factory are considered as sunk costs because, any change in the business activity say, depreciation on machinery and the resulting incremental costs will have to take these preceding costs as given. So, sunk costs are mostly related to fixed (establishment) costs only. These sunk costs are also called as Non-avoidable or Non-escapable costs.

Among these two costs, sunk costs are irrelevant for decision-making, as these costs do not vary with the changes made in the business activity. So, incremental costs are important for decision-making compared to sunk costs. It is important that, in most cases, incremental costs are variable, but not always. Say, for example, if the sugar factory desires to purchase tables, chairs, install electricity wiring etc., these incremental costs fall under fixed costs. So, depending upon the nature of the business activity, the incremental cost may be a fixed cost or variable cost.

11. Out-of-Pocket Costs and Book Costs

As explained earlier, Out-of-pocket costs are explicit costs incurred in the production programme by the entrepreneur. They include actual cash outlay to purchase the resources like raw material, machinery, technology, hired labour costs etc.

On the other hand, book costs are those imputed costs, which do not involve any cash outlay, but for them a provision is made in the account books to consider them in profit and loss statements and take tax advantages. They include depreciation on capital equipment, interest on the owned fixed capital of the business etc.

12. Original Costs and Replacement Costs

Original costs are the costs incurred on the capital assets, equipment and materials at the price originally paid for them, when such items are included first time in the business. They are also called as Historical costs. On the other hand, Replacement costs are the costs that are incurred, if the above items are to be replaced again in the business. The main difference between the Original costs and Replacement costs is that, there will be price changes over time. Suppose, an entrepreneur acquired a processing machinery for Rs. 1.00 lakh in 1995 and he wishes to replace it with a new machinery of same capacity and it costs about Rs. 1.20 lakhs in 2010, as the earlier machinery was depreciated due to wear and tear. Here, Rs. 1.00 lakh is the Original cost of the machinery and Rs. 1.20 lakhs is its Replacement cost. The difference of Rs. 20000 between the two costs has resulted because of the price change of the machine during this period of 15 years.

13. Controllable Costs and Non-Controllable Costs

Controllable costs are the costs, which can be regulated by the farmer, so as to ensure cost-effectiveness in the production programme. The farmer aims to achieve resource use efficiency in using the variable resources. So, variable costs in the production programme falls under the category of Controllable costs, as the level of these resources vary with the level of output and thereby, achieving resource use efficiency counts much in the profitability of the production programme. These costs include, use of fertilizers, pesticides, irrigation etc., in the production programme.

On the other hand, Non-controllable costs are those, which cannot be subjected to control or supervision on the part of the farmer, as they represent the fixed costs in the production programme. So, the farmer tries to ensure lower fixed cost per unit of output produced in the production programme through increasing the output from efficient use of variable resources. These costs include, depreciation, land revenue, interest on borrowed capital etc.

14. Shutdown Costs and Abandonment Costs

Shutdown costs are the costs, which the factory incurs, if it temporarily stops its operations. So, these costs can be saved or recovered by the factory, if it continues its business operations. These costs include, wages of permanent employees, land tax, the cost of sheltering plant, machinery and equipment etc.

Abandonment costs are the costs of retiring altogether a fixed asset from its usage. For example, with the advent of computers, the type writers are completely out of usage, thereby, they may not be used in the factory. Thus, abandonment costs involve the problem of the disposal of unwanted assets.

15. Urgent Costs and Postponable Costs

As the name indicates, Urgent costs are those costs that must be incurred in taking up the operations, whose postponement does affect the production programme adversely. For example, farmer must necessarily transplant paddy seedlings in the main field after 25-30 days after sowing, thereby, he cannot postpone this operation and hence, transplanting costs are urgent costs. On the other hand, if the farmer can postpone the farm operations that does not affect (at least for some time) the operational efficiency of the production programme, then the costs incurred on such operations are known as Postponable costs. Examples include, maintenance of machinery, godowns, implements etc.

16. Short-run Costs and Long-run Costs

The short-run is defined as a period in which the supply of at least one of the factors or resources cannot be varied by the farmer. For example, if the farmer cultivates paddy during kharif season in this year, certain factors like machinery, land, farm buildings etc., cannot be changed whenever it so desires. That means, it takes time on the part of the farmer to replace, add or dismantle them. So, in a short run production programme, some factors are fixed and some are variable. In terms of costs, therefore, two types of costs arise in the short run *viz.*, fixed costs and variable costs.

Long-run is defined as a period in which all the factors can be varied as desired by the farmer to increase the output. That means, in the available time span, the farmer can make all possible adjustments and changes in the level of factors of the farm, as he desires. So, in long run production programme, all the factors are variable and no factor is fixed. In terms of costs, there are no fixed costs in the long run and all costs are variable costs.

In the time of distinction between fixed costs and variable costs is a matter of degree, it all depends upon the contracts of a firm and the period of time under consideration. For example, if a farmer makes contract with the agricultural labour for a certain period, then the farmer has to bear the cost (wages) of the labour irrespective of the total output. Under such conditions, the wages paid to the agricultural labour will be classified as fixed cost and not variable cost. Secondly, when the period of time is short, the distinction between fixed cost and variable cost can be made rigid, but not in a longer period of time, as all fixed costs change into variable cost in the long run.

Cost Analysis in the Short Run and Long Run Production Programmes

It is important for a farmer to recover the costs incurred in short run (fixed costs and variable costs) and in long run (variable costs) production programmes. The following are the important aspects:

A. Recovery of Costs

As discussed earlier, in short run, the farmer incurs both fixed costs and variable costs. So, the farmer tries to recover these costs from the gross returns he derive from the business. But, in short run, due to price fluctuations, the gross returns may or may not recover the TC incurred by the farmer. At times of fall in market prices, the gross returns tend to be lower than TC incurred by the farmer. In such case, the distinction between fixed costs and variable costs in the short run production programme guides the farmer to plan the business. But, in long run, the farmer has to recover the TC incurred in the production programme, as all costs are variable costs in the long run. The following Table 15.1 highlights the importance of recovery of costs from the gross returns derived by the farmer both in short run and long run production programmes:

Table 15.1: Recovery of short run and long run costs in the production programme.

Period	*Absolute Terms*	*Per Unit Output Terms*	*Inferences*	*Recommendation to Farmer*
Short run production programme	Gross returns > Short Run Total Costs (SRTC)	Price > Short Run Average Total Cost (SRATC)	Gross returns from the output covers both Total Fixed Costs (TFC) and Short Run Total Variable Cost (SRTVC)	Continue the business
	SRTVC < Gross returns < SRTC	Short Run Average Variable Cost (SRAVC) < Price < SRATC	Gross returns of the output cover SRTVC and part of TFC. So, loss is than the amount of TFC	Continue the business
	Gross returns < SRTVC	Price<SRAVC	Gross returns of the output could not even cover SRTVC. So, loss is more than the amount of TFC	Stop the production programme temporarily.
Long run production programme	Gross returns > Long Run Total Costs (LRTC)	Price>LRATC	Gross returns of the output covers the LRTC	Continue the business
	Gross returns < LRTC	Price<LRATC	Gross returns of the output could not cover LRTC	Stop the business permanently and plan for an alternate enterprise.

B. Equilibrium Conditions of the Firm

Since, in short run production programme, some factors are fixed and some are variable, thereby, the best possible combinations of factors is not obtained. So, only the proportion of variable factors has to be changed to increase the output. Hence, for a short run production programme, the point of equilibrium output or maximum profits is given by, Short Run MC (SRMC)= MR.

In long run production programme, the point of optimality is Long Run MC (LRMC) = MR. It is important that, the SRMC is influenced by variable costs only, whereas LRMC is influenced by both fixed costs and variable costs, as all costs are variable costs in the long run.

17. Incremental Cost and Marginal Cost

These two costs are very closely related, as they deal with increments in output, but there lies few important differences. MC is the additional cost incurred to produce an additional unit of an output. That means, MC deals with unit-by-unit changes in output. But, incremental cost is additional cost incurred in relation to change in any number of units of output or even change in the quality of the output. That means, incremental cost is not restricted to a unit change in output, unlike MC. The concept of MC useful in the cases like LDR to decide the optimum level of output, profit maximization in a firm etc. However, the concept of incremental cost is useful in the case of linear cost functions, in comparing the two technical processes (though yielding same level of output) etc.

15.2. Cost Function

Cost function refers to the functional relationship between cost and output. Various resources and resource services determine the cost of output in the production of programme. So, we can write the cost function as,

C = f (fixed factors, variable factors, technology, output, etc)

But, all the factors *viz.*, fixed factors, variable factors and technology influence the output. Hence, finally we can write, C = f(output).

A TC curve, C_1 shown in the Figure 15.2 is a 'S' shaped curve. This is because, at the beginning of the production programme or initial quantities of output *i.e.*, OQ_1, there will be economics of scale, where the firm experiences increasing returns or decreasing costs. Hence, the TC curve C_1 rises slowly between A and B points. If the output is increased from OQ_1 to OQ_2, due to employment of more of variable resources, due to constant returns or constant costs, the cost curve rises at a constant rate between B and C points on the TC curve. If the output is further increased beyond OQ_2 by increasing the level of variable resources, due to operation of diminishing returns/increasing costs, the cost curve rises rapidly beyond point C on the TC curve thus, giving 'S' shape to TC curve.

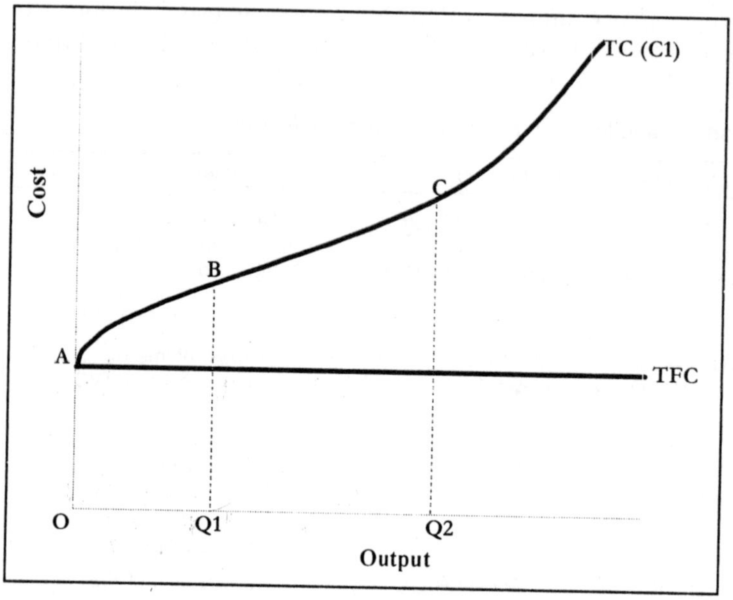

Figure 15.2: Cost function showing the relation between cost and output.

Cost function can also be explained from the concepts of both TFC and TVC in the short run production programme. We know, TC= TFC+TVC. But TFC remains constant at different levels of output. Since TVC is equal to total quantum of variable factors used multiplied by their price value, so,

$$TC = K + (L*W)$$

$$TC = K + LW \qquad \textit{Equation 15.1}$$

where,

K = Fixed factors

L = Total quantum of variable factors

W = Prices of variable factors

From the above Equation 15.1, it is clear that, the increase in TC of the production programme is due to increase in TVC, because TFC remains same at different levels of output. So, by increasing the variable factors *i.e.*, LW, output will increase and when output increases, TC also increases. So we can write the above Equation 15.1 as,

$$TC = f(Q) \qquad \textit{Equation 15.2}$$

where,

Q = Output.

Cost function can be linear or curvi-linear depending upon the response between output and costs incurred in the production programme. The following are the two forms of cost functions:

$$C = a + bQ \ (C = 4000 + 7Q) \qquad \textit{Equation 15.3}$$

$$C = a + bQ + cQ^2 \ (C = 4000 + 5Q + 2.5Q^2) \qquad \textit{Equation 15.4}$$

In the above examples, C = Cost incurred, Q = Output, 4000 is the TFC incurred in the production programme.

15.3. Theories of Costs

There are two theories to explain the cost analysis of the production programme *viz.*, Traditional theory and Modern theory.

I. Traditional Theory

According to this theory, there are seven basic costs *viz.*, Total Fixed Cost (TFC), Total Variable Cost (TVC), Total Cost (TC), Average Fixed Cost (AFC), Average Variabl Cost (AVC), Average Total Cost (ATC) and Marginal Cost (MC). This theory explains the behaviour of these costs both in the short run and long run production programmes.

A. Traditional Theory - Short Run Costs

They include both Absolute costs and Per unit costs.

1. Traditional Theory – Short Run Absolute Costs

They include TFC, SRTVC and SRTC.

(a) TFC

It is the cost incurred by the farmer in the production programme irrespective of the level of output. That means, whatever may be the level of output or even at zero output level, TFC remains same. So, when there is zero production, TC is equal to TFC. These fixed costs are the costs related to fixed resources only.

The distinction between fixed costs and variable costs will arise only in short run because, in the long run, all the resources are variable and hence, all costs are variable costs in the long run. But, the recovery of the fixed costs is considered important in the long run and not in the short run. Hence, fixed costs are also called as long run costs. To an Economist, the fixed costs are Overhead costs and to an Accountant, they are Indirect costs. Firms have no control over fixed costs in the short run. For this reason, fixed costs are sometimes called as Sunk costs. They are also called as On-costs or Supplementary costs. TFC curve is a straight line parallel to X-axis because, irrespective of the level of output, TFC remains same (Figure 15.3). When the output goes up, the fixed cost per unit of output comes down, as the TFC is then divided between larger number of units of output. Examples include, depreciation, interest on loan amount, land revenue etc.

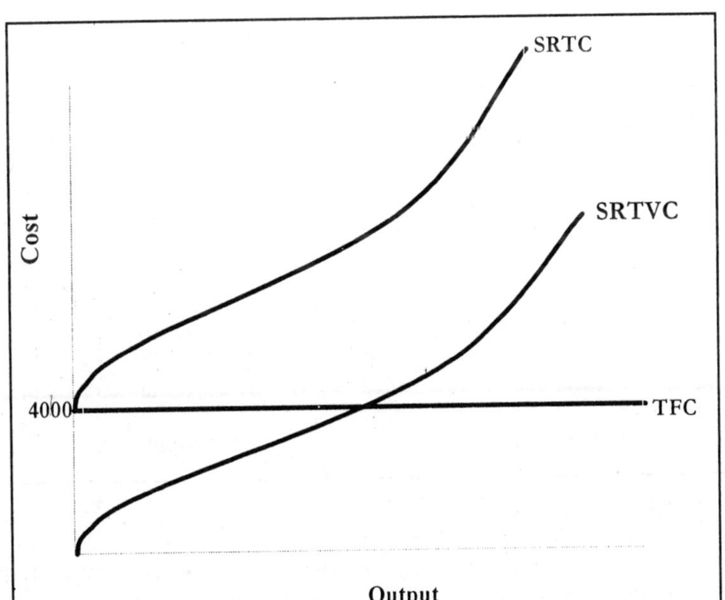

Figure 15.3: Traditional theory – Short run absolute cost curves in production of paddy.

(b) SRTVC

As the name indicates, this cost varies with the level of output. So, when output is zero, variable cost is also zero and when output increases, variable cost also increases. These costs are related to variable resources only and they increase, but not necessarily in the same proportion as the increase in output. In order to increase output, the farmer must increase the quantity of variable factors of production that he employs. Therefore, as farm output increases, the farm's variable costs must also increase. The degree of proportionality between the variable cost and output depends upon the utilization of fixed factors during the process of production.

The SRTVC curve is an inverted S shaped curve passing through the origin because, when output is zero, SRTVC is zero. At the beginning of production programme SRTVC curve rises rather slowly because at the initial levels of output, the production programme comprises of more of fixed factors and less of variable factors. So, when there is less amount of variable resource, they are properlv managed by the farm manager leading to higher resource use efficiency. So, due to internal economies of scale, increasing returns or decreasing costs is experienced, thereby, the curve rises slowly. But with the gradual increase in output, the proportion of variable factors in relation to fixed factors will be more leading to diseconomies of scale, which results in diminishing returns or increasing costs and hence, the curve rises steeply at higher levels of output thus, giving an S shape to the SRTVC curve.

SRTVC are considered in the short run because, in a short period, the output of the production programme can be increased by increasing the variable factors only and not of fixed factors. So, the farmer has to recover these costs from the returns in the short run production programme. So, variable costs are also called as short run costs. They are, otherwise, called as direct costs or prime costs. Examples include, pesticide cost, seed cost, fertilizer cost, human labour wages etc. But, if the period under consideration is long enough to allow the necessary adjustments in the capacity of the farm, the fixed costs no longer remain fixed and they can also be varied. That means, all the costs are variable in the long run, as all the factors are variable in the long run and no factor is fixed.

Since SRTVC contains at least one of the variable resources, SRTVC curve can be derived from the TPP curve of short run production function. This infers that, the concept of LDR helps to explain the shape and slope of SRTVC. From TPP, we can easily compute SRTC. The slopes of these two costs *i.e.*, SRTVC and SRTC indicates MC and hence, the concept of MC can also be explained from LDR concept. In deriving the SRTVC from TPP, we need not consider the III stage of production function and from the Stage I and Stage II of the production function, we can explain the shape of SRTVC curve. At the beginning of production programme or in the initial quantities of output, the SRTVC curve has a flat slope, as the production experiences increasing MRs implying Stage I of the production function. However, gradually, the SRTVC rises upward with a steep slope indicating the diminishing MRs operates in the Stage II of the production function. So, SRTVC and TPP are mirror images of each other (Figure 15.4).

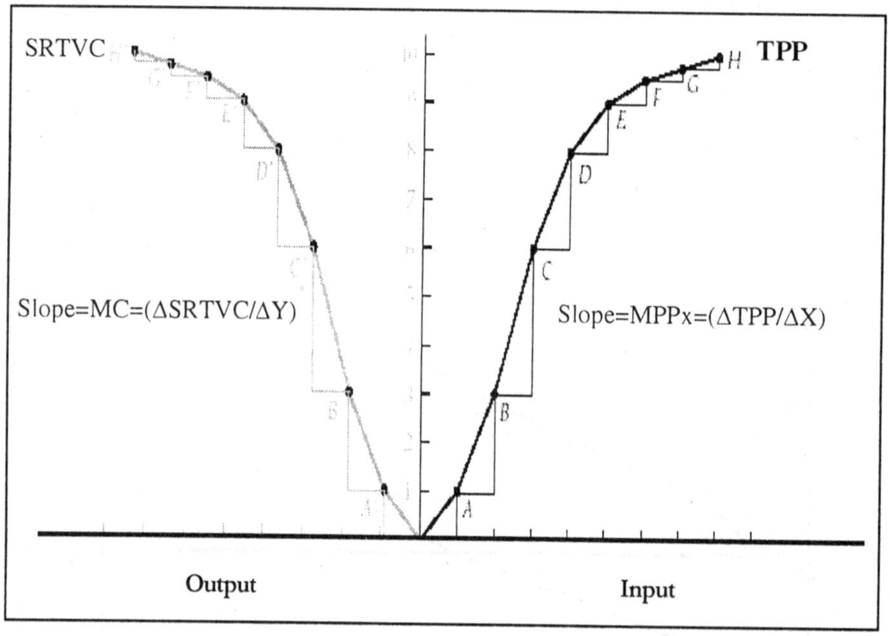

Figure 15.4: TPP and SRTVC curves are mirror image.

It is usually assumed by the Economists that, the variable costs continuously vary with the output. But, there are some cases, where costs remains fixed for each of range of output, but the movement from one range of output to another only, the costs will increase. This is common in case of electricity bills paid by the farmer, as upto a certain units of electricity consumption, the tariff remains same, but beyond certain units of electricity consumption, the tariff changes. Similar is the case of telephone bills, over time wage rates of workers in fertilizer companies etc. So, these costs consist of a fixed portion and a variable portion and are, therefore, known as Semi-variable costs.

In normal cases, as we see in the crop production programmes, the variable cost increases as output increases, but this increase need not be equally proportionate. Its proportion first declines, becomes constant and then starts rising. SRAVC also behaves in a similar way. Accordingly, the variable costs are of three types,

- ☆ ***Progressive variable costs:*** If the variable costs increase faster than the output is increased, it is referred as Progressive variable costs. Say, if output is increased by one per cent, variable costs increase by more than one per cent.

☆ ***Proportional variable costs:*** If the variable costs increase in the same proportion as output is increased, it is referred as Proportional variable costs. Say, if output is increased by one per cent, variable costs also increase by one per cent.

☆ ***Regressive variable costs:*** If the variable costs increase at slower pace than the output is increased, it is referred as Regressive variable costs. Say, if output is increased by one per cent, variable costs increase by less than one per cent.

Differences between TFC and SRTVC

The differences between the two types of factors *i.e.*, fixed and variable factors forms the basis for cost analysis and the laws of returns. If all the factors of production are perfectly divisible and variable, the cost of production would have increased in the exact proportion of the output. As this is not the case, a special cost analysis becomes important. The differences between these two costs are summarized through Table 15.2.

Table 15.2: Differences between TFC and SRTVC incurred in the production programme.

TFC	*SRTVC*
Total fixed costs do not vary with the level of output *i.e.*, they are independent of the level of output.	Total variable costs vary with the level of output *i.e.*, they are dependent of the level of output.
When output is nil, fixed costs are incurred in the production programme. So, SRTC=TFC.	When output is nil, no variable costs are incurred in the production programme.
They are considered as Long-run costs or Indirect costs or Sunken costs	They are considered as Short run costs or Direct costs
TFC are related to fixed resources only	SRTVC are related to variable resources only
Concept of fixed costs will arise only in the short run production programme.	Concept of variable costs will arise both in the short run and long run production programmes
The recovery of fixed costs assumes more importance in the long run	The recovery of variable costs assumes more importance both in short run and long run production programmes. (All costs are variable costs in the long run)
To continue the production programme in the long run, the gross returns must cover these fixed costs also (in addition to variable costs).	To continue the production programme in the short run, the gross returns must cover at least variable costs, if not fixed costs.
When the output goes up, TFC remains same, thereby, fixed cost per unit of output comes down.	When the output increases, variable costs also increases, but this increase need not be equally proportionate. Its proportion first declines, becomes constant and then starts rising. SRAVC also behaves in the similar way.
TFC curve is parallel to X-axis. So, the TFC curve starts above the point of origin on Y-axis.	SRTVC curve is 'S' shaped curve and it starts from the point of origin.
TFC include indivisible large units. For example, a machinery has to be employed in its full capacity, irrespective of the volume of the output.	SRTVC include divisible small units. For example, resources like fertilizers, pesticides, weedicides, human labour etc., can be employed in small or large units according to the varying quantity of output.
Fixed factors are supplementary in nature. For example, machinery makes production activity more convenient and efficient. However, even in their absence, some amount of output can be produced.	Variable factors are called Prime factors without which no output can be produced.
Examples include, land, machinery, managerial resources, depreciation, interest on loan amount, land revenue etc.	Examples include, hired labour wages, fertilizers, pesticides, weedicides etc.

(c) SRTC

It is also called as Production cost. It refers to the total expenditure incurred in the production programme to produce a certain level of output. SRTC is the sum of TFC and SRTVC *i.e.*, SRTC = TFC + SRTVC. SRTC curve will have the same shape (inverted 'S' shaped) and slope like SRTVC curve, as the behaviour of SRTC curve is similar to that of SRTVC curve. This is because, the variation in SRTC is due to the variation in SRTVC only because, TFC always remains constant at all levels of output and hence, they are placed at equi-distance at all levels of output. To illustrate the concepts of TFC, SRTVC and SRTC, consider the example of a single farm operating in the short-run with fixed factors *viz.*, land and machinery and variable factors as shown through Table 15.3. The same data are depicted graphically (Figure 15.3) and they are referred to as Absolute cost curves. As shown in the Figure 15.3, total output units have been measured along the X-axis and costs have been shown on the Y-axis. At a distance of Rs. 4000, we find TFC curve as a line parallel to X-axis. This shows that, even at zero output level, SRTC is equivalent to TFC. From here onwards, SRTC curve rises continuously and in the same proportion as SRTVC.

SRTVC increases with increase in output, but at the initial levels of output *i.e.*, upto 11 units of output, SRTVC increases at diminishing rate, but between 11 and 18 units of output, the SRTVC increases at constant rate and beyond 18 units of output, the SRTVC increases at increasing rate. Both SRTC and SRTVC curves have the same shape and slope, as the changes in SRTC is due to changes in SRTVC only and there is no change in TFC due to changes in output. So, both SRTVC and SRTC are parallel curves and SRTC curve is just the vertical summation of the TFC and SRTVC curves. The vertical distance between SRTC and SRTVC curves is the same at all output quantities. The reason, of course, is that, this vertical distance is TFC. Since, TFC is constant, the vertical distance is constant, thereby, the SRTVC and SRTC remain parallel and they do not intersect each other. In other words, the SRTC and SRTVC curves are essentially the same curves, but the SRTC curve will be higher than SRTVC curve, as it includes the TFC component. This further implies that, the slopes of both SRTC and SRTVC curves are identical at each and every output quantity. So, both SRTC and SRTVC indicate the SRMC of producing additional quantity of output. That means, the slopes of both SRTC and SRTVC curves indicates the additional cost incurred in the production programme to produce an additional unit of output (SRMC). This further indicates that, any quantity-induced change in SRTC must come from SRTVC only, as TFC is fixed at all levels of output.

Table 15.3: Short run absolute costs, per unit costs and MPP in the production of paddy.

Output, Y (Units)	*TFC*	*SRTVC*	*SRTC*	*AFC*	*SRAVC*	*SRATC*	*SRMC*	*Rate of Change of SRMC*	*MPP*	*Rate of Change of MPP*
0	400	0	400.00	–	0	–	–	–	–	–
4	400	250.00	650.00	100.00	62.50	162.50	250.00	Diminishing MC	4	Increasing MPP
11	400	392.86	792.86	36.36	35.71	72.08	142.86	Constant MC	7	Constant MPP
18	400	535.72	935.72	22.22	29.76	51.98	142.86		7	
23	400	735.71	1135.71	17.39	31.99	49.38	200.00	Increasing MC	5	Diminishing MPP
25	400	1235.71	1635.71	16.00	49.43	65.43	500.00		2	
26	400	2235.71	2635.71	15.38	85.99	101.37	1000.00		1	

2. *Traditional Theory – Short Run per Unit Costs*

They include, AFC, SRAVC, SRATC and SRMC. According to Traditional theory, the per unit cost curves both in short run and long run are 'U' shaped, but in long run, the curves are more flatter than in the short run. In Figure 15.5, short run per unit cost curves are shown considering varying output units on the X-axis and different cost amounts on the Y-axis. All the four per unit cost curves have been drawn on the basis of respective values in the Table 15.3. The column I shows output of the production programme. In column 5, we computed AFC, which is a ratio of TFC to total output. Column 6 reads SRAVC. It is a ratio of SRTVC to total output. In column 7, we computed SRATC for different units of output. It is a ratio of SRTC to total output or AFC + SRAVC. In column 8, SRMC is computed as a ratio between $\Delta SRTVC/\Delta Y$ or differences between successive SRTVC values or as a ratio between $\Delta SRTC/\Delta Y$ or differences between successive SRTC values. The behaviour of these per unit costs is an interesting and important part of cost analysis. From both Table 15.3 and Figure 15.5, we can better explain the behaviour of these short run per unit cost curves.

(a) AFC

The AFC is the fixed cost per unit of output. The average fixed cost obtained by dividing the TFC by the quantity of the commodity produced or output *i.e.*, AFC = TFC/Y. Alternatively, AFC can also be computed by subtracting SRAVC from SRATC, given by, AFC = SRATC – SRAVC.

It is generally referred as 'Spreading the over heads' because, TFC is same irrespective of the level of output and hence, AFC will spread over different levels of output in terms of TFC incurred per each unit level of output. From the above formula, it is clear that, since TFC remains same at different levels of output, the AFC decreases continuously with increase in level of output (Table 15.3). Hence, AFC is a continuous falling curve from left to right giving a shape of rectangular hyperbola (Figure 15.5) implying that, AFC decreases with increase in level of output. But, TFC curve is parallel to X-axis indicating that, TFC remains same irrespective of the level of the output (Figure 15.3). From the Table 15.3 and Figure 15.5, we can also have the following inferences:

- ☆ AFC is based on the values of TFC. AFC curve is seen to be continuously falling from Rs. 100 to Rs. 15.38. However, initially it falls sharply and then with a slower tempo because, at the beginning output rises fastly (due to increasing MRs) and later slowly (due to diminishing MRs).

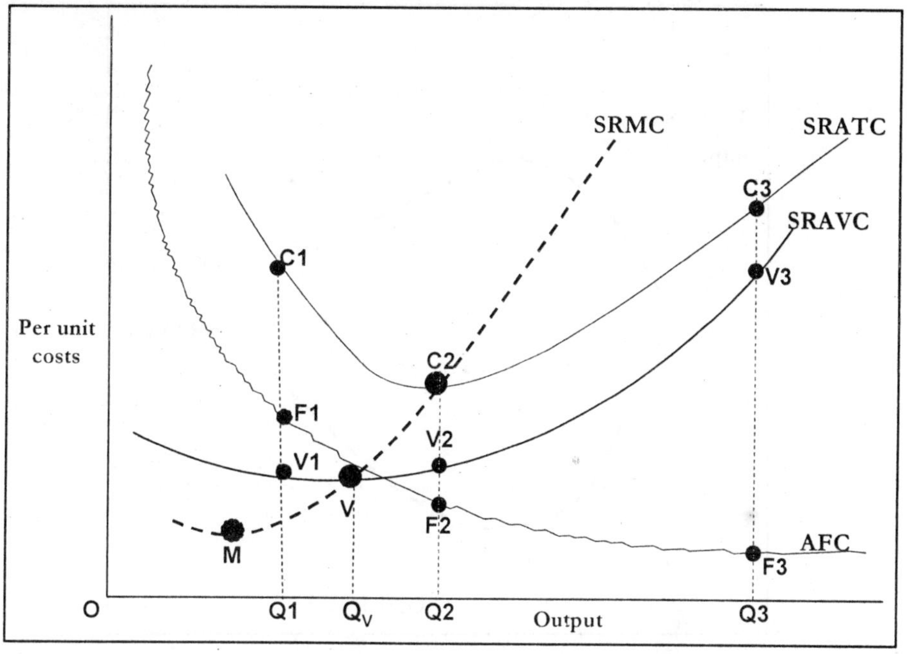

Figure 15.5: Traditional theory – Short run per unit cost curves in cultivation of paddy.

- ☆ AFC is high at relatively small output quantities and low at relatively large output quantities. That means, with increase in output, TFC is spread more thinly over a larger quantity.
- ☆ AFC remains positive, it never reaches a zero value and never turns negative. It is asymptotic *i.e.*, it won't touch either of the two axes. If AFC curve touches the X-axis, it implies that, for that given level of output, AFC is zero. But, this is not possible because, TFC will not be zero even at zero level of output. It wont touches the Y-axis also, as it implies, AFC is infinity (TFC/Y = TFC/0 = infinity). So, AFC curve will not touch either of the axes.
- ☆ AFC curve will have negative slope through out *i.e.*, slopes downward from left to right. This declining AFC curve is a major reason that the SRATC curve is negatively sloped for relatively small output quantities. In fact, firm that employs a lot of fixed factors relative to variable factors (TFC has a substantial share in SRTC), enjoys a downward falling SRATC.

We can also derive AFC from both SRATC and SRAVC curves. At the initial levels of output, the gap between the SRATC curve and the SRAVC curve is relatively more. But, as the firm expands the output, the gap between SRATC curve and SRAVC curve continuously decreases, indicating that, there is continuous decline in AFC with increase in output. So, from the gap between SRATC curve and SRAVC curve, we can easily compute the AFC. So, the AFC curve need not be indicated in the family of 'U' shaped per unit cost curves, as the gap between SRATC and SRAVC curves itself indicates the behaviour of AFC curve. The rise in SRAVC curve further suggests that, it is solely responsible for the positive slope of SRATC curve. On the other hand, the negative slope of SRATC is mainly contributed by the negative slope of AFC curve, though SRAVC also contributed to some extent.

AFC when compared with price of the commodity, indicates whether or not the farmer should shut down production in the short run. If price is greater than AFC, then the farmer is able to cover, at least, fixed expenses. Even though, the farmer might be incurring an economic loss, he will lose less by producing than by shutting down the production. However, if price is less than AFC, then it is better for the farmer to shut down the production.

(b) SRAVC

It is the variable cost per unit of output. So,

SRAVC = SRTVC/Y = (X.Px)/Y = Px/(Y/X) = Px/APP

From the above formula, we can say that, SRAVC is inversely related to APP (Figure 15.6). As shown in the Figure 15.6, when APP is rising, it implies technical efficiency of variable factor is increasing and thereby, SRAVC decreases. Similarly, when APP is maximum, SRAVC is minimum. When APP declines, it implies decline in technical efficiency of variable resource, thereby, SRAVC rises. An alternative specification for computation of SRAVC is by subtracting AFC from SRATC, given by, SRAVC = SRATC – AFC.

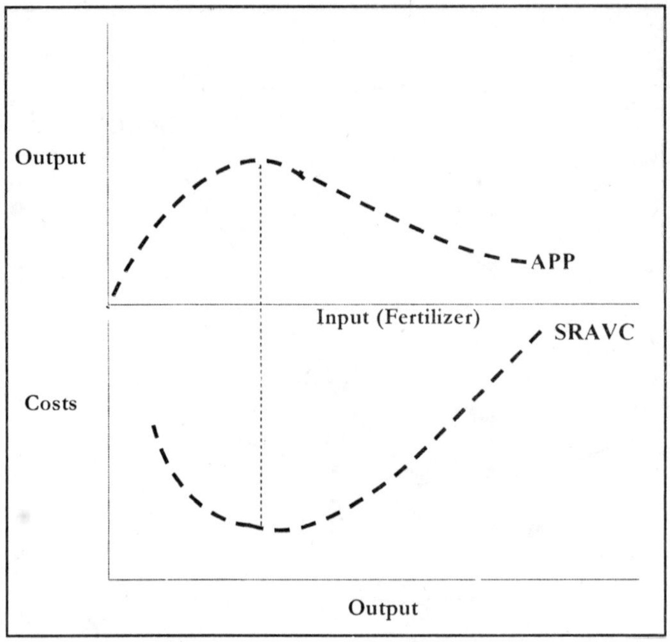

Figure 15.6: Relation between APP and SRAVC curves in short run production programme.

From the Figure 15.5 it is clear that, SRAVC curve first falls, reaches a minimum point 'V' and later rises. When it reaches the minimum point 'V', SRAVC = SRMC curve. That means, SRMC curve cuts the SRAVC curve at its minimum point. SRAVC is a U-shaped curve indicating that, it has a negative slope for small quantities of output, reaches a minimum value, then has a positive slope for larger quantities of output. This U-shape is indirectly attributable to the LDR. The negatively-sloped portion of SRAVC is attributable to increasing MRs and the positively-sloped portion is attributable to diminishing MRs. The SRAVC curve falls first because, at the beginning of production programme or when the output is small, there is more amount of fixed resources and less amount of variable resource. So, the farmer can efficiently manage this less amount of variable resource leading to higher resource use efficiency. So, due to internal economies of scale, increasing returns/diminishing costs are experienced leading to fall in SRAVC curve. This is up to 18 units of output and at 18th unit of output, the SRAVC reaches minimum (Rs. 29.76). But, beyond 18 units of output, the SRAVC curve rises at an increasing rate. This is because, at higher levels of output, the production programme comprises of more amount of variable resources and comparatively less amount of fixed resources. Hence the production programme experiences diseconomies of scale regarding management of labour and other variable resources, thereby, leading to diminishing returns or increasing costs. Hence, the SRAVC curve starts rising upwards, thus giving a U shape. It is also clear that, SRAVC remains positive, it never turns negative. It may be zero, when no production programme is taken up by the farmer because, SRTVC = 0. The behaviour of SRAVC depends on the behaviour of SRTVC *i.e.*, if SRTVC are progressive, SRAVC increases, if SRTVC are proportional, SRAVC is constant, if SRTVC are regressive, SRAVC decreases.

The SRAVC curve is most important to the analysis of a farmer's decision to shut down production in the short run. If price of output is greater than SRAVC, then the farmer may or may not be receiving an economic profit, but it is better to continue the production in the short run rather than shutting down the production. Shutting down production entails a loss equal to TFC. When price is greater than SRAVC, sufficient revenue is generated to pay all variable costs and some TFC, making the operating loss less than TFC. On the other hand, if price is less than SRAVC, then the farmer incurs a loss greater than TFC by producing the output. The farmer's loss includes both TFC plus part of the SRTVC not covered by the returns. As such, it is better for the farmer to shut down the production programme, if price falls below SRAVC and await for better times.

(c) SRATC

It is also called as Short Run Average Cost. SRATC is the ratio between SRTC to output or it is the sum of AFC and SRAVC. So,

SRATC = SRTC/Y = (TFC/Y + SRTVC/Y) = AFC + SRAVC

As discussed earlier, SRTC curve is the vertical summation of the SRTVC curve and the TFC curve. This also holds true for the SRATC curve, which is just the vertical summation of the SRAVC curve and the AFC curve. SRATC curve is a U- shaped curve. It implies that, the SRATC has a negative slope for small quantities of output, reaches a minimum value, then has a positive slope for larger quantities of output. This U-shape is indirectly attributable to the concept of

LDR. When the output level is low, there is more of fixed factor and less of variable factor in the production programme. Hence, the farmer can efficiently supervise the variable factor, thereby, the resource use efficiency is increased. This implies that, there is internal economies of scale leading to increasing returns/diminishing costs. Besides these, the fixed resources like machinery, equipment etc., are used efficiently. Hence, SRATC curve starts falling downwards. When the output is reached to a certain level, the SRATC curve will be at minimum. This is because of two reasons,

- ☆ There is a proper balance between the levels of both fixed and variable factors.
- ☆ The variable resources are used efficiently and fixed resources like equipment and machinery are used up to the normal capacity, thereby, the firm experiences constant returns/constant costs leading to minimum of SRATC. This minimum SRATC implies the optimum normal capacity of business firm or cost of production at minimum level.

With gradual increase in the level of output, beyond optimal normal capacity of the production of enterprise, there is more amount of variable factor compared to fixed factor. Hence, it leads to diseconomies of scale, thereby, leading to diminishing returns/increasing costs. Similarly, the fixed resources like equipment and machinery are over-utilized in the production programme (working for more number of hours), thereby, leading to diminishing returns/increasing costs and hence, SRATC rises upwards and thus giving a 'U' shape to SRATC curve. Thus, increasing MRs is indirectly responsible for negatively sloped portion of SRATC curve, the constant MRs is indirectly responsible for minimum portion of SRATC curve and diminishing MRs (which overwhelms the continuing decline of AFC curve) is indirectly responsible for the positively-sloped portion of the SRATC curve.

The behaviour of SRATC curve can also be explained from the behaviour of both AFC and SRAVC curves (Figure 15.5). At the beginning of production programme, the SRATC curve is initially at high level because, there is more of fixed cost and less of variable cost. As explained earlier, initially, SRAVC curve starts falling due to efficient utilization of variable resources. AFC also falls because, TFC remains same at all levels of output. Hence, SRATC curve starts falling downwards. But, with gradual increase in level of output, SRAVC curve starts rising (larger variable factor compared to fixed factor), but still AFC falls. The fall in AFC is more or outweighs the rise in SRAVC curve and hence, SRATC curve continues to fall. This is upto a certain level of output and beyond that, the SRAVC rises more than the fall in AFC. That means, now SRAVC out weighs AFC and hence, SRATC starts rising upward and thus, giving a 'U' shape to SRATC curve.

As shown through Figure 15.5, SRATC curve shows three distinct phases. Between C_1C_2, the SRATC curve slopes downward. At C_2 the curve reaches a minimum point. From C_2 to C_3 the curve starts continuously rising upwards. Thus, two components of SRATC are noticeable. At output level Q_1, the value of SRATC – C_1Q_1, which is equal to the sum of F_1Q_1, the fixed cost component and V_1Q_1, the variable cost component. So, $C_1Q_1 = F_1Q_1 + V_1Q_1$. The lowest point on SRATC is C_2. Around this point SRATC remains constant for a moment. At output level Q_2, the value of SRATC = C_2Q_2, which is equal to the sum of F_2Q_2, the fixed cost component and V_2Q_2, the variable cost component. So, $C_2Q_2 = F_2Q_2 + V_2Q_2$. At output level Q_3, the value of SRATC = C_3Q_3, which is equal to the sum of F_3Q_3, the fixed cost component and V_3Q_3, the variable cost component. So, $C_3Q_3 = F_3Q_3 + V_3Q_3$. So, to the left of C_2, SRATC falls (C_1C_2) and to its right SRATC rises (C_2C_3). SRATC, thus, shows the three phases of cost behaviour, namely, the diminishing, the minimum (constant) and the increasing cost phases. In the beginning of production programme, when there is diminishing MCs, the SRATC slopes downward *i.e.*, between C_1C_2. This diminishing MCs is due to efficient utilization of less quantity of variable resources and given fixed resources. So long as fixed and indivisible factors are not used to their fullest capacity, the SRATC keeps on falling. Once the fixed resources are utilized at their fullest capacity and there is proper combination with variable resources, the MCs are at minimum/constant, thereby, the SRATC reaches a minimum point (C_2). This is the point of optimum utilization of fixed resources. If the output continues to increase beyond this point (OQ_2), the fixed resources get over utilized and the variable resources are in excess (leading to increasing MCs), thereby, they have improper combination with less quantity of fixed resources and coupled with improper management cause SRATC to rise. From the Table 15.3 and Figure 15.5, we can also have the following inferences:

- ☆ SRATC remains positive, it never reaches a zero value and never turns negative.
- ☆ Each SRATC value consists of two average costs *viz.*, AFC and SRAVC. For example, when SRATC is Rs. 162.50, Rs. 100 is AFC and Rs. 62.50 is SRAVC at four units of output.
- ☆ At low levels of output, the share of AFC is more (compared to SRAVC) in SRATC, but at higher levels of output, the share of SRAVC is more (compared to AFC) in SRATC.

(d) SRMC

It is the additional cost incurred by the farmer to produce additional unit of output. SRMC includes only variable cost because, it is an additional cost made in the production programme. No fixed cost forms a component of SRMC because, it remains same at different levels of output.

$SRMC = \Delta SRTC/\Delta Y = \Delta SRTVC/\Delta Y = \Delta X.Px/\Delta Y = Px/MPP$ [since, $\Delta Y/\Delta X = MPP$]

In the above formula, it is clear that, SRMC is inversely related to MPP. That means, when MPP increases, SRMC decreases *vice-versa*. When MPP is maximum, SRMC is minimum. When MPP is zero, SRMC is maximum and it is vertical and lies parallel to Y-axis (Figure 15.7). SRMC can also be computed from the following formulae:

$SRMC = (SRTC_n - SRTC_{n-1})$ or $(SRTVC_n - SRTVC_{n-1})$

where,

$SRTC_n$ = SRTC at n[th] level of output, $SRTC_{n-1}$ = SRTC at n-1th level of output, $SRTVC_n$ = SRTVC at n[th] level of output, $SRTVC_{n-1}$ = SRTVC at n-1th level of output.

Opportunity Cost vs MC

The following points highlights the differences between opportunity cost and MC:

- Opportunity cost expresses the relationship between scarcity and choice, while MC represents the cost of producing an additional unit of output.
- MC always has a monetary value, while opportunity cost can have a monetary value or not.
- Opportunity cost includes the value of lost time, output, utility and the benefits that might have been enjoyed, if the other choice is made, while MC does not.
- MCs are visible, while opportunity costs are not.
- MC is the cost incurred during the production of a unit output, while opportunity cost is the cost incurred during the producer's or consumer's choice of which resource to use or product to buy respectively.

(i) Relation between SRMC and MPP in the Short Run Production Programme

From the Table 15.3, it is interesting that, SRMC behaviour shows three phases of change *i.e.*, diminishing SRMC, constant SRMC and increasing SRMC. Initially, upto 11 units of output, the SRMC decreases from Rs.250 to Rs. 143. Then, between 11 to 18 units of output, SRMC remains constant at Rs. 143. Finally, above 18 units of output, SRMC continuously increases from Rs.143 to Rs.1000. These three phases are marked as diminishing SRMC, constant SRMC and increasing SRMC respectively and they exactly correspond with increasing MPP or MRs, constant MPP or MRs and diminishing MPP or MRs on the output or returns side. This infers that, when returns diminish, costs increase and both are constant for the same units of output.

Regarding the shapes of SRMC and MPP curves, the U-shape of the SRMC curve is closely related to the hump-shape of the MPP curve (Figure 15.7). The increasing portion of the MPP curve corresponds with the decreasing portion of the SRMC curve. This is because, each additional unit of an input is increasingly more productive, thereby, a given

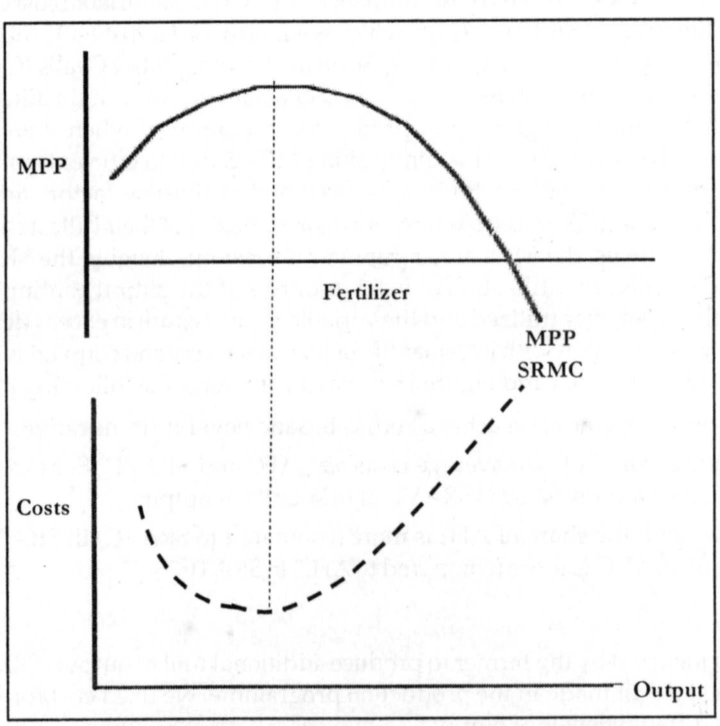

Figure 15.7: Relation between MPP and SRMC curves in short run production programme.

quantity of output can be produced with lesser quantity of variable input. The decreasing portion of the MPP curve corresponds with the increasing portion of the SRMC curve. This is because, each additional unit of an input is less and less productive, thereby, a given quantity of output can be produced only with higher quantity of variable input. The peak of the MPP curve corresponds with the minimum of the SRMC curve. So, the variable cost is associated with the cost of employing a variable unit in the short run production programme and this guides to form the relationship between MPP and SRMC curves. As depicted in Figure 15.7,

- The MPP curve in the Panel A has a distinctive hump-shape, with MPP rising, reaching a peak, then falling. The rising portion of the MPP curve is the result of law of increasing MRs. The falling portion is attributable to diminishing MRs.
- The SRMC curve in the Panel B has a distinctive U-shape, with SRMC initially falling, reaching a minimum and then rising. The falling portion of the SRMC curve is the result of increasing MRs. The rising portion is attributable to diminishing MRs.
- The quantity of output at which SRMC of additional input is minimum is the same quantity of output produced by variable input when the MPP of that variable input is at a maximum.
- Over the range of output, in which MPP increases, then the variable input experiences increasing MRs, thereby, SRMC declines.
- Over the range of output, in which MPP falls, then the variable input experiences diminishing MRs, thereby, SRMC rises.

The above relationships infer that, the correspondence between the MPP and SRMC of a variable input is guided by the concept of LDR.

(ii) Relation between SRMC and SRTC

As the shape of the SRMC curve is influenced by the concept of LDR, the same concept also influences the shape of the SRTC curve (as the slope of SRTC is SRMC). The flattening slope of the SRTC curve for small quantities of output is due to increasing MRs. However, with the onset of the diminishing MRs causes the SRTC curve to rise steeply. The following are the conclusions from both the Panels A and B of the Figure 15.8:

- If the SRTC curve has a positive slope, then SRMC curve also show positive slope.
- If the SRTC curve has a positive and increasingly steeper slope, then the SRMC curve is positive and rising.
- If the SRTC curve has a positive and increasingly flatter slope, then the SRMC curve is positive, but falling.

It is important that, the above relation between SRMC and SRTC also holds good between SRMC and SRTVC. This is because, any change in SRTVC is due to change in SRMC and any change in SRTC is due to change in SRTVC, as TFC is fixed at all levels of output. This also infers that, the slopes of both SRTVC and SRTC indicates SRMC.

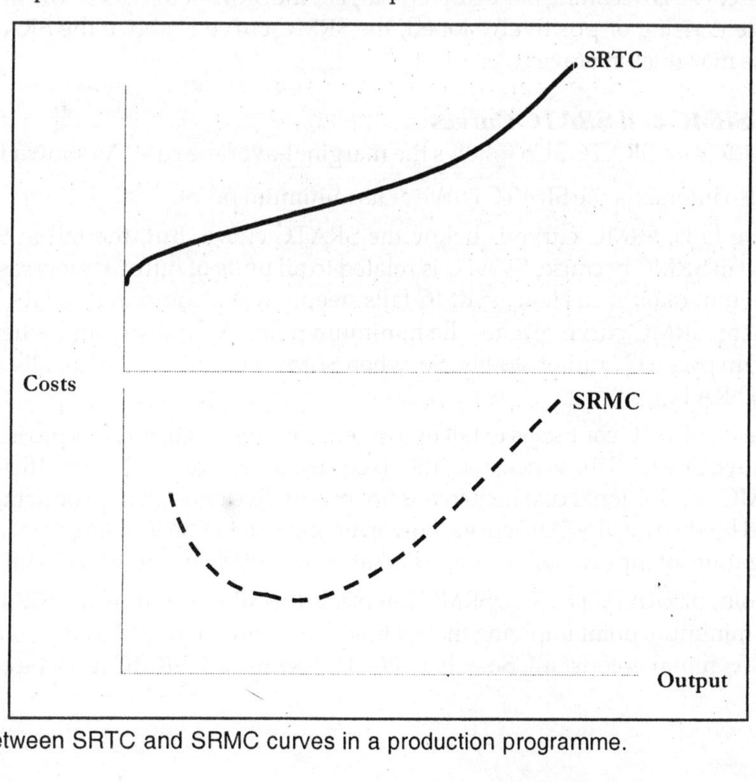

Figure 15.8: Relation between SRTC and SRMC curves in a production programme.

(iii) Relation between SRMC and SRAVC

We know, SRMC can also be expressed as SRMC = $SRTC_n - SRTC_{n-1}$ or $SRTVC_n - SRTVC_{n-1}$. Since no fixed cost form a component of SRMC, marginal fixed cost is equal to zero. This can be well explained from the equations given below:

$$SRMC = SRTC_n - SRTC_{n-1} \quad \textit{Equation 15.5}$$

$$SRTC_n = TFC + SRTVC_n \quad \textit{Equation 15.6}$$

$$SRTC_{n-1} = TFC + SRTVC_{n-1} \quad \textit{Equation 15.7}$$

Substituting the equations 15.6 and 15.7 in 15.5, we get

$$SRMC = (TFC + SRTVC_n) - (TFC + SRTVC_{n-1})$$

$$SRMC = TFC + SRTVC_n - TFC - SRTVC_{n-1}$$

$$SRMC = SRTVC_n - SRTVC_{n-1} \quad \textit{Equation 15.8}$$

So, from the above Equation 15.8, we can say that, SRMC includes only variable cost and no fixed cost. Thus, SRMC is independent of fixed cost and an increase in SRMC is due to increase in SRTVC.

The behaviour of SRMC curve resembles the behaviour of SRAVC curve. This is because, SRMC includes only variable cost. The SRMC curve is U-shaped. It is relatively high at small quantities of output, then as the output increases, SRMC declines, reaches a minimum value, then rises. So, SRMC curve is constructed to study the relation between SRMC and the level of output, holding other variables, like technology and factor prices constant. The 'U' shape of SRMC curve is directly attributable to increasing and then diminishing MRs of the concept of LDR. That means, as the production programme experiences increasing MRs, the SRMC falls and when the production programme experiences diminishing MRs, the SRMC rises., thus giving a 'U' shape to the SRMC curve. So, the negatively-sloping portion of the SRMC curve coincides exactly with increasing MRs of stage I of production function and the positively-sloping portion of the SRMC curve coincides exactly with diminishing MRs of stage II of production function. As shown in the Figure 15.5,

- ☆ Below the SRAVC curve, SRMC curve first falls and then rises.
- ☆ As long as SRMC curve is below SRAVC curve, SRMC curve falls rapidly because, SRMC is related to only one additional unit of output whereas, SRAVC is related to all the units of output.
- ☆ While rising, SRMC curve intersects SRAVC curve at its minimum point 'V' and beyond that point, SRAVC curve starts rising.
- ☆ When SRMC curve is above SRAVC, SRMC rises rapidly because, SRMC is related to only one additional unit of output whereas, SRAVC is related to all the units of output.
- ☆ Thus, when SRAVC curve is declining or negatively sloped, the SRMC curve is below the SRAVC curve. But, when SRAVC curve is rising or positively sloped, the SRMC curve is above the SRAVC curve. This is in accordance with the marginal-average rule.

(iv) Relationship between SRMC and SRATC Curves

The relation between SRMC and SRATC also follows the marginal-average rule. As shown in the Figure 15.5,

- ☆ The SRMC curve also intersects the SRATC curve at its minimum point.
- ☆ When SRATC curve falls, SRMC curve is below the SRATC curve. But, the fall in SRATC is less when compared to the fall in SRMC because, SRATC is related to all units of output, whereas, SRMC is related to only one additional unit of output. Hence, SRMC falls steeply when compared to SRATC. This is shown in the Figure 15.5 where, SRMC curve reaches the minimum point 'M' quickly and whereas, SRATC curves reaches the minimum point 'C_2' rather slowly. So, when SRMC falls, SRATC also falls and the fall in SRMC is faster than fall in SRATC.
- ☆ When SRMC increases, SRATC continues to fall upto a certain level of output *i.e.*, upto its minimum point 'C_2' and SRATC lies above SRMC. This is because, the fixed resources are used very efficiently to increase the output, but the SRMC or additional cost incurred is not productive enough in producing more of additional output, as indicated by the rise in SRMC curve. However, the rise in SRMC should not be above the SRATC of all the preceding units of inputs. So, this implies that, when SRMC is increased, still SRATC falls.
- ☆ At the minimum point of SRATC *i.e.*, 'C_2', SRMC intersects the SRATC curve. So, SRMC curve touches the SRATC curve at its minimum point implying that, at that minimum point of SRATC, both SRMC and SRATC are equal and SRATC remains constant. So, when SRATC is constant, SRMC must be constant.

- When SRMC rises above SRATC, SRATC also rises, but at a slower pace compared to SRMC *i.e.*, when SRMC is above SRATC, SRMC rises faster than SRATC. This is because, SRMC is related to only one additional unit of output whereas, SRATC is related to all the units of output. So, this implies that, SRMC rises at a faster pace, whereas, SRATC rises at slower pace.
- It is not necessary that, SRMC must necessarily rise to rise SRATC. Even if SRMC falls, SRATC can be increased, but the SRMC should be greater than SRATC of all the preceding units of inputs.
- Since MC is an additional cost incurred in the production programme, we can write, $SRTC_{n+1}$ = SRTCn + SRMC. Thus,
  - if the SRMC of (n+1)th unit is less than SRATCn (SRATC of all the previous 'n' units) then, SRATCn+1 will be smaller the SRATCn.
  - if the SRMC of (n+1)th unit is more than SRATCn (*i.e.*, SRATC of all the previous 'n' units), then SRATCn+1 will be more than SRATCn.

  So the final the relationship between SRATC and SRMC is shown through the following diagram:

  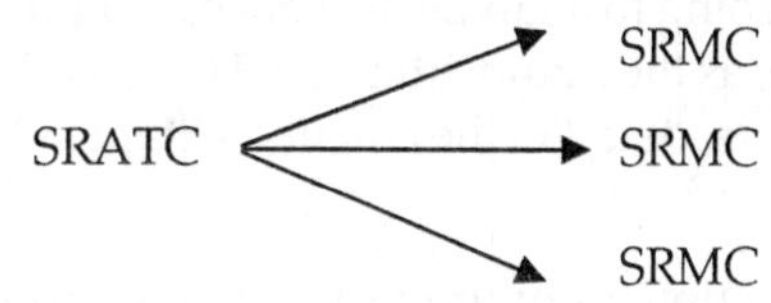

  - When SRMC is below SRATC, SRATC falls *i.e.*, SRMC curve pulls the SRATC curve downwards. As long as SRMC is below SRATC curve, a rise or fall in SRMC will lead to fall in SRATC.
  - When SRMC = SRATC, SRATC remains constant and SRATC is at minimum point. So, to have constant SRATC, we should have constant SRMC.
  - When SRMC is above SRATC, SRATC rises upward. That means, SRMC curve pulls SRATC curve upwards. As long as SRMC curve is above SRATC curve, a rise or fall in SRMC will lead to rise in SRATC.
- It is important that, the minimum values of the SRAVC curve and SRATC curve occur at different quantities of output Q_V and Q_2 respectively. This is because of the following reasons:
  - SRMC curve intersects each of the SRAVC and SRATC curve at their minimum points
  - At the point of intersection of SRMC curve with SRAVC curve, the SRMC curve has positive slope and this positive slope implies rise in SRMC, rise in SRTVC and thereby, the SRAVC quickly reaches the minimum compared to SRATC.
  - Since SRATC includes SRAVC + AFC, the rise in SRMC above SRAVC though contributes to rise in SRAVC due to increase in SRTVC, but the greater fall in AFC outweighs the rise in SRAVC, thereby, the SRATC still continue to slope downwards. So, after a certain level of output *i.e.*, Q_VQ_2, SRMC intersects the SRATC at its minimum C_2 due to downward fall of AFC at a faster pace compared to rise in SRAVC.

Thus, the family of short run per unit cost curves *viz.*, AFC, SRAVC, SRATC and SRMC have U-shapes. Each is U-shaped because, it begins with relatively high but falling cost for small quantities of output, reaches a minimum value, then has rising cost at large quantities of output. Although, the AFC curve is not U-shaped, it is occasionally included with the other three just for sake of completeness. The U-shapes of the SRAVC, SRATC and SRMC curves are directly or indirectly the result of increasing MRs for small quantities of output (Stage I of short run production function) followed by diminishing MRs for larger quantities of output (Stage II of short run production function). Among these three per unit cost curves, the SRMC curve is directly influenced by the concept of LDR. The U-shaped pattern for the SRMC curve that results from increasing and diminishing MRs is then indirectly responsible for creating the U-shape of the SRAVC and SRATC curves. So, between the three curves *i.e.*, SRAVC, SRATC and SRMC, the basic mathematical relation *viz.*, average-marginal relation exists. The advantage of analyzing the family of 'U' shaped cost curves is that, from these per unit cost curves, we can easily compute the total cost values *i.e.*, TFC, SRTVC and SRTC. This is because, SRTVC is the quantity times SRAVC, SRTC is the quantity times SRATC and TFC is the quantity times AFC. Otherwise, if we compute SRTC and SRTVC from SRATC and SRAVC curves, the difference between SRTC and SRTVC indicates TFC. These U-shaped per unit cost curves form the foundation for the cost and output analyses in short-run production programme, thereby, guide for profit-maximizing production in a firm.

The importance of SRMC curve is discussed in the ensuing chapters, as it takes a center stage in analyzing the point of optimality of short run production function and in the case of perfect competition, the portion of this SRMC curve above the SRAVC curve represents the supply curve of the firm in the short run (and the portion of the LRMC curve above LRATC curve represents supply curve of the firm in the long run). Besides these, it also indicates the efficient allocation of resources, as the decline in SRMC indicates efficient allocation of a variable resource and *vice versa*.

B. Traditional Theory – Long Run per Unit Cost Curves

1. LRATC

As discussed earlier, in the long run production programme, all the factors are variable and no or very few factors are fixed. So, to get the LRATC, the LRTC is divided by the quantity of output produced. So, LRATC is the per unit cost incurred by a firm in a long run production programme, when all the factors/inputs are variable. This LRATC is guided by the returns to scale. Since, all the factors are variable in the long run, there are no fixed costs and hence, no AFC in the long run. Hence, there is no need to distinguish between LRATC and long run AVC (LRAVC). So, in the long run, LRATC is merely LRAVC. In the absence of fixed factors (as all the factors are variable) the laws of returns *i.e.*, increasing, constant and diminishing MRs are not relevant to LRATC. But, there are two similar influences *viz.*, economies of scale (or increasing returns to scale) and diseconomies of scale (or diminishing returns to scale). In explaining the behaviour of SRATC, we noticed that, the negative sloping of SRATC is due to increasing MRs and positive sloping is due to diminishing MRs in the short run production programme. But, in long run production programme, since all the factors are variable, the MRs or concept of LDR will not guide the production programme. Instead, the LRATC is guided by the increasing and diminishing returns to scale, which translates into economies of scale and diseconomies of scale respectively. Before going to analyze the concept of LRATC, it is essential to highlight the concepts of economies and diseconomies of scale, as they guide the LRATC behaviour in the long run production programme. (However, these concepts are discussed in-detail in the ensuing Chapter 16)

Economies of Scale

It implies increasing returns to scale in the production programme, where the proportionate increase in output is more than proportionate increase in all the factors, thereby, leading to fall in LRATC. For relatively small levels of output, a firm tends to experience economies of scale and increasing returns to scale.

Diseconomies of Scale

It implies diminishing returns to scale in the production programme, where the proportionate increase in output is less than proportionate increase in all the factors, thereby, leading to rise in LRATC. For relatively higher levels of output, a firm tends to experience diseconomies of scale and diminishing returns to scale.

Derivation of LRATC

So far, we have studied cost analysis with reference to the short run production programme, where we assumed that, some factors are variable and some factors like machines, land, manager, farm building etc., are fixed. When a firm continues to produce for a sufficiently long number of years, some or all of its fixed factors can be replaced. So, in the long run, the distinction between fixed and variable factors becomes less important. So, 'long run' of a firm can be defined as, a period sufficiently long enough during which at least one or more of the fixed factors become variable and can be replaced. Thus, in the long run, the firm can change all the factors, thereby, enlarges the scale of operation and produces more output in the most efficient way. So, in the long run, all the factors are variable or very few factors are fixed. This implies, indivisible factors become divisible in the long run and therefore, they can be used more economically. But, it is important that, there is no fixed limit on the number of years to distinguish between short run and long run. For each variety of fixed factor, its long run depends upon its life span. Let us assume that, a firm employs three factors of production with their life span given as machinery (5 years), manager (10 years) and factory building (20 years). Then, for machinery replacement, six years is the long run, for replacement of manager 11 years is the long run and after 20 years all the three factors become perfectly variable or replaceable. With the passage of time, each of these factors can be altered and replaced. This permits the firm to change its scale of production. An old machine can be replaced by a new one, which has better capacity, a manager can be replaced at the end of the contract period with another more qualified and experienced person etc. Such changes make productive activities more efficient and hence, they contribute to economies of large scale. Since the long run goes on combining such individual advantages, it is also called as a 'chain of the short runs'.

In the short run, the firm is tied up with a given plant, as the firm size is relatively fixed in the production programme. The term 'plant' refers to the firm consisting of given capital equipment, machinery, land etc., which are fixed in the short run. So, in the short run, the firm has to tie up only on a single plant and given the plant size, the firm can increase the output only through varying the variable factors. However, in the long run, the firm can change the size of the plant, thereby, it moves from one plant to another with the main objective of increasing the scale of operation and to minimize the per unit costs. That means, long run is a period of time, sufficiently long enough, to bring all desirable changes in the plant (both fixed and variable resources) *i.e.*, changes in capital equipment, machinery land etc., so as to expand the output or reduce the output. Thus, in the long run production programme, all the factors are variable including the size of the plant and the firm tries to ensure least possible average cost of production to produce a given level of output and this cost of production is referred as LRATC. In the words of Leftwitch, *'the LRATC shows the least possible cost per unit of producing the various outputs when the firm has time to build any desired scale of plant'*. According to J.S, Bain, *'the LRATC curve shows for each possible output the lowest cost of producing that output in the long run'*. So, the

LRATC represents the per unit cost of producing a product or service in the long run, when all the factors are variable. So, in long run, the firm can move from one plant to another plant to expand or contract the output, so as to minimize the LRATC, unlike short run production programme, where the firm has to tie up only on a single plant.

We know, SRATC is derived by dividing the SRTC with the quantity of output. Similarly, LRATC is derived by dividing the LRTC with the level of output. This curve depicts least possible average cost of production of a firm in the long run. It is, otherwise, defined as the summation of SRATCs of different plants of the same firm. The SRATC curves are also called as plant curves, since in the short run, plant size is fixed and each of the SRATC curve correspond to a particular plant. The concept of LRATC can be better explained through the Figures 15.9 and 15.10.

Assume, there are three plants of a firm represented by their SRATC curves $SRATC_1$, $SRATC_2$ and $SRATC_3$ (Figure 15.9). Each SRATC corresponds to a particular size of the plant, which represents the average cost of producing the output in each plant. $SRATC_1$ depicts a lower plant size, while the movement to $SRATC_2$ and $SRATC_3$ shows the firms of larger size. From the Figure 15.9, the following inferences can be drawn:

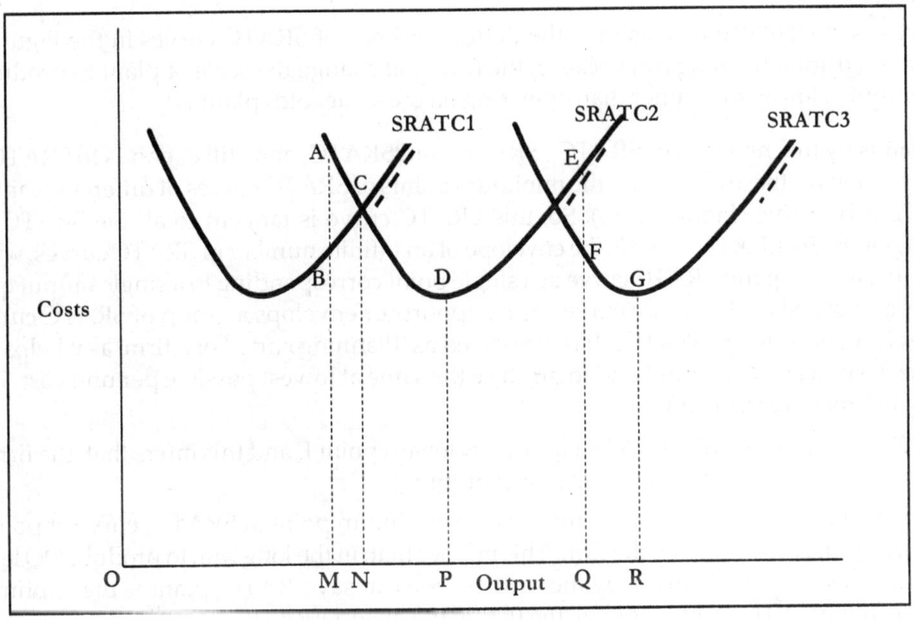

Figure 15.9: Traditional theory – Different plant sizes with different SRATCs in producing the output.

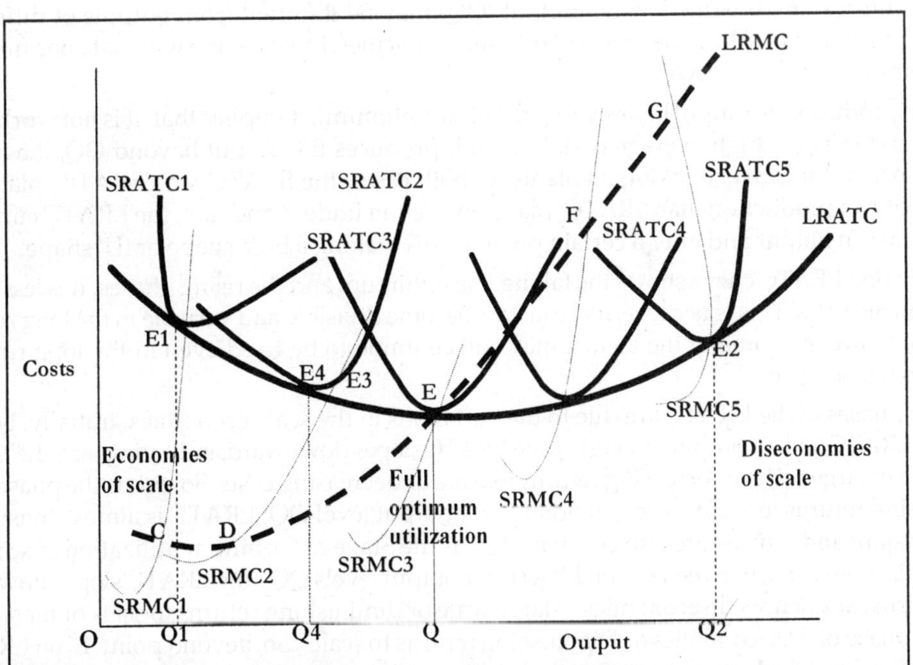

Figure 15.10: Traditional theory – Derivation of LRATC and LRMC curves.

- For producing OM level of output, both $SRATC_1$ and $SRATC_2$ plants can be used. However, it is economical to use $SRATC_1$ plant, as OM level of output can be produced with a lower average cost of BM compared to AM average cost of $SRATC_2$ plant.
- If the firm is to produce ON level of output, it can be produced by both $SRATC_1$ plant and $SRATC_2$ plant. But, it is advantageous to use $SRATC_2$ plant because, the same plant can be used to produce still a higher output OP, at lowest possible average cost of DP. If we employ $SRATC_1$ plant to achieve this OP level of output, a huge average cost has to be incurred.
- To produce a still higher level of output OQ, it is better to use $SRATC_3$ plant compared to $SRATC_2$ plant, as from $SRATC_3$ plant, a lower average cost of FQ is incurred compared to EQ of $SRATC_2$ plant.
- To produce a still higher level output OR, $SRATC_3$ plant is to be used to produce at lowest possible average cost of RG, but not from $SRATC_2$ plant.
- Thus, in the long run, to produce any given level of output, the firm uses that plant, which leads to minimum average cost.
- The above discussion also reveals that, the dotted portions of SRATC curves in the Figure 15.9 are of no consideration during the long run because, the firm will change the scale of plant to produce a still higher level of output at lower cost rather than operating on the same (old) plant.

If the firm expands by the three stages $SRATC_1$, $SRATC_2$ and $SRATC_3$ and still expands to $SRATC_4$ and $SRATC_5$, the thick wave like portions of these curves (*ie.*, the minimum points of SRATC curves of different plants), when joined together form the LRATC curve (Figure 15.10). So, this LRATC curve is tangent to all the SRATC curves at their minimum or lowest points. So, LRATC curve is the envelope of an infinite number of SRATC curves, with each SRATC curve tangent to or just touching the LRATC curve at a single point corresponding to a single output quantity. Hence, the LRATC curve is also called as 'Envelope curve', as it supports or envelops a group of SRATC curves of different plant sizes. It passes from outside the SRATCs. It is also called as 'Planning curve' of a firm, as it helps in choosing the plant to produce the desired level of output and to produce the same at lowest possible per unit cost. From the Figure 15.10, the following inferences can be drawn:

- The LRMC curve intersects the LRATC curve at its lowest point E and this infers that, the firm produces OQ equilibrium output by incurring QE unit cost of output.
- The lower most point of LRATC curve touches the minimum point of $SRATC_3$ curve at point E, where OQ quantity of output is produced by the firm. This implies that, in the long run, to produce OQ level of optimum output, QE is the LRATC incurred by the firm. So, we can say $SRATC_3$ plant is the 'Optimum plant size' needed to produce OQ level of output by the firm at the lowest LRATC.
- At the long run cost minimizing level of output (OQ), LRTC = SRTC, LRATC = $SRATC_3$ and LRMC = $SRMC_3$.
- The long run cost minimizing level of output (OQ) may be different from outputs of different minimum SRATCs, as the firm employs the plants of different capacities, but coincide with only one minimum SRATC of optimum plant size (*i.e.*, $SRATC_3$).
- If the firm produces the output below OQ (*i.e.*, below optimum), it implies that, it is not working upto its full capacity and it incurs high average cost. Even if it produces the output beyond OQ, it again incurs high average cost and it is overworking its plants. In both cases, the $SRATC_2$ and $SRATC_4$ plants have higher average costs of production than $SRATC_3$ plant. So, we can understand that, the LRATC curve initially falls with increase in output and after a certain point it rises making a boat shape or 'U' shape.
- The 'U' shaped LRATC curve shows the falling, the minimum and the rising phases. It is less pronounced or flatter than SRATCs. This is because, more factors become divisible and variable in the long run compared to short run. However, some of the factors may still continue to be fixed even in the long run and it makes LRATC to fall and rise.
- The three phases of the LRATC are due to the variations in the scale economies. Initially, between the E_1E points on LRATC and for output levels Q_1Q, the LRATC slopes downward, indicating that, the firm experiences economies of large scale or increasing returns to scale or decreasing costs. So, this is the phase of 'Economies or Increasing returns to scale'. Then, at point E, for output level OQ, LRATC is almost constant. Therefore, cost is constant and returns are also constant. This is the Stage of Optimum utilization of scale advantages. Beyond this point E, *i.e.*, between EE_2 on LRATC for output levels QQ_2, the LRATC slopes upward, indicating that, the firm experiences diseconomies of large scale or diminishing returns to scale or increasing costs. So, this is the phase of 'Diseconomies or Diminishing returns to scale'. So, beyond point 'E' on LRATC curve, no further output can be advantageously produced. Thus, economies of scale are at the disposal of a firm only in the long run, when fixed factors become variable and replaceable and that too, if the firm operates on an 'optimum plant size'.

☆ The minimum of a $SRATC_3$ occurs when the slope is zero. Thus, the points of tangency between the U-shaped LRATC curve and the minimum of the $SRATC_3$ curve would coincide only with that portion of the LRATC curve exhibiting constant economies of scale.

☆ At least one of the SRATCs is tangential to LRATC. This is essential because, the advantages available in the long run must be relevant to the respective short run plant size.

☆ None of the SRATCs can ever intersect LRATC. This is because of the fact that, if SRATC of a plant intersect LRATC, then some portion of SRATC will lie below LRATC. It would suggest some advantages that firm can enjoy in the short run that are not available to it in the long run. This is absurd since, LRATC is a chain of SRATCs.

☆ The U shape of SRATC is more pronounced than that of LRATC, as the diminishing MRs of a variable factor sets in very early compared to diminishing scale of all the variable factors due to expansion of output.

☆ LRATC curve will touch the $SRATC_3$ of optimum plant size at its minimum point and all other SRATC curves are touched by the LRATC either to the left or to the right of their minimum points. So, SRATC curve is tangent to LRATC at the long-run cost minimizing level of output. At the point of tangency LRATC = SRATC. At all other levels of output, SRATC will exceed LRATC. To the left of the point of tangency, the firm employs too much of fixed capital and less of variable capital. To the right of the point of tangency, the firm employs too much of variable capital relative to fixed capital, thereby, leading to diminishing returns in the production programme.

☆ The slope of the TC curves indicates MC. So, when $SRATC_3$ is tangent to $LRATC_3$, $SRMC_3 = LRMC_3$.

Differences between SRATC and LRATC

The following are the differences between SRATC and LRATC curves:

☆ SRATC curve represents the ATC of the output in short run with reference to a particular plant size. LRATC curve represents the ATC of the output in the long run, with reference to different plant sizes.

☆ LRATC curve cannot be higher than any SRATC curve of a different plant sizes and it lies tangent to all the SRATC curves

☆ LRATC curve cannot cut SRATC curve of any plant size and it envelopes all the SRATC curves

☆ LRATC curve can represent the optimum size of the firm, while SRATC curve can indicate only optimum level of output that can be produced in the short run.

☆ LRATC will be equal to the SRATC that produces the output at lowest ATC

☆ LRATC curve is flatter compared to SRATC

☆ LRATC curve indicates the lowest possible average cost at which the output can be produced in the long run and not in the short run.

2. Shapes of LRATC Curve

Depending upon length of the time period, LRATC curve will behave in one way or the other. If all the factors are perfectly variable and there are no fixed factors at all, then the LRATC curve will be a horizontal straight line parallel to X-axis (Figure 15.11). The straight line LRATC curve implies that, increasing levels of output is possible through incurring same LRATC. In case of straight line LRATC, when the plant size is expanded, the SRATC of each plant will remain same *i.e.*, they does not rise or fall and this is because of constant returns to scale in the firm. When LRATC is same across different levels of output, it implies LRMC also is same and coincides with LRATC. But, if some of the factors are variable, while a few other factors continue to be fixed even in the long run, then LRATC curve will be 'U' shaped, but flatter than SRATC (Figure 15.10). This 'U' shaped LRATC implies that, the firm experiences varying returns to scale. The downward slope of LRATC implies the firm is experiencing increasing returns to scale (economies of scale) and the upward slope of LRATC implies operation of diminishing returns to scale (diseconomies of scale). Thus, in case of U or boat shaped LRATC, when the plant size is expanded, the SRATC of each plant will first fall and then starts rising as the scale of production is increased, further implying that, the firm is experiencing varying returns to scale. For some sectors like electricity supply, water supply etc., the LRATC is 'L' shaped, thereby, the firm in the industry enjoys economies of scale indefinitely. Such sectors will enjoy natural monopoly in the market.

So, scale economies and diseconomies produce a U-shaped LRATC curve. For relatively small quantities of output, the curve is negatively sloped and for large quantities, the curve is positively sloped. Whiie the shape of the LRATC curve looks surprisingly like that of a SRATC curve, but the underlying forces are different. This U-shape is not the result of increasing and then diminishing MRs that surface in the short run when a variable input is added to a fixed input. The negatively-sloped portion of this LRATC curve reflects economies of scale and increasing returns to scale. The positively-sloped portion reflects diseconomies of scale or diminishing returns to scale. But, the LRATC curve is

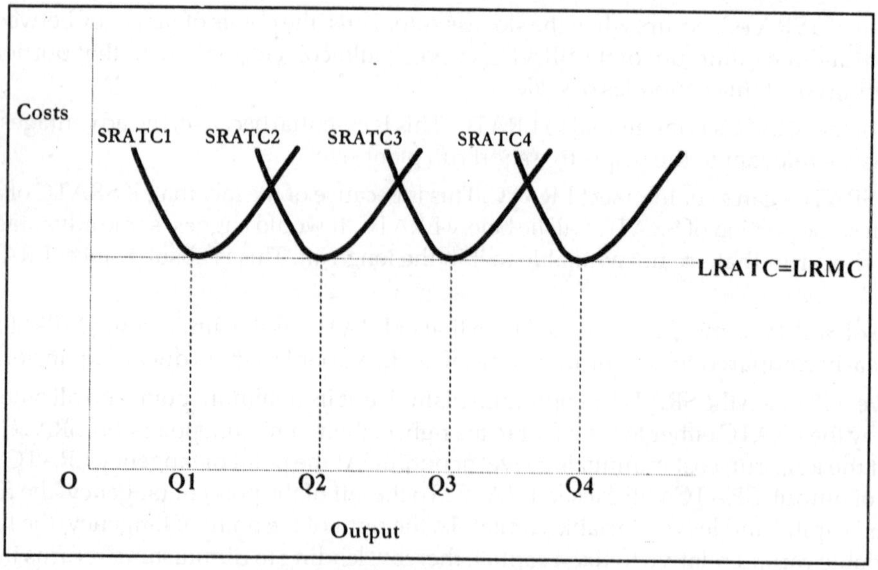

Figure 15.11: Traditional theory – Straight line LRATC curve under constant returns to scale.

flatter than SRATC curve (Figure 15.10). It means that, LRATC curve is also 'U' shaped curve like SRATC curve, but first falls slowly and then rises gradually after a minimum point is reached. Initially, LRATC curve slopes downward due to internal economies of scale such as labour economies, technical economies, managerial economies, bulk-buying economies, risk bearing economies, financial economies, welfare economies etc., thereby, helps to attain increasing returns to scale in the production programme. After reaching the minimum point, LRATC may flatter over a certain range of output. In such a situation, both economics and diseconomies of scale balance each other and thus, LRATC curve has a disc base at OQ level of output. However, with the further expansion of scale beyond minimum point of LRATC, the internal diseconomies like managerial diseconomies, technical diseconomies, bureaucracy, Government interventions etc., arise, which more than counterbalance the economies, so that, the LRATC curve begins to rise. To these internal diseconomies, external diseconomies also add like rise in factor prices, more spending on advertisement and propaganda, higher social costs, increased transportation costs etc., and as a result, the LRATC curve starts rising upward.

However, in terms of economies of scale and in terms of diseconomies of scale, the LRATC curve falls and rises comparatively slowly than SRATC curves because, in the long run, all costs are variable costs and very few costs are fixed and hence, the firm has more control on the variable factors. That means, in the long run, the existing factors can be worked fully and more efficiently with long term planning. Hence, LRATC curve will be more flatter compared to SRATC curve.

3. Derivation of LRMC

- ☆ Like LRATC curve, LRMC curve is also derived from the summation of SRMC curves of different plant sizes. LRMC curve connects the C, D, E, F, G points of the SRMC curves of the considered five plants (Figure 15.10). The LRMC curve intersects the LRATC curve at its lowest point of $SRATC_3$ plant at minimum point E, where equilibrium output OQ is produced by the firm. So, at equilibrium level of output OQ, $LRATC=LRMC=SRATC_3=SRMC_3$.
- ☆ Thus, there exists the usual relation between LRATC and LRMC curves. To the left of 'E', LRATC > LRMC and to its right, LRATC < LRMC and at optimum output, LRATC=LRMC.

4. Shape of LRMC Curve

LRMC curve is flatter compared to SRMC. In the long run, LRMC curve falls and rises slowly than SRMC curve because, in the long run, due to long run planning of production, the resources are used efficiently. The LRMC curve bears usual relation with the LRATC. LRMC curve first falls and is below the LRATC curve and gradually rises and cuts LRATC curve at its lowest point 'E' and later, it is above LRATC curve throughout its length (Figure 15.10). So, when LRMC is below LRATC, LRATC falls, when LRATC is constant LRMC is constant and when LRMC is above LRATC, LRATC rises.

5. Optimum Size Firm

In the Figure 15.10, it is evident that, the optimum output OQ is produced by employing plant 3 with $SRATC_3$ and $SRMC_3$ and it infers that, the plant 3 is of 'optimum size' to derive the optimum output in the long run production programme at least possible LRATC. So, the optimum size of firm refers to the ideal size of the firm, where the firm can produce the output at least possible LRATC. To be more specific, the optimum size of the firm is the ideal size of plant that the firm should use to produce the output at lowest possible LRATC. An extensive research on the concept of optimum size firm was done by E.A.G. Robinson and in his words, *'an optimum firm is the one which operates at the scale at which, in the existing conditions of technology and organizing ability, has the lowest average cost of production, when all those costs which must be covered in the long run are included'*. This means that, an optimum size of the firm is one, which operates at the lowest point of LRATC to produce the desired output. When the output is produced at the lowest point on the LRATC, it implies, the resources are used efficiently in the production programme. This also infers that, the firm has expanded to its maximum capacity, so as to ensure twin benefits *viz.*, efficient use of resources and producing desired level of output at lowest possible LRATC. It is clear from the Figure 15.10 that, the point 'E' on LRATC curve is the minimum point indicating that,

- ☆ OQ is the desirable output produced in the firm at lowest possible LRATC.
- ☆ The resources are used efficiently to produce OQ output.
- ☆ The firm employs plant 3 with $SRATC_3$ and $SRMC_3$ to produce the OQ output.
- ☆ Any level of output other than OQ in the long run production programme is not efficient from the LRATC point of view. That means, if the firm produces output less than OQ or more than OQ, the LRATC will rise and it leads to low profits for the firm.
- ☆ The point 'E' further infers that, the plant, whose minimum SRATC (less than compared to SRATC of all other plants in the firm) coincides with the minimum LRATC of the firm refers to optimum plant size. This plant size should be employed by the firm to derive the benefit of minimum LRATC and in the Figure 15.10, it refers to plant 3 with $SRATC_3$ and $SRMC_3$. Here, it is important to distinguish between 'optimum plant size' and 'optimum use of plant capacity'. The optimum plant size gives the optimum output OQ (Figure 15.10) *i.e.*, with reference to plant 3 at lowest LRATC. That means, when plant size is optimum, it implies, the SRATC of the plant lies tangent to LRATC at optimum level of output OQ. For any given line of production, there will be only one optimum level of output and it is produced at lowest point on LRATC and that corresponding plant size refers to 'optimum plant size'. Regarding 'optimum use of plant capacity', we can say that, a plant can be put into use at different levels of its capacity to produce different levels of output. For example, in the Figure 15.10, the optimum capacity of plant 3 is shown at point E, where its SRATC is minimum. For plant 2, the optimum capacity is at point E_3, but the plant is used less than its optimum capacity to produce OQ_4 output with reference to point E_4 on $SRATC_2$. So, for plant 3, the optimum rate of output is point E, where its SRATC is minimum. Similarly, for plant 2, the optimum rate of output is E_3. That means, the optimum plant size refers to volume of output, whereas, the optimum use of plant capacity of plant refers to rate of output. So, a firm in the long run, may have optimum plant size, but may or may not be using it to optimum capacity. This is shown in the Figure 15.10, as it is economical for the firm to employ plant 3 to produce output OQ rather than plant 2.
- ☆ The output produced by employing the optimum plant size is the least cost output or optimum output of the production programme, which is nothing but, OQ with reference to plant 3. So, an optimum plant size is one, which produces optimum output or least cost output in the long run production programme with lowest LRATC.

Factors Influencing Optimum Plant Size

The optimum plant size, slope and position of LRATC curve of a firm are influenced by both internal economies and diseconomies of scale and external economies and diseconomies of scale. As discussed earlier, the downward (negative) slope of LRATC is influenced by the economies of large scale through enjoying increasing returns to scale and the upward (positive) slope of LRATC is influenced by the diseconomies of large scale through experiencing diminishing returns to scale. As shown in the Panel A of the Figure 15.12,

- ☆ The movement from point A to point E infers that, the firm is experiencing internal economies of large scale through enjoying increasing returns to scale.
- ☆ At point E, where LRATC is relatively more flat, it implies, the firm is experiencing constant returns to scale.
- ☆ The movement from point E to point B infers that, the firm is influenced by internal diseconomies of large scale through experiencing diminishing returns to scale.
- ☆ When LRATC is at minimum *i.e.*, at E, it implies that, the firm has achieved optimum size in producing desired level of output through efficient use of resources through enjoying internal economies of large scale and further infers that, the internal diseconomies of scale have not yet started accruing to it.
- ☆ So, due to internal economies and diseconomies of scale, we move along the same LRATC curve.

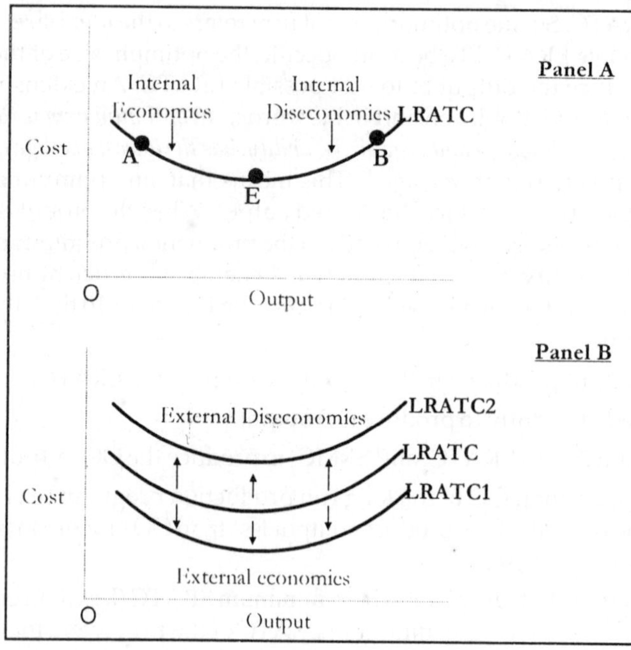

Figure 15.12: Economies and Diseconomies of scale and LRATC curve.

With reference to external economies and diseconomies of scale, the position of LRATC itself changes. When the firm enjoys external economies of scale, the LRATC curve of the firm shifts downward as $LRATC_1$ (Panel B of the Figure 15.12) indicating that, due to external economies, the LRATC of production of the product by the firm will fall implying more profits to the firm. Conversely, when the firm experiences external diseconomies of scale, the LRATC curve of the firm shifts upward as $LRATC_2$ indicating that, due to external diseconomies, the LRATC of production of the product by the firm will rise implying less profits to the firm.

It is interesting that, the optimum size of the firm varies across the industries. In case of agricultural sector, the optimum size of the firm will be smaller (Figure 15.13), as the diseconomies of scale will set in earlier in the firm compared to manufacturing sectors like computers, automobiles, steel industries etc. That means, at lower levels of output only, the minimum point of LRATC is reached in case of agricultural sector. However, in case of manufacturing industries, the minimum point of LRATC will be reached at comparatively higher levels of output (Figure 15.14).

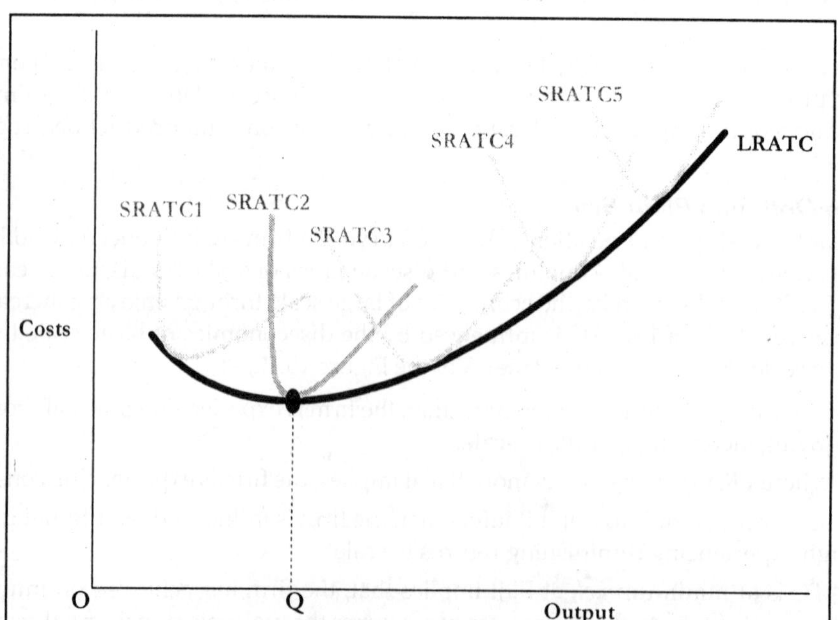

Figure 15.13: Firm operating under 'Agricultural sector' showing minimum LRATC at small scale output.

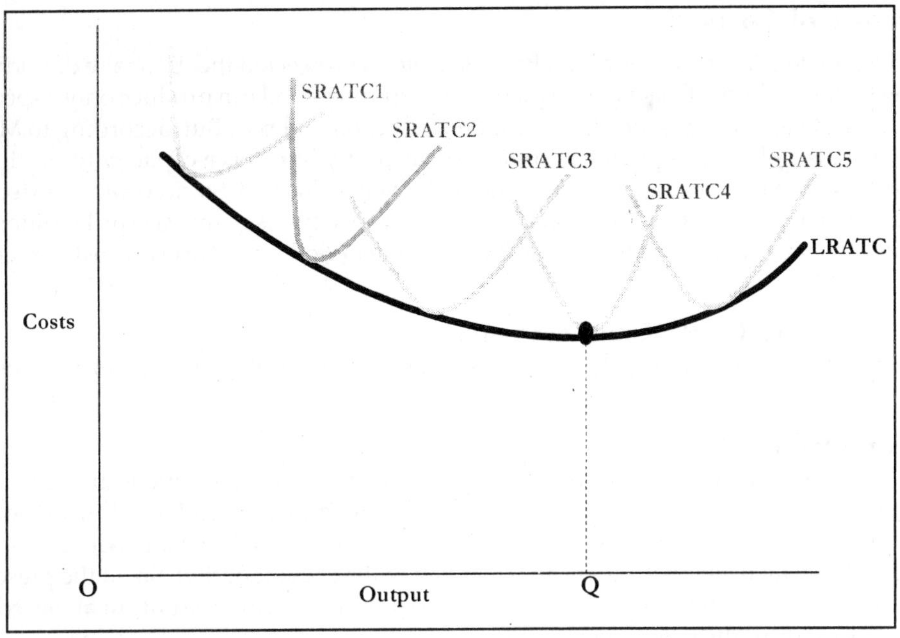

Figure 15.14: Firm operating under 'Manufacturing sector' showing minimum LRATC at large scale output.

6. Relation between LRATC and SRATC Curves

From the above discussion, we can formulate the following relations between LRATC and SRATC curves:

- ✰ The LRATC curve of the firm lies tangent to the SRATC curves of different plants and it cannot intersect any SRATC of any plant.
- ✰ When LRATC curve of the firm is U shaped, it will be tangent to minimum SRATC of an optimum plant size and for other plants, the LRATC will be tangent either at decreasing portion or increasing portion of SRATCs. As shown in the Figure 15.10, the LRATC is tangent to minimum $SRATC_3$ of plant 3 and for plant 2, the LRATC is tangent to $SRATC_2$ at its (LRATC) decreasing portion. This infers that, the output OQ can be produced most cheaply by installing plant 3 rather than plant 2.
- ✰ The SRATC curve can lie wholly 'above' the LRATC curve with no tangency point. That means, LRATC will always equal to or be less than SRATC.
- ✰ With fixed unit costs of inputs, if the production programme experiences constant returns to scale, then at the minimal level of the SRATC curve, we have SRATC = LRATC = SRMC = LRMC. So, the LRATC and SRATC will intersect at their common minimum values. If the production programme experiences increasing returns to scale, the minimum of the SRATC curve is to the right of the point of tangency between the LRATC and the SRATC curves, where LRTC = SRTC, LRATC = SRATC and LRMC = SRMC. That means, long run minimum will occur at a lower level of output than SRATC. This is because, there are economies of scale that have not been exploited. So, in the long run, a firm could always produce a quantity at a price lower than minimum SRATC simply by using a larger plant. If the production programme experiences diminishing returns to scale, the minimum of the SRATC curve is to the left of the point of tangency between LRATC and SRATC, where LRTC = SRTC, LRATC = SRATC and LRMC = SRMC. That means, minimum SRATC occurs at a lower output level than minimum LRATC because, a firm could reduce average costs by simply decreasing the size or its operations. This implies, if the production programme experiences increasing or diminishing returns to scale, then the minimum SRATC does not equal minimum LRATC.
- ✰ With fixed unit input costs, a firm that is experiencing increasing (diminishing) returns to scale and is producing at its minimum SRATC can always reduce LRATC by expanding (reducing) the use of the fixed input.

7. Relation between LRMC and SRMC Curves

Like LRATC and SRATC relationship, no such interesting relationship exists between LRMC and SRMCs of different plants. The only interesting aspect is that, when LRATC is minimum, the SRMC of the optimum plant size equals the LRMC.

It is essential to note that, the above relationships between short run and long run cost curves assume that, the firm is using the optimal level of capital for the quantity of output produced. If not, then the SRATC curve would lie 'wholly above' the LRATC and would not be tangent at any point.

II. Modern Theory of Costs

Economists like Stigler, Andrews, Florence, Freidman etc., has objected the 'U' shape of traditional per unit cost curves. According to the Traditional theory, each plant in a short run period can produce only a specific level of output. If more than this output has to be produced, the cost of production will rise. But, according to Modern Economists, whenever a firm installs a plant, this plant has some more capacity or reserve capacity to produce in addition to a specific level of output, such that, higher outputs are possible even at low cost. So, according to Modern theory, the per unit cost curves are explained based on the concept of reserve capacity. As in the case of Traditional theory, Modern theory also studies two types of cost curves on the basis of time period *viz.*, short run cost curves and long run cost curves.

1. Modern Theory – Short Run Per Unit Cost Curves

As in Traditional theory, the Modern theory also studies four types of short run per unit cost curves *viz.*, AFC, SRAVC, SRATC and SRMC.

(a) Modern Theory–AFC Curve

According to Modern theory, a firm's fixed costs consists of same items of expenditure as in Traditional theory such as, depreciation, interest on fixed capital, land tax, salaries of permanent staff etc. But, at the time of installation of plant or machinery, the firm takes into consideration the future demand of the product or output. So, considering the future demand, the firm will install the higher capacity plant in the present period. So, in the present period, the firm will have some reserve capacity and this facilitates the firm to produce more of the output at low cost. The firm should have some reserve capacity for the following reasons:

- ✰ To meet the seasonal and cyclical fluctuations in demand for the product. If demand increases, the reserve capacity can be utilized to produce more output at low cost (economies of scale).
- ✰ Reserve capacity will give greater flexibility to the firm at times of power failures, repairs etc., thereby, the production process will be ensured smoothly.
- ✰ The firm can compete out its rivals for earning significant share in the market.
- ✰ The firm can improve the quality and standards of the products suiting to the tastes and preferences of the customers.
- ✰ Having the machinery with sufficient reserve capacity is more desirable compared to large number of smaller machinery, as it saves more labour, power and other resources.

So, in view of changes in future demand, the firm will install a higher capacity plant than warranted by the present demand. This is so because, in the future, if the demand for the product increases, the firm can immediately rise to the occasion to increase the production of the product. So, the firm will maintain certain reserve capacity regarding its plant size. Of course, not only in plant size, but also in terms of land, buildings, managerial resources etc.

So, considering the reserve capacity of the plant, the AFC curve will be as shown in the Figure 15.15. Since, the firm installs the machinery with reserve capacity, it can increase the output beyond OB units. In normal cases of demand, it

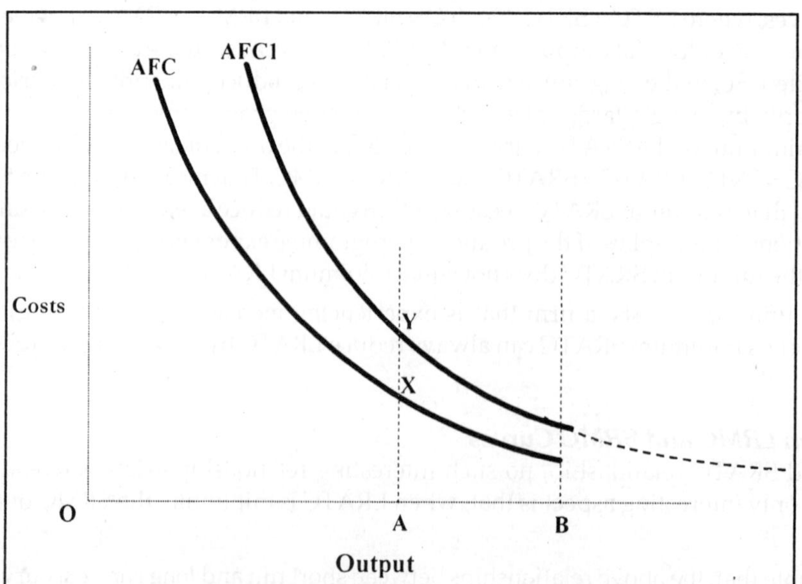

Figure 15.15: Modern theory – AFC curve with reserve capacity.

produces the normal output upto OB units by using the machinery with normal capacity. But, if the future demand for the product increases in the market, the firm will employ the machinery with certain reserve capacity, thereby, it increases the output beyond OB. It is important that, with the addition of new machinery having considerable reserve capacity, the firm's AFC increases from point X to Y on the boundary line A, but the important aspect is that, the firm can increase the production beyond OB to meet the anticipated increase in demand for the product. Though the AFC increases for the product initially, but with the increase in production beyond OB, the AFC finally decreases. This is shown by the new AFC curve AFC_1, which implies machinery with reserve capacity has been installed in the plant to meet the future demand. In general, the firm increases the output to meet the future demand in two ways *viz.*,

- ☆ By paying over time wage rates to the workers.
- ☆ By installing small machines and this is shown by AFC_1 curve (shifted towards right).

(b) Modern Theory–SRAVC and SRMC Curves

According to Modern theory, both SRAVC and SRMC curves have saucer shape or bowl shape (Figure 15.16) with a flat stretch over a range of output rather than a 'U' shape, as discussed under Traditional theory. The reasons for the saucer shaped or bowl shaped SRAVC and SRMC curves are as follows:

- ☆ SRAVC curve first falls upto point A and similarly, SRMC also first falls and later rises, but lies below the SRAVC upto point A, where they both are equal. The reasons for the fall of SRAVC is the reduction in costs due to better utilization of variable factors (say, labour) and increase in output due to better utilization of fixed factors. Due to better utilization of labour, the wastage of resources is greatly reduced, thereby, the whole plant is better utilized.

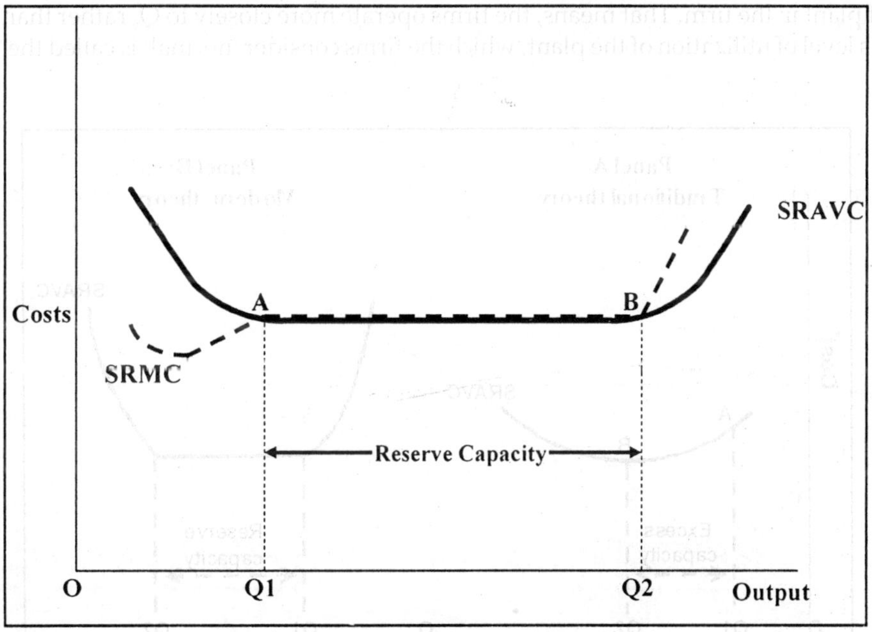

Figure 15.16: Modern theory – Saucer shape/Bowl shape SRAVC and SRMC curves.

- ☆ Before point 'A', though SRMC rises, still SRAVC falls upto point 'A' because, SRMC is related to only one additional unit of output, but SRAVC is related to all the units of output produced.
- ☆ Due to better utilization of whole plant for a wider range of output produced, the firm operates as per the law of constant returns. Hence, for a wide range of output Q_1Q_2, the SRAVC and SRMC do not fluctuate and they are equal and lie parallel to X- axis. So, the flat stretch corresponds to reserve capacity of the plant, where both SRAVC and SRMC are constant per unit of output.
- ☆ The another reason for the constancy of SRAVC and SRMC over a wide range of output Q_1Q_2 is that, during this period, the fixed factor is divisible and till the complete exhaustion of use of all the fixed factors, the SRAVC tend to remain same. This is because, there is no need for the firm to depart from the optimal combination of factors that are used in the production programme. Thus, for a long range of output Q_1Q_2, SRAVC remains same or constant. Since, there is no change in optimal combination of factors, during this large range of output Q_1Q_2, the SRMC also remains same. Hence, SRAVC and SRMC are equal and are constant per unit of output during the range Q_1Q_2 output. Thus, the firm continue to produce within the range of Q_1Q_2 output and it refers to 'Reserve capacity' of the plant.

☆ After point B, both SRAVC and SRMC curves start rising *i.e.*, when the firm departs from the normal or reserve capacity of the plant in order to obtain further higher levels of output beyond OQ_2, it leads to both higher SRAVC and SRMC. The increase in costs may be due to overtime operations of old and less efficient machinery leading to frequent breakdowns, wastage of raw materials, reduction in labour productivity, increase in labour costs due to overtime operations etc. In the rising portion of the curves beyond point B, the SRMC curve rises quickly compared to SRAVC curve, SRMC is related to only one additional unit of output, but SRAVC is related to all the units of output produced.

So, according to Modern Economists, both SRAVC and SRMC curves remains flat over a certain range of output. That means, both SRAVC and SRMC are equal and constant over a certain range of output (Q_1Q_2). This Q_1Q_2 reserve capacity gives more flexibility to the business firm. This 'Reserve capacity' concept of Modern theory is different from the 'Excess capacity' concept of Traditional theory in explaining the shape of SRAVC curve. The Traditional theory assumed only one point of optimality (lowest point) on the SRAVC curve to produce a single level of output and there is no flexibility concept with reference to SRAVC curve. As shown in the Panel A of the Figure 15.17, if the firm produces OQ_2 output with reference to point B on the SRAVC, it is the point of optimality, as point B is the minimum point of SRAVC curve. So, Q_1Q_2 is the excess capacity over OQ_1 with reference to point A on the SRAVC. That means, if the firm moves from point B to point A on the SRAVC, it increases the unit cost of producing output.

But, according to Modern theory, the range of output Q_1Q_2 represents 'Reserve capacity' of the firm and over this range of output, the SRAVC remains same, unlike 'Excess capacity' concept of Traditional theory. So, the firm can produce any amount of output in the given range Q_1Q_2, where the SRAVC remains constant (Panel B of the Figure 15.17). In general, the firms operate between two-thirds and three-fourths of the reserve capacity, so as to consider the normal utilization of plant in the firm. That means, the firms operate more closely to Q_2 rather than Q_1 of the range of reserve capacity. This level of utilization of the plant, which the firms consider 'normal' is called the 'load factor' of the plant.

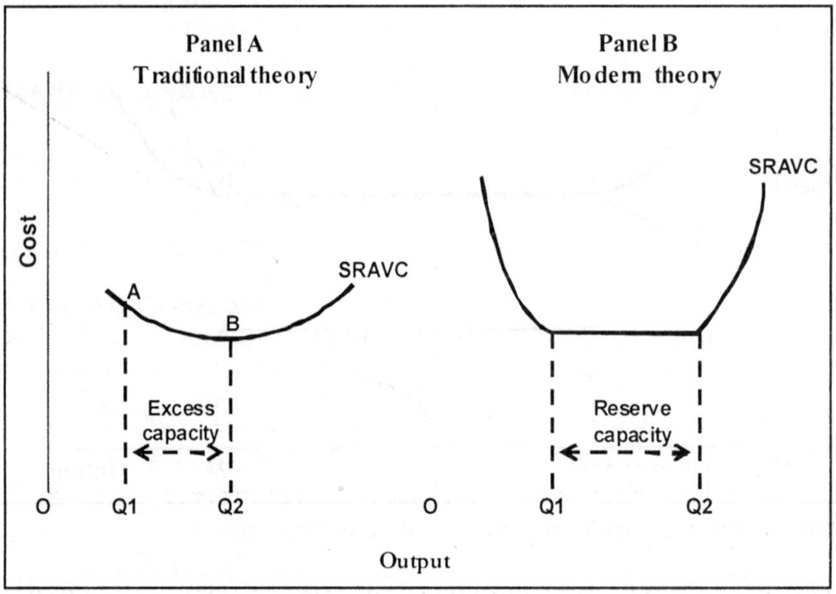

Figure 15.17: Reserve capacity (Modern theory) Vs Excess capacity (Traditional theory) of SRAVC curve.

(c) Modern Theory–SRATC Curve

The SRATC curve is obtained by adding SRAFC and SRAVC curves at each level of output. The SRATC curve falls continuously upto OQ_2 level of output, at which the reserve capacity of the plant is fully exhausted (Figure 15.18). Upto OQ_2 level of output, SRATC falls continuously due to the fact that, SRAFC is a rectangular hyperbola, which falls continuously and SRAVC first falls and then becomes horizontal within the range of reserve capacity of output. Hence, the continuous decline in SRAFC brings the downward trend to SRATC upto OQ_2 level of output. But, beyond OQ_2 level of output, SRATC curve rises upward and this is due to rise in SRAVC, which outweighs fall in SRAFC.

The SRMC curve intersects the SRATC at its lowest point M. This point 'M' will be to the right of point 'B' or Q_2, at which the flat stretch of SRAVC ends. From point M, the SRATC curve rises steeply because, SRAVC starts rising upward steeply from point B and the rise in SRAVC now outweighs slowly falling AFC.

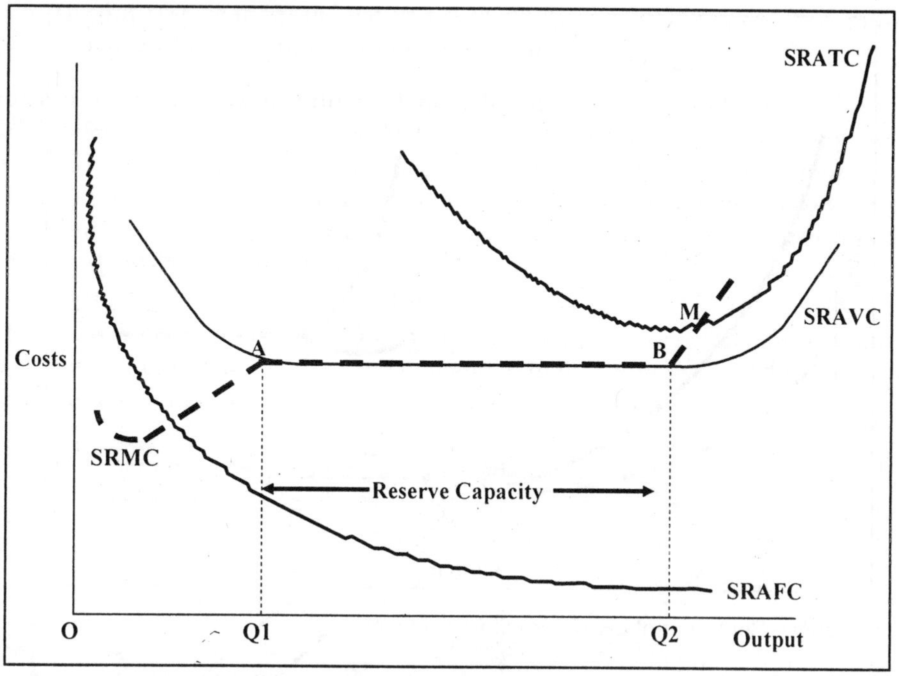

Figure 15.18: Modern theory – Long run per unit cost curves.

2. Modern Theory – Long Run Per Unit Cost Curves

According to Modern theory, the LRATC curve is 'L' shaped rather than 'U' shaped. In the beginning, the LRATC curve falls rapidly with increase in output, but after a point, the curve remains flat or may slope gently downwards towards right side. The following are the important reasons:

(a) Production and Managerial Costs

In the long run, since all the costs become variable, production and managerial costs should be studied to analyze their influence on output, so that, average cost can be worked out. As output increases, production cost/unit of output will fall, while managerial cost/unit of output may rise. But, the fall in production costs outweighs the rise in managerial costs, so that the LRATC falls with increase in output. This is discussed below.

☆ ***Production costs:*** If the firm increases its production, the production costs fall steeply at the beginning and then gradually. This is because, in the beginning, the firm enjoys substantial technical economies of large scale production, thereby, the production costs fall steeply. But, after a certain level of output, all or most of the economies have been achieved by the firm, thereby, the firm reaches the Minimum Optimal Scale or Minimum Efficient Scale (MES), beyond which, economies of scale will not be achieved at a significant level. The MES is best defined as the scale of production, where the internal economies of scale have been fully exploited. It corresponds to the lowest point on the LRATC curve and it refers to an output range over which a business achieves productive efficiency. So, the quantity of output that achieved at the lowest possible LRATC is termed as MES. This is shown as output OQ at lowest possible LRATC, QA in Panel B of the Figure 15.19. It is important that, the MES is not a single output level rather, it comprises a range of output levels, where the firm achieves constant returns to scale and has reached the lowest feasible cost per unit in the long run. But, with the given level of technology, sometimes, the firm continue to enjoy technical economies larger than MES due to the following reasons:

- Division of labour and improvement in skills of labour.
- Repairing of the machinery instead of purchasing new machinery (no huge extra costs).
- Self-production of raw materials and equipment instead of purchasing new items.

☆ ***Managerial costs:*** As the firm expands, new managerial techniques will be employed and separate divisions will be started and hence, employment increases. Upto a certain plant size, the managerial costs will decline, but at very large scale of output, the managerial costs will rise.

To sum up, production costs fall smoothly at very large scale, but managerial costs rise only at very large scales of output. According to Modern Economists, the fall in production costs outweighs the rise in managerial costs, so that, LRATC curve fall smoothly or becomes flat at very large scales of output, thereby, giving 'L' shape to the curve instead of 'U' shape.

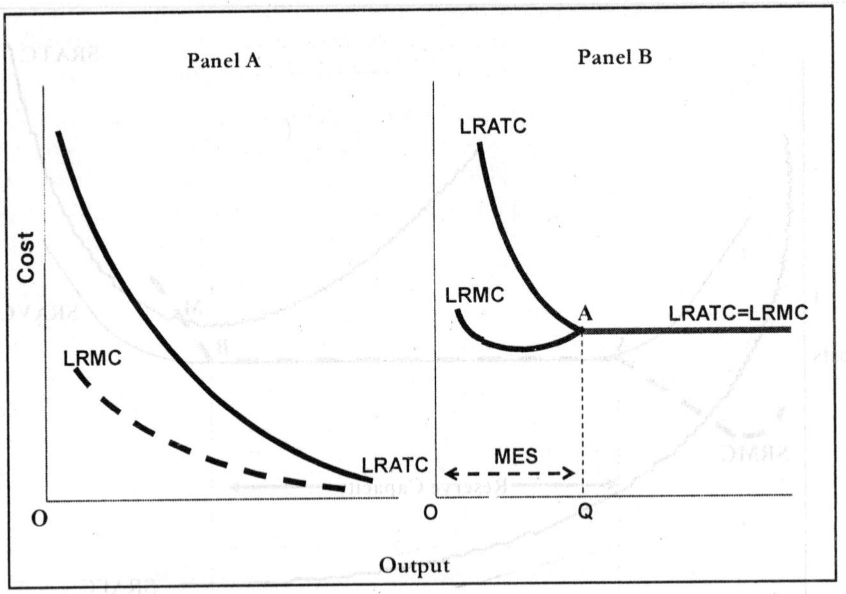

Figure 15.19: Modern theory – Behaviour of LRATC and LRMC curves.

So, to derive the LRATC curve, let us assume three SRATC curves *viz.*, $SRATC_1$, $SRATC_2$ and $SRATC_3$ representing three plants with the same technology (Figure 15.20). Each SRATC curve includes production costs, managerial costs, other fixed costs etc. Since fall in production costs outweighs rise in managerial costs at large scales of output, the SRATC curves of successive plants falls downward. The points A, B and C on successive SRATC curves represent the MES of output of each plant. By joining all such points A, B and C of the three SRATC curves, we get downward sloping LRATC curve towards right. The important differences between Modern theory LRATC curve and Traditional theory LRATC is that, as per Modern theory, the LRATC curve

- ✰ falls smoothly from left to right
- ✰ it does not rise upwards
- ✰ it does not envelope SRATC curves, but intersects them at the MES of output of each plant.

According to some Modern Economists, at all scales of output, LRMC curve lie below LRATC curve (Panel A of the Figure 15.19). If there is MES scale of plant, upto the point of MES, LRMC curve lie below LRATC curve. But, after MES of output, the LRATC remains constant and lies parallel to X-axis and LRMC curve coincides with LRATC curve (Panel B of the Figure 15.19). However, in Traditional theory, there is altogether a different picture with reference to the

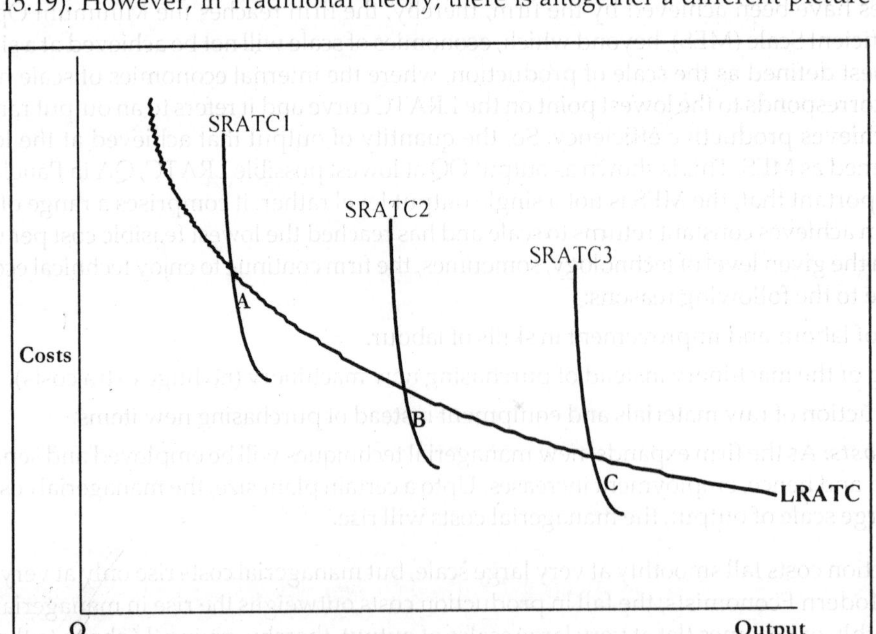

Figure 15.20: Modern theory – Declining production costs outweighs rising managerial costs giving 'L' shape to LRATC curve

behaviour of LRATC and LRMC *i.e.*, the LRATC is an envelope curve and LRMC curve intersects the LRATC curve at minimum point. But, empirical evidences proved that, the Modern theory of costs is more valid over Traditional theory.

(b) Technical Progress

The Traditional theory assumed the constancy of technology in explaining the behaviour of cost curves. But, technology changes in real life and hence, Modern theory considers that, with the improvement in technology or techniques of production, the cost of production declines in the firm and with the advent of new technology, the firm shifts from one plant to another plant, so as to enjoy the benefits of economies of large-scale. The same is explained through the Figure 15.21.

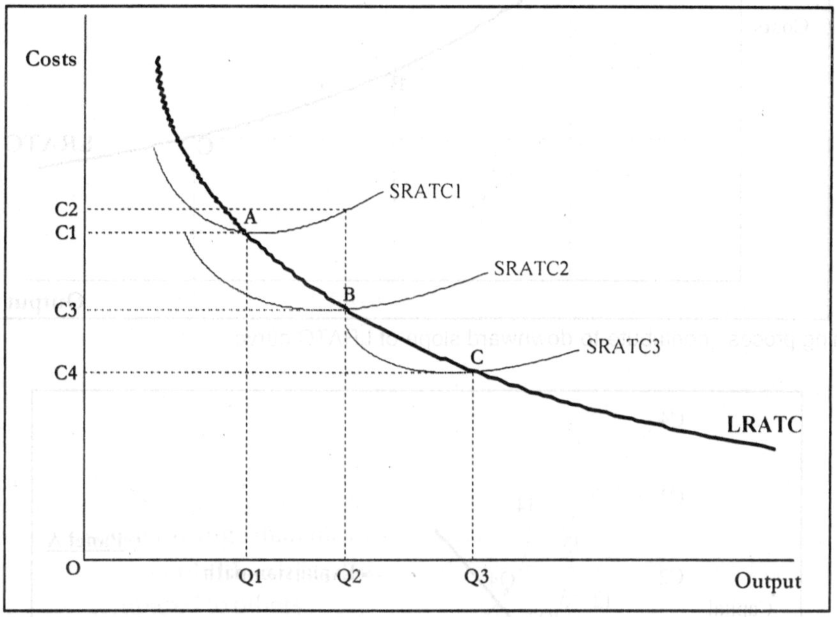

Figure 15.21: Technical Progress contributing to 'L' shape of LRATC curve.

Let us assume, the firm is operating on plant 1, where point A on $SRATC_1$ infers that, the firm is producing OQ_1 level of output at OC_1 cost. If the demand for the product is increased in the market, now the firm produces OQ_2 quantity of output on plant 1 at a higher cost of OC_2. But, with the advent of modern technology, the firm establishes new plant with $SRATC_2$, thereby, it enables the firm to produce same OQ_2 output at lower cost OC_3. That means, due to technological progress, the firm utilizes modern plant $SRATC_2$ to produce same output at low cost OC_3. Similarly, the firm utilizes plant size with $SRATC_3$ due to improvement in technology to produce still higher output OQ_3 at lower cost OC_4. So, by connecting the minimum points of $SRATC_1$, $SRATC_2$ and $SRATC_3$ *i.e.*, A, B and C respectively, we can derive the downward falling LRATC curve indicating that, the technological progress contributes much to the fall of LRATC in the firm. This LRATC will not rise upwards, as the evidences of diseconomies is much less certain.

(c) Learning Process

The experiences gained by the entrepreneur in the production programme enable him to formulate and take the business decisions in a scientific manner and this contributes to produce more output at low cost. If the manager produces an output a particular cost level, in the next instance or period, he tries to produce it at still lower cost and in lesser time *i.e.*, with more efficiency. Thus, he aims to maximize the resource use efficiency in the production activities through the learning by doing process. So, this process makes the LRATC curve to slope downwards from left to right direction and it is 'L' shaped and will not slope upward, indicating that, more output is produced at lower cost (Figure 15.22).

15.4. Derivation of LRTC Curve from Expansion Path

Having studied the concept of expansion path in Chapter 14, which indicates the points of optimal factor combinations at different output levels through employing isoquants and iso-cost lines, we can derive the LRTC from the expansion path. The same is explained through Figure 15.23, where Panel A shows the derivation of expansion path and Panel B shows the derivation of LRTC curve from the expansion path.

In Panel A, let us assume, the farmer employs two factors *viz.*, labour and capital in the production programme. Four isoquants I_1, I_2, I_3 and I_4 indicating different levels of output *viz.*, 100 units, 200 units, 300 units and 400 units respectively are drawn. Considering the different cash outlays, four iso-cost lines *viz.*, C_1L_1, C_2L_2, C_3L_3 and C_4L_4 are

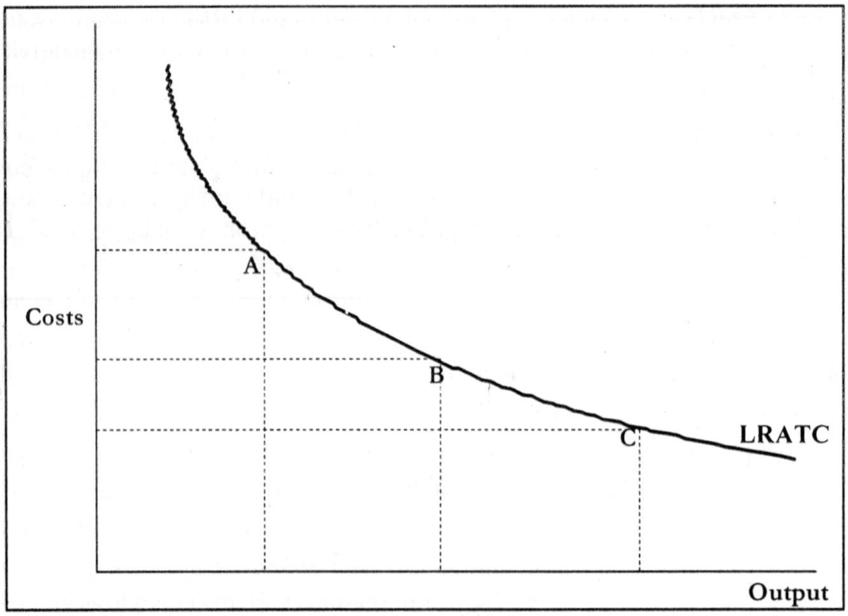

Figure 15.22: Learning process contribute to downward slope of LRATC curve.

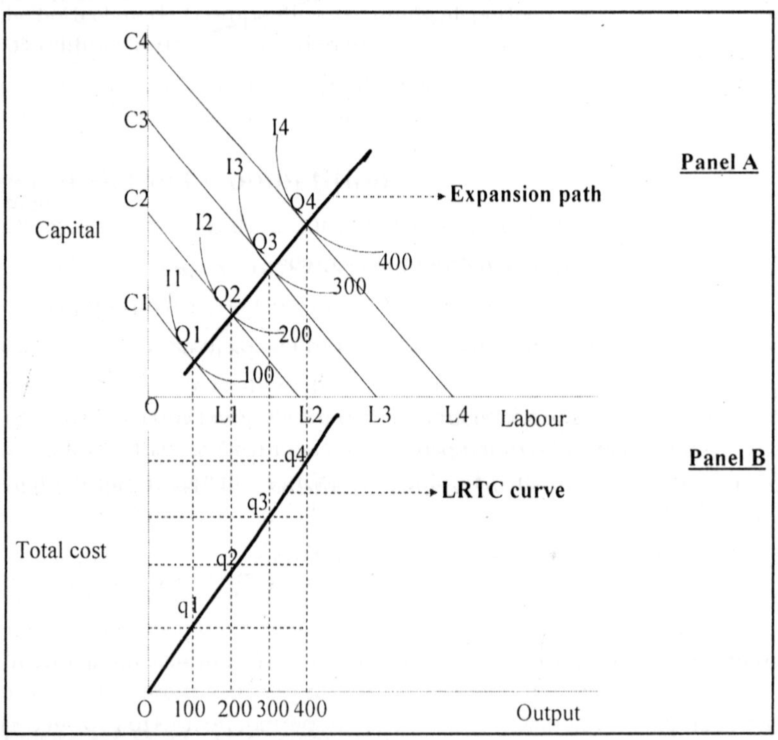

Figure 15.23: Derivation of LRTC curve from expansion path.

drawn and these iso-cost lines are tangent to different isoquants. The iso-cost lines are parallel to each other indicating that, the factor prices are constant. Considering the isoquant I_1 for 100 units of output, the point of optimum factor combination is at point Q_1, where the isoquant I_1 and iso-cost line C_1L_1 are tangent to each other. Similarly, with respect to isoquant I_2, I_3 and I_4 and iso-cost lines, C_2L_2, C_3L_3 and C_4L_4, the points of optimum factor combinations are Q_2, Q_3 and Q_4 respectively. By connecting these points of optimum factor combinations *i.e.*, Q_1, Q_2, Q_3 and Q_4, we can derive an Expansion path 'OE'. This expansion path infers that, by using different factor combinations at given factor prices, we can produce different levels of output. This guides us to derive LRTC from the expansion path.

To derive LRTC curve from the expansion path, in Panel B, we consider total cost (quantity of factors employed x prices of the factors) on the Y-axis and output levels on the X-axis. The points of optimum factor combination at different levels of output *i.e.*, Q_1, Q_2, Q_3 and Q_4 of Panel A are extended downwards to the panel B as points, q1, q2, q3 and q4 respectively and if we connect these points, we get an upward sloping LRTC. This infers that, with increase in output in the production programme, the LRTC also increases and *vice versa*.

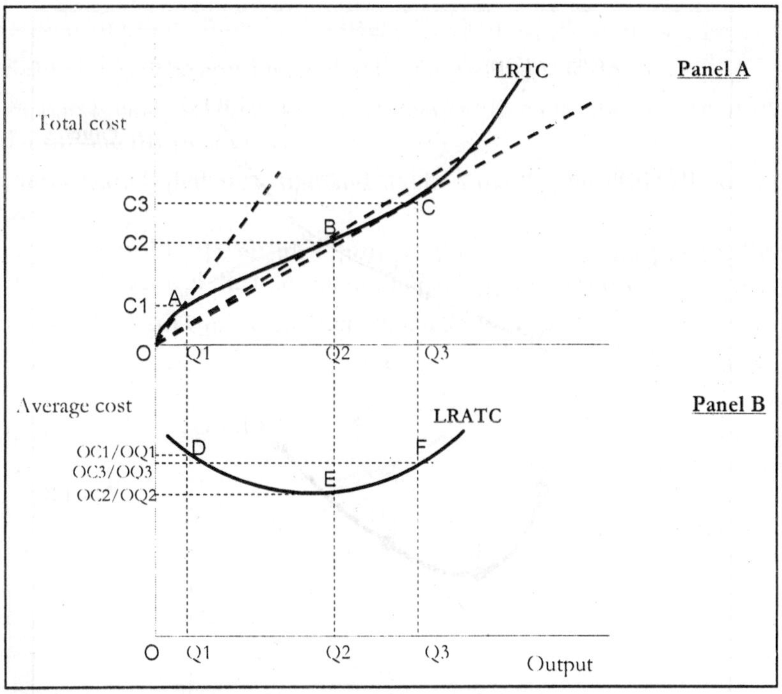

Figure 15.24: Derivation of LRATC curve from LRTC curve.

15.5. Derivation of LRATC Curve from LRTC Curve

As we can compute LRATC from LRTC, we can derive LRATC curve from LRTC curve and the same is shown through Figure 15.24. We know, LRATC is the LRTC divided by the output. 'LRTC' is the long-run total cost curve drawn in the Panel A and it starts from the origin, as all the costs are variable costs in the long run production programme. Let us assume three points A, B and C on the LRTC curve, by drawing tangents to these points from the point of origin. At point A on the LRTC curve, we can say that, to produce OQ_1 quantity of output, OC_1 is the total cost incurred in the production programme. So, the LRATC incurred on per unit of output with reference to point A is given by OC_1/OQ_1. Similarly, at point B on the LRATC, we can say that, to produce OQ_2 quantity of output, OC_2 is the total cost incurred in the production programme. So, the LRATC incurred on per unit of output with reference to point B is given by OC_2/OQ_2. At point C on the LRATC, we can say that, to produce OQ_3 quantity of output, OC_3 is the total cost incurred in the production programme. So, the LRATC incurred on per unit of output with reference to point C is given by OC_3/OQ_3.

To derive the LRATC curve for the production programme at different levels of output, consider Panel B. The Panel A helps to compute LRATC at different levels of output. So to derive the LRATC curve, consider the LRATC values as calculated above from Panel A on the Y-axis and output of the production programme on the X-axis. So, this helps us to plot the values of LRATC at different levels of output. For this, the tangency points A, B and C of Panel A were extended downwards to Panel B as D, E and F respectively and if they are connected, gives the 'U' shaped LRATC curve.

It is interesting that, the tangent drawn to point A on LRTC is more steeper followed by points C and B, thereby, LRATC is more with reference to point D followed by points F and E. So, the LRATC curve first falls, reaches minimum and then rises and the reasons for these are explained earlier under Figure 15.10.

15.6. Derivation of LRMC Curve from LRTC Curve

There are two ways of deriving the LRMC curve from LRTC curve.

Method 1: From the Slope of LRTS

As we can compute LRMC from LRTC, we can derive LRMC curve from LRTC curve and the same is shown through Figure 15.25. We know, the slope of the LRTC curve indicates the LRMC curve. 'LRTC' is the long-run total cost curve drawn in the Panel A and it starts from the origin, as all the costs are variable costs in the long run production programme. Let us assume four points A, B, C and D on the LRTC curve, by drawing tangents to these points from the point of origin. At point A on the LRTC curve for output level OQ, the slope of the tangent is steeper, which means that,

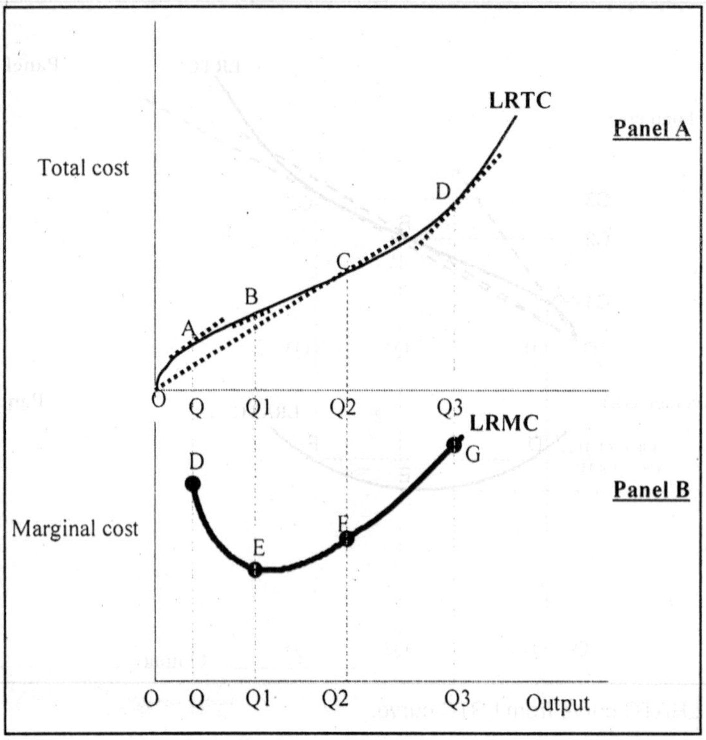

Figure 15.25: Derivation of LRMC curve from LRTC curve.

the LRMC to produce OQ output is high. But, the slope of the tangent at point B is the smallest, so that, the LRMC is the lowest at OQ_1 level of output. The slope of the tangent drawn at point C on the LRTC curve is steeper than the tangent at point B and it implies that, LRMC at output level OQ_2 is rising. A close examination of the slope at point C on the LRTC in this Figure 15.25 and point B of the Figure 15.24 reveals that, the slopes are equal, thereby, at these slopes, LRMC will be equal to or intersect LRATC at OQ_2 level of output. The slope of the tangent drawn at point D on the LRTC curve is very steep indicating that, LRMC curve rises steeply at OQ_3 output level.

To derive the LRMC curve, in Panel B, the LRMC computed from the slope of LRTC curve at points A, B, C and D are taken on Y-axis and output is taken on X-axis. The points A, B, C and D considered on LRTC curve of Panel A at which we computed LRMC values were extended downwards to Panel B as D, E, F and G and if they are joined, it gives a 'U' shaped LRMC curve.

It is interesting that, the tangent drawn to point D on LRTC (Panel A) is more steeper followed by point A and C, thereby, LRMC is more with reference to point G followed by points D and F. The slope of the tangent drawn at point B on LRTC is smallest, thereby, LRMC at point E is smallest, inferring it is the minimum point of LRMC curve. As we noticed at point F in this Figure 15.25, the slope of LRMC is equal to slope of LRATC at point E (Figure 15.24) with reference to OQ_2 level of output, thereby, below point F of LRMC (*i.e.*, below OQ_2 output level), the LRATC>LRMC and above point F of LRMC (above OQ_2 output level), the LRMC>LRATC.

Method 2: Summation of SRMCs

We know, MC curve is important for taking cost-output decisions both in short run and long run production programmes and this guides us to compute the LRMC from the SRMCs of different plants considered in short run production programme and the same is explained through the Figure 15.26.

We know, LRATC is derived from the SRATCs of different plants of a firm and it is the envelope of a number of SRATC curves of a firm. Assume point E_1 on the LRATC curve, where LRATC is tangent to $SRATC_1$ of plant 1. This infers that, OQ_1 is the output produced in the long run with reference to LRATC curve considering the plant size with $SRATC_1$ and $SRMC_1$. With reference to the point E_1 on LRATC curve tangent to $SRATC_1$, there is another point corresponding to E_1 on the $SRMC_1$ curve *i.e.*, C. This infers that, for the production of OQ_1 output in the long run, with reference to $SRATC_1$ and $SRMC_1$, the SRMC incurred is Q_1C. So, this gives the indication that, on the LRMC curve to be derived, this point C on the $SRMC_1$ should be considered.

Similarly, assume point E_2 on the LRATC curve, where LRATC is tangent to $SRATC_2$ of plant 2. This infers that, OQ_2 is the output produced in the long run with reference to LRATC curve considering the plant size with $SRATC_2$ and $SRMC_2$. This infers that, for the production of OQ_2 output in the long run, with reference to $SRATC_2$ and $SRMC_2$, the

SRMC incurred is Q_2E_2. So, this gives the indication that, on the LRMC curve to be derived, this point E_2 on the $SRMC_2$ should be considered.

With reference to the point E_3 on LRATC curve tangent to $SRATC_3$, there is another point corresponding to E_3 on the SRMC curve *i.e.*, F. This infers that, for the production of OQ_3 output in the long run, with reference to $SRATC_3$ and $SRMC_3$ of plant 3, the SRMC incurred is Q_3F. So, this gives the indication that, on the LRMC curve to be derived, this point F on the $SRMC_3$ should be considered.

So, by joining the points C, E_2 and F on the SRMCs of different plants producing different levels of output, we can derive the LRMC. It is clear from the Figure 15.26 that, LRMC so derived is more flatter compared to SRMCs and the following are the other interesting points:

- ☆ When LRMC curve is below the LRATC curve, the LRATC curve is falling.
- ☆ At the point of intersection of LRMC curve and LRATC curve, the LRATC curve is minimum.
- ☆ When LRMC curve is above the LRATC curve, the LRATC curve is rising.
- ☆ So, the relationship between LRMC curve and LRATC curve in long run production programme is similar to the relationship between SRMC curve and SRATC curve in short run production programme.

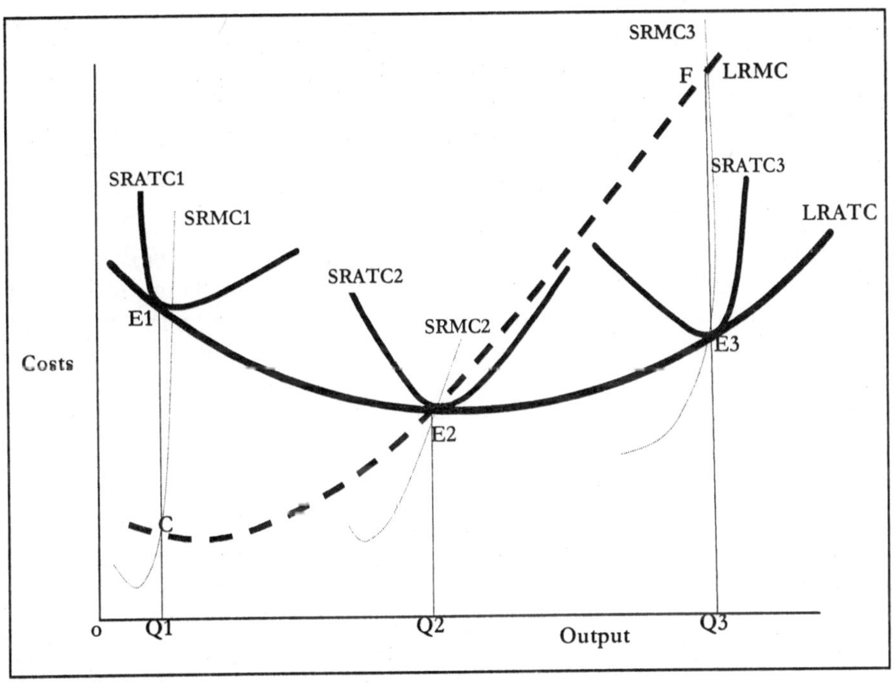

Figure 15.26: Traditional theory – Derivation of LRMC as ΣSRMCs.

15.7. Relation between Cost Function and Production Function

As studied in the Chapter 14, production function deals with the relationship output and factors employed in the production programme. Cost function deals with the functional relationship between cost and output. Since, output is a function of factors employed in the production programme, we can derive the cost function from the production function, provided, if we know the prices of the factors employed and quantity of output produced in the production programme. Say, for example, if 10 units of manure and 20 units of fertilizer are employed to produce 100 units of output and if the price of manure is Rs 50 per unit and price of fertilizer is Rs 100 per unit, then the TC of producing 100 units of output is given by (10x50) + (20x100) = Rs.2500. The following explanation clearly infers that, the production and cost functions are duals to each other.

1. TPP and SRTC

As discussed under Figure 14.4, TPP first increases at increasing rate, then at constant rate and then at diminishing rate. So, SRTC increases first at diminishing rate, then at constant rate and then at increasing rate (Figure 15.3).

2. APP and SRATC

As discussed under Figure 14.4, APP first rises, reaches maximum (at the end of zone I) and falls continuously. So SRATC first falls, reaches minimum and starts rising (Figure 15.5). So APP will have an inverted 'U' shape, where as SRATC will have 'U' shape.

3. MPP and SRMC

As discussed under Figure 14.4, MPP first increases, reaches maximum then falls, become zero and turns negative, if the dosage of a variable factor is increased indiscriminately in the production programme. So, SRMC first falls, reaches minimum and gradually rises (Figure 15.5). But, SRMC can never be zero and negative unlike MPP.

16

Economies and Diseconomies of Scale

In the earlier discussion (Chapter 14), we studied that, increasing and decreasing MRs in a long run production programme are influenced by economies of scale and diseconomies of scale respectively. Further, we also discussed that, the behaviour of cost curves in the long run are influenced by both economies and diseconomies of scales. So, it is high time now, to discuss these aspects of economies and diseconomies of scales, as it gives the true picture regarding the production behaviour of the firm or an industry as a whole.

16.1. Economies of Scale

We know, a farm-firm is an independently administered business unit (say, cattle farm) by the farmer, while an industry is made up of a number of firms producing broadly similar items or items that are connected to each other (dairy industry). Depending upon the level of output, firms and industries may be of small-scale or large-scale. As every firm or industry aims to maximize the output, they expand the business, thereby, reap profits and during this process, they are able to secure certain benefits that are not available to small firms or industries. These benefits or advantages are referred to as economies of large scale and they result in reduction of LRATC of production due to expanding the output in the production programme. So, an economies of scale refers to, when a larger output is associated with lower per unit cost of production (*i.e.*, low average cost of production) in a production process. Thus, economies of scale are the cost advantages exploited by expanding the scale of production by the firm in the long run. It is, otherwise, defined as the efficiency gained in a production process due to increase in the rate of production. It can also be defined as a state, when the number of goods or services produced goes up by a sizeable margin, but that hardly affects the cost of production and input costs remain the same. These definitions infer that, the growth of an individual firm or industry is possible only when economies of scale is attained, as it ensures two benefits *viz.*, increase in output and decline in average cost of production. In an era, where large scale private investments are flowing into the sectors, economies of scale are the goal and this will enable the firm to gain a competitive advantage in the market. This is exemplified through the Table 16.1 and Figure 16.1.

The Table 16.1 shows a simple example for a firm (sugar factory) enjoying economies of large scale. The rate of increase in output of sugar is more compared to the rise in total costs and this contributes to decline in LRATC of sugar production in the factory. As long as the LRATC is falling with increase in output of sugar, it implies, economies of scale are being exploited by the firm. This also infers that, if the firm shows increasing returns to scale, it enjoys economies of large scale production. So, as the firm increases the factor scale, total cost rises, but LRATC falls, as the percentage change in output is more than the percentage change in costs. The same is explained through Figure 16.1.

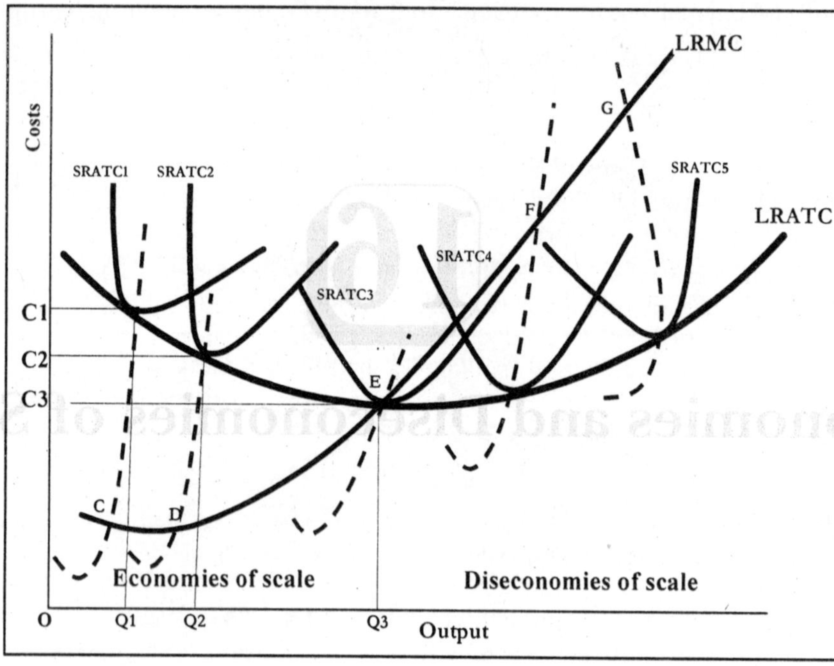

Figure 16.1: Economies of large scale – Decline in SRATC and LRATC with increase in scale.

If the sugar factory operates and expands from one scale of production to another, the output increases in the factory, as we move from $SRATC_1$ to $SRATC_2$ to $SRATC_3$. Each SRATC curve represents a given quantity of scale of factors. So, the movement from $SRATC_1$ to $SRATC_2$ to $SRATC_3$ reflect that, as the scale of factors increases, output increases and this is accompanied by the decline in SRATC. The LRATC curve is drawn enveloping the SRATC curves of different plants and it shows that, upto OQ_3 level of output, the LRATC falls with increase in scale of factors and at OQ_3 level of output, LRATC is the lowest and the firm enjoys economies of large scale.

Table 16.1: Economies of large scale in a sugar factory.

Scale	Total Cost (Rs)	Output of Sugar (Units)	LRATC (Rs/unit)
1 Capital + 1 Labour	250000	12500	20.00
2 Capital + 2 Labour	280000	16000	17.50
3 Capital + 3 Labour	368000	23000	16.00
4 Capital + 4 Labour	486400	32000	15.20
5 Capital + 5 Labour	518000	35000	14.80
6 Capital + 6 Labour	528000	44000	12.00
7 Capital + 7 Labour	533600	46000	11.60
8 Capital + 8 Labour	573750	51000	11.25
9 Capital + 9 Labour	590000	59000	10.00

The above example indicates that, if the firm experiences increasing returns to scale, it enjoys economies of large scale. This is especially true in case of large firms like Coromandel fertilizers limited, Nagarjuna fertilizers limited, Bayer India limited etc., which produce larger output at huge fixed costs. For these large scale firms, the striking aspect is, the MC involved in the production of an additional unit output is nearly zero. The lower MC will further reduce the LRATC and the firm gains more reputation from the farming community, as it can offer the product at lower price. Lower costs normally mean higher profits and increasing financial returns for the firm. From the above discussion, we can infer that, the economies of scale can be studied from the LRATC curve (Figure 16.2). If the LRATC curve is falling, it implies that, with increase in output, the LRATC of commodity is decreasing, thereby, the firm is experiencing economies of large scale. At OQ_1 level of output, the LRATC of sugar is OC_1. With increase in output to OQ_3, the LRATC of a sugar is OC_3, indicating that, at OQ_3 level of output, the LRATC is minimum for sugar, thereby, the factory is experiencing economies of scale between points A and B on LRATC curve. But, beyond OQ_3 level of output, LRATC curve rises indicating, diseconomies of scale sets in into the sugar factory.

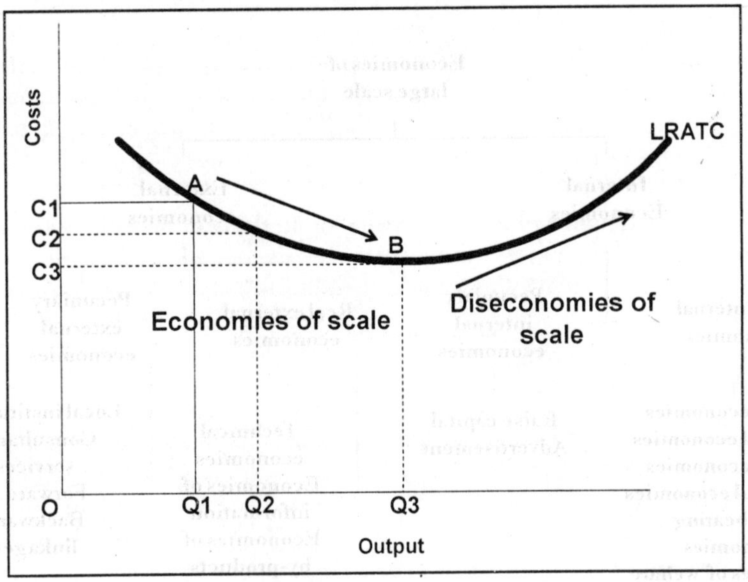

Figure 16.2: Economies and Diseconomies of scale in a sugar factory.

16.2. Importance of Economies of Large Scale

The following are the advantages for the business firm, if it enjoys economies of large scale:

- ☆ Large-scale business is possible at lower LRATC, thereby, market share of the business firm increases significantly.
- ☆ With the decline in LRATC due to economies of large scale, the firm need not cut the market price, rather, it can charge same price for the product and accept higher profit margins. For example, a sugar factory produces 1000 units of sugar at a cost of production of Rs. 500/unit. At a market price of Rs. 600/unit, the entrepreneur sells the product, thereby, he derives the profit margin of Rs. 100000. If the firm experiences increasing returns to scale, the factory produces 2000 units of sugar at a cost of production of Rs. 400. At the same market price of Rs. 600/unit, the entrepreneur sells the product, thereby he derives the profit margin of Rs. 400000. So, by charging same market price, the factory can maximize the profits through enjoying economies of large scale.
- ☆ In cases of cut-throat competition, the business firm enjoying economies of large scale can go for destroyer pricing, such that, the price quoted by the firm for the product will be lower than LRATC of other firms.
- ☆ The lower LRATC attained by the business firm is a measure of an improvement in productive efficiency and it can be passed on to the consumers in the form of lower market prices. This also gives the firm a competitive advantage both in the domestic and international markets. These advantages are recently seen in the soft ware sector. Just a few years back, high-performance personal computers are sold at very high prices, but with the experience of economies of large scale, the same are sold now at affordable prices.
- ☆ Economies of large scale serve as an incentive for business firm to grow.
- ☆ Because of reduced costs, business firms can price their products to be more profitable.
- ☆ Practicing economies of large scale allows business firms to be more competitive

16.3. Types of Economies of Large Scale

According to Alfred Marshall, there are two kinds of economies of scale. These include Internal economies of scale and External economies of scale (Figure 16.3). Internal economies of scale refers to the benefits or advantages to one particular firm, as it goes into large-scale production. External economies of scale refers to the benefits accrued to an entire industry that has been localized or concentrated in a particular area. They are discussed in-detail in the following pages.

I. Internal Economies of Large Scale

Internal economies of scale relate to the lower LRATC a single firm can obtain by growing in size itself. So, it is related to the economies of production, which accrue to the business firm through enjoying increasing returns to scale. This makes the business firm to expand the output, so that, the cost of production will fall considerably and hence, the firm will be in a better position to compete in the market effectively. So, internal economies of scale show a greater

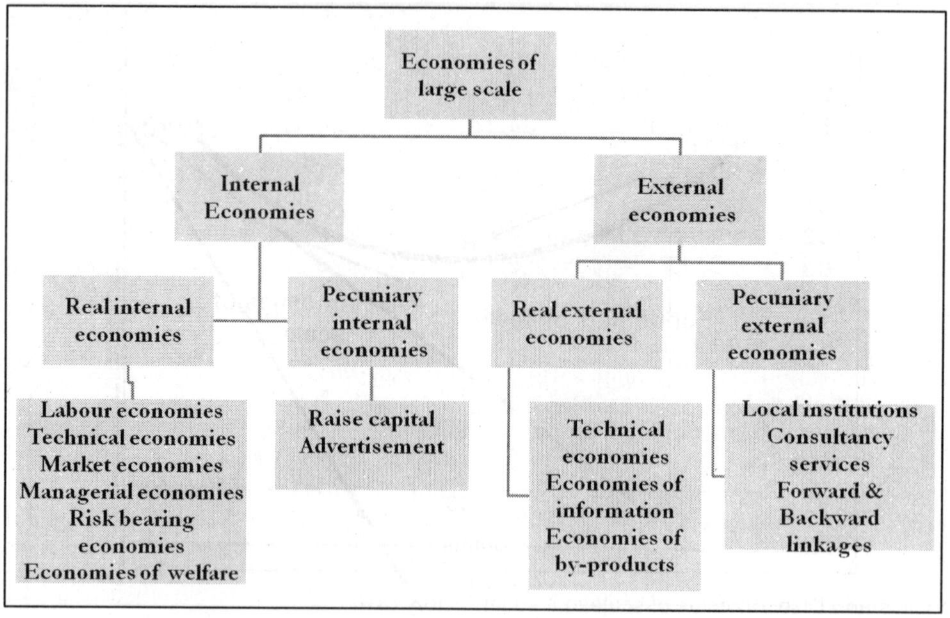

Figure 16.3: Different types of economies of scale.

potential impact on the cost-reduction, thereby, on the profitability of a firm. Thus, economies of scale of a firm refers to Internal economies. To put it in other way, internal economies of scale are the savings, which occur within a firm and it is independent of other firms. So, the factors responsible for this are purely within the establishment or endogenous or internal of the business firm. Thus, internal economies of scale are solely enjoyed by the individual firm when its scale of production changes, independent of the actions of other firms. Hence, the factors will benefit the firm only and not the entire industry in minimizing the LRATC of the product through enjoying economies of large scale. They are not the result of inventions of any kind, but are due to the uses of known methods of production, but more scientifically. These economies of scale will arise in the firm and they include,

(a) Labour economies

This is Adam Smith's idea of the division of labour. It can be achieved by specialization. In large firms, the complex production processes is splitted into separate tasks or activities and for accomplishing each activity, specialized skilled workers will be employed. Division of labour in mass production of motor vehicles, manufacturing of electronic products, fruit processing etc., are good examples. So, the labour or workers in a factory become more efficient at a particular job, when specialization occurs. It also means, no time is lost due to labour changing jobs in the middle of a shift. With people specializing in certain activities of the business, their levels of experience and expertise will also increase, thereby, resulting in more output through cost-effectiveness. This is because, with increased specialization, a less number of labour will be required and hence, the cost for hiring labour will subsequently decline. The quality of output also increases, if the firm employs specialists in every aspect of production.

(b) Technical Economies

These are often called as economies in the use of factors of production *i.e.*, they are related to the production process itself. Prof. Cairncross has given the following kinds of technical economies:

- ☆ Large-scale business firms employ expensive, specialist and modern machinery that contributes to higher output through cost-effectiveness. For example, if a firm manufacturing dairy products installs expensive machinery that ensures higher output through cost-effectiveness, it enjoys low LRATC and thus, it easily competes out other firms in the dairy industry.
- ☆ The 'Law of Increased Dimensions' or the 'Container Principle' also contributes to technical economies of scale. When a godown's length, breadth and height are doubled, eight times the quantity can be stored inside. This is an important economies of scale in areas, where bulky material are handled. The same law can be employed in increasing the size of processing firms, fertilizer, pesticide and other such firms to increase the output through cost-effectiveness. This is because, the size of machinery can be doubled in these firms, without doubling the labour and material costs.
- ☆ A large firm can also afford to invest more in Research and Development (R&D). This is because, as the firm expands, it can afford to install highly technical and expensive equipment needed for in-depth experiments. This leads to employment of highly skilled and technically qualified persons, who can experiment and develop modern techniques of production that ensures cost-effectiveness in large scale production. Further,

the modern production techniques employed by the firm are suitable for using cheaper raw materials. The products developed from such modern technologies enjoy low LRATC and thereby, the firm can easily compete out the competing firms. Further, the firm can enjoy higher margin of safety. The expenditure on R&D will not pose any burden on the part of the firm, as the firm produces larger output that too through cost-effectiveness. So, the returns will flow significantly into the business, as large quantities of output are sold at low price in the market and such higher returns into the business makes the R&D expenditure affordable on the part of the firm. If the Government feels the product produced by the firm will benefit the society, it may offer funding for encouraging R&D and will promote the firm. Likewise, private sector might also be willing to offer support to the firm by providing beneficial services such as transportation and communication to the growing industry.

- Effective utilization of by-products is possible in case of large scale production, unlike small scale production. For example, in a sugar factory (large size firm), the by-product, molasses can be used in the production of alcohol. This valuable by-product enable the entrepreneur to start a subsidiary industry or he can sell the by-product as raw material to other firms engaged in the production of alcohol. This will result in the saving in cost of production in the main (sugar) industry. (Compensate the cost of production in the main (sugar) industry and thereby, cost of production of sugar is minimized indirectly).
- Linking the activities of production will save expenses incurred on intermediaries, thereby, reduces the cost per unit of output. For example, a large sugar factory may have its own sugar cane production farms, manufacture sugar, packing, transportation and distribution network etc., and these contribute to economies of large scale.
- There is a growing interest in the concept of a network economy of scale. Some networks and services have huge potential for economies of scale. That is, as they are more widely used or adopted, they become more valuable to the business. The basic idea behind the net work economies of scale is that, the MC of adding one more user to the network is close to zero, but the MRs derived from each new user is very high and this contributes large economies of scale. The rapid expansion of e-commerce is a good example of the exploitation of network economies of scale. This type of economies of scale is linked more to the growth of demand for a product.
- The business firm enjoys internal economies of scale through the concept of 'learning-by-doing'. So, the entrepreneur uses all his experience in the production process and aims to minimize the costs, so that, LRATC of production will fall through increasing the output. The earlier studies conducted on different industries revealed that, 70 to 90 per cent of manufacturing costs are declined due to learning experiences of the entrepreneur. The same is explained through the Figure 16.4. By implementing the production strategies through the experiences gained in the past production processes, the entrepreneur could minimize the cost of production of sugar by C_1C_2.
- Because of indivisibility of capital equipment, only a single largest firm in the industry can afford to buy the heavy machinery at a high price. For example, blast furnace by a steel plant, thereby, it can easily compete out other small firms engaged in manufacturing steel.

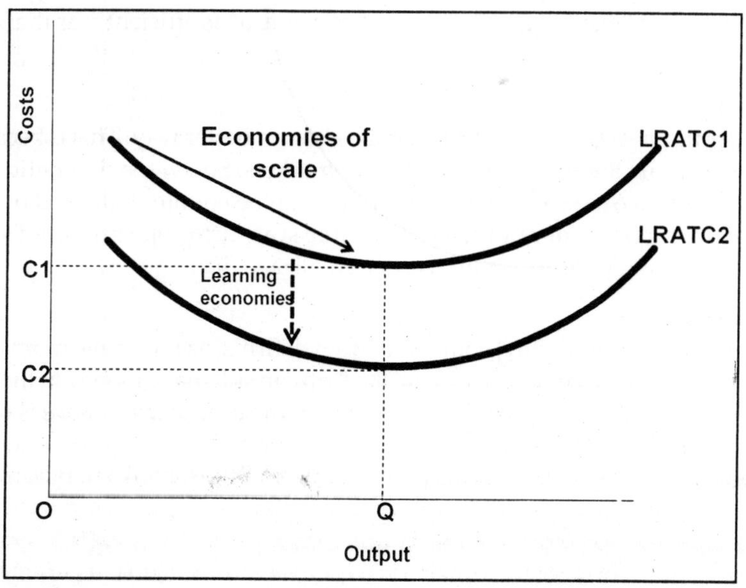

Figure 16.4: Learning economies in a sugar factory contributing to fall in LRATC curve.

(c) Market Economies

This gives the highest benefit to the firm engaged in large scale production. It includes buying economies and selling economies.

(i) Buying Economies

It includes:

- ✫ Larger firm will have the ability to purchase the raw materials in bulk at discounted prices, if it enjoys monopsony (buying) power in the market. For example, a fruit processing firm enters a contract with the fruit growers in a particular area regarding the supply of fruits to the firm after the harvest.
- ✫ As the firm expands the business, the demand increases for raw material. So, the firm's order value increases, thereby, it obtains more bargaining power with the suppliers. It may be able to purchase the raw materials in bulk quantities and obtain discounts and lower prices for the raw materials and this is referred as Bulk-buying economies. For example, a processing firm purchases potatoes from farmers in bulk to make French fries, thereby, it enjoys volume discounts.
- ✫ In terms of localization, proximity to its suppliers will enable the firm to reduce transportation costs for inputs. Also, a convenient location, where there is plenty of access to public transport, will minimize the distance for customers to have the product in large quantities in right time.

(ii) Selling Economies

It includes:

- ✫ Every part of marketing has cost, particularly promotional costs, such as advertising and employing and running a sales force. Many of these marketing costs are fixed costs and so as the business gets larger, it is able to spread the cost of marketing over a wider range of output and sales, thereby, cutting the average marketing cost per unit. The successful results of such promotional methods will be increased demand and brand loyalty towards the product and both will benefit the firm.
- ✫ For selling the output, the firm can plan for exporting the product to distant markets, where it enjoys more comparative advantage.
- ✫ The firm enjoys economies of freight, as it can maintain full cargo, thereby, the transportation cost per unit of product will be low.

(d) Managerial Economies

This is a form of division of labour. As the firm expands its operations, it need not expand its administrative staff to the same degree or percentage, as the expansion in terms of operations. So, the large-scale firm employ managers, who specialize in various fields like marketing management, human resource management, finance management, pricing management, materials management etc., and they supervise the production systems more efficiently, as they possess higher qualifications and expertise compared to one person in a small firm trying to perform all these roles. This 'functional specialization' contribute much to economies of large scale through better ways of organization, production, communication etc. These managerial economies reduce per unit cost of management because, with the expansion of firm, various department managers will manage large output as efficiently as they were managing small output at the same salary.

(e) Risk Bearing Economies

Compared to small scale firms, a large scale firm can handle risk in a better way. This is because, a large scale firm produces variety of products and sale them in different areas or markets. So, by the diversification of its production process, the large firm is able to reduce risk, as loss in one product can be counter balanced by returns from another product. Similar is the case of procurement of raw material, as large scale firm enjoys assured supply of raw material from several sources rather than a single source, as in case of a small size firm.

(f) Financial economies of scale

Large firms raise capital easier (credit worthy) compared to small firms, as they have more capital assets and high risk bearing ability. This is because, if one market goes down, the firm can still make money in the others. In this context, small firms find it hard to obtain finance and when they do obtain it, the cost of the finance is often quite high. Large firms, therefore, find it easier to find potential lenders and to raise money at lower interest rates. They are seen as less likely to become bankrupt and able to repay their loans in right time and thus, banks may actually compete for their accounts.

(g) Economies of Welfare

Large scale firm with its huge resources can provide better working facilities both inside and outside the factory. The firm can run subsidized canteens, resting rooms for the workers etc., within the factory premises and provide staff quarters, educational and medical facilities, provisional stores, recreation clubs etc., outside the factory. Though providing such facilities involves huge cost, yet the factory aims to increase the productive efficiency of the workers, through which it raises the production and cover these costs. The business expansion may allow the firm even to afford to sponsor sporting events and become popular among the customers. Such activities of a large firm also promote good customer relations and create a good impression in the eyes of the public. Ultimately, the firm benefit from increased sales.

II. External Economies of Large Scale

External economies of scale occur outside of a firm but within an industry. So, external economies of scale occur, when a firm benefits from lower unit costs as a result of the whole industry growing in size. That means, a firm may enjoy certain cost savings, not through its own expansion, but by being a part of a well-organized and large industry. So, external economies are shared by a number of firms, when the entire industry prospers. That means, the external economies are not monopolized by any single firm in the industry. So, the factors responsible for economies of large scale are external to firm and they not only influence the firm, but the whole industry. So, economies of scale of an industry refers to External economies. They accrue to a firm due to technological influences on the total output produced in the industry, which reduces per unit cost of production. So, all the firms in industry will develop due to technological interdependence of firms. They include:

- ☆ *Technical economies:* This is related to division of labour and specialization. The supply of skilled labour may locate in the vicinity of the industry and this will cause firms in the industry to have an easy access to the type of labour they need at low costs. This ensures specialization of labour among the firms in the industry, which contributes higher output to the industry. For example, in cotton textile industry, some firms may specialize in manufacturing threads, some in dyeing, some in long cloth, some in dhotis, some in shirting etc. This enhances the production efficiency of the firms thereby cost of production falls. Further, a network of suppliers or support industries may grow in size and locate close to the main industry and they specialize in various activities like preparing tools, implements, repairs and maintenance etc., thereby, lowering unit cost of production of all the firms. So, an individual firm has a greater chance of finding a high quality, yet affordable supplier close to its site. Agglomeration economies may also result due to clustering of similar businesses in a distinct geographical location.
- ☆ *Economies of information:* The firms exchange ideas and publish articles and magazines (Trade journals) that promote the spread of technical information throughout the industry. The industry will be in a better position capable enough to establish research laboratories, as it can pool large amount of resources. All these will help the firms in raising the production efficiency and reduction in per unit costs.
- ☆ *Economies of by-products:* When industry is localized, it turns out large quantities of waste materials, such as molasses in sugar industry, iron scrap in steel industry etc. New firms enter the industry to purchase these waste materials (For example, an alcohol firm will purchase molasses from sugar industry to prepare a spirit or alcohol) at reasonable prices and use them in manufacturing by-products. So, the firms in industry enjoy economies of scale in two ways *viz.*, no expenses are incurred in disposing off the waste material and revenue is earned through selling by-product to other manufacturers.
- ☆ *Transport and communication links will improve:* If an industry establishes itself and grows in a particular region, the Government will provide better transport and communication links to improve accessibility to the region. This will lower transportation costs for firms in the area and it attract more potential customers.
- ☆ *Training and education becomes more focused on the industry:* Universities and colleges will offer more courses suitable for a career in the industry, which has become dominant in a region or at national level. For example, considering the potentiality of processed food products in the export basket of the country, many Agricultural Universities of different states in India has started Food Science and Technology colleges. Similarly, new courses on Information Technology (IT) are started in the Engineering colleges, as the whole IT industry expanded in India. This will benefit the firms in the industry in getting a larger pool of skilled workers in accomplishing the technical duties. Besides educational facilities, research and training facilities also improves, thereby, all firms in the industry can benefit from such facilities.
- ☆ *Collective advertisement:* Industry can also benefit from collective advertisement, which is cheaper than each firm undertaking its own advertising. This is more commonly seen in case of tourism industry.
- ☆ *Accessibility to other services:* When the entire industry prospers, banks, insurance, catering, cleaning and other businesses will be located in the vicinity to the major industry, allowing them to save on costs. Normally, firms will have to pay more for such things, if they are not located near to them.

- *Centralized maintenance services:* Firms in the industry can benefit from centralized maintenance services, thereby, the cost of maintenance to each firm will be much less than, if each undertook its own maintenance.
- *Spill over effects:* The entire industry prospers due to spill over effects. That means, if one firm enjoys economies of large scale due to modern technology, the same will be spread across different firms in the industry, thereby, the entire industry prospers and it will also lead to the creation of standards within an industry.
- *Localized markets:* The expansion of an industry in a particular locality will lead to the development of well-organized markets and this ensures fair returns to all the firms in the industry.

16.4. Diseconomies of Large Scale

In an era, when large corporations rule the business landscape, economies of scale are the goal. Not all expansions, however, produces positive results. Diseconomies or disadvantages could result. The diseconomies of scale are the forces that cause large firms to produce goods and services at increased per-unit costs. But, they are less well-known than what Economists have long understood as economies of scale. So, when the firm experiences diseconomies of scale, LRATC do not continue to decline as business expands its output (Figure 16.1), but rises and this is the result of diminishing returns to scale. This implies that, there are inefficiencies within the firm or industry, thereby, resulting in rising LRATC. The causes of diseconomies of scale are numerous and may not be obvious without comprehensive analysis of individual cases. Basic causes, however, can still be defined. The diseconomies will be looked at under two headings *viz.*, internal diseconomies of scale (disadvantages to the firm of large-scale production) and external diseconomies of scale (disadvantages or drawbacks to the industry of expansion or localization).

I. Internal Diseconomies of Large Scale

This arises due to the following reasons:

- *Managerial diseconomies:* Managerial problems are often specific to diseconomies of scale. When the firm greatly expands, managers have a harder time for coordinating tasks and processes. That means, it becomes difficult on the part of the business managers to co-ordinate the complicated production processes across several plants in different locations. This results in a loss of competitive advantage due to inefficiencies or mismanagement or inexperience with scale. All these leads to slackness, waste, confusion, low morale of the workers etc., contributing to rise in costs more than proportionately compared to the rise in output. This inability to monitor the large scale production programme by the top administrator has been called *'Control loss'* by Oliver Williamson.
- *Marketing diseconomies:* When the firm expands greatly, raw material availability in right time and in right quantity may become a problem (may be due to scarcity). The demand for a product may fall, as a result of changes in tastes of the people. The organization may fail to foresee the changes in market conditions and hence, the sales of products might fall.
- *Technical diseconomies:* Specialization and massive investment in machinery and equipment though contributes to higher output, but may also lead to loss of flexibility necessary to respond quickly to changes in demand. The expansion of business firm beyond the optimum size, may lead to repeated breakdown in plants and hence, the firm may not function up to its installed capacity.
- *Diseconomies of risk taking:* As the scale of firm expands, risks also increase with it. An error of judgement, say regarding the price fixation of product, adversely affects the trade and it leads to greater loss.
- *Worker dissatisfaction:* Workers in large firms may feel a sense of alienation and subsequent loss of morale. If they do not consider themselves to be an integral part of the business, their productivity may fall leading to wastage of factors of production. The collective feeling and morale towards the organization are less in large firms than in small firms. Both the managers and workers of large firms feel more secured in the organization and have no fear of getting sacked. This makes them lethargic and devoid of entrepreneurship. The workers find it easier to form into groups such as, trade unions, workers associations, labour unions etc., and the restrictive practices of such groups may result in increased costs for firms.
- *Bureaucracy:* Diseconomies of scale are particularly likely in the public sector, as Government bureaucrats may be ill-equipped to manage large organizations. Nepotism and Red tapism are rampant and these coupled with elaborate and time-consuming procedures to perform the activities in these organizations makes the customers not to realize the benefits on par with their expectations and in right time, thereby they find the larger organizations are too impersonal and lack personal touch.
- *Government intervention:* As the firm expands, it may be subjected to Government intervention in the form of price control or other restrictions. This is especially true, if the expansion results in monopolies and oligopolies.

II. External Diseconomies of Large Scale

These arise due to the following reasons:

- *Scarcity of factors:* When an industry expands, the demand for factors like labour, capital equipments, raw materials etc., will increase. But, there may be shortage of these factors in the economy and they will be available at higher prices. Hence, cost per unit of output increases.
- *Increased competition among the firms:* As the number of firms in the industry increases, competition becomes stiffer, thereby, the firms have to invest more on advertising and propaganda about their products and this raises the cost per unit of the output.
- *Localization of firms:* Localization of firms in an industry leads to more social costs, which impacts negatively on the firms or the industry and on society at large.
- *Negative externalities:* If one or two firms in the industry suffer decline, all other firms in the industry may feel the same adverse effects.

All the above diseconomies are external to each firm in the industry because, these diseconomies are not caused by the expansion of any single firm, but are the consequences of the expansion of the whole industry.

16.5. Avoiding Diseconomies of Large Scale

The Economists believe that, effective management techniques and the appropriate incentives can do much to reduce the risk of rising LRATC. The following are the suggestive measures to overcome the diseconomies of scale in large-scale business:

- Developments in human resource management should be focussed on priority basis. By this, the productive efficiency of the workers can be greatly improved through streamlining the areas like worker recruitment, training, promotion, retention and support of faculty and staff. These counts much, especially, when the firm experiences shortage of skilled workers.
- Offering incentives to workers based on the work performance in the firm counts much in promoting the work culture thereby, leading to higher productivity. This also motivates the other staff in the firm to function towards the directed goals and objectives.
- Recently, large scale businesses are engaged in out-sourcing of production and distribution activities, as they aim to expand the sales even to ever-distant markets. By doing so, the costs can be greatly reduced, but at the same time, the firm can retain control over the production and distribution activities.

17

Supply and Elasticity of Supply

In Chapter 3, we studied that, business firms are one of the major economic players in an economic system. This is because, they supply the requisite goods and services suiting to the tastes and preferences and income levels of the consumers at their affordable prices. While demand explains the consumer's side of purchasing decisions, supply relates to the producer's desire to make a profit. This chapter focuses on the supply behaviour of the business firms and the price elasticity of supply (e_s).

The word 'supply' in Economics is used in a different sense compared to its normal use in everyday conversation. In normal sense, 'supply' is most often used as a verb to provide someone with something. But, in Economics, 'supply' refers to offering either good or service on payment basis to the buyer. To be clear, 'supply' refers to the amount of a good or a service that a producer is willing and able to offer across a range of prices. Supply at macro level represents, how much the overall market can offer to the buyers. Thus, 'supply' differs from 'stock' in the sense that, it is the actual quantity of total production that was offered for sale by the producer (farmer). The total production of the produce by the farmer in the current year together with previous years' retentions, if any, constitutes total stock of produce. Supply refers to the actual quantity of the total produce (stock) that was offered for sale by the farmer. Thus, stock refers to potential supply. So, supply of a commodity into the market may be less than stock or equal to stock. In case of perishable commodities like eggs, leafy vegetables, milk etc., supply equals to stock, as these commodities cannot be stocked. On the other hand, for commodities like rice, wheat, sugar, chilli powder etc., the supply may be less than the stock (if prices in the market are not remunerative for the farmer), as these commodities can be stored for a short period of time or if the prices are remunerative for the farmer, then supply equals the stock. The following are the popular definitions of supply given by the Economists:

'Supply is a schedule of amounts that will be offered for sale at different prices during any given time period, other facto remaining same'.

–(Bac

'Supply may be defined as a schedule, which shows the various amounts of a product which a producer is willing to and to produce and make available for sale in the market at each specific price in a set of possible prices during some given period'.

–(McConnell)

From the above definitions, we can infer the following important aspects of supply in an economic sense:

☆ Supply is always at a price. If we say, a farmer supplied a commodity with out referring the price, it will be meaningless.

- Supply is always with reference to a particular period of time, as time element plays a vital role in changing the supply of a commodity and in bringing the balance between demand and supply forces in the market.

17.1. Law of Supply

From the above definitions, we can infer that, the supply of a commodity is influenced by its price. It is the practical experience of every farmer that, with the rise in price of the commodity, the farmer will increase the supply of the commodity into the market, with a view to earn more profits. With the fall in price of the commodity, the farmer will stock the commodity (reduce supply into the market), so as to avoid losses. So, there lies direct relationship between price and quantity supplied of the commodity and this direct relationship is explained by the Law of supply. The statements of the Law of supply given by the famous Economists are as under.

'Ceteris paribus, the quantity of a commodity produced and offered for sale will increase as the price of the commodity rises and decreases as the price falls'

– (Lipsey)

'Other things remaining the same, higher the price, the greater the quantity supplied or lower the price, the smallest the quantity supplied'.

– (Dooley)

The above definitions infer that, price and quantity supplied of the commodity are positively related. To put it simply, when the price of the commodity increases, the quantity of commodity offered for sale by the farmer increases, *ceteris paribus'*. Of course, besides price, there are several other factors that influence the quantity supplied of a commodity and this functional relationship between various factors and quantity supplied of a commodity is called Supply function. It is represented in the form of an equation given below.

$$Q_P = f(P_P / I_P, P_E, T, \ldots\ldots) \qquad \textit{Equation 17.1}$$

Q_P = Quantity of paddy supplied into the market

where,

f = Functional relationship between quantity supplied and factors influencing it

P_P = Price of paddy in the market

I_P = Input prices

P_E = Price expectations

T = Technology

The 'bar' in the above equation denotes that, the variables I_P, P_E, T etc., are kept constant and the influence of price of the commodity (paddy) is studied on the quantity supplied for paddy. We can rewrite the above Equation 17.1 in terms of relationship between quantity supplied and price of paddy commodity (Supply function) and this is shown below.

$$Q_P = f(P_P), \textit{ ceteris paribus} \qquad \textit{Equation 17.2}$$

The above general functional form does not show, how much quantity of paddy is supplied by the farmer due to change in price of the paddy. So, to estimate the quantity supplied in numerical form, we generally consider a linear supply function given below,

$$Q_P = a + bP_P \qquad \textit{Equation 17.3}$$

In the above equation, 'a' implies intercept of the supply function, 'b' represents the regression coefficient or slope of the supply function. The positive sign in the above function implies positive relationship between the two parameters under consideration. To explain the linear supply function, assume the following function:

$$Q_P = 20 + 4P_P \qquad \textit{Equation 17.4}$$

This equation implies that, for every one unit increase in the price of the paddy, the quantity supplied of paddy also increases by four units.

Features of Law of Supply

From the law of supply, we can ascertain the following features:

- There exists direct relationship between price and quantity supplied of the commodity.
- Price of the commodity is the independent variable and quantity supplied is the dependent variable.
- It operates under *ceteris paribus* assumption.
- It helps to estimate the elasticity of supply for the commodity, based on which we can forecast the variables.

Assumptions of the Law of Supply or Determinants of Supply

The supply function between quantity supplied and price of the commodity is valid, when the following assumptions holds good:

- There is no change in number of firms or sellers producing a commodity
- There is no change in scale of production and rate of production.
- There is no change in techniques of production
- There is no change in the cost of production of the commodity
- There is no change in prices of related goods
- There is no change in price expectations of the goods in the future
- There is no change in goal of the firm
- There is no change in prices of factors of production
- There are no labour problems or lock-outs in the factory
- There is no change in Government policy
- There is no change in natural factors
- No discovery of new substitutes
- There should not be any change in the tastes and preferences of the consumers
- No change in population
- No change in psychology of the consumer
- No change in the credit availability to the consumer
- No change in the propensity to consume
- No change in trade cycles
- No change in advertisement and propaganda about the commodity

If the above assumptions do not hold good, the law of supply may not hold true. Say, for example, when the price of paddy in the market is Rs. 12/kg, the farmer supplies 100 kilograms. If the price of paddy is increased in the market, say to Rs. 15/kg, now the farmer supplies 120 kilograms. This indicates that, with increase in price of paddy, the quantity offered for sale by the farmer will increase, thereby, the law of supply holds good, as long as other factors *viz.*, no change in prices of related goods, no change in price expectations of the commodity in the future, no change in prices of factors of production etc., are valid. If the increase in price of the paddy in the market is accompanied with rise in prices of factors of production, then the law of supply may not hold good, as the farmer may shift to another enterprise instead of paddy. Similarly, the present supply of paddy will not increase due to increase in price of paddy in the market, if the farmer expects further rise in price of paddy in the near future. If price of substitute commodity *viz.*, wheat is decreased in the market, the traders will not increase supply of rice into the market, as the consumer will now prefer more of wheat to rice. From these examples, we can infer that, the supply of the commodity responds positively to the change in price of the commodity, only under *ceteris paribus* condition. If this *ceteris paribus* condition is not fulfilled, the law of supply may not hold true. Supply, thus, is a positive relationship between price and quantity, under *ceteris paribus* condition and thereby, the supply curve (between quantity supplied of a commodity and price) will have a positive slope.

Importance of Law of Supply

The concept of law of supply is very useful for the farmers, traders, Government etc.

- *Importance to farmers:* The law of supply guides the farmers towards realization of better price for their crop produce. If the prices of the produce in the market increases, the farmers will increase the supply of the produce into market. During peak marketing season, the concept of supply guides the farmers to stock the produce, as excessive supply will adversely affect the price of the commodity in the market. So, depending upon the supply situation in the market, the farmer will plan stocking or destocking the produce. At high price, the farmer will de-stock the produce, *ceteris paribus* and *vice versa*.
- *Importance to traders:* Through analyzing the production and supply scenario of the commodity, the traders will fix the price for the commodity. If there is scarcity of production of commodity due to adverse climatic conditions or pest and disease incidences, the traders exploit the situation and quote higher prices for the produce from the consumers. If there is bumper harvest of the crop, the traders purchase the produce at low prices and stock the produce during peak marketing season. At times of lean season, the traders will de-stock the produce and place the commodity for sale at higher prices.

- ☆ ***Importance to Government:*** This law will guide the Government in planning for maintenance of buffer stocks. At times of bumper harvest, when the prices fall at the farmers' level, the Government will then procure the produce at the announced MSP. This offers protection to the farmers against distress sales. Further, at low prices during bumper harvest of the crop, the Government plan for exports to earn significant amount of foreign exchange. So, depending upon the supply of commodity into the market, the Government plan its procurement and export strategies
- ☆ ***Basis for price fixation:*** The law of supply guides the entrepreneur to fix the price for the commodity based on drawing the market supply schedule. If the market supply is expected to increase, he will purposefully quote a low price, so as to improve his sales in the market. Conversely, if the supply of the commodity is low into the market, he will quote a high price to earn more profits.
- ☆ ***Planning the new enterprises:*** In an area, where there are specific market outlets and prevalence of more comparative advantage for the produce, it guides the farmer to take up new enterprises. This is because, the produce so harvested by the farmer can be supplied to the different market intermediaries in time.

17.1.1. Individual Supply

The individual supply of a commodity is the quantity of a commodity that farmer is able and willing to offer for sale in a market at a given price at a point of time. The individual supply of a commodity varies directly with the price of the commodity, *ceteris paribus.*

Individual Supply Schedule

We know, a farmer sells different quantities of a commodity at different prices. So, a schedule or a table showing such information for a commodity at a given point of time refers to Individual supply schedule. The Table 17.1 shows the hypothetical supply schedule of a farmer for paddy commodity at different prices. This infers that, as the price of the paddy increases in the market, more quantity of it is supplied by the farmer.

Table 17.1: Individual supply schedule for paddy in the market.

Price (Rs/kg)	*Quantity Supplied (kg)*
18	38
20	40
25	44
30	47
35	50

Individual Supply Curve

Supply curve is a graphical representation of the supply schedule. The individual supply schedule can be converted into an individual supply curve by measuring price on Y-axis and quantity of commodity supplied on X-axis, as shown in Figure 17.1. SS is the supply curve for paddy, which slopes upwards from left to right showing that, when price increases, the quantity supplied extends and *vice versa.* So, different points on the individual supply curve indicates different quantities of a commodity offered for sale by the consumer at its different prices. Here, it is often important to distinguish the 'supply' from the 'quantity supplied'. Typically, 'supply' refers to the entire supply curve

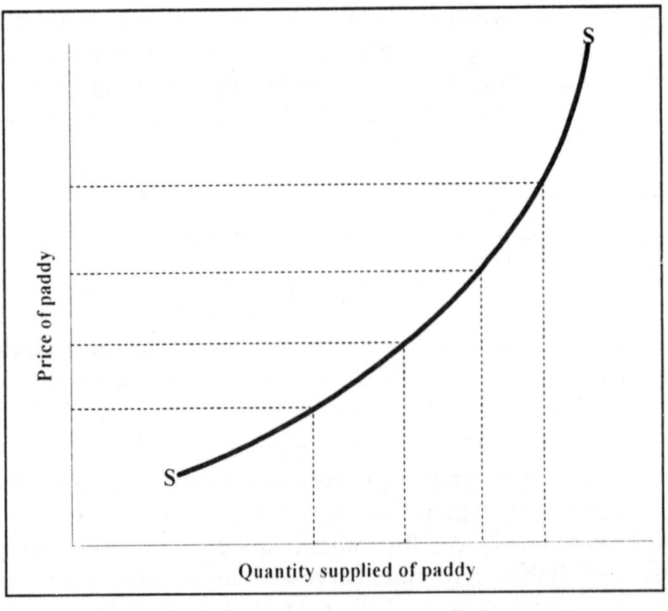

Figure 17.1: Upward (positive) slope of supply curve.

and not just one point on the supply curve (the entire relationship between price and quantity supplied), while 'quantity supplied' is a point on the supply curve. So, when the price of the commodity changes, the quantity supplied changes, but supply does not change. But, when cost of production changes, supply changes, thereby, the entire supply curve will shift.

Reasons for Upward Sloping of Supply Curve

As discussed earlier, the quantity supplied of a commodity varies directly with the price of the commodity, thereby, the supply curve will have a positive slope. That means, the two important variables *viz.*, price and quantity supplied work in the same direction and this gives a positive slope to the supply curve. So, upward sloping supply curve depicts the law of supply. The following are the main causes responsible for the upward sloping of supply curve:

- ☆ High price is an incentive to sellers, thereby, they offer more for sale at higher prices.
- ☆ In short run, due to changes in the level of variable factors in the production programme, the cost of cultivation increases and this increases the output. So, to cover the variable costs, the farmer desires to sell more output at higher prices.
- ☆ The increase in price for the commodity in the market attracts new farmers to cultivate the same crop, thereby, supply into the market increases with increase in price of the commodity.
- ☆ Marginal sellers do not sell the commodity at lower prices and they stock the produce in the godowns, as long as price of the commodity in the market rises considerably.

17.1.2. Exceptions to the Law of Supply

The Law of supply is a general statement explaining the direct relationship between quantity supplied and price of the commodity. However, there are certain peculiar cases in which the law of supply will not hold good. In such cases, more quantity of a commodity is supplied even at lower price and *vice versa* and hence, the supply curve will not have a positive slope. The following are such exceptions of law of supply:

- ☆ ***Low price expectations in the future:*** If the farmers or traders expect that, the prices of the commodities would still fall in the future, they increase the supply of the commodities in the present market even at low prices. Similarly, if the farmers or traders expects that, the prices of the commodities will further rise in the future, they decrease the supply of the commodities in the present market even at high prices.
- ☆ ***Backward bending slope of supply curve of labour:*** If we go on increasing the wage rate, the labour contributes more number of working hours to earn more income. But, it is interesting to say that, if wage rate is still increased, the number of working hours contributed by the labour will not increase, rather they decrease because, at one point of time, the labour prefers leisure hours to working hours, because of inelastic cash requirements. The same is explained through IDC analysis in Chapter 9 (Figure 9.59).
- ☆ ***Supply of agricultural products:*** Since the production of agricultural products is season bound, supply cannot be increased beyond a certain limit, even with the rise in their prices.
- ☆ ***Supply of artistic works:*** The supply of artistic works, creative paintings etc., cannot be increased with the rise in prices.
- ☆ ***Goods kept for auction:*** The goods that are meant for auction cannot be increased with increase in their prices.
- ☆ ***Perishability of agricultural produce:*** Some of the agricultural produce like milk, leafy vegetables, eggs etc., are highly perishable and hence, their entire stock represents supply. So, their supply cannot be increased with increase in prices.

17.1.3. Market Supply

It refers to the quantity of supply of the commodity in the entire market *i.e.*, various quantities of a commodity supplied by all the farmers at various prices in a market at a given point of time. The market supply of a commodity is obtained by adding up the quantities supplied by all the individual farmers at various prices at a given point of time in the market. It is, otherwise, described as the horizontal summation of all the individuals' supply of a commodity at various prices in the market. The market supply for a commodity also varies directly with the price of the commodity. Different farmers will sell different quantities of a commodity at different prices because, farmers incur different costs of production in producing the commodity. At a low price, only the most efficient farmers can make a profit, so only they produce. At a high price, even high cost producers can make a profit, so everyone produces.

Market Supply Schedule

Since, all the farmers offer different quantities of a commodity at different prices, the schedule or a table showing such information for a commodity at a given point of time refers to Market supply schedule. So, we can arrive at the market supply schedule by the horizontal summation of individual's supply schedules for a commodity. It is, otherwise, defined as the inverse relationship between price of the commodity and quantity supplied by all the farmers in the market. Table 17.2 shows the market supply schedule for paddy at different prices. It infers that,

- ☆ Market supply schedule is the horizontal summation of individual's supply schedules of paddy at different prices.
- ☆ Market supply schedule also obey the law of supply

Table 17.2: Market supply schedule for paddy in the market

Price (Rs/kg)	*Quantity Supplied (kgs)*				
	Farmer A	*Farmer B*	*Farmer C*	*Farmer D*	*Market Supply*
15	8	27	29	36	100
25	19	30	36	48	133
35	29	34	50	60	173

Market Supply Curve

The graphical representation of market supply schedule is the market supply curve for the commodity. Figure 17.2 shows the market supply curve for paddy in the market. The market supply schedule can be converted into market supply curve by measuring price on Y-axis and quantity of commodity supplied by all the farmers on X-axis. S_MS_M is the market supply curve, which slopes upward from left to right showing that, when price increases, the market supply for the commodity extends and *vice versa*. So, different points on the market supply curve indicates different quantities of a commodity offered for sale by all the farmers at its different prices in the market. With the rise in price of the commodity, new sellers enter the market and this will further increase the quantity supplied of the commodity. This will also contribute to upward sloping of market supply curve from left to right with the rise in price of the commodity.

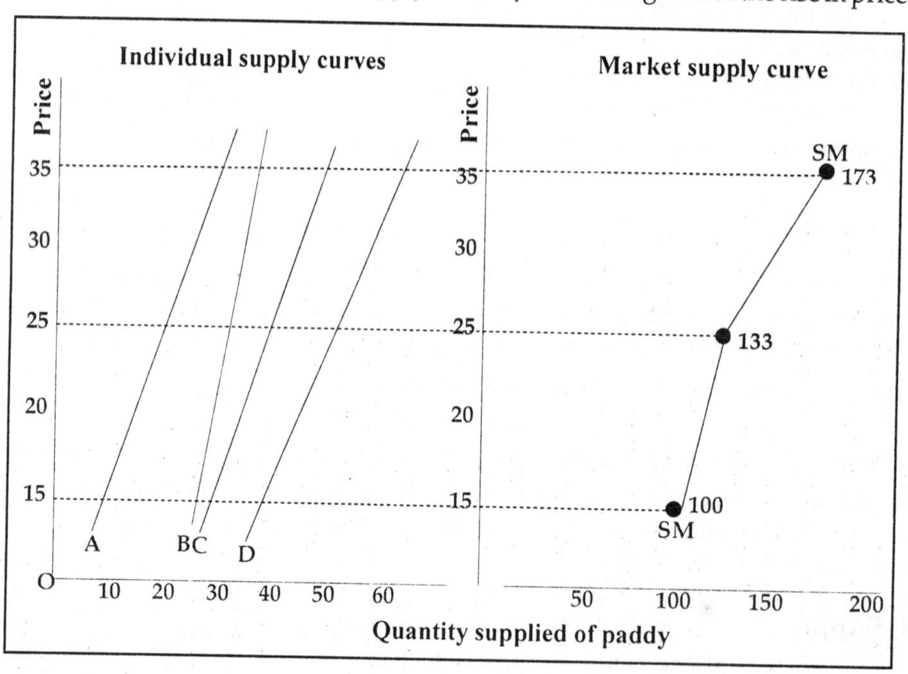

Figure 17.2: Derivation of Market supply curve from Individual's supply curves.

So, the above explanation infers that, if we have the information pertaining to individual supply schedule for a commodity, we can easily prepare individual's supply curves, market supply schedule and market supply curve for the commodity.

Individual Supply Schedule vs Market Supply Schedule

The important differences between individual supply schedule and market supply schedule are discussed here under.

- ☆ Individual supply schedule indicates the selling behaviour with reference to single seller (farmer), while market supply schedule indicates the selling behaviour of all the farmers in the market.

- ☆ Individual supply schedule forms the basis for market supply schedule, whereas, market supply schedule indicates the individual supply schedule.
- ☆ Individual supply schedule may obey the law of supply, but market supply schedule may not obey the law of supply and *vice versa*.
- ☆ Individual supply schedule may be more elastic, but market supply schedule may be less elastic and *vice versa*.
- ☆ Individual supply will not influence the price of the commodity (in perfect competitive market), but market supply influences the price of the commodity
- ☆ The individual supply schedule may involve discontinuity, as it is quite difficult to predict how the farmers behave at different prices of the commodity. But, there is continuity with reference to market supply schedule, as individual differences may disappear in the collective supply.

17.2. Supply Function

As discussed earlier, the functional relationship between quantity supplied of a commodity and the factors influencing it constitute supply function. Besides price of the commodity, there are several other factors, which influence the quantity supplied of a commodity. They are discussed here under through the following equation:

$$Q_P = f(P_P, C, P_W, P_E, D_R, L, T_E, T_A, S_U, T_P, A_G, W, S\,G, P_D, C_T, C_R \text{——})$$ ***Equation 17.5***

where,

Q_P = Quantity supplied of commodity, paddy
f = Functional relationship between quantity supplied and factors influencing it
P_P = Price of paddy commodity
C = Cost of production of commodity
P_W = Price of the related commodity wheat (substitute)
P_E = Price expectations in the future
D_R = Domestic retentions
L = Labour trouble
T_E = Change in technology
T_A = Taxation policy
S_U = Subsidy offered by the Government for the production of commodity
T_P = Tastes and preferences of the consumer
A_G = Agreement between sellers
W = Weather
S = Seasonality factor
G = Goal of the firm/seller
P_D = Political disturbances
C_T = Communication and transportation facilities
C_R = Credit availability

- ☆ ***Price of the commodity:*** Among various factors, this is the most important factor, which influences the quantity supplied of the commodity. There exists positive or direct relationship between quantity supplied and price of the commodity in the market. This is already explained through Table 17.1 and Figure 17.1.
- ☆ ***Cost of production of the commodity:*** With increase in cost of production of the commodity (say, paddy) due to rise in prices of factors of production, the farmer shifts to another enterprise (say, sericulture), from which he can derive more profits. So, supply of a commodity into the market decreases with increase in cost of production of the commodity. So, an increase in the cost of production will cause a leftward shift in the supply curve
- ☆ ***Change in prices of related goods:*** The change in prices of related goods like substitutes and complements also bring the changes in the quantity supplied for a commodity under consideration. In case of substitutes, the price of one good (wheat) and quantity supplied of other good (rice) will move in the opposite direction. However, in case of complements, the price of one good (bread) and quantity supplied of other good (jam) will move in the same direction.

- ☆ ***Price expectations:*** If the farmer expects that, the price of the commodity will further rise in the future, the present supply of the commodity into the market decreases and *vice versa*.
- ☆ ***Domestic retentions:*** If the domestic retentions of the farmer are more due to bigger family size, the supply of the commodity into the market decreases and *vice versa*. In general, for food grain crops like rice, wheat etc., the domestic retentions are more, thereby, supply into the market is less. Conversely, for commercial crops like cotton, turmeric, tobacco etc., the domestic retentions are less, thereby, supply into the market is more.
- ☆ ***Labour trouble:*** If the firm experiences labour trouble or strikes or lockouts, it leads to low production of the commodity/product and hence, less supply into the market.
- ☆ ***Change in technology:*** If the modern technology ensures higher production at cost-effectiveness, it leads to increase in supply of the commodity into the market.
- ☆ ***Taxation policy:*** If heavy sales tax is imposed on the commodity, it discourages production of that commodity, thereby, supply into the market decreases.
- ☆ ***Subsidies:*** If the Government offers more subsidies for the production of the commodity, it encourage more production, thereby, supply of the commodity increases into the market.
- ☆ ***Tastes and preferences of the consumer:*** They influence the supply behaviour of the producer. Generally, a producer produces and supplies the commodities suiting to the tastes and preferences of the consumer. If the consumers' tastes and preferences changes, the nature and quantity of the commodities produced and supplied by the producer will also changes. For example, if the consumers prefer to eat processed food products, then the producer instead of supplying fresh food into the market, will plan for processing the produce and supply the processed products into the market. So, the supply of fresh food into the market decreases and supply of processed food products increases into the market.
- ☆ ***Agreement between the sellers:*** This is commonly seen in oligopoly market competition. If few firms are engaged in the production of a commodity, they form an association (Cartel) and they supply the commodity only at a high (common) price.
- ☆ ***Weather:*** If the weather conditions are normal, it ensures higher production of the commodity, thereby, larger supply into the market. Further, some commodities are produced in certain specific climatic conditions, thereby, these conditions will influence their supply into the market. For example, in Southern parts of India, more of paddy is produced and supplied into the market, as the climatic conditions are suitable for paddy cultivation. Similarly, in Northern parts of India, the climatic conditions are more suitable for wheat cultivation, thereby, more of wheat is produced and supplied into the market.
- ☆ ***Seasonality factor:*** For some commodities like mangoes, sapota, pine apples, oranges, apples etc., seasonality factor assumes vital role in influencing the quantity supplied of these commodities. For example, during summer, the supply of mangoes will be more into the market and during winter, the supply of oranges will be more.
- ☆ ***Goal of the firm or seller:*** If the goal of the firm is to earn more profits, it will supply more of the commodity, when the prices of the commodity increase in the market. On the other hand, if the goal of the firm is to earn more image and popularity in the society (initially), it offers more of the commodity even at low price.
- ☆ ***Political disturbances:*** If the State experiences more political disturbances, the transportation, communication and other facilities gets affected or disturbed, thereby, supply of commodities into the market gets affected.
- ☆ ***Communication and transportation facilities:*** If these facilities improve in any area, the supply of the commodities will increase into the market, as the farmers and traders are quick responsive to changing price trends for agricultural commodities.
- ☆ ***Credit availability:*** If more credit is available to the farmers in taking up the enterprises, more output is realized, thereby, supply of the commodities into the market increases.

Importance of Supply Function

The supply function guides us in the following aspects:

- ☆ It helps us to study the elasticity of supply for the commodity
- ☆ Based on the quantities supplied of a commodity at various prices, it guides the farmers to plan the marketing strategies, such that, they offer the commodity for sale during lean marketing season and considering the principle of comparative advantage.
- ☆ It helps in forecasting of prices and quantity supplied of the commodity and thus, guides the farmers in taking output-pricing decisions.
- ☆ It further guides in analyzing the influence of various factors on the quantity supplied of a commodity.

17.3. Movement Vs. Shifts in the Supply Curve

So far, we have discusssed individual supply curve, market supply curve, law of supply etc., and it is noted that, the quantity supplied of a commodity varies directly with the price of the commodity. However, the law of supply operates only under *ceteris paribus* assumption. If other factors are not held constant, they also extend influence on the supply of the commodity. In this context, it is essential to differentiate between 'supply' and 'quantity supplied' of a commodity. 'Supply' refers to the entire supply schedule or supply curve of the commodity, which indicates various quantities of a commodity offered for sale at various prices of the commodity. The changes in 'supply' for a commodity is influenced by several factors other than its price, as discussed in the supply function earlier. But, 'quantity supplied' infers only the quantity of a commodity offered for sale at a particular price. That means, it explains about a particular point on the supply curve. So, the 'changes in quantity supplied' of a commodity is influenced by price of the same commodity only and other factors are held constant. So, both 'changes in quantity supplied' and 'changes in supply' of a commodity can be shown through the supply curve in two ways *viz.*, Movement along the supply curve and Shifts in the supply curve respectively.

1. Movement along the Supply Curve

As explained through supply function, the quantity supplied of the commodity is influenced by various factors, implying that, supply is a multivariable function. If the influence of the price alone was studied on the quantity supplied of the same commodity, keeping other influential factors like technology, prices of related goods, tastes and preferences of the consumers, weather etc., constant, then we move along the same supply curve. This movement along the same supply curve represents various price-quantity combinations and it has a positive slope. Thus, whenever there is a change in the quantity supplied of a commodity due to change in its price, there is a movement from one point price-quantity combination to another on the same supply curve. Such a movement from one point price-quantity combination to another along the same supply curve is known as Movement along the supply curve. But, the movement along the same supply curve implies that, the level of supply remains consistent. So, in case of movement along the same supply curve, there is no shift in the supply curve either left side or right side. So, due to the movement along the same supply curve in accordance with the influence of the price of the same commodity, technically, there are two categories of supply called as Extension of supply and Contraction of supply.

- ☆ ***Extension of supply:*** Other things remaining equal, when more quantity of commodity is supplied due to rise in its price, it is called Extension of supply. So, the farmer moves from bottom to top along the same supply curve. The increase in price of paddy from OP_{P1} to OP_{P2} leads to increase in quantity supplied from OQ_{P1} to OQ_{P2}. As a result of price rise, the farmer moves from point B to point A along the same supply curve (Figure 17.3).
- ☆ ***Contraction of supply:*** Other things remaining equal, when less quantity of commodity is supplied due to fall in its price, it is called Contraction of supply. So, the farmer moves from top to bottom along the same supply

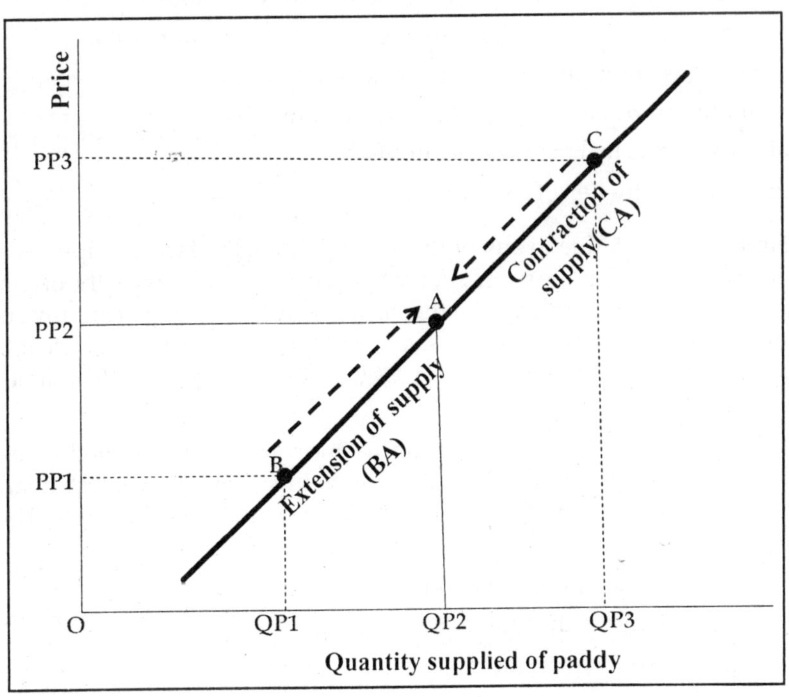

Figure 17.3: Movement along the supply curve – Extension and Contraction of supply.

curve. The decline in price from OP_{P3} to OP_{P2} leads to decrease in its quantity supplied from OQ_{P3} to OQ_{P2}. As a result of price fall, the consumer moves from point C to point A along the same supply curve (Figure 17.3).

The Figure 17.3 further reveals that, due to rise or fall in price of the commodity, the quantity supplied of the commodity extends and contracts respectively and the important aspect is that, the farmer moves along the same supply curve, but there is no shift in the position of supply curve. This movement along the supply curve is designated as 'change in quantity supplied'.

2. Change or Shift in the Supply Curve

If we study the influence of one or more factors other than price factor of the same commodity (*i.e.*, price of the commodity remains same) on the quantity supplied of the commodity, then there is a shift in supply curve from its original position. For example, if there is improvement in technology, decrease in price of substitute good, changes in tastes and preferences of the consumers favouring the commodity etc., there is an increase in supply of the commodity. This increase in supply shifts the supply curve from its original position to the right side or away from the origin. On the other hand, if the technology is outdated, increase in price of substitute good, changes in tastes and preferences of the consumers against the commodity etc., there is a reduction or decrease in supply of the commodity. This fall or decrease in supply shifts the supply curve from its original position to the left side or towards origin. So, the shift in supply curve implies that, the original supply relationship has changed for the commodity. That means, in this case, at original price of the commodity, the supply curve shifts its position either left side or right side depending upon the influence of other factor(s) on the quantity of commodity supplied. Since, we study the relationship between various quantities of a commodity supplied at the same price, but due to change in other factors, the supply curve has a positive slope. So, such a shift in supply curve from one position to another position is known as Shift in the supply curve or 'Change in supply'. In this case, there are two categories of supply called as Increase in supply and Decrease in supply.

- ☆ ***Increase in supply:*** When due to factors other than the price of the same commodity, more quantity of commodity is supplied at the same price, it is called as Increase in supply. Since, more quantity of a commodity is supplied at the same price, the supply curve shifts right side or away from the origin. The right side shift in supply curve may be due to:
  - ❖ Fall in the price of the substitute commodity
  - ❖ If price is expected to fall in the future
  - ❖ No labour strikes or lockouts in the factory
  - ❖ Improvements in technology
  - ❖ Less sales tax imposed on the commodity
  - ❖ More subsidies offered by the Government for the production of commodity
  - ❖ Tastes and preferences of the consumer in favour of the commodity
  - ❖ Agreement between the sellers to increase the supply of the commodity
  - ❖ Weather conditions favouring the production of the commodity
  - ❖ On-season for the production and supply of the commodity
  - ❖ If the goal of the firm is to maximize the profits
  - ❖ Decline in cost of production of commodity

As shown in the Figure 17.4, say, due to any of the above factors, the farmer supplies more quantity of paddy ($OQ_{P2}>OQ_{P1}$) at the same price OP. Hence, the supply curve S_1S_1 will move away from the origin as S_2S_2, indicating more quantity of paddy (OQ_{P2}) is supplied by the farmer at the same price OP. The higher supply curve shows that, farmers are able and willing to offer more of the commodity for sale at each price than before. It is clear from the Figure 17.4 that, increase in supply refers to 'more quantity offered for sale (OQ_{P2}) at the same price (OP) or same quantity offered for sale (OQ_{P1}) at lower price'.

- ☆ ***Decrease in supply:*** When due to factors other than the price of the same commodity, less quantity of commodity is supplied at the same price, it is called as Decrease in supply. Since, less quantity of a commodity is supplied at the same price, the supply curve shifts left side or towards the origin. The left side shift in supply curve may be due to:
  - ❖ Rise in the price of the substitute commodity
  - ❖ If price is expected to rise in the future
  - ❖ Labour strikes or lockouts in the factory
  - ❖ Outdated technology

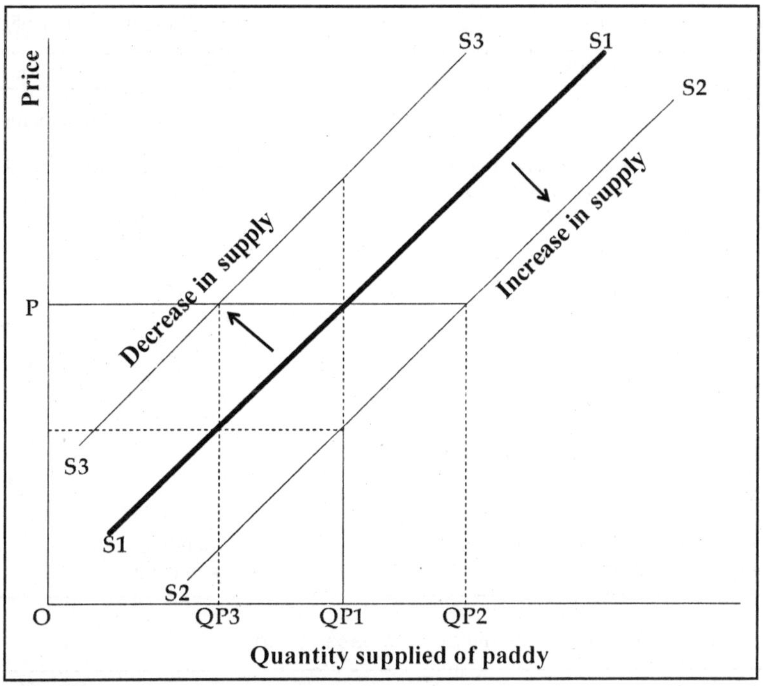

Figure 17.4: Shifts in the supply curve – Increase and Decrease in supply.

- ❖ More sales tax imposed on the commodity
- ❖ Less subsidies offered by the Government for the production of commodity
- ❖ Changes in tastes and preferences of the consumer against the commodity
- ❖ Weather conditions not favouring the production of the commodity
- ❖ Increase in cost of production of commodity

As shown in the Figure 17.4, say, due to any of the above factors, the farmer supplied less quantity of paddy ($OQ_{P3}<OQ_{P1}$) at the same price OP. Hence, the supply curve S_1S_1 will move towards the origin as S_3S_3, indicating less quantity of paddy (OQ_{P3}) is supplied by the farmer at the same price OP. The lower supply curve shows that, farmers are able and willing to offer less of the commodity at each price than before. It is clear from the Figure 17.4 that, decrease in supply refers to 'less quantity supplied (OQ_{P3}) at the same price (OP) or same quantity (OQ_{P1}) supplied at higher price'.

From the above discussion, we can infer the following:

- ☆ 'Changes in the quantity supplied' are represented by movement along the same supply curve.
- ☆ 'Extension in supply' refers to more quantity supplied due to increase in price of the same commodity.
- ☆ 'Contraction in supply' refers to less quantity supplied due to fall in the price of the same commodity.
- ☆ 'Changes in supply' are represented by the shifts in the supply curve.
- ☆ 'Increase in supply' refers to more quantity supplied due to changes in factors other than price of the same commodity.
- ☆ 'Decrease in supply' refers to less quantity supplied due to changes in factors other than price of the same commodity.

Supply vs Demand

Supply and demand have an inverse relationship with each other. If one is up, then the other will goes down. Supply has a direct relationship with price, thereby, supply curve slopes upward from left to right, while demand has an inverse relationship with price and the demand curve will slope downward from left to right. Both concepts have their own determinants. Supply's determinants reflect the firm, while determinants of demand reflect the consumers. There is no Total-Revenue test for price elasticity of supply (e_S), unlike E_P.

17.4. Elasticity of Supply

From the earlier discussion regarding the theory and basic concepts of supply, it is evident that, the law of supply simply explains the positive relationship between quantity supplied and price of the commodity. However, it does not explain the rate at which supply for a commodity changes due to change in price of the commodity. So, the law of

Table 17.3: Differences between Movement along the same supply curve and Shifts in the supply curve

Movement along the Same Supply Curve	*Shifts in the Supply Surve*
It implies changes in quantity supplied of the commodity	It implies changes in the level of supply itself for the commodity.
Other things remaining equal, the change in quantity supplied of a commodity due to change in the price of the same commodity refers to Movement along the same supply curve or change in quantity supplied.	When there is change in quantity supplied of a commodity due to changes in various factors other than price of the same commodity, it refers to shift in the supply curve or change in the supply.
The causal factor for the movement along the same supply curve is the change in price of the same commodity	The causal factors for shifts in the supply curve are changes in price of the related commodity, labour trouble, change in technology, taxation policy, subsidy offered by the Government for the production of commodity, tastes and preferences of the consumer, weather conditions, cost of production of commodity etc., other than price of the same commodity
There is no shift in the supply curve and the farmer moves up and down along the same supply curve	There is altogether shift in the supply curve either towards right side or left side.
There are two kinds of movements along the same supply curve *i.e.*, Extension of supply and Contraction of supply	There are two kinds of shifts in the supply curve *i.e.*, Increase in supply and Decrease in supply

Table 17.4: Differences between Extension of supply and Increase in supply.

Extension of Supply	*Increase in Supply*
It is a situation, where quantity supplied of a commodity varies due to change in price of the same commodity, other things remaining equal	It is a situation, where quantity supplied of a commodity varies due to factors other than price of the same commodity.
There is no shift in the supply curve and the farmer moves up along the same supply curve	There is altogether shift in the supply curve towards right side or away from the origin
The causal factor for extension of supply is increase in price of the same commodity	The causal factors for increase in supply are fall in the price of the substitute commodity, price is expected to fall in the future, no labour strikes or lockouts in the factory, improvements in technology, less sales tax imposed on the commodity, more subsidies offered by the Government for the production of commodity, tastes and preferences of the consumer in favour of the commodity, agreement between the sellers to increase the supply of the commodity, decrease in cost of production of commodity etc., other than price of the same commodity.

Table 17.5: Differences between Contraction of supply and Decrease in supply.

Contraction of Supply	*Decrease in Supply*
It is a situation, where quantity supplied of a commodity varies due to change in price of the same commodity, other things remaining equal	It is a situation, where quantity supplied of a commodity varies due to factors other than price of the same commodity.
There is no shift in the supply curve and the farmer moves down along the same supply curve	There is altogether shift in the supply curve towards left side or towards the origin
The causal factor for contraction of supply is decrease in price of the same commodity.	The causal factors for decrease in supply are rise in the price of the substitute commodity, price is expected to rise in the future, labour strikes or lockouts in the factory, outdated technology, more sales tax imposed on the commodity, less subsidies offered by the Government for the production of commodity, changes in tastes and preferences of the consumer against the commodity, increase in cost of production of commodity etc., other than price of the same commodity.

supply is straight forward, as it does not indicate how much responsive the supply is to a change in price. The impact of the price change is always not the same. Sometimes, the supply for a good changes considerably even for small price changes. On the other hand, there are some goods for which, the supply is not affected much by price changes. Supply, for some goods, are very responsive to price changes, while supply for certain others are not so responsive to price changes. So, the concept of elasticity of supply gains significance in measuring the rate of change in supply in response to change in price of the same commodity. According to Samuelson, *'elasticity of supply is the degree of the responsiveness of supply of a commodity to a change in its price'*. In the words of Prof. Bilas, *'elasticity of supply is a the percentage change in quantities supplied divided by the percentage change in price'*.

The quantity of a commodity supplied by the farmer per unit of time depends upon various factors *viz.*, price of a commodity, price of the related commodity, labour trouble, changes in technology, taxation policy, subsidies offered by the Government for the production of commodity, tastes and preferences of the consumer, weather conditions etc. While studying the elasticity of supply, we consider the influence of price of the same commodity (keeping other factors constant) on the responsiveness of various quantities of a commodity supplied by the farmer. So, when the relative responsiveness or sensitiveness of the quantity supplied is measured to changes in its price, the elasticity is said be e_S.

Price Elasticity of Supply (e_S)

The concept of e_S is commonly used in economic literature. It is the measure of degree of responsiveness of the quantity supplied of a commodity to changes in its price, *ceteris paribus*. Precisely, it is the ratio between proportionate or percentage change in quantity supplied of a commodity to a given proportionate or percentage change in the price of the same commodity. Symbolically, the e_S with reference to paddy commodity is given by

e_S = (% change in quantity supplied of paddy/% change in price of paddy) ***Equation 17.6***

So, to calculate the e_S, we need to know, what the percentage change in quantity supplied is and what the percentage change in price of the commodity is. It is best to calculate these one at a time.

Percentage change in quantity supplied: The formula used to calculate the percentage change in quantity supplied is:

[Quantity supplied(NEW) – Quantity supplied (OLD)] / Quantity supplied (OLD)

Percentage change in price: The formula used to calculate the percentage change in price is:

[Price(NEW) - Price(OLD)] / Price(OLD)

So, $e_S = (\Delta Q_P / Q_P) \div (\Delta P_P / P_P) = (\Delta Q_P / \Delta P_P) \times (P_P / Q_P)$ ***Equation 17.7***

where,

Q_P = Original quantity of paddy supplied

P_P = Original price of paddy

ΔQ_P = Change in quantity of paddy supplied

ΔP_P = Change in price of paddy.

Interpretation of e_S

The computation of e_S is helpful to see how sensitive the supply for a commodity is to its price change. The higher the price elasticity, the more sensitive farmers are to price changes. A very high price elasticity suggests that, when the price of a commodity increases, the farmers will offer a great deal more of it and when the price of that commodity goes down, the farmers will offer a great deal less of it. A very low price elasticity implies just the opposite that, changes in price have little influence on supply. The following are the rules of thumb to interpret the e_S:

- ☆ If $e_S > 1$ then, supply for the commodity is price elastic (Supply is sensitive to price changes)
- ☆ If $e_S = 1$ then, supply for the commodity is unit elastic
- ☆ If $e_S < 1$ then, supply for the commodity is price inelastic (Supply is not sensitive to price changes)

In computing the e_S, we get a positive sign because, the changes in quantity supplied of a commodity is in same direction to the change in its price. Thus, there are no minus signs to drop, as was necessary with E_P. Being a ratio in terms of percentages, the e_S will have no units for measurement.

Example 1

Let us suppose that, price of a mango falls from Rs. 15/unit to Rs.10/unit in a day. The decline in price leads to decrease in number of mangoes supplied by the farmer from 35 to 20. The e_S is computed using the above formula will be:

$e_S = (\Delta Q/\Delta P) \times (P/Q)$

Given, original quantity = 35, original price = Rs.15

ΔQ = 35-20 = 15, ΔP = 15-10= 5

e_S = (15/5)x(15/35)= 1.29

So, e_S is >1 and this implies, the commodity is price elastic and thus, supply is very sensitive to price changes.

Example 2

Suppose that, price of rice falls from Rs. 35/kg to Rs.20/kg in a day. The decline in price leads to decrease in the quantity supplied of rice from 25kgs to 20kgs. The e_S is computed using the above formula will be:

$e_S = (\Delta Q/\Delta P) \times (P/Q)$

Given, original quantity = 25, original price = Rs.35

ΔQ = 25-20 = 5, ΔP = 35-20 = 15

$e_s = (5/15) \times (35/25) = 0.47$

So, e_s is <1 and this implies, the commodity is price inelastic or less elastic and thus, supply is not very sensitive to price changes.

17.5. Degrees of e_s

As explained earlier, the supply of a commodity is sensitive or responsive to its price change. The variation in supply is, however, not uniform with a change in price across the commodities. For some commodities, a small change in price leads to a relatively larger change in quantity supplied. For example, a rise in price of the commodity by one per cent may extend the quantity supplied by 10 per cent. In some commodities, a greater change in price may bring about a very small change in the quantity of commodity supplied. For example, a rise in price of the commodity by 10 per cent may increase the quantity supplied by only one per cent. So, depending upon the degree of responsiveness and unresponsiveness of the quantity supplied to a change in price, the Economists categorized the degrees of e_s into five categories *viz.*, Perfect elastic or Infinite elastic supply, Perfect inelastic or Infinite inelastic supply, More elastic supply or More than unitary elastic supply, Unitary elastic supply and Less elastic supply or Less than unitary elastic or More inelastic supply and they are discussed here under.

(a) Perfect Elastic or Infinite Elastic Supply

When a very small change in price of the commodity brings about infinite change in quantity supplied of it, it refers to infinite elastic supply. This is an extreme degree of price elastic supply. But, this degree of price elasticity is not practically seen in the economy. The supply curve lies parallel to X-axis on which the quantity supplied for the commodity is taken, thereby, the e_s is equal to infinity (Panel A of the Figure 17.5). That means, when supply is perfectly elastic, the supplier (farmer) can supply any amount at the same price. This occurs, when the firm can supply at a constant price per unit and has no capacity limits to its production. A change in demand alters the equilibrium quantity, but not the market price.

(b) Perfect Inelastic or Infinite Inelastic Supply

When an infinite change in price of the commodity brings about no absolute change in quantity supplied for it, it refers to infinite inelastic supply. That means, say, a 50 per cent decrease or increase in the price leads to no change in the quantity supplied of the commodity. This is an extreme degree of price inelastic supply. But, this degree of elasticity is also not practically seen in the economy. The supply curve lies parallel to Y-axis on which the price of the commodity is considered, thereby, the e_s is equal to zero (Panel B of the Figure 17.5). When supply is perfectly inelastic, a shift in the demand curve has no effect on the equilibrium quantity supplied into the market, but only market price changes. Supply of rare books, stamps, coins etc., is perfectly inelastic. As these commodities are rare, their supply cannot be

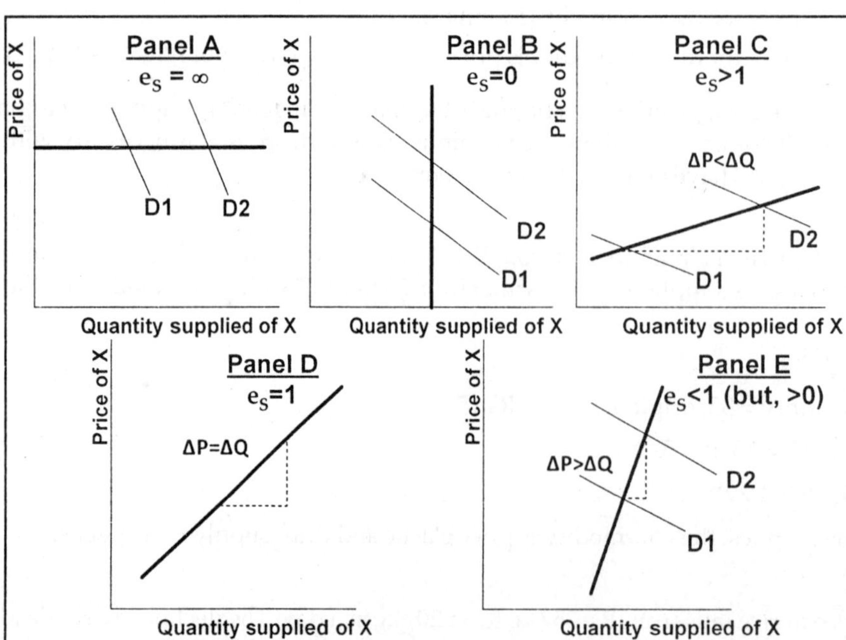

Figure 17.5: Different degrees of price elasticity of supply.

changed irrespective of the change in their prices. For example, if the supply of rare coins is 100, it will remain the same, whether the price is Rs. 1000/coin or Rs. 10000/coin in the market. Similarly, for the supply of tickets for movies, sports and supply of highly perishable agricultural produce in a very short period market, the e_s is equal to zero and the supply curve is vertical.

(*c*) More Elastic Supply or More than Unitary Elastic Supply

When a small change in price of the commodity brings about a greater or substantial change in quantity supplied, it implies, the supply for the commodity is more elastic. That means, when the percent change in quantity supplied of a commodity is greater than the percent change in its price, the supply is said to be more elastic. For example, one per cent change in price causes greater than a one per cent change in the quantity supplied of a commodity, thereby, $e_s = (\%\Delta Q/\%\Delta P) > 1$. The supply curve slopes upward with a flat slope indicating, $\Delta Q > \Delta P$ (Panel C of the Figure 17.5). So, when supply is relatively elastic, a change in demand influences the price less than the quantity supplied.

(*d*) Unitary Elastic Supply

When the percentage change in the quantity supplied of a commodity equals the percentage in its price, the e_s is said to have unitary elasticity. That means, $e_s = (\%\Delta Q/\%\Delta P) = 1$. For example, a one per cent change in price causes a response of exactly one percent change in the quantity supplied of a commodity. The supply curve slopes upward with a uniform slope, thereby, $\Delta Q = \Delta P$. As shown in the Panel D of the Figure 17.5, the percentage change in price brings about an exactly equal percentage in quantity at any point on the supply curve. This unitary elasticity is essentially a dividing line or boundary between the elastic and inelastic ranges of supply.

(*e*) Less Elastic Supply or Less than Unitary Elastic or More Inelastic Supply

When a greater or substantial change in price of the commodity brings only a small change in the quantity supplied, it implies, the supply for the commodity is less elastic or more inelastic. That means, when the per cent change in quantity supplied of a commodity is less than the per cent change in its price, the supply is said to be less elastic. For example, one per cent change in price causes less than a one per cent change in the quantity supplied of a commodity, thereby, $e_s = (\%\Delta Q/\%\Delta P) < 1$ (but greater than zero). The supply curve slopes upward with a steep slope, thereby, $\Delta Q < \Delta P$ (Panel E of the Figure 17.5). So, when supply is relatively less elastic, a change in demand influences the price more than the quantity supplied.

So, the above discussion infers that, the e_s ranges from zero to infinity across the commodities. For clarity purpose, different degrees of e_s and the degree of relationship of supply with price are explained through the Figure 17.6.

Degree of elasticity	e_s	Relationship of supply with price
Perfect elastic supply	Infinity	Infinite increase or decrease in supply for a very small change in price of the same commodity.
Perfect inelastic supply	Zero	No increase or decrease in supply even for an infinite change in price of the same commodity.
More elastic supply	Greater than one	% change in quantity supplied is more than % change in price of the same commodity.
Unitary elastic supply	Equal to one	% change in quantity supplied is equal to % change in price of the same commodity.
Less elastic supply	Less than one, but greater than zero	% change in quantity supplied is less than % change in price of the same commodity.

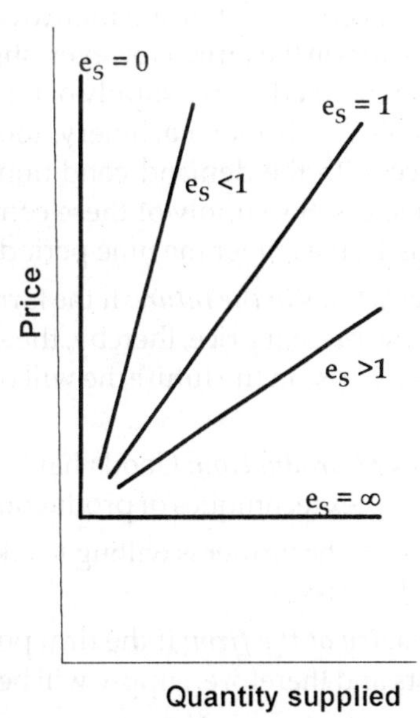

Figure 17.6: Different degrees of e_s and relationship of supply with price of the commodity.

17.6. Differences between Slope of Supply Curve and e_s

The slope of the supply curve differs from e_s in the following aspects:

- ☆ The slope of supply curve is given by the ratio between change in price of the commodity to change in quantity supplied of the commodity, given by,

  Slope = (ΔP)/(ΔQ).

  But, the e_S is the ratio between percentage change in quantity supplied of a commodity (paddy) to percentage change in price of the commodity (paddy), given by,

  $e_S = (\Delta Q_P / Q_P) \div (\Delta P_P / P_P) = (\Delta Q_P / \Delta P_P) \times (P_P / Q_P) = [(\Delta Q_P / \Delta P_P)] / [(Q_P / P_P)]$

  = Marginal change / Average change

- ☆ Price is in the numerator and quantity is in the denominator for slope. In contrast, quantity is in the numerator and price is in the denominator for e_S. So, slope is the inverse of e_S. When one is bigger, the other is smaller.
- ☆ Slope is calculated using the measurement units for price and quantity. In contrast, e_S is calculated using percentage changes. As such, slope includes the measurement units (such as Rs/kg or Rs/quintal etc), whereas, e_S is just a number with no measurement units. So, the value of slope changes, if the measurement units change (such as, Rupees versus Dollars, kilograms vs. pounds). However, the value of e_S remains same, as it is the ratio of percentages.

17.7. Factors Influencing e_S

The e_S is not the same for all the commodities. It may be high or low depending upon the number of factors influencing the quantity supplied of the commodity. The important factors which influence the e_S are discussed here under.

- ☆ *Nature of the commodity:* Durable commodities have elastic supply, as these commodities can be stored, if prices in the market are not favourable. If prices in the market turn favourable, the trader can offer more quantity for sale. Perishable commodities, on the other hand, have inelastic supply, as their supply cannot be varied even with increase in their price.
- ☆ *Changes in cost of production:* If the cost of production of a commodity increases due to rise in prices of factors of production, the farmers prefer to take up another enterprise, which is cost-effective. So, with increase in cost of production, the production of the crop gets affected, thereby, supply decreases in the market.
- ☆ *Element of time:* The time period plays a vital role in increasing the supply of the commodity into the market. In a very short period market dealing with perishable commodities like milk, leafy vegetables, eggs etc., their supply into the market is fixed, thereby, only demand plays a crucial role in influencing the price of the commodity. So, supply of these commodities is inelastic. In a short period market dealing with semi-durable goods like rice, wheat, pulses, oilseeds etc., their supply into the market can be slightly increased, as these can be stored for certain period and more over, their production can be slightly increased through increasing the variable factors in the farm. However, still demand plays a crucial role compared to supply in influencing the price of the commodity. So, supply of these commodities is elastic. However, in a long period market dealing with durable goods like machinery, tools, implements etc., their supply into the market can be varied in accordance with the demand conditions for these goods through changing the levels of both fixed and variable factors. So, supply of these commodities is highly elastic. This implies that, supply is likely to be highly elastic, the longer the time period a firm has to adjust its production.
- ☆ *Price expectations in the future:* If the farmer perceives that, the prices will fall in the future, he will offer more supply at the present price, thereby, the supply will be more elastic. Conversely, if the farmer perceives that, the prices will rise in the future, he will offer less supply at the present price, thereby, the supply will be less elastic.
- ☆ *Techniques of production:* Goods having simple techniques of production have elastic supply and goods having complex techniques of production, have inelastic supply.
- ☆ *Risk taking:* If the farmer is willing to take more risk in the production programme, the supply will be more elastic and *vice versa*.
- ☆ *Spare capacity of the firm:* If the firm possess more spare capacity, it is capable of increasing the output at lower costs and therefore, supply will be elastic.
- ☆ *Stocks with the farmer:* If the farmer possesses more stocks of produce in his godowns, he can respond quickly for offering large quantities for sale at times of higher prices, thereby, supply will be more elastic.
- ☆ *Elasticity of factor substitution:* If the E_{LC} between labour and capital is high, the e_S will be high and *vice versa*.
- ☆ *Stages of returns:* If the business experiences increasing returns, the supply is highly elastic, if the business experiences constant returns, the supply is elastic and if the business experiences diminishing returns, the supply is inelastic.

17.8. Methods of Measuring e_S

There are two methods of measuring the e_S *viz.*, Mid-point method and Point method or Geometric method. They are discussed in-detail here under.

1. Mid-point Method

For an accurate measurement of e_S, we employ this method. In this method, the initial and final data pertaining to both quantity supplied and price of the commodity should be taken into consideration to compute e_S. So, we have to take the average of initial and final data of both quantity supplied and price to compute the e_S and it is shown through the following equation:

$$e_S = \frac{(\text{New quantity} - \text{Old quantity})/[(\text{New quantity} + \text{Old quantity})/2]}{(\text{New Price} - \text{Old price})/[(\text{New price} + \text{Old price})/2]} \qquad \textit{Equation 17.8}$$

$$e_S = (\Delta Q)/[(Q_2+Q_1)/2]/(\Delta P)/[(P_2+P_1)/2]$$

Rearranging the terms, we get, $e_S = (\Delta Q/\Delta P) \times [(P_2+P_1)/(Q_2+Q_1)]$ *Equation 17.9*

Example 1

For example, if price of rice in the market is increased from Rs.20/kg to Rs. 35/kg, then the rice miller increased the supply from 50 kilograms to 100 kilograms. The e_S is given by,

$$e_S = [(\Delta Q/\Delta P)] \times [(P_2+P_1)/(Q_2+Q_1)] = [(50/15)] \times [(55)/(150)] = 1.22$$

Since $e_S > 1$, the supply of rice into the market is more elastic *i.e.*, it is more sensitive to price changes.

Example 2

For example, if price of rice in the market is increased from Rs.20/kg to Rs. 35/kg, then the rice miller increased the supply from 50 kilograms to 75 kilograms. The e_S is given by,

$$e_S = [(\Delta Q/\Delta P)] \times [(P_2+P_1)/(Q_2+Q_1)] = [(25/15)] \times [(55)/(125)] = 0.73$$

Since $e_S < 1$, the supply of rice into the market is less elastic *i.e.*, it is less sensitive to price changes.

2. Point Method or Geometric Method

The measurement of e_S can be illustrated with the help of Geometric method. As shown in the Panel A of the Figure 17.7, SS is the supply curve and assume that, we have to measure e_S at point M on the supply curve. In order to measure the e_S at point M, draw a tangent AB, such that, it touches X-axis at point A. So, e_S at point M is given by,

$$e_S = [(\Delta Q)/(\Delta P)] \times [(P)/(Q)] = [(Q_1Q_2)/(P_1P_2)] \times [(OP_1)/(OQ_1)]$$

Considering the triangle MCD, $Q_1Q_2 = MD$, $P_1P_2 = CD$. So,

$$e_S = [(MD)/(CD)] \times [(OP_1)/(OQ_1)] \qquad \textit{Equation 17.10}$$

Since, the triangles MAQ_1 and MCD are similar, the ratio of their sides are equal *i.e.*, $MD=AQ_1$, $CD=Q_1M=OP_1$.

Substituting above terms in Equation 5, we get,

$$e_S = [(MD)/(CD)] \times [(OP_1)/(OQ_1)] = [(AQ_1)/(OP_1)] \times [(OP_1)/(OQ_1)]$$

$$= (AQ_1)/(OQ_1) \qquad \textit{Equation 17.11}$$

Therefore, the value of $(AQ_1)/(OQ_1)$ gives the coefficient of e_S at point M on the supply curve SS. In Panel A, as $AQ_1<OQ_1$, e_S is less than one.

Similar methodology is followed to compute e_S at points M_1 (Panel B of the Figure 17.7) and M_2 (Panel C of the Figure 17.7) on the supply curves S_1S_1 and S_2S_2 respectively. At point M_1 on the supply curve S_1S_1 (Panel B), the e_S is more than unity, as point A touches the X-axis to the left of the point of origin. At point M_2 on the supply curve S_2S_2 (Panel C), the e_S is equal to unity, as the tangent touches the X-axis at the point of origin.

To conclude, e_S will be more than unity, if the tangent drawn to the point considered on supply curve touches the Y-axis. e_S will be equal to unity if the tangent drawn to the point considered on supply curve touches the X-axis at the point of origin and e_S will be less than unity, if the tangent drawn to the point considered on supply curve touches the X-axis.

The above analysis also helps to compare e_S at different points on the supply curve S_3S_3. As shown in Panel D of the Figure 17.7, since $[A_1Q_1/OQ_1]>[A_2Q_2/OQ_2]$, the e_S is more at point M compared to point N.

There is no Total-Revenue test for e_S. Supply curve shows a positive or direct relationship between price and quantity supplied, thereby, the supply curve is upward sloping. So, regardless of the degree of elasticity or inelasticity, price and TR always move together.

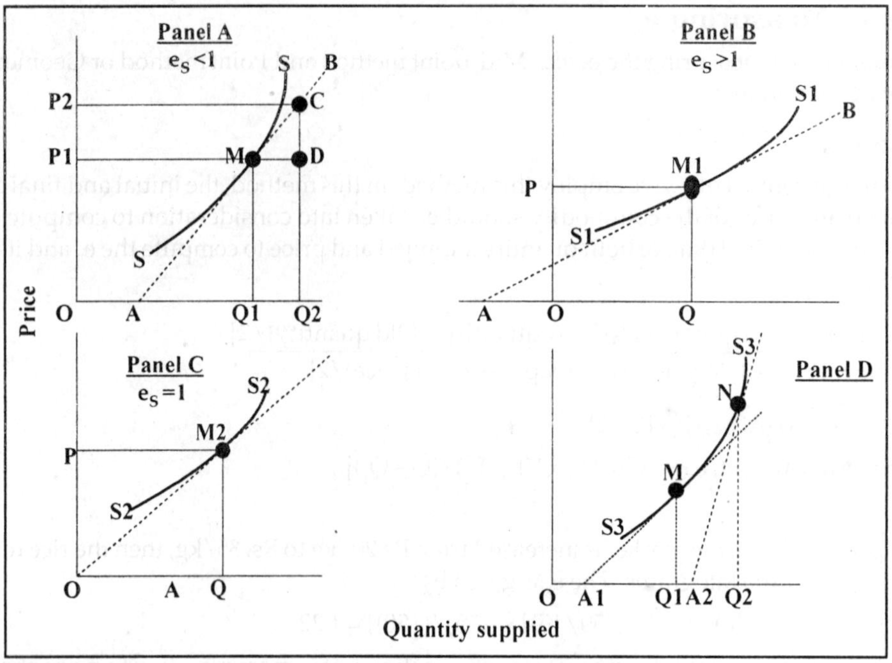

Figure 17.7: Point or Geometric method of measuring price elasticity of supply.

17.9. Applications of e_S

The idea of price elasticity of supply has widespread applicability, as suggested by the following examples:

Volatile Gold Prices

Price of gold is volatile, sometimes shooting upwards one moment and dramatically decreasing the next. The main sources of these fluctuations are due to increase or decrease in demand only, as the supply of gold is highly inelastic. This is because, gold production is a costly and time-consuming process. Moreover, the physical availability of gold is highly limited. So, increases in gold prices do not elicit substantial increases in quantity supplied. Conversely, gold mining is costly to shut down and existing gold bars are expensive to store. So, price decreases do not influences in the large drops in the quantity of gold supplied. This implies, the supply of gold is inelastic. However, gold is demanded for several purposes like jewelry, dental fillings, coins and for speculative financial investment. The demand for gold is further influenced by the fear general inflation or changes in exchange values of currencies across the countries etc. So, because of the inelastic supply of gold, even relatively small changes in demand bring about large changes in its price.

Elasticity of Labour Supply

The concept of e_s can be applied to labour to show, how the number of labour supplied responds to changes in their wages. What makes this case interesting is that, it has sometimes been found that, the measured elasticity is negative, that is, that an increase in the wage rate is associated with a reduction in the number of labour supplied. In most cases, labour supply curve have normal upward slope *i.e.*, higher wages induce labour to work more. For them, having the additional income from working more is preferable to having more leisure time. However, wage increases may lead some people in very highly paid jobs to cut back on the number of hours they work because, their incomes are already high and they would rather have more time for leisure activities. In this case, the labour supply curve would have a negative slope. The reasons for this phenomenon are already explained under Figure 9.59.

Antiques and Reproductions

Since, antiques are impossible to reproduce, these products have high inelastic supply. So, the high price of an antique results from strong demand and highly inelastic supply. The higher price might prompt the discovery of a few more of the remaining originals and thus add to the quantity available for sale, but this quantity response is usually quite small. So the supply of antiques and other collectibles tends to be inelastic. So, the influential factors like increase in population, higher income greater enthusiasm for collecting antiques etc., will increase the demand for antiques. Since, supply being highly inelastic, the increase in demand is the determining factor for boosting the prices of antiques.

17.10. Supply Curve Under Perfect Competition

The supply curve under perfect competitive market situation can be studied under short run and long run production programmes.

1. Supply Curve Under Perfect Competition in Short Run

As discussed earlier, in short run production programme, the farmer can vary only few factors (fertilizers, pesticides, hired labour etc) to influence the output and remaining factors (machinery, land etc) are fixed. In a perfect market situation,

- ☆ The price line, AR Curve (ARC) or demand curve, MR Curve (MRC) are equal and parallel to X-axis.
- ☆ The point of optimality of output in the production programme is SRMC=MR.
- ☆ The price of output of firm must at least cover SRAVC to continue the production programme.

(a) Supply Curve for Firm in Short Run

In the Figure 17.8, the SRMC curve cuts the SRAVC curve at point A, where the firm sells OQ_1 quantity of commodity at OP_1 price. Below this OQ_1 quantity, the SRMC curve lies below the SRAVC curve and the price will not cover the SRAVC. So, the firm will not operate when the price falls below the SRAVC. When the price of the commodity increases beyond SRAVC, the firm will increase the supply of the commodity beyond OQ, thereby, the SRMC curve will lie above the SRAVC. When the SRMC curve intersects the SRATC curve at point B, the price per unit of output is OP. At this price OP, the SRAVC is covered and this is due to more quantity of commodity offered for sale beyond OQ_1. So, the portion of SRMC curve beyond SRAVC refers to Short Run Supply (SRS) curve. So, at price OP, the firm is earning normal profits (discussed in Chapter 18) by selling OQ quantity of commodity. If the price of the commodity increases to OP_2, the firm still increases the supply of commodity even beyond OQ and earn super normal profits (discussed in Chapter 18) by selling OQ_2 quantity of commodity. This clearly infers that, the portion of SRMC curve, which lies above the SRAVC curve and to the right of point of intersection of SRMC curve and SRAVC curve constitute supply curve of firm in the short run.

(b) Supply Curve for Industry in the Short Run

In the above analysis, we analyzed the supply curve for an individual firm. The SRS curve for the industry is given by the summation of supply curves of all the firms operating under perfect competition producing the same commodity. In the Panel A of the Figure 17.8, the portion of SRMC curve above 'A' constitute SRS curve for an individual firm and it is clear that, at price OP_1, the quantity of commodity supplied by the firm is OQ_1, at OP price the quantity supplied by the firm is OQ and at OP_2 price the quantity supplied by the firm is OQ_2. Let us assume, there are 100000 firms in the industry operating under perfect competition producing the same commodity. So, for the 100000 firms, the total quantity of commodity supplied at OP_1 price is given by OM_1=OQ_1*100000. Similarly, the industry supply at OP price is given by, OM=OQ*100000 and at OP_2 price is given by OM_2=OQ_2*100000. So, now connecting the respective prices and

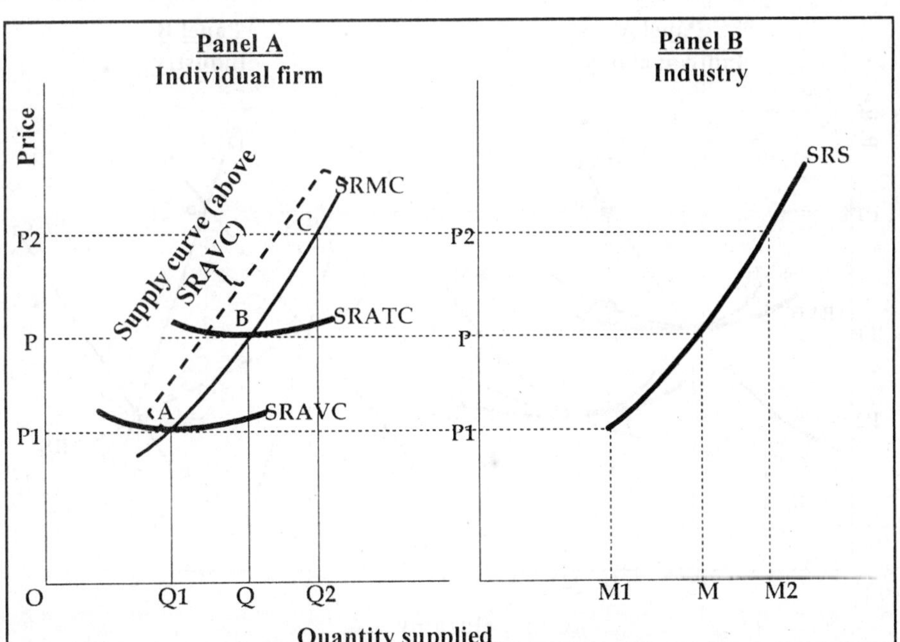

Figure 17.8: Short run supply curve of the firm and industry under perfect competition.

quantities supplied by the industry, we get the industry supply curve for the commodity and it is also an upward sloping SRS curve from left to right. Like an individual firm, the industry also stops supplyiiig the commodity below OP_1 price.

The upward sloping of the industry supply curve is due to upward sloping of SRMC curves of the firms in the industry. But, if the SRMC curves of some of the firms slopes downward, the industry supply curve also may slope downwards. But, such a situation is not possible because, for the point of firm's equilibrium or optimality of output, the SRMC curve should cut the MR curve from below and not from above.

2. Supply Curve Under Perfect Competition in the Long Run

In the long run production programme, all the factors are variable and no factor is fixed in the firm. So, the firm can change all the factors, thereby, the cost structure will be altered and hence, there will be shift in the cost curves of the individual firms. In the long run, under perfect competition, there is no possibility for an individual firm to enjoy super normal profits. This is because, if the profits are attractive in the industry in the short run, new firms will enter the industry and this will squeeze out the super normal profits in two ways *viz.*, because of entry of new firms, the super normal profits are shared or distributed among all the firms and with the entry of new firms, the demand for factors of production will increase leading to escalation of cost structure. So, the firms operating under perfect competition in the long run will enjoy only normal profits and super normal profits are not possible. So, the point of optimality when the firm enjoys normal profits in the long run is LRMC is equal to MR=LRATC. Since, due to variations in the cost structure, the cost curves shift from their original position and accordingly, the Long Run Supply (LRS) curve will have different shapes and they are discussed here under when the industry experiences increasing returns to scale or constant returns to scale or diminishing returns to scale in the production programme.

(a) Supply Curve when the Industry Experiences 'Increasing Returns' or 'Decreasing Costs' in the Long Run

Let us assume that, the industry is operating under increasing returns to scale or decreasing costs. These decreasing costs are due to economies of scale experienced by the industry. So, due to decrease in costs, the cost curves will shift downward from their original position (Panel A of the Figure 17.9).

In Panel B of the Figure 17.9, the industry supplies OX quantity of commodity at OP price with reference to equilibrium point A of demand (DD) and supply (SS) curves. At individual firm level (Panel A), the firm supplies OQ quantity at the same price OP, where the firm is in equilibrium, when price line (OP)=MR=SRMC=SRATC=LRATC. Let us suppose, the demand for the commodity increases in the market, thereby, the new demand curve is D_1D_1 (Panel B) and it intersects the industry supply curve SS at point B, where the price of the product is OP_1. At this price OP_1, the individual firm supplies OQ_1 quantity of commodity into the market, where the SRMC curve cuts the price line=MR at point M. Since this price OP_1 is above SRATC (by MN), the firm enjoys super normal profits in the business. So, new

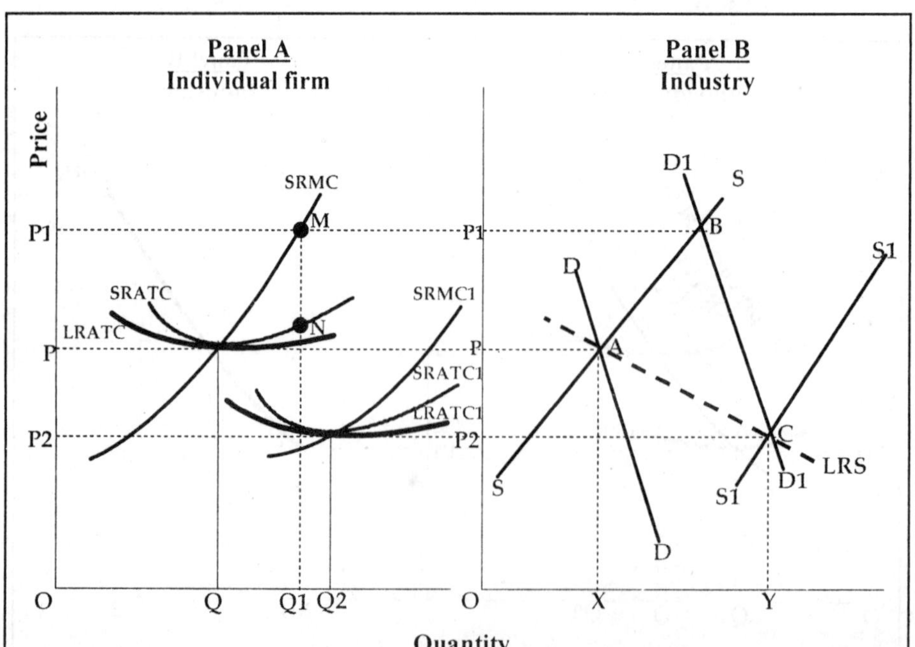

Figure 17.9: Long run supply curve of the industry under perfect competition experiencing increasing returns or decreasing costs.

firms gets attracted to these profits, enter the industry and since the industry is experiencing increasing returns or decreasing costs, the cost structure will fall from SRATC to $SRATC_1$, SRMC to $SRMC_1$ and LRATC to $LRATC_1$. So, when new firms enter the industry, the industry supply of the commodity into the market increases and the new supply curve for the industry is S_1S_1 (Panel B) and it intersects the new demand curve D_1D_1 at point 'C' and the equilibrium price is OP_2. At this price OP_2, the individual firm is in equilibrium, where price line (OP_2)=MR=$LRATC_1$=$SRMC_1$=$SRATC_1$. At this price OP_2, the firm supplies OQ_2 quantity of commodity and the industry supply is OY. Since, the price decreases from OP_1 to OP_2, some firms could not even recover their average costs, thereby, they leave the industry. So, the remaining firms, who enjoy price OP_2 greater than their average costs will contribute to expansion of output, as they experience increasing returns, and hence, the industry output increases from OX to OY. So, by connecting the initial and new equilibrium points of demand and supply of commodity in the industry, *i.e.*, A and C, we can derive the LRS curve of the industry. It is downward sloping as shown in Panel B of the Figure 17.9 indicating that, as industry expands, more quantity of commodity is offered for sale at lower factors costs due to increasing returns in the industry.

(*b*) Supply Curve when the Industry Experiences 'Constant Returns' or 'Constant Costs' in the Long Run

Let us assume that, the industry is operating under constant returns or constant costs. These constant costs are due to the simultaneous experience of economies of scale and diseconomies of scale and they both counter balance each other in the long run. That means, either when new firm enter the industry or leave the industry, there will not be any change in the factor prices. That means, there is no change in cost structure and hence, there is no shift in the position of cost curves (Panel A of the Figure 17.10).

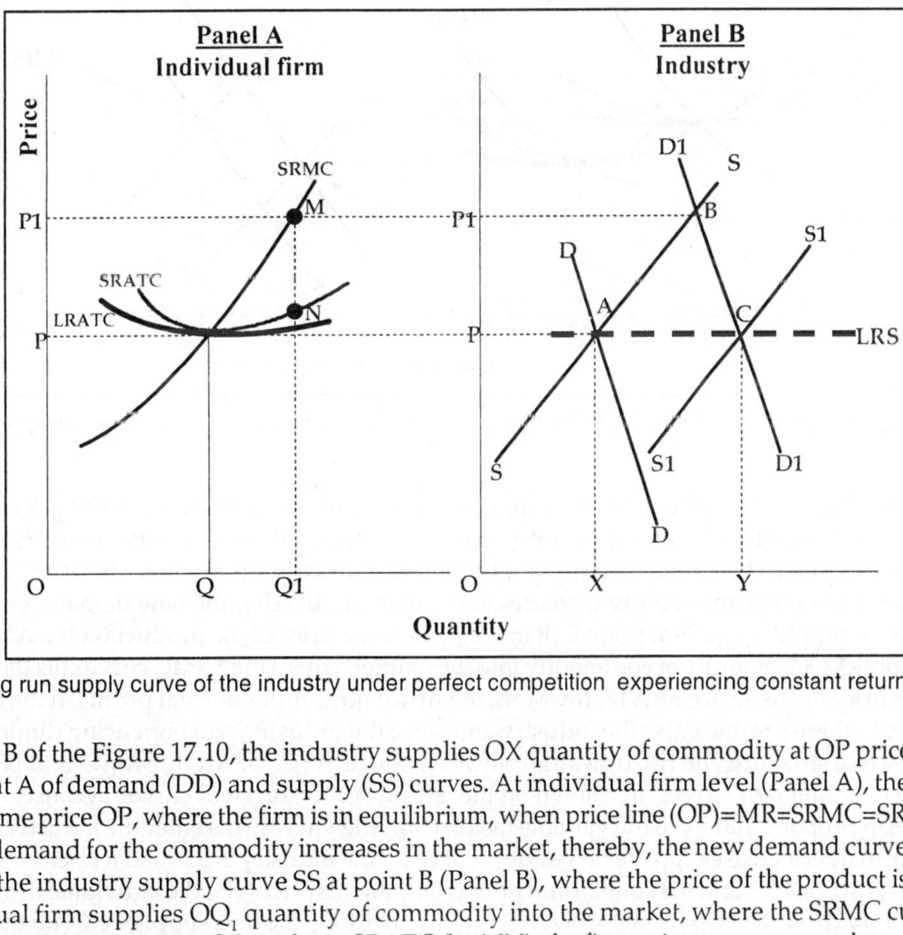

Figure 17.10: Long run supply curve of the industry under perfect competition experiencing constant returns or constant costs.

In the Panel B of the Figure 17.10, the industry supplies OX quantity of commodity at OP price with reference to equilibrium point A of demand (DD) and supply (SS) curves. At individual firm level (Panel A), the firm supplies OQ quantity at the same price OP, where the firm is in equilibrium, when price line (OP)=MR=SRMC=SRATC=LRATC. Let us suppose, the demand for the commodity increases in the market, thereby, the new demand curve is D_1D_1 (industry) and it intersects the industry supply curve SS at point B (Panel B), where the price of the product is OP_1. At this price OP_1, the individual firm supplies OQ_1 quantity of commodity into the market, where the SRMC curve cuts the price line=MR at point M. Since this price OP_1 is above SRATC (by MN), the firm enjoys super normal profits in the business. So, new firms gets attracted to these profits, enter the industry and since the industry is experiencing constant returns or constant costs, the cost structure will remains same for the firms. That means, there is no change in minimum point of LRATC curve. So, when new firms enter the industry, the supply of the commodity into the market increases and the new supply curve for the industry is S_1S_1 and it intersects the new demand curve D_1D_1 at point 'C' (Panel B) and the price falls back to OP. At this price OP, the individual firm will be at the original equilibrium position *i.e.*, price line (OP)=MR=SRMC=SRATC=LRATC, as there is no change in cost structure of the firm due to constancy of factor prices and factors are in perfect elastic supply. So, by connecting the initial and new equilibrium points of demand and supply of commodity in the industry, *i.e.*, A and C, we can derive the LRS curve of the industry. It is a straight line parallel to X-axis as shown in Panel B of the Figure 17.10 indicating that, as industry expands, more quantity of commodity is offered for sale at constant factors costs due to constant returns in the industry. It is further interesting to note that, the total output supplied by the industry increases from OX to OY, though the supply of output by each firm remains same at OQ. This is due to increase in number of firms in the industry, as some new firms gets attracted towards the super normal profits enjoyed by firms in the industry in the short run.

(c) Supply Curve when the Industry Experiences 'Diminishing Returns' or 'Increasing Costs' in the Long Run

Let us assume that, the industry is operating under diminishing returns to scale or increasing costs. This increasing costs is due to two reasons *viz.*, the entry of new firms into the industry leads to increase in demand for the factors of production, thus, increasing their prices and due to diseconomies of scale experienced by the industry. These contribute to shift in cost curves upward from their original position (Panel A of the Figure 17.11).

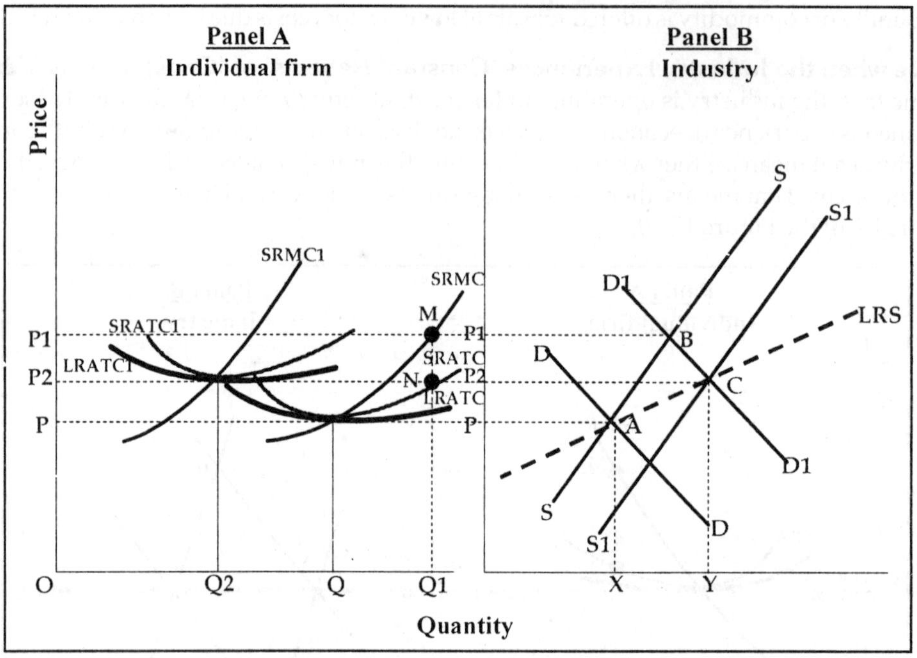

Figure 17.11: Long run supply curve of the industry under perfect competition experiencing diminishing returns or increasing costs.

In Panel B of the Figure 17.11, the industry supplies OX quantity of commodity at OP price with reference to equilibrium point A of demand (DD) and supply (SS) curves. At individual firm level (Panel A), the firm supplies OQ quantity at the same price OP, where the firm is in equilibrium, when price line (OP)=MR=SRMC=SRATC=LRATC. Let us suppose, the demand for the commodity increases in the market, thereby, the new demand curve is D_1D_1 and it intersects the industry supply curve SS at point B (Panel B), where the price of the product is OP_1. At this price OP_1, the individual firm supplies OQ_1 quantity of commodity into the market, where the SRMC curve cuts the price line=MR at point M. Since this price OP_1 is above SRATC (by MN), the firm enjoys super normal profits in the business. So, new firms gets attracted to these profits, enter the industry and since the industry is experiencing diminishing returns or increasing costs, the cost structure will rise upward from SRATC to $SRATC_1$, SRMC to $SRMC_1$ and LRATC to $LRATC_1$. These new (higher) per unit cost curves are shown on the left side of the original curves because, the prices of fixed factors rise more than proportionately to the variable factors. So, when new firms enter the industry, the supply of the commodity into the market increases and the new supply curve for the industry is S_1S_1 and it intersects the new demand curve D_1D_1 (industry) at point 'C' and the equilibrium price is OP_2. At this price OP_2, the individual firm is in equilibrium, where price line (OP_2)=MR=$LRATC_1$=$SRMC_1$=$SRATC_1$. At this price OP_2, the firm supplies OQ_2 quantity of commodity and the industry supply is OY. So, by connecting the initial and new equilibrium points of demand and supply of commodity in the industry, *i.e.*, A and C, we can derive the LRS curve of the industry. It is upward sloping as shown in Panel B of the Figure 17.11 indicating that, as industry expands, more quantity of commodity is offered for sale at higher factors costs due to diminishing returns in the industry.

17.11. Supply Curve Under Imperfect (Monopoly) Competition

Monopoly is a market situation, where there is a single seller for the commodity in the market. Since, he is the only seller for the commodity, he is a price maker, unlike in perfect competition and moreover, he can also decide the level of output of the commodity in the firm (industry). So, a monopolist can fix either price or determine the output as he wishes and hence, we cannot draw a unique supply curve for the monopolist. The same is explained through the following cases:

Case 1: Given Price is Associated with Two Different Levels of Output

Let us assume, the monopolist faces two AR curves AR_1 and AR_2 (Panel A of Figure 17.12). The given MC curve of the firm will intersect MR_1 curve at E_1, where the firm supplies OQ_1 quantity of commodity into the market at OP price. When the demand for the commodity is given by AR_2, the MC curve will intersect the MR_2 curve at E_2, where the firm supplies OQ_2 quantity of commodity into the market at the same OP price. So, the output supplied by the monopolist depends upon the demand conditions for the commodity, as at the same price OP, when demand is more, more quantity is offered for sale and when the demand is less, less quantity is offered for sale by the monopolist. So, we cannot draw a unique supply curve for the firm.

Case 2: Given Output is Associated with Two Different Prices

Let us assume, the monopolist faces two AR curves *viz.*, AR_1 and AR_2 (Panel B of Figure 17.12). The given MC curve for the firm will intersect MR_1 curve at E, where the firm supplies OQ quantity of commodity into the market at OP_1 price. When the demand for the commodity is given by AR_2, the MC curve will intersect the MR_2 curve at E, when the firm supplies the same quantity OQ, but charge a higher price OP_2. This is the case, when the monopolist experiences different elasticities of demand in different periods. In one period, when the demand for the product is more elastic, (*i.e.*, AR_1 curve), the monopolist charges OP_1 (lower) price and sells OQ quantity of commodity. In another period, when the demand for the product is less elastic (i.e AR_2 curve), the monopolist charges OP_2 (higher) price and sells the same OQ quantity of commodity. So, for selling the same output, the monopolist can charge different prices depending upon the E_P.

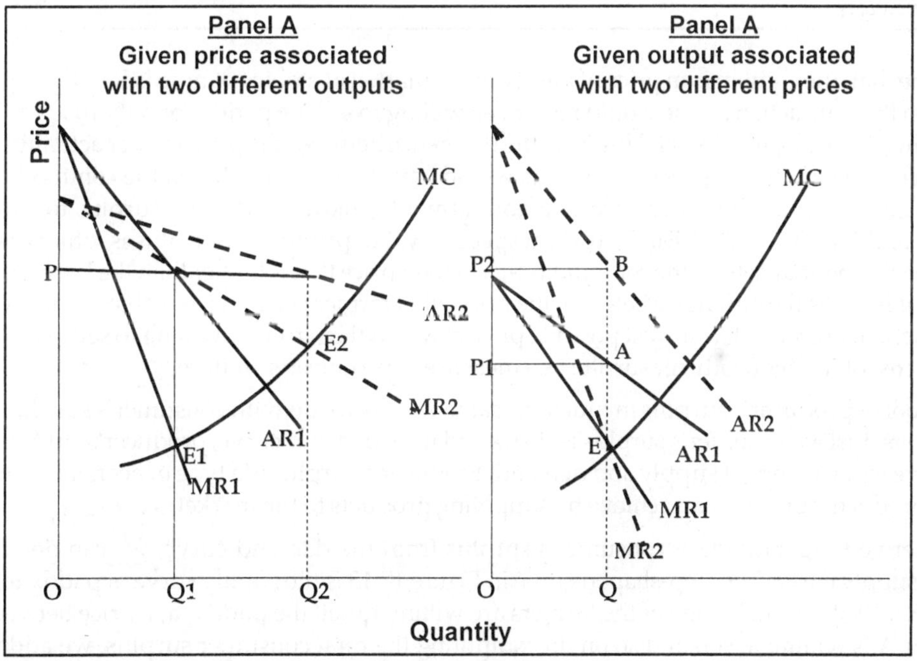

Figure 17.12: Output-Price determination under Monopoly competition.

So, it is evident from the above two cases that, we cannot draw a unique supply curve for the monopoly firm, as the supply of commodity into the market depends upon demand conditions and E_P for the product.

17.12. Producer Surplus

Having discussed the concept of supply, it is essential to study the concept of producer surplus. Just as buyers of a commodity would have been willing to pay more for their purchase than the price they actually pay in the market, the sellers of a commodity would have been willing to sell it for less than the price they actually receive. So, this guides us to focus on the concept of producer's surplus through analyzing the supply behaviour of the producer. This concept runs almost exactly parallel to that of consumer surplus of demand analysis.

We practically see, several producers or traders selling the products in the market. Let us assume, five farmers are selling one quintal of paddy each in the market. Since the cost of production of paddy is different across the farmers, they sell the same at different prices in the market. Table 17.6 shows the different prices at which the paddy is offered for sale by the five farmers in the market. Farmer A is willing to sell the paddy as long as he can get anything more than Rs. 1000/quintal, Farmer B will sell the paddy only if he gets a minimum of Rs 1050/quintal, Farmer C won't sell unless he gets at least Rs. 1075/quintal, Farmer D, unless he gets Rs. 1100/quintal and Farmer E, unless he gets Rs. 1200/quintal. These prices are also called as Seller's cost, as it is the lowest price at which a potential seller is willing to sell the commodity in the market. So, Farmer A's selling cost is Rs. 1000/quintal, Farmer B's selling cost is Rs. 1050/quintal and so on. It is interesting to note that, here, we mentioned the word 'cost', which people normally use the same referring to monetary cost of producing the paddy. The term 'cost' is used here in the sense of 'opportunity cost' of selling the paddy in the market. This is because, if the farmer sells a quintal of paddy, he won't have it later for his food consumption, seed purpose etc. So, the true measure of the cost of doing anything is always its opportunity cost *i.e.*, the real cost of something is what the seller must give up to get it. So, it is appropriate to say the minimum price at which the farmer will sell the paddy as the 'cost' of selling that commodity.

Table 17.6: Total producer surplus in transacting paddy by five farmers.

Farmers	*Cost/Price at which the Farmer is Willing to Sell Paddy (Rs/quintal)*	*Market Price/Price at which Farmer Actually Sold Paddy (Rs/quintal)*	*Individual Producer's Surplus (Rs)*
Farmer A	1000	1090	90
Farmer B	1050	1090	40
Farmer C	1075	1090	15
Farmer D	1100	—	—
Farmer E	1200	—	—
Total producer surplus			145

Let us assume that, the equilibrium price of paddy in the market at a point of time is Rs. 1090/quintal. So, Farmer A has gained from the transaction *i.e.*, he would have been willing to sell the paddy for only 1000/quintal, but actually sells at Rs. 1090, thereby, he gained Rs. 90. This gain, the difference between the price farmer actually gets (Rs.1090) and his cost (Rs.1000) *i.e.*, the minimum price at which he would have been willing to sell the commodity is known as his Individual producer's surplus. Similarly, farmer B and farmer C enjoys producer's surplus to a tune of Rs. 40 (*i.e.*, Rs.1090-Rs.1050) and Rs. 15 (*i.e.*, Rs.1090-Rs.1075) respectively. So, producer's surplus is, otherwise, defined as the amount that farmers benefit by selling the commodity at a market price that is higher than the lowest price he would sell for. Similarly speaking, it is the difference between the price what farmer is willing and able to supply a commodity for and the price he actually receive. The lowest possible price at which the farmer is willing to sell a commodity is usually the average total cost of it. This producer surplus is a measure of producer's welfare.

We use the tool of producer's surplus in much the same way as we do with consumer's surplus in measuring the well-being of buyers. Just as consumer's surplus is abstracted from demand curve, producer's surplus is closely related to the supply curve in the theory of supply and demand. Producer's surplus is a type of economic surplus. It relates to the welfare that producers or firms can achieve by supplying products to the market.

Just as we derived the individual consumer's surplus from the demand curve, we can derive the producer's surplus from the supply curve. The step-shaped curve in Figure 17.13 is the supply curve of paddy for five farmers. At a price less than Rs. 1000/quintal, none of the farmers are willing to sell the paddy, at a price between Rs. 1000 to Rs. 1050, only Farmer A is willing to sell and so on. In computing the total consumer surplus, we added the individual consumer's surpluses. So, in the same way, we can calculate the total producer surplus, by adding the total gains of all the individual producer's surpluses in the market.

Table 17.6, shows the net gain to each of the farmers (producer's surplus), who sold paddy at the equilibrium market price (Rs.1090/quintal) *i.e.*, Rs. 90 for Farmer A, Rs. 40 for Farmer B and Rs. 15 for Farmer C. The total producer surplus is Rs.90 + Rs.40 + Rs. 15 = Rs. 145. Each step in the supply curve (Figure 17.13) is one quintal quantity of paddy wide and represents one seller. The height of Farmer A's step is Rs. 1000 *i.e.*, his cost. This forms the bottom of a rectangle, with Rs. 1090, the price he actually receives for his one quintal of paddy, forming the top. The area of this rectangle, (Rs.1090 - Rs.1000) × 1 = Rs. 90, is his producer's surplus. So the producer's surplus, the Farmer A gains from selling his one quintal of paddy is the area of the rectangle ABCD, shown in the Figure 17.13. Likewise, the individual producer's surpluses of Farmer B and Farmer C will be Rs. 40 and Rs. 15 respectively, represented by the rectangles DEFG and GHIJ respectively. So, the above discussion infers that, the total producer surplus from sales of a paddy at a given price is the area above the supply curve, but below the equilibrium market price *i.e.*, AJC. This is exemplified more clearly through the Figure 17.14, where in, a large number of farmers transacted paddy in the market to a tune of 10 million tonnes at an equilibrium market price of Rs. 1090/quintal. So, the total producer surplus is ABC *i.e.*, the

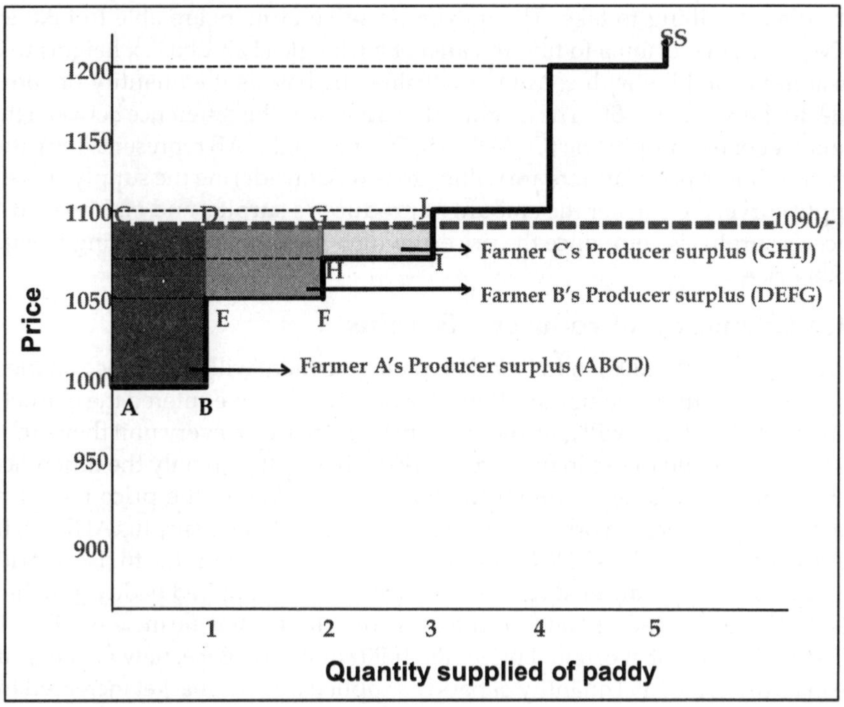

Figure 17.13: Individual producer's surplus in selling paddy .

triangular area above the supply curve, but below the equilibrium market price, OB. This is only part of the TR earned by selling paddy in the market, which is indicated by the rectangle OBCQ. So, mathematically,

Producer surplus = Amount received by sellers – Cost to sellers

1. Measurement of Producer's Surplus

Producer's surplus measures the gain the farmers receive when they can sell their commodity at a higher price than the one they were willing to sell. The methodology employed to calculate the producer's surplus is similar to the concept of consumer's surplus. In the Figure 17.14, the triangle ABC represent producer's surplus of the farmers selling paddy. That means, producer surplus's is the triangle made up of the area above the supply curve, but under the equilibrium price and to the right of the vertical axis. The vertical intercept of the supply line (OA) indicates the

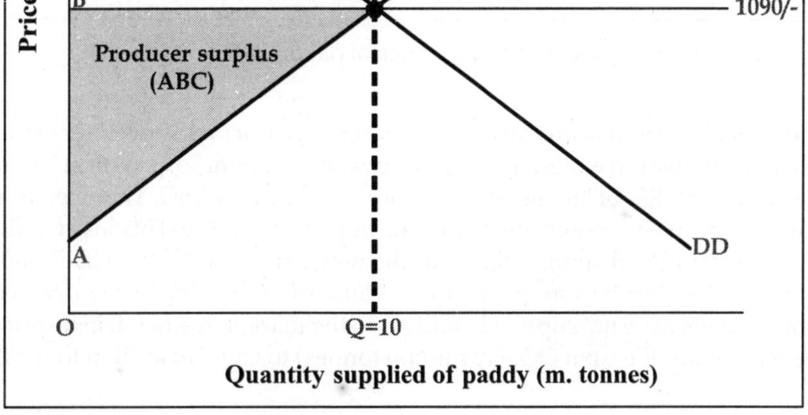

Figure 17.14: Total producer surplus in selling paddy.

minimum price the farmers are willing to take. The maximum price farmers are able to take is illustrated by the equilibrium price line. Now, use the formula to find the area of a triangle (1/2 x base x height) to calculate producer surplus. The ½ is a constant factor. Like with consumer surplus, the base is the quantity of commodity sold at the equilibrium price parallel to the X-axis *i.e.*, BC. The height of the triangle is the difference between the equilibrium price (OB) and the vertical intercept of the supply line (OA) *i.e.*, AB. So, the height, AB represents the difference between the equilibrium price and the minimum price farmers are willing to take. Considering the supply curve in the Figure 17.14, the more elastic the supply curve, the smaller the amount of producer's surplus and *vice versa*. If the supply curve is perfectly elastic, producer's surplus is zero, since the price at which the farmers are willing to supply their output is equal to the ruling market price.

2. Influence of Price Changes on Producer's Surplus

If the price of paddy increases (Rs.1300) in the market, paddy farmers will experience an increase in producer's surplus, though not all farmers gain the same amount. Some farmers would have entered the market even at the original price of paddy (Rs. 1090/quintal) and they will gain the entire price increase on every unit they produce. Other farmers will supply paddy into the market due to rise in its price, thereby, they will gain only the difference between the new price and their cost. The same is explained through the Figure 17.15. Before the price rise of paddy, ABC is the producer's surplus. But, with the increase in price of paddy, the total producer's surplus ADF. So, the new producer's surplus gained due to rise in price of paddy is BDCF. This new producer's surplus due to rise in price of paddy (BDCF) involves two components *viz.*, rise in producer surplus to the farmers who supplied paddy into the market even when the price of paddy is Rs. 1090/quintal (BCDE) and producer surplus to the new farmers (CEF), who supplied paddy only at a higher price (Rs. 1300), but not at original price (Rs. 1090/quintal). Since, new farmers supplied paddy into the market with the rise in its price, the total quantity of paddy supplied into the market increased from OQ_1 (10 million tonnes) to OQ_2 (12 million tonnes).

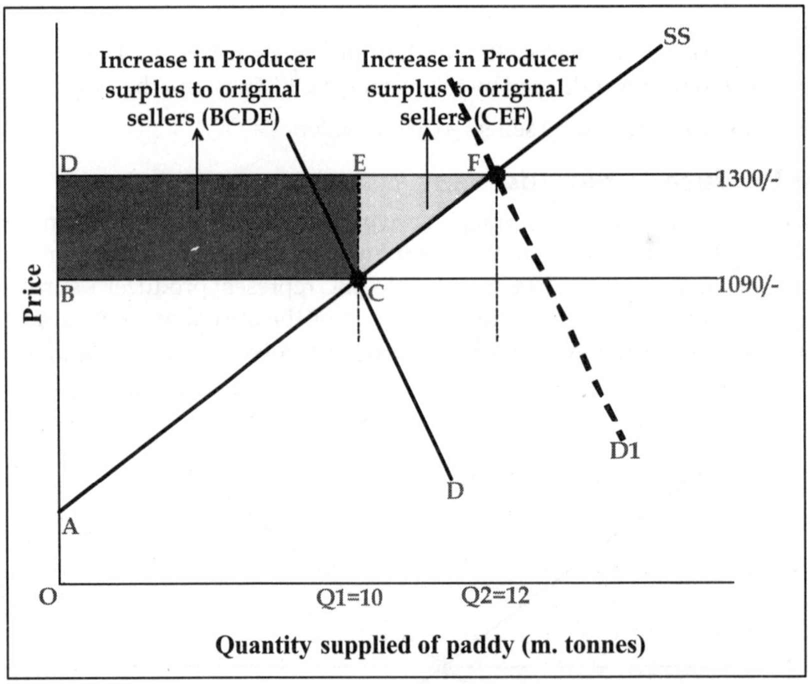

Figure 17.15: Changes in producer surplus due to change in price of paddy.

Due to implementation of cost-effective production technology, price of paddy declines in the market from Rs. 1300 to Rs. 1090, thereby, the whole discussion would run in the opposite direction. So, producer's surplus, at OD (Rs. 1300) is ADF. With the fall in price to OB (Rs. 1090), the new producer's surplus is ABC. That means, the whole area BDCF would now be the decline in producer surplus due to decline in price of paddy. This loss would consist of two parts *viz.*, the loss to farmers who would still supply paddy at the new price, Rs. 1090 (BCDE) and the loss of producer surplus (CEF) to farmers, who decide not to supply paddy because of the decline in its price in the market to Rs. 1090 from Rs. 1300. Since, some farmers have not supplied paddy into the market due to fall in its price, the total quantity of paddy supplied into the market decline from OQ_2 (12 million tonnes) to OQ_1 (10 million tonnes).

3. Influence of Changes in Demand and Supply on Producer Surplus

The changes in demand and supply of a commodity also influence the producer surplus. Let us assume first, the demand for the commodity is increased in the market (say, due to increase in consumers' income) and this causes the demand curve to shift away from the origin (Figure 17.16). This outward shift in the demand curve (D_1D_1) causes a rise both in the equilibrium quantity (OQ to OQ_1) and price (OP to OP_1), thereby, the producer surplus increases from SPE to SP_1E_1 to all the suppliers. This increase in producer surplus is due to selling more units of the commodity at a higher price.

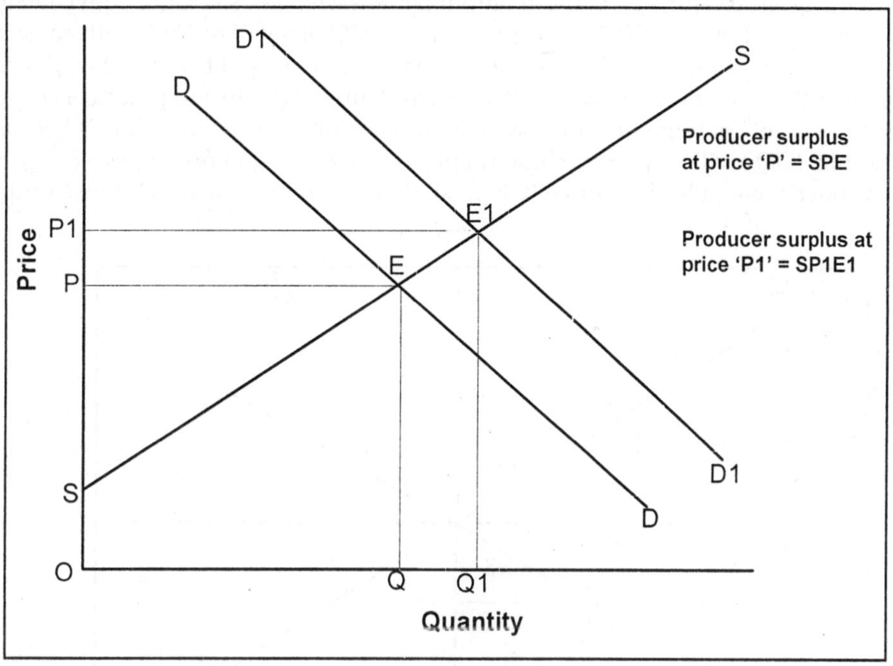

Figure 17.16: Change in producer surplus due to change (increase) in demand for the commodity.

On the contrary, when the supply of the commodity is increased in the market due to improvement in production technology or a fall in the cost of raw materials, the supply curve will shift towards right side (Figure 17.17). This outward shift in the supply curve (S_1S_1) causes a rise in the equilibrium quantity (OQ to OQ_1), but a fall in equilibrium price (OP to OP_1). At the initial supply of the commodity, producer surplus is SPE, but after increase in supply, the

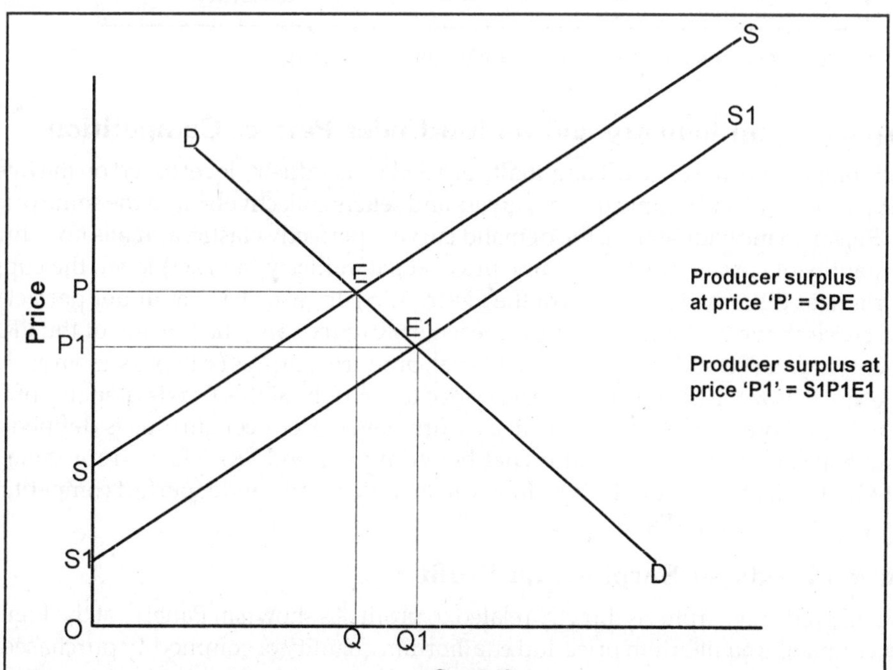

Figure 17.17: Change in producer surplus due to change (increase) in supply of the commodity.

producer supply changes to $S_1P_1E_1$. However, the extent of change in producer surplus depends upon the e_S. The more elastic the supply curve, the smaller the amount of producer surplus. If the supply curve is perfectly elastic, producer surplus is zero, since the price at which the firm is willing to supply its output is also (equal to) the ruling market price. If the e_S is completely inelastic, the producer surplus is infinite.

4. Sizes of the Consumer and Producer Surpluses at Various Output Levels

Depending upon the level or quantity of output transacted in the market, the sizes of producer and consumer surpluses vary. As shown in the Figure 17.18, at OQ_1 level of output transacted in the market, P_1DA_1 is the consumer surplus and $SP_1A_1B_1$ is the producer surplus. If more output is transacted *i.e.*, say OQ_2, both consumer surplus and producer surplus increases to P_2DA_2 and $SP_2A_2B_2$ respectively. At OQ_3 level of output, both consumer surplus and producer surplus further increases and this will continue up to the point of equilibrium. At equilibrium output level (OQ) and equilibrium price (OP) of the commodity, both consumer surplus (PDE) and producer surplus (PSE) reaches maximum. However, we cannot continue this process beyond the equilibrium point. This is because, output levels greater than the equilibrium (OQ_5) will only be purchased at prices below the equilibrium price (OP_B), but they will only be produced at prices above the equilibrium price (OP_A). So, there is no price at which those output levels will be produced and sold.

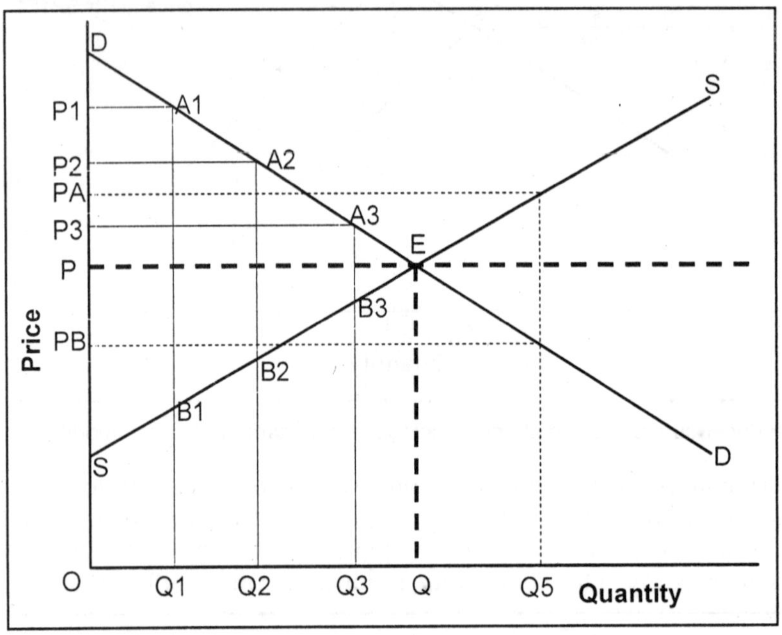

Figure 17.18: Sizes of producer and consumer surpluses at various output levels.

5. Producer Surplus for an Industry and a Firm Under Perfect Competition

In perfect market competition, the price of commodity is decided at industry level based on the interaction between two market forces *viz.*, of demand and supply (*i.e.*, by buyers and sellers collectively) and the same price is passed on to the individual firms. So, for an individual firm, the demand curve is perfectly elastic, as it has no control over the price and has to sell any quantity of commodity at the same price. So, at industry (market) level, the supply curve slopes upward from left to right (as discussed in Panel B of the Figure 17.8). In case of a firm under perfect competition, the portion of the SRMC curve above the SRAVC curve represents SRS curve and the portion of the LRMC curve above LRATC curve represents LRS curve. At industry (market) level, producer surplus (ABC) is shown as the area of triangle above the supply line and below the equilibrium market price to the right of the Y- axis (Panel A of the Figure 17.19). This same price OB is passed onto the individual firms. At firm level, producer surplus is defined as the difference between the firm's revenue and its SRTVC or difference between price and SRAVC *i.e.*, represented by the shaded rectangular area PABC (Panel B of the Figure 17.19). However, in the long run, under perfect competition, the firm earns no producer surplus, as price=LRATC (Figure 17.22).

6. Relation between Producer Surplus and Profit

It is interesting that, producer surplus is directly related to profit. As shown in Panel C of the Figure 17.19, the area below the supply curve upto the equilibrium price and equilibrium quantity of commodity purchased represents TVC. We discussed earlier, producer surplus is the area above the supply curve upto the equilibrium price to the right of the vertical axis. We know, TR is equal to price times the equilibrium output. So, we can say that,

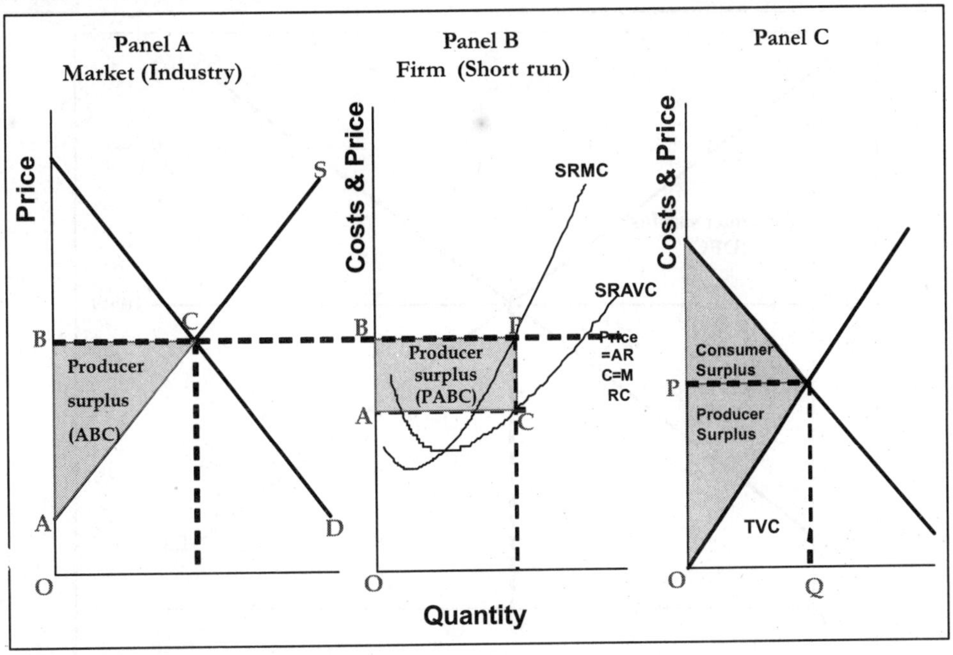

Figure 17.19: Producer surplus in a perfect competitive market (industry) and for a firm (short run) and its relation to profit

OP * OQ = TR = (Producer surplus + TVC)

We also know that, profit is the TR minus TC. So, we can write,

Profit = TR – TC = TR – (TFC + TVC)

= (Producer surplus + TVC) – (TFC + TVC)

Profits = Producer surplus – TFC

As discussed in Chapter 15, fixed costs exist only in the short run and in the long run, all costs are variable costs. So, in the short run, profit equals to producer surplus less of TFC. However, in the long run production programme, as there are no fixed costs, profit and producer surplus are equal to each other.

17.13. Social Surplus (SS) or Social Welfare or Economic Surplus or Total Surplus

So far, we studied about consumer's surplus in the theory of demand and producer's surplus in the theory of supply. Economists use the terms consumer surplus and producer surplus to illustrate the monetary gains realized by the consumers and producers respectively.

Consumer surplus is the difference between the total amount that consumers are willing and able to pay for a commodity or service (indicated by the demand curve) and the total amount that they actually pay (equilibrium market price) or it is the difference between value to buyers and amount paid by buyers. If the market price of the commodity is less than the maximum price the consumer is willing to pay, the consumer gains consumer's surplus. So, consumer's surplus depends on the maximum price a consumer is willing to pay for a particular commodity above the equilibrium price. The amount of consumer's surplus is shown by the area below the demand curve and above the equilibrium market price to the right of Y-axis.

Producer surplus is the difference between the equilibrium market price the producers receive and the minimum price at which they are willing and able to supply a commodity in the market or it is the difference between amount received by producers and cost to sellers. If the equilibrium market price of the commodity is more than the minimum price at which the producer is willing to offer the commodity, the producer gains producer's surplus. So, producer's surplus depends on the minimum price at which producers are willing to sell their goods below the equilibrium market price. That means, the producer gains producer's surplus when consumers are willing to pay more than the producer's minimum price. The level of producer's surplus is shown by the area above the supply curve and below the equilibrium market price to the right of Y-axis.

In the Figure 17.20, consumer surplus is the area of the triangle below the demand curve line and above the equilibrium price line to the right of the Y-axis. The vertical intercept of the demand curve on the Y-axis shows the maximum price the consumers are willing to pay. The minimum price consumers are willing to pay is indicated by the equilibrium price line. So, the consumer surplus can be calculated by using the formula of a triangle *i.e.*, (½ x base x

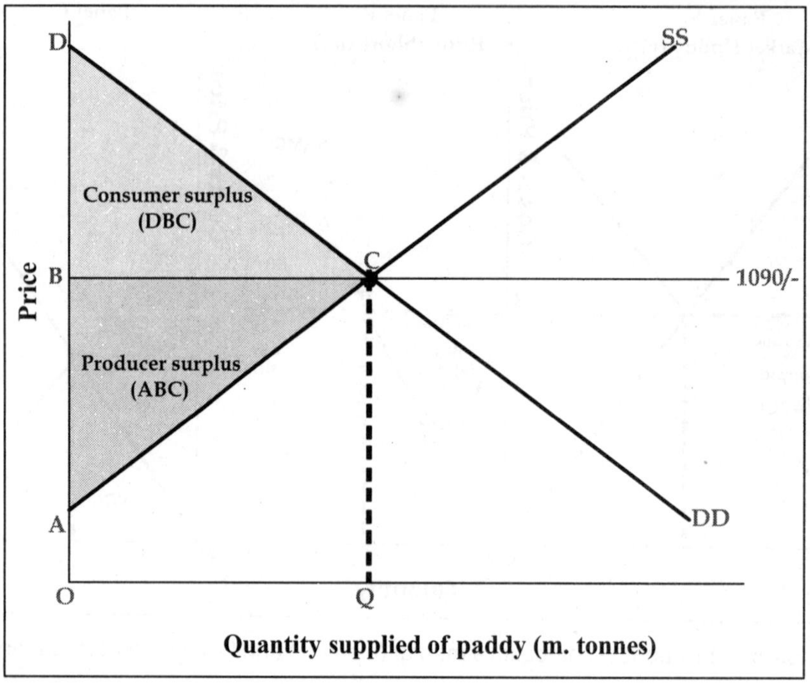

Figure 17.20: Total surplus in selling and buying paddy.

height). The ½ is a constant number and base is the quantity of commodity purchased at equilibrium market price (BC) and height is the difference between vertical intercept (maximum price the consumer is willing to pay) and the equilibrium market price *i.e.,* BD.

Regarding producer surplus, it is the area of triangle above the supply line and below the equilibrium market price to the right of the Y- axis (Figure 17.20). The vertical intercept of the supply line (OA) indicates the minimum price producers are willing to take. The maximum price producers are able to take is indicated by the equilibrium price line (OB). So, again use the area of a triangle formula (½ x base x height) to calculate producer surplus. The ½ is a constant factor, base is the quantity of commodity offered for sale at equilibrium market price (BC) along the X-axis and height of the triangle (AB) is the difference between equilibrium market price (OB) and the minimum price producers are willing to take (OA). This discussion further infers that, we use the tool of producer surplus in much the same way as we do with consumer surplus.

In Figure 17.20, we bring together the demand and supply curves to show the equilibrium price and quantity and the previously described regions of consumer and producer surpluses. All markets that have downward-sloping demand curves and upward-sloping supply curves yield consumer and producer surpluses. The equilibrium quantity OQ in Figure 17.20 reflects economic efficiency, which consists of productive efficiency and allocative efficiency. Productive efficiency is achieved because, competition forces the farmers to use the best techniques of production and combinations of resources in growing and selling paddy. Production costs of each level of output are minimized. Allocative efficiency is achieved because, the correct quantity of output OQ is produced relative to other goods and services. Points on the demand curve measure the MR of paddy at each level of output. Points on the supply curve measure the MC of paddy at each output level. The demand and supply curves intersect at the equilibrium output OQ, indicating that, MR=MC. (The significance of the MR=MC equality for efficiency is discussed under Chapter 18). This analysis of consumer and producer surpluses provides another way of thinking about efficiency. Each point on a demand curve identifies not only the MR of the corresponding unit of output, but also the maximum willingness to pay for it. Willingness to pay derives from the benefit that a product provides. Similarly, each point on the supply curve identifies not only the MC of a commodity, but also the minimum acceptable price for the commodity. To stay profitable, sellers must receive minimum prices that cover their MC.

In Figure 17.20, the maximum willingness to pay for each unit of paddy upto OQ exceeds the corresponding minimum acceptable price. So, each of these units adds a positive amount (= maximum willingness to pay minus minimum acceptable price) to the total of consumer and producer surpluses. Only at the equilibrium price OP, where maximum willingness to pay for the last unit equals minimum acceptable price for that unit, does society exhaust all the opportunities to add to combined consumer and producer surpluses. So, allocative efficiency occurs, where the triangle representing 'consumer surplus + producer surplus' is at its maximum size (ACD). Other things remaining equal, competitive market produce equilibrium price and quantity that maximize the sum of consumer and producer surpluses. Allocative efficiency occurs at quantity levels, where the following three conditions exist:

- ✰ MR=MC
- ✰ Maximum willingness to pay = Minimum acceptable price.
- ✰ Combined consumer and producer surpluses are at a maximum.

Social Surplus (SS) is the sum of consumer surplus and producer surplus. We know,

- ✰ Consumer surplus = Value to buyers - Amount paid by buyers
- ✰ Producer surplus = Amount received by producers - Cost to sellers
- ✰ SS = Consumer surplus + Producer surplus
  = (Value to buyers - Amount paid by buyers) + (Amount received by sellers – Cost to sellers)
- ✰ But since, amount paid and received are the same, SS is given by,
  SS = Value to buyers - Cost to sellers

Example 1

As shown in the Figure 17.21, SS is calculated as the sum of consumer surplus and producer surplus. A close look at the Figures 17.20 and 17.21 further reveals that, the SS or social welfare is maximized at the competitive equilibrium.

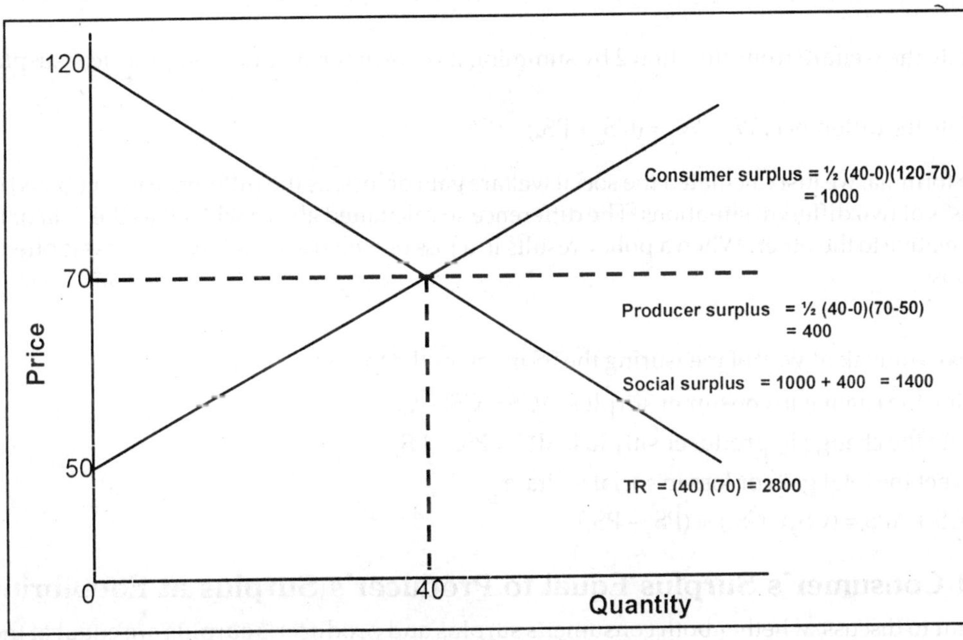

Figure 17.21: Computation of Social Surplus in transacting the commodity.

Example 2

Given information that,

Supply Equation: Ps = 40 + 4Qs

Demand Equation: Pd = 100 – 2Qd

We know, at equilibrium Ps=Pd, thereby,

40+4Qs=100-2Qd or 40+4Q=100-2Q (since, Qs=Qd).

So, equilibrium quantity is given by, 6Q=60 = 10 units

Now, to find the equilibrium price, Ps=40+ 4(10) = Rs.80 or Pd*=100-2(10) = Rs.80.

So, to graph, we should know at what price the quantity supplied is zero and what price the quantity demanded is zero. This is given as,

Ps = 40 + 4Qs = 40 + 4(0)= Rs.40

Pd = 100 – 2Qd = 100 – 2(0) = Rs.100

The above calculated values *i.e.*, Rs.40 and Rs.100 are the intercepts on the Y-axis for supply and demand curves.

To calculate the intercept on the X-axis *i.e.*, how much is the quantity demanded when the price is zero and it is given by,

$0 = 100 - 2Qd$ or $2Qd=100$ or $Qd=100/2=50$ units

There is no need to calculate the X intercept for the supply equation. From the above findings, we can easily compute the SS and TR as below.

Consumer surplus = ½ (10-0)(100-80) = Rs. 100

Producer surplus = ½ (10-0)(80-40) = Rs. 200

So, SS = 100 + 200 = Rs.300

TR = Equilibrium quantity x Equilibrium price = (10)(80) = Rs. 800

Comparison of SS at Two Different Situations

The following methodology is employed to measure the difference in SS under the given two different situations:

Method 1

This method involves the following steps:

- ☆ Calculate the welfare from Situation 1 by summing its consumer surplus and producer surplus: $W_1 = CS_1 + PS_1$.
- ☆ Calculate the welfare from situation 2 by summing its consumer surplus and producer surplus: $W_2 = CS_2 + PS_2$.
- ☆ Calculate the difference: $W_2 - W_1 = (CS_2 + PS_2) - (CS_1 + PS_1)$.

In the above formula, we just calculated the social welfare gain or loss, as the difference in combined consumer and producer surpluses of two different situations. The difference so calculated above will tell us the gain or loss of welfare of one situation relative to the other. When a policy results in a loss of welfare to society, that loss is often referred to as the deadweight loss.

Method 2

An alternative equivalent way of measuring the SS involves the following steps:

- ☆ Calculate the change in consumer surplus: $\Delta CS = CS_2 - CS_1$.
- ☆ Calculate the change in producer surplus: $\Delta PS = PS_2 - PS_1$.
- ☆ Add to get the total gain or loss in social welfare:
  $SS = \Delta CS + \Delta PS = (CS_2 - CS_1) + (PS_2 - PS_1)$

17.14. Must Consumer's Surplus Equal to Producer's Surplus at Equilibrium Price?

It is important to discuss, whether both consumer's surplus and producer's surplus are equal at the equilibrium price of the commodity. The consumer's surplus need not necessarily be equal to producer's surplus at the equilibrium price. We can even exemplify the case, where consumer's surplus is positive, but producer's surplus is zero. Let us assume long run equilibrium under perfect competition. The point of optimality is when LRMC=MR. But, in long run equilibrium, when LRMC=MR, price is equal to LRATC, as no firm under perfect competition in the long run earn super normal profits and only normal profits are possible (Figure 17.22). So, in such case, the producer's surplus enjoyed by the firm is zero, but the consumer enjoy consumer's surplus in accordance with the concept of LDMU *i.e.*, except at the point of consumer's equilibrium, where MU derived in terms of money = Price of the commodity (as per the equation 6.2), and for the remaining units of commodity he enjoy consumer's surplus, as MU derived in terms of money > Price of the commodity.

17.15. Inter-related Supply

Thee following are the different types of inter-related supply of commodities:

1. Joint Supply

When two or more commodities have a common source of supply, it refers to joint supply. Examples include, paddy grain + straw, wheat grain + straw, cotton seed + lint, sheep wool + mutton etc. These commodities are produced and supplied jointly through a single production programme. The farmer cannot produce a single commodity in the absence of other. Hence, these products are also called as joint products. However, with the recent advances in the field

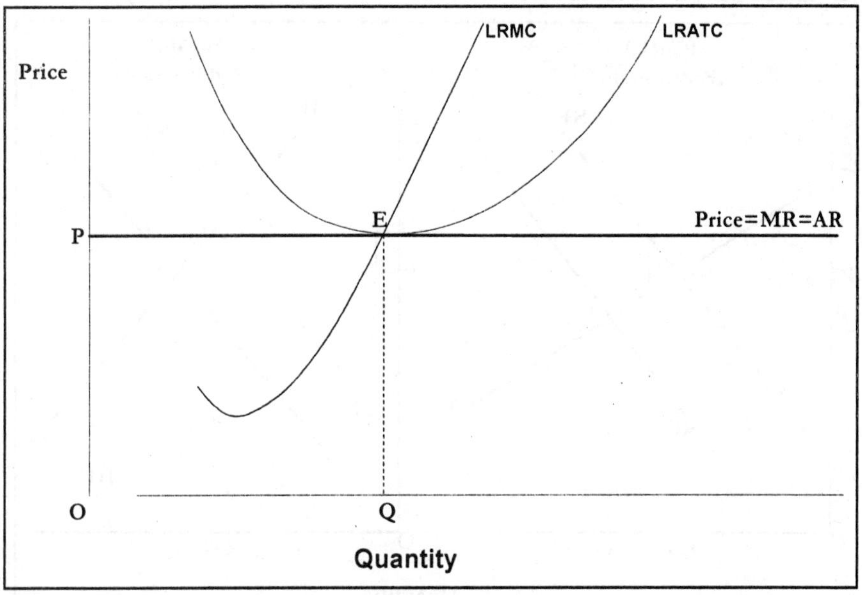

Figure 17.22: Consumer's surplus is positive and Producer's surplus is zero at equilibrium price under perfect competition in the long run.

of Bio-technology and Genetic Engineering, we are capable of changing the proportions of the quantities of the commodities supplied jointly. Examples include, seedless grapes, seedless papaya, changing the proportions of wool and mutton components of sheep etc. So, we can analyze the influences of changes in demand and supply on the price of the commodities considering the following two cases:

(a) Joint Supply of Commodities with Fixed Proportions

Let us assume the joint supply of commodities *viz.*, paddy grain and straw and the production of these two commodities have a fixed proportion. In Panel A of the Figure 17.23, when the demand for paddy grain increases in the market from DD to D_1D_1, it leads to increase in price of paddy from OP to OP_1. As a result, the supply of paddy straw will increase from SS to S_1S_1 (Panel B of the Figure 17.23), thereby, it leads to fall in price of straw from OP to OP_1. Thus, due to increase in demand for paddy grain in the market, the price of paddy grain and price of the straw moves in the opposite direction in the market.

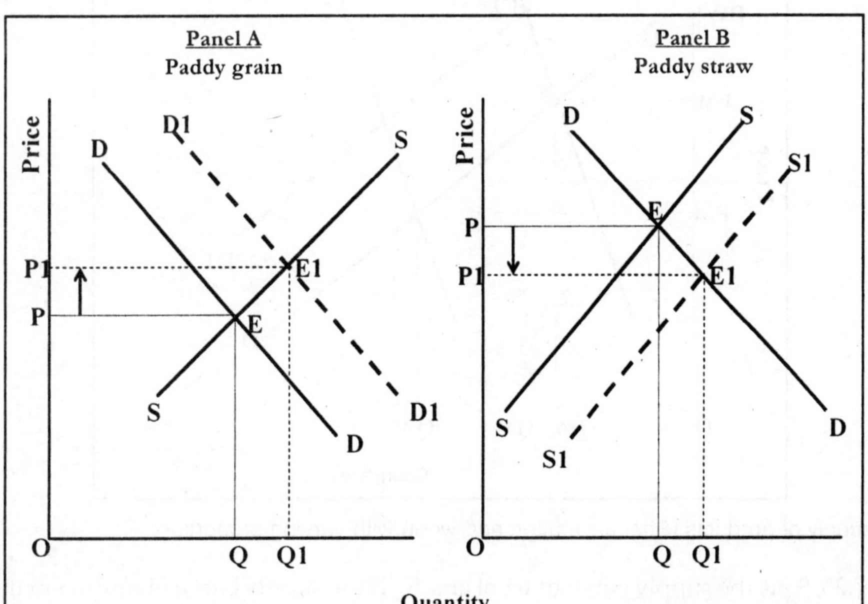

Figure 17.23: Joint supply of products with fixed proportion.

When there is decrease in supply of paddy grain in the market due to adverse climatic conditions *i.e.*, from SS to S_1S_1 (Panel A of the Figure 17.24), it leads to rise in price of paddy grain from OP to OP_1 thereby, the quantity demanded for paddy grain will fall from OQ to OQ_1, as the consumers prefer substitute goods like wheat, jowar etc. Accordingly,

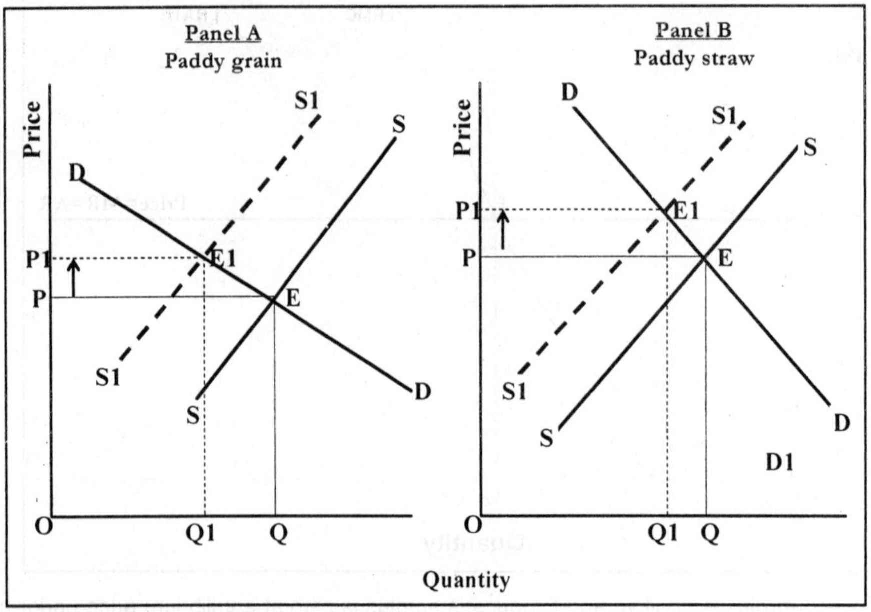

Figure 17.24: Joint supply of commodities (increase in price of rice).

the supply of straw into the market also falls from OQ to OQ_1 (Panel B of the Figure 17.24). So, the price of straw also increases from OP to OP_1 due to decline in its supply in the market.

(*b*) Joint Supply of Commodities with Varying Proportions

As explained earlier, due to advances in Bio-technology, Genetic Engineering and due to other breeding methods, the proportions of the combinations of the commodities can be varied. For example, sheep yields wool and mutton and the proportions of wool and mutton can be varied by different breeding methods. The inter-relationship of the prices of such jointly supplied commodities under varying proportions is discussed through Figure 17.25.

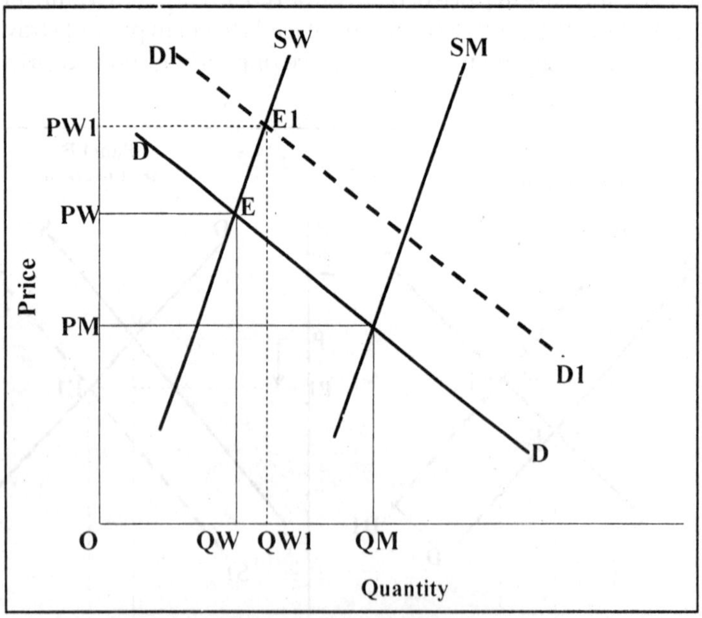

Figure 17.25: Joint supply of products (Sheep's mutton and wool) with varying proportion.

In the Figure 17.25, S_W is the supply curve of wool and S_M is the supply curve of mutton in the market. DD is the demand curve for these two products. It is clear from the figure that, the supply of mutton in the market is more than wool. Accordingly, the price of wool is (OP_W) is higher than price of mutton (OP_M). If the demand for wool is increased in the market, the demand curve (DD) will shift towards right side as D_1D_1 and thereby, the price for wool will increase in the market to OP_{W1} from OP_W and accordingly supply of wool increases to OQ_{W1} from OQ_W. But, this increase in supply of wool may not lead to increase in supply of mutton into the market in the same proportion. It all depends upon the degree of proportion of the production of these two products in the production process and their relative elasticities of demand in the market.

2. Composite Supply

When two or more commodities are supplied to meet the same purpose or need, it is said to be composite supply. Examples include, sugar and jaggery, tea and coffee, rice and wheat etc. In case of factors of production, family labour and hired labour, labour and machinery etc., are in composite supply. So, the commodities or factors which have composite supply are substitutes for each other. That means, these commodities will compete each other to fulfill a human want. The same is explained through Figure 17.26.

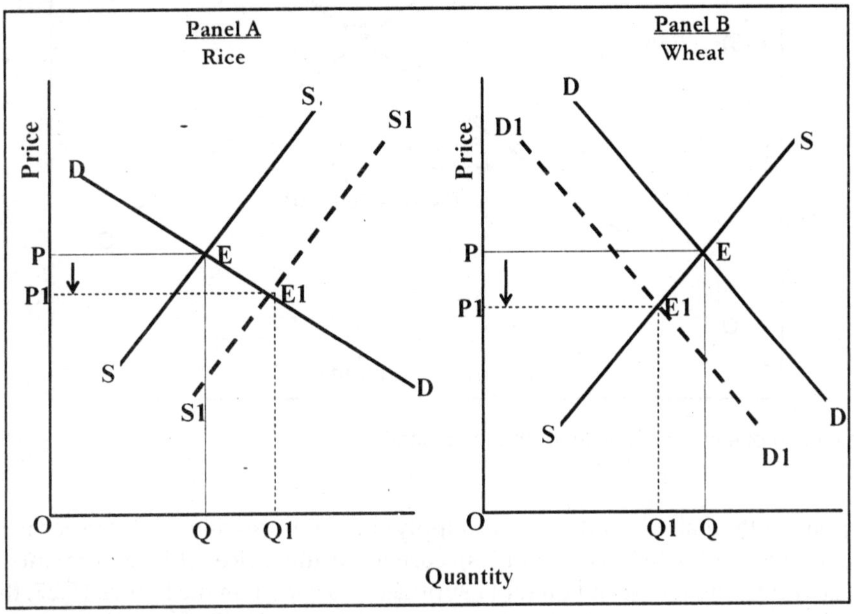

Figure 17.26: Composite supply for competitive goods.

In Panel A of the Figure 17.26, the supply of rice is increased in the market from SS to S_1S_1, thereby, the price of rice in the market falls from OP to OP_1. When there is fall in price of rice in the market, the demand for the substitute commodity (wheat) will fall from DD to D_1D_1 (Panel B of the Figure 17.26). As a result, the price of wheat in the market decline from OP to OP_1. Thus, when the two commodities have composite supply, their prices will move in the same direction.

17.16. Equilibrium Price

Having studied the concepts of demand and supply, it is essential to analyze and discuss about the equilibrium price. An equilibrium price is a price, which is influenced by these two market forces *i.e.*, demand and supply. That means, equilibrium price is the price that exists when a market is in equilibrium. It is the price that equates the quantity demanded and quantity supplied in a market. At equilibrium price, the quantity of the commodity that the buyers are willing and able to buy exactly balances the quantity that the sellers are willing and able to sell. The equilibrium price is also called as Market clearing price because, at this price, buyers have bought all they want to buy and sellers have sold all they want to sell.

As shown in the Figure 17.27, the equilibrium price is found at the intersection of the demand curve and the supply curve. So, equilibrium price is the price that achieves as a result of market balance. Since quantity demanded and quantity supplied are equal at the equilibrium price, there is neither shortage nor surplus in the market, which means that, neither buyers nor sellers are inclined to change the price, which is the primary condition for equilibrium. When demand exceeds the supply, the equilibrium price will be raised and when the supply exceeds the demand, the equilibrium price will fall, thereby, in both the cases, the market is said to be in dis-equilibrium. In dis-equilibrium situation, again the market forces operate in such a way that, the equilibrium price will be ensured. Say, if price of the commodity is increased in the market above the equilibrium price, then, there will be excess supply over demand. So, some of the sellers could not sell a portion of commodity. In such case, they even try to dispose the excess stock at lower price. Thus, the price will gradually decrease up to the level of equilibrium price. On the other hand, if price is decreased in the market below the equilibrium price, then the demand extends and supply contracts. So, some of the buyers could not get the commodity or desired quantity of commodity because, demand exceeds supply. Hence, to get the desired quantity, the buyers try to purchase the commodity even at a higher price, thereby, price gradually increases and equals to equilibrium price. So, to derive an equilibrium price, two curves *viz.*, demand curve and supply curve has to be drawn and as discussed earlier, the demand curve is negatively sloped in accordance with the Law of demand (*i.e.*,

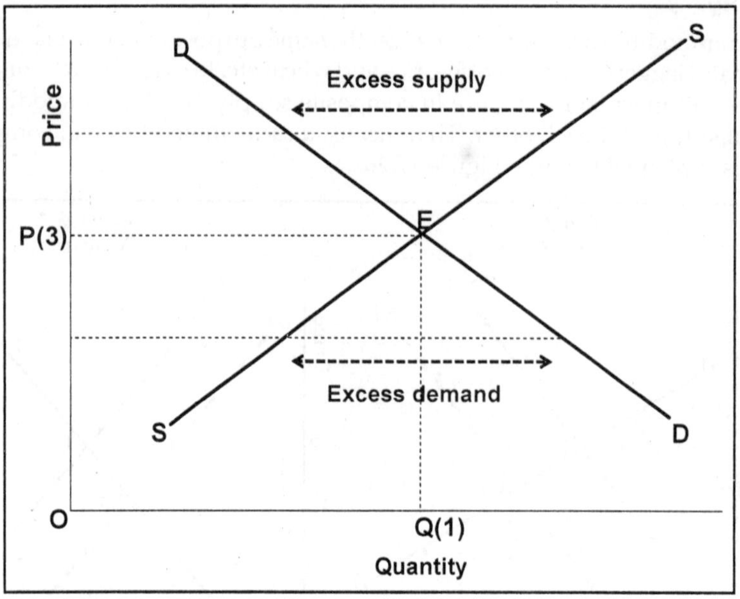

Figure 17.27: Determination of equilibrium price for the commodity.

higher prices correspond with smaller quantities) and supply curve is positively sloped in accordance with the Law of supply (higher prices correspond with large quantities). So, equilibrium price achieves a balance in the market, which is equality between quantity demanded and quantity supplied. As shown in the Figure 17.27, the demand curve DD and supply curve SS intersects at point E, which indicates that, OQ quantity of commodity is supplied and demanded in the market at OP equilibrium price. Any shortage (or excess) of either demand or supply will bring back the price back to this equilibrium price OP. So, at equilibrium price, the quantity demanded is equal to the quantity supplied. The buyers can buy all that they want, so there is no shortage. The sellers can sell all that they want, so there is no surplus. In other words, equilibrium price represents that price at which buyers are willing to buy the good and sellers are willing to sell it. This is the point of satisfaction for both the groups. Neither buyers nor sellers are motivated to change the price. The forces of demand and supply are in balance unless or until any external force intervenes. The quantity of commodity transacted at the equilibrium price is said to be equilibrium quantity. So, equilibrium quantity is simultaneously equal to both the quantity demanded and quantity supplied. In the Figure 17.27, the equilibrium quantity is OQ found at the intersection of the demand curve and the supply curve.

Let us illustrate the establishment of equilibrium with the help of demand and supply functions. We begin with two equations *viz.*, $Q_D = 10 - 3P$ and $Q_S = -5 + 2P$. So, at equilibrium, quantity demanded and supplied are equal, thereby,

$$Q_D = Q_S = (10 - 3P) = (-5 + 2P)$$

$$15 = 5P$$

So, equilibrium price, P = Rs.3/unit

On substituting the value of price in demand and supply functions, we get,

$$Q_D = 10 - 3P$$

$$Q_D = 10 - 3(3) = 10 - 9 = 1\text{unit}$$

$$Q_S = -5 + 2P$$

$$Q_S = -5 + 2(3) = 1\text{unit}$$

Hence, equilibrium price is Rs.3/unit, at which both quantity demanded and supplied are equal to 1 unit each. The same is represented through Figure 17.27 geometrically, where both demand and supply curves intersect at point E (equilibrium point) at which price P = Rs.3/unit and quantity demanded and supplied are both equal (Q_D=Q_S= 1).

Changes in Equilibrium Price

There are three possible cases to discuss about the disequilibrium in the market economy. They include:

1. Change in demand and no change in supply
2. Change in supply and no change in demand
3. Changes in both demand and supply

1. Change in Demand and No Change in Supply

Let us assume, there is increase in demand for wheat due to changes in consumers' tastes and preferences. As a result, the demand curve (DD) shifts towards right side or away from the origin as D_1D_1. As assumed in this case, there is no change in supply of wheat in the market (SS). So, this event causes increase in both equilibrium quantity (OQ to OQ_1) and equilibrium price (OP to OP_1) of the commodity (Panel A of the Figure 17.28). In case, when demand for wheat decreases in the market, the new demand curve (D_2D_2) shifts towards left side, thereby, there is decline in both equilibrium quantity (OQ to OQ_2) and equilibrium price (OP to OP_2) of the commodity (Panel A of the Figure 17.28).

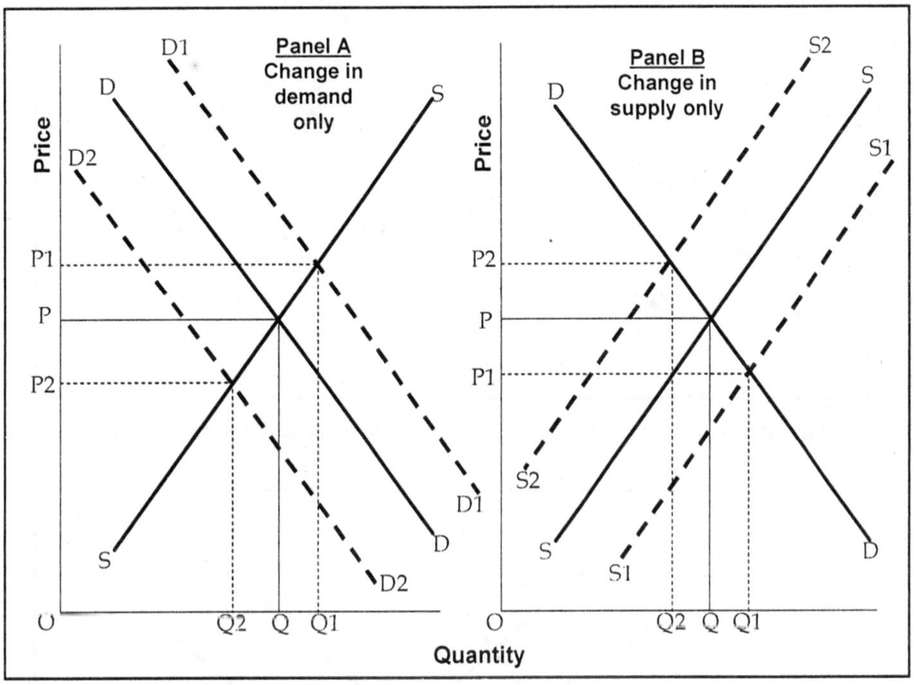

Figure 17.28: Changes in equilibrium quantity and equilibrium price due to changes in demand or supply.

2. Change in Supply and No Change in Demand

Let us assume, there is increase in supply of wheat in the market due to its bumper harvest. As a result, the supply curve (SS) shifts towards right side or away from the origin as S_1S_1. As assumed in this case, there is no change in demand for wheat in the market (DD). So, this event causes increase in equilibrium quantity (OQ to OQ_1), but fall in equilibrium price (OP to OP_1) of the commodity (Panel B of the Figure 17.28). In case, when supply of wheat decreases in the market, the new supply curve (S_2S_2) shifts towards left side, thereby, there is decline in equilibrium quantity (OQ to OQ_2), but increase in equilibrium price (OP to OP_2) of the commodity (Panel B of the Figure 17.28).

3. Changes in Both Demand and Supply

These influences can be studied considering the following four cases:

Case 1: Simultaneous Increase in Both Demand and Supply

Let us assume, both demand and supply of wheat increases in the market. As a result, demand curve (DD) shifts towards right side or away from the origin as D_1D_1 and supply curve (SS) also shifts towards right side or away from the origin as S_1S_1 (Figure 17.29) Further, also consider another supply curve (S_2S_2) in the analysis, which is more elastic compared to SS and S_1S_1 curves. So, due to simultaneous increase in both demand and supply of the commodity, the equilibrium quantity definitely increases (OQ to OQ_1). But, equilibrium price may not necessarily change. As long as there is no change in elasticities of both demand and supply curves, there is no change in equilibrium price due to simultaneous increase in both demand and supply of the commodity. As shown in the Figure 17.29, since the supply curve S_2S_2 is more elastic compared to SS, the simultaneous increase in both demand and supply of the commodity leads to decline in equilibrium price from OP to OP_2. But, the equilibrium quantity increases from OQ to OQ_2.

Case 2: Simultaneous Decrease in Both Demand and Supply

Let us assume, both demand and supply of wheat decreases in the market. As a result, demand curve (DD) shifts towards left side or towards the origin as D_1D_1 and supply curve (SS) also shifts towards left side or towards the origin as S_1S_1 (Figure 17.30). Further, also consider another supply curve (S_2S_2) in the analysis, which is more inelastic compared to SS and S_1S_1 curves. So, due to simultaneous decrease in both demand and supply of the commodity, the

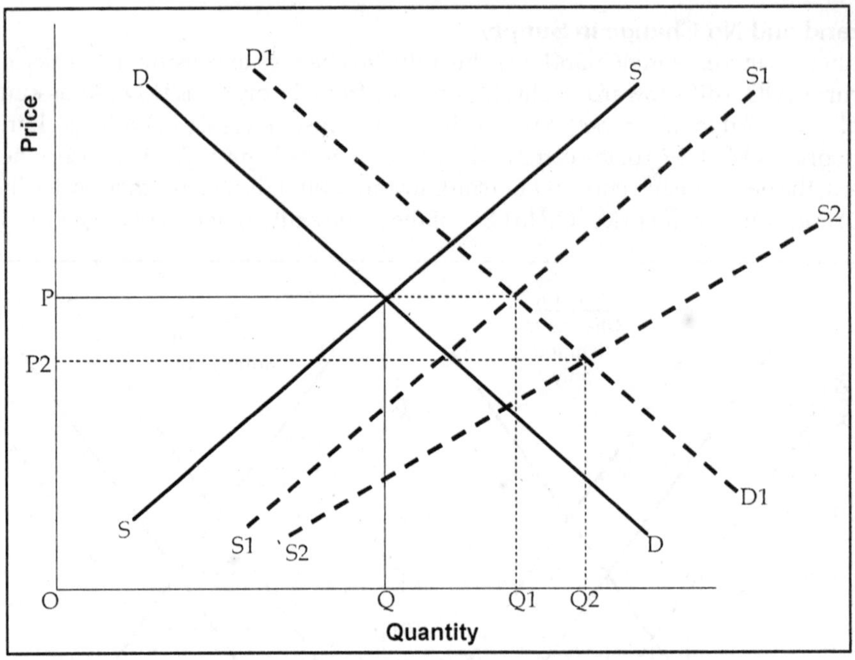

Figure 17.29: Changes in equilibrium quantity and equilibrium price due to simultaneous increase in both demand and supply.

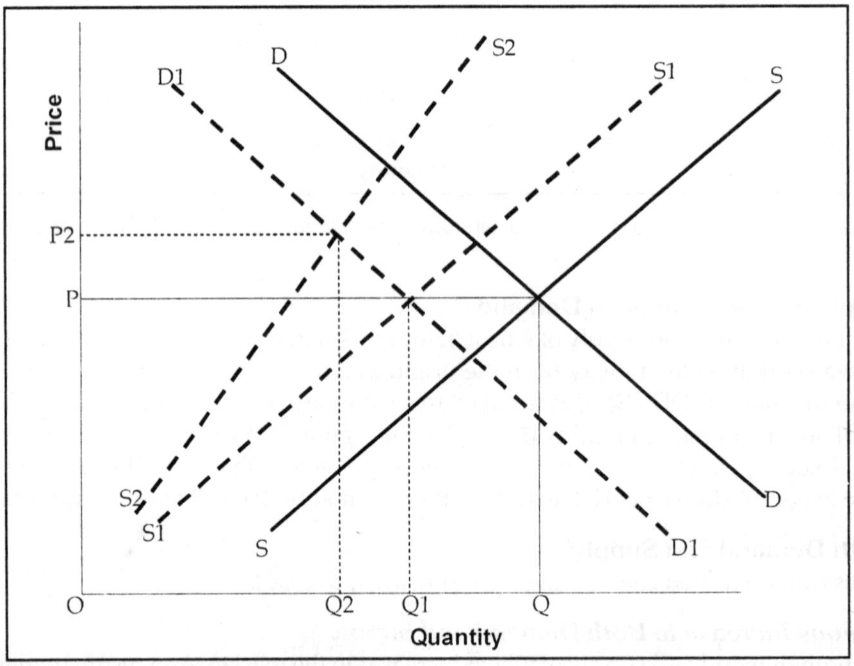

Figure 17.30: Changes in equilibrium quantity and equilibrium price due to simultaneous decrease in both demand and supply.

equilibrium quantity definitely decreases (OQ to OQ_1). But, equilibrium price may not necessarily change. As long as there is no change in elasticities of both demand and supply curves, there is no change in equilibrium price due to simultaneous decrease in both demand and supply of the commodity. As shown in the Figure 17.30, since the supply curve S_2S_2 is more inelastic compared to SS, the simultaneous decrease in both demand and supply of the commodity leads to increase in equilibrium price from OP to OP_2. But, the equilibrium quantity decreases from OQ to OQ_2.

Case 3: Increase in Demand and Decrease in Supply

Let us assume, supply of the wheat decreases in the market (due to decline in production) and demand for the same increases (due to changes in consumers' tastes and preferences) in the market. As a result, demand curve (DD) shifts towards right side or away from the origin as D_2D_2 and supply curve (SS) shifts towards left side or towards the origin as S_1S_1 (Figure 17.31). Further, also consider another demand curve (D_1D_1) in the analysis, which is more

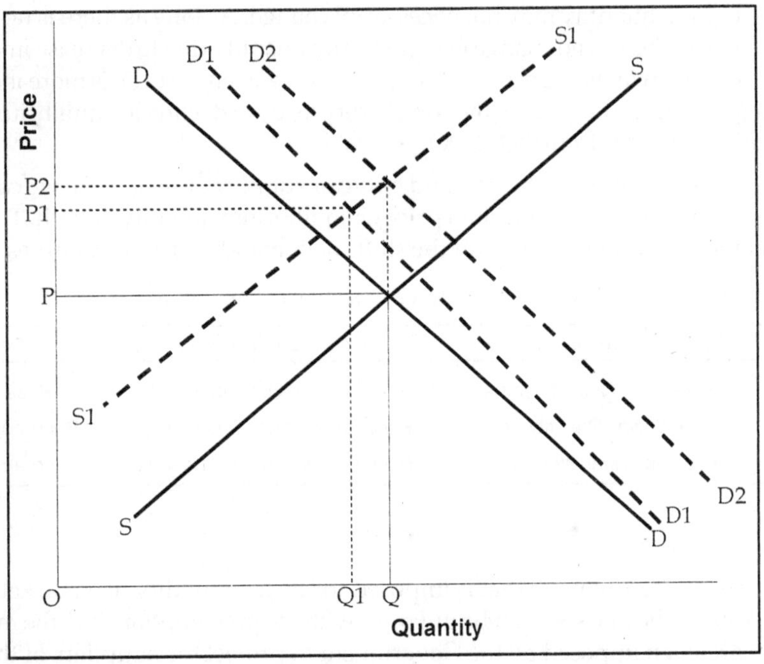

Figure 17.31: Changes in equilibrium quantity and equilibrium price due to increase in demand and decrease in supply.

inelastic compared to DD and D_2D_2 curves. So, due to increase in demand and decrease in supply of the commodity, the equilibrium price definitely increases (OP to OP_2). But, equilibrium quantity may not necessarily change. As long as there is no change in elasticities of both demand and supply curves, there is no change in equilibrium quantity due to increase in demand and decrease in supply of the commodity. As shown in the Figure 17.31, since the demand curve D_1D_1 is more inelastic compared to DD, the increase in demand and decrease in supply of the commodity leads to increase in equilibrium price from OP to OP_1. But, the equilibrium quantity decreases from OQ to OQ_1.

Case 4: Decrease in Demand and Increase in Supply

Let us assume, supply of the wheat increases in the market (due to bumper harvest) and demand for the same decreases (due to changes in consumers' tastes and preferences) in the market. As a result, demand curve (DD) shifts towards left side or towards the origin as D_1D_1 and supply curve (SS) shifts towards right side or away from the origin as S_1S_1 (Figure 17.32). Further, also consider another demand curve (D_2D_2), which is more inelastic compared to DD and D_1D_1 curves. So, due to decrease in demand and increase in supply of the commodity, the equilibrium price definitely

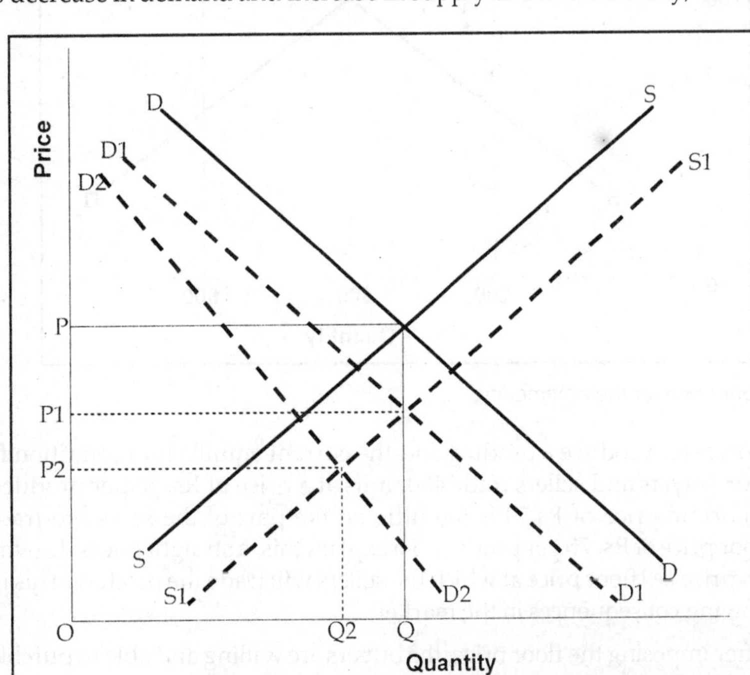

Figure 17.32: Changes in equilibrium quantity and equilibrium price due to decrease in demand and increase in supply.

falls (OP to OP_1). But, equilibrium quantity may not necessarily change. As long as there is no change in elasticities of both demand and supply curves, there is no change in equilibrium quantity due to decrease in demand and increase in supply of the commodity. As shown in the Figure 17.32, since the demand curve D_2D_2 is more inelastic compared to DD, the decrease in demand and increase in supply of the commodity leads to decline in equilibrium price from OP to OP_2. But, the equilibrium quantity decreases from OQ to OQ_2.

It is clear that, whenever an event shifts the demand curve or supply curve or both, we can employ these tools to predict to what extent there is a change in equilibrium price or equilibrium quantity or both. To summarize, Table 17.7 shows the predicted outcome for any combination of the shifts of demand and supply curves.

Table 17.7: Predicted outcomes due to shifts in demand curve or supply curve or both.

	No Shift in Demand Curve	*Increase in Demand*	*Decrease in Demand*
No shift in supply curve	Price same, Quantity same	Price up, Quantity up	Price down, Quantity down
Increase in supply	Price down, Quantity up	Price ambiguous, Quantity up	Price down, Quantity ambiguous
Decrease in supply	Price up, Quantity down	Price up, Quantity ambiguous	Price ambiguous, Quantity down

17.17. Floor Price

It is a legally established minimum price that is imposed for the commodity in a market above the equilibrium price. It is generally placed above the market equilibrium price with the presumption that, the market equilibrium price is too low. So, price floors are often imposed by the Government to protect the suppliers of the product or service in question. That means, a floor price places a downward limit on the price of good. But, if floor price is imposed, it will create a surplus or excess supply. Buyers and sellers can exchange the good at prices more than the floor price, but not less than the floor price. So, fixation of floor price will keep the price above the equilibrium price and not below it, thereby, it helps the sellers to generate more revenue. So, price floors are much important for the sellers, who face drastic fall in the prices of the commodities like agricultural produce, labour etc. The concept of price floor is explained through Figure 17.33.

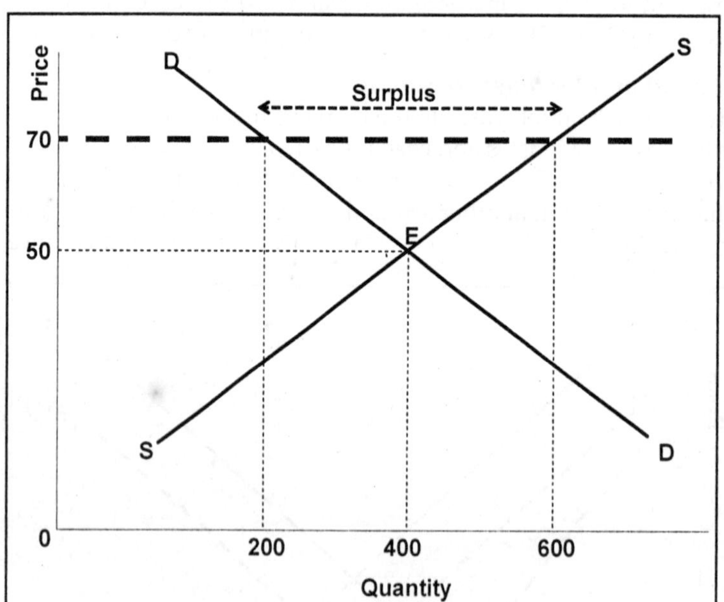

Figure 17.33: Fixation of Floor price for the commodity.

For example, fruit processors produce a product and the current equilibrium condition for this product reveals that, without the price floor, buyers and sellers trade 400 units at a price of Rs. 50 per product. But, the Government considered that, the equilibrium price of Rs.50 is too little on the part of the sellers to trade the product. Hence, Government imposed a floor price of Rs. 70 per product. To explain this, a straight line is drawn parallel to X-axis at Rs. 70 indicating the minimum price *i.e.*, floor price at which the sellers will trade the product. This imposition of floor price (Rs. 70) will show the following consequences in the market:

- ☆ ***Demand side***: After imposing the floor price, the buyers are willing and able to purchase only 200 units of the product. So, the quantity demanded by the buyers at floor price decline compared to equilibrium price. That means, they are unwilling and unable to purchase more of the product.

☆ *Supply side*: The sellers, on the other hand, are willing and able to offer upto 600 units of the product at the announced floor price. This is because, at the announced floor price, they can cover the production cost of 600 units of the product.

☆ *Surplus*: The combination of the quantity demanded and the quantity supplied of the product reveals that, at the announced floor price, the market has a surplus of 400 units of the product. That means, the sellers want to sell 400 units of the product more than the buyers' demand.

The above example clearly infers that, the main objective of fixing floor price above equilibrium price is to generate an adequate amount of revenue for the sellers. So, at Rs.70/unit of the product, the seller receives an extra of Rs.20 on each unit of the product. However, from the buyers point of view, fixing a higher floor price is a loss for them, as they have to pay Rs. 20 extra on each unit of the product they purchase. So, they demand only less quantity of the product *i.e.*, only 200 units from the sellers. That means, at the floor price, the total quantity demanded by the buyers is 200 units, but the total quantity of product offered for sale by the sellers is 600, thereby, it gives the following important conclusions:

☆ At higher (floor) price, only few quantity of the product is sold in the market (200 units).

☆ The supply into the market (600 units) is more than quantity demanded (200 units). So, some number of sellers sell less quantity or none at all, thereby, they receive less revenue or none at all. So, there is a loss on the part of these sellers too.

☆ The total surplus of the product in the market increases with increase in floor price

However, it is important that, the extent of surplus of product left over in the market after fixing the floor price depends upon the elasticities of demand and supply curves. If the demand and supply curves are more steeply sloped or more inelastic, the surplus left over will be less. That means, if the demand and supply of the product are relatively inelastic, then the goal of generating revenue to sellers, without excluding a lot of sellers from the market can be more easily achieved.

Assuming normal conditions with reference to elasticities of demand and supply for the product, more surplus will be left over with the sellers (400 units) with the announcement of floor price (as demand<supply). So, the sellers resort to illegal exchange of the product, thereby, violating the legally mandated floor price. So, to clear off the excess surplus left over, the sellers will offer the same to the buyers at prices below the legal maximum (floor price). But, they are not legally allowed to do so with a floor price. However, they resort for this illegal trade to clear off their surplus and from the buyers side, they are always inclined to pay less than the maximum floor price for the product. That means, if the sellers offers the product at less than Rs. 70/unit, the buyers readily accept the product. But, the extent of price reduction below floor price at which the seller will offer the product to the buyer depends upon the impact of price rise. If the floor price fixed above equilibrium price is too high (*i.e.*, huge difference between floor price and equilibrium price), then the seller has to lower the price to a greater extent below floor price, so as to clear off the extra surplus and *vice versa*.

For example, in the United States, the Government has imposed price floors on sugar. This means that, the sugar producers are guaranteed a minimum price for their sugar, regardless of what the market will pay for them. But, if the market pays the price less than the floor price (say, due to less demand for sugar in the market), then the Government will pay the difference between floor price and market price to the suppliers. Further, the surplus quantity of sugar left over with the farmers due to fixation of floor price will be procured by the Government. In India, we see the taxis run at minimum fare (the fares begin at a certain level above zero) and it is nothing but a floor price. Similarly, labour wages in agriculture are fixed at minimum, so as to safeguard their income levels.

17.18. Price ceiling (Maximum Price)

It is a legally established maximum price that is imposed for the commodity in a market below the equilibrium price. It is generally place below the market equilibrium price with the presumption that, the market equilibrium price is too high. So, price ceilings are often imposed by the Government to protect the markets from failure, as a price ceiling creates excess demand. But, if price ceiling is imposed, it will create a shortage of supply of the product. That means, a ceiling price places an upward limit on the price of product. Buyers and sellers can exchange the product at prices below the ceiling price, but not greater than the ceiling price. So, fixation of ceiling price will keep the price below the equilibrium price and not above it, thereby, it helps the buyers to purchase the product at affordable prices. So, the concept of ceiling price is more important for the buyers, who face drastic rise in the prices of the basic necessaries, apartments, gasoline, natural gas, food, prescription drugs etc. This is explained through Figure 17.34.

For example, fruit processors produce a product and the current equilibrium condition for this product reveals that, without the ceiling price, buyers and sellers trade 400 units at a price of 50 per product. But, the Government considered that, the equilibrium price of Rs.50 is too high on the part of the buyers to purchase the product. Hence, Government imposed a ceiling price of Rs. 30 per product. To explain this, a straight line is drawn parallel to X-axis at

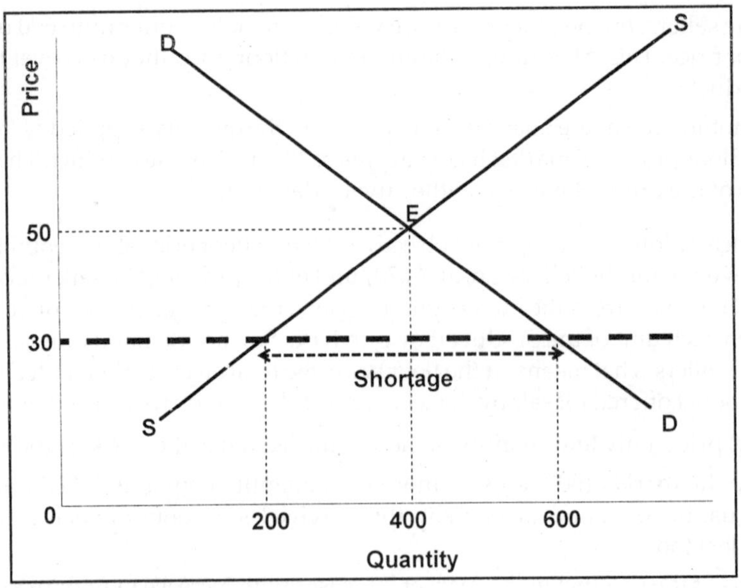

Figure 17.34: Fixation of Ceiling price for the commodity.

Rs. 30 indicating the maximum price *i.e.*, ceiling price at which the sellers must sell the product. This imposition of ceiling price will show the following consequences in the market:

- ✰ ***Demand side***: After imposing the ceiling price, the buyers are willing and able to purchase 600 units of the product. So, the quantity demanded by the buyers at ceiling price will increase compared to equilibrium price. That means, the buyers are willing and able to purchase more of the product.
- ✰ ***Supply side***: The sellers, on the other hand, are willing and able to offer only 200 units of the product at the announced ceiling price. This is because, at the announced ceiling price, they can cover the production cost of only 200 units of the product and they are unwilling and unable to sell more.
- ✰ ***Deficit or Shortage***: The combination of the quantity demanded and the quantity supplied of the product reveals that, at the announced ceiling price, the market has a shortage of 400 units of the product. That means, the buyers want to buy 400 units of the product more than the seller's supply.

The above example clearly infers that, the main objective of fixing the ceiling price below equilibrium price is to offer the product at an affordable price to the consumers. So, at Rs.30/unit of the product, the buyer receives a benefit of Rs.20 on each unit of the product. However, from the sellers point of view, fixing a ceiling price lower than equilibrium market price, is a loss for them, as they will receive less of Rs. 20 on each unit of the product. So, they offer less quantity of the product for sale *i.e.*, only 200 units. That means, at the ceiling price, the total quantity offered by the sellers is 200 units, but the total quantity of product demanded by the buyers is 600, thereby, it gives the following important conclusions:

- ✰ At lower (ceiling) price, large quantity of the product is demanded in the market (600 units).
- ✰ The supply into the market (200 units) is less than quantity demanded (600 units). So, some number of buyers cannot purchase the product at low (ceiling) price. So, there is a loss on the part of these buyers too.
- ✰ The total shortage of the product in the market increases with the fall in ceiling price

However, it is important that, the extent of shortage of product left over in the market after fixing the ceiling price depends upon the elasticities of demand and supply curves. If the demand and supply curves are more steeply sloped or more inelastic, the shortage will be less. That means, if the demand and supply of the product are relatively inelastic, then the goal of providing the product at affordable price, without excluding a lot of buyers from the market can be more easily achieved.

Assuming normal conditions with reference to elasticities of demand and supply for the product, more shortage (400 units) will be there with the announcement of ceiling price (as demand>supply). So, the buyers resort to illegal purchase of the product, thereby, violating the legally mandated ceiling price. So, the buyers will even offer more than Rs. 30 to have the units of the product through entering a black market exchange. If the buyers offer more than Rs.30 per unit of the product, the sellers listen to them. But, the extent of price increase above the ceiling price at which the buyer is ready to purchase the product depends upon the impact of price fall. If the ceiling price fixed below equilibrium price is too low (*i.e.*, huge difference between equilibrium price and ceiling price), then the buyer has to pay higher price in the black market above ceiling price, so as to purchase the product at times of shortages and *vice versa*.

17.19. Welfare Analysis

Having discussed the concepts of consumer surplus, producer surplus and SS, we can analyze the effects of market interventions (say, price ceiling) and over and under productions of a commodity on the changes in SS and such analysis is called as Welfare analysis.

(a) Imposition of Price Ceiling

In the Figure 17.35, OP is the equilibrium price of the commodity at which OQ is the equilibrium quantity demanded and supplied in the market. Assuming that, the market equilibrium price (OP) is too high, the Government imposed price ceiling (OP_1) below the equilibrium price. With the imposition of ceiling price, the market price of the commodity decreases from OP to OP_1 and the quantity from OQ to OQ_1.

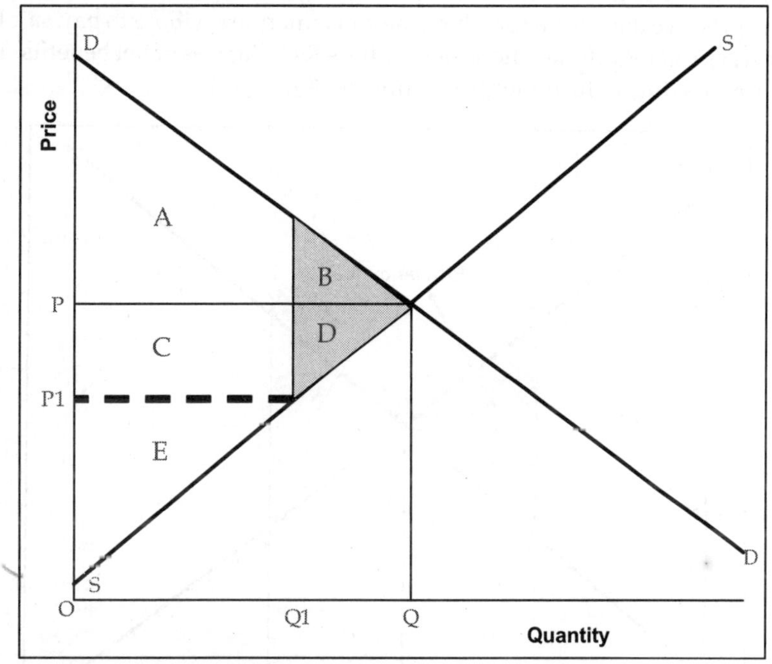

Figure 17.35: Welfare analysis due to imposition of price ceiling.

Before the imposition of ceiling price or at the equilibrium price (OP):

Consumer surplus = A + B

Producer surplus = C + D + E

SS = A + B + C + D + E

After the imposition of ceiling price (OP_1):

Consumer surplus = A

Producer surplus = C + E

SS = A + C + E

Changes in consumer surplus and producer surplus due to imposition of price ceiling:

- ☆ ***Changes in consumer surplus:*** Consumers have lost 'B', but gained 'C' from the producers.
- ☆ ***Changes in producer surplus:*** Producers have lost 'C' and 'D' and no gain to them.
- ☆ ***Changes in SS:*** This is given by,

= SS before imposition of ceiling price – SS after imposing ceiling price.

= (A + B + C + D + E) – (A + C + E)

= **B + D**

So, with the imposition of price ceiling on the commodity, the producers are definitely adversely affected due to loss in producer surplus by 'C' and 'D' components. To know, whether consumers are better off or worse off with the imposition of price ceiling, it depends upon the relative sizes of the areas 'C' and 'B'. If area 'C' is larger than area 'B', then the consumers are safe guarded with the market intervention and *vice versa*. However, SS is always minimized from '(A + B + C + D + E)' to '(A + C + E)' with the imposition of price ceiling in a perfectly competitive market. This amount of loss in SS *i.e.*, '(B + D)' is called the deadweight loss. So, we can define deadweight loss as the loss in welfare due to market imperfections in the economy.

(b) Over and Under Productions of a Commodity

Figure 17.36 demonstrates the efficiency losses *i.e.*, reductions of combined consumer and producer surpluses due to over and under productions of paddy. Suppose that, the farmers produced OQ_1 output (under production) rather than the efficient level OQ. So, the sum of consumer and producer surpluses (SS), previously DES falls to DABS. So, the SS declines by the amount of the triangle AEB to the left of OQ output and this represents an efficiency loss to buyers and sellers. And since buyers and sellers are members of the society, it represents an efficiency loss or deadweight loss to the society. For output levels from OQ_1 to OQ, the maximum willingness to pay by the consumers (as reflected by points on the demand curve) exceeds the minimum acceptable price of farmers or sellers (as reflected by points on the supply curve). By failing to produce the output for which the consumers are willing to pay, say, Rs.25/kg and for which the farmers are willing to accept say, Rs.20/kg, the society suffers Rs.5/kg loss of net benefits. The triangle AEB shows the total loss of such net benefits due to the underproduction at OQ_1.

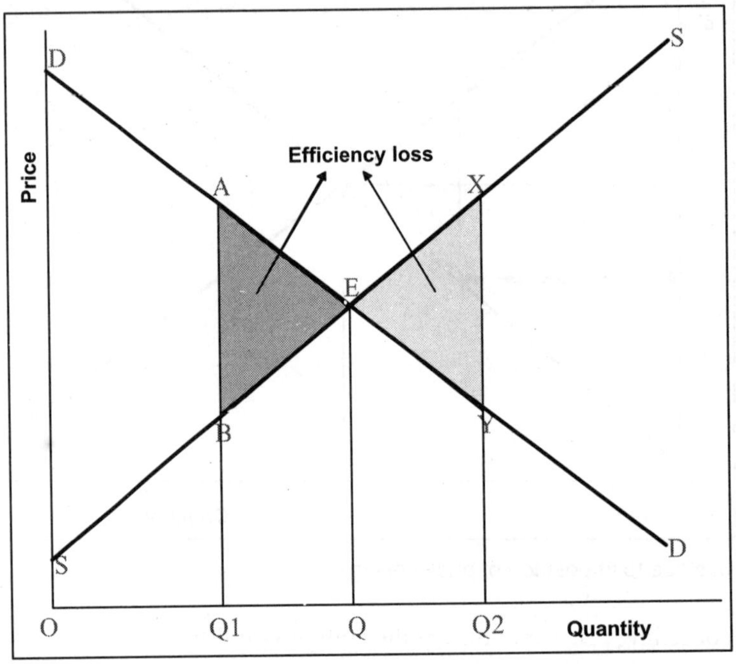

Figure 17.36: Welfare loss due to over and under productions of paddy.

In contrast, suppose that, there is over production of paddy in the economy and the output produced is OQ_2 rather than the efficient level of OQ. So, the SS declines by EXY, the triangle to the right of OQ. This triangle subtracts from the SS of DES that would occur, if the quantity of output had been OQ. For all units beyond OQ, the consumers' maximum willingness to pay is less than the farmers' minimum acceptable price. Producing an item for which the maximum willingness to pay is, say, Rs.30 and the minimum acceptable price is Rs.35 subtracts Rs.5 from society's net benefits. Such over production is uneconomical and creates an efficiency loss or deadweight loss for society, as shown by the triangle EXY to the right of OQ. So, it is essential to note that, too much or too little output compared to optimum leads to loss in societal welfare.

18

Morphology of Markets and Analysis of Firm's Equilibrium

In an ordinary language, the word 'market' implies a particular place, where the buyers and sellers interact with each other for the transaction of commodities. In other words, an area, large or small, can be considered as a market, where buyers and sellers are in easy contact with one another. So, the term 'market' thus, indicates a geographical location. However, in economic sense, market implies a contact or a network of dealings between buyers and sellers. In the modern days of business, with the development of communications and banking, the markets have widened and dealings in some commodities are now world-wide. Therefore, the essential feature of a market is that, buyers should be able to strike bargains with sellers. In the words of Wicksteed, '*market is the characteristic phenomenon of economic life and the constitution of markets and market prices is the central problem of Economics*'. P.Tailor suggests that '*Marketing is not about providing products or services it is essentially about providing changing benefits to the changing needs and demands of the customer*'. In the words of Joan Violet Robinson '*market, a means by which the exchange of goods and services takes place as a result of buyers and sellers being in contact with one another, either directly or through mediating agents or institutions*'. There are several criteria based on which the markets may be broadly classified, but the classification relevant for the present study is based on the extent of competition prevailing in the market and it is studied in-detail in the ensuing chapters. The present chapter mainly focuses on the morphology of markets, concepts of revenue across the markets and various approaches to determine the firm's equilibrium. Considering the market competition as the basic criterion, the markets are classified on the basis of:

- ☆ The number of firms in the industry
- ☆ The nature of the product produced
- ☆ The degree of control the firm has on output and price policies
- ☆ Profit levels
- ☆ Non-price competition
- ☆ The extent of barriers to entry and exit

18.1. Classification of Market Structure or Competition

The detailed classification of markets based on market competition is presented through Figure 18.1. From the Figure 18.1, it is clear that, the markets are broadly classified into two types *viz.*, Perfect markets and Imperfect markets. Under imperfect market structure, there are different forms of market competition *viz.*, Monopoly, Duopoly, Oligopoly and Monopolistic competition. The basic features of these different types of market structures are presented through Table 18.1.

Table 18.1: Basic features of different types of market structures

Market Structure	*Number of Firms*	*Degree of Product Differentiation*	*Firms Degree of Control Over Price*	*Examples*
I. Perfect competition	Infinite number of firms	Homogenous nature	No control over price	Agricultural commodities, Internet service
II. Imperfect competition				
a. Monopoly	One firm	Product with no close substitutes (homogenous)	Considerable control over price	Microsoft is held to have monopolized market for computer operating systems, Public utilities like railways, electricity, irrigation etc.
b. Duopoly (A pure duopoly occurs very rarely in practical world and the more common is two dominate firms hold majority of the market share)				
❖ Perfect Duopoly	Two firms	Homogenous	Some control over price	Visa and MasterCard control a large proportion of the electronic payment processing market, Hindustan Petroleum vs Bharat Petroleum
❖ Imperfect Duopoly	Two firms	Heterogeneous	Some control over price	Pepsi vs Coca-Cola in soft drink market, Gillette vs Wilkinson Sword in razor blade market, Kodak vs Fujifilm in the color 135 film market
c. Oligopoly				
❖ Perfect Oligopoly	Few firms	Homogenous	Some control over price	Organization of Petroleum Exporting Countries (OPEC), Steel industry in USA, Cement.
❖ Imperfect Oligopoly	Few firms	Heterogeneous	Some control over price	Automobile industry in USA, Computers, Electrical appliances, Television etc.
d. Monopolistic competition	Many firms (but less than perfect competition)	Heterogeneous	Some control over price	Retail outlets of fertilizers, pesticides, restaurants, clothing, shoes etc.

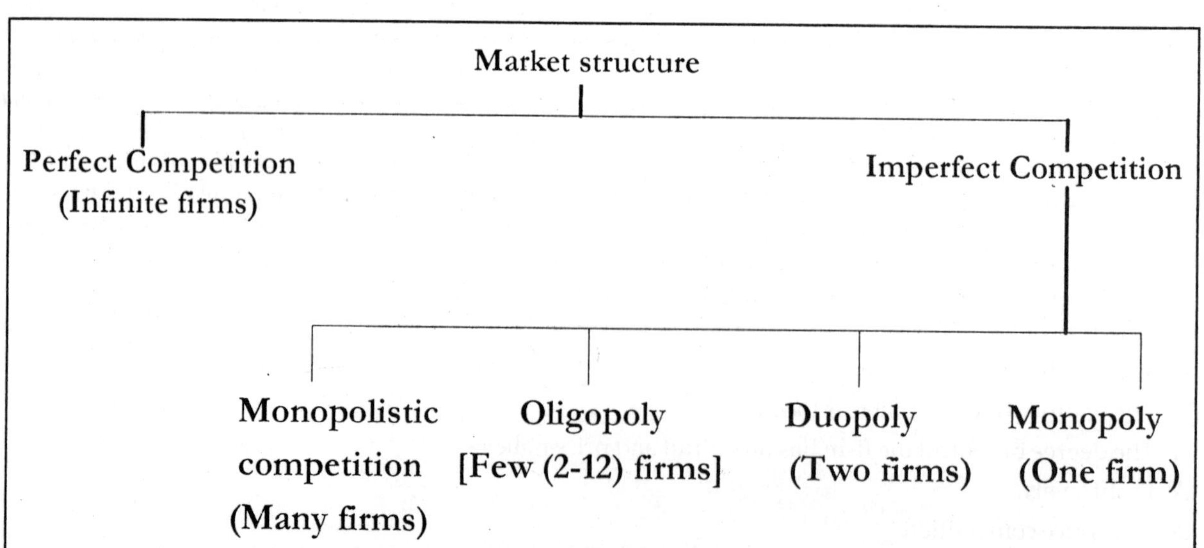

Figure 18.1: Different types of market structures based on competition.

The above classification of markets into different structures is with reference to sellers' side (firm's side). But, with reference to buyers' side, the following is the classification corresponding to sellers' view.

Classification of markets

(Sellers' view)

- **Perfect competition**
- **Imperfect competition**
  - **Monopoly**
  - **Duopoly**
  - **Oligopoly**
  - **Monopolistic competition**

Classification of markets

(Buyers' view)

- **Perfect competition**
- **Imperfect competition**
  - **Monopsony** (Ex. Cotton Corporation of India procuring cotton from farmers, Employment of nurses etc)
  - **Duopsony** (Ex.Two firms who dominate in selling the refined petrol also likely to dominate the input market for buying crude oil)
  - **Oligopsony:** (Ex. Sugarcane market, Jute market, Few firms who dominate in manufacturing and selling the computers also likely to dominate the input market for buying computer components, such as silicon chips, hard disk drives, programmers etc.)
  - **Monopsonistic competition:** (Ex. Firms engaged in hiring skilled labour in agricultural sector)

1. Perfect Competition

A market is said to be perfect, when all the potential buyers and sellers of the product have complete knowledge about the prices of the commodities of homogenous nature at which transactions takes place. Under such circumstances, same price will prevail for the same commodity at the same time. No single buyer or seller has much effect on the going market price. A seller cannot charge more than the going price because, buyers can obtain as much they need at the going price. In a perfectly competitive market, marketing research, product development, pricing, advertising and sales promotion play little or no role. Thus, sellers in these markets do not spend much time on the marketing strategy. The following are the important features of perfect competition:

☆ Infinite number of buyers and sellers

☆ The products are homogeneous products and they are perfect substitutes

☆ Free entry and exit of firms in the industry

☆ There is perfect knowledge about the product for buyers and sellers

☆ Firms are 'price takers' and each firm face a perfectly elastic demand curve for their product

☆ All buyers and sellers are completely aware of the price of product

☆ No individual firm or seller can influence the price of the product

☆ Same price will prevail for the same commodity at the same time

☆ No Government regulation on prices of the products

☆ Perfect mobility of factors of production

☆ ARC or demand curve is parallel to X-axis, and MRC coincides with ARC for a perfect competitive firm

☆ Complete absence of transportation costs

Pure Competition and Perfect Competition

Economists like Chamberlin and others often make a distinction between Pure competition and Perfect competition. The term 'Pure competition' (coined by Chamberlin) is used in a restricted sense. In order that competition be 'pure', it requires the fulfillment of first three features of perfect competition *viz.*, infinite number of buyers and sellers, homogenous nature of the product and free entry and exit of firms. These conditions together mean that, no individual firm can exert any influence over the market price. But, the term perfect competition is a wider concept, in the sense, that it includes the features of pure competition and some additional features as mentioned earlier. This means, perfect competition requires that, there should be no imperfections in the market. Such imperfections arise due to imperfect knowledge or immobility of the factors of production. In fact, pure competition is a part and parcel of perfect competition. American

Economists prefer to use the term pure competition, while the English Economists prefer the term perfect competition. However, both the terms are used interchangeably to analyze the features of perfect markets.

2. Imperfect Competition

As discussed earlier, perfect competition describes a market structure with extreme assumptions, thereby, it is highly unlikely to exist in the practical world. This is because, in reality, most markets are imperfectly competitive. So, any deviation from the conditions of perfect competition in a market leads to the existence of imperfect competition *i.e.*, less than infinite number of sellers, homogenous or heterogeneous nature of the product, absence of price information, restrictions on free entry and exit of firms, existence of transportation costs etc. Depending upon the degree of imperfections, the different market forms include, monopoly, duopoly, oligopoly and monopolistic competitions.

(a) Monopoly

Monopoly is a form of imperfect competition in which, a single firm produces the product with no close substitutes. That means, there are no other firms in the economy that make a similar product (or service) in the market, thereby, the single firm represents an industry and such monopoly is called as 'Pure monopoly'. Such 'Pure monopolies' are relatively rare, but they do exist. Local electric and water utilities are good examples for pure monopolies, as they produce a product for which there are no close substitutes. It is important that, a firm's degree of monopoly may vary among the markets. For example, an electric company enjoys monopoly status in the production of electricity meant for lighting purposes, but for heating a commodity or product, the firm has to face stiff competition from gas, coal and fuel oil supplying industries. So, the electric company may consequently be a 'pure monopoly' in the lighting market, but do not enjoy monopoly status in the heating market. Similarly, railway transport do not enjoy monopoly status, as there are other means of transport for the customers to move to distant places such as air lines, buses etc. To generalize, by law, a monopoly occurs, if a firm has a market share of more than 25 per cent.

The fundamental cause of monopoly is due to the existence of strong barriers for entry of new firms into the industry. The barriers may be technological or economical that raise the cost for the new firms to enter the industry above the cost of a firm enjoying monopoly status in the market. If the barriers to entry prevent others firms from entering the market, there is no competition and the monopolist remains the only seller for the product in the market. So, the monopolist can either decide the output or price for the product. Here lies the major difference between a monopoly and a perfect competition. A monopolist has the ability to influence the price of the product (price maker), whereas, the firm under perfect competition is simply a price taker rather than a price maker. This is because, under perfect competition, the competitive firm enjoys a very small share in the total market, thereby, the price of the product is determined by market conditions (industry) and this price is passed on to each individual firm. On the other hand, under monopoly competition, a single firm is the sole producer of the product in the market, thereby, it can often alter the price of the product by adjusting the quantity of supply of output into the market. The following are the important features of monopoly:

- ☆ A single firm or seller selling all output in a market *i.e.*, the product have only single control
- ☆ A unique product, thereby, the E_C for the product is zero.
- ☆ A monopoly often acquires and generally maintains single seller status due to restrictions on the entry of other firms into the market.
- ☆ The monopolist is either a price maker or output decider.
- ☆ The distinction between firm and industry will disappear in monopoly.
- ☆ Power to influence price is the essence of monopoly.
- ☆ There are absolutely no rivals, as monopoly firm represents the industry.
- ☆ A monopoly firm often possesses information not available to others. This specialized information comes in the form of legally-established patents, copyrights or trademarks etc.
- ☆ Because of the presence of only seller, the firm faces a downward-sloping demand curve (*i.e.*, industry demand curve) and this implies that, the monopolist sells more of the output at lower price.

(b) Oligopoly

The word 'Oligopoly' is derived from the Greek word '*oligos*' which means few. When the whole market structure of few firms controls the major share of market sales, then the resulting structure is referred as Oligopoly. In this form of competition, the entire market is dominated by as few as two firms or as many as twelve firms. These few firms are interdependent on each other with respect to investment, output and pricing decisions. That means, each firm is aware of the actions of the rivalry firms. Normally, an oligopoly market exists, when the top five firms in the market account for more than 60 per cent of the total market sales. With fewer than two firms, the market structure is Monopoly. The simplest form of oligopoly is Duopoly, where there are only two firms producing and selling the product. For example,

Pepsi vs Coca-Cola in soft drink market, Gillette vs Wilkinson Sword in razor blade market, Kodak vs Fujifilm in the color 135 film market etc., are the good examples for Duopoly market. But, as the number of firms is many, then Oligopoly becomes Monopolistic competition.

Oligopoly competition dominates the modern economic landscape, accounting for about half of all output produced in the economy. Oligopolistic industries deals with a wide range of products like breakfast cereals, cars, computers, television broadcasting, pharmaceuticals, petroleum etc. OPEC is an example of Oligopoly since few countries control the production of oil, the steel and the automobile industry in United States of America is another example. These industries tend to be a prime source of innovations that promote technological advances and economic growth. The oligopolies may be Perfect oligopolies and Imperfect oligopolies with respect to differentiation of products. In a 'perfect oligopoly' such as the oil industry, there is little or no differentiation in terms of the product sold. But, in an 'imperfect oligopoly' like the IT and automotive industry, there's lot of differentiation between the products. The pricing strategies followed in perfect oligopolies vary greatly compared to imperfect oligopolies. However, the differentiations across the products are often short-lived, as the market is attuned to replicating originality, thereby, imperfect oligopolies becomes more perfect in due course of time. So, oligopoly is a market structure characterized by a small number of large size firms that dominate the market, selling either identical or differentiated products, with significant barriers to entry into the industry. The following are the important features of oligopoly:

- ☆ Presence of few sellers in a market situation for a commodity.
- ☆ Interdependency among the firms for decision-making.
- ☆ The individual firm faces an indeterminate demand curve, as the price elasticity depends on the rival's reaction to changes in investment, output and price.
- ☆ Firms sell branded products, which are close substitutes of each other.
- ☆ High selling costs is the unique feature of this competition. So, the non-price competition is a consistent feature of the competitive strategies of oligopolistic firms. Examples of non-price competition includes *viz.*, free deliveries and installation, extended warranties for consumers and credit facilities, longer opening hours (supermarkets and petrol stations), branding of products and heavy spending on advertising and marketing, extensive after-sales service etc.
- ☆ The existence of both the co-operative and competitive trends among the business firms.
- ☆ Entry barriers for the other firms are high and the barriers can be due to patents, copyrights, Government rules and regulations, ownership of scare resources etc.
- ☆ Lack of uniformity in size of firms
- ☆ Products may be homogenous (standardized) or heterogeneous (differentiated).
- ☆ The sellers are the price makers and not price takers, since the few sellers mutually dominate the pricing decisions.
- ☆ The sellers can achieve super normal profits even in the long run.
- ☆ The sellers can achieve economies of scale, since for the large producers, as the level of production rises, the cost per unit of product decreases, thus ensuring higher profits.
- ☆ No unique pattern of price behaviour.
- ☆ The firms tend to keep the prices at the same level or rigid.

(c) Monopolistic Competition

Monopolistic competition refers to a market competition, in which large number of firms sells differentiated products. Since the products produced by these firms are close, but not perfect substitutes to each other, there exists stiff competition among the firms. It is important that, though the number of firms are larger in monopolistic competition, but they are not as large as in perfect competition. This form of market competition is extremely common. Perhaps, half of the economy's total production comes from monopolistically competitive firms. Almost all retail operations in the economy are in this form of market competition. For example, retail outlets of fertilizers, pesticides, restaurants, clothing, shoes etc., fall under this category. Under this competition, starting a business is relatively easy, but staying in the business is quite difficult, as the firm requires an ability to convince the customers that the product is different and better than that of competitors. Considering the above features, it infers that, the market, which consists of large number of buyers and sellers who trade over a range of prices rather than a single market price is called monopolistic competition. A range of prices occurs because, sellers can differentiate their products offered to the buyers. This product differentiation is usually achieved in one of three ways *viz.*, physical differences (differences in size, shape, colour of the product), perceived differences (supplying the products in tins, bottles, packets etc) and support services (repairs and maintenance services, offering discounts etc). This product differentiation is the primary reason that each firm operating in a monopolistically competitive market is able to create a little monopoly all to itself.

Perfect competition and Monopoly are the two extreme forms of market competition, the former being extreme form of perfection and the latter being extreme form of imperfection. However, these two competitions are purely theoretical and not practically seen. Between these two extremes, there is a monopolistic competition, which is more common and practically seen in the economy. So, in effect, monopolistic competition is something of a hybrid between perfect competition and monopoly and this combines the features of both perfect competition and monopoly. Monopolistic competition is the best approximation of perfect competition that the real world offers. Edward Hastings Chamberlin is regarded as the 'Founding Father' for studying the concepts of monopolistic competition in his famous work, '*Theory of Monopolistic Competition* (1933)'. The following are the important features of monopolistic competition:

- ☆ There are large number of sellers producing differentiated products and no firm has total control over the market price.
- ☆ Product differentiation is the essence of monopolistic competition *i.e.*, every producer tries to keep his product dissimilar than his rival's product in order to maintain his separate identity.
- ☆ There is more freedom of entry and exit of firms into the market. This leads to stiff competition and helps to maintain normal profits in the market for a longer span of time.
- ☆ Due to product differentiation, every firm has to incur some additional expenditure in the form of selling cost. This cost includes sales promotion expenses, advertisement expenses, salaries of marketing staff etc.
- ☆ The firms produce very close substitutes, thereby, E_C is very high.
- ☆ Large numbers of firms are different in their size. Each firm has its own production and marketing policies. So, no firm is influenced by other firm and each firm is independent in its output-pricing policies.
- ☆ There is two dimensional competition in this market *viz.*, price competition (firms compete with each other on the basis of price) and non-price competition (firms compete on the basis of brand, product quality, advertisement etc).
- ☆ The firm faces a downward sloping demand curve *i.e.*, elastic demand curve. It means, the firm can sell more output at lower price and *vice versa*. The demand curve in monopolistic competition is more elastic compared to monopoly.
- ☆ Each firm produces similar, but not identical goods. So, each firm is an industry itself. Since products are not homogeneous, there is no industry, as in the case of perfect competition. So, Chamberlin introduced the term 'group' under monopolistic competition. A group is a cluster of firms producing very closely related, but differentiated products. That means, the collection of firms that produce same type of products with high E_C will constitute a 'group' or a 'product group'.
- ☆ The firm under monopolistic competition is a price maker.

So, the features of a monopolistically competition are almost the same as in perfect competition, but the difference is, in monopolistic competition, the products are heterogeneous and thereby, involves a great deal of non-price competition. Further, the individual firm's demand curve is downward sloping in monopolistic competition, in contrast to perfect competition, which is perfectly elastic.

Competitive Structure vis-à-vis Competitive Behaviour

A perfect competitive market structure is characterized by the presence of infinite number of firms producing and selling a homogenous commodity. But, each individual firm is unable to influence the price of the commodity in the market, as they are simply price takers. The price of the commodity is determined by the demand and supply forces at the industry level and the same is passed on to the individual firms at which, each individual firm has to transact any amount of output. So, each firm does not need to compete with other firms to obtain a market for its produce. This is clearly the opposite of the meaning of what is commonly understood by the competitive behaviour. For example, in case of imperfect competitive market structure, say duopoly, where two giant firms like Bayer and Syngenta compete with each other in a variety of ways to achieve a higher level of sales or profits. Thus, intense competitive behaviour exists between these two giant firms to gain significant share in the market. Conversely, we do not find individual farmers competing among themselves to sell a larger amount of produce. This is because, both Bayer and Syngenta firms possess the market power to influence the price of pesticides, while an individual farmer does not. Thus, competitive market structure and competitive behaviour are, in general, inversely related. That is, the more competitive the market structure, the less competitive is the behaviour among the firms, as in case of perfect competition. On the other hand, the less competitive the market structure, the more competitive is the behaviour of firms towards each other for gaining significant market share. However, pure monopoly is the most visible exception.

18.2. Analysis of Firm's Equilibrium

1. Concepts of Revenue

In modern days of business, every firm, either small or large tend to produce and market the commodities with the aim of maximization of profits. The amount of money which the firm earns by selling the output is called Revenue. So, every firm aims to earn more of the revenue, so as to cover the TC incurred in the production programme. So, in the business, the firms play dual role *viz.*, as a producer and as a seller. As a producer, it aims at minimization of costs and as a seller, it aims at maximization of profits. In this process of selling the output in the market, the firm earns revenue to sustain in the business. There are different concepts of revenue *viz.*, TR, AR and MR and these concepts are widely useful for analyzing the firm's equilibrium with reference to optimization of output or maximization of profits. So, before discussing the different approaches to arrive at the firm's equilibrium, it is essential to study different concepts of revenue under different forms of market competition.

(*a*) TR

TR is the amount of money or income that the firm gets from the sale of its product in the market. This is derived by multiplying the quantity of product or output sold by the firm in the market with the price at which it is transacted to the buyer. The concept of TR helps to derive both AR and MR of the firm. The curve, which depicts the TR of the firm refers to TRC.

(*b*) AR

In simpler terms, the term AR was quoted as 'Price' received by the firm in selling the product in the market. Using the term AR often provides a connection to other related terms, especially TR and MR. We know, TR refers to the total income or returns earned by the farmer or firm in selling the product in the market. So, AR refers to the ratio between TR to the quantity of output sold in the market. It is given by AR = TR/Y. This infers that, TR relates to total income earned by the firm from the total output transacted in the market, whereas, AR refers to the revenue (price) earned per unit of the product. In the concepts of SRATC and LRATC, we studied the cost per unit of output in the short run and long run production programmes respectively and in the concept of revenue, AR refers to the revenue or price earned by the seller per unit of output transacted in the market. The concept of AR is of vital importance for the firm, as it indicates the firm's profits and its future prospects of expanding the business in the long run production programme. The concept of AR is represented by ARC. The ARC is also called as Demand curve because, the demand curve indicates the various quantities of a commodity purchased by the buyer at different prices, from which, we can compute the price or AR at which each unit of the commodity is transacted by the seller. The price paid by the buyer constitute the revenue to the seller.

AR Vs. Price

In Economic sense, we generally use the terms AR and price interchangeably. But, in all cases, they both are not the same. This is because, if the farmer sells various quantities of a commodity at the same price, AR is equal to price. But, if the seller offers different quantities of a commodity at different prices, the AR is not the same as price. Say, for example, if the farmer sells 100 quintals of paddy to rice miller at Rs. 1000/quintal and 50 quintals of paddy to wholesaler at Rs. 1000/quintal, the TR obtained by the farmer is Rs. 1.50 lakhs. The AR obtained by the farmer is Rs. 1000/qtl, which is same as price/quintal of paddy. Let us consider another case, where the farmer sells 100 quintals of paddy to rice miller at Rs. 1000/quintal in Andhra Pradesh and 100 quintals of paddy to another rice miller in Tamil Nadu at Rs. 1200/quintal. Then, the TR obtained by the farmer is Rs. 2.20 lakhs and the AR obtained by the farmer is Rs. 1100/qtl, which is not equal to the price/quintal of paddy. This infers that, when the farmer or firm discriminates the price across the markets, the AR he receives is not equal to price of the commodity. This is the case normally we see, when a monopoly firm discriminates the markets. But, in general, the farmer transacts the large quantities of commodities at the same price, thereby, AR equals to price and hence, we use these two terms interchangeably.

(*c*) MR

The concept of AR only indicates the average price per unit of output sold by the seller to the buyer. But, it won't indicate the change in TR due to additional unit of output transacted by the seller to the buyer and the same is indicated by MR. That means, MR indicates, how much extra revenue the firm receives by selling an extra unit of output. We ca derive this MR as the slope of the TR curve. So, MR is the net revenue added to the seller's revenue, when an extra unit of output was transacted in the market. This word 'net' clearly infers that, the MR is not equal to the price (AR) of the commodity, unless the firm operates under perfect competition. This is because, in perfect competition, as discussed earlier, the firms are simply price takers and a single firm cannot influence the price of the commodity. Further, the commodity being homogeneous, the firm gets same additional revenue for an additional unit of output transacted in the market. Hence, MR equals to AR under perfect market competition. However, in other (imperfect) markets, different quantities of output are sold at different prices in accordance with the law of demand, thereby, MR is not equal to the price (AR) of the commodity. The concept of MR is represented by MRC.

Measurement of MR

There are three methods of computation of MR. For example, a farmer transacted 10 quintals of paddy at Rs. 1000/qunital. Due to decrease in price to Rs. 950, the farmer could transact 11 quintals of paddy.

Method 1

MR is computed as the difference between TR of quantity of output sold at new price and TR of quantity of output sold at old price. This is given by the formula : $MR = TR_{n+1} - TR_n$ So, MR = (11 x 950) – (10 x 1000) = 450. This infers that, by selling additional unit (quintal) of paddy, the farmer gets additional revenue or net revenue or MR of Rs. 450.

Method 2

MR is, otherwise, explained as the ratio between change in TR to the change in quantity of output sold by the farmer in the market. So, the MR is given by, $\Delta TR/\Delta Y$. This implies that, the slope of TR indicates MR. In the above example, the change in TR is Rs. 10450 – Rs. 10000 = Rs. 450 and change in quantity of output sold by the farmer is one quintal, thereby, MR = Rs. 450.

Method 3

The above concepts can be studied in a different way. When the price of paddy is decreased in the market from Rs. 1000 to Rs. 950, it implies, the farmer is losing Rs. 50 on each quintal he transacted earlier, thereby, the total loss on the previous 10 units will be Rs. 500. But, for the 11th unit (quintal) of paddy, the farmer is getting the price Rs. 950. So, the net gain (MR) to the farmer by selling 11th unit of paddy is 950 minus total loss incurred by the farmer on the previous 10 units due to fall in price of paddy *i.e.*, MR = Rs.950-Rs.500 = Rs. 450. So, we can derive the formula for computation of MR and it is given by,

MR = Price of additional unit of output – Loss in revenue on previous units sold due to price fall

The above discussion infers that,

- ☆ When different quantities of commodities are offered at different prices by the farmer or firm, the MR is not equal to the price.
- ☆ MR of transacting additional unit of paddy (Rs. 450/quintal) is not equal to price of paddy (Rs. 950) at which the 11th quintal of paddy is sold. That means, when additional unit of output is sold at less new price (Rs. 950) than the old price (Rs. 1000), then MR (Rs. 450) is less than the new price. But, when additional unit of output is sold even when there is no change in price, then MR is equal to the price.
- ☆ From Method 2, we can also infer that, the slope of TR indicates MR. So, given a TR curve, we can easily compute the MR of each and every additional unit of output transacted by the firm in the market.

As discussed under the relationship between SRMC and SRAVC curves, there exists the same marginal-average relation between MRC and ARC. This is commonly we see in imperfect market competition. That means, when AR falls, MR also falls and lies below AR. When AR is constant, MR is also constant and is equal to the AR. This is commonly we see in perfect market competition. Among these three concepts of TR, AR and MR, the concepts of AR and MR are important to draw comparative picture across different market structures.

2. ARC and MRC in Different Forms of Market Competition

The concepts of AR and MR are represented in the form of ARC and MRC respectively. The relation between these two curves in relation to output depends upon the market structure. For a perfectly competitive firm, AR is not only equal to price, but also equal to MR. So, for a perfectly competitive firm, the ARC is a horizontal line parallel to X-axis or perfectly elastic (as the firm is only a price taker) and MRC coincides with ARC. But, in case of imperfect market competitions *viz.*, monopoly, oligopoly and monopolistic competitions, for a firm, the AR is greater than MR and they both fall when larger quantities of output are transacted. So, for monopoly, oligopoly and monopolistic competitive firms, both ARC and MRC are negatively sloped. In each case, the ARC is above the MRC, but with different elasticities. As explained earlier, the ARC is also the demand curve facing the firm. But, both the constant or decreasing nature of ARC and MRC are the indicators of the market control of a firm. The point of optimality in the production programme is indicated by the equality between MC=MR. If MR>MC, the firm can still increase the output till the equality MR=MC is ensured. On the other hand, if MR<MC, the firm has to reduce the output till the equality MR=MC is ensured. The ARC indicates the price at which the optimal output has to be transacted to maximize the profits. Hence, ARC is otherwise termed as Demand curve of the firm. Let us discuss the pattern of ARC and MRC across different market competitions.

(a) ARC and MRC Under Perfect Market Competition

We know, perfect competition is a market structure with an infinite number of small firms, each selling a homogenous good. All the perfect competitive firms have perfect knowledge about the market and perfect mobility into and out of the market. The firms under perfect competition are simply price takers, thereby, they have no market control and receive the same or going market price for all the output sold. So, at the same price, the firms have to transact all the output and this implies, the demand curve or ARC are perfectly elastic. Since, the firm has no control over the price, it has to sell the additional units of output at the same price, thereby, MRC is also constant and it is equal to ARC. So, for a perfect competitive firm, both ARC and MRC are equal and when ARC is constant, MRC must be constant. The same concept is explained through Table 18.2 and Figures 18.2 and 18.3.

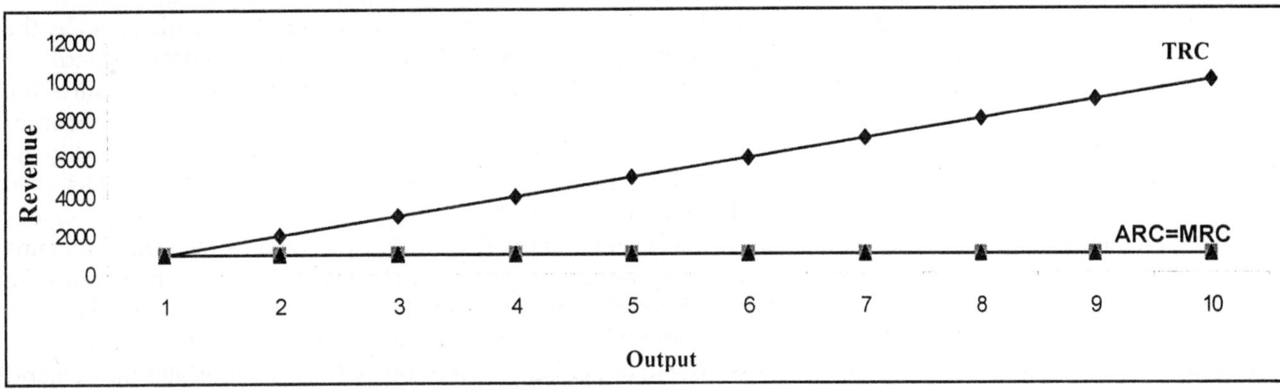

Figure 18.2: TR, AR and MR curves of a firm under perfect market competition.

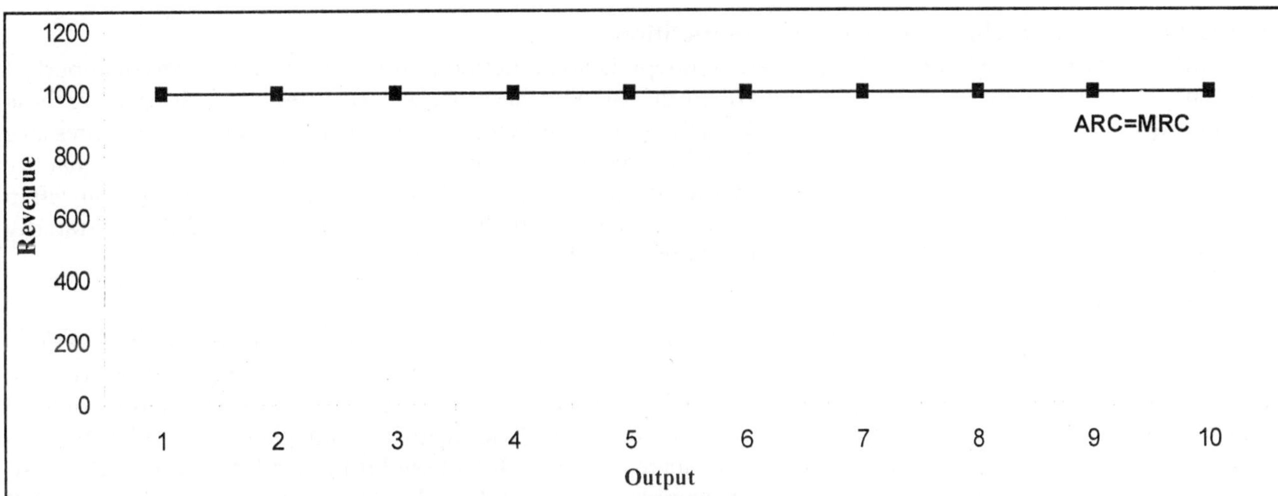

Figure 18.3: ARC and MRC of a firm under perfect market competition.

Table 18.2: TR, AR and MR of a perfect competitive firm transacting paddy.

Output (Quintals)	*Price (Rs/quintal)*	*TR (Rs.)*	*AR (Rs/quintal)*	*MR (Rs/quintal)*
0	1000	0	–	–
1	1000	1000	1000	1000
2	1000	2000	1000	1000
3	1000	3000	1000	1000
4	1000	4000	1000	1000
5	1000	5000	1000	1000
6	1000	6000	1000	1000
7	1000	7000	1000	1000
8	1000	8000	1000	1000
9	1000	9000	1000	1000
10	1000	10000	1000	1000

In the Table 18.2, it was shown that, the farmer transacted various quantities of paddy at the same price, as under perfect market competition, same price will prevail for the same commodity at the same time. From the quantity-price data, the TR, AR and MR are computed. TR is computed in the third column as the product between quantity of paddy transacted and price received. AR is computed in the fourth column by dividing the TR with quantity of paddy transacted. The obvious point here is that, AR is equal to Rs. 1000 for each and every unit (quintal) of paddy sold. The reason is that, price is constant at Rs. 1000 for each and every unit (quintal) of paddy sold. So, AR is price or price is AR and these two are one and the same. MR is calculated by dividing the change in TR by the change in quantity of paddy sold. For example, when the farmer sells one quintal of paddy, he receives Rs. 1000 MR and if he sells 10th quintal of paddy, still he receives Rs. 1000 MR. So, the MR is equal to Rs. 1000 for every extra quintal of paddy sold. This results because, price is constant at Rs. 1000 for every quintal of paddy sold. So, for every extra quintal of paddy sold by the farmer, he receives an extra Rs. 1000, which is the going market price. So, MR is equal to price and AR for a perfectly competitive firm. The same results are shown through Figures 18.2 and 18.3, where the vertical axis measures the revenue and the horizontal axis measures the quantity of output (paddy) sold by the farmer. The ARC and MRC curves are perfectly elastic and hence, they lie parallel to X-axis. That means, if the farmer sells one quintal of paddy, he receives Rs. 1000 AR and Rs. 1000 MR. If the farmer sells 10 quintals of paddy, still he receives Rs. 1000 AR and Rs. 1000 MR (for 10th quintal), but the TR increases from Rs. 1000 to Rs. 10000. This implies that, TR continuously slopes upwards with increase in quantity of paddy sold by the farmer, but the AR and MR received by the farmer remains same irrespective of the quantity of paddy transacted by the farmer. So, MRC coincides with ARC. Another interesting aspect here is, under perfect competition, the AR and MR coincides with price of the output. So, AR is just another term for price and this ARC is actually the demand curve. Under the perfect market competition, the MRC of the farmer is also his demand curve, as MRC coincides with ARC. But in imperfect competition, ARC alone indicates the demand curve for the farmer, as MRC is negatively sloped and lies below the ARC (demand curve) and the same is discussed in the ensuing pages.

(*b*) ARC and MRC Under Imperfect Market Competition

As discussed earlier, monopoly, oligopoly and monopolistic competitions are the important forms of imperfect market competition. While conceptually AR and MR are the same for these market structures, as the firms under these competitions are price makers rather than price takers, there are a few key differences. All these market structures face negatively-sloped demand curve (ARC), thereby, the price received by the firm is not fixed, but depends on the quantity of output sold in the market. When ARC slopes downward from left to right, MRC also slopes downward from left to right and lies below ARC. This is because, if ARC falls, MRC must fall and lies below ARC, (*i.e.*, MRC falls more rapidly than ARC), as MRC is related to one additional unit of output, but ARC is related to all the units of output.

(i) Monopoly Competition

Let us assume, a monopoly firm enjoys patent for marketing Bt cotton seed in the market and it offers various quantities of seed for sale at different prices and the same are shown through Table 18.3. From the quantity-price data, the TR, AR and MR are computed. TR is computed in the third column as the product between quantity of seed transacted and price received. AR is computed in the fourth column by dividing the TR with quantity of seed transacted. The obvious point here is that, AR decreases with the increase in quantity of seed transacted. Moreover, AR is also equal to the price of seed for each quantity sold. For example, when the firm sells one unit of seed, it receives Rs. 1000 as AR and it is equal to the price of the seed *i.e.*, Rs. 1000. Similarly, if the firm sells 10 units of seed, it receives Rs. 550/ unit as AR and it is equal to the price of the seed *i.e.*, Rs. 550/unit. So, AR is price or price is AR and these two are one and the same.

MR is calculated by dividing the change in TR by the change in quantity of seed sold. For example, when the firm sells one unit of Bt cotton seed, it receives Rs. 1000 MR and if it sells 10th unit of seed, it receives Rs. 100 MR. So, the MR decreases with the increase in quantity of seed sold by the firm. Further, the MR is less than the price (AR) of seed for each quantity sold. This is because, any decline in an average, which is most certainly happening to AR, is associated with a marginal that is less than the average. That means, when AR is decreasing, MR must be less than AR. For exampie, the price at which three units of seed is sold is Rs. 900/unit, but the MR of the firm is only Rs. 800/unit. Similarly, the price at which ten units of seed is sold is Rs. 550/unit, but the MR of the firm is only Rs. 100/unit. And because, price is AR, MR is also less than AR. As shown through the Table 18.3:

- ☆ If the firm sets the price at Rs. 850/unit, then buyers (farmers) are willing to purchase four units of seed.
- ☆ If the firm wants to increase the quantity of sale from four units to five units, then it must lower the price from Rs. 850 to Rs. 800 per unit. But, the important aspect here is, the firm must lower the price for all the five units now and not just the extra unit *i.e.*, 5th unit of seed.
- ☆ By decreasing the price to Rs. 800, now two forces influences the TR *viz.*, the revenue gained by selling extra unit of seed (*i.e.*, 5th unit only) and the revenue lost by lowering the price from earlier four units of seed. We have to work out these these two forces separately, as MR is the net result of both.
    - ❖ First, by lowering the price, the firm increases the quantity of seed sold from four to five units. This extra 5th unit generates an extra of Rs. 800 of revenue, *i.e.*, the price of the fifth unit of seed. This Rs. 800 of extra revenue the firm did not have at the higher price (*i.e.*, when the price/unit is Rs.850).

Table 18.3: TR, AR and MR of a monopoly firm selling *Bt* cotton seed

Quantity (Units)	*Price (Rs/unit)*	*TR (Rs)*	*AR (Rs/unit)*	*MR (Rs/unit)*
0	1050	0	–	–
1	1000	1000	1000	1000
2	950	1900	950	900
3	900	2700	900	800
4	850	3400	850	700
5	800	4000	800	600
6	750	4500	750	500
7	700	4900	700	400
8	650	5200	650	300
9	600	5400	600	200
10	550	5500	550	100
11	500	5500	500	0
12	450	5400	450	-100

- ❖ Second, by lowering the price, the firm collects less revenue from the earlier four units of seed. For the earlier four units of seed, the firm would have collected Rs. 850 per unit of seed for a total of Rs. 3400, before the fall in price. But, with the lower price of Rs. 800, now the firm collects only Rs. 3200 for the earlier four units of seed *i.e.*, a reduction of TR by Rs. 200 or Rs. 50 per unit of seed.

☆ That means, Rs. 800 gained by selling the extra fifth unit of seed is partially offset by the Rs. 200 lost from lowering the price for earlier four units of seed. On net, TR increases by only Rs.800-Rs.200 = Rs.600. So, the total loss of revenue on the earlier four units of seed is the key reason that, MR is less than price.

The Table 18.3 further infers that, the rate of fall in MR is more compared to price or AR, and this gives more steepness to the slope of MRC compared to ARC (Figures 18.4 and 18.5). So, both AR and MR decreases with increase in quantity of seed sold by the firm, thereby, both ARC (demand curve) and MRC are negatively sloped. As the slope of MRC is steeper compared to ARC, it implies, ARC is more elastic compared to MRC. That means, if the price of seed falls in the market, MR falls more rapidly than AR, but the TR of the firm increases continuously and reaches maximum, when MR is equal to zero. Thus, ARC and MRC for the firm enjoying monopoly status look different when compared to perfect competition. For a monopoly firm, both ARC and MRC are negatively-sloped and the MRC lies below the ARC, which means that, MR is less than AR (and price) for any given quantity and larger quantities of output are sold only at lower prices, whereas under perfect market competition, MRC coincides with ARC and price and they are perfectly elastic.

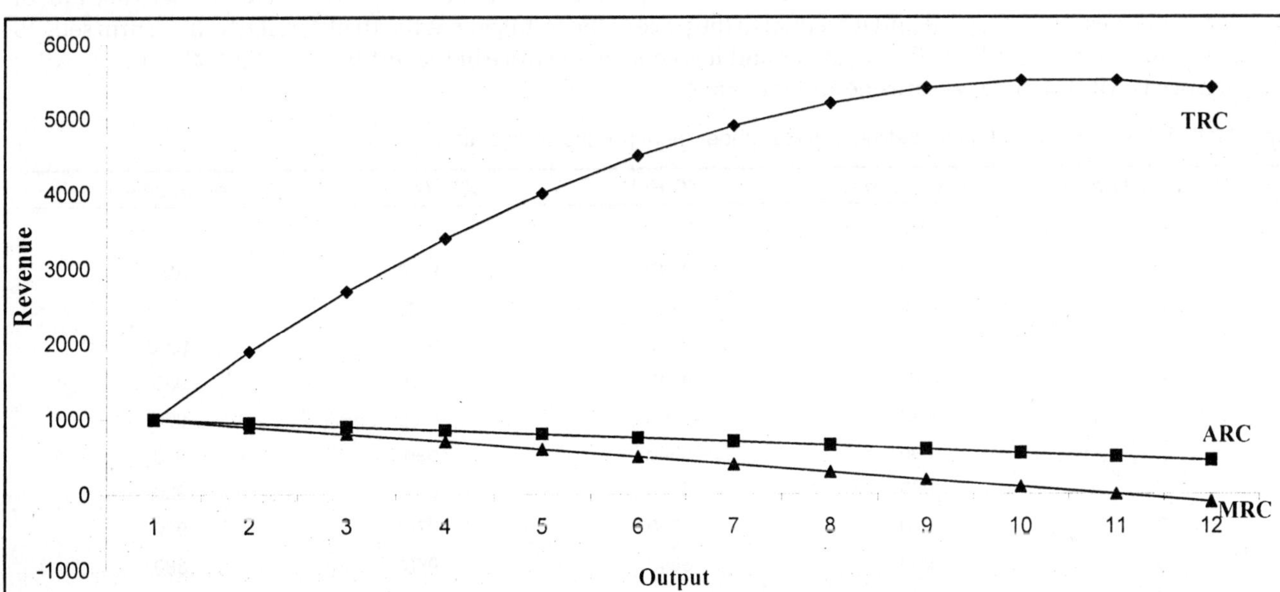

Figure 18.4: TR, AR and MR curves of a monopoly firm selling *Bt* cotton seed .

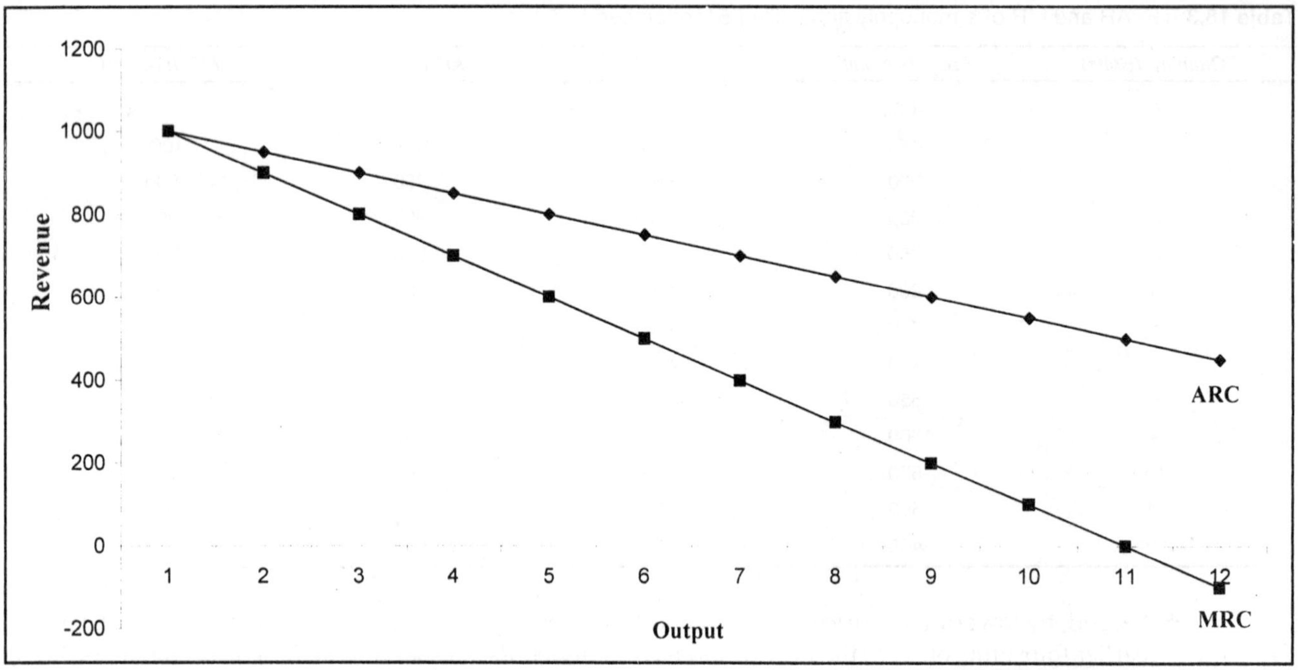

Figure 18.5: ARC and MRC of a monopoly firm selling *Bt* cotton seed.

(ii) Oligopoly Competition

Even in oligopoly competition, the ARC and MRC slopes downward from left to right and MRC lies below ARC. But, the interesting aspect is that, the ARC or demand curve will bear a kink such that, above kink 'K', the ARC is more elastic for price rise and below kink, the ARC is less elastic for price fall (Figure 18.6). Due to differences in elasticities along the ARC, the MRC will show discontinuity below the kink. These concepts are discussed in-detail in the Chapter 22.

(iii) Monopolistic Competition

Let us assume, a retail firm operating under monopolistic competition selling fertilizer in the market and it offers various quantities of fertilizer for sale at different prices and the same are shown through Table 18.4. From the quantity-price data, the TR, AR and MR are computed. TR is computed in the third column as the product between quantity of fertilizer transacted and price received. AR is computed in the fourth column by dividing the TR with quantity of fertilizer transacted. The obvious point here is that, AR decreases with the increase in quantity of fertilizer transacted. Moreover, AR is also equal to the price of fertilizer for each quantity sold. For example, when the firm sells one unit of fertilizer, it receives Rs. 1040 as AR and it is equal to the price of the fertilizer *i.e.*, Rs. 1040. Similarly, if the firm sells 10 units of fertilizer, it receives Rs. 950/unit as AR and it is equal to the price of the fertilizer *i.e.*, Rs. 950/unit. So, AR is price or price is AR and these two are one and the same.

Table 18.4: TR, AR and MR of a firm selling fertilizer under monopolistic competition.

Quantity (Units)	*Price (Rs/unit)*	*TR(Rs)*	*AR (Rs/unit)*	*MR (Rs/unit)*
0	1050	0	–	–
1	1040	1040	1040	1040
2	1030	2060	1030	1020
3	1020	3060	1020	1000
4	1010	4040	1010	980
5	1000	5000	1000	960
6	990	5940	990	940
7	980	6860	980	920
8	970	7760	970	900
9	960	8640	960	880
10	950	9500	950	860

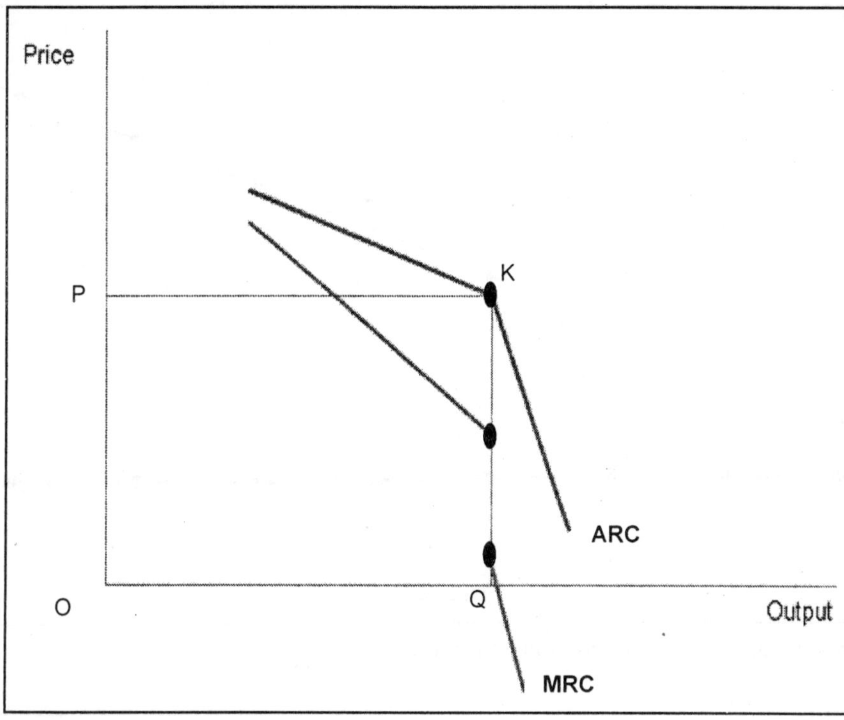

Figure 18.6: ARC and MRC of an oligopoly firm.

MR is calculated by dividing the change in TR by the change in quantity of fertilizer sold. For example, when the firm sells one unit of fertilizer, it receives Rs. 1040 MR and if it sells 10th unit of fertilizer, it receives Rs. 860 MR. So, the MR decreases with the increase in quantity of fertilizer sold by the firm. Further, the MR is less than the price (AR) of fertilizer for each quantity sold. This is because, any decline in an average, which is most certainly happening to AR, is associated with a marginal that is less than the average. That means, when AR is decreasing, MR must be less than AR. For example, the price at which three units of fertilizer is sold is Rs. 1020/unit, but the MR of the firm is only Rs. 1000/unit. Similarly, the price at which ten units of fertilizer is sold is Rs. 950/unit, but the MR of the firm is only Rs. 860/unit. And because, price is AR, MR is also less than AR. As shown through the Table 18.4,

- ✰ If the firm sets the price at Rs. 1010/unit, then buyers (farmers) are willing to purchase four units of fertilizer.
- ✰ If the firm wants to increase the quantity of sale from four units to five units, then it must lower the price from Rs. 1010 to Rs. 1000 per unit. But, the important aspect here is, the firm must lower the price for all the five units now and not just the extra unit *i.e.*, 5th unit of fertilizer.
- ✰ By decreasing the price to Rs. 1000, now two forces influences the TR *viz.*, the revenue gained by selling extra unit of fertilizer (*i.e.*, 5th unit only) and the revenue lost by lowering the price from earlier four units of fertilizer. We have to work out these two forces separately, as MR is the net result of both.
  - ❖ First, by lowering the price, the firm increases the quantity of fertilizer sold from four to five units. This extra 5th unit generates an extra of Rs. 1000 of revenue, *i.e.*, the price of the fifth unit of fertilizer. This Rs. 1000 of extra revenue the firm did not have at the higher price (*i.e.*, when the price/unit is Rs.1010).
  - ❖ Second, by lowering the price, the firm collects less revenue from the earlier four units of fertilizer. For the earlier four units of fertilizer, the firm would have collected Rs. 1010 per unit of fertilizer for a total of Rs. 4040, before the fall in price. But, with the lower price of Rs. 1000, now the firm collects only Rs. 4000 for the earlier four units of fertilizer *i.e.*, a reduction of TR by Rs. 40 or Rs. 10 per unit of fertilizer.
- ✰ That means, Rs. 1000 gained by selling the extra fifth unit of fertilizer is partially offset by the Rs.40 lost from lowering the price on earlier four units of fertilizer. On net, TR increases by only Rs.1000-Rs.40 = Rs.960. So, the total loss of revenue on the earlier four units of fertilizer is the key reason that, MR is less than price.

The Table 18.4 further infers that, the rate of fall in MR is more compared to price or AR, and this gives more steepness to the slope of MRC compared to ARC (Figures 18.7 and 18.8). So, both AR and MR decreases with increase in quantity of fertilizer sold by the firm, thereby, both ARC (demand curve) and MRC are negatively sloped. As the slope of MRC is steeper compared to ARC, it implies, ARC is more elastic compared to MRC. That means, if the price of fertilizer falls in the market, MR falls more rapidly than AR, but the TR of the firm increases continuously and reaches

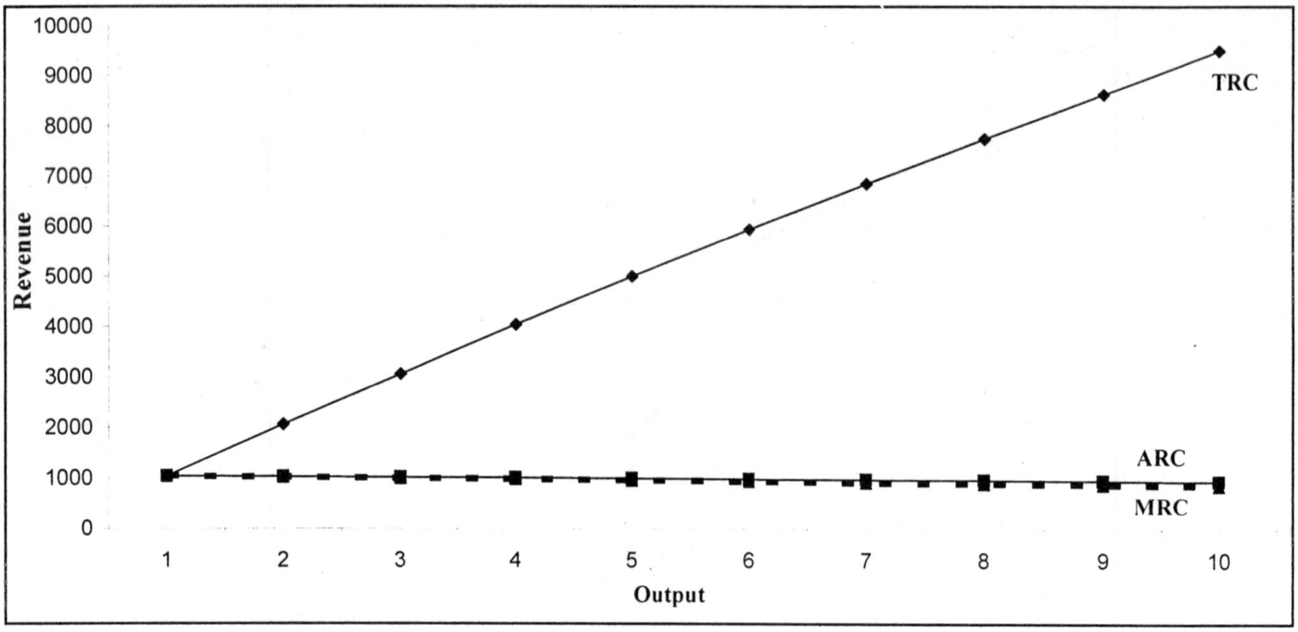

Figure 18.7: TR, AR and MR of a firm selling fertilizer under monopolistic competition.

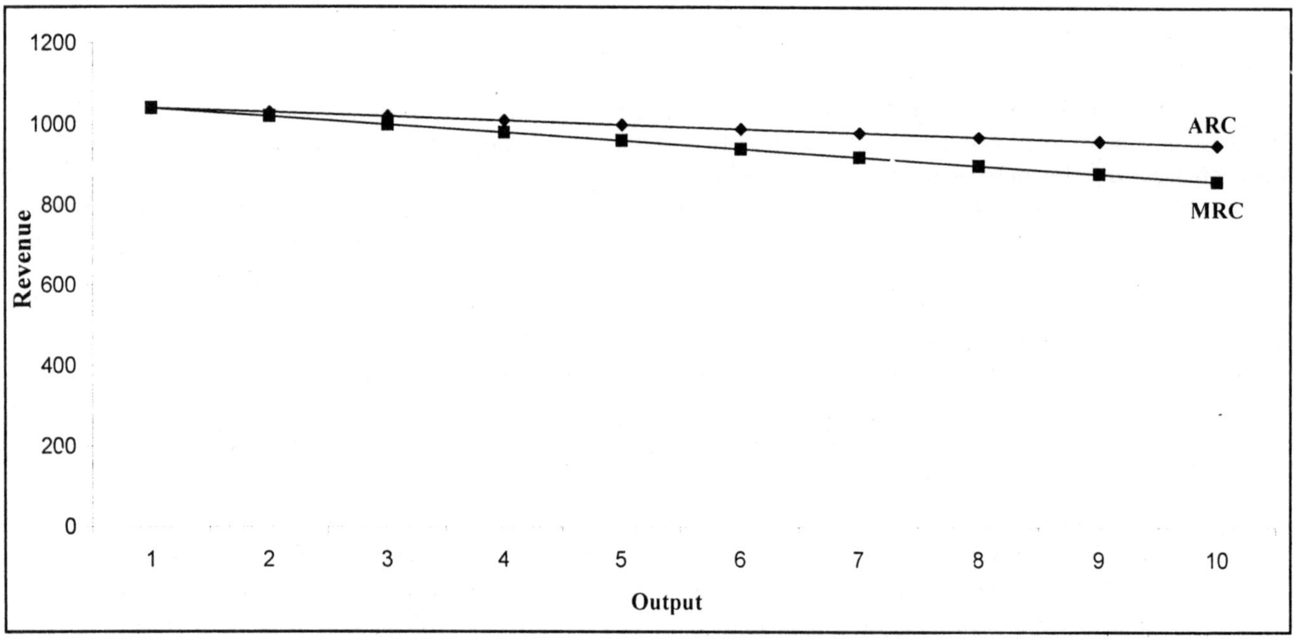

Figure 18.8: ARC and MRC of a firm selling fertilizer under monopolistic competition.

maximum, when MR is equal to zero. Thus, ARC and MRC for the firm under monopolistic competition look different when compared to perfect competition and monopoly. For a firm operating under monopolistic competition, both ARC and MRC are negatively-sloped and the MRC lies below the ARC, which means that, MR is less than AR (and price) for any given quantity and larger quantities of output are sold only at lower prices. The same is the case under monopoly, but the AR and MR curves are more steeper compared to monopolistic competition. As the firm itself represent the industry under monopoly, the steepness of the slopes fetch the firm higher price for the product, if the firm wishes to maximize the profits through increasing the price of the product. However, under perfect market competition, MRC coincides with ARC and price and they are perfectly elastic.

3. Derivation of ARC and MRC from TRC

From the TRC, we can easily derive the ARC and MRC and the same is discussed through the Figure 18.9. On X-axis, quantity of output and on the Y-axis, revenue of the firm are taken. TRC rises upward from left to right from the point of origin, but it continues to flatter as it rises *i.e.*, with the increase in the quantity of output sold. That means, TR increases at diminishing rate with increase in quantity of output sold in the market.

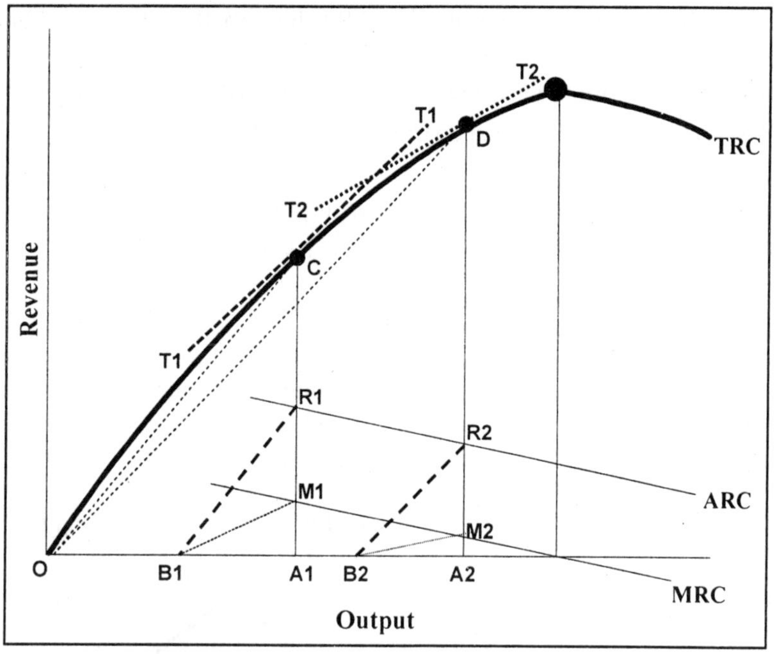

Figure 18.9: Derivation of ARC and MRC from TRC.

Derivation of AR

To derive ARC from TRC, consider a point C on the TRC. To this point draw a straight line (dotted line) OC from the origin and the slope of the straight line represent the AR. This is because, with reference to point C on TRC, TR is A_1C and total output is OA_1. So, A_1C/OA_1 gives the AR with reference to point C. So, the slope of the straight line OC represent the AR. Now, to derive the ARC, consider one unit output level from point A_1 on X-axis and it is B_1. From this point B_1 draw a line parallel to OC i.e, B_1R_1 and it touches the A_1C line at R_1. Since, the point B_1 is at one unit of output distance from point A_1, the slope of this line B_1R_1 represents AR per unit of output (as B_1R_1 is parallel to OC). So, with reference B_1R_1 line, the AR is equal to A_1R_1/B_1A_1. Since, B_1A_1 represents one unit of output, the AR is equal to A_1R_1.

Derivation of MR

We know, the slope of TRC indicates MR of the firm. So, a tangent T_1T_1 is drawn with reference to point C on the TRC. Again, considering one unit output length from A_1 *i.e.*, B_1A_1, draw a line B_1M_1 parallel to T_1T_1 tangent and the line touches the A_1C line at point M_1. Since, the two lines *i.e.*, T_1T_1 and B_1M_1 are parallel to each other, the slope of the line B_1M_1 also indicates the MR for an output B_1A_1. So, slope of line B_1M_1 *i.e.*, MR is given by A_1M_1/B_1A_1. Since, B_1A_1 represents one unit of output, the MR is equal to A_1M_1.

The same procedure for computation of AR and MR is repeated again with reference to point D on the TRC and accordingly we derive AR as A_2R_2 and MR as A_2M_2. So, when the points R_1R_2 are connected by a line, it represents ARC of the firm and the line connecting the points M_1M_2 represents MRC of the firm. Also notice that, when MR is equal to zero, TR is maximum.

4. Relation Between TR, AR and MR Curves

The following are the interesting relationships we can have between TR, AR and MR curves:

- ✰ When MRC is positive TRC increases, when MRC is zero TRC is maximum and when MRC is negative TRC falls (Figure 18.9).
- ✰ In case of perfect market competition, ARC is a straight line parallel to X-axis and thereby, MRC is also a straight line and coincides with ARC. That means, when ARC is constant, MRC must be constant (Figure 18.3).
- ✰ When ARC is a straight line and slope downwards from left to right as under imperfect competition, the MRC also is a straight line slopes downwards from left to right and lies below the ARC (Panel A of the Figure 18.10). The perpendicular drawn at any point from the ARC to the Y-axis will cut the MRC exactly at the mid point of the perpendicular line. From point A on the ARC, a perpendicular line AY is drawn to Y-axis. This perpendicular line AY will cross the MRC at point E, thereby, MRC will lie at mid-point on the perpendicular line.

From Panel A of the Figure 18.10, we can also prove that, MRC lies exactly at half way between ARC and Y-axis. From Panel A,

- ☆ The TR of the firm can be computed in two ways *viz.*, from the AR and from the MR.
  - ❖ ***From the AR:*** With reference to point A on the ARC, the TR can be computed as, TR=AR*output sold.
    So, TR = QA*OQ=OYAQ *Equation 18.1*
  - ❖ ***From the MR:*** The summation of MRs at the given output gives the TR. So, the area below the MRC with reference to OQ level of output gives TR. This is given by area, OREMQ *Equation 18.2*
- ☆ So, we can write, Area OYAQ = Area OREMQ *Equation 18.3*
- ☆ In the above two areas, the area OYEMQ is common. So, let us deduct this area OYEMQ from the above two areas *i.e.*, OYAQ and OREMQ. By doing so, we will have Area REY= Area MEA *Equation 18.4*

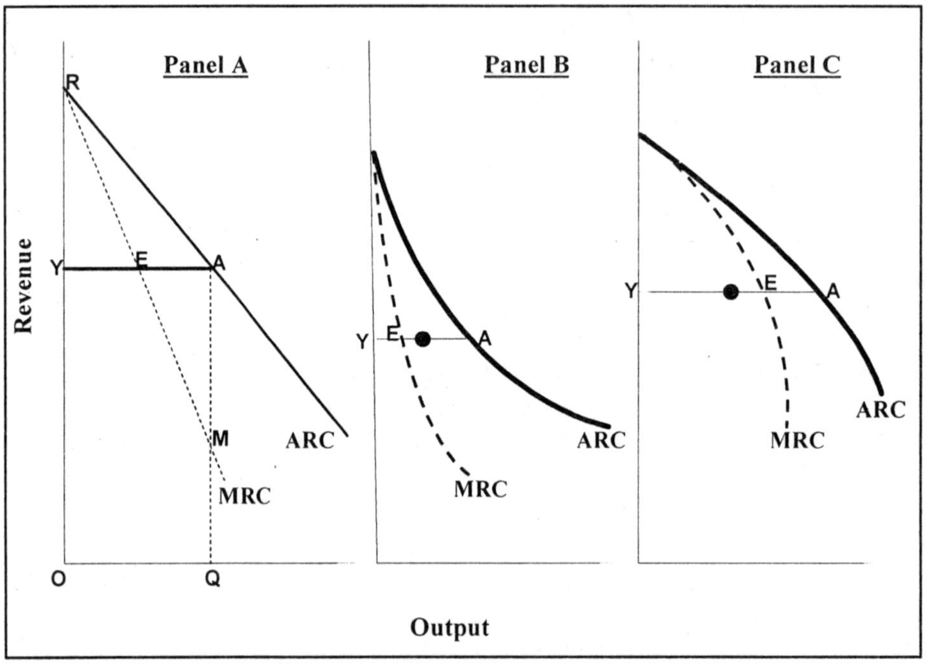

Figure 18.10: Geometric relationships between ARC and MRC under imperfect competition.

- ☆ Since, the two areas REY and MEA are equal and similar [because, angle REY = angle MEA (being vertically opposite angles)], we can conclude that, when ARC is a straight line and downward sloping, MRC is also a straight line and downward sloping and lies below ARC at exactly half way distance between ARC and Y-axis, and as shown in the Panel A, YE=EA.
- ☆ When ARC is a curve, the corresponding MRC also traces a curve and lies below ARC.
  - ❖ If ARC is convex (Panel B of Figure 18.10), MRC is also convex in shape. When ARC is convex to the origin, the perpendicular drawn from ARC to the Y-axis cuts the MRC at E *i.e.*, to the left of the mid-point of the perpendicular (AY) drawn.
  - ❖ If the ARC is concave (Panel C of Figure 18.10), MRC is also concave in shape. When ARC is concave to the origin, the perpendicular drawn from the ARC to the Y-axis cuts the MRC at E *i.e.*, to the right of the mid-point of the perpendicular (AY) drawn.
- ☆ When ARC is more elastic, MRC is also elastic (but, comparatively less elastic) as in case of monopolistic competition (Figure 18.8). But, when the ARC is less elastic, MRC is still less elastic, as in the case of monopoly firm (Figure 18.4). This infers that, both MRC and ARC are more elastic in monopolistic competition compared to monopoly competition.
- ☆ When the slope of ARC shows different elasticities as in case of oligopoly market competition (Figure 18.6), the MRC shows discontinuity at the point, where the elasticity on the ARC curve changes.

The above discussion also infers an interesting aspect that, in the output–price analysis of the firm, we commonly use AR and MR curves and therefore TRC is superfluous for the purpose, as MR=MC is the optimality criterion and ARC indicates the price of the commodity.

5. Relationship between TR, AR, MR and E_P

This concept of useful relationships between TR, AR, MR and E_P was extensively studied by Joan Robinson in her work, *'The Economics of Imperfect Competition'*. In the Figure 18.11, DD' is the ARC or demand curve of the firm under imperfect market competition. So, the E_P at point 'A' on the DD' curve is given by (Point method),

$E_P = AD'/AD$ *Equation 18.5*

So, with this measure of E_P, we can formulate the relationship between TR, AR, MR and E_P as explained below.

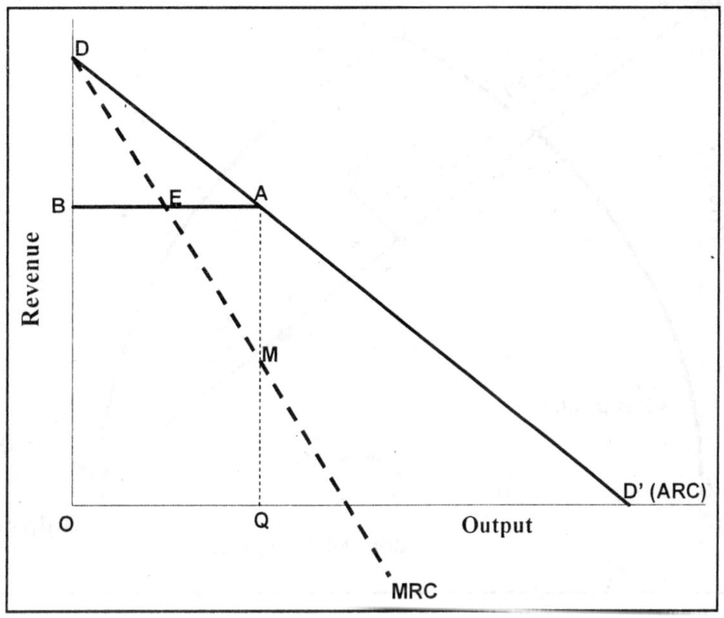

Figure 18.11: Relation between AR and MR curves of a firm under imperfect competition.

- When ARC or demand curve slopes downward, MRC also slopes downward from left to right and lies below the ARC.
- From the point 'A' considered on the ARC or demand curve, draw a perpendicular to Y-axis as AB.
- The triangles AQD' and ABD are similar, as angle AD'Q and angle DAB are corresponding angles and angle AQD' and angle DBA are right angles. We know, the corresponding sides of these two similar triangles will have the same ratio, thereby, AD'/AD =AQ/DB. But, as per the Equation 18.5, E_P =AD'/AD. Therefore,

  E_P =AD'/AD= AQ/DB *Equation 18.6*
- In the Figure 18.11, the triangle DBE is equal/similar to triangle EAM in all aspects as per the Equation 18.4. So, DB=AM. Therefore, substitute DB as AM in Equation 18.6 above, we get,

  E_P= AD'/AD = AQ/DB = AQ/AM *Equation 18.7.*
- In the Figure 18.11, AM =AQ-MQ. So, the above Equation 18.7 can be written as,

  E_P=AQ/(AQ-MQ) *Equation 18.8*
- E_P=Average Revenue/(Average Revenue-Marginal Revenue)

  = AR/(AR-MR) *Equation 18.9*
- The above Equation 18.9, can be rewritten to compute AR as,

  $E_P(AR - MR) = AR$

  $(E_P AR) - (E_P MR) = AR$

  $(E_P AR) - AR = E_P MR$

  $AR(E_P - 1) = E_P MR$

  Therefore, $AR = E_P MR/(E_P - 1) = MR[E_P/(E_P-1)]$ *Equation 18.10*
- From Equation 18.10, we can write, $MR = AR(E_P-1)/E_P = P(E_P-1)/E_P$ *Equation 18.11*
- The above Equations 18.10 and 18.11 infers that, given E_P, we can estimate the MR and AR relationship.

Case 1

Assume E_P of the firm's ARC is equal to one, then MR is given by, as per the Equation 18.11, $MR = AR(E_P-1)/E_P = AR(1-1)/1 = 0$.

The above case is represented through Figure 18.12 with reference to point A on the ARC. So, when E_p of the firm's ARC is equal to one, MR is zero, *i.e.*, MRC touches the X-axis.

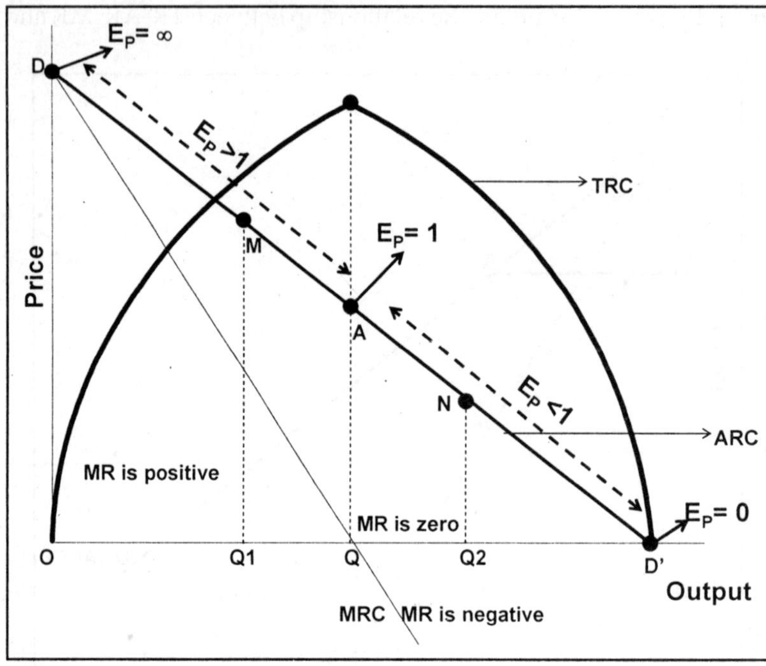

Figure 18.12: Relation between TR, AR, MR and E_P.

Case 2

Assume E_P of the firm's ARC is equal to 2, then MR is given by, as per the Equation 18.11, $MR = AR(E_P-1)/E_P = AR(2-1)/2 = 0.5AR$.

The above case is represented through Figure 18.12, with reference to point M on the ARC. So, when E_P of the firm's ARC is equal to two, it implies, MR will be half of the AR. Further, it also infers that, when E_P of the firm's ARC is more than one, MR is positive.

Case 3

Assume E_P of the firm's ARC is equal to 0.5, then MR is given by, as per the Equation 18.11, $MR = AR(E_P-1)/E_P = AR\ (0.5-1/0.5) = -1AR$.

The above case is represented through Figure 18.12, with reference to point N on the ARC. So, when E_P of the firm's ARC is less than one, MR is negative.

To sum up, as long as the E_P on the ARC is more than one, the MR of the firm will be positive at any output level. If the E_P on the ARC is equal to one/unitary, then MR of the firm will be zero. If the E_P on the ARC is less than one, the MR of the firm will be negative at any output level. So, the above methodology of estimating the relationship between MR and AR at different levels of output based on E_P of the firm's ARC or demand curve guides the firm in making profits under different types of market situations.

So far, we derived the relation between AR, MR and E_p of the firm's output. Now, we are in a position to formulate the relationship between TR, AR, MR and E_P and the same is explained through Figure 18.12. The Figure 18.12 shows the usual total-average-marginal relation as discussed under the concept of LDR. The inferences derived from the Figure 18.12 are summarized through Table 18.5.

Table 18.5: Relationship between TR, AR, MR and E_P

MR	*E_P of the firm's ARC*	*TR and AR*
MR is Positive	>1	TR rises upward, AR is positive
MR is zero	=1	TR is maximum, AR is positive
MR is less than zero	<1	TR slopes downward, AR is positive. AR will be zero only, when TR is zero.

6. Determination of Firm's Equilibrium

After analyzing the concepts of TR, AR and MR under different forms of market competition, we now study different approaches to determine the firm's equilibrium. In Economics, a firm refers to a technical and managerial unit of production. In other words, it is an enterprise engaged in the production of a commodity. Considering the market conditions, the firm takes up the production activity to produce the commodity at lowest cost possible. In the words of Prof. Watson, *'a firm is a unit engaged in production for a sale at a profit and with the objective of maximizing the profit'*. So, a firm is considered as an income earning unit and it tries to maximize the income in the business. Once the point of profit maximization is reached, the firm will not disturb the production activity *i.e.*, it will neither expand nor contract the output. Such a stage in the production programme is referred as 'Equilibrium point'. So, in determining the firm's equilibrium, it is assumed that, the firm tries to attain profit maximization and this forms the basis for analyzing the equilibrium output in the production programme. The profits to the entrepreneur of the business firm comprises of three important elements *viz.*, normal profits, super-normal profits and sub-normal profits. In general, profits refer to the excess of TR over TC in the production programme. Symbolically, P = R – C, where, P is the Profits, R is the TR and C is the TC incurred in the production programme. When R>C, then the firm is said to be earning profits and when R<C, then the firm is said to be incurring losses. But, this is the accounting sense of interpreting the concept of 'profit'. But, in economic sense, the total cost includes both explicit costs and implicit costs. So, Economists use three different notions of profits *viz.*, normal profits, super-normal profits and sub-normal profits.

(a) Normal Profits

It is the amount of profits the business firm earns that is sufficient for the firm to stay in the industry. So, normal profits are the minimum reasonable level of profits that the entrepreneur must earn in the long run, so that, he is induced to continue the production programme in the present form. It is, otherwise, defined as the minimum level of profits required to keep the factors of production in their current use in the long run.

Normal profit is always regarded as a part of TC. Say, for example, farm manager labour, farm family labour are the important factor services employed in the crop production programme. The price paid for these factor services is the normal profit, but they are to be included as implicit costs while computing the TC. So, TC includes explicit costs + implicit cost (normal profit). That means, in economic sense, TC include normal profit. So, in computing firm's equilibrium, when we say, TR=TC, it implies, in economic sense, there is no pure business profit, but there is normal profit, which is already included as implicit cost in the TC. That means, normal profits occur when TR = TC or when the firm experiences break even point.

It is important to note that, the entrepreneur desires to have a fixed amount of normal profits, irrespective of the level of output. This is because, normal profits includes a fixed implicit cost element. So, when the firm expands the output, like TFC, the implicit cost or normal profit will spread over a wide range of output. This inclusion of normal profit also has a bearing on the ATC curve of the firm. The same concept is explained through Figure 18.13.

ATC is the ATC curve excluding normal profit and ATC + NP is the ATC curve including normal profit/implicit cost of the firm. Consider two points A and B on ATC + NP curve and the movement from A to B indicates the gap

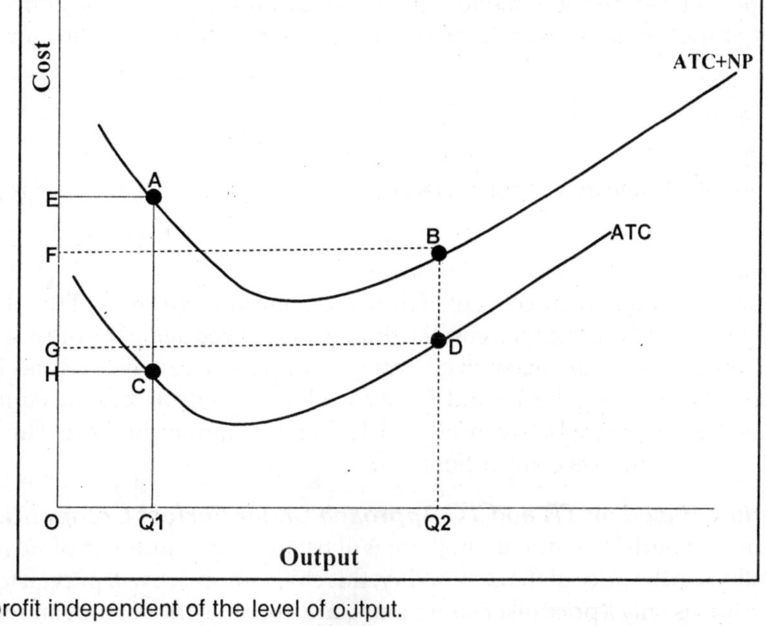

Figure 18.13: Normal profit independent of the level of output.

between ATC and ATC + NP curve tends to become narrow. This implies that, as the output increases, the normal profits/implicit costs are spread over larger output, thereby, the gap between the two curves (normal profits or implicit costs) gradually decreases. But, the total profit at all levels of output remains same. This is because,

- At OQ_1 level of output, the average normal profit is AC. So, the total profit is givén by total output * average normal profit = (OQ_1 = EA)*AC = EACH.
- At OQ_2 level of output, the average normal profit is BD. So, the total profit is given by, total output*average normal profit = (OQ_2 = FB)*BD=FBDG.
- So, at output OQ_1, the total profit is EACH and at output OQ_2, the total profit is FBDG and EACH=FBDG.

It is important that, in economic theory, when ever ATC curve is drawn, it implies ATC + normal profit/implicit cost. Hence, ATC curve represents ATC + normal profit. So, when the firm enjoys normal profits in the business (price=ATC), it implies, there are no pure business profits and the normal profits include only the implicit costs of the firm. Thus, when there are only normal profits, no new firms enter the industry and existing firms in the industry show no tendency to leave the industry.

(b) Super-normal Profits

A business is said to be earning super-normal profits, when the TR > TC. That means, the super-normal profits are over and above the normal profits. Since, normal profits are included in the TC, when TR exceeds the TC, the business firm is said to be earning excess profits over normal profits. So, super-normal profits are also called as 'Pure business profits' or 'Excess profits' or 'Abnormal profits' or 'Economic profits'. When a business firm is earning super-normal profits, it implies, the entrepreneur has successfully managed all the risks and uncertainties and the resources are exploited efficiently. Since, super normal profits are over and above the normal profits, and if the firm enjoys super-normal profits, new firms get attracted to these profits and enter the industry.

(c) Sub-Normal Profits

This is another important concept of profits of the business firm. As the name implies, the profits earned by the business firm are below normal profits. That means, the TR earned by the business firm fully covers the explicit costs plus only a part of the implicit costs, thereby, the profits are sub-normal profits. They are also called as 'Economic loss'.

6.1. Assumptions to Determine Firm's Equilibrium

The concept of the firm's equilibrium rests upon the following assumptions:

- The entrepreneur acts rationally and tries to maximise the profits in the production programme. If losses in the production programme are inevitable, he tries to minimise the losses.
- The firm produces only one commodity and its price is stable.
- The factor prices remains constant, thereby, the firm purchases the desired quantity of factors at a given price.

6.2. Methods to Determine the Firm's Equilibrium

As explained earlier, a firm is said to be in equilibrium, when it shows no inclination to change the output from the optimum level, thereby, it can maximize the profits or minimize the losses. This concept is applicable to all types of market competitions *i.e.*, perfect competition, monopoly, oligopoly and monopolistic competition. There are two complementary approaches to determine the conditions of profit maximization or equilibrium output of the firm and they include,

- TR and TC approach
- MR and MC approach

Both the approaches are applicable in cases of perfect competition, pure monopoly, oligopoly and monopolistic competition.

6.2.1. TR and TC Approach

This methodology is simple to explain the concept of firm's equilibrium. We know, TR is the total income the firm earns through selling the output and TC is the total cost the firm incurs in producing the output. Profit is the difference between TR and TC. For a firm, the profits are maximized, when the difference between TR and TC is maximum. So, we have the firm's equilibrium by comparing the TR and TC at each level of output and the equilibrium output can be determined, where the positive difference between TR and TC is maximum or highest. The same principle can be studied under perfect and imperfect market competitions.

6.2.1.1. Firm's Equilibrium Based on TR and TC Approach Under Perfect Competition

As explained earlier, under perfect competition, there will be an infinite number of buyers and sellers and no single seller or buyer can influence the price of the commodity, thereby, same price will prevail for the same commodity at the same time. Hence, the firm is only a price taker, as it cannot influence the price of the commodity. In the Table 18.6, at the prevailing price of paddy (Rs.1000/quintal), the firm has to sell any quantity of paddy. The TC of producing the paddy increases with increase in the quantity of output and TR from the commodity increases with increase in the quantity transacted at the prevailing market price. It is evident that, by producing eight quintals of output, the firm can maximize the profits (Rs. 3300), as the difference between TR and TC is maximum at this level of output. The firm also maximizes the profits (Rs. 3300) even at seven quintals of output, but this indeterminacy is due to assumption of discrete data. If continuous data are considered, there is no such indeterminacy and only specific output level will be obtained, which contributes to maximum profit. The same concept is represented graphically (Figure 18.14), where TRC and TC curve are shown as continuous upward sloping curves from left to right. The following are the interesting points from the Figure 18.14:

Table 18.6: TR, TC and Profit of a perfect competitive firm transacting paddy

Output (quintals)	*Price (Rs/quintal)*	*TR (Rs.)*	*TC (Rs.)*	*Profit (Rs.)*	*MC (Rs/quintal)*	*MR (Rs/quintal)*
0	1000	0	1000	-1000	0	—
1	1000	1000	1600	-600	600	1000
2	1000	2000	2000	0	400	1000
3	1000	3000	2100	900	100	1000
4	1000	4000	2200	1800	100	1000
5	1000	5000	2500	2500	300	1000
6	1000	6000	3000	3000	500	1000
7	1000	7000	3700	3300	700	1000
8	**1000**	**8000**	**4700**	**3300**	**1000**	**1000**
9	1000	9000	6100	2900	1400	1000
10	1000	10000	8100	1900	2000	1000

- ☆ The TRC curve rises from the origin indicating that, at zero output level, there is no revenue to the firm. The rise in TRC as a straight line indicates that, the TR of the firm increases with increase in quantity of paddy transacted in the same ratio, as price of paddy remains same.
- ☆ The TC curve of the firm also rises upward, but not as a straight line because, it rises rather slowly at the initial levels of output, as the production programme comprises of more of fixed factors and less of variable factors. So, when there is less amount of variable resource, they are properly managed by the farm manager leading to higher resource use efficiency. So due to internal economies of scale, increasing returns or decreasing costs is experienced, thereby, the curve rises slowly. But with the gradual increase in output, the proportion of variable factors in relation to fixed factors will be more leading to diseconomies of scale, which results in diminishing returns or increasing costs and hence the curve rises steeply at higher levels of output, thus giving an S shape to the TC curve. Further, TC curve originates above the origin, as TC includes TFC even at zero level of output.
- ☆ At 8 units of output (OQ), the gap between TR and TC is maximum (EF), thereby, the firm maximizes the profits (Rs. 3300).
- ☆ Below 8 units (<OQ) of output, two distinct phases are seen with reference to the behaviour of TR and TC curves.
    - ❖ Upto OQ_1 level of output, the TC curve is higher than TRC indicating that, TC incurred by the firm are more than TR earned by the firm, thereby, the firm incurs losses. This is shown by the profit curve (dotted line curve) and its portion upto OQ_1 level of output lies below the X-axis indicating that, the firm incurs losses.
    - ❖ At OQ_1 level of output, TC curve intersects TRC at point E_1, thereby, TR=TC and hence, it is the point of zero profits to the firm. It is otherwise called as 'Break even point' of the business.
- ☆ Beyond OQ_1 level of output, TRC is higher than TC curve upto OQ_2 level of output. Between OQ_1 and OQ_2 levels of output *i.e.*, at OQ, the gap between TRC and TC curve is maximum by EF, thereby, at this OQ output level, the firm earns maximum profits, QP. So, if the farmer continues the production programme beyond OQ_1 (*i.e.*, after break even output), he can maximize the profits at OQ level of output.
- ☆ If the farmer still continues the production programme beyond OQ level of output, again the point of break even will be reached at OQ_2 level of output, where TR=TC at point E_2, but production beyond OQ level of output will reduce the profits.

- ☆ At the break even points *i.e.*, at OQ_1 and OQ_2 output levels, TR covers all TC. So, there will be no economic profits and the farmer enjoys only normal profits. Between these two break even points, the farmer enjoys economic profits and it is maximum at the maximum gap between TR and TC curves *i.e.*, at OQ output level.
- ☆ So, it is evident that, to maximize the profits in the production programme the firm should produce and market the output such that, the gap between TR and TC is maximum. Geometrically, this is obtained when the slopes of both TRC and TC curve are equal. As shown in the Figure 18.14, the tangent drawn TT at point F on TC curve is parallel to TRC, indicating that, their slopes are equal, thereby, between E and F with reference to output OQ, the profits are maximum in the production programme. So, at output OQ, TR is QE

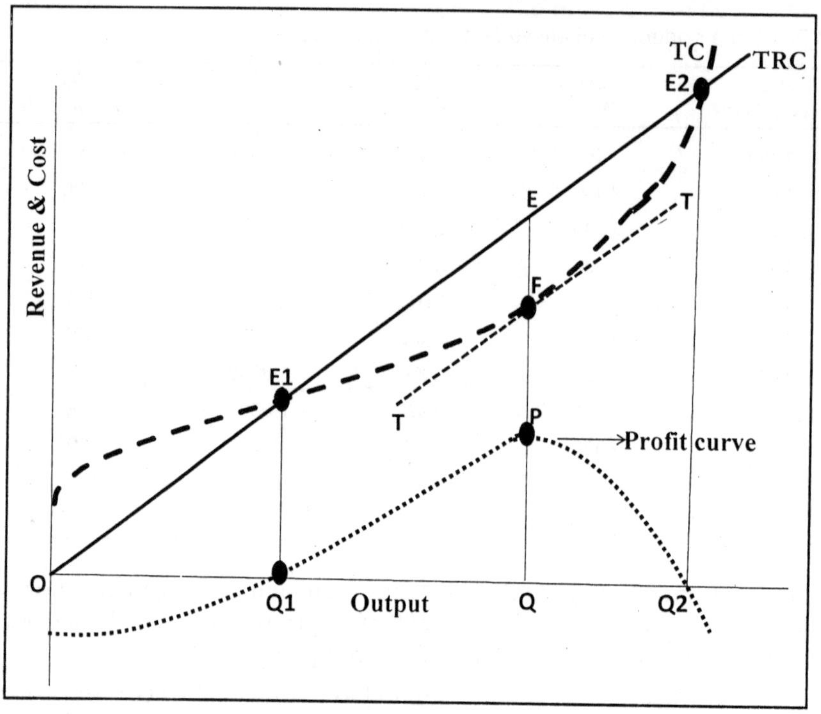

Figure 18.14: TR – TC approach to determine firm's equilibrium under perfect competition.

and TC is QF, thereby, the profits is QE-QF=EF=QP. So, at this output OQ, the total profit is maximum, as indicated by QP with reference to the profit curve.

- ☆ At first break even point *i.e.*, at OQ_1 output, the farmer should produce more. The farmer should maintain the output, where there is maximum vertical distance between TR and TC curves *i.e.*, OQ, as he enjoys maximum economic profits. At second break even point *i.e.*, at OQ_2, the farmer should stop producing and reduce the quantity of output till OQ.
- ☆ But, the demerit of this TR and TC approach is that, the exact point at which profit is maximized is difficult to pinpoint from inspection, which is why the MR=MC method is superior. In MR and MC approach, the reason, why a profit-max point can be found is due to the law of diminishing returns or increasing marginal costs and the same is discussed in the ensuing pages.

6.2.1.2. Firm's Equilibrium Based on TR and TC Approach Under Imperfect Competition

1. Case of Profit Maximization

Unlike perfect competition, in imperfect competition, the firm has control on the price of the commodity. So, the entrepreneur is not a price taker, but rather a price maker. In accordance with the law of demand, the firm can sell large quantities of output only at lower prices. So, under imperfect competition, the TRC also rises from left to right with increase in quantity of output sold, but the rate of rise of TRC falls with increase in quantity of output sold. Hence, TRC is not a straight line, as with the price fall only, the firm can transact large quantities of output in the market. As usual, the TC curve behaves as discussed under the perfect competition. The concept of profit maximization or firm's equilibrium as discussed under the perfect competition also holds good even under imperfect competition, but the only difference is that, the TRC is not a straight line. So, as shown in the Figure 18.15, the firm maximizes the profits (QP) at OQ level of output, where the gap between TR and TC curves is maximum. Geometrically, the tangents drawn at the two points F and E are exactly parallel to each other indicating that, the slopes of TR and TC curves are equal, thereby,

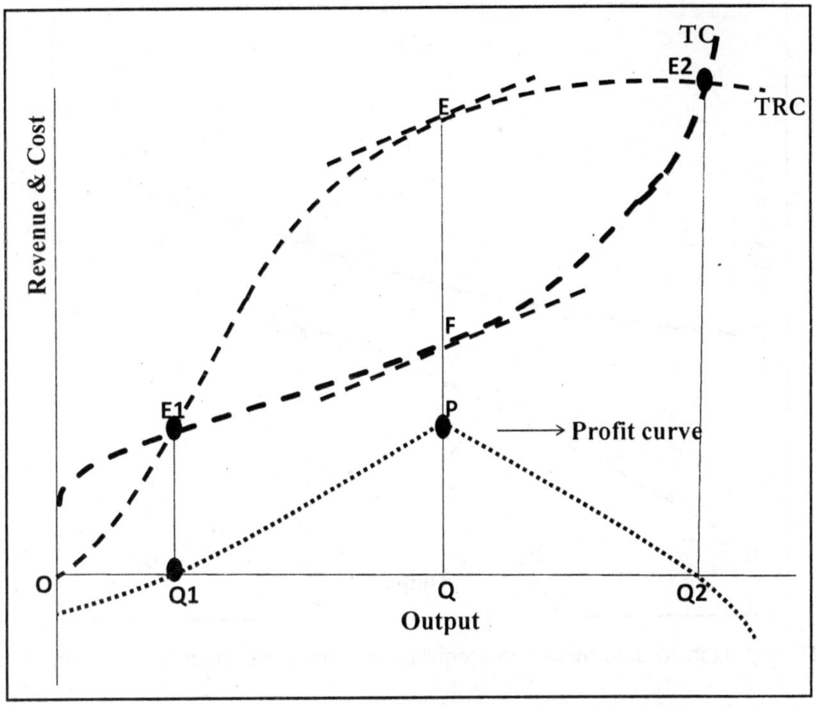

Figure 18.15: TR – TC approach to determine firm's equilibrium under imperfect competition – Case of maximization of profits.

the firm maximizes the profits QP, as indicated by the total profits curve. As shown in the Table 18.7, the firm maximizes the profits at six units of output, where the gap between TR and TC is maximum *i.e.*, Rs. 1500. The firm also maximizes the profits (Rs. 1500) even at five units of output, but this indeterminacy is due to assumption of discrete data. If continuous data are considered, there is no such indeterminacy and only specific output level will be obtained, which contributes to maximum profit.

Table 18.7: TR, TC and Profit of a firm selling the product under imperfect competition

Output (Units)	*Price (Rs/unit)*	*TR (Rs.)*	*TC (Rs.)*	*Profit (Rs.)*	*MC (Rs/unit)*	*MR (Rs/unit)*
0	1050	0	1000	-1000	0	–
1	1000	1000	1600	-600	600	1000
2	950	1900	2000	-100	400	900
3	900	2700	2100	600	100	800
4	850	3400	2200	1200	100	700
5	800	4000	2500	1500	300	600
6	**750**	**4500**	**3000**	**1500**	**500**	**500**
7	700	4900	3600	1300	600	400
8	650	5200	4300	900	700	300
9	600	5400	5100	300	800	200
10	550	5500	6000	–500	900	100
11	500	5500	7000	–1500	1000	0
12	450	5400	8100	–2700	1100	–100

2. *Case of Minimization of Losses*

Sometimes, the firm may even undergo losses either due to fall in prices of output in the market or due to rise in prices of factors. In such case, a rational producer will try to minimize the losses to the extent possible. So, the case of firm's equilibrium will be such that, the vertical gap between TR and TC curves should be minimum. In the Figure 18.16. as the firm is incurring losses TC curve is higher over TRC at each level of output. The gap between TC curve and TRC is minimum between C and R, thereby, producing at least OQ level of output, the firm can minimize the losses. Geometrically, the tangents drawn at the two points C and R are exactly parallel to each other indicating that, the

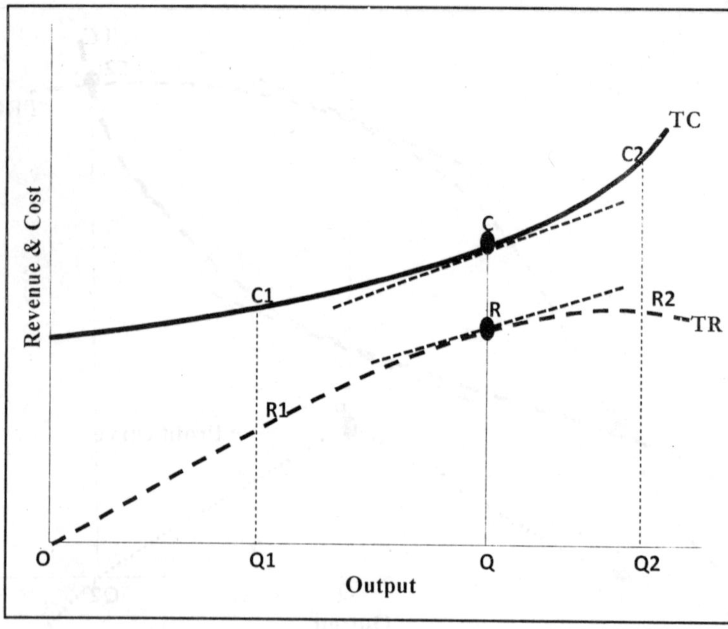

Figure 18.16: TR and TC approach to determine firm's equilibrium under imperfect competition – Case of minimization of losses.

slopes of TR and TC curves are equal, thereby, the firm minimizes the losses CR at OQ level of output. If the firm produces the output less than OQ or more than OQ, the gap between TC curve and TRC is widened, thereby, the losses in the firm will increase by C_1R_1 and C_2R_2 respectively.

Limitations of TR and TC Approach

Though the TR and TC approach is the simplest method to study the firm's equilibrium, it suffers from the following limitations:

- ☆ It is difficult to locate the maximum gap between TR and TC curves at a glance from the figure.
- ☆ Many tangents have to be drawn to locate the highest points of TR and TC curves.
- ☆ This approach will not directly indicate the cost per unit of output (MC) and revenue per unit of output (MR). This is a serious limitation because, if the firm enjoys economies of large scale, even the TC may be high, but cost per unit of output may be low due to higher levels of output. Similarly, even at higher levels of TR, revenue per unit of output may be low due to transacting large quantities of output. So, the maximum gap between TR and TC curves may not give the true picture of firm's equilibrium without considering MR and MC of the output.

So, considering the above limitations, the business firm generally employ MR and MC approach, as this method directly reveals the additional cost incurred in producing the additional unit of output and additional revenue the firm gets in selling additional unit of output, thereby, the firm can easily analyze optimum output and whether it is profitable for the firm to produce and sell additional unit of output.

6.2.2. MR and MC Approach

This concept was proposed by Mrs. Joan Robinson. This approach is more informative and superior in computing the firm's equilibrium when compared to the earlier approach. As discussed earlier, the slopes of TR and TC curves will give MR and MC respectively and these two concepts *viz.*, MR and MC will guide the firm to optimize the output or maximize the profits. MR indicates the additional revenue the firm gets in selling additional unit of output and MC indicates the additional cost incurred by the firm in producing the additional unit of output. The rules of thumb are:

- ☆ The firm attains equilibrium, when MR=MC.
- ☆ When MR>MC, the firm should continue the production programme to increase the output, so that MC rises and thereby, the equality between MR and MC is ensured.
- ☆ When MR<MC, the firm has to reduce the output, so that MC falls and thereby, the equality between MR and MC is ensured.

The above rules of firm's equilibrium implies that, the firm will not produce an additional unit of output, which involves more MC compared to MR earned from the same. In other words, the firm will produce an additional unit of output. only when the MR earned from that unit is more than MC incurred in producing the same. So, the firm continues to produce more units of output and attains optimum level of output at the point, where MR = MC. According to Stonier and Hague, *'a firm will be in equilibrium when it is earning maximum money profits. But, the money profits of a firm will always be maximized when its marginal revenue equals the marginal cost'*. In the graphical illustration of the MR and MC approach, the equality between MRC and MC curve will not be the only sufficient condition for the firm's equilibrium, but the firm must also satisfy the second order condition *i.e.*, the MC curve should cut the MR curve from below, while MC curve is rising upward. Thus, the two important conditions for the firm's equilibrium include,

- ☆ MR= MC
- ☆ MC curve should cut the MR curve from below, while rising upward.

This approach is applicable to study the firm's equilibrium both under perfect and imperfect market competitions.

6.2.2.1. Firm's Equilibrium Based on MR and MC Approach Under Perfect Competition

As mentioned earlier, the MRC of the firm under perfect competition lies parallel to X-axis and coincides with ARC. As usual, the MC curve will first fall, reaches minimum and rises upward with gradual increase in the output. In the Figure 18.17, the MC curve intersects the MRC at two points *viz.*, E_1 and E_2. Considering the above two essential conditions of optimality, the firm's equilibrium will be at OQ_2 level of output, where the MR=MC at E_2. Even though, MR=MC at E_1, it is not the point of the firm's equilibrium because,

- ☆ At E_1, the MC curve cuts the MRC from above and not from below, thereby, the second-order condition is not fulfilled.
- ☆ If the firm expands the output beyond OQ_1, the MC is lower than MRC thereby, there is still scope for the firm to expand the output till MR=MC is ensured and this is at E_2, where OQ_2 is the optimum output produced by the firm.
- ☆ If the firm still expands the output beyond OQ_2, now the MC curve rises over MRC, thereby, the firm incurs losses.

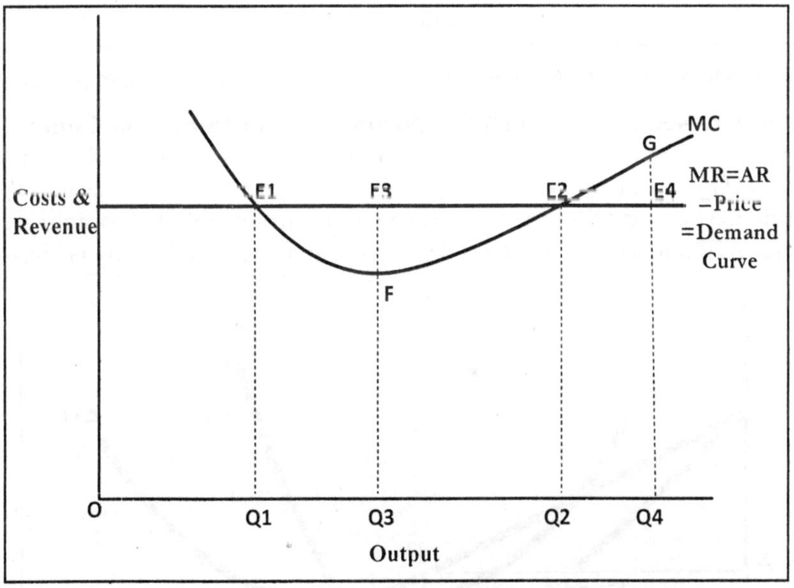

Figure 18.17: MR and MC approach to determine firm's equilibrium under perfect competition.

So, we infer that, when the two essential conditions *viz.*, MR = MC and MC curve should cut the MRC from below while rising are fulfilled, the firm attains equilibrium and it is at point E_2 with reference to OQ_2 level of output. Diagrammatically, the area under the MRC indicates the TR of the output and the area underlying the MC curve measures the TC incurred in producing the output. With reference to OQ_1 and OQ_2 levels of output, the area between MR and MC curves *i.e.*, $E_1E_3FE_2$ refers the profits gained by the firm due to increase in output from OQ_1 to OQ_2 and the profits are maximum at OQ_2 output. Say, if the firm increases the output from OQ_1 to OQ_3, the firm earns the profits E_1E_3F area, as MR>MC. Since, MR>MC, the firm can still increase the output from OQ_3 to OQ_2, where MR=MC, thereby, the profits are increased from E_1E_3F to $E_1E_3FE_2$ and are maximum at OQ_2 level of output. But, if the firm still increases the output beyond OQ_2, *i.e.*, upto OQ_4, now the firm experience losses to the extent of E_2GE_4, thereby, the firm has to reduce the output back to E_2 to ensure firm's equilibrium.

Even in imperfect competition, the second necessary condition *i.e.,* MC curve should cut the MRC from below, is considered essential to ensure firm's equilibrium and its importance is shown graphically through Panel A and Panel B of the Figure 18.18. In the Panel A, MC intersects MR at point 'E_1', but this is not the condition for firm' equilibrium because, here MC intersects MR from above rather than from below and MC is below MR beyond point E_1. But in Panel B, the necessary condition of firm' equilibrium is established at point 'E_2', as MC intersects MR from below and beyond E_2, MC curve is above MRC. So, OQ_2 is the optimum output produced by the firm to realize maximum profits in the business.

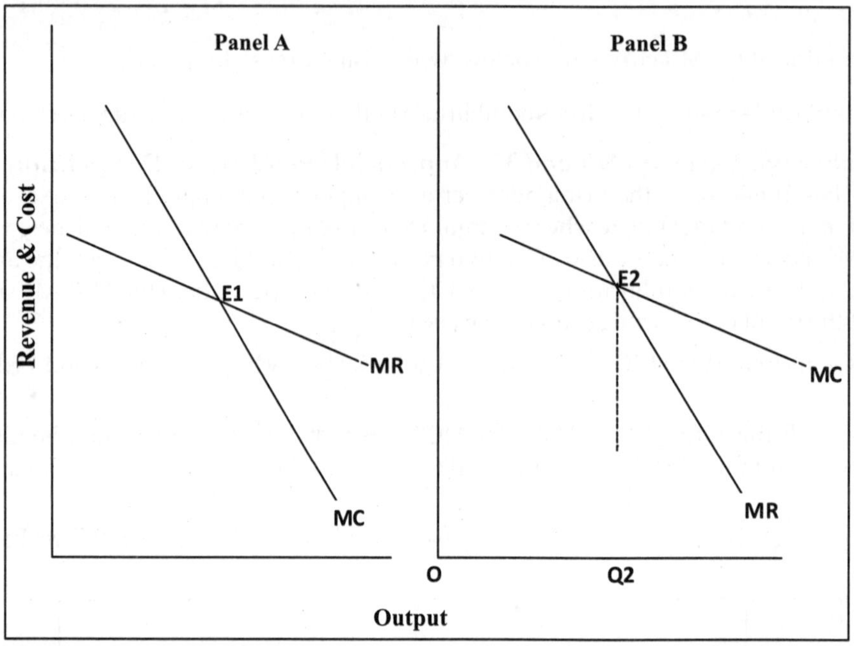

Figure 18.18: Second-order condition of MR and MC approach to determine firm's equilibrium (Imperfect competition).

6.2.2.2. Firm's Equilibrium Based on MR and MC Approach Under Imperfect Competition

As discussed earlier, under monopoly, oligopoly and monopolistic competitions, ARC and MRC are not equal, as the firm sells more quantities of the output only at lower prices because, the firm is a price maker and not simply price taker as in perfect competition. ARC and MRC are downward falling curves from left to right and MRC lies below the ARC (Figure 18.19). As usual, ATC and MC are 'U' shaped curves. The point of optimality is MR=MC and in the Figure

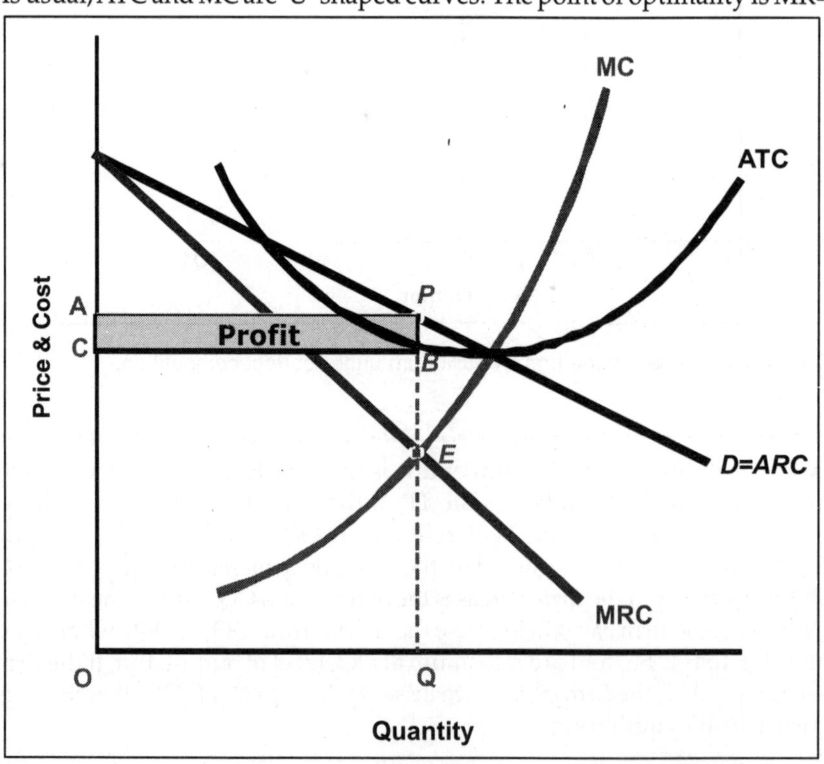

Figure 18.19: MR and MC approach to determine firm's equilibrium under imperfect competition.

18.19. it is at 'E', with reference to OQ level of output and OP level of price. So, TR to the firm will be total output * average price= OQ*QP=OQPA. The TC of producing OQ output is = total output*average cost= OQ*QB=OQCB. So, total profits of the firm is TR-TC, *i.e.*, OQPA- OQCB=PABC.

In the Figure 18.19, since the ARC lies above the ATC curve, the firm is enjoying profits (super-normal profits) *i.e.*, price is above the ATC. But, when the price < ATC or ARC< ATC curve, the firm experience losses in the business. The conditions for shut down of the business either temporarily or permanently will be in accordance with the relationship between price, ATC and AVC and they are already discussed through Table 15.1 of Chapter 15. The graphical representations for these cases are studied in-detail under each type of market structure in the ensuing pages.

So, the explanation offered under MR and MC approach highlights the following aspects:

- ☆ It helps to compute TR of the firm in selling the optimum output.
- ☆ It helps to compute TC of the firm in producing the optimum output.
- ☆ It helps to compute profits of the firm in producing the optimum output.

So, the conclusions we derived from the MR and MC approach leads to the conclusions of earlier TR and TC approach, thereby, MR and MC approach of determining firm's equilibrium is a more superior approach.

6.2.2.3. MR=MC Analysis in Markets with Linear Demand

We know, the equation for a linear demand curve is given by,

$$P = a - bQ$$

where, P is price of the commodity (paddy), Q is output, 'a' is the vertical intercept of the demand curve and 'b' is the slope of the demand curve. As studied earlier, the firm's equilibrium point is MR=MC. Since, we assumed linear demand curve (ARC), MRC is also linear and has the same vertical intercept as the demand curve, but twice the slope of the demand curve. This is given by,

$$MR = a - 2bQ$$

As per the Panel A of the Figure 18.10, we already proved that, when demand curve (ARC) is a straight line and downward sloping, MRC is also a straight line and downward sloping and lies below ARC at exactly half way distance between demand curve and Y- axis. The following is another approach to prove the above statement:

We know, TR is the product between price and output and it is given by,

$$TR = PQ = aQ - bQ^2$$

Assume, output increases from Q to Q + ΔQ *i.e.*, by an amount ΔQ. So, the new TR is given by,

$$TR = a(Q + \Delta Q) - b(Q + \Delta Q)^2$$

The change in TR due to change in output is given by,

$$\Delta TR = a(Q + \Delta Q) - b(Q + \Delta Q)^2 - (aQ - bQ)^2$$

$$= a\Delta Q - 2bQ\Delta Q - b(\Delta Q)^2$$

We know that, MR is the change in TR due to per unit change in output. So,

$$MR = \frac{\Delta TR}{\Delta Q} = \frac{[a\Delta Q - 2bQ\Delta Q - b(\Delta Q)^2]}{\Delta Q} = a - 2bQ - b\Delta Q$$

Since, output quantities can vary continuously, we will be interested in evaluating MR for very small changes in output. Assuming a very small change in output, then ΔQ tends to be zero. So, MR is given by,

$$MR = a - 2bQ$$

To find the profit-maximising output level, we should know MC of producing the output. Assume, MC is constant in producing additional unit of output *i.e.*, 'c' then, the following is true:

$$a - 2bQ = c$$

$$2bQ = a - c$$

$$Q = \frac{a - c}{2b}$$

To estimate the firm's best price for profit maximization, substitute the above output equation in the demand curve P = a – bQ, gives a price

$$P = a - b\left(\frac{a-c}{2b}\right)$$

Example 1

Suppose, a firm faces a linear demand curve of P = 2500 – 100Q, where 'Q' represents quintals of output and 'P' represents price/quintal. The firm faces a constant MC of Rs. 300. Estimate the firm's profit-maximising output and price.

Answer

We know, when demand curve (ARC) is linear (P = 2500 – 100Q), MRC will have the same vertical intercept as the demand curve, but with twice the slope.

So, MR is given by, MR = 2500 – 200Q.

The profit maximizing level of output is given by, MR = MC. So, given MC as Rs. 300, we can write,

2500 – 200Q = 300

So, firm's profit maximizing level of output, Q = 11quintals.

Inserting above Q in the demand equation, P = 2500 – 100Q, we can find firm's best price for profit maximization given by,

P = 2500 – 100(11) = Rs. 1400/quintal.

Example 2

Assume, the firm faces the same demand curve as above, P = 2500 – 100Q. The cost function of the firm is given by, C(Q) = 300Q + 2000. Estimate the firm's profit-maximising output and price.

Answer

From the above cost function, it implies that, MC continues to be equal to Rs.300/quintal and the firm incurs a fixed cost equal to Rs. 2000. Following the same procedure as followed in the above example, we have firm's profit maximizing output level = 11quintals and firm's best price = Rs.1400/quintal. This infers that, even in the presence of fixed costs, there has been no change in the firm's profit maximizing output level and firm's best price.

6.3. Importance of AR and MR Curves

The AR and MR curves form significant tools for economic analysis. They include:

- ***Profit determinants:*** As discussed earlier, ARC is the price line for the firm in all market structures. By studying the relationship between ARC and ATC of the firm, we can ascertain whether the firm is earning supernormal or normal profits or incurring losses. If the ARC is above the ATC curve at the point of equilibrium (MR=MC), it implies, the firm is earning super normal profits (Figure 19.10 and Figure 21.4). If the ARC is tangent to the ATC curve at the point of equilibrium (MR=MC), it implies, the firm is earning normal profits (Figure 19.10 and Figure 21.5). If the ARC is below the ATC curve at the point of equilibrium (MR=MC), it implies, the firm is incurring losses in the production programme (Figure 19.10 and Figure 21.7)
- ***Determination of full capacity:*** The concept of tangency between ARC and ATC indicates, whether the firm is producing at is full capacity or under capacity. If the ARC is tangent to the ATC curve at its minimum point, as in the case of perfect competition in the long run (Panel A of the Figure 23.11), it implies, the firm produces at its full capacity. If ARC is tangent to the downward falling ATC, as in case of monopolistic competition in the long run (Panel B of the Figure 23.11), it implies, the firm posses idle or reserve capacity.
- ***Determination of point of equilibrium:*** In all market structures, the point of equality of MR and MC curves determines the equilibrium position of the firm. Their point of intersection determines price, output and profit or loss of a firm.
- ***Determination of factor pricing:*** The same concepts of MR=MC and ARC also helps in determining the factor prices through describing the equilibrium of the firm under different market conditions. But, in factor pricing, MRC becomes the Marginal Revenue Productivity (MRP), MC becomes the Marginal Factor Cost (MFC) and ARC becomes the Average Revenue Productivity (ARP). It is essential to note that, in factor pricing, MRP and ARP are 'inverted U shaped' curves, unlike MR and AR curve.

19

Perfect Competition

In the preceding chapter, we studied about different types of market structures, concepts of revenue and firm's equilibrium. The spectrum of competition ranges from perfectly competitive market, an extreme form of perfection to a pure monopoly, an extreme form of imperfection among different market structures. In the concept of firm's equilibrium, we analyzed that, the revenue of a firm depends upon the quantity of output transacted in the market and we discussed MC=MR approach (meritorious over TR and TC approach) to maximize the profits in the firm. This general rule applies to all firms, but the way in which revenue is related to output varies across the firms depending upon the nature of market competition. From this chapter, we begin with a detailed explanation and analysis of different market structures and this chapter focuses on perfect market competition.

In Economics, the term competition refers to the competitive market structure. The competitiveness of market structure refers to the extent to which an individual firm have the power to influence the market price. The less is the power of the firm to influence the market price, the more competitive is the market structure. In the extreme case of zero market power of the firm, the market structure is said to be perfectly competitive. So, in a perfect competition, no single firm can influence the price of the commodity and they set the price based on the market (industry) equilibrium. That means, the price of the product is determined at the industry level based on the demand and supply forces and the same price is passed on to the individual firms to transact their produce in the market (Figure 19.1). In perfect competition, all sellers produce homogeneous product and this ensures the existence of single price in the market. So, perfect competition refers to a market in which there are infinite numbers of buyers and sellers selling homogeneous product. So, with infinite number of firms and homogeneous product under perfect competition, no individual firm in it is in a position to influence the price of the commodity in the market. The notion of competition is very widely used in Economics in general and in Microeconomics in particular. This term 'competition' is used in the sense of rivalry between the firms. But, perfect market competition is characterized by the complete absence of rivalry among the firms. The following are the popular definitions of perfect competition given by the Economists:

'Perfectly competitive market is a situation where large number of buyers and sellers are engaged in the purchase and sale of identically similar commodities, who are in close contact with one another and who buy and sell freely among themselves.'

– ***(Boulding)***

'Perfect competition, as a condition of market in which there will be fluidity and mobility of factors of production, so that the number of firms and the size of firms can freely increase and decrease'

– ***(Frank Knight)***

'The perfect competition is characterized by the presence of many firms, they sell identically the same product. The seller is a price- taker.'

– **(Bilas)**

'Perfect competition prevails when the demand for the output of each product is perfectly elastic.'

– **(Joan Robinson)**

'Perfect competition describes a market in which there is a complete absence of direct competition among economic groups.'

– **(Ferguson)**

19.1. Features or Assumptions of Perfect Competition

Traditionally perfect competition is considered as an ideal form of market. For classical writers, perfect competition is a golden rule. The following are the important features of perfect competition:

1. Infinite Number of Buyers and Sellers

The market is said to be Atomistic market (Atomicity) because, there will be infinite number of buyers and sellers and their actions in the market are independent *i.e.,* they have no impact on others. So, many sellers form the total market supply of the commodity. Individually, seller is a firm and collectively, it is an industry. The price of commodity is decided at industry level by market forces of demand and supply *i.e.,* by buyers and sellers collectively and the same price is passed on to the individual firms. As shown in the Panel A of the Figure 19.1, at industry level, equilibrium price OP is determined based on the interaction between demand and supply forces and the same price OP is accepted by the individual firm (Panel B of the Figure 19.1). Hence, no individual seller is in a position to change the price by controlling supply because, individual seller's supply constitute an insignificant percentage of total market supply. So, if that seller alone raises the price, his product will become costlier than others and automatically, he will be out of the market. Hence, that seller has to accept the price in the market and he has no control over the price of the product. This ensures single price in the market and in this way, seller becomes price taker and quantity adjuster and not a price maker. So, the individual firm cannot have the price policy of its own. The sellers have to accept the price and adjust the quantity of output to be sold, so as to maximize their profits.

Similarly, the market comprises of infinite number of buyers for the product and no individual buyer can influence the price by changing or controlling the demand *i.e.,* there is no monopsony power. This is because, individual buyer's demand is insignificant of total demand or market demand. So, every buyer has to accept the price decided by market forces of demand and supply. In this way, like sellers, all buyers are price takers and not price makers. This also ensures existence of single price in the market. The buyers should not have any organization or union and they compete for the market demand on an individual basis.

2. Homogenous Nature of the Product

All sellers produce and sell homogeneous *i.e.,* perfectly identical products. The products are perfectly same in terms of size, shape, taste, colour, ingredients, quality etc. So, the products are perfect substitutes, thereby, the E_C is infinite. This ensures each firm being passive 'price taker' and the existence of a uniform price in the market. So, non-price competition will not influence the buyers to purchase the product and the buyers are simply attached to the sellers only on the basis of prevailing uniform market price.

From the above two features, we can infer that, in a perfect market competition, the ARC (demand curve) of the firm will be horizontal and parallel to X-axis, as no single seller and buyer can influence the price of the commodity and commodity being homogenous, the buyers will not pay differential prices for the same product. So, due to absence of price differences, ARC remains same at different quantities of output sold and MRC also remains same and coincides with ARC (Panel B of the Figure 19.1).

3. Free Entry and Exit of Firms

There are no restrictions regarding the entry and exit of firms into an industry. During short run, when the existing firms enjoy super-normal profits, new firms enter the industry. As a result, the supply of the product into the market increases and it leads to fall in price. Further, due to entry of new firms into the industry, the demand for factors and factor services will increase and this escalates the cost of production of product. So, these influences reduce profits from abnormal to normal level. Similarly, when the existing firms suffer from losses, some firms will leave the industry. This leads to fall in supply, thereby, price will shoot up in the market and the profits of the firms will increase. So, again this ensures normal level of profits. This free entry and free exit of firms are possible only in the long run. This is because, the size of the plant cannot be changed in the short run. This feature of free entry and exit of firms under perfect competition ensures stiff competition among the firms, thereby, excess profits are completely ruled out and only normal profits prevail in the long run. The ease of entry and exit of firms is also facilitated by the absence of sunk costs on the part of the firms. Of course, the new entrants into the industry produce similar products.

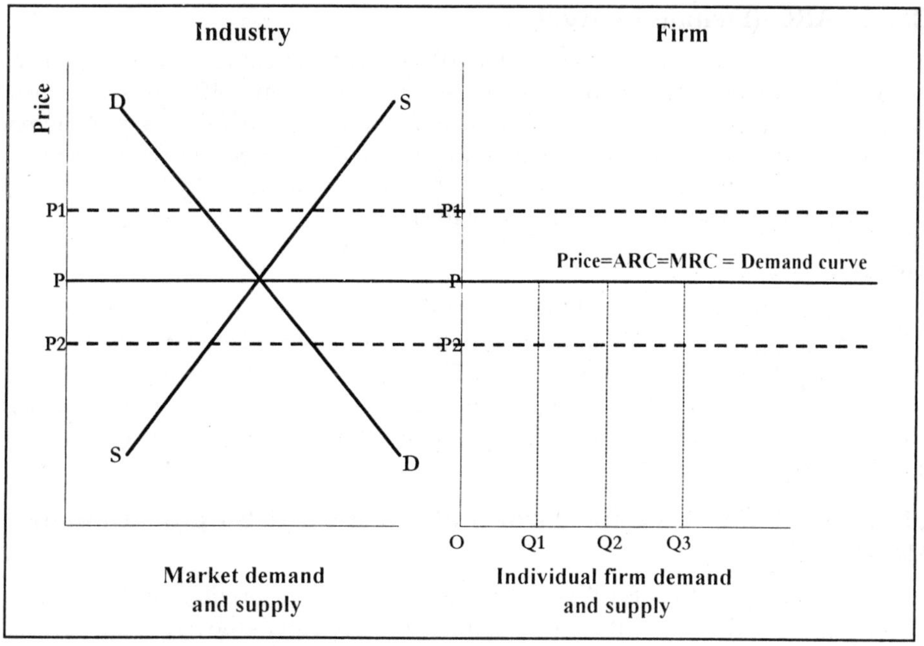

Figure 19.1: Price, AR, MR and Demand Curve of Perfectly Competitive Firm.

If the above three features alone are fulfilled, then the market structure refers to Pure competition. This concept of Pure competition was developed by E. H. Chamberlin. Of the above three elements, the third element *i.e.*, 'free entry and exit of firms' implies complete absence of monopoly element in the perfect competition. In addition to the above three features of Pure competition, the following features are also attached to the Perfect competition:

4. Perfect Knowledge

All the buyers and sellers have readily available information about prices and products from competing suppliers. So, no buyer will pay a price higher than market price and no seller will accept a price less than the one prevailing in the market. The seller must know the price at which the product is offered for sale by the other sellers in the market. Similarly, buyers must know the price being charged by the different sellers. So, this feature facilitates to have a single price for the product at the same time and no buyer will purchase the product from other sellers who quote a higher price for the product. So, this perfect knowledge about the output, prices etc., among the sellers and buyers ensures prevalence of uniform price for the product in the market.

5. No Government Intervention

There is complete absence of Government regulation regarding the supply, prices and pricing mechanism of the products. There is also no collusion among sellers or buyers. The Government does not interfere in economic activities of the firms and price mechanism is influenced only by the two market forces *viz.*, demand and supply.

6. Perfect Mobility of Factors of Production

The factors of production should be free to move from one use to another or from one industry to another easily to get better remuneration. No one should have monopoly power to control the factors of production. The assumption of perfect mobility of factors is essential to fulfill the first condition namely large number of sellers in the market. This feature is also essential to keep the supply of the product at par with the demand conditions in the market. When the demand for a product of a particular industry increases in the market, then it facilitates easy movement of factors of production across the industries, thereby, it leads to increased production of that product. So, this feature helps to maintain balance between demand and supply conditions in the market, thereby, it ensures single price for the product in the market.

7. Complete Absence of Transportation Costs

It is assumed that, buyers and sellers are close to the market, such that, there is complete absence of transportation costs in the transaction of commodities. This also ensures prevalence of same price for the same product at the same time in the market. If transportation costs are incurred, the firms nearer to the market will charge a low price than the firms far away. Hence, it is assumed that, there are no transportation costs. This feature also ensures that, the market price charged by different sellers does not differ even due to differences in locations of the sellers in the market.

8. Perfectly Elastic ARC (Demand curve)

Since, same price prevails for the same commodity, the ARC for a perfect competitive firm both in the short run and long run is perfectly elastic *i.e.*, it lies parallel to X-axis. Since, ARC is constant, MRC must be constant and coincides with ARC (Table 18.2). This is so because, an individual firm under perfect competition cannot influence the price of the product and hence, additional units are also sold at the same price as in the case of previous units. So, under perfect market competition, Price=ARC=MRC=Demand curve (Panel B of the Figure 19.1). However, at industry level, the demand curve slopes as usual from left to right (Panel A of the Figure 19.1).

9. Uniform Prices

Under perfect market competition, an individual firm cannot influence the price of the commodity, thereby, it has to take the market price as given, constant and uniform in nature. As shown in Panel B of the Figure 19.1, the firm sells OQ_1, OQ_2 and OQ_3 quantities of paddy at the same price OP. The firm cannot charge a higher or lower price than the OP. If it attempts to charge a some what higher price like OP_1, the firm will be able to sell nothing because, the buyers prefer the same commodity from other firms at price OP. On the other hand, if the firm charges a somewhat lower price OP_2, the firm will unnecessarily suffer losses.

10. Horizontal Demand Curve of the Firm and Downward Sloping Demand Curve of the Industry are Related

For an individual firm, the demand curve is perfectly elastic, as the firm has no control over the price and it has to sell any quantity of commodity at the same price. So, demand curve also equals the ARC and MRC of a firm (Panel B of the Figure 19.1). Earlier, we studied that, the demand curve for a commodity slopes downward from left to right. In a perfect competitive market as well, there is a downward sloping demand curve, but it is applicable to the entire market or the industry. Such a demand curve has been shown as DD in Panel A of the Figure 19.1.

The relation between a firm's demand curve and market demand curve can be explained from the Figure 19.2. When there are infinite number of firms under perfect competition, each firm's demand quantity appears as a dot or a point (q1, q2, q3, q4, q5 etc) on the market demand curve. With each addition of a firm to the market, the industry demand curve gets more flexible and thereby, it becomes more flatter. Gradually, it starts moving in the direction of DD_1, DD_2, DD_3, DD_4, DD_5 etc., and ultimately assumes the form of a horizontal straight line parallel to X-axis as DD. So, there is no contradiction between the horizontal demand curve of a firm and the downward sloping demand curve of a market or industry under perfect market competition.

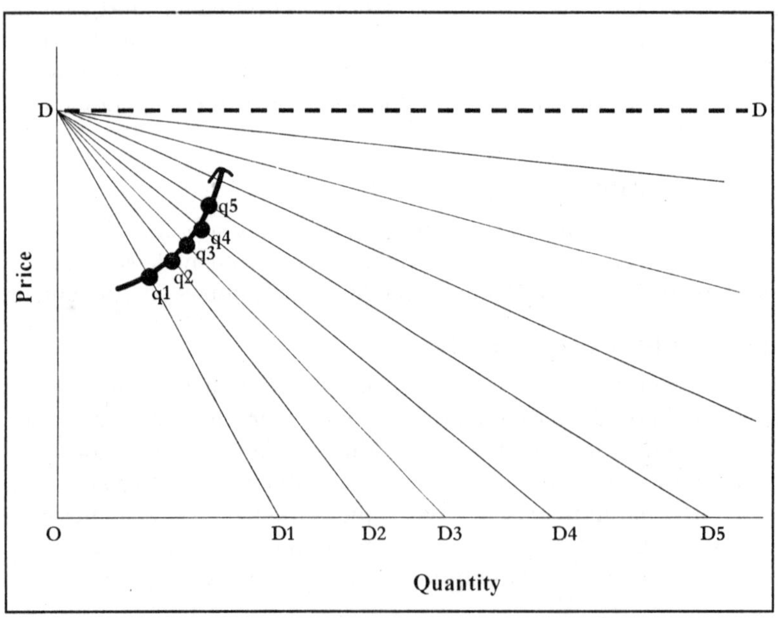

Figure 19.2: Relation between Market demand curve and Firm's demand curve.

11. Supply Curve

Supply curve of a perfect competitive firm varies in short run and long run production programmes. In short run, the portion of the SRMC curve above SRAVC curve represents SRS curve of a perfect competitive firm (Figures 19.3). In long run, the portion of the LRMC curve above LRATC curve represents LRS curve of a perfect competitive firm (Figures 19.4).

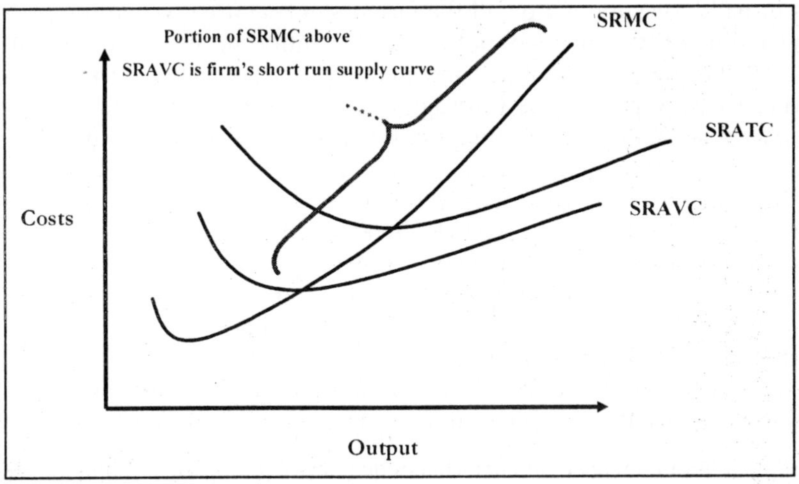

Figure 19.3: Short run supply curve of a firm under perfect competition.

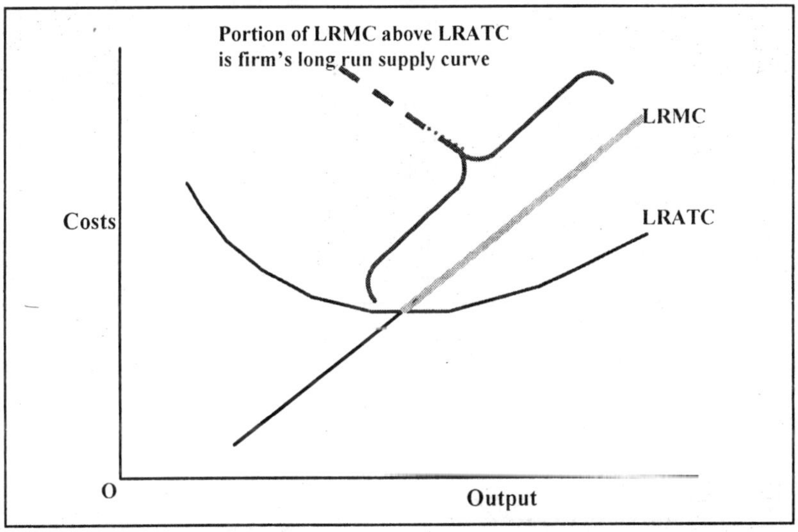

Figure 19.4: Long run supply curve of a firm under perfect competition.

12. Absence of Non-Price Competition

Since all the firms deal with homogenous and standardized products with respect to quality, taste, shape, colour etc., there is no scope for product differentiation. More over, since buyers have perfect knowledge about the features of products and their prices in the market, there is no need for the firms to incur any expenditure on advertisement and other selling costs.

13. No Formation of Cartel

There is no agreement between the firms to collude in the market.

19.2. Price Determination Under Perfect Competition

As discussed above, under perfect competition, there are infinite number of sellers and buyers in dealing with a homogenous product. But, the actions of either individual seller or individual buyer will not influence the price of the product in the market. So, the individual firms are simply 'price takers' and not 'price makers' and they have to take the prevailing market price for granted for transacting their output. Similarly, each buyer has to decide his cash outlay based on the prevailing market price. Though individual's actions with reference to both buyers and sellers could not influence the price, but their aggregate influences in terms of total demand and total supply can influence the price. So, price of the product under perfect competition is influenced by two market forces *viz.*, demand and supply. The demand side is governed by law of demand and supply side is influenced by the cost of production of the commodity, thereby, the law of supply will operate.

In accordance with the law of demand and law of supply, the quantity demanded and supplied of paddy is shown through Table 19.1. When the price of paddy is Rs. 1000/qtl, 1200 quintals of paddy is demanded by the rice millers, but only 200 quintals of paddy is supplied by all the farmers in that area. When the price of paddy is gradually increased in the market, the quantity demanded contracts and quantity supplied extends. When the price of the paddy is at Rs. 1500/quintal, both the quantity demanded and quantity supplied in the market are equal at 500 quintals. The price so determined when the aggregate demand and aggregate supply forces are equal is called equilibrium price and the quantity of paddy transacted at this equilibrium price refers to equilibrium quantity. Once the equilibrium price is arrived in the market, it shows less tendency for change. If the price of paddy falls or rises compared to equilibrium price (Rs. 1500/quintal), again the demand and supply forces will react to bring the price back to the equilibrium price. Say, for example, if price of the paddy is increased to Rs. 1600/qtl, then supply exceeds the demand in the market. In such case, some farmers cannot dispose their produce in the market and thereby, they even try to sell their output even at low price. This leads to gradual fall in the price from Rs. 1600/quintal back to equilibrium price Rs. 1500/quintal. On the contrary, when the price of the paddy falls to Rs. 1400/quintal, then demand exceeds the supply in the market, thereby, there is a shortfall of supply of paddy into the market. In such case, some rice millers could not get the produce. So, they even try to pay more to have the commodity and this gradually escalate the price of paddy from Rs. 1400/ quintal to back to equilibrium price Rs. 1500/quintal. The same concept is explained through Figure 19.5.

Table 19.1: Determination of equilibrium price and equilibrium quantity of paddy under perfect market competition.

Price (Rs/qtl)	*Quantity Demanded (Qtls)*	*Quantity Supplied (Qtls)*
1000	1200	200
1150	1000	350
1275	800	410
1400	650	465
1500	**500**	**500**
1600	430	725
1720	320	800
1800	200	1000

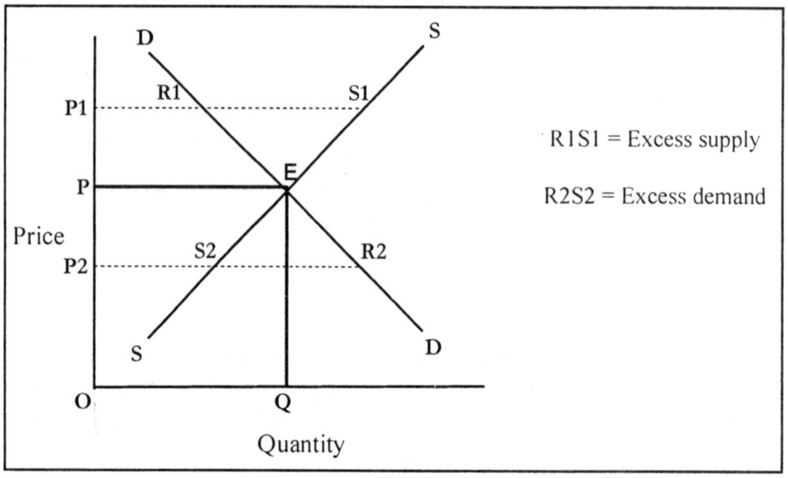

Figure 19.5: Price determination of paddy under perfect market competition.

On X-axis, quantity demanded and supplied and on Y-axis, price of the commodity (paddy) are taken. DD is the demand curve and SS is the supply curve drawn for paddy in the market (industry). DD and SS curves intersect at point 'E', which indicates that, OQ equilibrium quantity of paddy is supplied and demanded in the market at OP equilibrium price.

Case 1

If price of the commodity is increased to OP_1 in the market above the equilibrium price OP, the demand is only P_1R_1 (contracts) and supply is P_1S_1 (extends). The excess supply over demand is R_1S_1. So, some of the sellers could not sell a portion of commodity. In that case, they will try to dispose the unsold stock at lower price. Thus, the price will decrease and equals with equilibrium price OP. So, the sellers will dispose the unsold stock at OP level of price instead of OP_1 and hence, price gradually decreases to equilibrium price OP.

Case 2

If price is decreased to OP_2 from OP, then, the demand extends to P_2R_2 and supply contracts to P_2S_2. Therefore, the excess of demand over supply is S_2R_2. So, some of the buyers will not get the commodity or desired quantity of commodity because, demand exceeds supply. Hence, to have the commodity or to get the desired quantity of commodity, the buyers even purchase the commodity at a higher price, thereby, price gradually increases and equals to equilibrium price OP.

Changes in Equilibrium Price

In the above discussion, we moved along the same demand curve and same supply curve due to changes in price of the commodity around the equilibrium price, but there is no shift in the demand and supply curves. However, changes in equilibrium price may also be due to increase or decrease in demand, increase or decrease in supply, increase or decrease of both demand and supply, differences in elasticities of demand and supply etc., and these cases are discussed here under.

1. Change in Either Demand or Supply of the Commodity

Case 1 - Increase in Demand from the Buyers and there is No Change in Supply of the Commodity

As shown in the Panel A of the Figure 19.6, DD is the demand curve, SS is the supply curve and 'E' is the equilibrium point, where OQ quantity of paddy is supplied and demanded in the market at OP equilibrium price. Let us suppose, due to increase in demand from the rice millers, the demand curve DD shifted towards right as D_1D_1 and there is no change in supply of paddy in the market. The new demand curve D_1D_1 intersects the original supply curve SS at E_1, where OQ_1 quantity of paddy is supplied and demanded in the market at OP_1 price. So, due to increase in demand at constancy of supply of paddy, the equilibrium quantity of paddy demanded and supplied is increased from OQ to OQ_1 and the equilibrium price is also increased from OP to OP_1. So, the new equilibrium price is OP_1, which is higher than original equilibrium price OP.

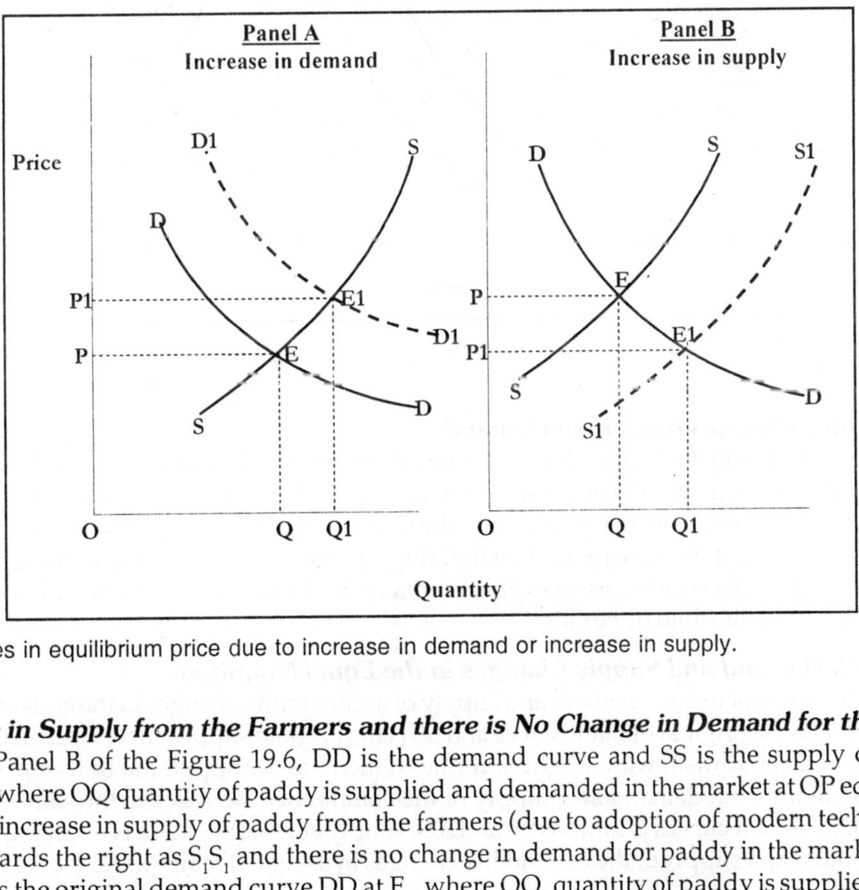

Figure 19.6: Changes in equilibrium price due to increase in demand or increase in supply.

Case 2 - Increase in Supply from the Farmers and there is No Change in Demand for the Commodity

As shown in Panel B of the Figure 19.6, DD is the demand curve and SS is the supply curve and 'E' is the equilibrium point, where OQ quantity of paddy is supplied and demanded in the market at OP equilibrium price. Let us suppose, due to increase in supply of paddy from the farmers (due to adoption of modern technology), the supply curve SS shifts towards the right as S_1S_1 and there is no change in demand for paddy in the market. The new supply curve S_1S_1 intersects the original demand curve DD at E_1, where OQ_1 quantity of paddy is supplied and demanded in the market. So, due to increase in supply of paddy at constancy of demand, the equilibrium price falls from OP to OP_1 and the quantity demanded and supplied increases from OQ to OQ_1. So, the new equilibrium price is OP_1 which is lower than original equilibrium price OP.

2. Changes in Both Demand and Supply of the Commodity

In the above two cases, we studied how the equilibrium price of the paddy changes due to change either in demand or supply of the commodity. Now, let us analyze, how the equilibrium price changes due to changes in both demand and supply of the commodity. When both demand and supply of the commodity changes, the extent of influence on the original equilibrium price depends upon their relative magnitude change *i.e.*, whether demand changes more than supply of the commodity, whether supply changes more than demand of the commodity and when demand and supply changes in the equal magnitude. Let us examine each case here under.

Case 1 - When Demand Changes more than Supply

Let us assume the case, when the demand changes (increases) more than supply change (increases). In the Panel A of the Figure 19.7, DD is the original demand curve and SS is the original supply curve. Due to increase in demand, the new demand curve D_1D_1 shifts towards right and due to increase in supply, the new supply curve also shifts towards right as S_1S_1. Since the increase in demand is more than increase in supply, the gap between DD and D_1D_1 curves is more than the gap between SS and S_1S_1 curves. Since, increase in demand is greater than increase in supply, the new equilibrium price OP_1 is higher than original equilibrium price OP.

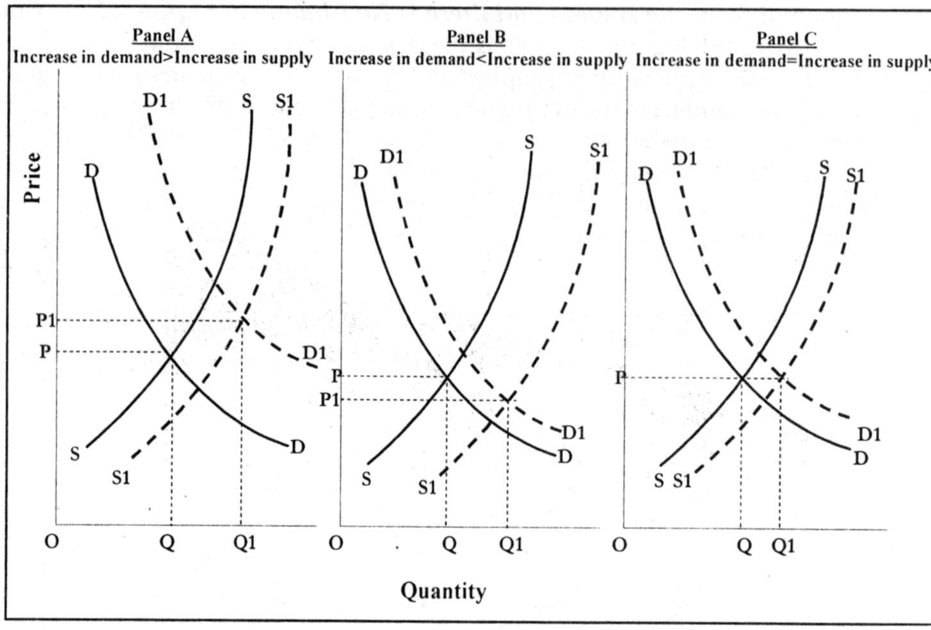

Figure 19.7: Changes in equilibrium price due to changes (increase) in both demand and supply.

Case 2 - When Supply Changes more than Demand

Here, let us assume that, supply changes (increase) more than demand (increase). In Panel B of the Figure 19.7, DD is the original demand curve and SS is the original supply curve. Due to increase in demand, the new demand curve D_1D_1 shifts towards right and due to increase in supply, the new supply curve also shifts towards right as S_1S_1. Since, the increase in supply is more than increase in demand, the gap between SS and S_1S_1 curves is more than the gap between DD and D_1D_1 curves. Since, increase in supply is greater than increase in demand, the new equilibrium price OP_1 is lower than original equilibrium price OP.

Case 3 - When Both Demand and Supply Changes in the Equal Magnitude

Let us assume the case, where both demand and supply of a commodity changes in the equal magnitude. In Panel C of the Figure 19.7, DD is the original demand curve and SS is the original supply curve. Due to increase in demand, the new demand curve D_1D_1 shifts towards right and due to increase in supply, the new supply curve also shifts towards right as S_1S_1. Since, both demand and supply of the commodity increases in the same magnitude, the gap between DD and D_1D_1 demand curves is same as SS and S_1S_1 supply curves. So, there is no change in the equilibrium price OP, as both demand and supply of the commodity changes in the equal magnitude.

3. When the Market Faces with Different Elasticities of Demand and Supply

When there is change in demand or supply of a commodity, it does not always cause the same change in equilibrium price. Two possible cases can be discussed here *viz.*, when the market faces with two different elasticities of demand and when the market faces with two different elasticities of supply. Let us discuss one by one.

Case 1 - When the Market Faces with Two Different Elasticities of Demand

Let us assume, two demand curves DD and D_1D_1 with different elasticities (Panel A of the Figure 19.8), the former being less elastic (more steeper) and the latter being comparatively more elastic (more flatter). Let us study the influence of increase in supply of paddy in the market with reference to differences in elasticity of demand curves. Initially, when the demand is DD and supply is SS, the equilibrium price is OP with reference to equilibrium point 'E', where OQ

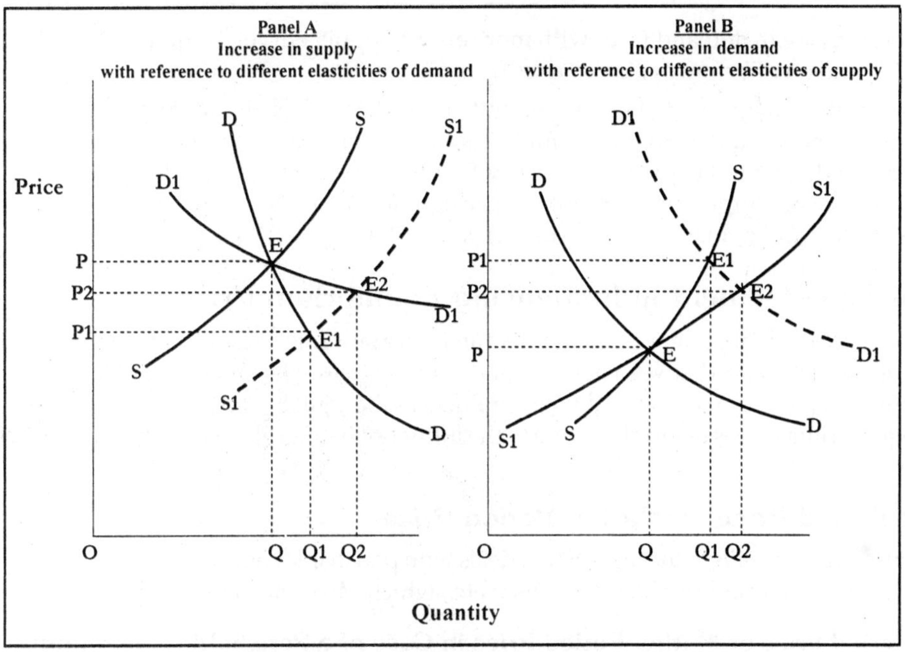

Figure 19.8: Changes in equilibrium price when the market faces with different elasticities of demand and supply.

quantity of paddy is supplied and demanded in the market. Let us suppose, the supply of the commodity increases in the market, thereby, S_1S_1 is the new supply curve and the effects of this increase in supply can be studied with reference to two demand curves DD and D_1D_1 of different elasticities. If the market faces with less elastic demand curve DD, then the rise in supply *i.e.*, S_1S_1 intersects the demand curve DD at E_1, where OQ_1 quantity of paddy is supplied and demanded in the market at OP_1 equilibrium price.

If the market faces with more elastic demand curve, D_1D_1, the new supply curve S_1S_1 intersects D_1D_1 demand curve at E_2, where OQ_2 quantity of paddy is supplied and demanded in the market at OP_2 equilibrium price.

The above discussion infers the following interesting points:

- When the increase in supply faces with less elastic demand curve, the quantity demanded increases from OQ to OQ_1 and equilibrium price decreases from OP to OP_1.
- When the increase in supply faces with more elastic demand curve, the quantity demanded increases from OQ to OQ_2 and equilibrium price decreases from OP to OP_2.
- The comparative analysis of the above two points infer that, when the increase in supply faces with a less elastic demand curve, the fall in equilibrium price is more compared to increase in quantity of paddy supplied and demanded in the market, *i.e.*, $(OQ_1\text{-}OQ) < (OP\text{-}OP_1)$. But, when the increase in supply faces with a more elastic demand curve, the fall in equilibrium price is less compared to increase in quantity of paddy supplied and demanded in the market, *i.e.*, $(OQ_2\text{-}OQ) > (OP\text{-}OP_2)$.

Case 2 - When the Market Faces with Two Different Elasticities of Supply

Let us assume, two supply curves SS and S_1S_1 are with different elasticities (Panel B of the Figure 19.8), the former being less elastic (more steeper) and the latter being comparatively more elastic (more flatter). Let us study the influence of increase in demand for paddy in the market with reference to differences in elasticity of supply curves. Initially, when the demand is DD and supply is SS, the equilibrium price is OP with reference to equilibrium point 'E', where OQ quantity of paddy is supplied and demanded in the market. Let us suppose, the demand for the commodity increases in the market, thereby, D_1D_1 is the new demand curve and the effects of this increase in demand can be studied with reference to two supply curves SS and S_1S_1 of different elasticities. If the market faces with less elastic supply curve SS, then the rise in demand *i.e.*, D_1D_1 intersects the supply curve SS at E_1 where OQ_1 quantity of paddy is supplied and demanded in the market at OP_1 equilibrium price.

If the market faces with more elastic supply curve S_1S_1, the new demand curve D_1D_1 intersects S_1S_1 supply curve at E_2, where OQ_2 quantity of paddy is supplied and demanded in the market at OP_2 equilibrium price.

The above discussion infers the following interesting points:

- When the increase in demand faces with less elastic supply curve, the quantity demanded increases from OQ to OQ_1 and equilibrium price also rises from OP to OP_1.

- When the increase in demand faces with more elastic supply curve, the quantity demanded increases from OQ to OQ_2 and equilibrium price also rises from OP to OP_2.
- The comparative analysis of above two points infer that, when the increase in demand faces with a less elastic supply curve, the increase in equilibrium price is more compared to increase in quantity of paddy supplied and demanded in the market, *i.e.*, $(OP_1\text{-}OP) > (OQ_1\text{-}OQ)$. But, when the increase in demand faces with a more elastic supply curve, the increase in equilibrium price is less compared to increase in quantity of paddy supplied and demanded in the market, *i.e.*, $(OP_2\text{-}OP) < (OQ_2\text{-}OQ)$.

19.3. Role of Time Element in Influencing Equilibrium Price

According to Alfred Marshall, time element plays an important role in determining the equilibrium price of the commodity through the interaction between demand and supply forces. The time element influences the changes in the supply of the commodity into the market and thereby, influences the equilibrium price of the commodity. The following are the different equilibrium prices determined due to the influence of time element in the interaction between demand and supply forces:

1. Very Short Period Price or Market Period Price

This equilibrium price prevails in the market that deals with perishable commodities like fish, milk, leafy vegetables, eggs and slightly durable commodities like rice, wheat etc., which can be stocked for a limited period.

(a) Very Short Period Price or Market Period Price in Case of a Perishable Commodity

In case of a perishable commodity, supply into the market is fixed and cannot be changed at all. Since supply is fixed, demand plays a decisive role in determining the price of the commodity and it is explained through Panel A of the Figure 19.9. The Figure 19.9 reveals that,

- On X-axis, quantity demanded and supplied of milk and on Y-axis, price of the milk are taken.
- SS is the supply curve of commodity (milk) and DD is initial demand curve of the commodity (milk). Since, the supply of milk cannot be increased in the market in a very short period, the supply remains fixed and thereby, the supply curve SS remains parallel to Y-axis.
- These two curves, DD and SS intersect at point E, which indicates that, OQ quantity of milk is supplied and demanded at OP equilibrium price.
- If the demand for milk is increased, D_1D_1 is the new demand curve (shifts upwards) and it intersects the SS supply curve at E_1, thereby, the equilibrium price is increased to OP_1 (*i.e.*, same quantity purchased at higher price).
- If the demand for milk is decreased, D_2D_2 is the new demand curve (shifts downward) and it intersects the SS supply curve at E_2, thereby, the equilibrium price is decreased to OP_2 (*i.e.*, same quantity purchased at lower price).
- Thus, demand plays a decisive role in determining the price of milk in a very short period market.

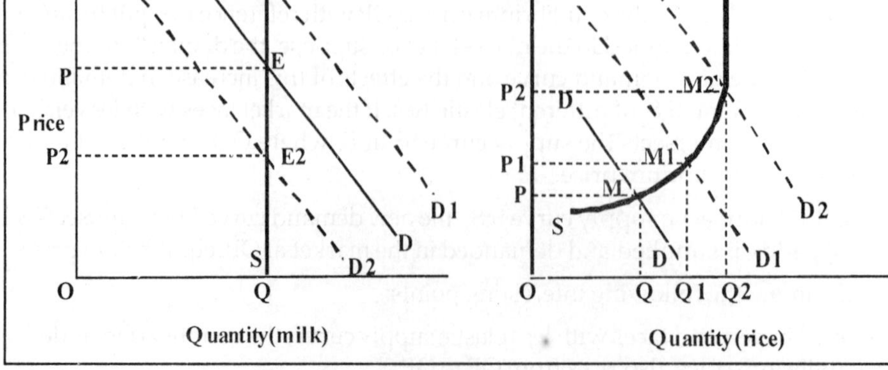

Figure 19.9: Market period (Very short period) price determination under perfect market competition.

(b) Very Short Period Price or Market Period Price in Case of a Semi-durable Commodity

In this case, the commodity is not perishable, but semi-durable and hence, it can be stocked for a certain period. So, supply of the commodity can be slightly varied in the market depending upon the demand situations, thereby, supply has little influence on the price of the commodity. This market deals with the commodities like rice, wheat, pulses etc. Since, the commodity is semi-durable, the sellers can stock or withhold the commodity, if the market price is not sufficiently high. However, demand still plays a dominant role in influencing the price of the commodity in the market. There is no question of distress sale of the commodity (price<cost of production of commodity), as the seller can stock the commodity at times of fall in prices. The seller can fix a minimum price below which he will not sell the commodity and that minimum price is called Reserve price or Floor price. If the seller anticipates increase in demand for the commodity in the future, he fixes higher reserve price for the commodity and *vice versa*. This is explained through Panel B of the Figure 19.9. It is clear from the Figure 19.9 that,

- ☆ On X-axis, quantity demanded and supplied and on Y-axis, price of the commodity (rice) are taken.
- ☆ SS_1 is the supply curve and DD is initial demand curve of the commodity (rice). The SS_1 supply curve initially slopes upwards indicating that, upto a certain quantity or output of the commodity (OQ_2), the supply can be increased in the market in response to rise in price of the commodity. But, beyond OQ_2 quantity of rice, what ever may be the increase in demand for the commodity, the supply of rice into the market remains fixed, as it cannot be increased in the very short period.
- ☆ Initially, at OP level of equilibrium price, OQ is the quantity demanded and supplied. This OP is the reserve price below which the farmer is not willing to sell the commodity. So, if the demand for the commodity is decreased below DD, price will not fall much and it will be around reserve price only.
- ☆ If the demand for the commodity is increased, D_1D_1 is the new demand curve (shifts towards right). Farmer will increase the supply from OQ to OQ_1 and the new equilibrium price is OP_1. So, the price increases from OP to OP_1.
- ☆ If the demand is still further increased, the new demand curve is D_2D_2 (shifts towards right). The farmer further increase the supply of the commodity to OQ_2 and the new equilibrium price is OP_2. Beyond OQ_2 supply, the farmer cannot increase the supply, as it is relatively fixed now. Beyond this point *i.e.*, at OQ_2 quantity of supply, if demand increases, price increases greatly, as supply has no influence on price. So, the supply curve (SS_1) has three phases:
  - ❖ *Minimum price phase:* SM phase of the curve.
  - ❖ *Maximum price phase:* S_1M_2 phase of the curve.
  - ❖ *Change in price due to changes in demand and supply of commodity phase:* MM_2 phase of the curve.

So, we can infer that, the farmer sells OQ quantity of rice at minimum price OP and withhold the stock QQ_2 in anticipation of higher prices in the future. At higher price level, OP_2 the farmer disposes all the stocked portion of the commodity (QQ_2).

Factors Influencing Reserve Price

The following are the factors that influence the fixation of reserve price for the commodity:

- ☆ ***Durability of the commodity:*** The more durable the commodity is, the higher will be its reserve price.
- ☆ ***Prices in future:*** If the seller expects that, prices rise in future, he will fix a higher reserve price.
- ☆ ***Future cost of production:*** If the seller expects that cost of production rise in future, he will fix a higher reserve price.
- ☆ ***Expenses on storage:*** The reserve price also depends upon the time and expenses involved in storage of the commodity. The greater the time and expenses incurred in storing the commodity, the lower is the reserve price. This is so because, the seller would like to dispose of the commodity at the earliest, so as to avoid the expenses of storage.
- ☆ ***Liquidity preference:*** The higher the liquidity preference by the seller, the lower is the reserve price.
- ☆ ***Demand in future:*** If the seller expects demand for product rises in the future, he will fix high reserve price.

2. Short Period Price

The short period relates to a few months period in which supply can be changed in accordance with demand. This is possible by changing the variable factors only in the production programme. In the short period, it is not possible to change the fixed factors, the scale of production and organization. Therefore, supply of output of durable good (cloth) can be increased or decreased to match the increase or decrease in demand by changing the variable factors alone.

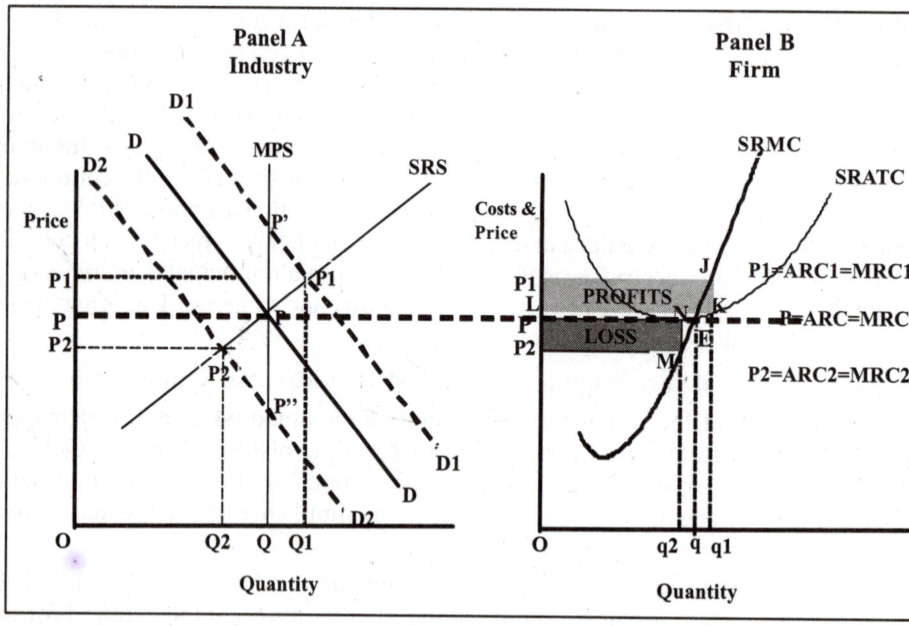

Figure 19.10: Short period price determination for a durable good under perfect market competition.

In short period, price is determined by the forces of demand and supply. The SRS curve slopes upward from left to right like the ordinary supply curve. It establishes short-run equilibrium price when it is intersected by the demand curve. The short-run price is also known as the short-run normal price. Its determination is shown in Figure 19.10.

- In Panel A, at industry level, DD is the original demand curve and MPS is the Market Period Supply (MPS) curve, which lies parallel to Y-axis, as discussed earlier (Panel A of the Figure 19.9). Both the curves intersect at P and establishes OP=PQ equilibrium price and OQ equilibrium quantity.
- Suppose the demand (for cloth) rises, as indicated by the D_1D_1 curve. The immediate effect will be for the market price to increase from PQ to P'Q. This is because, since supply is fixed in the market period, it is not possible to increase the supply beyond OQ.
- But, in short period, it is possible to increase the supply with the help of existing plants and machinery by employing extra labour, raw material and other variable factors. Thus, by increasing the variable factors, the supply can be increased along the SRS curve. This SRS curve will intersect the new demand curve D_1D_1 at P_1. In this way, the short-run normal price $OP_1=P_1Q_1$ is determined at which OQ_1 quantity of commodity is bought and sold. This short-run price ($OP_1=P_1Q_1$) is higher than the original market price (OP=PQ), but it is lower than the new market period price OP'=P'Q, which is prevailed immediately after the increase in demand.
- Suppose, the demand for cloth falls in the market. The demand curve falls from DD to D_2D_2. The market period price immediately falls from PQ to PΔQ. So, all the firms in the industry will reduce the output by reducing the amount of variable factors like labour, raw materials etc., in the production programme and also reduce supply by stocking the goods, thereby, the new supply is decreased along SRS. As a result, the new equilibrium is established at P_2, where the SRS and D_2D_2 intersect. Now, OQ_2 quantity is bought and sold at $OP_2=P_2Q_2$ price. This short-run price $OP_2=P_2Q_2$ is less than the original market price OP=PQ, but higher than the new market price OPΔ=PΔQ established after fall in demand.
- Thus, supply is more important than demand in the short period because, supply can be increased or decreased to match the rise or fall in demand by changing the variable factors in the production programme.
- As explained earlier, the same prices QP=OP, $Q_1P_1=OP_1$ and $Q_2P_2=OP_2$ determined in the industry due to changes in the supply of the commodity in the short run, will be passed on to the individual firms and as shown in the Panel B, they are horizontal lines parallel to X-axis (as individual firms are price takers) and these price lines coincides with respective ARC and MRC of the individual firms under perfect market competition.
- With the original price OP=QP, we will compare SRATC and SRMC curves. This SRMC curve will cut P=ARC=MRC curve from below and hence, the firm reaches the equilibrium at 'E' and the corresponding point on X-axis represent the Oq quantity of output supplied and demanded in the market at which the firm earns normal profits, as price=SRATC.

- ☆ At $OP_1=P_1Q_1$ price, for the firm, $OP_1=ARC_1=MRC_1$. This MRC_1 is intersected by SRMC at J, which implies that, the firm is again at equilibrium at point J and the corresponding point on X-axis represent Oq_1 level of output supplied and demanded in the market at OP_1 price, where the firm earns super-normal profits, P_1JKL because, price is above SRATC by JK. So, rise in price above SRATC enables the firm to earn super-normal profits.
- ☆ At OP_2 price, for the firm, $OP_2=ARC_2=MRC_2$. This SRMC intersects the MRC_2 curve at M, which implies that, the firm is in equilibrium at point M and the corresponding point on X-axis represent Oq_2 level of output supplied and demanded in the market at OP_2 price, where the firm incurs losses, PP_2NM because, price OP_2 is less than SRATC by NM.

3. Long Period Price or Normal Price

The long period relates to many years in which supply can be changed or fully adjusted to changes in demand. This is possible by changing both fixed and variable factors of production. During this period, the old machines, equipments and plants can be replaced by new ones. There will be free entry and exit of firms depending upon the prospects in the industry. The scale of production, organisation and management can also be changed. Thus, the supply of output of durable good can be fully adjusted to changes in demand in every possible way, in the long run. The price that prevails for the commodity in the long run is called Long run normal price or Normal price. It represents the normal or natural value of the commodity in the long run based on the interaction between permanent demand and supply forces. This is explained through the Figure 19.11.

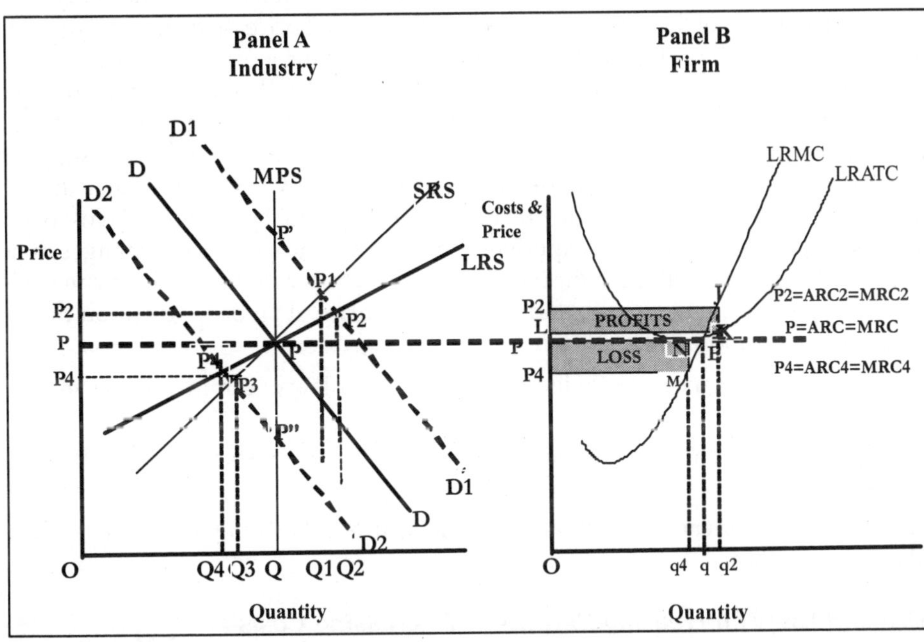

Figure 19.11: Long run price determination for a durable good under perfect market competition.

- ☆ In Panel A, at industry level, the demand curve DD intersects the MPS curve at P, where OQ quantity of commodity (Tractor, Implements) is demanded and supplied at OP level of price.
- ☆ When there is increase in demand, D_1D_1 is the new demand curve, thereby, the market period price is OP', the short period price is OP_1 and the long period price is OP_2. Hence, a rise in demand in the market, increases the price of the commodity from OP to OP_2 in the long run.
- ☆ If the demand for the commodity falls, D_2D_2 is the new demand curve and hence, the market period price falls to OP'', the short period price is OP_3 and the long period price is OP_4. Hence, a fall in demand in the market, leads to fall in price of the commodity from OP to OP_4 in the long run.
- ☆ The same prices OP, OP_2 and OP_4 determined in the industry due to changes in the supply of the commodity in the long run will be passed on to the individual firms and as shown in the Panel B, they are horizontal lines parallel to X-axis (as individual firms are price takers) and these price lines coincides with their respective ARCs and MRCs of the individual firms under perfect market competition.
- ☆ With the original price OP, we will compare LRATC and LRMC curves. This LRMC curve will cut P=ARC=MRC curve from below and hence, the firm reaches the equilibrium at 'E', where the firm sells Oq quantity of output at price OP and the firm earns normal profits, as price=LRATC.

- ☆ At OP_2 price, for the firm, OP_2=ARC_2=MRC_2. This LRMC intersects the MRC_2 at J, which implies that, the firm is again at equilibrium at point J, thereby, it sells Oq_2 output at OP_2 price and earns extra (super-normal) profits, P_2JKL. But, under perfect market competition, the firms enjoy only normal profits in the long run instead of super-normal profits. This is because, if the existing firms enjoy extra (super-normal), it will attract new firms to enter the industry, thereby, the demand for factors increases and hence, their prices will increase in the market. Further, the existing market of the industry will be shared by more number of firms. So, the entry of new firms will increase supply of the commodity into the market, thereby, the price falls from OP_2 to original price OP. So, again the firms earn normal profits in the long run.
- ☆ At OP_4 price, for the firm, OP_4=ARC_4=MRC_4. The LRMC intersects the MRC_4 at M, which implies that, the firm is again at equilibrium at point M, thereby, it sells Oq_4 output at OP_4 price and incurs losses PP_4MN. These losses make some firms to leave the industry. So, this reduces the supply of the commodity into the market, thereby, the price gradually rises from OP_4 to original price OP. So, again the firms earn normal profits in the long run.

Long Run Price or Normal Price and the Laws of Returns

From the above discussion, we can infer that, long period in the industry is the period, that is sufficient enough to bring significant changes both in the fixed and variable factors to suit the changes in the demand for the product. The price ruling in the long period on the basis of LRS is called Long period price or Normal price. In the long period, the supply of the commodity into the market depends upon the cost of production of the commodity in the industry. So, in the long run, the cost of production of the commodity decides the normal price. If the normal price is above the cost of production, profits will be there in the industry, thereby, new firms enter the industry and the existing firms expand the output. So, due to increase in supply of output in the market, the price will fall and gets equal to the cost of production, thereby, the firms experience only normal profits in the long run. On the contrary, if the normal price is lower than cost of production, the industry will experience losses, thereby, some firms will leave the industry, supply of the commodity into the market decreases and gradually price increases and equals the cost of production and again the firms experience normal profits in the long run. So, the cost of production influences the normal price and this cost of production, in turn, depends upon the laws of returns, the industry experiences. So, we have to discuss the changes in normal price in the long run as influenced by the cost conditions in the industry. That means, whether the long run price is lower or same or higher than very short period (market) price and short period price, it all depends upon the laws of returns the business firm experiences. It is important that,

- ☆ if the business firm experiences law of increasing returns or law of diminishing costs, the long run price is less than market price.
- ☆ if the business firm experiences law of constant returns or law of constant costs, the long run price is equal to market price.
- ☆ if the business firm experiences law of diminishing returns or law of increasing costs, the long run price is higher than market price.

(a) Long Run Price and Law of Increasing Returns or Decreasing Costs

In case of law of increasing returns or decreasing costs, LRS curve falls downward from left to right. This is because, in the initial stage of development of an industry, there are external economies and hence, per unit cost of production of output will decrease. Hence, LRS curve slopes downward from left to right indicating that, more supplies can be made at a lower price because, as the cost of production decreases, more output can be produced. In Panel A of the Figure 19.12,

- ☆ At industry level, MPS is the MPS curve, SRS is the SRS curve and LRS is the LRS curve and it is downward sloping from left to right, when the firm experiences increasing returns or decreasing costs.
- ☆ DD is the demand curve and it intersects the MPS, SRS and LRS curves at P. This implies that, at OP=QP price, OQ is the quantity demanded and supplied in the industry. This OP=QP is the normal price, which is passed on to the individual firm (Panel B), where the firm earns normal profits, because price=LRATC.
- ☆ Suppose, the demand for the commodity is increased in the market, thereby, D_1D_1 is the new demand curve. This D_1D_1 demand curve intersects the MPS at P′, thereby, market period price rises sharply to QP′.
- ☆ In the short run, the firm can increase supply due to increased use of variable factors and hence, due to increase in supply (OQ_1), price gradually falls from QP′ to Q_1P_1. At this price Q_1P_1, the firm will earn super-normal profits, as price>SRATC. Hence, more firms will enter the industry and it leads to normal profits in the long run. Due to expansion of industry with the entry of new firms, external economies (increasing returns to scale) will prevail, thereby, the cost of production decreases.

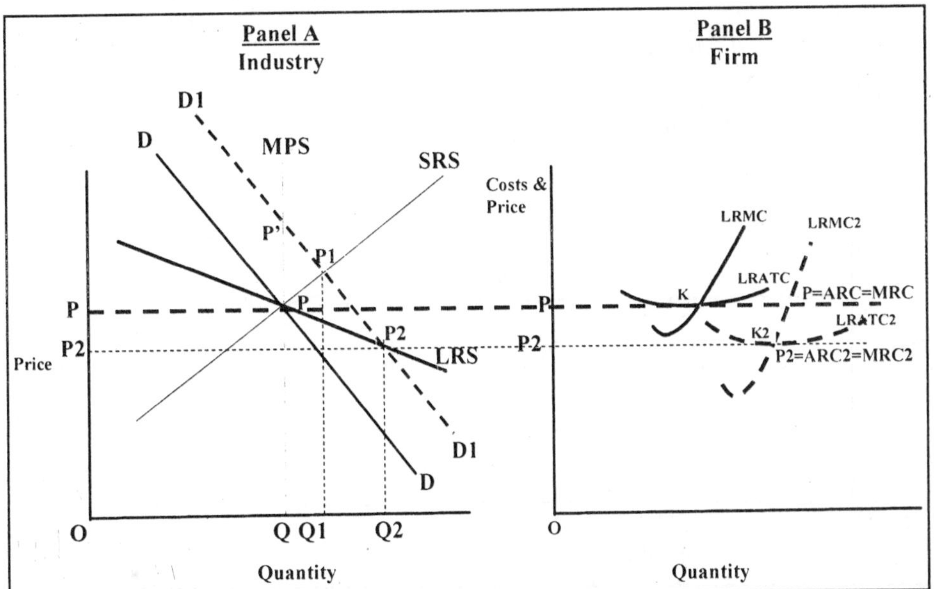

Figure 19.12: Long run price determination under perfect market competition and Law of increasing returns or decreasing costs.

- In the long run, the firm still increases the supply at low cost of production and hence, the LRS curve slopes downward from left to right.
- The D_1D_1 curve cuts the LRS curve at P_2, *i.e.*, at price Q_2P_2. That means, due to increase in supply (OQ_2) in the long run, the price falls from Q_1P_1 to Q_2P_2. This Q_2P_2 is also the normal price like QP, but due to expansion of output with the operation of increasing returns to scale and increase in number of firms in the long run. So, this normal price Q_2P_2 is below the original normal price QP.
- The prices OP and OP_2 determined in the industry due to changes in the supply and demand of the commodity during long run will be passed on to the individual firms and as shown in the Panel B, they are horizontal lines parallel to X-axis and these price lines coincides with their respective ARCs and MRCs of the individual firms.
- With the original price OP, we compare LRATC and LRMC curves. This LRMC curve will cut P=ARC=MRC from below at the point K, where the price of output is OP at which the firm earns only normal profits in the long run, as price OP=LRATC.
- Due to the operation of law of increasing returns or decreasing costs, the per unit cost curves will fall (due to increase in output) for a firm. The price Q_2P_2 is equal to the ARC_2=MRC_2 and it lies below the original P=ARC=MRC curve. The $LRMC_2$ curve intersects this new MRC_2 curve from below at point K_2, where the price of output is OP_2 at which the firm earns only normal profits in the long run, as price OP_2=$LRATC_2$.
- This implies that, when an industry is working under increasing returns or decreasing costs, an increase in demand for the commodity in the long run, leads to fall in long run normal price below the original normal price, but leads to increase in quantity supplied by the industry into the market due to increase in number of firms. That means, in case of decreasing cost industry, more quantity of output comes into the market at a lower price. The extent to which the long run normal price differs from original (normal) price depends upon the extent of costs decreased due to expansion of the industry.

(b) Long Run Price and Law of Constant Returns or Constant Costs

In case of law of constant returns or constant costs of an industry, LRS curve will be parallel to X-axis. This is because, external economies and external diseconomies counter balance each other, so that, the firms of an enlarged industry will not experience shift in cost curves. In Panel A of the Figure 19.13,

- At industry level, MPS is the MPS curve, SRS is the SRS curve and LRS is the LRS curve and it lies parallel to X-axis, when the firm experiences constant returns or constant costs.
- DD is the demand curve and it intersects the MPS, SRS and LRS curves at P. This implies that, at OP=QP price, OQ is the quantity demanded and supplied in the industry. This OP=QP is the normal price which is passed on to the individual firm (Panel B), where the firm earns normal profits, because price=LRATC.
- Suppose, the demand for the commodity is increased in the market, thereby, D_1D_1 is the new demand curve. This D_1D_1 demand curve intersects the MPS at P′, thereby, market period price rises sharply to QP′.

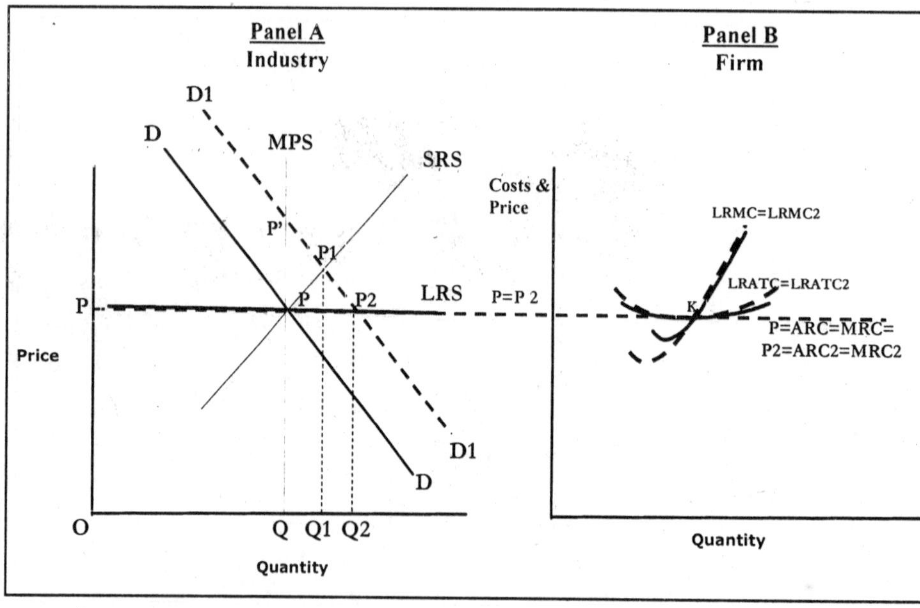

Figure 19.13: Long run price determination under perfect market competition and Law of constant returns or constant costs.

- ☆ In the short run, the firm can increase supply due to increased use of variable factors and hence, due to increase in supply (OQ_1), price gradually falls from QP′ to Q_1P_1. At this price Q_1P_1, the firm will earn super-normal profits, as price>SRATC. Hence, more firms will enter the industry and it leads to normal profits in the long run. Due to expansion of industry with the entry of new firms, the output supply increases in the long run. But, since the firms experience constant returns, LRS curve is a straight line parallel to the X-axis, thereby, cost of production remains same.
- ☆ The D_1D_1 curve cuts the LRS curve at P_2, *i.e.*, at price Q_2P_2. This Q_2P_2 normal price is equal to original normal price QP and it is due to operation of law of constant returns, but, the output increases from OQ to OQ_2 due to increase in number of firms in the long run.
- ☆ The same prices $OP=Q_2P_2$ will be passed on to the individual firms and as shown in the Panel B, it is a horizontal line parallel to X-axis and this price lines coincides with ARC and MRC of the individual firms.
- ☆ With this price $OP=Q_2P_2$, we compare LRATC and LRMC curves. This LRMC curve will cut P=ARC=MRC or $P_2=ARC_2=MRC_2$ curve from below at point K, where the firm earns only normal profits in the long run, as price Q_2P_2 (=OP)=$LRATC_2$ (LRATC).
- ☆ Due to operation of law of constant returns or constant costs, the per unit cost curves of the firms will remain on the same price line and there is no shift in per unit cost curves. But, the total output from the industry increases due to entry of new firms into the industry. As there is no change in cost structure, the firm produces the product at same price, thereby, $QP=Q_2P_2$, $ARC=ARC_2$ or $MRC=MRC_2$.
- ☆ This implies that, when an industry is working under constant returns or constant costs, an increase in demand for the commodity in the long run, leads to constant level of normal price even in the long run same as original normal price, but leads to increase in quantity supplied by the industry into the market due to increase in number of firms. That means, if the industry experiences constant returns or constant costs in the long run, simply the output from the industry increases due to increase in number of firms, but there is no effect on price, as the cost structure is not disturbed.

(c) Long Run Price and LDR or Increasing Costs

In case of LDR or increasing costs, LRS curve slopes upwards from left to right. This is because, when a well sized industry expands, as a result of the increased demand for the product in the long run, it experiences certain external economies and diseconomies. But, in the case of increasing cost conditions, diseconomies will outweigh economies and hence, the cost curves will shift upward. When the industry expands, the costs rises due to greater increase in the wages of specialized labour, costs of machinery and prices of raw materials due to the entry of new firms into the industry. So, in practical competitive situation, we will meet only diminishing returns or increasing costs condition. Hence, LRS curve slopes upward from left to right indicating that, large supplies can be made only at increasing costs (prices) because, the firm will experience diminishing returns or increasing costs. In Panel A of the Figure 19.14,

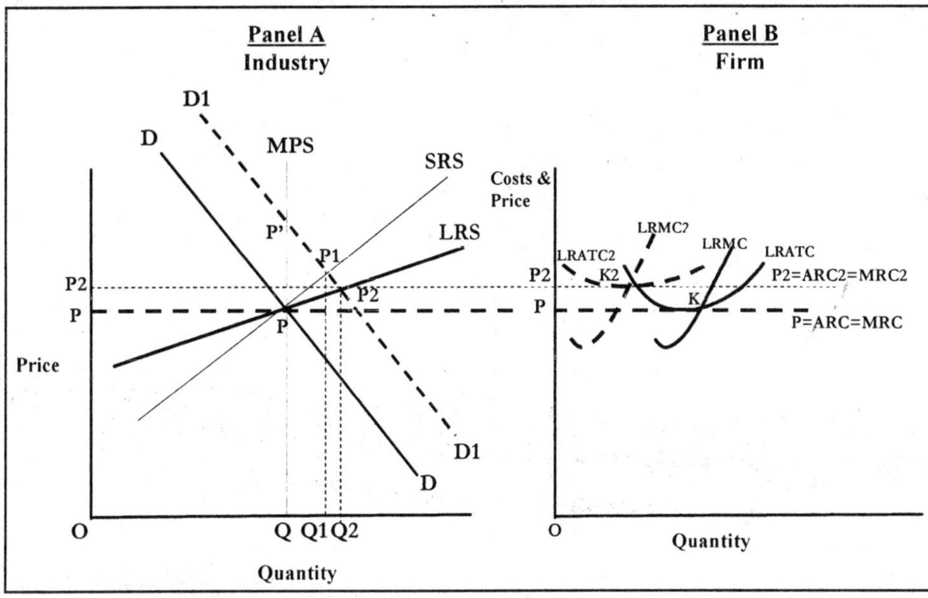

Figure 19.14: Long run price determination under perfect market competition and Law of diminishing returns or increasing costs.

- At industry level, MPS is the MPS curve, SRS is the SRS curve and LRS is the LRS curve and it is upward sloping from left to right, when the firm experiences diminishing returns or increasing costs.
- DD is the demand curve and it intersects the MPS, SRS and LRS curves at P. This implies that, at OP=QP price, OQ is the quantity demanded and supplied in the industry. This OP=QP is the normal price, which is passed on to the individual firm (Panel B), where the firm earns normal profits, because price=LRATC.
- Suppose, the demand for the commodity is increased in the market, thereby, D_1D_1 is the new demand curve. This D_1D_1 demand curve intersects the MPS at P', thereby, market period price rises sharply to QP'.
- In the short run, the firm can increase supply due to increased use of variable factors and hence, due to increase in supply (OQ_1), price gradually falls from QP' to Q_1P_1. At this price Q_1P_1, the firm will earn super-normal profits, as price>SRATC. Hence, more firms will enter the industry and it leads to normal profits in the long run. Due to expansion of industry with the entry of new firms, the firms will employ increased use of machinery and equipments and more employment of skilled labour. This will increase the costs and hence, the cost curves will shift higher, as the firm is working under increasing costs.
- In the long run, the increase in number of firms leads to increase in supply (at higher costs or prices) and hence, LRS curve will slope upwards from left to right.
- The D_1D_1 curve cuts the LRS curve at P_2, *i.e.*, at price Q_2P_2. That means, due to increase in supply (OQ_2) in the long run, the price falls from Q_1P_1 to Q_2P_2. This Q_2P_2 is also the normal price like QP, but due to expansion of output with the entry of new firms in the long run. So, this normal price Q_2P_2 is higher than the original normal price QP.
- The prices OP and OP_2 determined in the industry due to changes in the supply and demand of the commodity during long run will be passed on to the individual firms and as shown in the Panel B, they are horizontal lines parallel to X-axis and these price lines coincides with their respective ARCs and MRCs of the individual firms.
- With the original price OP, we compare LRATC and LRMC curves. This LRMC curve will cut P=ARC=MRC from below at the point K, where the price of output is OP at which the firm earns only normal profits in the long run, as price OP=LRATC.
- Due to the operation of law of diminishing returns or increasing costs, the per unit cost curves will rise (due to increase in cost) for a firm. The price Q_2P_2 is equal to the $ARC_2=MRC_2$ and it lies above the original P=ARC=MRC. The $LRMC_2$ curve intersects this new MRC_2 curve from below at point K_2, where the price of output is OP_2 at which the firm earns only normal profits in the long run, as price $OP_2=LRATC_2$.
- This implies that, when an industry is working under diminishing returns or increasing costs, an increase in demand for the commodity, in the long run, leads to rise in long run normal price above the original normal price, but leads to increase in quantity supplied by the industry into the market due to increase in number of firms. That means, in case of increasing cost industry, more quantity of output comes into the market only at

a higher price. The extent to which the long run normal price differs from original normal price depends upon the extent of costs increased due to expansion of the industry.

4. Secular Period

According to Alfred Marshall, secular period is a period, which is more than ten years, thereby, the firm can fully adjust its supply to changes in demand conditions. However, during this lengthy period, is quite difficult to predict the changes in demand conditions due to various factors, and hence, Marshall did not analyze the pricing mechanism under the secular period.

The above discussion reveals that, time element plays an important role in analysing which of the two forces *viz.*, demand and supply is important in price determination of the product. In general, the shorter the time, the greater the influence of demand on price and the longer the time period, the greater will be the influence of supply on price.

Comparison between Market Price and Normal Price

Having discussed the concepts of market price and normal price under the price determination of a commodity in very short period market and long period market respectively, it is essential to discuss the differences between these two and they are presented through Table 19.2.

Table 19.2: Differences between Market price and Normal price.

Market Price	*Normal Price*
Market price is the actual price that prevails in the market.	Normal price is only an imaginary or expected price when supply adjusts itself to changes in demand conditions for the product in the long run.
Market price is determined by the temporary equilibrium between the demand and supply forces at a particular point of time.	Normal price is determined by the permanent equilibrium between the demand and supply forces, when supply conditions are fully adjusted to the changes in demand conditions in the long run.
It is quite easy to estimate market price of a commodity, as it involves temporary adjustment of demand and supply forces.	It is quite difficult to estimate normal price of commodity, as it involves adjustment of supply to changes in demand over a long period of time.
Market price frequently fluctuates, as it is governed by temporary fluctuations in demand and supply forces.	Normal price is relatively stable, as it is governed by permanent causes in demand and supply forces. So, normal price is the centre around which the market prices fluctuate.
It is a very short period price.	It is a long period price.
It is more influenced by demand.	It is largely determined by supply and cost of production of commodity.
Market price may be higher or less than cost of production.	Normal price is equal to cost of production.
All commodities *viz.*, perishable, durable, reproducible and non-reproducible have a market price.	Only durable and reproducible commodities alone have a normal price. If commodities like pictures of old masters, unique diamonds, old manuscripts etc., that cannot be reproduced, they do not have normal price, but have only market price.
Market price has nothing to do with average cost, as the seller is prepared to sell the commodity even at a market price less than average cost.	Normal price is based on the average cost. Under perfect competition, in the long run, the seller will sell the commodity at a price equal to the LRATC and under imperfect competition (monopoly), the normal price is above the LRATC.
Market price may give loss or super normal profits to the seller.	Normal price always includes normal profits. That means, normal price do not include the element of 'loss'.

19.4. Equilibrium of the Firm and Industry Under the Perfect Market Competition

In the earlier Chapter 18, we discussed about the firm and the different approaches of determining the firm's equilibrium. Among the two approaches discussed earlier, MR and MC approach is found to be more superior to determine the firm's equilibrium. In this chapter, let us study in-detail about the equilibrium of both the firm and industry under perfect competition. The concept of equilibrium of the industry will arise only under perfect market competition, as there are infinite number of firms producing a homogenous good, thereby, all the firms can represent an industry. However, in case of imperfect market competition, with reference to monopoly, the concept of firm's equilibrium will be same as industry equilibrium, as there will be only firm producing the product and thereby, E_c is zero. Regarding oligopoly and monopolistic market competitions, the firms produce differentiated products, thereby the concept of 'industry equilibrium' will not arise in such competitions.

We discussed earlier, a firm refers to a single production unit, where resources and resource services are employed to produce an output for sale in the market. Industry refers to a group of firms producing a homogenous good. A firm is said to be in equilibrium, when it shows no tendency to change the output level *i.e.*, the firm neither expand the output nor contract the output. But, the industry is said to be in equilibrium, when all the firms in the industry are in equilibrium. That means, no new firm has the tendency to enter the industry or no existing firm will leave the industry, as all the existing firms in the industry are earning normal profits.

In case of perfect market competition, the concepts of firm's equilibrium and industry equilibrium can be studied both in short run and long run production programmes. So, we can divide our analysis into two components *viz.*,

19.4.1. Equilibrium of firm and industry under perfect competition in the short run.

19.4.2. Equilibrium of firm and industry under perfect competition in the long run.

19.4.1. Equilibrium of Firm and Industry Under Perfect Competition in the Short Run

We know, in a short run production programme, some factors are fixed and some are variable. If the firm wishes to increase the production or output, it can vary only the variable factors and the quantity of fixed factors cannot be changed. The firm can increase the output by allocating more of variable resources, overworking of the plant or machinery etc. No new machinery can be installed in the production programme. New firms cannot enter the industry in the short period. The price of the commodity is determined by total demand and total supply at the industry level and this price is passed on to the individual firms. So, the two market forces, which determine the price of the commodity under perfect competition are demand and supply at the industry level. So, under perfect competition, each firm is a price taker and not a price maker. On this given price, the firm plans its output policy considering its cost conditions. We can discuss the equilibrium of the firm and industry under perfect competition in the short run considering two cases *viz.*, identical cost conditions and varying cost conditions.

1. Equilibrium of the Firm and Industry in the Short Run Under Identical Cost Conditions

When we assume identical cost conditions, it implies, all the factors of production are homogenous, thereby, all the firms will incur same costs in producing the given output. So, there will not be any shift in the cost curves with respect to all the firms. As discussed earlier, each firm under perfect market competition is only a price taker, thereby, first the price is determined based on total demand and total supply forces at the industry level and the same price is passed on to the individual firms.

Case 1

In panel A of the Figure 19.15, DD is the initial demand curve for paddy in the industry, which slopes downward from left to right and SRS is the SRS curve of the industry representing the total quantity of paddy supplied into the market in short run production programme. The DD demand curve intersects the SRS supply curve at point 'E', which indicates OQ is the equilibrium quantity of paddy supplied and demanded in the market at OP equilibrium price. This equilibrium price OP is passed on to the individual firms, as they are simply price takers in the market.

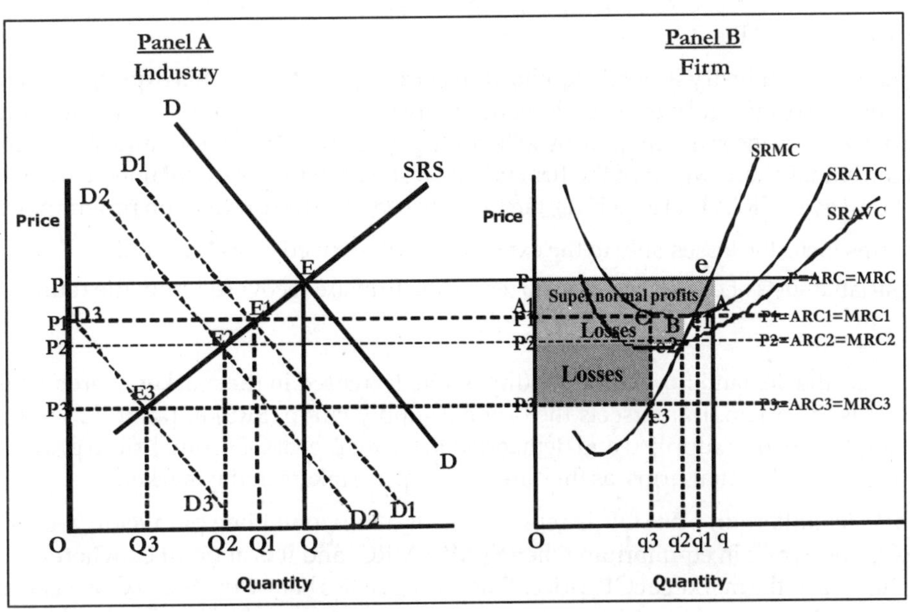

Figure 19.15: Equilibrium of firm and industry under perfect competition in the short run under identical cost conditions.

In Panel B of the Figure 19.15, a single firm is taken as a representative, as cost structure remains same across the firms under identical cost conditions. So, for the firm, the price of paddy is OP. We know, for a firm under perfect competition, the AR and MR curves are parallel to X-axis and MRC coincides with ARC. As usual, the per unit cost curves *viz.*, SRAVC, SRATC and SRMC are 'U' shaped and there is no shift in these cost curves, as cost structure remains same across the firms. The price OP passed on from the industry to the individual firm is represented as a horizontal line parallel to X-axis and under perfect competition, price line coincides with AR and MR curves.

The firm is in equilibrium when SRMC=MR and it is at point e, where Oq quantity of paddy is transacted in the market at OP price. This price OP is higher than SRATC curve by eA, thereby, the firm earns eA profits per unit of output. So, total profits = total output sold by the firm x average profit per unit of output = Oq x eA = $PeAA_1$

As the firm earns price more than SRATC, $PeAA_1$ refers to super-normal profits realized by the firm. So, in the short run, the firm can earn super normal profits under perfect competition. These super normal profits are earned by all the firms in the industry, each by producing and selling Oq quantity of output at OP level of equilibrium price. Attracting to these super-normal profits, new firms show the tendency to enter the industry in the long run, thereby in the long run, these super-normal profits gets eroded and it is discussed in the ensuing pages.

Case 2

Let us suppose, the demand for the commodity is decreased in the market, thereby, the new demand curve D_1D_1 move towards the origin (Panel A). It intersects the original supply curve (SRS) at point E_1, which indicates OQ_1 is the equilibrium quantity of paddy supplied and demanded in the market at OP_1 equilibrium price. This equilibrium price OP_1 is passed on to the individual firms, as they are simply price takers in the market.

In Panel B, the equilibrium price OP_1 is represented as a horizontal line parallel to X-axis and it coincides with ARC_1 and MRC_1. The firm is in equilibrium when $SRMC=MRC_1$ and it is at point e_1, where Oq_1 quantity of paddy is transacted by the firm in the market at OP_1 price. This price OP_1 is equal to SRATC, thereby, the firm earns only normal profits. In this situation, the industry is also in equilibrium, as no new firms enter the industry and the existing firms show no tendency to leave the industry.

Case 3

Let us suppose, the demand for the commodity is still decreased in the market, thereby, the new demand curve D_2D_2 move towards the origin. It intersects the original supply curve (SRS) at point E_2, which indicates OQ_2 is the equilibrium quantity of paddy supplied and demanded in the market at OP_2 equilibrium price. This equilibrium price OP_2 is passed on to the individual firms, as they are simply price takers in the market.

In Panel B, the equilibrium price OP_2 is represented as a horizontal line parallel to X-axis and it coincides with ARC_2 and MRC_2. The firm is in equilibrium when $SRMC=MRC_2$ and it is at point e_2, where Oq_2 quantity of paddy is transacted by the firm in the market at OP_2 price. This price OP_2 lies below the SRATC, but above SRAVC inferring that, the firm is incurring losses to the extent of less than fixed costs incurred in the business. This is shown as an area between SRATC and SRAVC curves *i.e.*, $P_1P_2e_2B$. Thus, total loss $P_1P_2e_2B$ is given by,

Total loss = total output sold by the firm x average loss per unit of output

$$= Oq_2 \times e_2B = P_1P_2e_2B$$

In this situation, the industry is not in equilibrium because, when the firms experience losses, some of the firms will quit the industry. So, the total supply of the commodity by the industry into the market decreases, thereby, price will rise in the market. So, the remaining firms will gradually recover their losses and start earning normal profits at OP_1 price and the equilibrium output of the firm will be restored at Oq_1. As explained earlier, when the price of the commodity at least covers SRAVC (Table 15.1), the firm can continue the production programme temporarily because,

- ☆ The firms incur the losses only to the extent of less than fixed costs incurred
- ☆ The variable costs or explicit costs incurred by the firm are recovered from the returns of the business firm.

Case 4

Let us suppose, the demand for the commodity is still decreased in the market, thereby, the new demand curve D_3D_3 move towards the origin. It intersects the original supply curve (SRS) at point E_3, which indicates OQ_3 is the equilibrium quantity of paddy supplied and demanded in the market at OP_3 equilibrium price. This equilibrium price OP_3 is passed on to the individual firms, as they are simply price takers in the market.

In Panel B, the equilibrium price OP_3 is represented as a horizontal line parallel to X-axis and it coincides with ARC_3 and MRC_3. The firm is in equilibrium when $SRMC=MRC_3$ and it is at point e_3, where Oq_3 quantity of paddy is transacted by the firm in the market at OP_3 price. This OP_3 price lies below the SRATC at a gap e_3C, inferring that, the firm is incurring huge losses to the extent of TFC and even some of the SRTVC. This total loss is shown by the area, $P_3e_3CP_1$ and it is given by,

Total loss = total output sold by the firm x average loss per unit of output

$= Oq_3 \times Ce_3 = P_3e_3CP_1$

In this situation, the firms will shut down their business, as even their SRTVC are not recovered besides TFC from the returns of the business.

2. Equilibrium of the Firm in the Short Run Under Varying Cost Conditions

In majority of the cases, the factors of production are heterogenous, thereby, the cost of production vary across the firms. We know, under perfect competition, price is determined by the industry and the same is passed on to the individual firms, as they are simply price takers. That means, the firms have to sell their output at the given price. As a result, the firms, which produce the output at low cost of production (price>SRATC) enjoy super-normal profits, some firms for which the price equals to SRATC enjoy only normal profits and some firms for which price<SRATC and price<SRAVC, suffers from losses. The same concept is explained through Figure 19.16, where the firms with different cost structures are shown due to heterogenous nature of factors of production and assume OP is the price determined for paddy in the industry and the same is passed on to the individual firms. The Figure 19.16 highlights the comparative picture with reference to the extent of profits earned and extent of losses experienced by the different firms at a given price.

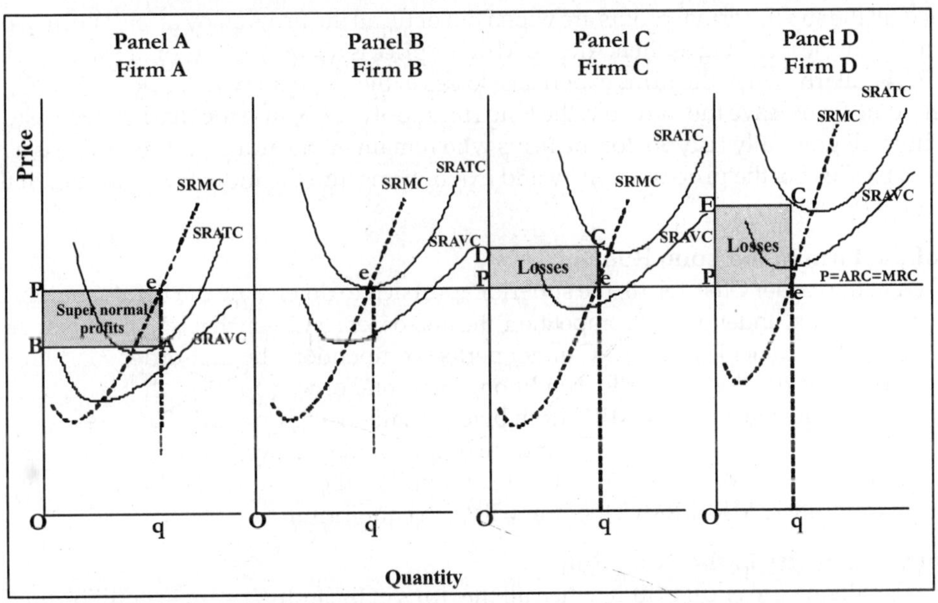

Figure 19.16: Equilibrium of firms under perfect competition in the short run under varying cost conditions.

In Panel A, the firm A is in equilibrium, when SRMC=MR *i.e.*, at 'e', which indicates that, the firm transacts 'Oq' quantity of paddy at OP level of price. As the price OP is above the SRATC by eA, the firm earns super normal profits given by,

Total profits = total output sold by the firm x average profit per unit of output

= Oq x eA = PeAB

In Panel B, the firm 'B' is in equilibrium, when SRMC=MR *i.e.*, at 'e', which indicates that, the firm transacts 'Oq' quantity of paddy at 'OP' price. A close examination of the Panel B reveals that, its cost structure is higher compared to firm 'A'. As the price OP is equal to the SRATC, the firm B earns only normal profits.

In Panel C, the firm C is in equilibrium, when SRMC=MR, *i.e.*, at 'e' which indicates that, the firm transacts 'Oq' quantity of paddy at 'OP' price. On comparison, the cost structure of the firm 'C' is higher than firm B and firm A. As the price OP is less than SRATC, but higher than SRAVC, the firm can continue the business temporarily, as the extent of losses to the firm are less than TFC. The total loss is given by,

Total loss = total output sold by the firm x average loss per unit of output

= Oq x eC = PeCD

Though the firm experience losses, since the price OP covers at least SRAVC, the firm can continue production programme temporarily.

In Panel D, the firm D is in equilibrium, when SRMC=MR, *i.e.*, at 'e' which indicates that, the firm transacts 'Oq' quantity of paddy at 'OP' price. The cost structure of firm D is higher compared to earlier three firms A, B and C, thereby, firm 'D' is the least efficient firm. As the price OP is less than SRAVC, the firm experience greater losses (compared to firm C) *i.e.*, PeCE, as it could not recover TFC and even some of the SRTVC. So, the firm will shut down the production programme and wait for better times.

The above discussion clearly infers that, since the cost structure varies across the firms, some firms will enjoy profits and some firms experience losses in the industry. So, the industry under different cost conditions will not be in stable equilibrium in the short run and the industry will be in stable equilibrium only in the long run. In the words of Ferguson and Kreps, *'in the short run, the number of firms in an industry cannot be changed. Each firm attempting either to maximize profit or minimize loss, produces that output, for which marginal cost equals the market equilibrium price. At this point, the firm may earn a pure economic profit, it may earn a pure economic loss or it may happen to break even'.*

19.4.2. Equilibrium of Firm and Industry Under Perfect Competition in the Long Run

As discussed earlier, in the long run, all the factors are variable and no factor is fixed in the production programme. That means, if the firm wishes to expand or contract the output, it can vary all the factors and the firm can move from one plant to another plant. When the firms earn super-normal profits, as discussed under Panel B of the Figure 19.15, new firms gets attracted to these super normal profits and enter the industry. The entry of new firms into the industry will be continued till the super-normal profits are wiped out or till all the firms enjoy at least normal profits. So, the firms will continue in the business as long as their ARC or MRC or price is above LRATC or at least equal to LRATC, but not less than LRATC. Similarly, when the firms experience losses in the long run (when price<LRATC), they will leave the industry. When some firms leave the industry, the long run supply (LRS) into the market decreases, thereby, the price of the commodity will gradually rise. So, for the firms who remain in the industry, their price is equal to LRATC and they earn normal profits. So, the process of entry and exit of firms into the industry continues in the long run till the price equals to LRATC.

Equilibrium of the Firm in the Long Run

As discussed in the earlier Chapter 18, the superior approach to determine the firm's equilibrium is MR and MC approach. So, in the long run, under perfect competition, the first-order condition that is required for the firm's equilibrium is LRMC=MR. Since, MRC coincides with ARC under perfect competition, the first-order condition can also be written as LRMC=MRC=ARC. Besides this first-order condition, the second order condition that is necessary for the firm to be in equilibrium is the LRMC must cut the MRC from below, while rising upward. That means, the LRMC should be rising at the point of determination of firm's equilibrium output. If MR>LRMC, it implies, the firm can still expand the output till the point of MR=LRMC and if MR<LRMC, it implies, the firm must reduce the output till the point of MR=LRMC and at the point of MR=LRMC, the firm will be in equilibrium.

Equilibrium of the Industry in the Long Run

An industry is said to be in equilibrium, when all the firms in the industry are in equilibrium. That means, at the point of equilibrium in the industry, no new firms show the tendency to enter the industry and no firms in the industry show the tendency to leave the industry. This is because, at the point of equilibrium in the industry in the long run, all the firms earn only normal profits. That means, the possibility of earning super-normal profits in the long run is completely wiped out and all the firms experience only normal profits. This is because, when the firms experience super-normal profits in the short run, new firms gets attracted to these super-normal profits and enter the industry, thereby,

- ✰ Competition increases for the employment of factors of production in the long run, thereby, factor prices will increase and this will reduce profits for all the firms in the long run.
- ✰ When large number of firms enters the industry, the existing market is to be shared among all the firms in the industry, thereby, profits share will decline and hence, only normal profits are experienced instead of super-normal profits.
- ✰ With the entry of new firms, the supply of the product into the market increases, thereby, price decreases and hence, all the firms will enjoy only normal profits instead of super normal profits.

Thus, in the long run, all the firms in the industry experience normal profits only instead of super-normal profits and thereby, at the point of equilibrium, no new firms shows the tendency to enter the industry and existing firms in the industry show no tendency to leave the industry. So, normal profits are the profits for a firm in the industry, which are just sufficient to induce the firm to continue in the industry and do not induce new firms outside the industry to enter the industry. So, when all the firms in the industry experience normal profits only in the long run, the point of equilibrium is LRMC = MR = AR = Price = LRATC.

As in short run, we can study the equilibrium of the firm and industry under perfect competition in the long run under two conditions *viz.*, identical cost conditions and varying cost conditions.

1. Long Run Equilibrium of the Firm and Industry Under Identical Cost Conditions

As discussed earlier, when we assumed identical cost conditions, it implies all the factors of production are homogenous, thereby, all the firms will incur same costs in the industry to produce a given output. So, there will not be any shift in the cost curves with respect to all the firms. We know, each firm under perfect market competition is only a price taker, thereby, first the price is determined based on total demand and total supply of paddy in the industry and the same price is passed on to the individual firms.

In Panel A of the Figure 19.17, DD is the initial demand curve of the industry for paddy and SS is the initial supply of paddy in the industry in the long run production programme. The demand curve DD intersects the supply curve SS at point E, where OQ quantity of paddy is supplied and demanded in the industry at OP long run equilibrium price. This price OP is passed on to the individual firms, as they are simply price takers in the market.

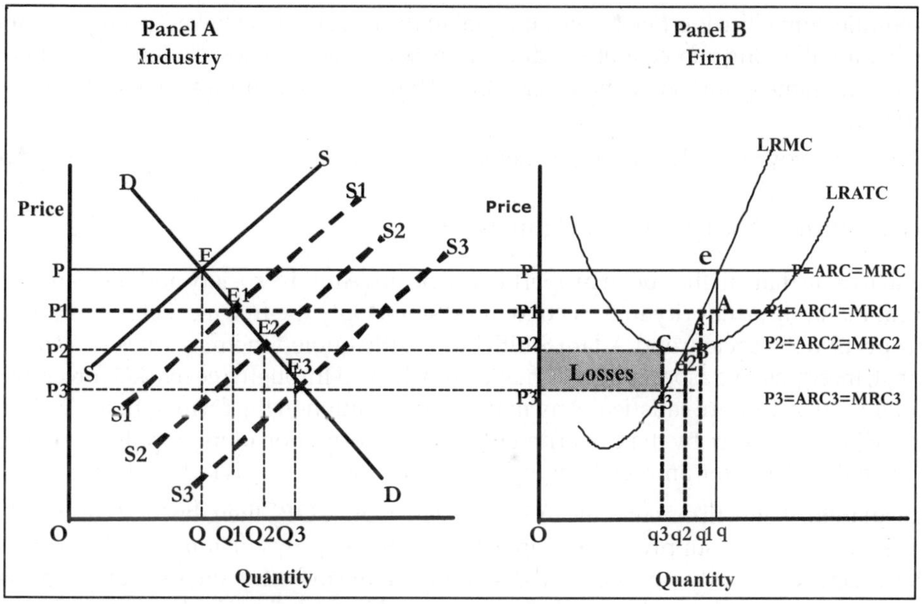

Figure 19.17: Equilibrium of firm and industry under perfect competition in the long run under identical cost conditions.

In Panel B of the Figure 19.17, a single firm is taken as a representative, as cost structure remains same across the firms under identical cost conditions. So, for the firm, the price of paddy is OP. We know, for a firm under perfect competition, MRC and ARC curves are parallel to X-axis and MRC coincides with ARC. As usual, the per unit cost curves LRATC and LRMC are 'U' shaped and there is no shift in these cost curves, as cost structure remains same across the firms. The price OP passed on from the industry to each individual firm is represented as horizontal line parallel to X-axis, as under perfect competition, price line coincides with ARC and MRC. The firm will be in equilibrium when the LRMC=MRC(=ARC=OP) and it is at point 'e', where Oq is the quantity of paddy transacted by the firm at OP long run equilibrium price. This price OP is higher than LRATC by eA, thereby, each firm will earn super-normal profits, $eAPP_1$. But, this is not the point of stable equilibrium. This is because, attracted to these super-normal profits, new firms will enter the industry and thereby, the total supply of paddy into the market by all the firms will increase. Accordingly S_1S_1 is the new (increased) supply curve for paddy in the market with the entry of new firms (Panel A).

With increase in supply of paddy in the industry, the new S_1S_1 supply curve will intersect the original demand curve DD at E_1, where, OQ_1 quantity of paddy supplied and demanded in the industry at OP_1 long run equilibrium price. This price OP_1 is passed on to the individual firms, as they are simply price takers in the market. The OP_1 price is shown as a horizontal line parallel to X-axis and it coincides with MRC_1 and ARC_1 for the firm under perfect competition. The point of equilibrium is when $LRMC=MRC_1(=ARC_1=OP_1)$ and it is at point e_1, where Oq_1 is the quantity of paddy transacted by the firm at OP_1 long run equilibrium price. Again, this price OP_1 is higher than LRATC by e_1B, thereby each firm will earn super-normal profits, $e_1BP_1P_2$. But again, this is not the point of stable equilibrium, as new firms gets attracted to these super normal profits and enter the industry, thereby, the total supply of paddy into the market by all the firms will increase from S_1S_1 to S_2S_2 (Panel A).

So, with increase in supply, the new supply curve S_2S_2 shifts towards right side and it will intersect the original demand curve DD at E_2, where OQ_2 quantity of paddy supplied and demanded in the industry at OP_2 long run equilibrium price. This price OP_2 is passed on to the individual firms, as they are simply price takers in the market. The OP_2 price is shown as a horizontal line parallel to X-axis and it coincides with MRC_2 and ARC_2 for the firm under perfect competition. The point of equilibrium is when LRMC=MRC_2(=ARC_2=OP_2) and it is at point e_2, where Oq_2 quantity of paddy is transacted by the firm at OP_2 long run equilibrium price. This price OP_2 is equal to the minimum of LRATC of the firm, which implies each firm will now earn only normal profits in the long run. So, at the point of firm's equilibrium *i.e.*, 'e_2' in the long run, when all the firms earn normal profits, the equilibrium condition is LRMC=MRC_2=ARC_2=OP_2=LRATC. At this equilibrium condition, all the firms will earn only normal profits only and the industry will also be in stable equilibrium. So, to generalize, the long run (stable) equilibrium for the firm (when all the firms experience normal profits only) is given by, LRMC=MRC=ARC=Price=LRATC. This general equilibrium condition in the long run prevails only, when the price equals the LRATC of the firm. So, the firm in the long run must produce the output at minimum LRATC.

Several central properties can be derived from this perfect competitive long run equilibrium (price=LRATC) and they are important from the normative economics point of view. They include,

- ☆ This equilibrium price is efficient, as it is equal to the LRATC at its lowest point. That means, at this point of equilibrium, the long run cost of production of paddy is at its lowest level and there is no other way to produce the same commodity that is cheaper. That is, there is no wastage of resources in the production programme.
- ☆ All the firms enjoy normal profits only, thereby, consumer welfare is maximized, as they pay only production costs.
- ☆ SS is maximized (Discussed under Figure 17.18).

It is essential to note that, all the above properties are positive for the society and they are not achieved due to the actions of Central Planning Authority in the economic system. Only the free market forces are solely responsible for these positive aspects in the society and Adam Smith has rightly quoted them as *'Invisible Hand'*. Of course, it is to be borne in mind that, these positive aspects are automatically achieved through free market forces, only if the assumptions formulated under Section 19.1 are satisfied. Any market intervention in the form of fixing ceiling price (as discussed under Figure 17.35) or floor price by the Government or by the formation of monopolies or collusive oligopolies will lead to reduction in the SS and the amount by which the SS or total welfare is reduced refers to deadweight loss.

If by miscalculation, if new firms enter the industry even at price OP_2, then the total supply of paddy in the market increases and it is shown by S_3S_3 supply curve (Panel A). This S_3S_3 supply will intersect the original demand curve DD at E_3, where OQ_3 quantity of paddy supplied and demanded in the industry and OP_3 long run equilibrium price. This price OP_3 is passed on to the individual firms, as they are simply price takers in the market. The price OP_3 is shown as a horizontal line parallel to X-axis and it coincides with MRC_3 and ARC_3 for the firm under perfect competition. The point of equilibrium is when LRMC=MRC_3(=ARC_3=OP_3) and it is at point e_3, where Oq_3 quantity of paddy is transacted by the firm at OP_3 long run equilibrium price. This price OP_3 is less than LRATC, thereby, the firm will incurr loss to the extent of e_3C per unit of output, thereby, it will shut down the production programme permanently. We know, no firm can stay in the production programme in the long run, if it makes losses. So, the new firms will not enter the industry, when the industry is in equilibrium, where OP_2 price is equal to minimum of LRATC for all the firms *i.e.*, when all the firms enjoy the normal profits only.

2. Long Run Equilibrium of the Firms Under Varying Cost Conditions

In general, the cost of production varies across the firms, as the factors of production are heterogenous. We know, price of the commodity is determined by the industry based on the interaction between total demand and total supply of the commodity and this equilibrium price is passed on to the firms, as they are simply price takers. So, the firms can determine the level of output in the long run considering the price offered by the industry. Depending upon the level of price offered by the industry, the firm can achieve either super-normal profits or normal profits in the long run depending upon its cost structure. That means, if the firms enjoy low cost structure, they can experience super-normal profits based on the price offered by the industry, *i.e.*, when price>LRATC of the firms. On the contrary, if the price offered by the industry equals the LRATC of the firm, the firms experience normal profits only. It is essential to note that, the firms experiencing losses (price<LRATC) in the long run, will quit the industry.

In Panel A of the Figure 19.18, let us assume, OP is the price offered by the industry and hence, firm A accepts this price. This price OP is represented as a horizontal straight line parallel to X-axis, as for the firm under perfect competition, the price=ARC=MRC. As usual, the LRATC and LRMC are 'U' shaped curves and the point of firm's equilibrium is LRMC=MR, which is at point 'e', where Oq quantity of paddy is transacted by the firm in the market at OP long run equilibrium price. Since, OP price>LRATC, the firm experiences super-normal profits PeAB. That means, if the firm maintains lower cost structure through efficient management, it can enjoy super normal profits.

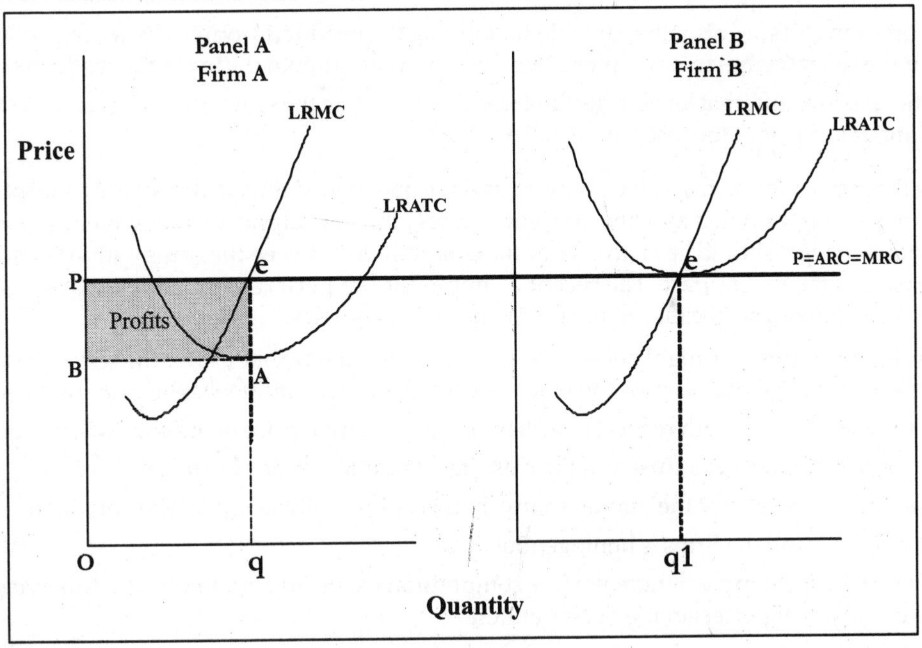

Figure 19.18: Equilibrium of firm and industry under perfect competition in the long run under varying cost conditions.

In Panel B, the cost structure of firm B is higher compared to firm A. So, at the same long run equilibrium price OP, the firm B experiences only normal profits, as LRMC=MRC=ARC=Price=LRATC.

19.5. Is Perfect Competition a 'Reality' or a 'Myth'

From the concepts discussed above, we can infer that, the perfect competition assumes extreme perfections in all aspects pertaining to the market behaviour of the firms and industry. But, in the practical world, these assumptions or features are violated in one form or the other. The different violations of the features of perfect competition are as follows:

- In real life, no two products are alike. That means, the same product offered by different firms is not homogenous in all aspects. For example, paddy is considered as a homogenous commodity under perfect competition, but paddy grains differ in their size, shape, colour, moisture content etc., across the varieties cultivated by different farmers. So, no product is homogenous across the firms and there is 'differentiation' across the products. In many industries, even firms go to great efforts to differentiate their products from others and this facilitate the firms to attract different customers with different tastes and preferences and able to charge higher prices.
- Many industries have significant barriers to entry such as, high start up or establishment costs, Government licensing etc., which limit the ability of firms to enter and exit such industries.
- The agri-business probably comes closest to exhibit the features of perfect competition, but still in the field of marketing, the buyers are well informed about the future trends in prices when compared to large chunk of small and marginal farmers.
- Government intervention in the fields of production and marketing of agricultural commodities indispensable, especially in case of developing countries and least developed countries, thereby, Governn regulates demand and supply of commodities.
- Perfect mobility of factors of production across the enterprises is not so easy, as it involves various obsta like, a labour performing skilled work in a jaggery processing industry cannot perform with same efficie in a fruit processing plant, family attachments, language problem, suitability of climate in a new area of etc.
- A single seller can influence the price of the commodity, at times of scarce production of the commodity similarly, a single buyer can also exert influence on the price of the commodity due to changes in his inco tastes and preferences etc.
- The output-price policy decisions of one firm is not known to other firms, if the firms are operating in distant areas and that too, pricing policy of a firm depends upon its ATC rather than the price offered by other fi Similarly, no buyer knows the price at which other buyers have purchased the product. So, this informat gap is indispensable, as the market involves infinite number of buyers and sellers.

- ☆ Transportation costs are also unavoidable for moving the products from one place to another and they count much, if the seller wishes to have inter-state trade or international trade for his product(s).
- ☆ As the firms adopt different levels of technologies in producing the same product, their cost structure definitely differs and it influences the price of the product.

Considering the above violations, many Economists have questioned the validity of the concept of perfect market competition and they strongly opined that, this concept is purely theoretical and it has no practical relevance. However, they still consider this concept should be analyzed, as the competition between the firms within an industry of different sectors has increased in the recent past. The recent example for the perfect competitive market is 'internet service providers' and it is considered so because of the following characteristics:

- ☆ The service providers face highly elastic demand curves, thereby, they offer the service at very low price. As the number of service providers is too many, each seller cannot influence the price of the service.
- ☆ The E_C for one's service with respect to a change in the relative price of another will be very high.
- ☆ New firms freely enter the industry, if the existing firms make super-normal profits.
- ☆ Since the firms compete for the market share, their economic efficiency will be very high.
- ☆ Service offered by all the firms is homogenous.

Besides above striking example, where perfect competition exists on a real note, the following reasons further recommend for the analysis of perfect market competition:

- ☆ The analysis of perfect market competition serves as a backdrop representing an ideal marketing situation
- ☆ Analyzing the perfect market competition assuming all perfections in a market guides us to clearly understand the complicated imperfections involved in imperfect market structures.
- ☆ The analysis of an imperfect market competition helps us to compare the extent of imperfections and degree of inefficiency under imperfect market structures compared to perfect market competition.

19.6. Advantages of Perfect Competition

Perfect competition is considered desirable for the society, as it offers the following advantages:

- ☆ There is consumer's sovereignty in a perfect competitive market.
- ☆ In the perfectly competitive market, the price=MC, thereby, it leads to consumer's welfare.
- ☆ As the products are homogenous, there is no need for the firm to incur huge expenditure on non-price competition.
- ☆ In the long run, the perfectly competitive firm produces the optimum output, *i.e.*, at the minimum point of LRATC and this infers that, there is no idle or unused or excess capacity.
- ☆ Due to lack of transportation costs, the product offered by the firm is at low cost and this ensures consumer's welfare.
- ☆ Due to free entry and exit of firms into the industry, there exists stiff competition among the firms and this leads to low price for the product at the consumer's level.
- ☆ As there is perfect knowledge with respect to products and price information among the sellers and buyers, it ensures prevalence of stable price for the product in the market and this benefits both the traders and consumers.

19.7. Disadvantages of Perfect Competition

The following are the demerits of free or perfect competition:

- ☆ There is no scope for economies of scale, as there is infinite number of firms, each producing insignificant share in total output.
- ☆ Industries with huge amounts of fixed or sunk costs would be particularly unsuitable under perfect competition.
- ☆ Lack of product variety offers little or no choice to consumers, as they generally prefer differentiated products suiting to their tastes and preferences.
- ☆ In the long run, firms earn normal profits only and supernormal profits are not possible, thereby, makes R&D investments less attractive in the industry.
- ☆ Because of perfect knowledge among the firms, there is no incentive to develop new technology.
- ☆ There will be insufficient profits for investment

- ☆ The market system allocates more goods and services to those consumers, who have more money than others
- ☆ Factors of production will be employed, if only, it is profitable to do so.
- ☆ Certain public goods such as roads, bridges, street lighting etc., may not be produced because, companies may not find it sufficiently profitable.
- ☆ Firms may produce goods that are in demand and they may be harmful to the customers or society as a whole (Ex. Pesticides)
- ☆ In order to maximize profits, the business firms might not consider the potentially negative social impacts of their products, Ex. Misuses of the internet for illegal or illicit purposes.
- ☆ Suffers from instability and leads to greater income inequality.

20

Imperfect Competition

In the earlier Chapter 19, we discussed in-detail about the perfect market competition, where infinite number of buyers and sellers characterize the market and no single seller or buyer has the power to influence the market price. Clearly the assumptions of perfect competition do not hold good in the vast majority of markets. This is because, some firms may exert some control over market supply and seek to exploit their monopoly power in the market. On the demand-side, some consumers may have monopsony power against suppliers because, they purchase a significant percentage of total demand. Further, they have imperfect market information and their preferences and choices can be influenced by the advertisement and other promotional measures of the firms. In every industry, there is always asymmetric information, where the sellers know more about the product and its features than the buyers. So, in the practical world, there are several imperfections in the product market and this may extend to related markets such as, the market for essential raw materials, labour and capital goods. So, perfect markets are far from reality and imperfect markets are more common. A market is said to be imperfect, if the actions of one or more sellers or buyers have an influence in determining the equilibrium price. So, the basic features of imperfect competition greatly differ from the perfect competition and they are listed below.·

- ☆ Firm exercises some influence over the price of the product. This influence is present when either the number of firms is one (monopoly) or few (oligopoly) or many (monopolistic).
- ☆ Products produced by the firms are differentiated or heterogenous. However, in monopoly, the product is unique and there are no close substitutes for the product.
- ☆ Demand curve or ARC facing an individual firm is downward sloping. This implies that, when a firm is operating under imperfect competition, it sells more quantity of its output only at lower prices and when it sells less quantity, its price will rise.
- ☆ Restrictions on free entry and exit of firms.
- ☆ Existence of transportation costs.
- ☆ Lack of perfect knowledge on the part of sellers and buyers regarding the price information.

So, any deviation from the conditions of perfect competition in a market leads to the existence of imperfect competition. This imperfect competition takes several forms generally based on the number of firms or sellers in the market and the important forms include,

☆ *Monopoly:* It is a market with only one seller and a large number of buyers.

☆ *Oligopoly:* It is a market characterized by few sellers, who are mutually interdependent. The product may be homogeneous or differentiated.

☆ *Monopolistic competition:* It is a market with many sellers, who compete to sell differentiated products and where free entry and exit is possible.

Let us discuss each type of imperfect market structure in-detail and in the ensuing chapter, the monopoly market is analyzed for its features and output-price determination.

21

Monopoly

As discussed earlier, all markets are 'competitive' to one degree or another, but the vast majority of markets are characterized as 'imperfectly competitive'. In Economics, a monopoly (from Greek words, *monos* means single + *polein* means to sell) exists, when a single firm has significant control over a particular product or service in the market. So, a monopoly firm is characterized by a lack of economic competition for the product or service from other firms, as the product or service offered by the firm have zero E_C. It is an extreme form of imperfect competition, where single seller have total control on the market through regulating the supply or price of the commodity. So, the single seller can dictate the price of the commodity to the buyers. In the UK, a firm is said to have Monopoly status, if it has more than 25 per cent of market share The barriers of entry for new firms into the monopoly industry are extremely strong, thereby, no competition exists in the monopoly industry and the single firm itself represents an industry. The number of buyers is assumed to be large for the product and their individual actions will not influence the price of the product. Some of the popular definitions of monopoly are given below.

'Pure monopoly is a market situation in which a single firm sells a product for which there is no good substitute'

– ***(Leftwhitch)***

'Under pure monopoly there is a single seller in the market. The monopolist's demand is market's demand. The monopolist is a price maker. Pure monopoly suggests a non-substitute situation'.

– ***(A.J.Braff)***

'Monopoly is market structure in which there is a single seller, there are no close substitutes for the commodity it produces and there are barriers to entry'.

– ***(Koutsoyiannis)***

'Pure or Absolute monopoly exists when a single firm is the sole producer of a product for which there are no close substitutes'.

– ***(McConnell)***

21.1. Sources of Monopoly Power

The different barriers that are responsible for the firm to enjoy the monopoly status in the market economy include,

☆ ***Natural resources:*** Monopoly arises due to naturally available resources in a particular area. So, the use and control of natural resources that are critical to the production of a final product ensures monopoly status to the firm. Examples include, South Africa has the monopoly of diamonds, Canada enjoys monopoly status with reference to nickel element, Middle East on the oil resources etc.

- ***Technological superiority***: Sometimes, monopoly arises due to technical reasons and such technicalities in the production of a product are not enjoyed by other firms. For example, the firm producing Coco Cola soft drink enjoys monopoly status for the formula involved in the product.
- ***Legal protection:*** The firms can even claim for legal protection in the form of Patents, Copy Rights, Trade Secrets etc., for the products, so as to enjoy the monopoly gains in the market. Through having legal protection, monopolist can gain exclusive control over the production and selling of certain products. For some firms, they enjoy legal protection over the materials or resources necessary to produce a product. For example, Monsanto Company enjoys patent protection with respect to *Bt* Cotton seed.
- ***Natural barriers to entry:*** For some products, the production processes involve huge amount of capital requirements, R&D costs, substantial sunk costs, which the small and medium firms could not afford to invest. In such cases, the large firm that is capable of producing that product enjoys monopoly status in the industry, as it is difficult for small and medium firms to enter the industry and expand. Examples include, public utilities like railway, electricity etc., and these are undertaken by the State. Economists, generally, refer to these monopolies as Natural monopolies.
- ***Deliberate actions:*** Some firms through their deliberate actions eliminate market competition. For example, the firms may collude, so as to enjoy monopoly gains in the market.
- ***Location advantages***: Due to specific geographic locations or agro-climatic conditions or special skills of the people in a particular location, the product may have uniqueness in the market. For example, Gadwal sarees from Mahboobnagar district, Nirmal toys from Adilabad district in Andhra Pradesh enjoy trade protection for their uniqueness in the market. A popcorn shop in a cinema theatre is another example.
- ***Quality advantage:*** A firm may produce the product with superior quality may be due to quality resources, superior technology, skills of the labour etc., thereby, they compete out all rivalry firms from the market and enjoy monopoly status.
- ***Economies of scale:*** Monopoly firm enjoy economies of large scale and this leads to decline in cost of production of the product. It contributes to decline in price of the product. The monopoly firm gains advantage over the new entrants in fixing the price of the product below their AVC incurred in producing the same product. This infers that, a monopoly firm generally goes for destroyer pricing, so as to eliminate the rivalry firms entering the industry.
- ***Control over raw materials:*** If a firm gains control over the purchase and use of raw materials in producing a product, it eliminates other firms in employing the same raw material, thereby, the firm enjoys monopoly status in producing the product.
- ***Brand loyalty:*** This is one important source that provides monopoly power to the business firms. If the customers become loyal for a particular brand of the product of a firm, it makes difficult for other existing firms to survive in the competition, thereby, they leave the industry. Similarly, if new firms enter the industry, they also find it difficult to compete with the well established firm. For example, with reference to computer software, Microsoft is presently enjoying brand loyalty from the customers.
- ***Network externalities:*** We know, the use of a product by a person influences the value of that product to other people. This influence or effect is called network effect. There is a direct relationship between the proportion of people using a product and the demand for that product. That means, when more number of people are using a product, there is a greater probability of any individual starting to use the product. This leads to both spill in and spill over effects of the use of the product across the customers. This leads to the development or acquisition of market power. A classical example is the market dominance of the Microsoft operating system in personal computers.
- ***No substitute goods:*** The absence of substitute goods makes the demand for the good relatively inelastic, thereby, enable the monopolist to extract positive profits.
- ***Cost advantages:*** If the monopolist enjoys the cheaper way of producing the product, it can push its competitors out of the market.
- ***Limit pricing:*** Sometimes the monopolist sets the price of the product so low such that, it becomes unattractive for the competitors to enter the market. Thus, Limit pricing refers to the situation in which a monopoly firm charges a price lower than the monopoly price (and earns less than the monopoly profit) to keep potential competitors out of the market.

21.2. Features of Monopoly

Monopoly represents the extreme form of imperfection exactly opposite to the perfect market competition and its features are discussed here under.

1. Single Seller

The essence of a monopoly is a market controlled by a single seller. The 'mono' part of monopoly means single and the 'poly' part of monopoly means to sell. So, the word 'monopoly' implies a single seller. This single-seller status gives monopoly firm an extensive market control. He is either price maker or quantity adjuster in the market and not a price taker, as in perfect market competition. But, the monopolist cannot control demand for the product, as there are many buyers.

2. No close Substitutes

There are no close substitutes for the product produced by the monopoly firm. That means, the E_C is zero. The buyers have no alternatives or choices for having the product. Either they have to purchase the product or go without it. The absence of substitutes makes the demand for the good relatively inelastic enabling monopolist to extract positive profits. If there are close substitutes, the monopoly power is lost because, the buyers can choose the substitute commodity produced. So, to enjoy the monopoly status, the product produced by the monopolist must be unique. The output-price policy of a monopoly firm does not influence firms of other industries nor it is affected by others, as the commodity produced by a monopolist will have no close substitutes. In the words of Prof. Boulding, *'a pure monopolist, therefore, is a firm producing a product which has no effective substitutes among the products of any other firm'.*

3. Strong Barriers to Entry

The existing seller will have full control on the market and hence, there is no freedom for other firms to enter the market. Some of the key barriers are discussed earlier and they ensure monopoly status to the firm. Not only for other firms, the monopolist himself face barriers in the production of a product. If the product or service produced by the monopolist falls under public utilities, then the Government may impose certain restrictions on the industry in the interests of the public. Examples include, electricity services, natural gas distribution companies, railway service etc.

4. Monopoly Power

Monopoly power is the ability of the firm to affect the terms and conditions of exchange. The monopoly firm may use its power in any manner in order to maximize its profits. Though a monopolist can control both output and price of the product, he cannot adopt both simultaneously. This is because, if he increases the price of commodity, he has to sell less quantity and *vice versa.* Normally, he decides the output supplies and allows the price to be decided in the market based on the demand by the customers. So, he can do either of the two things *i.e.,* either deciding output or fixing the price, thereby, the monopolist is either a price maker or output decider. Although a monopoly's market power is high, it is not absolute. This is because, a monopolist faces a negatively sloped demand curve (ARC) and it is not a perfectly inelastic curve. Consequently, any price increase will result in the loss of some customers.

5. No Distinction between Firm and Industry

Since, there is only one seller, the distinction between firm and industry disappears in monopoly. That means, in case of monopoly, both firm and industry are the same. Hence, the monopolist faces the industry demand curve (ARC).

6. Downward Sloping ARC or Demand Curve

As there is no distinction between firm and industry under monopoly, the firm's demand represents the industry or market demand. In monopoly, there is a single firm producing the product and there are large number of buyers for the product. We know, the individual's demand curve for a product slopes downward from left to right. So, the aggregate demand curve of all the buyers for the product also slopes downward from left to right, as the demand curve for the monopolist is the summation of the demand curves of all the buyers for the product. Thus, the industry or firm's demand curve or ARC is downward sloping from left to right indicating that, monopolist sells large quantities of a commodity or output at lower prices or price reductions (Figure 21.1).

7. Downward Sloping MRC

MRC also slopes downward from left to right and remain below ARC (Figure 21.1). The numerical slope of MRC is twice as steep as ARC (proved as per Panel A of the Figure 18.10). That means, the slope of MRC is more steeper than the slope of ARC, as MRC is related to one additional unit of output and ARC is related to all the units of output. Since, ARC is falling, the additional or extra units sold also fetch less and less revenue (MRC) in the market. When MRC is positive TRC rises, when MRC is zero TRC is maximum and when MRC is negative TRC starts falling. But, ARC falls and never becomes negative and when TRC is zero only, ARC is zero (Figure 21.2).

8. Price Maker

The monopolist is a price maker *i.e.,* he can set price for the commodity for getting maximum revenue. So, power to influence the price is the essence of monopoly. Hence, monopolist is a price maker and buyers are price takers.

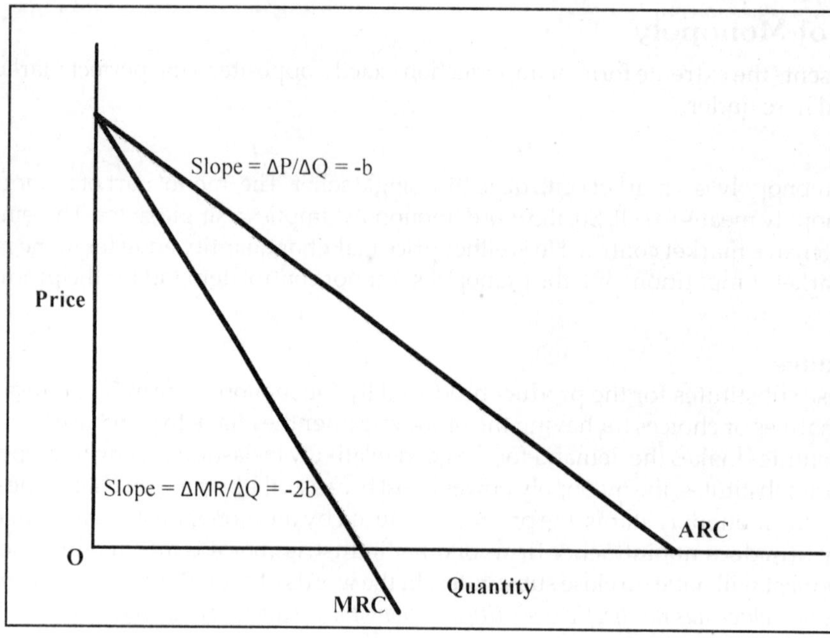

Figure 21.1: Downward sloping ARC and MRC of a monopoly firm.

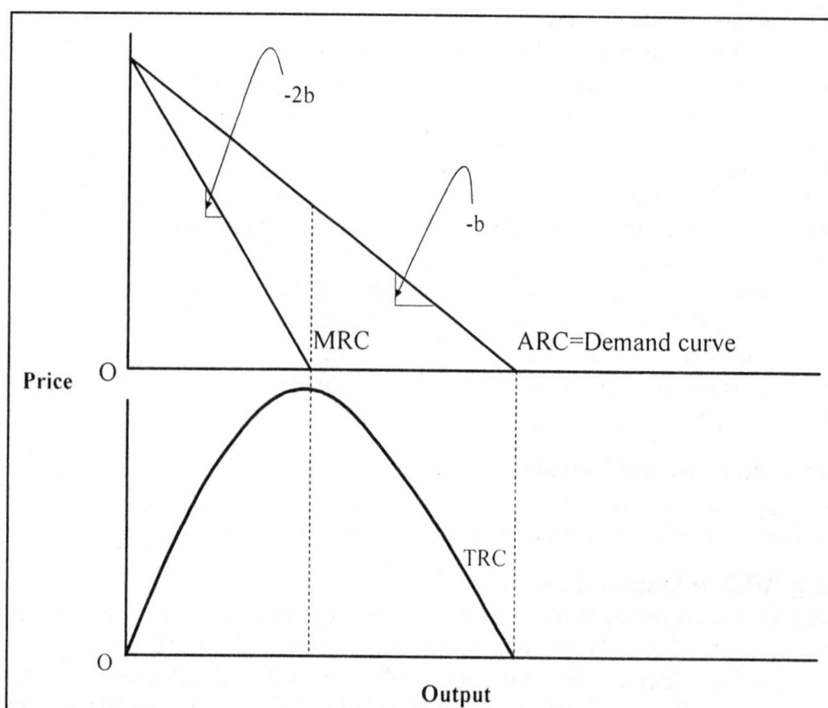

Figure 21.2: TR, AR and MR curves faced by a monopoly firm.

9. No Business Rivals

There are no rivals for the monopolist in the business, as monopoly firm represents the industry.

10. Specialized Information

The technology or process of production of a product is maintained secrecy, as it ensures trade protection for the monopoly firm. For example, the formula of Coca cola soft drink enjoys Trade Secret protection, as the firm enjoys unique formula in the preparation of soft drink.

11. Price Discrimination

Considering different elasticities of demand for the product in different markets, the monopolist sometimes may adopt price discrimination. He may even fix different prices for the same product among different categories of customers in the same market segment.

12. Economies of Scale

Monopolies are characterized by economies of large scale production (Natural monopolies). The firm enjoys a monopoly status due to declining costs over a relatively large range of output coupled with large establishment costs. The monopoly firm is often in a position to cut price of the product below the new entrants' AVC in producing the product and drive them out of the industry (Destroyer pricing).

21.3. Kinds of Monopoly

The monopoly is of different types and they include,

- ☆ ***Private monopoly:*** When monopoly control exists in private sector, it is called Private monopoly. That means, when production is owned, controlled and managed by the individual or private body or private organization, it refers to Private monopoly. This type of monopoly is profit oriented. For example, Monsanto Company enjoys monopoly status for *Bt* Cotton seed.
- ☆ ***Public monopoly***: If State controls the production and pricing of the commodity, it is called Public or State monopoly. It is welfare and service oriented, thereby, it is also called as 'Welfare monopoly'. For example, Railway industry and Defence in India falls under Public monopoly.
- ☆ ***Simple monopoly:*** Simple monopoly firm charges a uniform price or single price from all the customers. It operates in a single market.
- ☆ ***Discriminating monopoly***: In this form of monopoly, the firm charges different prices for the same product from different buyers in the same market or in different markets. For example, dumping is an example for price discrimination.
- ☆ ***Pure monopoly***: A pure monopoly is a firm that satisfies the following conditions:
    - ❖ The E_C with respect to every other product is zero.
    - ❖ The firm is the sole supplier of the product and there are absolutely no rivals.
    - ❖ The output-price policy of the firm does not influence firms of other industries nor this firm is affected by other firms.
    - ❖ The pure monopolist is so powerful that, he is always able to take the whole of all consumer's income *i.e.*, he won't leave any consumer's surplus. This is because, the ARC is a rectangular hyperbola with a unitary elasticity. Since, the elasticity of ARC is equal to one, the total outlay of firm's product will be the same at every price.
    - ❖ The pure monopolist can fix either price or output in the production programme. If he fixes OP_1 price, then the level of output OQ_1 to be sold is determined by the buyers or consumers (Figure 21.3). If he fixes OQ_1 level of output, then the price OP_1 at which the product is to be sold is decided by the buyers. Thus, even a pure monopolist with no rivals cannot fix both price and output at the same time.

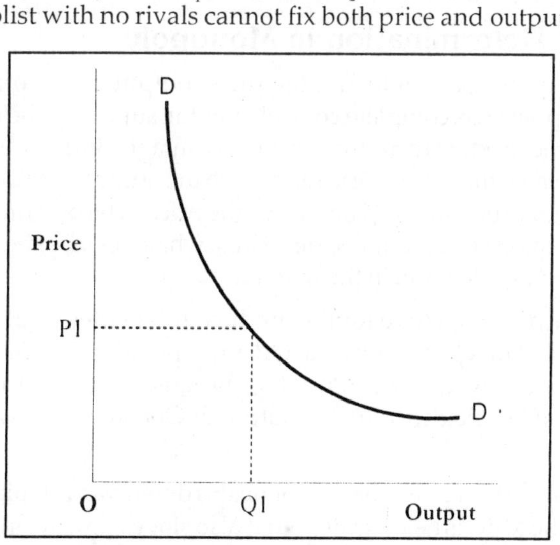

Figure 21.3: Pure monopolist – Price maker or Output decider.

In practice, pure monopolies are very rare. For example, a super market may enjoy monopoly status in a particular town, but if it raises its prices and earns too much of a profit, rivalry firms may enter the industry. So, the threat of entry of rival firms forces the super market to act conscionably and differently from how it would act otherwise. Similarly, buses may be the only carrier in a particular town, but if taxis are also available in that area, it implies, there exist a close substitute for buses.

- *Natural monopoly*: A natural monopoly occurs, when a single firm can produce and supply a product or service to an entire market at extreme economies of scale when compared to other firms. That means, the existing firm can produce and supply the product at a far lower cost than any other competitor. Examples include Columbia Gas, American Electric Power, a toll bridge across a river etc. Once the bridge is built, the natural monopoly does not fear entrants into the market. If a second bridge is produced, average costs would nearly double as the two producers split the market. Having just one bridge is more efficient. Natural monopolies also emerges as a result of natural advantages like good location, abundant mineral resources etc. For example, Gulf countries are having monopoly in crude oil exploration activities because of plenty of natural oil resources. Natural monopolies exist far more frequently than pure monopolies mainly because, the requirements are not as stringent. To produce some products, we need huge amount of establishment costs, R&D costs etc., and this makes the largest firm alone to produce the product, as it is difficult for other small firms to enter the industry. Another classical example for natural monopoly is the distribution of water in a community area. To provide water to residents in a locality, a firm must incur huge amount of fixed costs, say, installation of a network of pipes throughout the locality and houses. If two or more firms were to compete in providing the water distribution, each would have to pay the fixed cost of building a distribution network and in such case, the ATC of water distribution is lowest, when only one firm serves the entire locality. The other examples like, utilities and software industries fall under natural monopolies.
- *Perfect monopoly:* It is also called as Absolute monopoly. In this case, there is only a single seller of product having no close substitute, not even remote one. There is absolutely zero level of competition. Such monopoly is practically very rare.
- *Imperfect monopoly:* It is also called as Relative monopoly or Simple or Limited monopoly. It refers to a single seller market having no close substitute. It means in this market, a product may have a remote substitute. So, there is fear of competition to some extent *e.g.* Mobile (cell phone) industry (Vodaphone is having competition from landline phone service industry say, BSNL).
- *Legal monopoly:* When monopoly exists on account of trade marks, patents, copy rights, statutory regulation of Government etc., it is called Legal monopoly. Music industry is an example of legal monopoly.
- *Technological monopoly:* It emerges as a result of economies of large scale production, use of capital goods, new production methods etc. Examples include, engineering goods industry, automobile industry, software industry etc.
- *Joint monopoly:* When a number of business firms acquire monopoly position through amalgamation, cartels, syndicates etc., it refers to Joint monopoly. For example, pizza making firm and burger making firm are competitors of each other in a fast food industry. But, when they collude with each other in their business, it leads to reduction in competition. So, they can enjoy monopoly power in market.

21.4. Price and Output Determination in Monopoly

As monopoly firm itself represent the industry, the price, output and profits are determined by the forces of demand and supply. The monopolist has complete control over the supply of the product. Likewise, he is also a price maker, who can set the price of the product to his maximum advantage. But, he cannot fix the price and decide output simultaneously. That means, he can either fix the price and leave the output to be determined by the buyers' demand at that price. Or, he can fix the output to be produced and leave the price to be determined by the buyers' demand for that output. So, monopolist is a price maker or output decider. Thus, whatever price he fixes or whatever output he decides to produce, it is influenced by buyers' demand in the market.

In the Figure 21.3, DD is the demand curve for the product. It is evident that, the monopolist cannot fix both *i.e.*, selling his output at OQ_1 and expect to sell this output at OP_1 price per unit. This is because, if he decides to fix the price at OP_1, the output will be automatically fixed by him at OQ_1 because, at OP_1 price he can only sell OQ_1 units and not more than OQ_1 units. Similarly, if he decides to fix the output at OQ_1, the price will be automatically decided by the buyers at OP_1.

As explained earlier, the demand curve for the monopolist is downward sloping from left to right and it is definite. The demand curve is his ARC. The MRC is below ARC and it also slopes downward from left to right (Figure 21.1). But, the manner and extent to which the monopolist can influence the price and output will depend upon the laws of returns the firm experiences and E_P for the product. If the firm is producing under increasing returns or decreasing costs, he can produce more quantity of output at low cost and sell it at low price. On the contrary, if the firm is producing under diminishing returns or increasing costs, he produces less quantity of output and sells it at high price. If the demand for the product is highly elastic, he can sell more quantity of product by fixing lower price for the product. On the other hand, if the demand for the product is less elastic, he can get more profits by quoting higher price of the product.

As discussed earlier, the monopoly firm's equilibrium can be studied through two approaches,

- ☆ ***TR and TC approach:*** The maximum distance between TR and TC curves indicate the profit maximization of a monopoly firm.
- ☆ ***MR and MC approach:*** The point of equality between MR=MC determines the point of profit maximization of a monopoly firm. This approach is proved meritorious over TR and TC approach (Chapter 18).

The point of intersection between MR and MC curves and the perpendicular drawn from this intersection of MR and MC curves onto the ARC, determines the price of the product (as ARC is nothing but, demand curve). It implies that, in monopoly, price > MC=MR, as MRC lies below ARC.

We know, AR or monopoly price=MR($E_P/(E_P$-1)), as per the Equation 18.10. Since, in monopoly, the point of equilibrium is MC=MR, thereby, we can replace MR by MC in the above Equation 18.10. So,

AR or monopoly price = MC($E_P/(E_P$-1)) *Equation 21.1*

Hence, monopoly price is a function of MC and the E_P

21.4.1. Price Determination of a Monopoly Firm in Short Run and Long Run

The price determination of a monopoly firm both in short run and long run production programmes is based on the following assumptions:

- ☆ There is one firm in the production of a product
- ☆ The product is homogenous with out substitutes
- ☆ Monopolist is rational with an objective of profit maximization or cost minimization.
- ☆ Prices of factors and product remain same.
- ☆ The monopolist exercises full market power in fixing the market price and block the entry of rivalry firms.

I. Profitability Decisions of a Monopoly Firm in Short-Run or Short-Run Monopoly Equilibrium

Case 1: TR>SRTC or Price>SRATC

In the Figure 21.4, on X-axis, output and on Y-axis, revenue and costs are taken. ARC and MRC are downward sloping curves, the ARC is a higher level sloping down curve and MRC is a lower level sloping down curve. SRMC curve cuts the MRC at point 'E'. So, the monopolist is in equilibrium at 'E' and produces OQ quantity of output by fixing the price of the product at OC=QP. At this price and output, the monopolist realizes super-normal profits shown by the shaded rectangle PABC. This is because, at OQ level of output, QP=OC price is above SRATC curve by PA portion between ARC and SRATC curve. Beyond OQ level of output, MR<SRMC and below OQ level of output, MR>SRMC. So, to derive maximum profits, the monopolist has to produce the output OQ, where MR=SRMC.

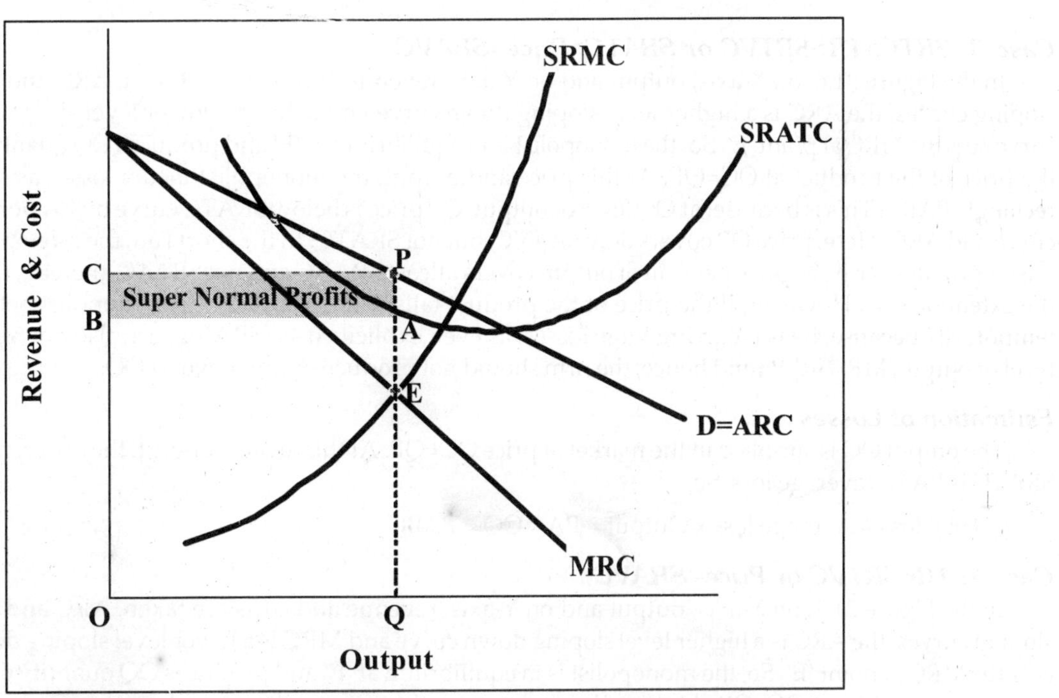

Figure 21.4: Price determination of monopoly firm in short run (Case: TR>SRTC or Price > SRATC).

Estimation of Profits

The output OQ is supplied in the market at price OC=QP. At this level of output, the difference between ARC and SRATC is PA *i.e.*, average profit. So,

Total profit = Average profit x Output = PA x OQ = PABC

Case 2: TR=SRTC or Price=SRATC

In the Figure 21.5, on X-axis, output and on Y-axis, revenue and costs are taken. ARC and MRC are downward sloping curves, the ARC is a higher level sloping down curve and MRC is a lower level sloping down curve. SRMC curve cuts the MRC at point 'E'. So, the monopolist is in equilibrium at 'E' and produces OQ quantity of output by fixing the price of the product at OC=QP. At this price and output, the monopolist realizes normal profits or the firm becomes break even, since price=SRATC. Beyond OQ level of output, MR<SRMC and below OQ level of output, MR>SRMC and at OQ output, MR=SRMC and the firm derives only normal profits, as price=SRATC.

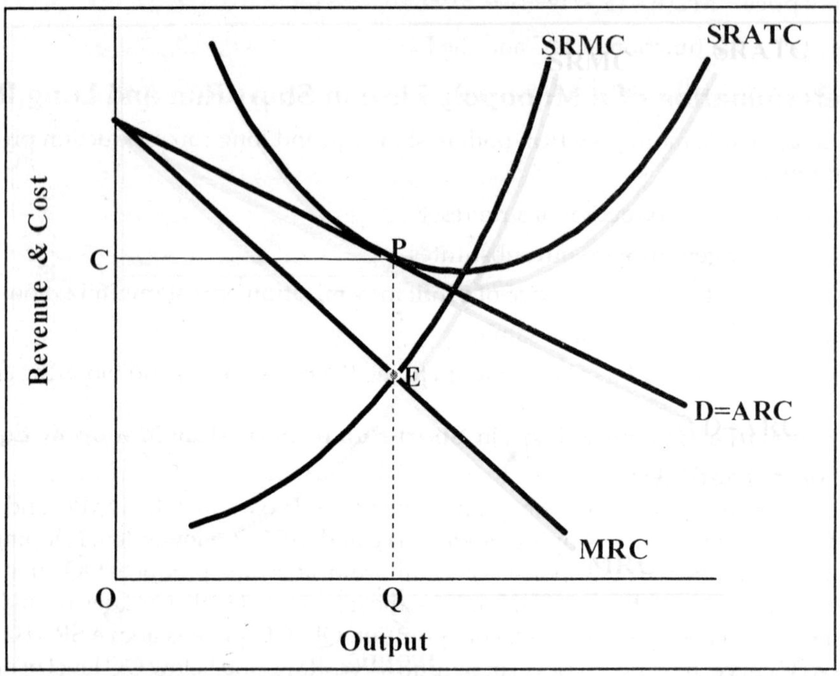

Figure 21.5: Price determination of monopoly firm in short run (Case: TR=SRTC or Price = SRATC).

Case 3: SRTC>TR>SRTVC or SRATC>Price>SRAVC

In the Figure 21.6, on X-axis, output and on Y-axis, revenue and costs are taken. ARC and MRC are downward sloping curves, the ARC is a higher level sloping down curve and MRC is a lower level sloping down curve. SRMC curve cuts the MRC at point 'E'. So, the monopolist is in equilibrium at 'E' and produces OQ quantity of output by fixing the price of the product at OC=QP. At this price and output, the monopolist incurs losses as shown by the shaded rectangle PABC. This is because, at OQ level of output, QP price is below SRATC curve by PA portion between SRATC curve and ARC. Here, price QP covers only SRAVC, but not SRATC. In the short run, the monopolist can continue the business temporarily, as price per unit of output covers atleast SRAVC and not SRATC, thereby, the losses are less than the extent of TFC. However, if the price of the product falls below SRAVC, the monopolist better stop the business temporarily because, he is incurring huge losses, as even explicit costs (SRAVC) are also not recovered. Beyond OQ level of output, MR<SRMC and hence, the firm should not produce output beyond OQ.

Estimation of Losses

The output OQ is supplied in the market at price OC=QP. At this level of output, the difference between ARC and SRATC is PA *i.e.*, average loss. So,

Total loss = Average loss x Output = PA x OQ = PABC

Case 4: TR<SRTVC or Price<SRAVC

In the Figure 21.7, on X-axis, output and on Y-axis, revenue and costs are taken. ARC and MRC are downward sloping curves, the ARC is a higher level sloping down curve and MRC is a lower level sloping down curve. MC curve cuts the MRC at point 'E'. So, the monopolist is in equilibrium at 'E' and produces OQ quantity of output by fixing the price of the product at OC=QP. At this price and output, the monopolist incurs huge losses, as shown by the shaded

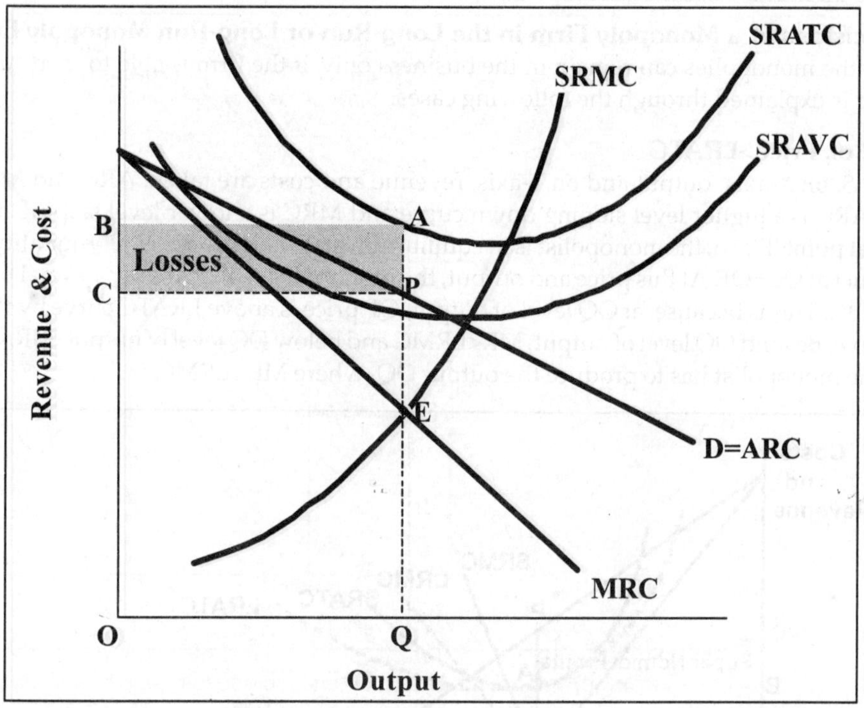

Figure 21.6: Price determination of monopoly firm in short run (Case: SRTC>TR>SRTVC or SRATC>Price>SRAVC).

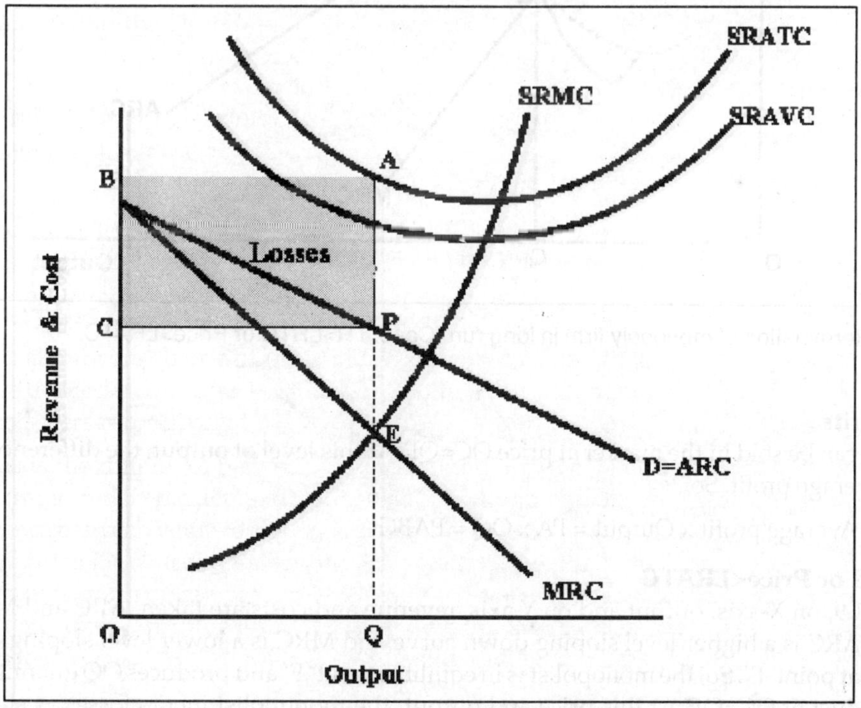

Figure 21.7: Price determination of monopoly firm in short run (Case: TR<SRTVC or Price<SRAVC).

rectangle PABC. This is because, at OQ level of output, QP price is below SRAVC curve. So, the price QP could not cover even SRAVC and hence, the monopolist better stop the business temporarily because, even explicit costs are also not recovered besides TFC.

Estimation of Losses

The output OQ is supplied in the market at price OC=QP. At this level of output, the difference between ARC and SRATC is PA *i.e.*, average loss. So,

Total loss = Average loss x Output = PA x OQ = PABC

II. Profitability Decisions of a Monopoly Firm in the Long-Run or Long-Run Monopoly Equilibrium

In the long run, the monopolies can remain in the business only, if the firm is able to earn super normal profits (Price>LRATC). This is explained through the following cases:

Case 1: TR>LRTC or Price>LRATC

In the Figure 21.8, on X-axis, output and on Y-axis, revenue and costs are taken. ARC and MRC are downward sloping curves, the ARC is a higher level sloping down curve and MRC is a lower level sloping down curve. LRMC curve cuts the MRC at point 'E'. So, the monopolist is in equilibrium at 'E' and produces OQ quantity of output by fixing the price of the product at OC=QP. At this price and output, the monopolist realizes super-normal profits shown by the shaded rectangle PABC. This is because, at OQ level of output, QP price is above LRATC curve by PA portion between ARC and LRATC curve. Beyond OQ level of output, MR<LRMC and below OQ level of output, MR>LRMC. So, to derive maximum profits, the monopolist has to produce the output OQ, where MR=LRMC.

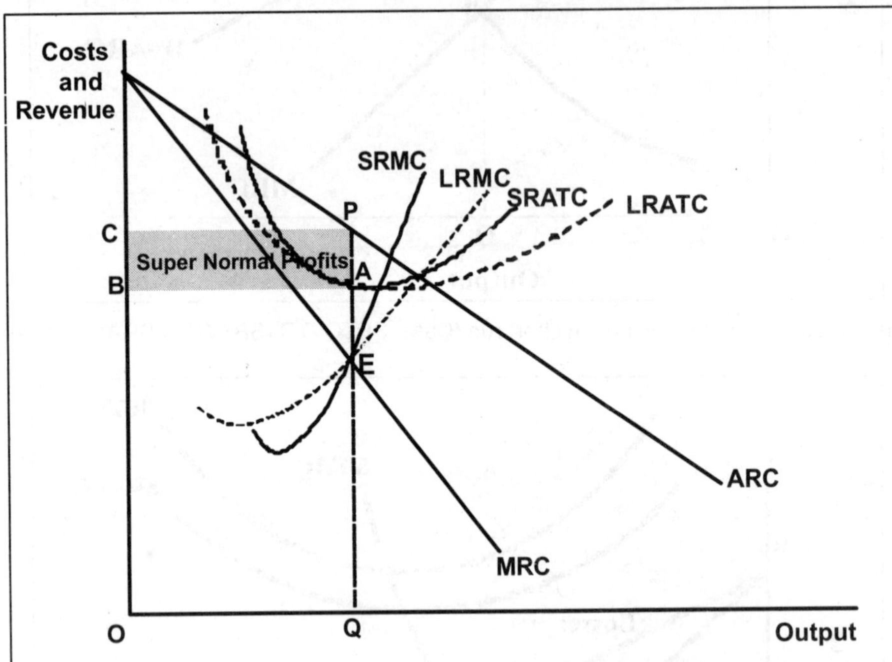

Figure 21.8: Price determination of monopoly firm in long run (Case: TR>LRTC or Price>LRATC.

Estimation of Profits

The output OQ can be sold in the market at price OC=QP. At this level of output, the difference between ARC and LRATC is PA *i.e.*, average profit. So,

Total profit = Average profit x Output = PA x OQ = PABC

Case 2: TR<LRTC or Price<LRATC

In the Figure 21.9, on X-axis, output and on Y-axis, revenue and costs are taken. ARC and MRC are downward sloping curves, the ARC is a higher level sloping down curve and MRC is a lower level sloping down curve. LRMC curve cuts the MRC at point 'E'. So, the monopolist is in equilibrium at 'E' and produces OQ quantity of output by fixing the price of the product at OC=QP. At this price and output, the monopolist incurs losses as shown by the shaded rectangle PABC. This is because, at OQ level of output, QP price is below LRATC curve by PA. So, the price QP could not cover LRATC and hence, the monopolist better stop the business permanently because, he is incurring huge losses.

Estimation of Losses

The output OQ is supplied in the market at price OC=QP. At this level of output, the difference between ARC and LRATC is PA *i.e.*, average loss. So,

Total loss = Average loss x Output = PA x OQ = PABC

In the earlier Figures 21.6 and 21.7, since price could not cover SRATC and even SRAVC respectively in the short run, the monopolist incurs losses and when the price falls below SRAVC, the monopolist has to stop the production programme temporarily. But, in the long run, the monopolist has enough time to make changes in the existing plant, so as to overcome the losses and to realize profits. With the entry of new firms is completely ruled out in monopoly market

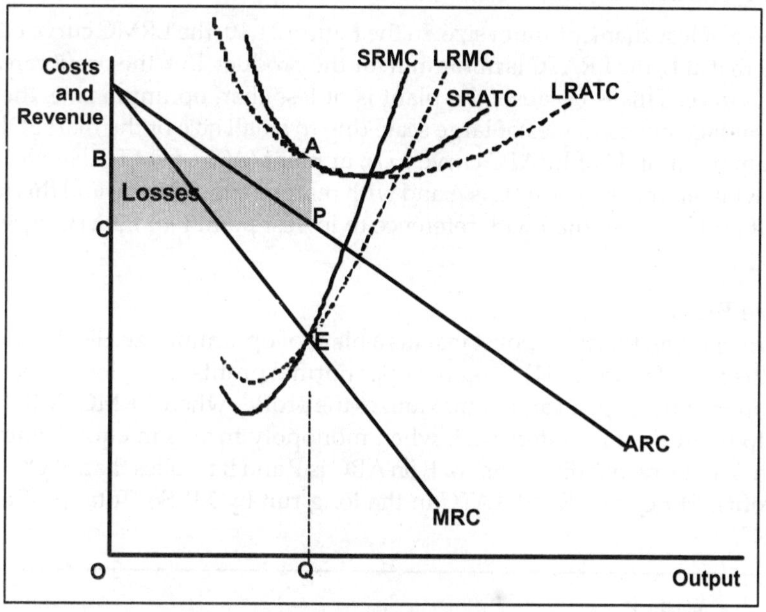

Figure 21.9: Price determination of monopoly firm in long run (Case: TR<LRTC or Price<LRATC).

situation, the monopolist can install a smaller than optimum scale plant or an optimum scale plant or a larger than optimum scale plant in the long run, depending upon the size of the market for the product he wishes to produce. The following conditions must be satisfied to ensure the profitable level of output in the case of each plant:

- ☆ LRMC curve=MRC and SRMC curve must pass through this point, and
- ☆ LRATC curve must lie tangent to the SRATC curve at this level of output.

Case 1: Smaller than Optimum Size Plant

If the size of the marker is small, the monopolist installs a plant of smaller than optimum size. SRMC and SRATC are the short run per unit cost curves of the firm (Figure 21.10). To earn super normal profits in the long run, if the size of market is small, the monopolist installs smaller than optimum size plant and the monopolist maximizes the profits, where LRMC=MRC=SRMC and LRATC is tangent to SRATC with reference to that output level. This is shown at point E and it infers that, when monopoly firm is in equilibrium in long run, it is also in equilibrium in short run. The corresponding point of E on ARC is P and it implies that, at OC=QP level of price, the firm earns super-normal profits, PABC, as ARC>LRATC in the long run by AP. So,

Total profit = Average profit x Total output =PA x OQ =PABC

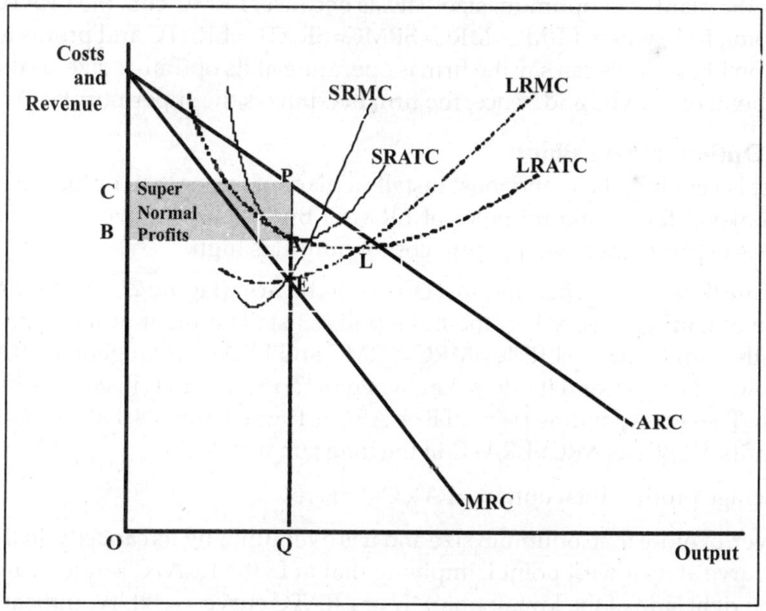

Figure 21.10: Price determination of monopoly firm in long run (Case: Smaller than Optimum size plant).

The plant is, however, of less than optimum size. In the Figure 21.10, the LRMC curve cuts the LRATC curve at its lowest point L implying that at L, the LRATC is minimum for the product. But, the profits are not studied at this lowest point 'L' on the LRATC curve. This is because, the plant is of less than optimum size, thereby, the firm is not in a position to take full advantage of economies of large scale due to small size of the market. Hence, profits are studied even before the minimum point of 'L' of LRATC curve., *i.e.*, at point 'A' on LRATC. So, the corresponding point of E (LRMC=MRC=SRMC) with reference to LRATC is A and with reference to ARC is P and this infers the firm earns super normal profits PABC. (If profits are studied with reference to lowest point L of LRATC curve, then the profits will be much higher).

Case 2: Optimum Size Plant

If the size of the market is big, the monopolist installs a plant of optimum size. SRMC and SRATC are the short run per unit cost curves of the firm (Figure 21.11). To earn super normal profits in the long run, if the size of market is big, the monopolist installs optimum size plant and he maximizes the profits, where LRMC=MRC=SRMC=SRATC=LRATC. This is with reference to point E and it infers that, when monopoly firm is in equilibrium in long run, it is also in equilibrium in short run. The corresponding point of E on ARC is P and it implies that, at OC=QP level of price, the firm earns super-normal profits, PEBC, as ARC>LRATC in the long run by EP. So, Total profit = Average profit x Total output =PE x OQ =PEBC.

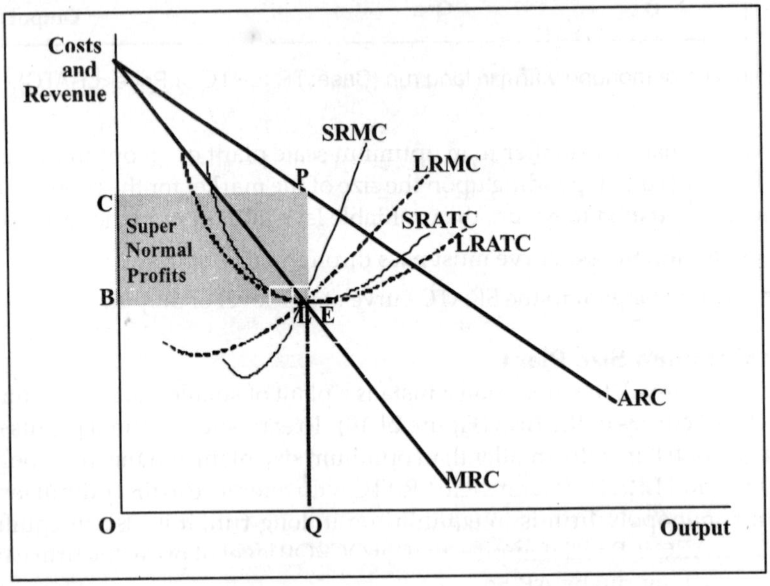

Figure 21.11: Price determination of monopoly firm in long run (Case: Optimum size plant).

In the Figure 21.11, the plant is of optimum size. This is because, LRMC cuts the LRATC at its minimum point L and at this minimum point, L=E, where LRMC=MRC=SRMC=SRATC=LRATC and profits are studied with reference to minimum point L=E on LRATC. We can say the firm is operating at its optimum size, as the profits are studied with reference to the lowest point of LRATC and hence, the firm maximizes the super-normal profits, PEBC.

Case 3. Larger than Optimum Size Plant

If size of the market is very big, the monopolist installs a plant of larger than optimum size. He will earn more by producing the output beyond the minimum point of LRATC. But it leads to over utilization of the plant, thereby, bringing in diseconomies of production, *i.e.*, per unit cost of output is high.

SRMC and SRATC are the short run per unit cost curves of the firm (Figure 21.12). To earn super normal profits in the long run, if the size of market is very big, the monopolist installs a plant of larger than optimum size and the monopolist maximizes the profits, where LRMC=MRC=SRMC and LRATC is tangent to SRATC with reference to that output level. This is shown at point E and it infers that, when monopoly firm is in equilibrium in long run, it is also in equilibrium in short run. The corresponding point of E on ARC is P and it implies that, at OC=QP level of price, the firm earns super-normal profits, PABC, as ARC>LRATC in the long run by PA. So,

Total profit = Average profit x Total output =PA x OQ =PABC

The plant is, however, of more than optimum size and it is over utilizing its capacity. In the Figure 21.12, the LRMC curve cuts the LRATC curve at its lowest point L implying that at L, the LRATC is minimum for the product. But, the profits are studied to the right side of the lowest point 'L' of LRATC curve *i.e.*, at 'A', indicating that, costs are slightly more at point A when compared to at point 'L'. This is because, since the plant is more than optimum size, to meet the

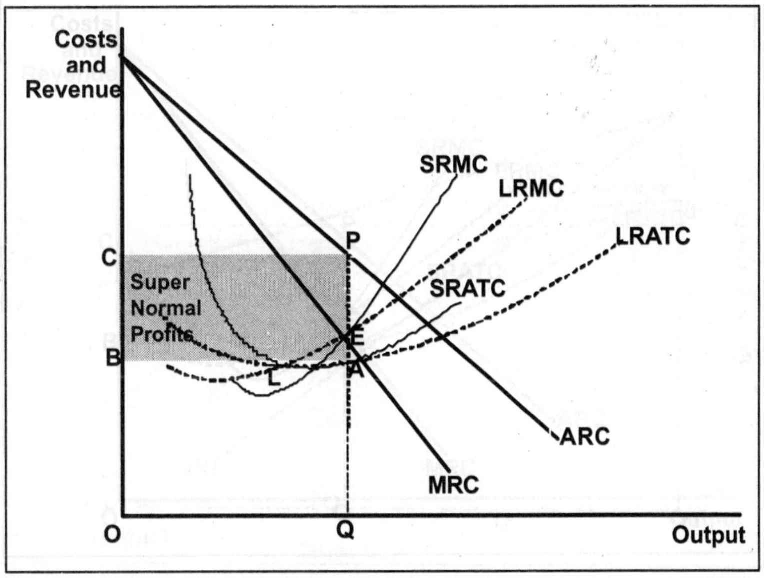

Figure 21.12: Price determination of monopoly firm in long run (Case: Larger than optimum size plant).

needs of big size market, it is over utilized above its capacity. This will leads to diseconomies of large scale, thereby, the profits are declined compared to the firm operating at optimum size. That means, had the profits studied at lowest point L on the LRATC curve, the profits will be much higher, as we studied in earlier Case 2 (Figure 21.11), when the plant is of optimum size. So, the corresponding point of E (LRMC=MRC=SRMC) with reference to LRATC is A and with reference to ARC is P and this infers that, the firm earns super-normal profits PABC.

To sum up, in the long run production programme, with the entry blocked to other firms in monopoly market situation, the firm meets the market demand by installing less than optimum or optimum or more than optimum plant sizes in accordance with the market sizes and try to attain super-normal profits. But, in case of perfect competition, in the long run, the firm must operate an optimum plant size at the lowest point of LRATC and derives only normal profits.

21.4.2. Laws of Returns and Monopoly Price Determination in the Long Run

In the long run production programme, the monopoly price is influenced by the laws of returns the firm experiences. So, a monopolist working under different laws of returns, has to adjust the output depending upon the equilibrium conditions. In general, the monopoly firm experiences three laws of returns *viz.*, the law of increasing returns or decreasing costs, the law of constant returns or constant costs and the law of diminishing returns or increasing costs. The following three Figures 21.13 to 21.15 illustrate the influences of these three laws of returns on the monopoly equilibrium:

1. Law of Increasing Returns and Monopoly Price

This law states that, the production programme experiences decreasing costs. In the Figure 21.13, LRATC and LRMC are sloping downward from left to right indicating decreasing costs in the production programme. For a monopoly firm, the ARC and MRC are downward sloping curves and MRC lies below the ARC. The point of equilibrium is where, LRMC intersects the MRC at point 'E', which indicates that, the monopoly firm produces OQ level of output and sell the same at QP price per unit of product and the firm enjoys super normal profits, PABC.

2. Law of Constant Returns and Monopoly Price

This law states that, the production programme experiences constant costs. In the Figure 21.14, LRMC coincides with LRATC and they both lie parallel to X-axis, as the firm experiences constant costs in the production programme. For a monopoly firm, the ARC and MRC are downward sloping curves and MRC lies below the ARC. The point of equilibrium is where, LRMC intersects the MRC at point 'E', which indicates that, the monopoly firm produces OQ level of output and sell the same at QP price per unit of product and firm benefitted by super normal profits PEBC.

3. LDR and Monopoly Price

This law states that, the production programme experiences increasing costs. In the Figure 21.15, LRATC and LRMC are upward sloping from left to right indicating increasing costs in the production programme. For a monopoly firm, the ARC and MRC are downward sloping curves from left to right and MRC lies below the ARC. The point of equilibrium is where, LRMC intersects the MRC at point 'E', which indicates that, the monopoly firm produces OQ level of output and sell the same at QP price per unit of product and the firm enjoys super-normal profits, PABC. Here, the monopolist enjoys super-normal profits by producing lower quantity of output than that under perfect competition.

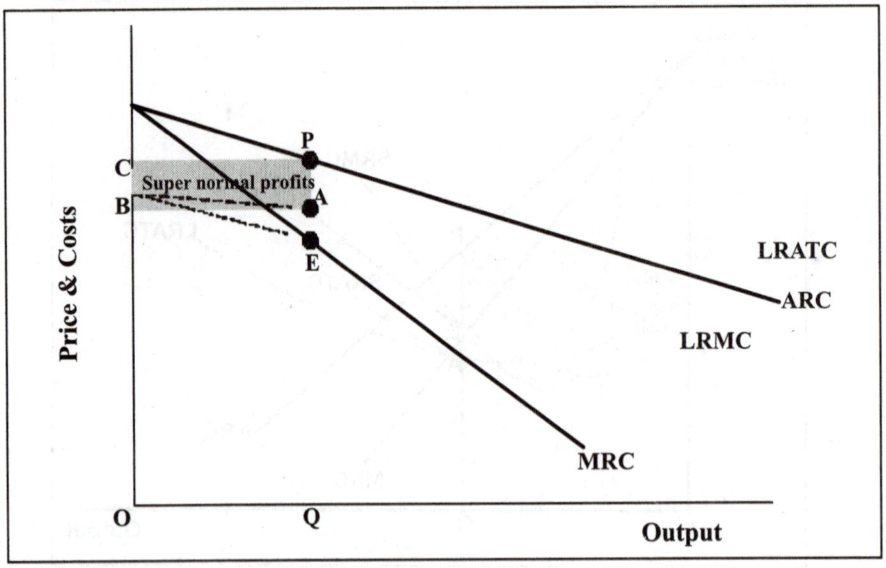

Figure 21.13: Monopoly price and Law of increasing returns.

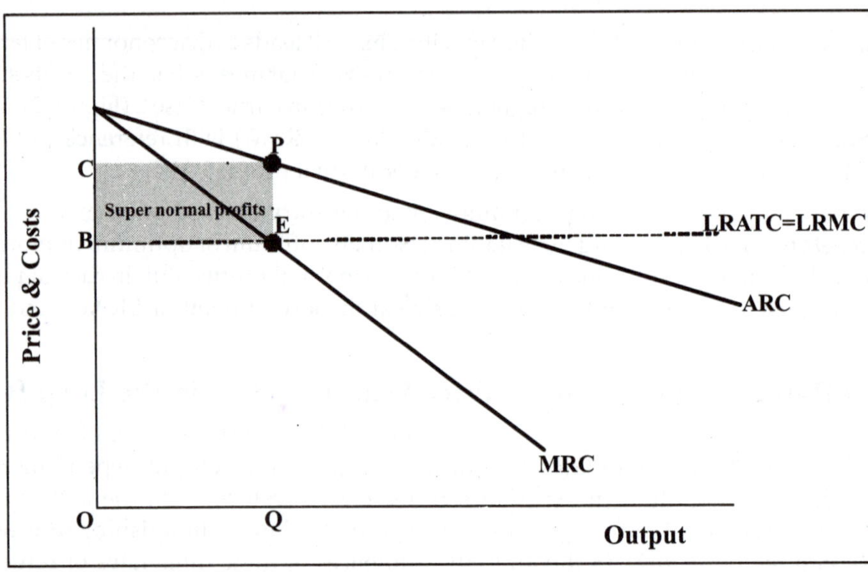

Figure 21.14: Monopoly price and Law of Constant returns.

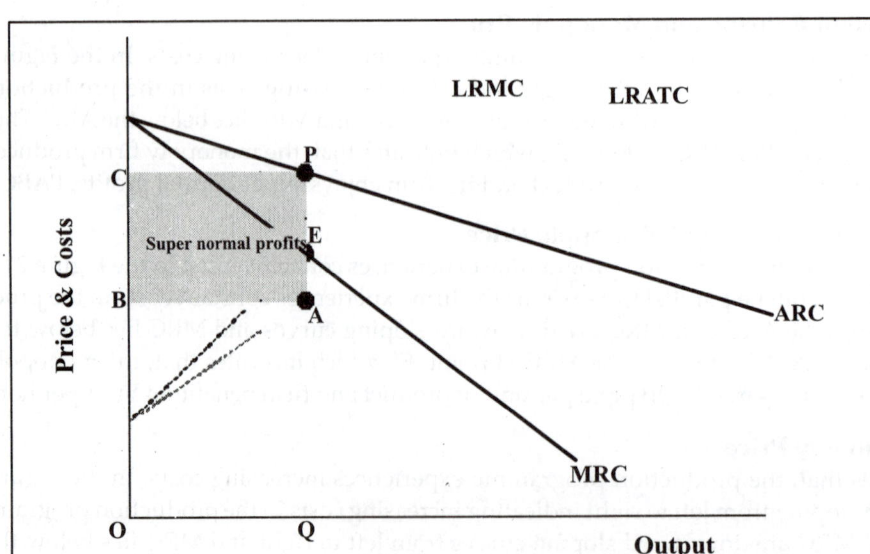

Figure 21.15: Monopoly price and Law of Diminishing returns.

21.4.3. Comparison of Price Determination Under Perfect Competition and Monopoly

Having studied the concepts of output-price determination both under perfect competition and monopoly, it is interesting to draw the comparative analysis between these two forms of market competition with reference to price determination and the same is presented through Table 21.1.

Table 21.1: Differences between Perfect market competition and Monopoly with reference to price determination.

Perfect Competition	*Monopoly*
The demand curve or ARC is perfectly elastic and it is a horizontal straight line parallel to X-axis, as the firm has no control over the price and it has to sell any quantity of commodity at the same price.	The demand curve or ARC is relatively less elastic and slopes downward from left to right indicating that, the monopolist sells large quantities of output at low prices.
The firm is in equilibrium at the level of output, where MC is equal to MR. Since in perfect competition, MR is equal to AR or price, therefore, when MC is equal to MR, it is also equal to AR or price at that equilibrium position, i.e., MC=MR=AR (Price). At the point of equilibrium, MC curve is rising and it cuts the horizontal MRC from below.	The firm is in equilibrium at the level of output, where MC is equal to MR. At the point of equilibrium, MC curve is rising and it cuts the downward falling MRC from below.
In equilibrium position, the price charged by the firm equals to MC, as ARC is equal to MRC.	In equilibrium position, the price charged by the firm is above MC, as ARC is above MRC.
In the long run, the firm earns only normal profit. The super normal profits enjoyed by the firms in the short run will be swept away in the long run, as new firms enter the industry.	The firm can earn super-normal profits even in the long run, as there are no competitors for the firm producing the same product.
The firm is in long-run equilibrium at the minimum point of the LRATC curve.	The firm is in long-run equilibrium at the point, where LRATC curve is still declining and has not reached the minimum point.
Under perfect competition, output is higher, but price is lower.	Under monopoly, output is lower, but price is higher.
In case of perfect competition, in long run, the firm must operate an optimum plant size at the lowest point of LRATC and derives only normal profits.	In the long run production programme, with the entry blocked to other firms in monopoly market situation, the firm meets the market demand by installing less than optimum or optimum or more than optimum plant sizes in accordance with the market sizes and try to attain super-normal profits.

21.5. Monopoly Price and E_p

We discussed earlier, when the E_p for the product is less (E_p<1), then the firm will quote a high price for the product. This is because, whatever may be the price, the customer has to purchase the requisite quantity of the product, when the demand for the product is less elastic. On the contrary, if the E_p is more, (E_p>1), then the monopolist will quote a low price for the commodity, as the buyers purchase more of the commodity at low price. So, the concept of E_p will influence the monopoly price and the same is explained through Figure 21.16.

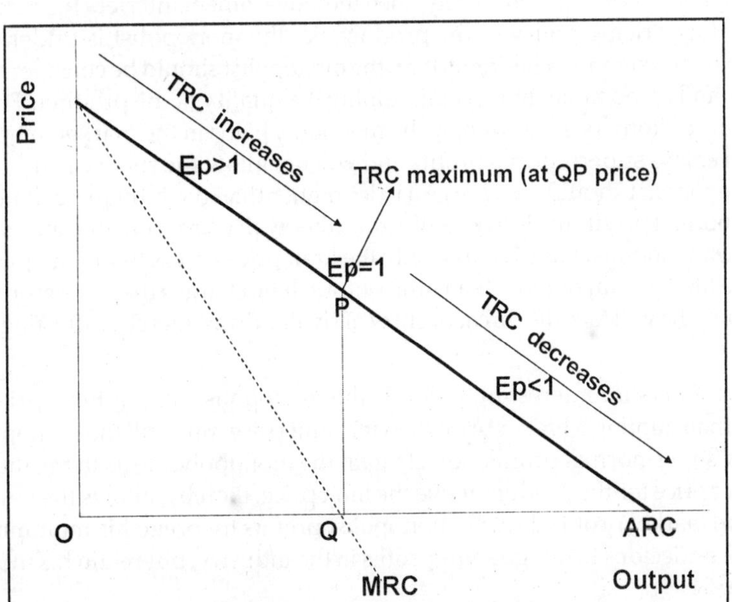

Figure 21.16: Monopoly price and E_p.

ARC and MRC are the AR and MR curves of the monopolist and they are downward sloping curves from left to right and MRC lies below the ARC. We know, ARC is also called as demand curve and as per the point elasticity method, at mid point of the demand (AR) curve *i.e.*, at 'P', the $E_p = 1$, below the mid point the E_p is less than unity and above the mid point the E_p is more than unity (Figure 12.12). The MRC of the firm infers that, at $E_p = 1$, MRC is zero, when MRC is negative E_p is less than unity and when MRC is positive, E_p is more than unity. From the Figure 21.16, we can have the following inferences:

- ☆ The monopolist will never fix the output, where E_p is less than unity *i.e.*, beyond OQ because, at that level of output, MRC is negative and this will adversely affect the TR of the firm. Moreover, since MC can never be negative, the firm will not be in equilibrium at output level, where MRC is negative.
- ☆ So, the monopoly firm will be in equilibrium, when MRC is positive (though falling downwards) because, MC is always positive. This infers that, the monopolist will be in equilibrium at the level of output, where the E_p is more than unity (*i.e.*, below OQ level of output).
- ☆ When there is no MC for the monopoly firm in dealing with the naturally or freely available commodities like mineral spring, the monopolist fix the output, where E_p is unitary at which MR is zero and thereby, TR is maximum. In such case, the monopolist fixes OQ quantity of output and sells the commodity at QP price to derive maximum TR.

The above analysis further infers the following interesting points:

- ☆ Monopolist never fix the output at a price where demand is less elastic ($E_p < 1$) or MR is negative (*i.e.*, beyond OQ level of output) because, MC cannot be negative.
- ☆ Monopolist fix the output, where $E_p = 1$ or MR = 0, when no MC is incurred in producing the output (*i.e.*, at OQ level of output).
- ☆ Monopolist fix the output, where $E_p > 1$ or MR is positive, as MC is always positive. That means, when MC = MR, MR must be positive (*i.e.*, below OQ level of output).

This concept of monopoly price in relation with E_p can be better analyzed and explained, while dealing with price discrimination. In the third degree price discrimination (Dumping), the monopolist fix a high price for the product in the domestic market, as the ARC is less elastic and in the foreign market he quotes a low price for the product, as the ARC is perfectly elastic. So, the concept of E_p guides the monopolist in influencing price-output policies.

21.6. Monopoly Pricing with Threat of Entry

If the monopolist earns significant level of super normal profits, new firms try to enter the industry and produce good substitutes for the product. So, earning super-normal profits through fixing very high price by the monopolist is a threat for himself, as there is a danger of entering rivalry firms into the industry. Besides fixing high price, even the technological advances that contribute to cost-effective production also pose a threat to the monopolist, as such technology attract rivalry firms to produce good substitutes to the existing product. Above all, if the monopolist exploit the market by charging higher prices indiscriminately, then the Government interfere the market and impose restrictions on the monopolist regarding pricing policy of the product. So, the monopolist is under continuous check to enjoy monopoly status in the market economy. This infers that, the monopolist should be considerative in charging reasonable price from the customers and at the same time in maintaining the quality of the product. The reasonable price that the monopolist charge from the customers, so as to enjoy the monopoly gains in the market refers to 'Limit price' and at this price, the monopoly firm enjoys super-normal profits and will not allow the rivalry firms to enter and compete in the same industry. So, the monopolist should not charge a price higher than the limit price. If he charges higher price than the limit price, the super-normal profits he derives will attract the rivalry firms into the industry and thereby, his market share gets affected. So, the monopolist has to charge only the limit price or less than limit price and enjoy super normal profits through selling the limit output or more than limit output. If he charges the price more than limit price or sells the output less than limit output, he will face the threat of entry of rivalry firms, thereby, affecting the monopoly status of the firm in the market.

The above discussion infers an interesting point. If the monopolist fixes a limit price for the product, he will definitely get less profits than quoting a higher price than the limit price. But, still the monopolist prefers to sacrifice the loss in earning monopoly super-normal profits, so as to gain the monopoly status in the market. The other option left before him is to fix a higher price for the product above the limit price, thereby, allows the new entrants (rival firms) into the industry and share the market profits. But, no monopolist prefers to forego his monopoly status in producing the product in the market. So, he decides to forego some profits in the industry, but retain his monopoly status. The same is explained through Figure 21.17.

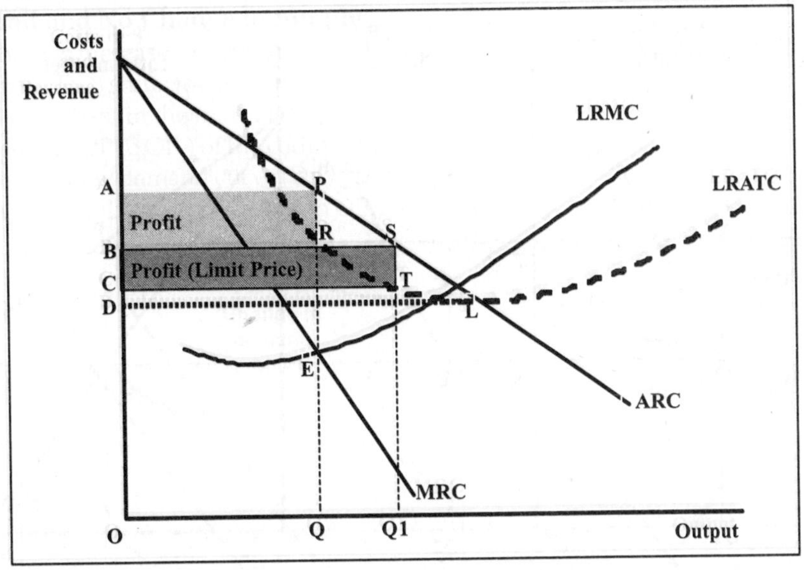

Figure 21.17: Monopoly pricing with threat of entry of rivalry firms.

In the Figure 21.17, on X-axis, output and on Y-axis, costs and revenue are taken. ARC and MRC are downward sloping curves, the ARC is a higher level sloping down curve and MRC is a lower level sloping down curve. LRMC curve cuts the MRC at point 'E'. So, the monopolist is in equilibrium at 'E' and produces OQ quantity of output by fixing the price of the product at OA=QP. At this price and output, the monopolist earns maximum super-normal profits, as shown by the shaded rectangle PRAB.

Assume, this price OA=QP is a very high price, thereby, the super-normal profits gained by the monopolist (PRAB) will attract other (rivalry) firms to enter the industry to produce the same product. This infers that, the monopolist should quote a price less than OA=QP, so that, he can eliminate the rivalry firms to enter the industry and he can enjoy monopoly status for the product in the market. So, now the monopolist should quote a price less than OA=QP and assume that, it is OB=Q_1S and it is the limit price. At this limit price, he sells more quantity of product (OQ_1>OQ) and still he derives super-normal profits BCST. So, the decline in super-normal profits due to decline in price from QP to Q_1S is slightly compensated by increase in quantity of product transacted at limit price *i.e.*, OQ_1>OQ. So, this price OB=Q_1S is the limit price, which fetches the monopolist two important advantages *viz.*, he can still reap super-normal profits BCST and there is no threat of the entry of rivalry firms into the industry. So, the monopolist should not charge a price above OB and should not increase or decrease output from OQ_1. If the monopolist raise the price above OB, the rivalry firms enter the industry and the present firm will lose the monopoly status. In fact, the limit price can be any where between P and L on the ARC. If the limit price is fixed at OD, it equals the lowest point of LRATC 'L', then the monopolist will earn only normal profits in the long run, though he can increase the quantum of sales of the product. So, there is little chance on the part of the monopolist to stick on to this price OD and hence, he quote OB price for the product, where he earns super-normal profits and also enjoys monopoly status in the market.

It is important that, the fixation of limit price is purely a guess work and based on the market experience and possibility of threat from the rivalry firms, the entrepreneur will decide the limit price for the product. If the modern technology is far superior over the present technology in terms of both economic viability and technical feasibility, then the rivalry firms may enter the market and produce close substitute goods. In such a case, the monopolist fixes a higher price and tolerates one or two competitors, as he still gets super normal profits. But, this monopoly is now designated as Weak monopoly, as when rivalry firms enter the industry, it leads to destruction of original monopoly power.

21.7. Multi-plant Monopoly Firm

Sometimes, instead of operating on a single plant, monopolist may employ plants of different capacities to produce the same product. Let us assume, the monopolist operates on two plants, Plant 1 and Plant 2 of different capacities and he employ these two plants to meet the market demand. In Panel C of the Figure 21.18, ARC and MRC are downward sloping curves for the entire market, the ARC is a higher level sloping down curve and MRC is a lower level sloping down curve. ÓMC is the cumulative MC of both the plants and it intersects the MRC at point E. So, the monopolist is in equilibrium at 'E' and produces OQ quantity of output by fixing the price of the product at OA=QP. At this price QP, the firm earns super-normal profits.

The monopolist has to allocate this output OQ between two plants in such a way that, the monopoly market MRC equals the MC curves of two plants *i.e.*, MC_1 and MC_2. This is done by drawing a line from E to M parallel to X-axis. This

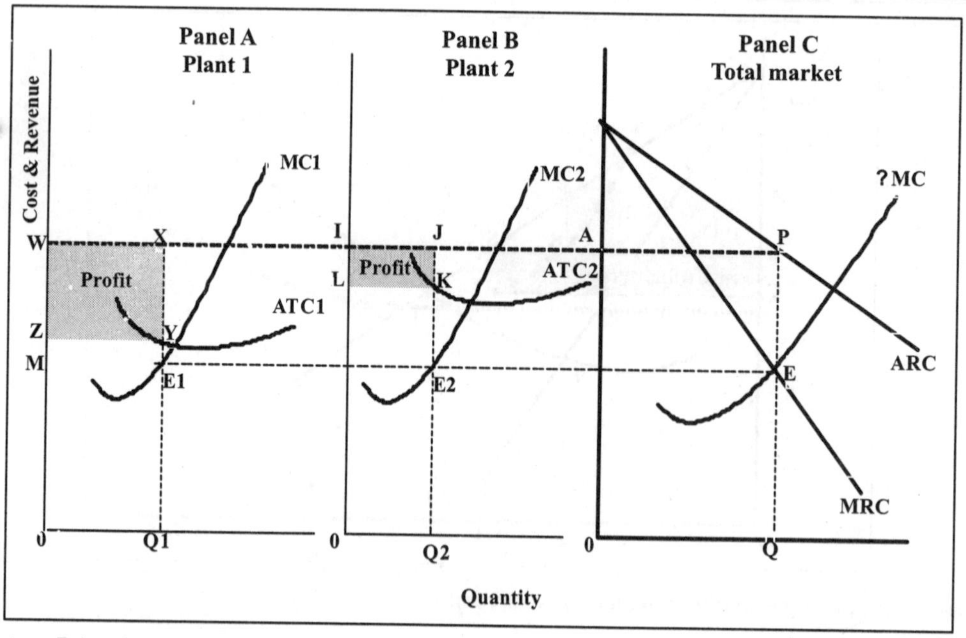

Figure 21.18: Output-Price determination in Multi-Plant Monopoly firm.

line touches the MC_1 of Plant 1 at E_1 where, it produces OQ_1 level of output and it also touches the MC_2 of Plant 2 at E_2 where, it produces OQ_2 level of output. Thus, monopolist produces OQ_1 level of output in Plant 1 and OQ_2 level of output in Plant 2, such that, total output $OQ=OQ_1+OQ_2$. He sells these outputs OQ_1 and OQ_2 at QP level of price, as decided at the industry or market level. As the cost of production of product is lower in Plant 1, he produces more output in Plant 1 *i.e.*, OQ_1 compared to Plant 2 *i.e.*, OQ_2 ($OQ_1>OQ_2$). Hence, he earns more super normal profits in Plant 1 *i.e.*, WXYZ compared to Plant 2 *i.e.*, IJKL (WXYZ > IJKL).

21.8. Price Discrimination

After going through the concepts of short run equilibrium and long run equilibrium of the monopoly firm, it is essential to discuss one of the important features of monopoly *i.e.*, price discrimination. In the earlier pages, we have seen the case, where the monopolist charges same price from all the customers and this is regarded as the single price monopoly case. But, the interesting aspect is that, the monopolist can charge different prices for the same product from different customers and this type of monopoly is called Price discriminating monopoly. The main objective of price discrimination is that, it may be more profitable than charging a single price. It occurs, when a firm charges different prices from different customers for an identical product or service, for reasons not associated with costs. It is, otherwise, defined as 'the sale of technically similar products at prices, which are not proportional to MC'. That means, the price variation of the product across the customers is not related to the cost of the product produced. In the words of Joan Robinson, *'the act of selling the same article, produced under single control at different prices to different customers is known as price discrimination'*. According to J.S. Bain, *'Price discrimination refers strictly to the practice by a seller of charging of different prices from different customers for the same good'*. Price discrimination is also called as Discriminating monopoly or Yield management and the monopolist practicing price discrimination is called Discriminating monopolist. The classification of customers into different groups for whom the product is offered at different prices refers to Market segmentation. This price discrimination is a strategy that is employed by almost every industry that has some power to determine prices. Theoretically, it is an important feature of monopoly markets, but in practice, it occurs with oligopolies such as, the airlines and even in fully competitive retail or industrial markets. This price discrimination is possible, when the monopolist sells the product in different markets in such a way that, it is not possible to transfer the product from cheaper market to the dearer market. It is, however, not possible under perfect competition, even if the two markets could be kept separate. This is because, it is based on the assumption that, same price will prevail for the same commodity at the same time. Moreover, the firm's demand curve is perfectly elastic, and hence, every seller would try to sell in the market as much as he can at the given price. While discussing the concept of price discrimination, we generally assume that, all the units of the product or products sold by the monopolist to different customers are homogeneous. Moreover, the costs of production faced by the monopolist against the product sold to different customers are also the same. The objectives of price discrimination include,

- Price discrimination helps to convert consumer surplus to producer surplus.
- Price discrimination increases the TR and profits of the monopolist

- Price discrimination helps the monopoly firm to off-load excess capacity
- Price discrimination can also be used as a technique to take market share away from rival firms.

21.8.1. Conditions of Price Discrimination

Essentially, the following conditions are required for practicing price discrimination by the monopolist:

- ***Monopoly power:*** The firm must enjoy monopoly power for the product in the market, thereby, the monopolist can gain firm control over the price of the product.
- ***Different elasticities of demand:*** The markets in which the monopolist sells the product must have different elasticities of demand, as it facilitates the monopolist to charge different prices for the product in different markets. In the market, where the demand for the product is less elastic, the monopolist charges high price and *vice versa.* According to Mrs. Joan Robinson, *'the sub-markets will be arranged in ascending order of their elasticities, the highest price being charged in the least elastic market'*. The demand of the customers must not be transferable from the high priced market to the low priced market.
- ***Different markets and customers:*** From the above condition, we can infer that, the markets and customers must be different, so as to facilitate the monopolist to charge different prices for the same product. This feature is important because of two reasons *viz.*, customers who purchase the product in low price market should not resell it in a high price market and the customers in the dearer market should not transfer themselves into the cheaper market in order to benefit from the low price. So, the monopoly firm must be able to prevent 'market seepage' or 'customer switching'.
- ***Price delusion:*** Sometimes, the customers purchase the products with an irrational feeling that, they are paying higher prices for the product because of its superior quality, although, actually it may be of the same quality. Sometimes, the price differences may be so small that the customers do not consider it worthwhile to bother about such differences.
- ***Taxes imposed on the commodity:*** In inter-state trade and international trade, certain duties are imposed on the movement of products, thereby, the monopolist will charge different prices in different markets.
- ***Ignorance:*** Ignorance of the customers would also enable the monopolist to practice price discrimination. If the customers of the product in the dearer market are ignorant of the cheaper price at which the same product is offered in another market, it enables the monopolist to practice price discrimination.
- ***Artificial differences in the product:*** Apparent differences in the nature of products and services will enable the monopolist to practice price discrimination. That means, by making artificial differences between products, *i.e.*, a monopolist may present the same commodity under different names, labels, packaging, forms etc., he can practice price discrimination. In the words of Mrs. Joan Robinson, *'various brands of a certain article which in fact are almost exactly alike may be sold as different qualities under names and labels which induce rich and snobbish customers to divide themselves from poor customers and in this way the market is split up and the monopolist can sell what is substantially the same thing at several prices'.*
- ***Legal sanctions:*** In some cases, the Government itself provides legal sanctions for the public utilities like electricity, railways etc., to charge different prices from different customers.

21.8.2. Types of Price Discrimination

The following are the important types of price discrimination:

- ***Personal price discrimination:*** If the monopolist charges different prices from different customers on the basis of their ability to pay, it refers to Personal price discrimination. Rich customers will be asked to pay more and poor customers to pay less. Ex:-Doctor service.
- ***Place or Geographical price discrimination:*** In this, the monopolist will charge different prices from the customers of different countries, states, regions or localities for the same commodity. Ex:-Dumping
- ***Trade price discrimination:*** It is also called as Use discrimination. In this, the monopolist will charge different prices for the same commodity for different types of uses. For example, electricity charges vary between industrial, domestic and agricultural uses. Similarly, accessories like springs, bolts, wires etc., will be charged higher price for cars and low prices for cycles and domestic use.
- ***Based on the nature of the product***: Unbranded products like open tea will be sold at lower price than the branded products like Brooke Bond or Lipton tea. In the case of services too, such price discrimination is practiced. During off-season, tourism department charges low and during season, they charge more from the customers.

- ☆ ***Based on age, sex and status of the customer:*** Price discrimination is also due to these factors. For example, barbers charge less for children's haircuts than adults. In military canteens, for employees, the products are offered at concessional prices.
- ☆ ***Based on the time of service:*** Depending upon the time of service offered to the customers, the firm charges different prices. For example, rates of ISD calls are different at night from those during the day, less on Sundays than on other days and some cinema houses at certain places in New Delhi, charge less for morning and matinee shows and charge more for first and second shows.

21.8.3. Degrees of Price Discrimination

So far, we understood price discrimination is the strategy followed by the monopolist to gain more revenue by charging different prices from different customers in accordance with the differences in E_P. The extent to which a monopolist can charge different prices for a product from his customers refers to Degree of price discrimination. There are three degrees of price discrimination explained by Arthur Cecil Pigou in his work, *'Economics of Welfare'*. They include,

1. Price discrimination of First degree
2. Price discrimination of Second degree
3. Price discrimination of Third degree

These three types all involve additional effort on the part of the monopolist to determine the preferences of different customers and their willingness to pay. These efforts are justified by a greater level of profits relative to that which can be earned by charging a single or uniform price.

1. Price Discrimination of First Degree

It is, otherwise, called as Personalized pricing or Optimal Pricing. According to Joan Robinson, it is also called as Perfect price discrimination. In this type of price discrimination, the monopolist charges from each individual the price that they are willing and able to pay based on their perceived E_P for the product. So, this price discrimination captures all the consumer surplus, thereby, the monopolist exploits the customers to the maximum extent. Thus, a firm engaged in first degree price discrimination will attempt to extract the entire consumer surplus from its customers as profits. Here, the monopolist charges different prices for each and every unit of the same commodity, such that, the price quoted for the commodity is equal to the demand price of the customer (*i.e.*, what the customer is willing and able to pay), thereby, leaving no consumer's surplus to the customer. For this, two conditions are required *viz.*, the customers are separated from each other and monopolist should deal with each customer on a take it or leave it basis. In fact, the first degree price discrimination manifests itself in the form of as many prices as many customers. For the most part, this type of price discrimination is considered unattainable and it is seen in case of rare products like diamonds, jewels, precious stones etc. The common practical examples of this degree of price discrimination include the prices charged by the doctors, lawyers etc.

Price	Units of commodity	TR	MR	TR_{D1}	MR_{D1}
11	0	---	---	---	---
10	1	10	10	10	10
9	2	18	8	19	9
8	3	24	6	27	8
7	4	28	4	34	7
6	5	30	2	40	6
5	6	30	0	45	5
4	7	28	-2	49	4
3	8	24	-4	52	3
2	9	18	-6	54	2
1	10	10	-8	55	1

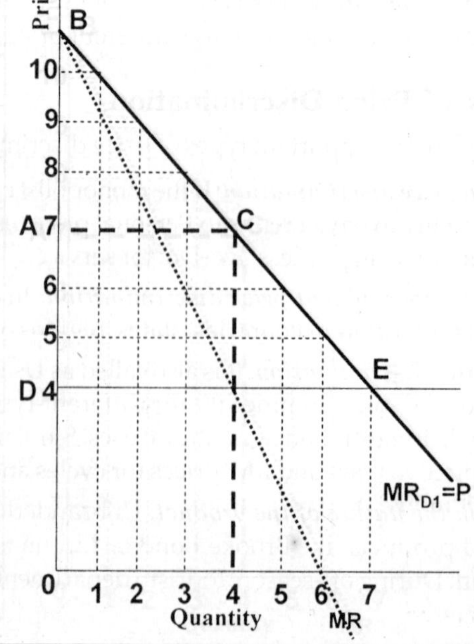

Figure 21.19: Price discrimination of First degree.

- ☆ The concept of first degree price discrimination is explained through Figure 21.19. When there is no price discrimination, the monopolist charges a uniform price for all the units of the commodity. When the price of the commodity is Rs. 10, one unit is demanded and the TR is Rs. 10. When the price is Rs. 9, two units are demanded and the TR is Rs. 18. So, the MR of the additional unit is Rs. 8. Notice that, when we have a single price monopoly, demand has the general form P = A – BQ, then the MR = A – 2BQ.
- ☆ Since the customers are simply price takers, the monopolist wishes to go for price discrimination of the product. In the Figure (table), TR_{D1} and MR_{D1} are the TR and MR of the firm in the case of adopting first degree price discrimination and it is interesting that, Price=MR_{D1} *i.e.*, demand curve and MRC are same. That means, under first degree price discrimination, if demand is P = A – BQ, then MR = A – BQ. So, in first degree price discrimination, the monopolist knows what the individual is willing to pay for each unit and thereby, he is able to extract that amount.
- ☆ The monopolist knows the individual will pay 10 for the first unit. Since two units are demanded at a price of Rs.9, the monopolist knows the individual is willing to pay Rs. 9 for the second unit. So, on the two units, the monopolist can charge Rs.10 for the first one and Rs. 9 for the second unit. If the individual wishes to purchase three units of the commodity, the monopolist charges Rs.10 for the first unit, Rs. 9 for the second unit and Rs. 8 for the third unit. This is just like a quantity discount idea. That is, pay Rs. 10 for one or get two for Rs. 19. So, the TR_{D1} for two units is thus Rs. 19 and the MR_{D1} for the second unit is Rs. 9. We follow the same idea, the rest of the way down the columns.
- ☆ We know the point of optimality is where, MR = MC.
- ☆ In the single price monopoly case, if MC is Rs. 4 all the time, the monopolist will produce four units of output (because, MC=MR) and charge Rs. 7 for each unit of the output. So, the TR = Rs. 28. But, in case of first degree price discrimination, if the MC is Rs. 4, the monopolist can sell seven units of output *i.e.*, first unit for Rs.10, second unit for Rs.9 and so on down to seventh unit for Rs. 4. So, the TR = Rs. 49.
- ☆ Considering the single price monopoly case, the consumer surplus is the small triangle with the horizontal line at Rs. 7 as the base of the triangle *i.e.*, ABC. But, under first degree price discrimination, the consumer surplus is the large triangle formed by the vertical axis, the horizontal line a Rs. 4 and the demand line *i.e.*, DBE. So, under first degree price discrimination, the monopolist will take away the entire consumer surplus that would have existed under single price monopoly case. So, by price discrimination, the monopolist will charge Rs. 10 for the first unit, Rs. 9 for the second unit, on down to Rs. 4 for the 7th unit of the commodity. He will not sell eight or more units because, the MR_{D1} on those units is less than the MC and thus, reduces the profit. Thus, under first degree price discrimination, the monopolist
    - ❖ sells more output than under single price monopoly case
    - ❖ charges a different price on each unit and the last unit has Price = MC
    - ❖ the entire consumer surplus is pocketed by the discriminating monopolist, as he charges different prices for different units of the same product thereby, the revenue of monopolist increases.

2. Price Discrimination of Second Degree

In this degree of price discrimination, the monopolist will classify the customers into different groups or slabs and charges different prices from each group. The monopolist charges each group the highest price which the marginal customer belonging to that group is willing to pay thereby, the product is accessible for other customers of the same group. That means, those customers who are affordable or prepared to pay more for the product will get higher consumer surplus, if they purchase the product at low price. This type of price discrimination also involves the establishment of a pricing structure for a particular good based on the number of units sold. In this case, the monopolist charges a higher per-unit price for fewer units sold and a lower per-unit price for larger quantities sold. So, here the monopolist will attempt to extract some of the consumer surplus as 'profits' with residual surplus remaining with the consumer over and above the actual price paid. The classical examples for this type of price discrimination is seen in public utilities like railways, electricity supply etc. Supermarkets also give discounts, if large quantities are bought. We commonly see, Electricity Corporation supplies electricity for domestic usage under different tariff rates in accordance with different units (kilowatts) of electricity consumed. The Corporation charges higher price for the first slab of kilowatts of electricity consumed and as more units of electricity consumed, the rate falls with subsequent slabs.

- ☆ In the Figure 21.20, DD_1 is the demand curve for electricity by the customers in a town. The Electricity Corporation charges Q_1P_1 rate per kilo watt upto OQ_1 units of power consumption (first slab), for consuming next Q_1Q_2 units, the rate is lowered to Q_2P_2 (second slab) and for consuming Q_2Q_3 units, the rate is further lowered to Q_3P_3 units (third slab). Q_3P_3 is the lowest rate, which will be charged even if the consumer consumes more than OQ_3 units of electricity. Since, the Corporation is charging different prices for different slabs of electricity consumed, the TR is given by $(OQ_1 \times P_1Q_1) + (OQ_2 \times P_2Q_2) + (OQ_3 \times P_3Q_3)$.

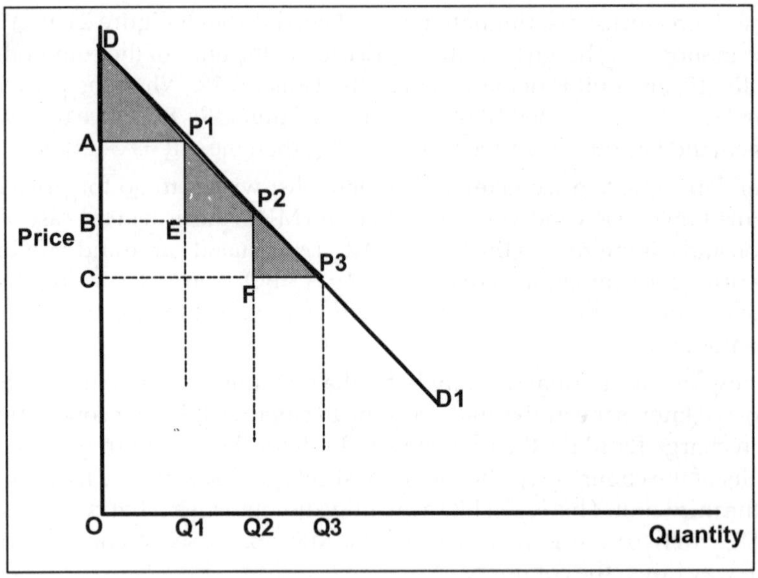

Figure 21.20: Price discrimination of Second degree.

- ☆ Suppose, if there is no price discrimination and if the Electricity Corporation charges uniform price Q_3P_3 from all the customers, irrespective of different units of electricity consumed, then the company will get a TR of OQ_3P_3C, thereby, leaving a consumer surplus of DCP_3.
- ☆ From the above two cases, due to price discrimination, the TR is increased for the Electricity Corporation given by,
  $[(OQ_1 \times P_1Q_1)+(OQ_2 \times P_2Q_2)+(OQ_3 \times P_3Q_3)] - [OQ_3P_3C]$.
- ☆ As a result, due to price discrimination, some of consumer's surplus is taken away by the Corporation *i.e.*, $ABEP_1 + BCFP_2$ (*i.e.*, of DCP_3). That means, consumer's surplus is decreased when compared to, if all the customers are provided electricity at uniform price Q_3P_3 (*i.e.*, DCP_3).
- ☆ But, this method of price discrimination allows some consumer's surplus for different categories of customers purchasing electricity at different prices, as here, for a group of units of electricity *i.e.*, slab-wise, the price is fixed and not for each and every unit of electricity consumed, as in the case of price discrimination of first degree. The amount of consumer's surplus allowed under this price discrimination include,
  - ❖ DAP_1 is the consumer's surplus gained by the customers consuming OQ_1 units of electricity
  - ❖ P_1EP_2 is the consumer's surplus gained by the customers consuming Q_1Q_2 units of electricity
  - ❖ P_2FP_3 is the consumer's surplus gained by the customers consuming Q_2Q_3 units of electricity

From the above discussion, we can differentiate between price discrimination of first degree and price discrimination of second degree (Table 21.2).

Table 21.2: Differences between First degree and Second degree price discriminations.

First Degree Price Discrimination	*Second Degree Price Discrimination*
The situation in which the monopolist sets different prices for each customer and for each unit purchased by each customer depending on their willingness and ability to pay. This attempts to capture the entire consumer surplus.	Monopolist charges different prices, when individuals purchases good or service in different quantities *i.e.*, monopolist sell different quantities at different prices
The customers are not differentiated into different groups or slabs.	The customers are differentiated according to different groups or slabs.
The monopolist charges a different price for each individual unit of the product. That means, price of the product differs from person to person.	The price differs across the units of the good, but not across the people. For example, in case of electricity consumption, the monopolist charges a uniform price for a number of units in one slab/group and as the slab size increases, the price will be lowered.
No consumer's surplus is left with the customer.	The monopolist takes away only a portion of the total consumer's surplus and the remaining portion is left with the customers

3. Price Discrimination of Third Degree

This is the most commonly practiced price discrimination by the monopolist. In this price discrimination, the market is divided into many sub-markets and in each sub-market, the monopolist charges different prices from the customers depending upon the demand for the product in that market. With the quantum of total output fixed on one hand, the monopolist will charge different prices for the same product from each sub-market based on demand for the product in each sub-market. The division here is among classes of customers and not among individual customers. This degree of price discrimination is possible only, if the classes of customers can be kept separate. Secondly, the various groups of customers must have different elasticities of demand for the product. The segment with a less elastic demand pays a higher price for the product than the segment with a more elastic demand. The customer faces a single price in each category of customers. He can purchase as much as desired at that price. Examples of third degree price discrimination include business vs tourist airfares, business vs residential telephone service, ticket prices in movie theaters etc.

Assume the monopolist sells his product in sub-two markets, M_1 and M_2 [Figure 21.21 (A)]. In sub-market M_1, there is high E_D and in sub-market M_2, there is less E_D for the product. The ARC for sub-market M_1 is ARC_1 with a flat slope (more elastic demand) and the ARC for sub-market M_2 is ARC_2 with a steep slope (less elastic demand). The corresponding MRCs in M_1 and M_2 sub-markets are MRC_1 and MRC_2 respectively.

In Panel C of the Figure 21.21 (A), MRC_T and ARC_T are downward sloping curves for the entire market, drawn by lateral summation of MRC_1 and MRC_2 and ARC_1 and ARC_2 of M_1 and M_2 sub-markets respectively. ΣMC is the cumulative MC of both the plants employed in both the sub-markets and it intersects the MRC_T at point E. So, the monopolist is at equilibrium at 'E' and produces OQ_T quantity of output by fixing the price of the product at $OA=Q_TP$. The monopolist must divide this total output OQ_T between the two sub-markets. So, a horizontal line is drawn from E to M parallel to X-axis, which passes through equilibrium points of MRC_1 and MC_1 of sub-market M_1 and MRC_2 and MC_2 of sub-market M_2 at points E_1 and E_2 respectively, representing OQ_1 and OQ_2 outputs respectively. Hence, $OQ_T = OQ_1 + OQ_2$.

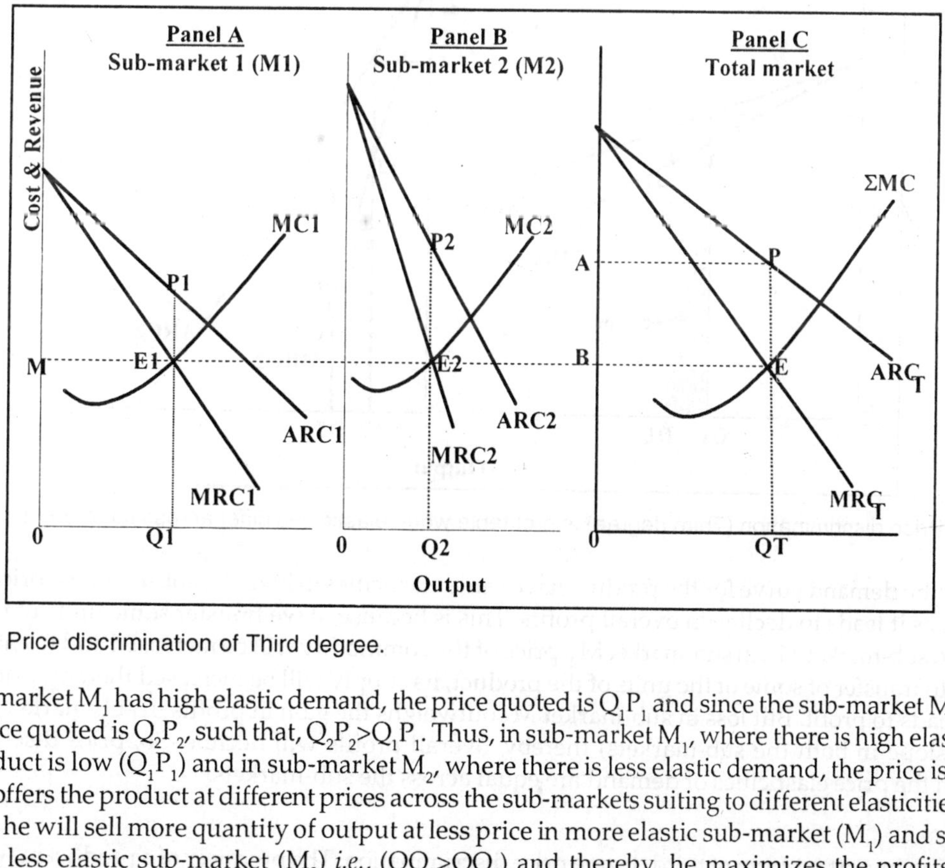

Figure 21.21(A): Price discrimination of Third degree.

Since, sub-market M_1 has high elastic demand, the price quoted is Q_1P_1 and since the sub-market M_2 has less elastic demand, the price quoted is Q_2P_2, such that, $Q_2P_2>Q_1P_1$. Thus, in sub-market M_1, where there is high elastic demand, the price of the product is low (Q_1P_1) and in sub-market M_2, where there is less elastic demand, the price is high (Q_2P_2). So, the monopolist offers the product at different prices across the sub-markets suiting to different elasticities of demand for the product. So, he will sell more quantity of output at less price in more elastic sub-market (M_1) and sell less quantity at high price in less elastic sub-market (M_2) *i.e.*, ($OQ_1>OQ_2$) and thereby, he maximizes the profits. If he sells the commodity at uniform price in both the sub-markets (say at high price), the demand is low in high elastic sub-market (M_1) and if he sells the commodity at low price in both the markets, he may leave large amount of consumer's surplus in less elastic sub-market (M_2). Hence, he goes for price discrimination and tries to maximize the profits. From the above discussion, we can conclude the following interesting aspects:

- ✰ The conditions essential for the discriminating monopolist to practice price discrimination include, $MRC_T=\Sigma MC$ and $MRC_1=MRC_2=\Sigma MRC_T$

- ☆ The greater the E_D for the product in a market, the greater the proportion of output sold in that market and *vice versa*.
- ☆ So, in third degree price discrimination, different prices are charged from different group of customers depending upon the E_D for the product. The greater the E_D for the product in a market, the less is the price quoted for the product and *vice versa*.

It is also essential to note that, (third degree) price discrimination is possible, when demand curves have different elasticities in the sub-markets and this the only way to maximize the profits. It is even worthwhile on the part of the monopolist to increase the profits by transferring the some units of the product from the sub-market M_2, with less elastic demand to the sub-market M_1, with high elastic demand. As shown in the Figure 21.21 (B), let us assume that, C_2D_2 units of product from sub-market M_2 (less elastic market) is transferred to sub-market M_1 (more elastic market), as C_1D_1. Due to this, the TR is sub-market M_2 is declined by $A_2B_2C_2D_2$ but due to transfer of units of product to sub-market M_1, there the TR is increased by $A_1B_1C_1D_1$. So, the net increase in TR or profits is more and this contributes to higher profits to the monopolist due to increase in sales of the product in more elastic sub-market M_1. So, it is obviously worthwhile for the monopolist to transfer some of the units of the product from less elastic sub-market to more elastic sub-market, as due to transfer of some units of the product, there will be rise in price of the product in less elastic sub-market and there will fall in price of product in more elastic sub-market. This shifting of units of the product from less elastic sub-market to more elastic sub-market can be continued, till the MRs of both markets are equalized, where the profits then attains maximum.

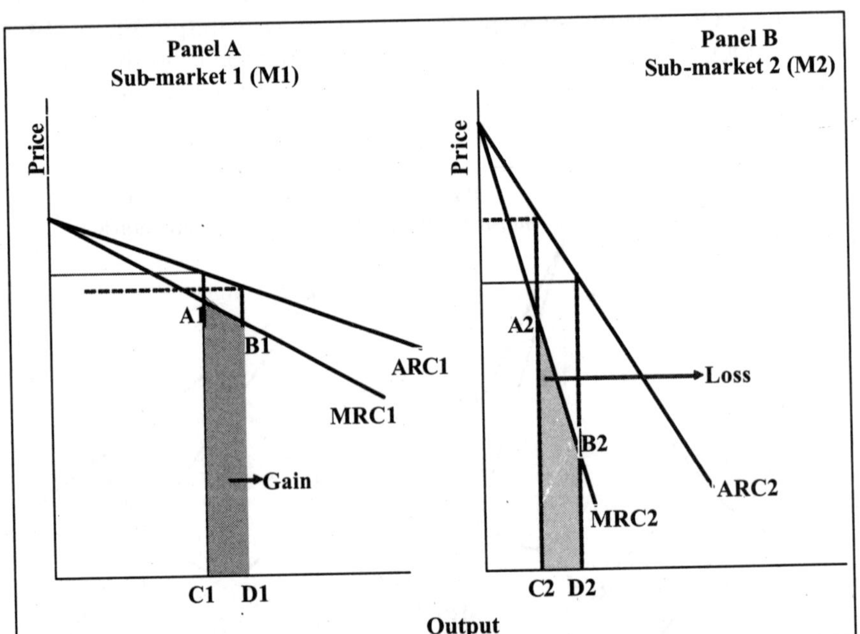

Figure 21.21(B): Price discrimination (Third degree) is profitable when elasticities differ across the markets.

However, if the demand curve for the product have same elasticities in both the sub-markets, price discrimination is not advisable, as it leads to decline in overall profits. This is because, if we transfer some units of the product from sub-market M_2 to sub-market M_1, in sub-market M_2, price of the commodity will be increased and it leads to loss. In sub-market M_1, due to transfer of some of the units of the product, its supply will be increased thereby, price of the product will fall and it leads to profit. But loss in sub-market M_2 outweighs the gain in profits in sub-market M_1 (as the ARCs have the same slope in both the sub-markets) thereby, overall profits will decline. So, price discrimination is not profitable, when the price elasticities of demand are equal across the sub-markets.

Dumping

Dumping is an important form of third degree price discrimination. The producer of a product may enjoy monopoly power in the domestic market, but he may not have the same in the world or international market, as he has to face lot of competition from infinite number of firms (*i.e.*, he has to face perfect competition). So, the monopolist has to discriminate between domestic and world markets. In the world market, he has to keep the price decided on the basis of perfect competition (perfectly elastic market) and he can keep the price at higher level in the domestic market (less elastic market). So, if the monopolist sells the product at lower price in the world market than in the domestic market, he is said to be dumping in the world market.

- In the Figure 21.22, ARC_W is the ARC in world market and in case of perfect market competition it is equal to MRC_W = Price of commodity in world market (P_W)= Demand curve for commodity in world market (D_W) and hence, it lies parallel to X-axis.

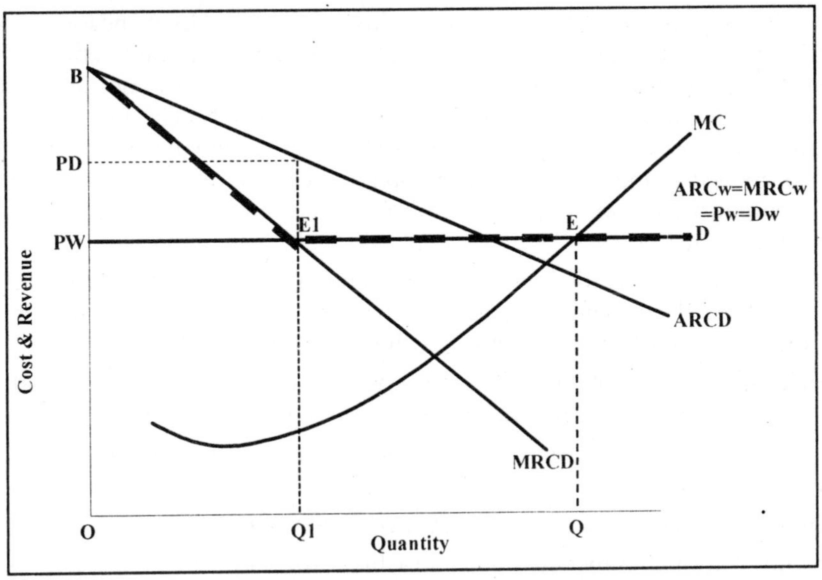

Figure 21.22: Price discrimination of Third degree – Dumping.

- ARC_D is the ARC in domestic market and MRC_D is the MRC in domestic market, which are downward falling curves, as in case of monopoly competition.
- BE_1ED is the Aggregate MRC, which is the summation of MRC_D and MRC_W.
- The MC curve of the firm cuts the Aggregate MRC *i.e.*, BE_1ED at point E, which is the equilibrium position, where total output, OQ is produced by the monopolist. Now, this output OQ has to be supplied to both domestic and world markets.
- The equilibrium position E on BE_1ED when extended to MRC_D cuts at point E_1 and it refers to equilibrium output OQ_1 in the domestic market. So, of the total output produced OQ by the monopolist, he sells OQ_1 quantity in the domestic market and remaining Q_1Q quantity in the world market.
- The monopolist sells the commodity at OP_W price in world market (lower price), as the market is more elastic and he sells the commodity at higher price OP_D in domestic market, as the domestic market is more inelastic or less elastic.
- This practice of selling the commodity at low price in world market and at higher price in domestic market is called dumping.

21.8.4. Justification of Price Discrimination

The question arises, 'whether charging different prices for the same product from different customers is really justifiable'? To answer this question, we have to analyze both the merits and demerits of price discrimination and the same is discussed here under.

1. Merits of Price Discrimination

Price discrimination is beneficial both to the firms and to the society and it is evident from the following points:

- Because of price discrimination, for poor sections of the society, the commodity may be offered at low price and for rich sections of the society, the commodity will be offered at high price (with the basic idea of the leaving no consumer's surplus or keeping the consumer's surplus as low as possible), such that, for all sections of the society, the commodity can be sold or offered. If no price discrimination is followed, the poor customers may not have access for the product. For example, in India, the electricity charges are low for farmers compared to industrial and household sectors.
- The concept of price discrimination generally favours the community to whom the product is offered at low price and disfavours the community to whom the product is offered at higher price. But, the net effect of price discrimination is important *i.e.*, how much percent of population is benefitted and if large scale poor sections of the society are favoured, then we can justify price discrimination.

- Through price discrimination, the firm maximizes the revenue. This will enable the firm to stay in the business, who otherwise would have made a loss.
- In case of geographical price discrimination (dumping), when the commodity is sold at high price in the domestic market (less elastic market) and commodity sold at low price in the foreign market (perfectly elastic market), price discrimination is clearly detrimental to the home community. But, if the firm experiences law of increasing returns or decreasing costs, even the less elastic market (domestic market) will get benefitted in case of price discrimination when compared to simple monopoly. This geographical price discrimination can also be justified because, if a country sells a product in foreign market (perfectly elastic market) at low price, it means more demand, more earning of foreign exchange, larger use of resources in the country, more expansion of output, more employment and income to the community etc. So, price discrimination encourages exports from the country.
- Price discrimination is also justified, as it helps in promoting economic welfare. Government, usually, charges high prices from the rich customers for the product or service offered by it and the funds so collected from rich sections of the society will be used to subsidize the products meant for the poor.
- Price discrimination is also beneficial to the society, as it helps in reducing inequalities in personal incomes, when higher prices are charged from rich people than from the poor people.
- The increased revenue for the firm through price discrimination can be used for R&D, which benefits the customers.
- Price discrimination increases the monopoly power of the producer.
- Price discrimination ensures the monopolist to achieve higher profits.
- The total output produced by the monopolist practicing price discrimination is larger than under simple monopoly.

2. Demerits of Price Discrimination

Though price discrimination ensures significant benefits to the monopolies and to the society, it suffers from the following limitations:

- There is decline in consumer's surplus especially, for the customers who purchase the product a high prices.
- In international trade, if a country resorts to dumping and export the products to other country at low prices, the producers in the importing country are adversely affected.
- There may be administration costs in separating the markets.
- Some times, the monopolist goes for price discrimination with a view to the exploit the customers and to maximize profits. For example, a seed producing firm enjoying monopoly status, charges very high price for the seed from the farmers during sowing season of the crop.
- Profits from price discrimination could be used to finance predatory pricing.
- Those who pay higher prices for the product may not be the richest. For example, if the monopolist charges higher prices from the adults compared to children, adults could be unemployed.

From the above discussion, we can infer that, the merits of price discrimination outweigh the demerits and thereby, we can justify the price discrimination followed by the monopolist. However, the justification we concluded, all depends upon the net positive effects of the price discrimination on the societal welfare. But, if the price discrimination is followed with the sole aim of profit maximization by the firm, but at the cost of customers' welfare, then it is considered socially unjustified.

21.8.5. Differences between Simple Monopoly and Discriminating Monopoly

The following points highlight the differences between simple monopoly and discriminating monopoly:

- In simple monopoly, the monopolist charges a uniform price for the product from different customers. However, in case of discriminating monopoly, the monopolist charges different prices from different customers for the same product.
- In discriminating monopoly, the monopolist is benefited by expanding the output, TR and profits beyond those which the simple monopolist would realize.
- In discriminating monopoly, the monopolist charges lower price from the poor customers and higher price from the rich customers for same product, unlike simple monopolist, where he charges uniform price from all the customers for the same product. So, discriminating monopoly ensures social welfare by safeguarding the interests of both rich and poor customers. It also facilitates redistribution of income in favour of the poor. So, discriminating monopoly is more socially desirable.

- ☆ In simple monopoly, the rich customers enjoy considerable consumer surplus in having the commodity whereas, in discriminating monopoly, the monopolist exploits the consumer surplus of the buyers, as he quotes different prices for different customers keeping in view the ability of the customer in willing to have that commodity.
- ☆ In case of public utilities, if the monopoly firm charges uniform price, it fails to recover LRATC thereby, it forces to withdraw even though the utilities or services are essential for societal welfare. In such cases, discriminating monopoly will fetch higher revenue and it covers LRTC thereby, ensures survival of the firms.

21.9. Monopoly Power

The degree of control, which the monopolist can exercise on the price and output policies of his product is called Monopoly power. If the monopolist has uncontrolled monopoly power, it is undesirable and then the Government must know the extent or Degree of Monopoly Power (DMP), so as to plan the strategies to limit or eliminate it. The following are the different methods suggested by different Economists to measure the DMP:

1. Lerner's Index

According to A. P. Lerner, the DMP is measured by formulating the relationship between price (P) and MC and it is given by

$$DMP = [(P - MC)/P] \qquad \textit{Equation 21.2}$$

We know, to maximize the profits under monopoly market competition, the criterion is MC=MR. So, we can write the above Equation 21.2 in terms of MR as,

$$DMP = [(P - MR)/P] \qquad \textit{Equation 21.3}$$

We know, $MR = P((E_P\text{-}1)/E_P)$, as per the Equation 18.11. So, the DMP is given by,

$$DMP = [(P-(P((E_P-1)/E_P))/P] = (E_PP - E_PP+P)/E_PP = 1/E_P \qquad \textit{Equation 21.4}$$

Hence, the DMP or Lerner's index is the reciprocal of E_P. The lower the E_P for the product, the greater is the monopoly power and *vice versa*.

From the above two Equations 21.2 and 21.4, we can infer that, [(P – MC)/P] is the mark-up over MC as a proportion of price. That means, the mark-up over price is inversely equal to the E_P for the product. This is shown below as,

$$DMP = [(P - MC)/P] = 1/E_P \qquad \textit{Equation 21.5}$$

From the above equation 21.5, we can conclude that, the lower the E_P, the greater the mark-up over MC as a proportion of price and hence, higher level of price is fixed for the product indicating the DMP is stronger and *viceversa*. So, the Lerner's Index provides a measure of how much a firm mark-up the price over MC. The higher the Lerner's Index, the greater the firm's markup and greater is the DMP. The concept of Lerner's measure is explained through the Figure 21.23. In Panel A, at OQ level of equilibrium output, the E_P is more (ARC is more flat), thereby, the power of the

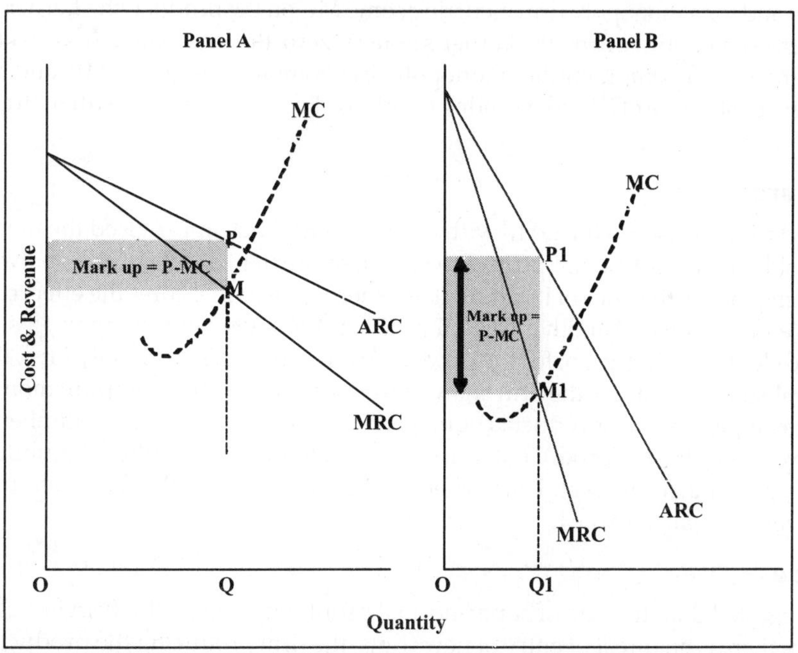

Figure 21.23: Lerner's measure of monopoly power – Relation between price mark up and E_P.

producer to raise the price above the MC is less. So, the mark up *i.e.*, P-MC is less. On the contrary, at OQ_1 level of output (Panel B of the Figure 21.23), the E_P is less (ARC is more steep), thereby, the power of the producer to raise the price above the MC is more. So, the mark up *i.e.*, P-MC is more.

Monopoly Power Under Perfect Market Competition

We know, in perfect market competition, since, MC=MR=Price, the DMP is zero and this is given by,

DMP in perfect competition = [(P–MC)/P] = [(P – P)/P] = 0 (zero).

Monopoly Power when Monopolist Incurs Zero MC

When the monopolist incurs no cost in the production of product (say, free good like mineral springs), the MC is zero, thereby, the DMP of the monopoly firm is given by,

DMP in monopoly (free good) = [(P–MC)/P] = [(P – 0)/P] = 1.

So, Lerner's index ranges from zero to unity. The greater is the value of Lerner's index, the greater is the monopoly power exercized by the monopolist.

Limitations of Lerner's Measure

Though this concept of Lerner's measure is very interesting in explaining the DMP, it suffers from the following limitations:

- ☆ This measure solely depends upon the gap between price and MC of the product. This gap may be high even when the demand for the product is less elastic.
- ☆ The gap between price and MC of the product also depends upon the supply of the product into the market. If the supply of the product into the market is greatly restricted by the monopoly firm, price of the product will shoot and thereby, gap increases between price and MC. But, restriction of market supply is not the true indicator of higher DMP because, a firm may reduce supply due to increase in factor costs or under-utilization of plants and machinery, under-investment by the business firm etc.

2. Kleczka's Measure

Kleczka's measure of monopoly power is slightly different from that of Lerner's measure. Just like Lerner's measure, even Kleczka's index of monopoly power is also based on the divergence between price and MC. But, here, the ratio between price (P) and MC directly yields the DMP of the firm. So,

Kleczka measure of DMP = (Price/MC).

Since, price is invariably higher than MC under monopoly (because, ARC is higher than MRC), the value of Kleczka's index is greater than one under monopoly. But, in perfect competition, as ARC=MRC=Price, the Kleczka's index is equal to one under perfect competition. So, higher the value of Kleczka's index, higher is the DMP.

So, according to Lerner's index, when DMP is equal to one, it implies the monopoly firm is enjoying highest DMP (if MC is equal to zero) and for a monopoly firm incurring some MC on the product, the Lerner's index ranges between zero to unity. For a perfect competitive firm, the Lerner's index is zero. But, according to Kleczka's measure, if the value of Kleczka's index is greater than one, it implies monopoly firm is enjoying higher DMP, and if the firm incurs no MC in producing the product, the value of Kleczka's index is infinity. For a perfect competitive firm, the Kleczka's index is unity.

3. Triffin's Measure

N. Kaldor and R. Triffin measured the DMP with the help of E_C. They improvised the measurement of DMP over Lerner's measure considering E_C instead of E_P. This E_C estimates the extent of substitution between the products of two firms, when one of them changes the price of its product. In other words, it measures the effect on demand for a product, as a result of a change in the price of another related product. We know, in monopoly, a single firm produces the product, which has no close substitutes. So, for the product produced by the monopoly firm, the E_C is zero, because of complete absence of substitutes, thereby, the firm enjoys absolute monopoly power (pure monopoly). On the contrary, under perfect market competition, the product is homogenous and there are infinite number of firms producing the same product, thereby, the E_C for the product of a firm is equal to infinity. So, the firm under consideration cannot exercise any monopoly power in the market. Thus, when E_C is low or zero, the DMP is maximum and *vice-versa*. In other words, the DMP is inversely related to the E_C.

4. Bain's Measure

Prof. J.S. Bain suggested that, the size of super-normal profits enjoyed by the firm indicates the DMP. In case of monopoly firm, where there is no threat of entry of new firms, the firm producing the product enjoys highest possible super-normal profits, thereby, it enjoys absolute DMP. We know, under pure monopoly, the divergence between price and ATC, *i.e.*, profit is the maximum and the monopolist enjoys absolute monopoly power. Further, due to complete absence of possible threat of entry of rivalry firms, these profits will not be competed away. But, as the threat of entry of new firms increases, the DMP of the firm decreases. In case of perfect market competition, where there is free entry and exit of firms, in the long run, the super-normal profits enjoyed by the firms in short run, are completely competed away, thereby, the DMP becomes equal to zero. According to Bain, a firm's profits over and above the level of normal profits is due to monopoly power. So, greater is the divergence between price and ATC, greater is the DMP. But, this measure too suffers from the following limitations:

- It is difficult to measure true monopoly power or profit because, it also depends upon the extent of amortization of fixed costs from gross income earned by the business firm.
- All profits accruing to firm are not monopoly profits because, it is influenced by windfall prices, demand, changes in cost conditions etc. So, these influences should be deducted from total profit, so as to analyze true monopoly power.

5. Rothschild's Measure

Rothschild has defined the DMP, as the ratio of the slope of the firm's demand curve and the slope of the industry's demand curve. Under pure monopoly, firm's demand curve coincides with the industry's demand curve, thereby, Rothschild's index is equal to one and it implies, there is absolute DMP for the firm. However, under perfect market competition, the DMP is equal to zero, since the firm's demand curve is a horizontal straight line with zero slope. Thus, according to Rothschild, the DMP lies between zero and one.

In the Figure 21.24, DD is the industry demand curve, 'dd' is the firm's demand curve. Firm's demand curve 'dd' is more elastic (flat slope) than industry demand curve 'DD' (steep slope). So,

DMP= Slope of firm's demand curve/Slope of industry demand curve

$= (BC/CE)/(AC/CE) = BC/AC$

Since, BC is less than AC, the firm enjoys less DMP. On the contrary, if the firm's demand curve is more steep, it enjoys higher DMP. But, the major short coming of this method is that, it is solely based on the demand conditions and neglected the supply conditions of the firms, which also influences the prices.

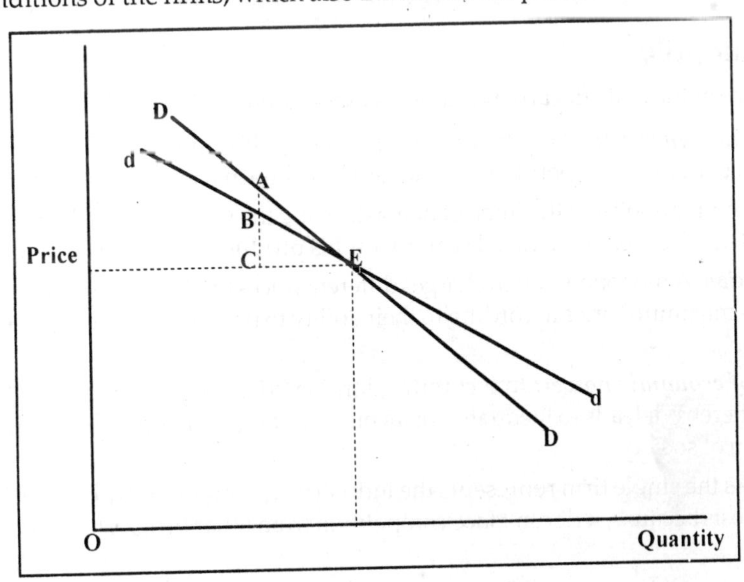

Figure 21.24: Monopoly Power- Rothschild's measure.

The different measures employed for studying the DMP was summarized and shown through Table 21.3.

21.10. Merits of Monopoly

The following are the merits of monopoly competition:

- Monopoly firms with their large-scale production possibilities can enjoy both internal and external economies of scale. This will result in the reduction of cost of production, thereby, price of the product. This is beneficial to the customers.
- Monopoly firms with their vast financial resources can invest on R&D and this enables the firms to innovate quickly. Ex. Pharmaceuticals

Table 21.3: Different measures of DMP.

Measure of DMP	*Formula*	*Monopoly*		*Perfect Market Competition*
		For Commodity MC=0	*For Commodity MC≠0*	
Lerner's index	(P-MC)/P	1	0 to 1	0
Kleczka's index	P/MC	Infinity	>1	1
Triffin's measure	$1/E_c$	0		Infinity
Rothschild's measure	(Slope of the firm's demand curve)/(Slope of the industry's demand curve)	1		0

- ☆ Monopolies with their huge financial strength can face the business depression phase more effectively.
- ☆ When a number of uneconomic firms producing same product combine together as a monopoly, they can manage the business more efficiently and enjoys economies of large scale.
- ☆ The operation of monopolies is indispensable in case of public utilities like electricity, railways, water supply etc. If there is competition in such public utilities, the fixed costs will rise and thereby, the price of the product or service at the customer level will be increased.
- ☆ By practicing price discrimination, the monopolist can expand the business in different markets.
- ☆ A monopoly firm compared to a large number of small firms can handle the foreign competition in a better way.
- ☆ Monopoly avoids duplication and hence, no wastage of resources.
- ☆ Monopolies may use price discrimination, which benefits the economically weaker sections of the society. For example, Indian railways provide discounts to students travelling through its network.
- ☆ Monopolies can afford to invest in latest technology and machinery in order to be efficient and to avoid competition.

21.11. Evils of Monopoly

The following are some of the evil effects of monopoly competition:

- ☆ ***Lower output and higher prices:*** Under monopoly, the equilibrium price is high and equilibrium output is lower compared to perfect competition. This leads to decline in consumer's surplus.
- ☆ ***Restrict output:*** A monopolist in the interest of maximizing the profits, may restrict the output and raise price of the product. Thus, he creates artificial scarcity for his product in the market.
- ☆ ***Price discrimination:*** A monopolist often charges different prices for the same product from different customers. Thus, he extracts maximum price according to their ability to pay for the product, thereby, leaving no surplus to the customers.
- ☆ ***Concentration of economic power:*** In a country dominated by monopolies, wealth is concentrated in the hands of a few, thereby, it leads to inequality of incomes among the people. This is against the principle of the socialistic pattern of society.
- ☆ ***X-Inefficiency:*** As the single firm represents the industry under monopoly competition, the firm shows less incentive to cut costs because, it doesn't face competition from other firms. Therefore, the ATC curve is higher than it should be.
- ☆ ***Productive inefficiency:*** A monopoly firm is productively inefficient because, it is not producing the output at the lowest point on the LRATC curve.
- ☆ ***Lower prices to suppliers:*** A monopoly firm may use its market power and pay lower prices to its suppliers in purchasing the raw materials.
- ☆ ***No technical progress:*** Lack of competition may not induce the monopolist to plan for improved techniques of production. This is because, a monopolist prefers to maintain or enjoy monopoly gains in the market without any threat of competition from the rivalry firms. So, he is supposed to be very conservative with regard to technological inventions and innovations. He even prefers to claim legal protection for the process or for the product, so as to block the entry of new firms into the industry.
- ☆ ***Loss of consumer's welfare:*** The consumer is exploited to a greater extent in monopoly compared to perfect market competition, as the monopolist can regulate either the output supply or price in the market. When the

monopolist regulates the supply of the product in the market, it leads to inefficient allocation of resources in the industry.

- *Wastage of resources:* The monopolist will not increase the output upto the minimum point of LRATC curve, as if he increases the output, the price will fall and his profits are affected. So, the monopoly firm will not reach the optimum size. This infers that, even though the monopolist can increase the output with the same plant and equipment, he does not. So, there is some excess or reserve capacity of plant under monopoly, thereby, leading to wastage of resources. Another type of wastage occurs, if the monopoly incurs costs to maintain its monopoly position. These resources could instead be used for productive purposes. For example, the monopoly could be wasting resources in order to lobby government officials for favorable legislation or contracts.
- *Price discrimination is harmful:* If the monopolist intends to practice price discrimination without increasing the output, it is harmful to the society, as the monopolist exploit the customers in charging very high prices for the same product.
- *Unfair business practices:* The monopolist may resort for unfair business practices like destroyer pricing, bulk purchasing of raw material at low prices etc., thereby, aim to enjoy monopoly status and try to exploit the customers.
- *Encourages political corruption:* Some monopolies to attract special favours from the Government, will contribute large sums to the political parties and try to influence the economic and trading policies suiting to their business prospects.
- *Super-normal profits:* Even in the long run production programme, the monopolist is capable of reaping super normal profits in the business and this indicates loss in consumer's welfare.
- *Income inequalities:* Greater exploitation of the customers often leads to unequal income distribution in the economy.
- *Dead weight loss:* Monopoly creates a deadweight loss due to the fact that, the monopoly restricts supply below the socially efficient quantity.
- *Outdated goods:* Lack of competition may lead to low quality and out dated goods and services and may not lead to improved product innovation.
- *Higher prices to suppliers:* A monopoly may use its market power and pay lower prices to its suppliers. Ex. Supermarkets have been criticized for paying low prices to farmers.
- *Diseconomies of scale:* It is possible that, if a monopoly gets too big, it may experience diseconomies of scale *i.e.*, higher LRATC

21.12. Regulation of Monopoly

When compared to the perfect market competition, monopoly competition leads to more evil effects than benefits from the stand point of society as a whole with reference to output, price of the product, income distribution, distribution of economic power, overall economic development etc. These adverse aspects due to the monopoly competition clearly reflects in aggravating the problems of unemployment and under employment, under allocation of resources, poverty etc. So, it calls for regulating the monopoly, so as to overcome the above evil effects on the economy. The following are the different measures followed to regulate the monopoly competition:

1. *Promotion of competition:* To check the monopoly power, the best way is to promote competition in the monopoly industries. When the Government liberalizes the licensing system, more and more firms start entering the industries and this encourages competition. This leads to technological inventions and innovations and also production of good substitute products. This acts as a restraint on the part of the monopolist to exercise monopoly power in the market.
2. *Nationalization of monopolies:* When a monopoly firm charges higher prices for the product indiscriminately, the Government, in the interest of the public, nationalizes the firm.
3. *Resistance from customers:* When the monopolist charges higher prices indiscriminately, the customers organize themselves as a union and should avoid purchasing that product. However, this regulation is somewhat difficult to execute because,
   - the customers of the product are scattered widely throughout the country, thereby, organizing them is quite complicated.
   - there being no substitutes for the product under monopoly, it compels the customers to purchase the product.

4. ***Price ceilings:*** If there is only one firm in the industry because of some artificial restrictions, then we can easily over come the monopoly power by simply removing these artificial restrictions, so that, the new firms enter the industry and the outcome is better than the original monopoly power. On the other hand, if there is a natural monopoly, then even lifting the restrictions will not induce entry of new firms, as new firms cannot make a profit in the industry. In such a case, the Government impose price-ceiling on the firm's product, such that, monopoly price equals the competitive price. The Government appoints a State Regulatory Authority and it recommends the fixation of a ceiling price for the product of a monopoly firm, below the monopoly price. This facilitates to increase the output from the firm's side and thereby, lowers the price at the customer's level. In the Figure 21.25,
   - Before imposing price ceiling, the monopolist is making PABC super-normal profits, *i.e.*, by selling OQ level of output at QP=OC level of price.
   - Suppose, the State Regulatory Authority imposed price ceiling at Q_1P_1=OD price limit for the product, suiting the present competitive conditions.
   - At this new ceiling price, Q_1P_1, the new demand curve or ARC is DP_1D and new MRC is DP_1GMRC.
   - Now the monopolist behaves as a perfectly competitive producer (because price is fixed at OD). The MC curve cuts the new MRC *i.e.*, DP_1GMRC from below at P_1, where the monopolist produces and sells OQ_1 level of output at Q_1P_1 new ceiling price.
   - So, due to price regulation, the monopolist sells more output (OQ_1) when compared to earlier case of selling at monopoly price OQ, *i.e.*, OQ_1>OQ. He still makes super normal profits *i.e.*, (P_1FBD), but this is smaller than monopoly profits PABC.

The above discussion infers that, in the presence of the ceiling price, where the MRC has a discontinuity, the firm's optimal output is OQ_1 and it earns super-normal profits (P_1FBD). The outcome of the regulation is that, the price falls from QP to Q_1P_1, but output increases from OQ to OQ_1. For smaller outputs (less than OQ_1), MRC (DP_1GMRC) exceeds MC and for larger outputs (more than OQ_1), MRC (DP_1GMRC) is less than MC. It is to be noted that, if the ceiling price is exactly the competitive price, then if the firm makes a profit, it is induced to produce the efficient output. If the ceiling price does not make a profit to the firm at this output, then a subsidy, in addition to a price ceiling, is necessary to induce the firm to produce the efficient output. So, in order to set the price ceiling exactly at the level that induces the firm to produce the efficient output, the regulator has to know the firm's MC curve, as it decides the efficient output.

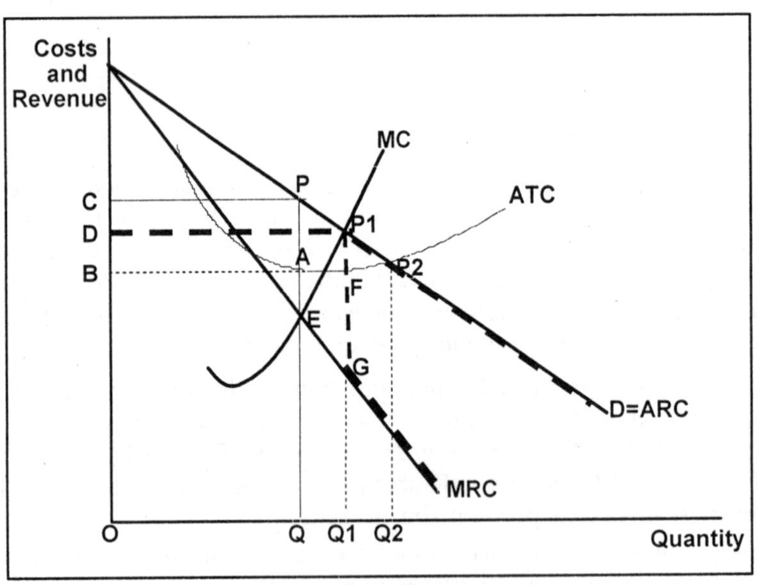

Figure 21.25: Regulation of monopoly power – Fixation of ceiling price.

5. ***Average cost pricing:*** If the State Regulatory Authority fixes the monopoly price Q_2P_2, which is equal to the ATC, then the firm will sells a larger quantity of output OQ_2 in the market (Figure 21.25). That means, it requires for the firm to choose an output-price combination at which ATC is equal to ARC. This regulation eliminates super-normal profits, but does not necessarily lead to an efficient output. This is explained through the Figure 21.26. In Panel A, the output OQ_1 under the price regulation is larger than the efficient output OQ, while in the Panel B, the output OQ_1 under the price regulation is smaller than the efficient output OQ.

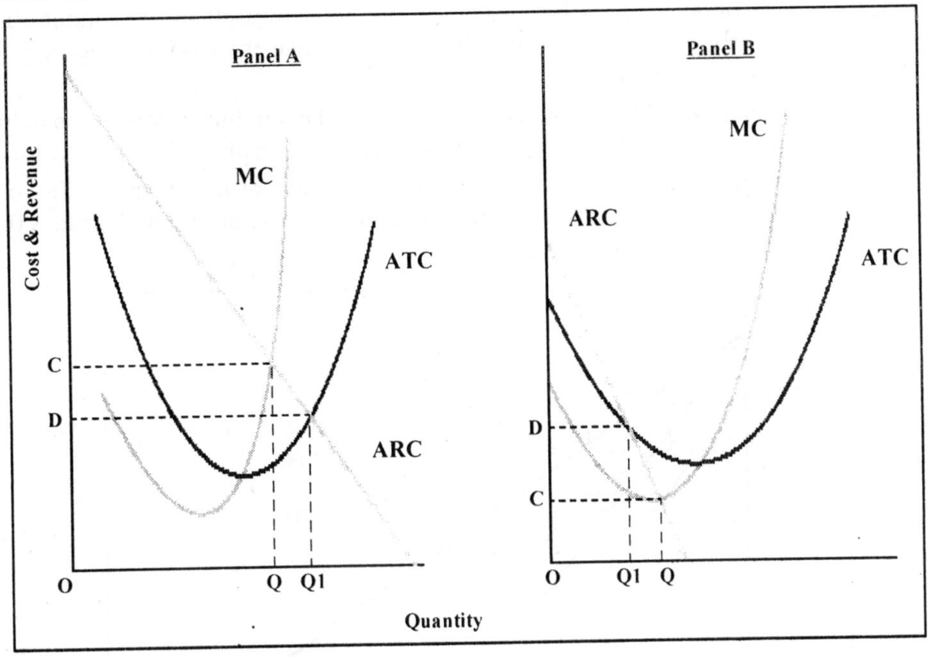

Figure 21.26: Regulation of monopoly power – Average cost pricing.

6. ***Fiscal measures:*** The excess or super-normal profits under monopoly often lead to inequality of income distribution and concentration of economic power. So, the Government attempts to regulate these excessive profits through fiscal means. The tax may be lumpsum without any regard to the level of output or it may be specific or proportional to output, and the amount of tax rises with increase in output. Let us discuss each case here under.

   (a) ***Lumpsum tax***: By imposing lumpsum tax, the Government can reduce the monopoly power without affecting the output and price policy of the firm. In the Figure 21.27, ATC and MC are the average and marginal cost curves before the lumpsum tax is imposed on the firm. Before imposition of tax, the firm earns super-normal monopoly profits of PABC, by selling output OQ at QP=OC level of price.

   ✫ Let us suppose, to check the monopoly power, Government imposed lumpsum tax. This is a fixed cost on the monopoly firm and it is independent of the level of output. Therefore, ATC of the firm rises above and hence, ATC_1 is the new average cost curve. The proportion of tax is indicated by AA_1 or BD. But, the MC of the firm remains same, as it is only an additional or variable cost to the firm.

   ✫ Due to imposition of lumpsum tax, now the profits of the firm will be declined (due to increase in ATC) to PA_1DC.

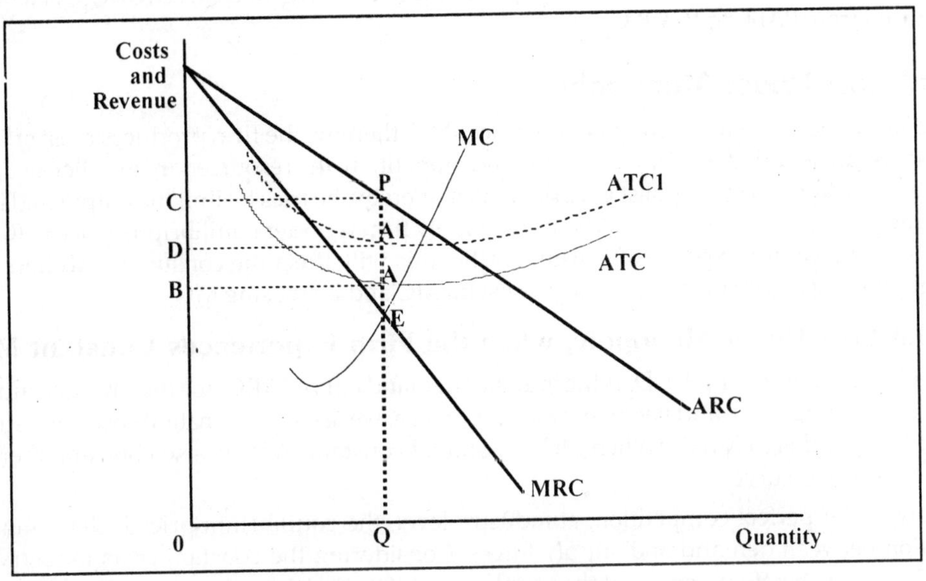

Figure 21.27: Regulation of monopoly Power – Imposition of Lumpsum tax.

- ☆ This, entire burden of tax has to be borne by the monopolist himself and he cannot shift the same to the customer by raising the price or by reducing the output. Hence, there is no change in price of the product (QP=OC) and no change in output of the firm, OQ.
- ☆ Since, there is no change in MC and MR of the monopolist even due to imposition of lumpsum tax, any change in price-output combination may lead to losses, because MC=MR equilibrium gets disturbed.

(b) ***Specific tax***: The Government can also reduce monopoly power by levying a specific tax, *i.e.*, tax imposed on per unit of product. Due to imposition of tax on per unit of product, both MC and ATC curves will shift upward and this is shown through Figure 21.28.

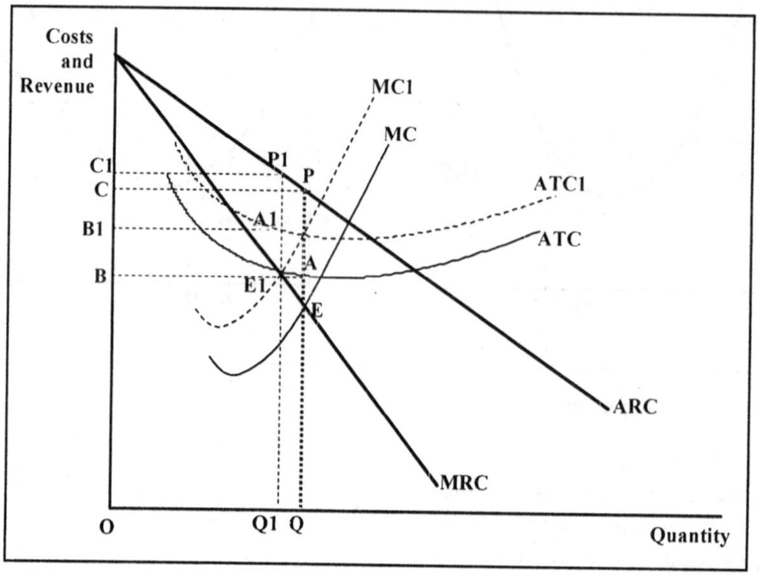

Figure 21.28: Regulation of Monopoly Power – Imposition of Specific tax.

- ☆ Before the imposition of tax, ATC and MC are the average cost and marginal cost curves respectively. Before the imposition of tax, the firm enjoys supernormal profits of PABC, by selling OQ output at QP=OC level of price.
- ☆ Let us suppose, specific tax was imposed on the product. Therefore, both ATC and MC curves rise upward as ATC_1 and MC_1 respectively, as specific tax refers to variable cost to the firm.
- ☆ So, the new equilibrium is at E_1 where, now the firm sells only OQ_1 quantity (OQ_1<OQ) at $Q_1P_1=OC_1$ level of price (QP_1>QP) and the firm earns only $P_1A_1B_1C_1$ super normal profits ($P_1A_1B_1C_1$<PABC profits). The proportion of tax is indicated by AA_1 or BB_1.
- ☆ The monopolist shifts this tax burden to the customer by rising the price of the product and he himself also bear the burden through decreasing the output from OQ to OQ_1, *i.e.*, by Q_1Q. Hence, his profits are reduced from PABC to $P_1A_1B_1C_1$.

21.13. Social Cost Under Monopoly

As discussed under perfect market competition, price=MC, thereby, the firm produces most efficient output and consumer's welfare is safe guarded. But, under monopoly competition, the resources are misallocated thereby, leads to loss of social welfare. We know, a monopolist is a price maker, thereby, he tries to limit the output and increase the price of the product, so as to maximize the profits. When a monopoly firm is under equilibrium, price>MC and this implies that, the monopolist gains at the expense of consumers. So, this will affect the consumer's welfare and the same is discussed through the cases given below assuming constant MC and increasing MC.

Case 1 – Social Cost Under Monopoly, when the Firm Experiences Constant MC

- ☆ As shown in the Figure 21.29, DD_1 is the industry demand curve (ARC) for the product and MRC is also the downward sloping curve and lies below the ARC. Since, we assumed constant costs industry, MC curve is a straight line parallel to X-axis. When MC is assumed constant, ATC is also constant, thereby, ATC curve coincides with MC curve.
- ☆ We know under perfect competition, at industry level, the equilibrium price is determined based on the interaction between demand and supply forces. Considering the constant costs industry, the horizontal MC=ATC curve, therefore, represent the supply curve of the industry.

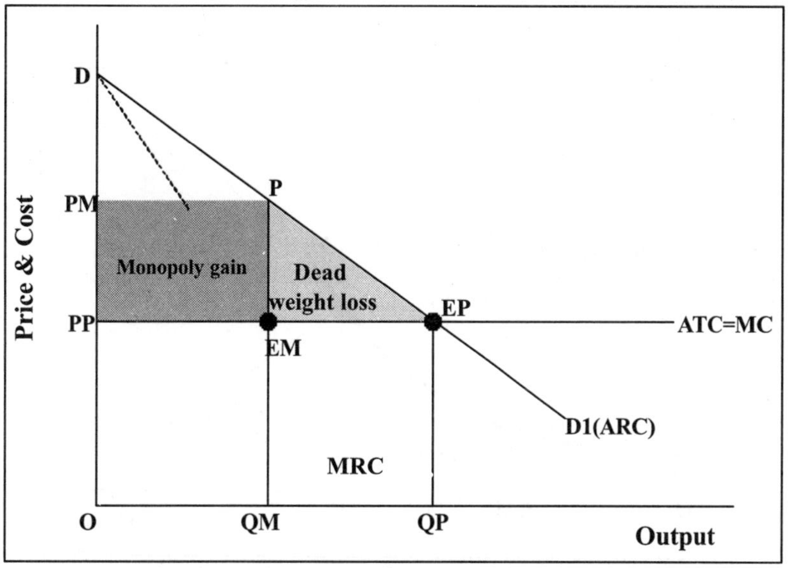

Figure 21.29: Social cost under monopoly assuming constant MC.

- Under perfect competition, the point of equilibrium is 'E', where demand curve DD_1 intersects horizontal supply curve (ATC=MC), thereby, the equilibrium output is OQ_P and the equilibrium price is OP_P.
- At this equilibrium price OP_P, the consumer's surplus is given by DP_PE_P and this implies the social welfare.
- Considering monopoly competition, the point of equilibrium is E_M, where MRC=MC (horizontal line) thereby, the equilibrium output is OQ_M and the equilibrium price is OP_M.
- So, when compared to perfect competition, the monopolist restrict the output OQ_M *i.e.*, $OQ_M<OQ_P$ and raise the price of the product to OP_M (*i.e.*, $OP_M>OP_P$).
- Due to rise in price of the product, the monopolist makes the profit equal to $P_MPE_MP_P$.
- Due to rise in price of the product, the consumer surplus declines to DPP_M from DP_PE_P. So, the loss in consumer's surplus is given by DP_PE_P-DPP_M = $P_MPE_PP_P$. This infers that, there is redistribution of consumer's surplus to the monopolist, as there is decline in consumer's surplus on one hand and rise in profits of the monopolist on the other. But, the interesting aspect is that, the loss in consumer's surplus *i.e.*, $P_MPE_PP_P$ is greater than monopoly profits $P_MPE_MP_P$.
- The loss in consumer's surplus *i.e.*, $P_MPE_PP_P$ can be split into two components;
- The first component refers to area $P_MPE_MP_P$ *i.e.*, profits made by the monopolist. This component refers to the loss of consumer's surplus to the consumers who are still purchasing the product even after rise in price of the product from OP_P to OP_M.
- The second component refers to area PE_ME_P and this component refers to the decline in consumer's surplus due to less output produced by the monopolist from OQ_P to OQ_M. So, the loss of consumer's surplus of the area PE_ME_P is due to allocative inefficiency caused by the monopolist. This is because, the monopolist will not produce the output beyond OQ_M as MC curve will be higher than MRC.
- So, from the above two components of total loss in consumer's surplus ($P_MPE_PP_P$) due to rise in price of the product from OP_P to OP_M, one component ($P_MPE_MP_P$) goes to the monopolist as a profit and other component (PE_ME_P) is a loss from the consumer's point of view and even the monopolist cannot pocket this consumer's surplus.
- This loss of component PE_ME_P cannot be eliminated because, it is due to misallocation of resources by the monopolist. As a result, less output is produced leading to rise in price of the product. Consequently, there will be net loss in consumer's welfare and since the loss cannot be eliminated it is rightly called as 'Dead-weight loss of welfare' or 'social cost' of monopoly.
- Thus, when the firm experiences constant MC, the loss in social welfare includes loss in consumer's surplus only.

Case 2 – Social Cost Under Monopoly, when the Firm Experiences Increasing MC

When the firm experiences increasing MC, the loss in social welfare includes both loss in consumer's surplus and loss in producer's surplus. The same is explained through Figure 21.30.

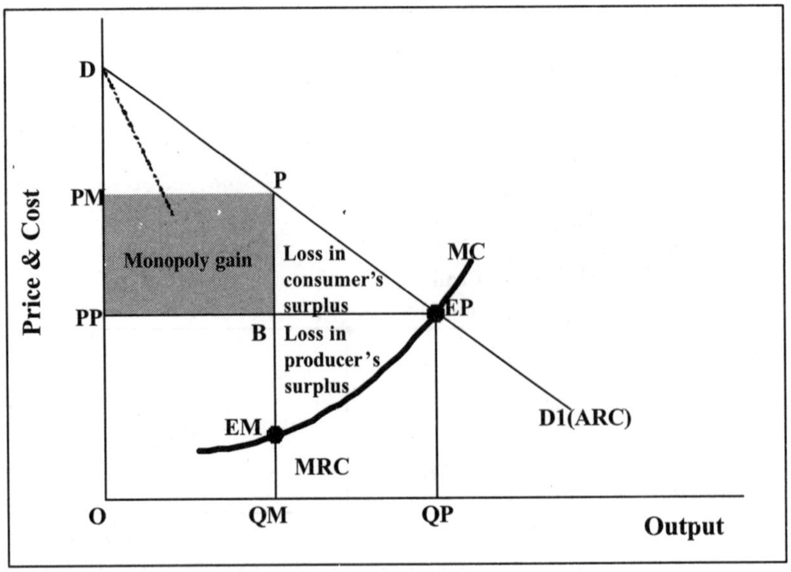

Figure 21.30: Social cost under monopoly assuming increasing MC.

- DD_1 is the industry demand curve (ARC) for the product and MRC is also the downward sloping curve and lies below the ARC.
- MC curve is upward sloping from left to right, as we assumed increasing MC industry.
- We know, under perfect competition at industry level, the equilibrium price is determined based on the interaction between demand and supply forces. Considering the increasing costs industry, the upward sloping MC curve, therefore, represent the supply curve of the industry.
- Under perfect competition, the point of equilibrium is E_p, where demand curve DD_1 intersects upward sloping supply curve (MC), thereby, the equilibrium output is OQ_P and the equilibrium price is OP_P.
- At the equilibrium price OP_P, the consumer's surplus is given by DP_PE_P and this implies the social welfare.
- Considering monopoly competition, the point of equilibrium is E_M, where, MC=MRC, thereby, the equilibrium output is OQ_M and the equilibrium price is OP_M.
- So, when compared to perfect competition the monopolist restrict the output to OQ_M and raise the price of the product to OP_M.
- Due to rise in price of the product, the monopolist makes the profit or rise in producer's surplus equal to P_MPBP_P.
- Due to rise in price of the product, the consumer's surplus declines to DP_MP from DP_PE_P. So, the loss in consumer's surplus is given by DP_PE_P-DP_MP = $P_MPE_PP_P$. This infers that, there is the redistribution of consumer's surplus to the monopolist, as there is decline in consumer's surplus on one hand and rise in profits of the monopolist on the other. But, the interesting aspect is that, the loss in consumer's surplus *i.e.*, $P_MPE_PP_P$ is greater than monopoly profits or rise in producer's surplus P_MPBP_P. So, the net loss in consumer's surplus is equal to PBE_P.
- It is interesting that, since the MC curve is rising, besides loss in consumer's surplus, there is also loss in producer's surplus. This is because under monopoly, the firm restrict the output to OQ_M and raise the price to OP_M when compared to perfect competition.
- Under perfect competition, with reference to equilibrium point E_P, the equilibrium output is OQ_P and equilibrium price is OP_P, thereby, the industry earns extra profits BE_ME_P when compared to monopoly situation with reference to equilibrium point E_M. But, under monopoly situation, when the firm restricts the output to OQ_M, the extra profits BE_ME_P are lost. So, under monopoly situation, due to restriction of output, there will be decline in producer's surplus by BE_ME_P.
- So, under monopoly situation, the net loss in consumer's surplus is PBE_p and net loss in producer's surplus is BE_ME_P and these losses are due to restriction of output to OQ_M by the monopolist. So, the total loss of welfare *i.e.*, loss of consumer's surplus + loss of producer's surplus is PE_ME_P.
- Since, this total loss of welfare PE_ME_P is due to misallocation of resources and cannot be eliminated, thereby, it is regarded as 'Dead weight loss of welfare' or 'Social cost of monopoly'.

Table 21.4: Differences between Perfect competition and Monopoly.

Perfect Competition	*Monopoly*
Number of firms: There will be infinite number of firms producing a product and their actions in the market are independent.	***Single firm:*** There will be a single firm producing the product, thereby, single firm represents the industry.
Firm is a price taker: Price of the product is decided by the market forces of demand and supply at the industry level and the same price is passed on to the individual firms. Hence, no individual firm is in a position to change the price by controlling supply because, individual seller's supply constitute an insignificant percentage of total market supply. This ensures each firm is only a price taker and quantity adjuster and not a price maker.	***Firm is a price maker:*** The single-firm status gives the monopolist an extensive market control, thereby, he is either price maker or quantity adjuster in the market and not a price taker.
Products are perfect substitutes: All firms produce homogenous products, which are perfect substitutes.	***No close substitutes:*** There is a single firm that produces a unique product, with out any close substitutes.
E_c is infinite: All firms produce and sell homogeneous products, which are perfectly same in terms of size, shape, taste, colour, ingredients, quality etc. So, the products are perfect substitutes, thereby, the E_C is infinite.	***E_c is zero:*** There are no close substitutes for the product produced by the monopoly firm, thereby, the E_C is zero.
Free entry and exit of firms: There are no restrictions regarding the entry and exit of firms into an industry. During short run, when the existing firms enjoy super-normal profits, new firms enter the industry. When the existing firms in the industry suffer from losses, they leave the industry.	***Strong barriers to entry:*** The existing firm will have firm control on the market and hence, there are strong barriers for other firms to enter the industry. Output-pricing strategies, control over the raw material, legal rights in the form of patents, copy rights, trade secrets etc., acts as strong barriers for other firms to enter the industry.
Perfect knowledge: All the buyers and sellers have perfect information about prices and products from competing firms. So, no buyer will pay a price higher than market price and no seller will accept a price less than the one prevailing in the market. The price determined at the industry level will be passed on to the individual firms, thereby, an individual seller must know the price at which the product is offered for sale by the other sellers in the market. Similarly, buyers must know the price being charged by the different sellers. So, this perfect knowledge among the sellers and buyers ensures prevalence of uniform price for the product in the market.	***Price maker or Output decider:*** The monopolist is either a price maker or output decider for the product. That means, he can either fix the price and leave the output to be determined by the buyers' demand at that price or he can fix the output to be produced and leave the price to be determined by the buyers' demand for that output.
No Government intervention: There is complete absence of Government regulation regarding the output, supply and pricing mechanism of the products. There is also no collusion among buyers or sellers.	***Regulation by Government:*** Government regulation is indispensable, especially when the monopolist exploits the customers indiscriminately in charging higher prices for the product.
Perfectly elastic ARC or demand curve: For the industry as a whole, the demand curve slopes downward from left to right. But, for an individual firm, the demand curve or ARC, both in the short run and long run is 'perceived' perfectly elastic *i.e.*, it lies parallel to X-axis.	***Downward sloping ARC or demand curve:*** The industry or firm's demand curve or ARC is downward sloping from left to right indicating that, monopolist sells large quantities of a commodity or output at lower prices or price reductions.
MRC coincides with ARC: Since, ARC is constant, MRC must be constant and coincides with ARC. This is so because, an individual firm in perfect competition cannot influence the price of the product and hence, additional units are also sold at the same price as in case of previous units. So, under perfect market competition, price=ARC=MRC=Demand curve.	***Downward sloping MRC:*** MRC slopes downward from left to right and remain below ARC and the numerical slope of MRC is twice as steep as ARC. Since, ARC is falling, the additional or extra units sold also fetch less and less revenue (MRC) in the market.
Same price for the product: The price determined at the industry level will be passed on to the individual firms, thereby, individual firm is only a price taker and it sells any amount of the commodity at the same price.	***Price discrimination:*** Considering different elasticities of demand for the product in different markets, the monopolist sometimes may adopt price discrimination. He may even fix different prices for the same product among different categories of customers in the same market segment.
No market power: Perfectly competitive firms have zero market power when it comes to setting price for the product, as they are simply price takers for the price settled at industry level. If one firm raises the price, its product will become costlier than others and automatically, it will incur losses. So, the seller has to accept the price, which is determined by two market forces namely the demand for and the supply of the product and he has no control over the price of the product.	***Significant market power:*** A monopolist has considerable, although not unlimited market power. He is a price maker, as he can set the price of the product to his maximum advantage. Likewise, the monopolist has complete control over the supply of the product. But, he cannot fix both the price and decide output simultaneously, thereby, he is either a price maker or output decider. The two primary factors determining monopoly market power are the firm's demand curve and its cost structure.

Contd...

Table 21.4–Contd...

Perfect Competition	*Monopoly*
Zero product differentiation: The product is perfectly homogenous in terms of size, shape, taste, colour, ingredients, quality etc. So, the products are perfect substitutes, thereby, there is zero product differentiation.	***Absolute product differentiation:*** The product produced by a monopoly firm is unique and with out close substitutes. So, in monopoly, there is high to absolute product differentiation, in the sense that, there is no available substitute for a monopolized product.
Supply curve: In short run, the portion of the SRMC curve above SRAVC curve represents supply curve of a perfect competitive firm. In long run, the portion of the LRMC curve above LRATC curve represents supply curve of a perfect competitive firm. Thus, for a perfect competitive firm, there is a well defined supply function.	***Supply curve:*** Since, monopolist is either price maker or output decider, we cannot draw a unique supply curve for the monopoly firm. Moreover, the monopolist's output decision depends not only on MC, but also on the demand curve. So, there is no one-to-one relationship between price and the quantity supplied.
Resource allocation: The resource allocation is optimum and hence, social welfare is maximum.	***Resource allocation:*** The resources are misallocated, thereby, causing a loss of social welfare.
Firm's equilibrium: The firm is in equilibrium at the level of output, where MC is equal to MR. Since, under perfect competition, MR is equal to AR or price, therefore, when MC is equal to MR, it is also equal to AR or price at that equilibrium position, i.e., MC=MR=AR=Price. At the point of equilibrium, MC curve is rising and it cuts the MRC from below. MC curve can neither be falling nor constant and it should be rising and then only, the firm under perfect competition is under equilibrium. That means, the firm will be in equilibrium only when the MC curve cuts the MRC from below, while MC curve is rising. Since the MRC is a horizontal straight line parallel to X-axis, the second order condition of firm's equilibrium *i.e.,* MC=MR is satisfied only when MC curve is rising.	***Firm's equilibrium:*** The firm is in equilibrium at the level of output, where MC is equal to MR. So, at the point of equilibrium MC=MR, the MC curve cuts the downward falling MRC. The MC curve may be falling (if the firm experiences increasing returns or decreasing costs) or rising (if the firm experiences diminishing returns or increasing costs) or constant (if the firm experiences constant returns or constant costs), but for stability of the equilibrium, when MC curve is falling, it should fall at slow pace than MRC. If this fails, there can be no equilibrium what so ever. The second order condition of firm's equilibrium *i.e.,* MC curve should cut MRC from below at the equilibrium point, is satisfied in all the three cases *i.e.,* when MC curve is rising, remaining constant or falling, as MRC is a continuous falling curve from left to right.
Price=MC: In equilibrium position, the price charged by the firm equals to MC, as ARC is equal to MRC.	***Price>MC:*** In equilibrium position, the price charged by the firm is above MC, as ARC is above MRC.
MRC and Price: In equilibrium position, the MRC equals the price, as MRC coincides with ARC.	***MRC and Price:*** In equilibrium position, the MRC is less than price, as MRC is lower than ARC.
Price and output: When compared to monopoly, under perfect competition, output is higher, but price is lower (Figure 21.31). A competitive firm can sell all the output it desires at the market price.	***Price and output:*** When compared to perfect competition, under monopoly, output is lower, but price is higher (Figure 21.31). For a monopoly, to increase sales it must reduce price.
Relation between TR and output: In perfect competition, the firm faces a perfectly elastic demand curve, thereby, TR is proportional to output. Thus, the TR curve for a competitive firm is a ray with a slope equal to the market price.	***Relation between TR and output:*** For a monopoly to increase sales, it must reduce price. Thus, the TR curve for a monopoly is a parabola *i.e.,* it begins at the origin, rises continuously, reaches maximum (when MR=0) and then falls continuously (when MR turns negative). So, the revenue maximizing quantity and price occur when MR = 0.
Profit maximization: A perfect competitive firm maximizes profits by producing output, where price equals MC. So, at the point of profit maximization, MC=MRC=ARC=Demand=Price.	***Profit maximization:*** A monopoly firm maximizes profits by producing output, where MRC equals MC. So, at the point of profit maximization, MC = MRC.
Normal profits in the long run: In the long run, the firm earns only normal profits. The super-normal profits enjoyed by the firms in the short run will be wiped out in the long run, as new firms enter the industry.	***Super normal profits even in the long run:*** The firm can earn supernormal profits even in the long run, as there are no competitors for the firm producing the same product and there are strong barriers for the entry of new firms into the industry.
Firm's equilibrium in long run: The firm is in long-run equilibrium at the minimum point of the LRATC curve. That means, a firm under perfect competition operates at its full capacity in the long run. So, no excess capacity is left over, since the firm under perfect competition is producing up to the point where MC=minimum point of LRATC. That means, the firm employs optimum plant size in the long run to achieve equilibrium.	***Firm's equilibrium in long run:*** The firm is in long-run equilibrium at the level of output, where LRATC curve is still declining and has not reached the minimum point. So, the firm operates at less than full capacity in the long run. So, some excess capacity or reserve capacity is left over, since the firm is not operating up to the minimum point of LRATC. That means, the firm employs less than full optimum plant size in the long run to achieve equilibrium.
Consumer's surplus: Under perfect competition, equilibrium output is higher and equilibrium price is lower, thereby the consumer's surplus is more	***Consumer's surplus:*** In monopoly, equilibrium output is lower and equilibrium price is higher, thereby the consumer's surplus is less.

Contd...

Table 21.4–Contd...

Perfect Competition	Monopoly
Consumer's welfare: It is considered to be more under perfect market competition. This is because, a firm under perfect competition ✰ Produces homogenous product ✰ Products are good substitutes ✰ Operates up to optimum plant size ✰ Can increase the profits by increasing the output rather than increasing the price of the product ✰ Firm operates at the minimum point of LRATC to produce equilibrium output.	***Consumer's welfare:*** Monopoly is equally beneficial in promoting consumer's welfare. But, there is a mis-conception that, monopolist exploits the customers by charging higher prices. The major difference between monopoly and perfect market competition is that, under monopoly, the firm produces sub-optimum output, thereby, price of product under monopoly is greater than price of product under perfect competition. From the Figure 21.31, it is clear that, equilibrium output is lower and equilibrium price is higher under monopoly when compared to perfect competition. Under perfect competition, the price of the product is determined at the industry level based on the interaction between two market forces *viz.*, demand and supply and the same price is passed on to the individual firms. The individual firms are simply price takers and the output produced by each individual firm constitute only an insignificant portion of total output of the industry, thereby, each firm cannot influence the price of the product in the market. On the contrary, in case of monopoly competition, the firm represents the industry, thereby, the firm has firm control over the supply of the product. So, the monopoly firm can influence the price of the product and this could be the major reason for the higher price under monopoly compared to perfect competition. But, the monopoly price is not always higher than the price under perfect competition, but generally it is. Even it is possible that, the monopoly price is lower than the price under perfect competition and following are the possible reasons. ✰ Because of large scale production of the product, the monopoly firm enjoys economies of large scale. But under perfect competition, each firm contributes an insignificant portion o the total output of the industry, thereby, the firm may not enjoy economies of large scale. ✰ A monopoly firm can plan for R&D investment, thereby, it can employ cost-effective techniques of production and lower the cost of production of product. Further, it can also plan for improving the quality of the product. ✰ A monopoly firm can plan for bulk purchasing of raw material at lower cost, thereby, it can reduce the cost of production of the product and hence, price. ✰ Having a firm control on the supply of the product into the market, the monopoly firm can plan for efficient marketing strategies of the product, which helps to reduce the price of product in the market. ✰ Government may impose restrictions on the monopoly firms in charging a lower price of the product.

21.14. Differences between Perfect Competition and Monopoly

In the earlier chapter, we discussed in-detail about the perfect market competition and in this chapter, we discussed in-detail about different concepts of monopoly and this guides us to draw comparative picture between these two market structures. It is interesting because, perfect competition represents extreme form of perfections in the market and monopoly competition represents the extreme form of imperfections in the market and this comparative analysis guides us in better understanding of these two concepts. There are some points of similarity between these two market structures *viz.*, the cost functions are the same, both monopolies and perfect competitive firms aim at cost minimization and profit maximization, shutdown decisions are same, both the structures are assumed to face perfectly competitive factors markets, for both the structures the profit determining equilibrium level of output and price is at the point of MC=MR etc. However, there are certain points of difference between perfect competition and monopoly (Table 21.4).

In the Figure 21.31, in perfect market competition, DD is the demand curve or ARC and 'S' is the supply curve or MC curve of the industry. Under perfect competition, the point of equilibrium in the industry is at 'E', where the two market forces *viz.*, demand and supply intersect each other, thereby, producing OQ_P equilibrium output and OP_P is the equilibrium price. Let us assume, all the firms merge together to form a monopoly in producing the same product. The monopoly firm or industry will be in equilibrium when the MC curve intersects the MRC from below and it is point E_1, which indicates that, the monopoly firm now produces OQ_M, output and the equilibrium price is OP_M. This infers that, the equilibrium output under monopoly (OQ_M) is lower than equilibrium output under perfect competition (OQ_P), but the equilibrium price under monopoly (OP_M) is higher than equilibrium price under perfect competition (OP_P).

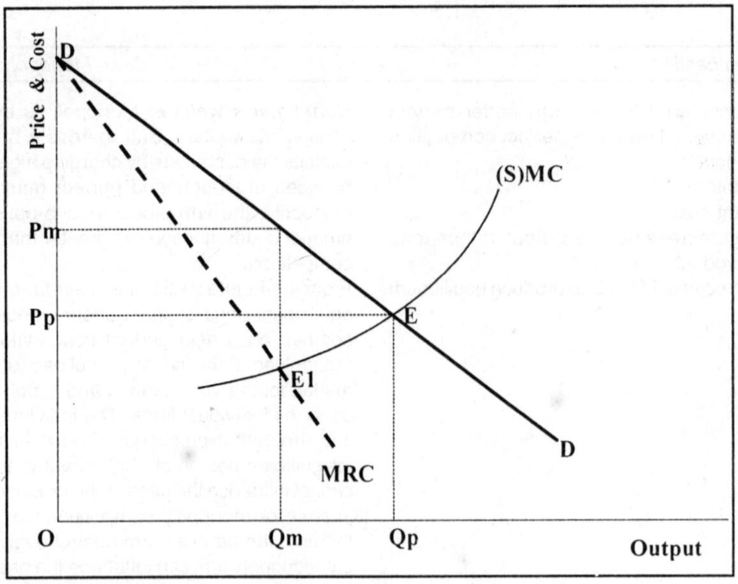

Figure 21.31: Monopoly price higher and output lower compared to Perfect competition.

21.15. Desirability of Monopoly Competition

An industry characterized by a monopoly market structure produces less output and charges higher prices than under perfect competition and in this context, it is less desirable socially. However, a natural monopoly is generally considered desirable, if the monopolist's price behaviour is regulated.

22

Oligopoly

Oligopoly is a common market form, in which a market or industry is dominated by a few number of sellers (oligopolists). The word *'oligopoly'* is derived from two Greek words, *'oligoi'* means *'few'* + *'polein'* means 'to sell'. So, oligopoly is otherwise referred as 'Competition among the few', as few firms are producing the products and competing in the market. In this form of market competition, the few sellers sell either homogeneous or heterogeneous goods. Because of the presence of few sellers, each oligopolist is likely to be aware of the actions of the others. The decisions of one firm influence and are influenced by the decisions of other firms. So, these firms require strategic planning to consider the reactions of other firms existing in the market. Since, perfect competition and monopoly are the exceptions and not the rule, in reality, it is the oligopoly market which exists, with a high degree of market concentration. That means, a huge percentage of the industry's market share is occupied by the top four firms. Oligopoly implies a blend of greater monopoly power and less competition. So, oligopoly is often referred to as 'monopoly of the few'. This small number of firms, if they collude, they could act like a monopoly. It is popularly defined as a market in which control over the supply of a commodity is in the hands of a small number of producers and each one can influence prices and affect competitors. In the words of George J. Stigler, *'that situation in which a firm bases its market policy in part on the expected behaviour of a few close rivals'*.

Firms within an oligopoly market structure have strong interdependence between them. That means, each firm must take into account the likely reactions of other firms in the market, while making investment, output and pricing decisions. So, considering this characteristic feature of mutual interdependence, the oligopoly competition can lead to a wide range of outcomes. Some times, the firms collude each other and restrict the output into the market, with a view to enjoy monopoly profits in the business. When the firms have a formal agreement, it is known as a Cartel. The classical example of such a cartel is OPEC, which has a profound influence on the international price of oil. In some cases, there will not be any formal agreement, but there may be an acknowledged market leader which informally sets prices to which other firms will respond, known as price leadership. Sometimes, the competition between the few firms is so severe that leads to drastic fall in prices and increased production, thereby, approaching the features of perfect market competition. Because of this diversified possible outcomes from the oligopoly market structure, there is no single, unified theory to explain its price-output mechanism. So, when compared to other forms of market competition, oligopoly is the least understood market structure. In this chapter, we deal about the features and price-output decisions under oligopoly according to different models. This chapter concludes with a brief look at the Theory of Contestable markets.

22.1. Different Forms of Oligopoly

Based on several criteria, oligopoly competition takes several forms and they are given here under.

1. Perfect and Imperfect Oligopoly

Perfect Oligopoly

If in a market competition, few sellers are selling a homogeneous product, it is said to be perfect or pure oligopoly. *e.g.*, Producers of industrial products such as, aluminum, cement, steel, zinc, copper etc.

Imperfect Oligopoly

If in a market competition, few sellers are selling a heterogeneous or differentiated product, it is said to be imperfect oligopoly. *e.g.*, Producers of consumer goods like automobiles, cigarettes, soaps and detergents, TVs, rubber tyres, refrigerators, typewriters, computers etc.

2. Collusive and Competitive Oligopoly

Collusive Oligopoly

When few firms of an oligopolistic industry come to a common understanding or act in collusion with each other in fixing price or output and both, it is a Collusive oligopoly

Competitive Oligopoly

A competitive oligopoly denotes lack of understanding between the firms regarding their output and pricing policies and they may be competing each other, thereby, making the market more competitive.

3. Partial and Full Oligopoly

Partial Oligopoly

Oligopoly is partial, when industry is dominated by one large firm, which is considered or looked upon as the leader of the group. The dominating firm will be the price leader. The rest of the firms will follow the leader regarding investment output, pricing, selling activities etc.

Full Oligopoly

In full oligopoly, the market will be conspicuous by the absence of price leadership by a single firm.

4. Syndicated and Organized Oligopoly

Based on the extent and degree of co-ordination between the firms, we can have,

Syndicated Oligopoly

Syndicated oligopoly refers to that situation, where firms sell their product through centralized syndicate.

Organized Oligopoly

Organized oligopoly refers to the situation, where firms organize themselves into central association for fixing prices, output, quotas etc.

5. Open and Closed Oligopoly

Open Oligopoly

In open oligopoly, new firms can enter the industry and compete with existing firms.

Closed Oligopoly

In closed oligopoly, new firms cannot enter the industry due to the output-price policies followed by the existing firms.

6. Duopoly

Duopoly (*'dyo'* (two)+*'polein'* (to sell)) is a specific type of Oligopoly, where only two firms or producers exist in one market. A pure duopoly situation occurs very rarely in the practical conditions and this definition is generally used, where two dominant firms hold the majority of the market share. Duopoly is an ideal model for analyzing oligopoly behaviour, as it is the simplest form of oligopoly. Examples include, Pepsi and Coke in a soft drink market, Gillette vs Wilkinson Sword in razor blade market, Airbus vs Boeing in commercial large jet aircraft market, Kodak vs Fujifilm in motion picture film stock and color 135 film markets, Canon Vs. Nikon in high end imaging optics, film and digital cameras etc. The following are the important features of duopoly competition:

- ☆ Two firms in the industry
- ☆ Strong control over price

- ☆ Uses non-price competition to compete between the firms
- ☆ Very strong barriers to entry
- ☆ Interdependent actions between the two firms.

7. Oligopsony

It is a form of market situation, where there are few buyers and large number of sellers. The other characteristics are same as oligopoly.

8. Bilateral Oligopoly

A market with a few sellers (Oligopoly) and a few buyers (Oligopsony) is referred as Bilateral oligopoly.

22.2. Causes of Oligopoly

The following are the important sources of oligopoly competition:

- ☆ *Economies of scale:* The firms in the industry will employ huge capital investments and improved technology, thereby, enjoy economies of scale in production, promotion and sales, which is difficult for the new firms to achieve, if entered the industry.
- ☆ *Merger:* If the few firms in the industry predict in advance the threat of entry of new firms into the industry, then the existing firms merge and formulate a joint policy regarding the investment, output and pricing of the product. This joint action of the few big firms discourages the entry of new firms into the industry.
- ☆ *Barriers to entry:* They are very strong in the oligopoly competition and the most important barriers include, economies of scale, patents, access to expensive technology, existing firms gained control over essential inputs, heavy capital investments due to huge plant costs, substantial advertising budgets, traditional brand loyalty etc., and these effectively check the entry of new firms into the industry.
- ☆ *Mutual interdependence:* As the number of firms is small in an oligopolistic industry, each firm keep a strict watch on the price charged by the rival firms in the industry. The firms generally avoid price wars and try to create conditions of mutual interdependence.

22.3. Features of Oligopoly

The features of oligopoly competition are discussed here under.

1. Few Sellers

Oligopoly market is dominated by a small number of firms, who are able to collectively exert control over supply of output and market prices.

2. Interdependence among the Firms

The distinctive feature of an oligopoly competition is mutual interdependence among the firms. Oligopolies are typically composed of a few large firms and each firm is so large that its actions affect market conditions. If one firm individually decides its output and pricing policies, other firms counter-moves to protect themselves in the market. So, the competing firms will have a strict watch on the rival firms' market actions and will respond appropriately. So, the decision making in a firm is connected with the output and price policies of the other firms. Hence, in oligopolistic market, a single firm cannot act independently in deciding the output and price of the commodity. So, all the firms will have mutual interdependence among themselves in deciding the level of total output and the price fixation for the product, so as to ensure mutual benefit. The importance of interdependence is that, it leads to strategic behaviour. Strategic behaviour is the behaviour that occurs, when what is best for firm A depends upon what firm B does, and what is best for firm B depends upon what firm A does.

3. Indeterminate Demand Curve (ARC)

An oligopolist faces a downward sloping demand curve, but its price elasticity depends upon the rivals' reaction to changes in its investment, output and price. So, the demand curve of a firm under oligopoly is indeterminate because, any strategy formulated by the rival firms, may affect or change the demand curve. It is more elastic than under simple monopoly and not perfectly elastic as under perfect market competition. This implies that, no single firm in oligopoly market situation can exactly determine the demand for its product by formulating its own output and price policies. This is because, the firm's output and pricing policies will definitely influence the fortune of other firms and hence, they counter move the strategies to survive in the market. That means, when one firm decides and implements a course of action, there will be a quick and equal reaction from the rival firms. Hence, by just changing the output and pricing policies, no firm can determine exactly their influence on demand or sales of the product. For example, if a firm reduces the price of a product, other firms will also reduce the price for the product. Hence, demand for the product may not increase as expected by the business firm, which reduces the price of product initially in the market. If the firm increases the price of the product, other firms will not increase the price, thereby, the firm which increases the price will suffer from huge losses. Hence, the demand curve or ARC is indeterminate, which is the characteristic feature of oligopoly. This is mainly due to interdependence among the firms.

According to the Kinked demand curve theory proposed by Paul M. Sweezy, for an oligopolist, the demand curve is popularly called as Kinked demand curve and it is shown through the Figure 22.1, as D_1KD_2. Under normal competitive conditions, D_1K is the more elastic demand curve (flat slope), KD_2 is the less elastic demand curve (steep slope). The demand curve is kinked at K because, if one seller reduces the present price of product from 'OP', the demand will not increase greatly because, other rivalry firms also reduce the price. Hence, the seller cannot increase his sales as expected. So, for this seller, demand will be less elastic below the price 'OP' *i.e.*, whatever may the price fall, sales will not increase much for the seller, as other rivalry firms also reduce the price. This indicates that, for kinked demand curve D_1KD_2, below 'K', *i.e.*, between K and D_2, the demand curve is less elastic. But, if the seller increase the price above 'OP', other firms will not increase the price and hence, for this seller the sales will fall drastically, indicating that, for the kinked demand curve D_1KD_2 above K *i.e.*, between D_1 and K, the demand curve is more elastic. Thus, demand curve in a oligopolistic market situation will have a kink at present or current price and this kinked demand curve is more elastic for price increase and less elastic for price fall or decrease.

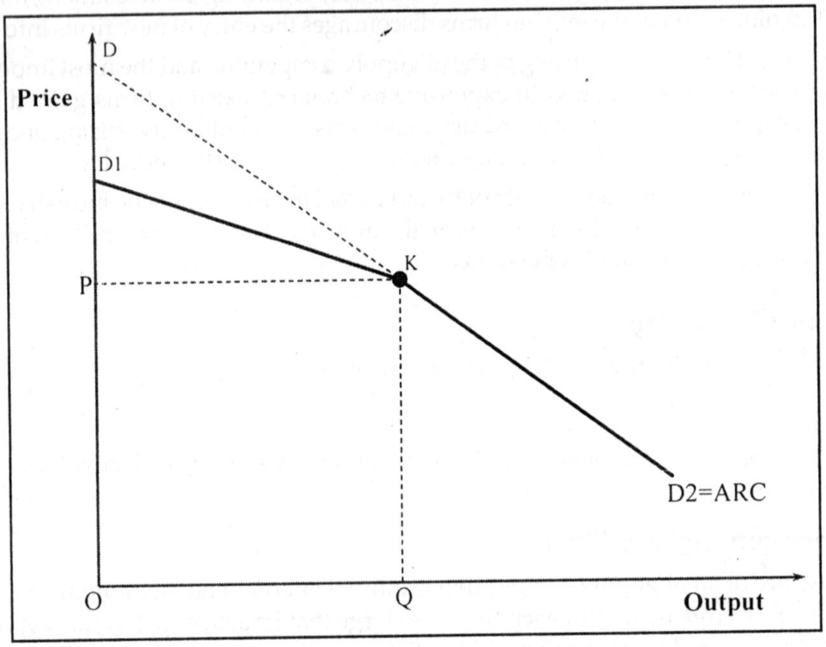

Figure 22.1: Indeterminate (Kinked) demand curve in oligopoly competition.

The concept of kinked demand curve is explained through the Figures 22.2 and 22.3. If a firm increases the price of the product from OP to OP_1, other firms will not increase the price of the product, thereby, the firm which increase the price of the product will suffer from the net loss of TR and as indicated in the Figure 22.2, the gain in TR (dark area) is less than loss in TR (light area).

If a firm decreases the price of the product from OP to OP_1, other firms will also decrease the price of the product, thereby, the firm which initially reduces the price of the product will suffer from the net loss of TR and as indicated in the Figure 22.3, the gain in TR (dark area) is less than loss in TR (light area).

So, no firm will unnecessarily disturb the price of the product and it is maintained stable. This leads to lot of non-price competition within the market. So, the demand curve in oligopoly competition will have a kink at the current price, as the demand curve has two different price elasticities. Above the kink, the demand curve is more elastic for price rise and below the kink, the demand curve is less elastic for a price fall. These two parts of the kinked demand curve have different MRCs associated with them, so we have the following Figures 22.4 and 22.5 to illustrate the indeterminate kinked demand curve and an MRC with a discontinuity. As shown in the Figure 22.5, the MRC (MRC_1MRC_2) will have a discontinuity JL below the kink of the ARC.

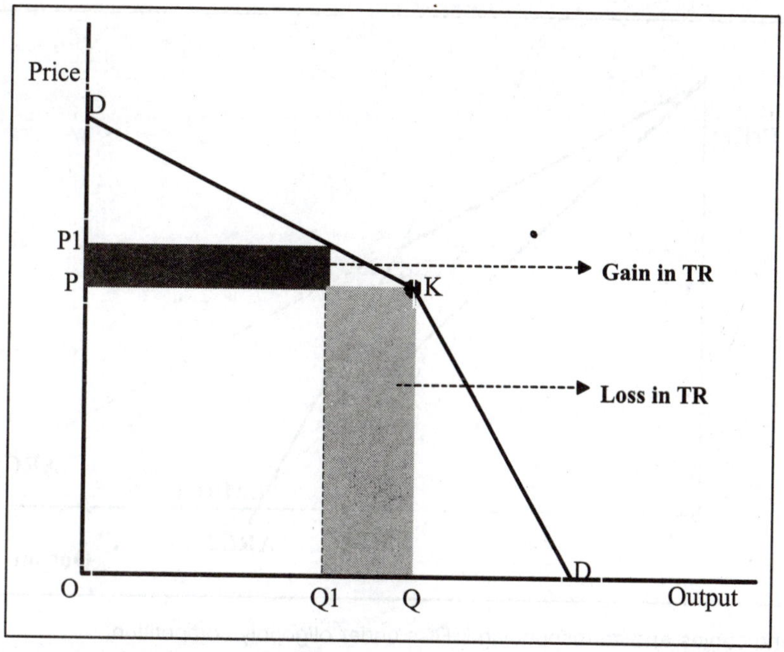

Figure 22.2: Net loss in TR due to increase in price by a single firm under oligopoly competition.

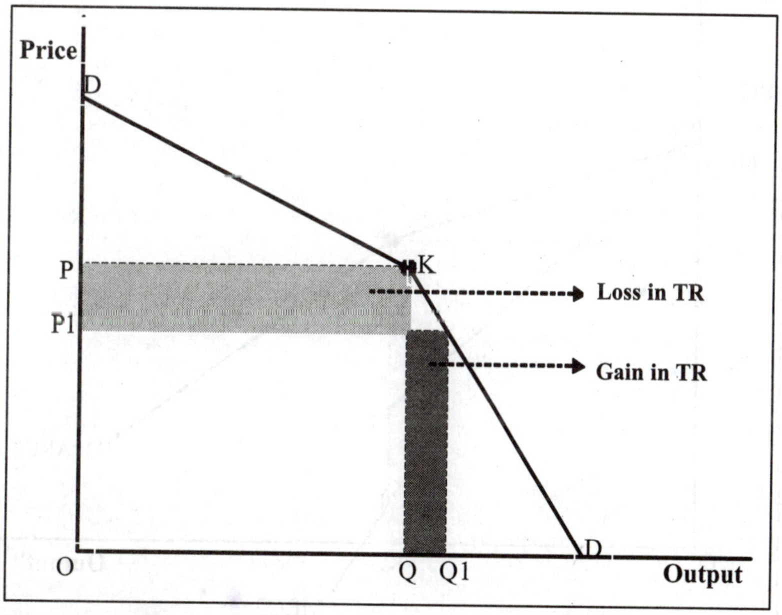

Figure 22.3: Net loss in TR due to price fall by a single firm under oligopoly competition.

4. MRC with Discontinuity

The MRC of the firm (MRC_1MRC_2) will have a gap or discontinuity below the point of kink of ARC (Figure 22.5). This is due to the change in the E_D *i.e.*, from the more elastic nature of the demand (above kink) to more inelastic nature of demand (below kink).

5. Non-price Competition

Firms compete for market share and to increase the demand from customers for their products in lots of ways. Before discussing the non-price competition, we make an important distinction between price competition and non-price competition. In price competition, the firm offers discounts on the price of the product (or a range of products) to increase demand. But, under non-price competition, the firm focuses on various strategies other than price discounts for increasing market share. The important examples of non-price competition include:

- Free deliveries and installation
- Extended warranties for customers

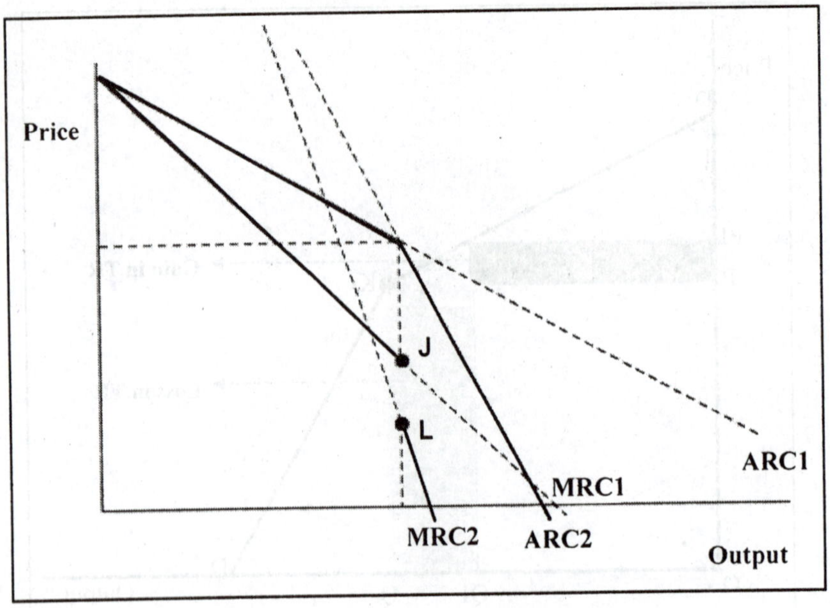

Figure 22.4: Two demand curves and corresponding MRCs under oligopoly competition.

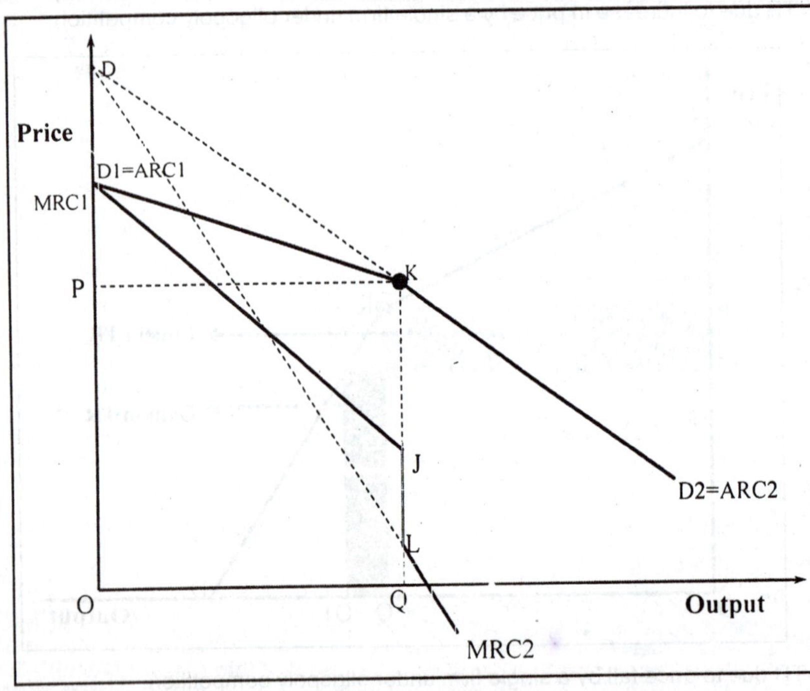

Figure 22.5: Indeterminate (Kinked) demand curve and MRC with discontinuity under oligopoly competition.

- ☆ Provision of credit facilities
- ☆ Offering discounts on the product
- ☆ Branding of products
- ☆ Heavy spending on advertisements
- ☆ Provision of repairs and maintenance services to the customers
- ☆ Home delivery facility
- ☆ Extension of opening hours (24 hours shopping in many stores)
- ☆ Internet shopping for customers

Considering the price rigidity, the only way the firms can enhance the sales of the product is by resorting to non-price competition. So, in order to increase the market share, the oligopolistic firms engage in aggressive and extensive sales promotion efforts by means of advertisement and by changing the design and improving the quality of their products, as these aim at stimulating and motivating the customers to purchase the products. One main objective of making higher selling costs or advertising costs is to make the demand curve more decisive and also favourable. Under monopolistic competition also, advertisement plays a considerable role because, the product is highly differentiated and it is necessary on the part of the firms to educate the customers about the features of the product. But, in oligopoly, advertisement still plays a bigger role, even when compared to monopolistic competition, mainly to make the output and price policies effective and to make the demand curve decisive or deterministic, as there is mutual interdependence among the firms regarding their investment, output and pricing policies. So, in oligopoly, non-price competition becomes a life and death matter for the survival of the firms.

6. Collusion (Group Behaviour)

The important feature of oligopoly is the existence of both the co-operative and competitive trends among the business firms. Understanding the interdependency and indeterminate demand curve behaviour of the firms, they generally resort for a mutual co-operation or collusion to their common or best advantage such that, all the business firms will prosper. So, they tend to collude in terms of output and price policies to grab the maximum from the market and all will mutually benefit. But, at times of rivalry, they compete each other for gaining maximum profit share and try to capture other firms' market by increasing the selling costs and by formulating suitable output and price policies. Such collusion or group behaviour is not seen in perfect competition, monopoly and monopolistic competition.

7. True Competition

Since there are few sellers, an action from one seller regarding output and pricing policies will be counteracted by his rivalries with an equal reaction. This is the case of a true competition, which is seen only in oligopoly and not in other forms of market competition.

8. Element of Monopoly

If all the few sellers produce a homogenous product, it is the case of perfect oligopoly and there will be true competition among the firms. But, if these few firms produce differentiated products, each firm should be considered as a monopoly firm, as no other firm is producing the same product. So, by producing the differentiated products by all the oligopoly firms, there will be more independence rather than the interdependence.

9. Barriers to Entry of Firms

As there is keen competition among the firms in oligopoly market situation, there are no barriers regarding entry of firms into the industry. However, in the long run, there will be strong barriers regarding the entry of new firms into the industry. They include:

- The existing firms gained control over essential and specialized inputs.
- High capital requirement due to huge plant costs.
- Exclusive patent rights on the raw materials used in the production programme or products produced.
- Less unused capacity, which makes the industry less attractive.
- Economies of scale enjoyed by only few large firms.
- Heavy advertising expenditure by the firms acts as a financial barrier for the new firms to enter the industry.
- Existing firms join as a group and go for destroyer pricing, which is not economical for the new firms.

10. Lack of Uniformity in the Size of Firms

This is one of the important characteristics of oligopolistic market or industry. The firms considerably differ in their sizes, *i.e.*, asymmetrical size situation is more common in oligopoly.

11. No Unique Pattern of Price Behaviour

Since, there exists mutual interdependence among the firms and the price-output policies of one firm influences the fortunes of other firms, there will not be any unique pattern of price behaviour for each firm in the oligopoly market situation. Sometimes, all the firms join as a group and each firm wishes to co-operate with its rivals on price-output policies. They may even recognize one firm as a leader of their group and under that leader's directions, other firms will plan their price-output policies for their mutualism. In such case, individual seller's demand curve is a part of industry demand curve. In case of conflicting situation, each firm plan its own price-output policies and try to gain maximum share from the market. Taking into consideration the possible behavioural pattern of firms, it is not possible to predict any unique pattern of pricing behaviour in the oligopoly market situation.

12. Profit Maximization Condition

An oligopoly firm maximizes its profits by producing the output, where MR equals MC.

13. Price Makers

Oligopolies are price makers rather than price takers, as the few sellers mutually take the pricing decisions.

14. Branded Products

Few firms sell branded products, which are close substitutes of each other. Some oligopolistic industries produce homogenous or standardized products (steel, zinc, copper, cement etc), whereas, others produce differentiated products (automobiles, detergents, greeting cards etc).

15. Perfect Knowledge

Oligopoly firms have perfect knowledge about their own cost structures and demand functions, but their inter-firm information may be incomplete. Buyers have only imperfect knowledge about cost, price and product quality.

16. Long Run Super Normal Profits

Since barriers to entry for new firms are very strong in oligopoly competition, the existing firms can earn super-normal profits even in the long run.

17. Price Rigidity

Price rigidity is a characteristic feature of oligopoly market. It implies that, in oligopoly market situation, the firms tend to keep the price of the product at the same level. This is because, if price is increased by one seller, other firms will not increase the price and thereby, the firm which increased the price will lose. If one seller reduces the price, other firms also reduce the price and thereby, the firm which reduced the price initially will not get the expected profits. Hence, the firms tend to keep the prices at the same level and this is called as price rigidity.

18. Market Concentration

It is one of the important features of oligopoly. Of all the four types of market competitions *viz.*, perfect competition, monopoly, oligopoly and monopolistic competition, the feature of 'market concentration' is an important issue under oligopoly. This is because, under monopoly, where there is a single firm and it represents the entire industry, the concept of market concentration is not relevant. In both, perfect competition and monopolistic competition, there are large number of firms, thereby, studying market concentration has no meaning. So, studying the concept of market concentration under oligopoly competition has special significance, as it influences the collusion, non-price competition etc., among the firms in the industry. This market concentration is measured in one of the following three ways:

Market Share Method

In this method, the market share of each firm is measured as a fraction of total production or total sales in the industry. For example, if a firm A sell Rs.10 crores of soft drinks each year and the total soft drink sales in the market are Rs.30 crores, then the market share of firm A is 33.3 percent.

Concentration Ratio Method

In this method, we combine the market shares of top four firms in the industry and popularly, it is called as Four-firm concentration ratio. So, this ratio measures the percentage of total product sales in the industry accounted by the top four firms. A low concentration ratio indicates a high degree of competition and *vice versa*. The range of concentration ratio is from almost zero for perfect competition to 100 percent for monopoly. If the percentage is fairly high (over 40 per cent), then we would say that, the industry is concentrated. A ratio of less than 40 percent indicates monopolistic competition. So, concentration ratio falls somewhere in the range of 0 to 100 percent. For example, there are 25 firms in an industry accounting for Rs. 450 millions worth of annual sales of the product. If top four firms of this industry account for Rs. 400 worth of annual sales, then the Concentration ratio will be 88.9 per cent and this infers the market structure is oligopoly.

Herfindahl-Hirschman Index (HHI) Method

In Herfindahl-Hirschman Index method, we consider the market shares of all firms in the market. In this method, instead of simply adding the market shares, we square each market share before adding. So, we place greater weightage on the large market shares. The value of Herfindahl-Hirschman Index ranges between 0 to 10,000. A high Herfindahl index number indicates a high degree of concentration by one or two firms. In perfect competition, the Herfindahl-Hirschman Index is small. In monopoly, the Herfindahl-Hirschman Index is 10,000 (100 squared).

22.4. Output and Price Determination

The output and price behaviour of the firms operating under duopolistic or oligopolistic market condition can be studied under the following sub-heads:

I. Output and Price Determination Under Duopoly

It depends upon the nature of the product the two firms produce in the industry.

- ☆ If the products produced by the two firms are homogenous and each having half of the total market share, the price-output policy of each firm is likely to affect the other firm appreciably. So, to maximize the profits, both firms decided to go for collusion and they may agree upon the price or share the total market or fix output quota or merge themselves into one unit to form a monopoly etc.
- ☆ If stiff competition arises between the two firms in terms of quoting a competitive price for the homogenous product, the firm with lower cost structure and better good will among the customers will establish in the industry and it drive the other firm out of the industry.
- ☆ If the products produced by the two firms are heterogenous, then each firm will have a close watch on the investment, output and pricing policies of the rival firm. The firm that produces the product with superior quality through cost-effective technologies will earn super-normal profits. Each firm will fix the price of the product and expand output in accordance with the demand for the product in the market.

II. Output and Price Determination Under Oligopoly

It depends upon the nature of the product the few firms produce in the industry.

- ☆ If the firms produce homogenous product and having powerful influence on the total market, the investment, output and pricing policy of each will adversely influence the others. So, to maximize the profits, the firms to tend to collude and plan the price-output policy of the industry.
- ☆ If the firms produce heterogenous products, then each firm can plan its own price-output policy without any fear of losing the customers or of immediate reactions from its rivals.

There is no single model that satisfactorily explains the oligopolistic price-output behaviour due to the wide diversity of oligopolistic situations and mutual interdependence that makes predictions about pricing and output quantity precarious. There are set of theories like Cournot duopoly model, Bertrand duopoly model, the Edgeworth duopoly model, Chamberlin Model, the Kinked Demand Curve Model, the Cartel model, Price leadership model, Game theory etc., which have been developed on particular set of assumptions about the reactions of other firms to the action of the firm under study. It is important that, in oligopoly, there are more than two firms and thereby, the analysis will be multi-dimensional. So, the best way to restrict the analysis to two-dimensional is to study the simplest analytical form of oligopoly *i.e.*, duopoly, where there are only two firms engaged in the production of a homogenous product. These models are discussed in-detail in the ensuing pages.

1. The Cournot Duopoly Model

This model was named after Antoine Augustin Cournot (1801-1877), a French Economist, published his theory of duopoly in 1838. This model assumes that that,

- ☆ There are two independent firms and they compete on the basis of quantity rather than on price.
- ☆ Each firm makes an output decision assuming the other firm's behaviour is fixed.
- ☆ Both the firms produce an homogenous product, 'mineral water' and there is no product differentiation.
- ☆ The number of firms are fixed and the number of buyers are large.
- ☆ There is no collusion among the firms.
- ☆ Neither of the single firm fixes the price of the product and each firm accept the market demand price.
- ☆ Each firm's output decision affects the product price.
- ☆ There is no cost incurred in the production of the product.
- ☆ Since, there is no cost of production for the product, each firm competes in terms of quantities of the product and choose quantities simultaneously.
- ☆ Both the firms are rational and each act strategically to maximize the profits given its competitor's decisions.

Given these assumptions, consider two firms A and B engaged in exploiting two mineral springs. As assumed above, 'there is no cost of production of mineral water' *i.e.*, each firm simply collect the mineral water coming from these springs, it is bottled and sold in the market. Since, there is no cost incurred in the production of mineral water, this analysis assumes to be simple, as we have to deal the concept only from the demand side. As shown in the Figure 22.6.

- ☆ DD_1 is the market demand curve for the product (mineral water).
- ☆ Since, the cost of production of mineral water is assumed to be zero for both the firms, the MCs of both the firms will be zero and the MC curves of both the firms coincide with the X-axis.

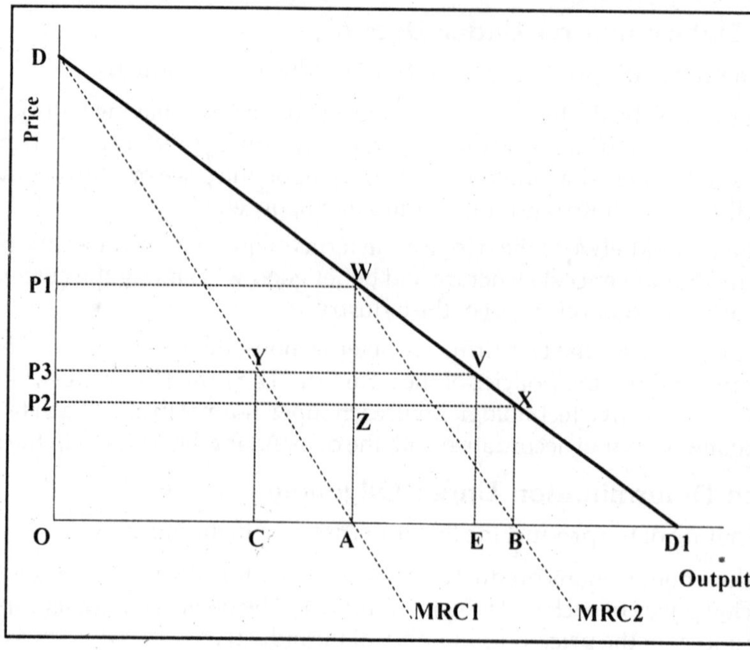

Figure 22.6: Cournot solution of duopoly competition.

- Suppose, the firm A is temporarily a monopolist and he takes up the business first.
- MRC_1 is the MRC of firm A and it is a downward sloping curve from left to right. This MRC_1 intersects its MC curve at point A and this infers that, of the total quantity of mineral water that can be produced by the two mineral springs *i.e.*, OD_1, firm A produces and supplies OA quantity of mineral water. So, this firm A produces 1/2 OD_1 output.
- For this OA quantity of mineral water, the firm A charges OP_1=AW price and thereby, the firm A earns OP_1WA monopoly profits.
- Assume, now firm B has entered the business. As per the assumption, firm B will expect that, firm A will not change its output level OA.
- Of the total market demand curve DD_1, DW is the market demand curve of firm A and thereby, WD_1 segment is the market demand curve for firm B.
- So, firm B's MRC is MRC_2 starts from point W of its market demand curve and it intersects its MC curve at B and this infers that, of the remaining quantity of mineral water that can be supplied into the market *i.e.*, AD_1, firm B supplies only AB *i.e.*, 1/2 AD_1 quantity of mineral water.
- Had the firm B supplies the remaining quantity of mineral water *i.e.*, AD_1 into the market, total supply of mineral water in the market is OD_1 ($OA+AD_1$) and thereby, the product will not fetch any price in the market or price is zero.
- So, by supplying only AB quantity of mineral water, the firm B charges OP_2=BX price and thereby, the firm B earns ABXZ profits.
- So, when firm B also entered the market, total supply of mineral water in the market increases to OB units *i.e.*, OA+AB. This increased supply of mineral water into the market brings down the price to OP_2 and thereby, profits of both the firms is OP_2XB, of which, firm A's profits are OP_2AZ and firm B's profits are ABXZ.
- So, with the entry of firm B into the market, now firm A must reconsider its situation in the business, as its profits levels were declined from OP_1WA to OP_2AZ.
- As per the assumption, firm A considers that, firm B will not change its output, AB=BD_1, and thereby, firm A considers OB is the portion of the market available to it, as BD_1 is the portion of the market of firm B (AB=BD_1).
- The best way that firm A can increase the profits is by producing 1/2 OB. So, by reducing the output from OA, the firm A can increase the profits, as it can increase the price of the product.
- Now, when firm A reduces the output and increases the profits, firm B will reappraise its position. Assuming firm A will hold its output constant, the best way that firm B can do is to increase its output to 1/2(OD_1-1/2OB), thereby, the price will fall and firm B can offer more quantity for sale.

- ☆ So, in this fashion, firm A will try to reduce the output and increase the price of the product, whereas firm B will react in opposite direction, *i.e.*, it increases the output and cause the price to decline in the market and ultimately an equilibrium price OP_3 will be attained.
- ☆ At this equilibrium price OP_3, the total output or supply of the mineral water into the market is OE.
- ☆ Out of this total supply OE, each firm supply 1/2 OE *i.e.*, firm A supplies OC and firm B supplies CE. Considering the total potential supply that is possible by both the firms *i.e.*, OD_1, each firm supplies $1/3OD_1$ and both the firms together supplies $2/3\ OD_1$.
- ☆ So, firm A supplies OC quantity of mineral water at OP_3 price and thereby, firm A earns $OCYP_3$ profits. Firm B supplies CE quantity of mineral water at OP_3 price, thereby, firm B earns CEVY profits.
- ☆ From the above discussion, we can infer that, the two firms produce $2/3OD_1$ output and supply the same into the market. This can be written as $[2/(2+1)]OD_1$. So, if there are two firms, the output that is actually supplied into the market as a proportion of total potential output that can be supplied by the two firms is given by (2)/(2)+1. So, if there are 'n' number of firms that are involved in supplying the mineral water, the quantity of mineral water actually supplied into the market will be $[[n/(n+1)].OD_1]$.

Comparison of Cournot Solution with Monopoly Solution

Let us assume both the firms A and B formed a coalition and they collectively produced OA units of mineral water *i.e.*, $1/2\ OD_1$. At this level of output OA, the price of the product is OP_1, thereby, the two firms get the maximum possible profits OP_1WA. Both the firms can share this profit equally. This coalition profit OP_1WA is considerably greater than the actual equilibrium profits $OCYP_3$ of firm A and CEVY of firm B. This infers that, when the two firms compete each other, both the price and profits are declined, as output supplied into the market is increased (OE) when compared to coalition of firms as monopoly (OA). So, when the firms compete each other, price of the product is declined in the market (OP_3), thereby, consumer's welfare is promoted. Algebraically, in the Cournot solution, the output will be $4/3^{rd}$ of the monopoly output (*i.e.*, larger than monopoly output).

Comparison of Cournot Solution with Perfect Competitive Solution

Considering perfect competition, the two firms produce OD_1 quantity of mineral water at zero MC. So, price will be zero. But, in Cournot solution, the price is OP_3 and it is higher than zero MC and price under perfect competition. But, the output under Cournot solution is only OE and it is lower compared to perfect competitive solution *i.e.*, OD_1. Algebraically, the output under Cournot solution will be $2/3^{rd}$ of the perfect competitive output. So, customers are worse off under Cournot solution compared to perfectly competitive solution, as output is reduced and price is higher.

Cournot equilibrium is an example of Nash equilibrium. The concept of Nash equilibrium was proposed by John Forbes Nash. It is a solution concept of a game involving two or more players, in which each player is assumed to know the equilibrium strategies of the other players and no player has anything to gain by changing only his or her own strategy (*i.e.*, by changing unilaterally). If each player has chosen a strategy and no player can benefit by changing his or her strategy, while other players keep theirs unchanged, then the current set of strategy choices and the corresponding payoffs constitute a Nash equilibrium. In Cournot solution, since each firm desires to produce a particular output level that maximizes its profits, given the profit maximizing output level of the other firm, Cournot duopoly is, otherwise, called as Cournot-Nash duopoly equilibrium.

Explanation of Cournot Solution with the Help of Reaction Rurves

The Cournot solution of duopoly can be better explained with the help of reaction curves. The reaction curves are output reaction curves, as the two firms under Cournot solution compete on the basis of quantity of output produced. It is important that, the reaction curves of the firms represent the firm's own reactions in terms of output to the moves of the rivalry firm. In the Figure 22.7,

- ☆ On X-axis, firm A's output and on Y-axis firm B's output are taken.
- ☆ AA' is the firm A's reaction curve and BB' is the firm B's reaction curve. The output reaction curve AA' of firm A indicates, how firm A will react to any change in output of firm B. Similarly, output reaction curve BB' of firm B indicates, how firm B will react to any change in output of firm A.
- ☆ In the Figure 22.7, we have drawn the two reaction curves as straight lines, as in the Figure 22.6 we assumed market demand curve DD_1 is a straight line and no MC is incurred in producing the product.
- ☆ Assume, firm A produces OA' output and this is the monopoly output, thereby, firm B's output is zero.
- ☆ Let firm B also enters the market and this firm B assumes that, firm A's output is constant at OA'. Considering firm B's reaction curve, firm B produces OB_1 quantity of output.
- ☆ When firm B is producing output OB_1, firm A assumes that, firm B's output is constant at OB_1 or continue production at OB_1 output, thereby, considering its reaction curve, firm A will now produce OA_1 output in reaction to firm B's output OB_1.

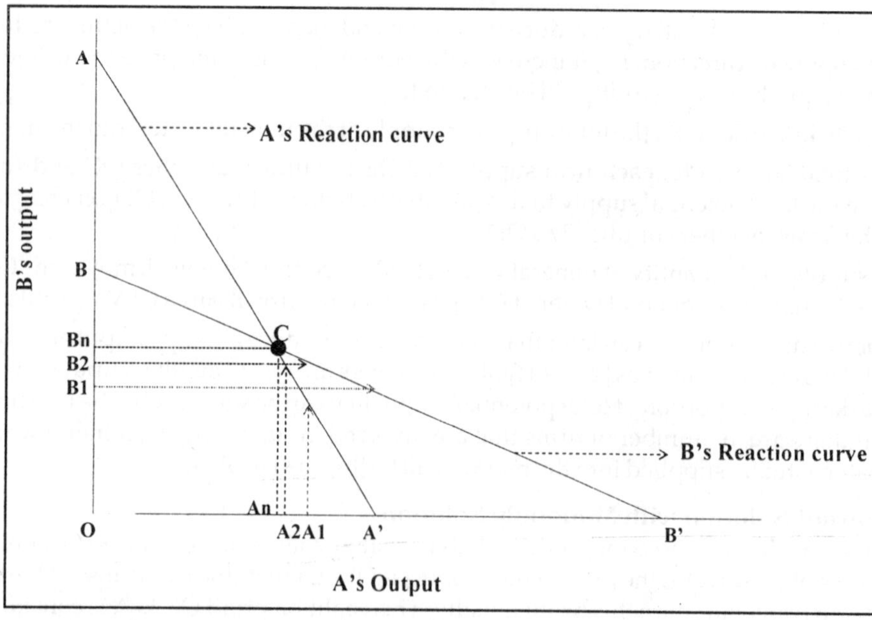

Figure 22.7: Cournot solution of duopoly in terms of reaction curves.

- Now, firm B will react to firm A's OA_1 output. That means, when firm A is producing OA_1 output, firm B assumes that, firm A's output is constant at OA_1, thereby, considering its reaction curve, firm B will produce OB_2 output in reaction to firm A's output OA_1.
- Now, firm A will react to OB_2 output of firm B and firm A produces OA_2 output considering its reaction curve.
- This process of reactions of firms with reference to their respective outputs will continue till both the firms reach the Cournot point 'C', where both the firms produce equal output. At this Cournot point 'C', firm A produces OA_n output and firm B produces OB_n output.
- At this Cournot point, no firm will show the tendency to change its output, thereby, Cournot solution will ensure stable and determinate equilibrium under duopoly. If any one firm changes the output, again the price war starts between the two firms.

Criticisms of Cournot Solution

Though Cournot solution has offered sufficient explanation for a stable equilibrium of pricing and output under duopoly, it suffers from the following criticisms:

- It is a static model. That means, while deciding the output policy, each firm will assume that, its rival firm will not change its output from the present level, what ever output it decides to produce. So, it is unrealistic.
- This model makes an unrealistic assumption that, MC of the product is zero.
- With reference to the production of output, the firms actions are independent. But, this is against the basic assumptions of mutual interdependency among the firms under oligopoly (duopoly) competition.
- In this model, one firm reacts in the production of output, only after the rival firm decides its level of output. This is unrealistic.

2. The Bertrand Model

The reasoning offered by the Cournot solution for a stable and determinate equilibrium under duopoly was criticized by Joseph Louis Francois Bertrand, a French Mathematician. In Cournot solution, a strong assumption was made that, each firm believes its rival firm will keep its output level constant, whatever output the firm wishes to produce. But, this assumption was subjected to severe criticism and thereby, in Bertrand solution, it is assumed that, each firm expects that, its rival firm will keep the price level constant. This model is based on the following assumptions:

- There are two firms producing a homogenous product, thereby, the products of two firms are perfect substitutes.
- There is no collusion among the firms.
- Each firm assumes its rival firm will keep the price level constant.
- There is no cost of production of the product.
- Both the firms compete solely on the price.

☆ Both the firms are rational and each act strategically to maximize the profits.

☆ Entry is blocked for other firms.

☆ Demand is linear.

Given these assumptions, assume there are two firms A and B engaged in the production of a homogenous product. As shown in the Figure 22.8,

☆ OP is the price axis of both the firms, thereby, AP and BP are the demand curves of firm A and firm B respectively.

☆ OA is the maximum quantity of output firm A sells and OB is the maximum quantity of output firm B sells.

☆ So, the total output the two firms can produce is OA+OB. If this total output OA+OB is supplied into the market, then the product fetches zero competitive price.

☆ As assumed earlier, there is no collusion between the firms. Let firm A first enters the market and it produces OC quantity of output and sells the same at OP_1 price.

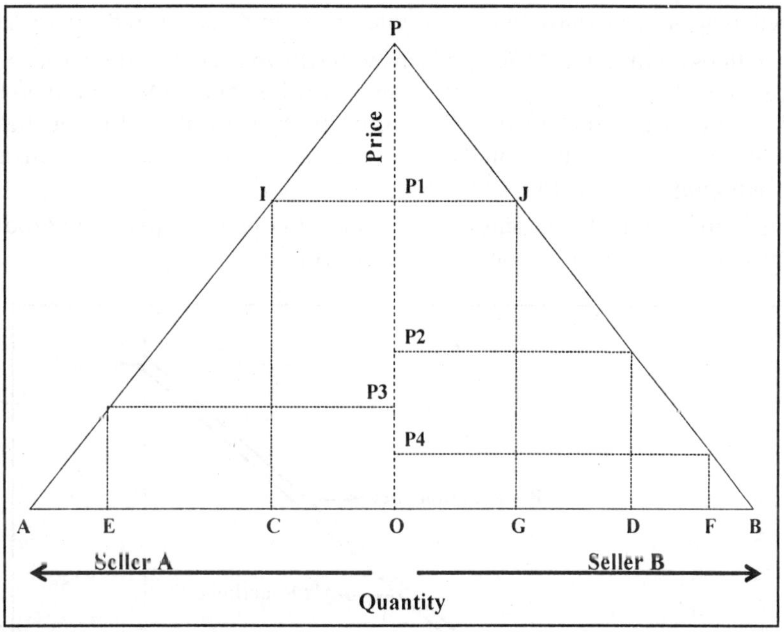

Figure 22.8: Bertrand solution of duopoly competition.

☆ Now, firm B enters the market and as per the assumption, it considers firm A will not change its price, thereby, it charges OP_2 price for the product by producing OD quantity of the output.

☆ By charging lower price OP_2 by firm B compared to OP_1 price of firm A, firm B attracts some of the customers of firm A. As OP_2 price of firm B is lower than OP_1 price of firm A, firm B could sell large quantity of output, OD than firm A, OC (*i.e.*, OD>OC)

☆ Now, firm A will reappraise its situation and to attract the customers of firm B, it will reduce the price to OP_3 by producing more output OE, considering firm B will not change its price. So, now firm A's price OP_3 is lower than firm B's price OP_2

☆ Again firm B will react to this situation and it will increase the output to OF and charges OP_4 price, assuming that, firm A will not change its price. So, now firm B's price OP_4 is lower than firm A's price OP_3.

☆ This process of adjustments and readjustments will continue in terms of price and both the firms A and B will finally sell their entire outputs OA and OB respectively at zero competitive price and stable equilibrium is attained.

☆ So, the stable equilibrium under Bertrand solution is $P_A=P_B=MC$, where P_A=price of product of firm A and P_B=price of product of firm B.

☆ At this equilibrium, neither of the two firms will raise or lower the price. If a firm raises the price above the competitive price, it will lose market to his rival, as the rival will continue to charge the same competitive price. If the firm charges the price below the competitive price, it will incur losses as price<MC. So, no firm will disturb the competitive price, thereby, the equilibrium is stable and determinate at $P_A=P_B=MC$. So, each firm will produce, where price=MC and there will be zero economic profits.

- ☆ If both firms set equal prices above MC, firms would get half the market at a price higher than MC. However, by lowering the prices slightly, they could gain the whole market and they lower the prices till each firm's price is equal to MC.
- ☆ The above discussion also infers an interesting point that, the two firms compete in terms of price, it means, the firms can easily change the quantity they supply, but once the firm choose a certain price, it is very hard to change it.

Explanation of Bertrand Solution with the Help of Reaction Curves

Like Cournot solution, Bertrand solution of duopoly equilibrium can also be explained with the aid of reaction curves. In the Figure 22.9,

- ☆ On X-axis, price of firm B's product and on Y-axis, price of firm A's product are taken. 45^0 line is drawn between the reaction curves of firm A and firm B to show the equality of prices charged by both the firms.
- ☆ Assume, when firm A entered the market as monopolist, it quotes a very high price, *i.e.*, monopoly price P_M.
- ☆ When firm B entered the market, to attract the consumer, it quotes a low price less than P_M.
- ☆ Now, firm A will reappraise its situation and quote a price less than firm B's price to increase profits.
- ☆ This chain of reactions continues till both the firms settle down at competitive price equal to MC. Say, if firm B fixes the price below the MC, then firm A maintains the price equal to MC and thereby, firm B incurs losses. So, no firm will reduce the price below MC. When firm B fix the price above MC, but below monopoly price P_M, firm A will quote the price below the firm B price by increasing the output. When firm B fixes the price above P_M (*i.e.*, above monopoly price), firm A charges monopoly price (P_M).
- ☆ Thus, both firms settle down at the point E, where for both the firms, price the product is equal to MC, *i.e.*, P_A=P_B=MC. This point E refers to the stable equilibrium of duopoly.

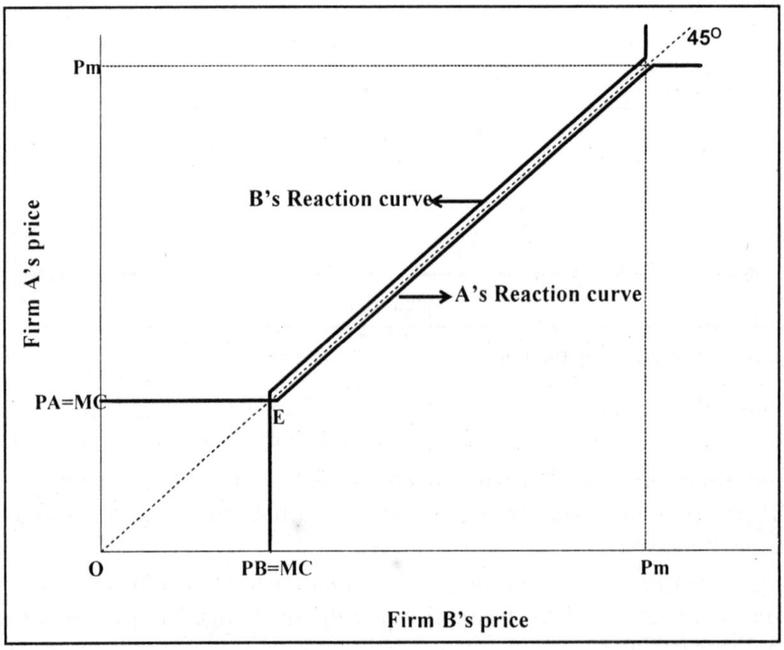

Figure 22.9: Bertrand solution of duopoly in terms of reaction curves.

Criticisms of Bertrand Solution

Bertrand solution too suffers from the following limitations:

- ☆ It is a static model. That means, while deciding the pricing policy, each firm will assume that, its rival firm will not change the price of the product.
- ☆ This model makes an unrealistic assumption that, cost of production of the product is zero.
- ☆ In this model, one firm reacts in fixing the price of the product, only after its rival firm decides the price policy of the product. This is unrealistic.
- ☆ This model considers only price competition between the firms, ignoring non-price competition.
- ☆ This model also ignores capacity constraints. That means, if a firm does not have the capacity to supply the whole market, then the equilibrium price=MC may not hold good.

Bertrand Solution versus Cournot Solution

Though both the above models have similar assumptions they differ in their implications. They include,

- Bertrand solution states that, price of the product under duopoly can be pushed upto MC level, thereby, duopoly ensures the benefits of perfect competition. According to Cournot solution, the price of the product is higher than MC. So, the output under Cournot solution is less compared to Bertrand solution.
- If capacity and output can be easily changed, Bertrand model is generally a better model of duopoly competition. But, if capacity and output cannot be easily changed, then Cournot model is better for the duopoly competition.
- In Bertrand solution, each firm expects that, its rival will keep its price constant. But in Cournot solution, each firm expects that its rival will keep its output constant.
- When the number of firms under Cournot model goes to infinity, even Cournot model gives the same result as in Bertrand model, *i.e.*, price is pushed upto the MC level.

3. The Edgeworth Model

F.Y.Edgeworth criticized the Cournot duopoly model and assumed that, each firm expects that, its rival firm will keep the price level constant. All other assumptions of Cournot solution are considered here in Edgeworth model except that the output of the rival firm remains constant. But, in addition to Cournot model assumptions, it further assumes that, the two firms have limited productive capacity and the total market is divided equally between the two firms. In the Figure 22.10,

- OP is the price axis of both the firms, thereby AP and BP are the demand curves of firm A and firm B respectively.
- OA+OB is the total market demand for the product produced by the two firms. But as assumed earlier, due to limited productive capacity, firm A can produce only OA_2 quantity of product and firm B can produce only OB_2 quantity of product.
- As assumed, there is no collusion between the two firms.
- Let firm A enters the market first and sells OA_1 output at OP_1 price, thereby, it earns OP_1CA_1 monopoly profits.
- Noticing the profits earned by firm A, firm B will enter the market, produce more output and fix a price lower than OP_1, so as to attract the customers and earn more profits, assuming that, firm A will not change its price.
- Now firm A will re-appraise its price-output policy and to attract the customers it will reduce the price below the firm B's price by increasing the output, assuming that, firm B will not change its price.
- Again, firm B will react to this situation and it will increase the output and reduce the price below firm A's price, assuming that, firm A will not change the price.

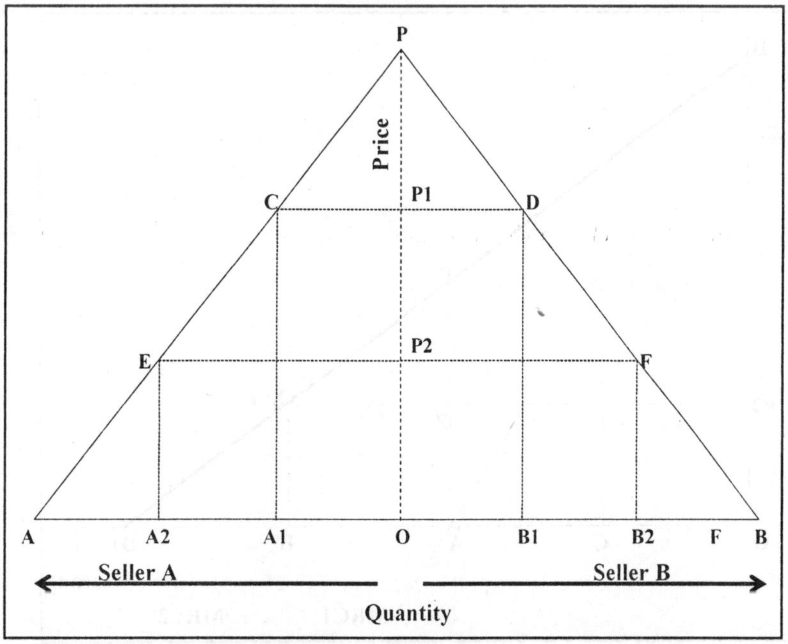

Figure 22.10: Edgeworth solution of duopoly competition.

- This price war will continue between the two firms such that, both firm A and firm B will sell the product at price OP_2, where firm A sells its capacity output OA_2 and firm B sells its capacity output OB_2. So, firm A will earn OP_2EA_2 profits and firm B will earn OP_2FB_2 profits.
- According to Edgeworth, this price OP_2 is not an equilibrium price. This is because, if firm A increases the price from OP_2 to OP_1, it leads to increase in profits of firm A from OP_2EA_2 to OP_1CA_1. So, firm A makes the price increase knowing that, firm B cannot attract any of the OA_1 units of sale, as it cannot supply them. So, firm B will also follow the suit and raise its price a little above OP_2 and earn more profits. So, this price war never ends. This makes the Edgeworth solution as an 'indeterminate and perpetual disequilibrium', constantly varying between competitive to the monopoly levels.
- Compared to the Cournot model, Edgeworth model showed that, price factor influences the duopoly solution rather than output, though the equilibrium is indeterminate.

4. The Chamberlin's Model

In the earlier models, the chief logic error is that, neither of the firms recognizes their mutual interdependence and this leads to different responses from both the firms in terms of their respective price-output policies. This is where, Chamberlin criticized the earlier models and hence, he framed an assumption that, both the firms recognizes their mutual interdependence. Considering mutual interdependence between the two firms, we can explain the Chamberlin solution in terms of both output adjustment and price adjustment. This is because, according to Chamberlin, when the two firms act independently, each firm will decide its own output-price policy assuming the rivalry firm will keep is output-price policy stable. But he opined that, oligopolists recognize their mutual interdependence and try to adjust their output-price policies, so that, they mutually benefit in the trade. This ensures the firms to attain stable and determinate equilibrium, enjoy monopoly output and price, even though the firms act in a non-collusive manner. In the Figure 22.11,

- Consider two firms A and B engaged in producing a homogenous product.
- DD_1 is the market demand curve for the product.
- Since, the cost of production of the product is assumed to be zero for both the firms, the MCs of both the firms will be zero and their MC curves coincide with X-axis.
- Suppose, the firm A is temporarily a monopolist and it takes up the business first.
- MRC_1 is the MRC of firm A and it is a downward sloping curve from left to right. This MRC curve intersects its MC curve at point A and this infers that, of the total quantity of product that can be produced by the two firms *i.e.*, OD_1, firm A produces and supplies OA quantity of product. Thus, firm A produces 1/2 OD_1 output.
- For this OA quantity of output, the firm A charges OP_1=AW price and thereby, the firm A earns OP_1WA monopoly profits.

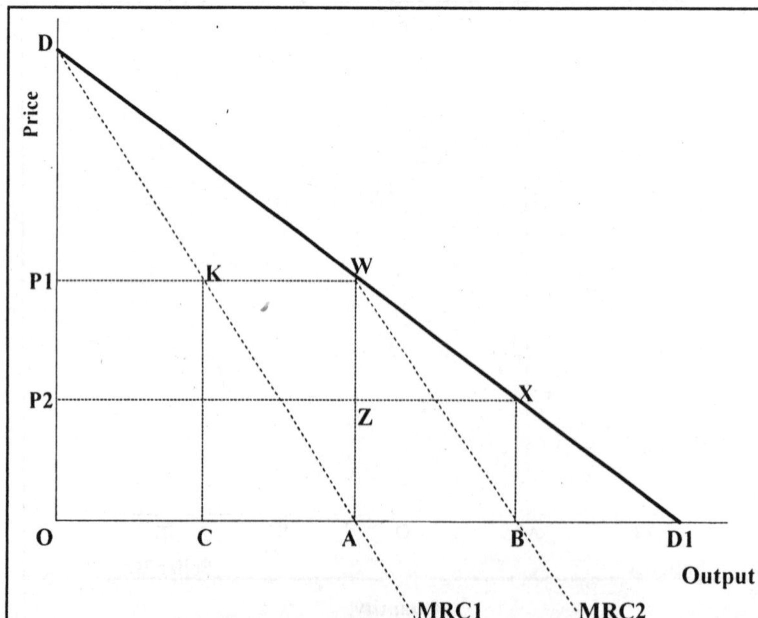

Figure 22.11: Chamberlin solution of duopoly competition.

- ☆ Assume, now firm B has entered the business. As discussed under Cournot model, firm B will expect that, firm A will not change its output level OA.
- ☆ Of the total market demand curve DD_1, DW is the market demand curve of firm A and thereby, WD_1 segment is the market demand curve for firm B.
- ☆ So, firm B's MRC *i.e.*, MRC_2 starts from point W of its market demand curve and it intersects its MC curve at B and this infers that, of the remaining quantity of output that can be supplied into the market *i.e.*, AD_1, firm B supplies only AB *i.e.*, 1/2 AD_1 quantity of output.
- ☆ Had the firm B supplies remaining quantity of output, *i.e.*, AD_1 into the market, total supply of output in the market is OD_1 (*i.e.*, $OA+AD_1$) and thereby, the product will not fetch any price in the market or price is zero.
- ☆ So, by supplying only AB quantity of output, the firm B charges OP_2=BX price and thereby, firm B earns ABXZ profits.
- ☆ So, when firm B also entered the market, total supply of product in the market increases to OB units *i.e.*, OA+AB. This increased supply of product into the market brings down the price to OP_2 and thereby, total profits of both the firms is OP_2XB, of which, firm A's profits are OP_2AZ and firm B's profits are ABXZ.
- ☆ So, with the entry of firm B into the market, the firm A profit level was declined from OP_1WA to OP_2AZ.
- ☆ From this juncture, Chamberlin's solution differs from Cournot solution. In case of Cournot solution, the firm A will reappraise its situation and readjust its output assuming firm B will not change its output. But, in case of Chamberlin solution, where mutual interdependence of firms is recognized, here firm A will decide to produce output OC, which is equal to output AB of firm B, such that the total output of both the firms is equal to monopoly output OA *i.e.*, OA=[(OC output of firm A)+(CA (=AB) output of firm B)].
- ☆ Since, output produced is monopoly output *i.e.*, OA, again the price will rise from OP_2 to OP_1. Thus, both the firms, given identical costs, share the total monopoly profits OP_1WA.
- ☆ Thus, under Chamberlin's solution, each firm produces half of the monopoly output and result in maximization of joint profits, though they act in a non-collusive manner.
- ☆ Thus, Chamberlin's solution ensures stable and determinate equilibrium under duopoly.

Criticisms of Chamberlin Solution

Though Chamberlin's solution gave valid explanation for the equilibrium under duopoly considering mutual interdependence between the firms, it too suffers from the following limitations:

- ☆ This model ignores the entry of new firms into the industry. This is because, when both the firms maximizes the monopoly profits, it will attract the new firms into the industry and with the entry of new firms, stable equilibrium is unlikely to occur.
- ☆ Arriving at monopoly profits through recognizing mutual interdependence is possible only when both the firms have identical cost structures and demands. But, in practicality, cost structure and demands varies across the firms, thereby, arriving at monopoly output and monopoly profits are not easily feasible.

5. Non-collusive Oligopoly Model – Kinked Demand Curve Theory

As mentioned earlier, according to the Kinked demand curve theory, each firm will face two market demand curves for its product. At high prices, the firm faces the relatively elastic market demand curve and at low prices, the firm faces the relatively inelastic market demand curve. This model aims to explain price stability or rigidity by the use of these two demand curves. This model seeks to explain that, even when there is no collusion among the firms under oligopoly, price of the product tends to be stable or rigid. It also illustrates the high degree of interdependency that exists among the firms, as the market demand curve that each oligopolist faces is determined by the output and price decisions of the other firms in the industry.

The concept of kinked demand curve analysis was explained by Paul M. Sweezy. This concept explains the price rigidity in oligopoly market situation and it illustrates the short run production activity of an oligopoly firm. This kinked demand curve represents a situation in which, the firm receives no incentives, either by increasing or decreasing the price of the product, but tends to keep the price at particular or same level. This model assumes that, if an oligopoly firm lowers the price, its rivals (other firms) also react in the same way and lower the price, so as to avoid losing the customers. Thus, the firm, by decreasing the price, will not be able to increase the demand for the product as expected because, other firms also decrease the price. This portion of the demand curve is less elastic or relatively inelastic. On the other hand, if an oligopoly firm increase the price of the product, other firms will not increase the price and thereby, the firm is not able to get higher revenue, as other firms offer the same product at lower price and customers prefer low priced product. So, the quantity demanded for the product offered by a firm at high price will fall considerably. This portion of the demand curve is more elastic or relatively elastic. Under such a situation, the demand curve of a firm will have a kink at the present price and this is explained below through the Figure 22.12.

In the Figure 22.12, the demand curve D_1D_2 with a kink at point 'K' is shown. OP is the price at which, the firm is selling the product by producing OQ units. When a firm increases the price above OP, other firms will not increase the price because, they will lose the market share or profit. This makes the portion of demand curve D_1K (above OP price) more elastic, *i.e.*, a small increase in price of the product will adversely affect the demand for the product.

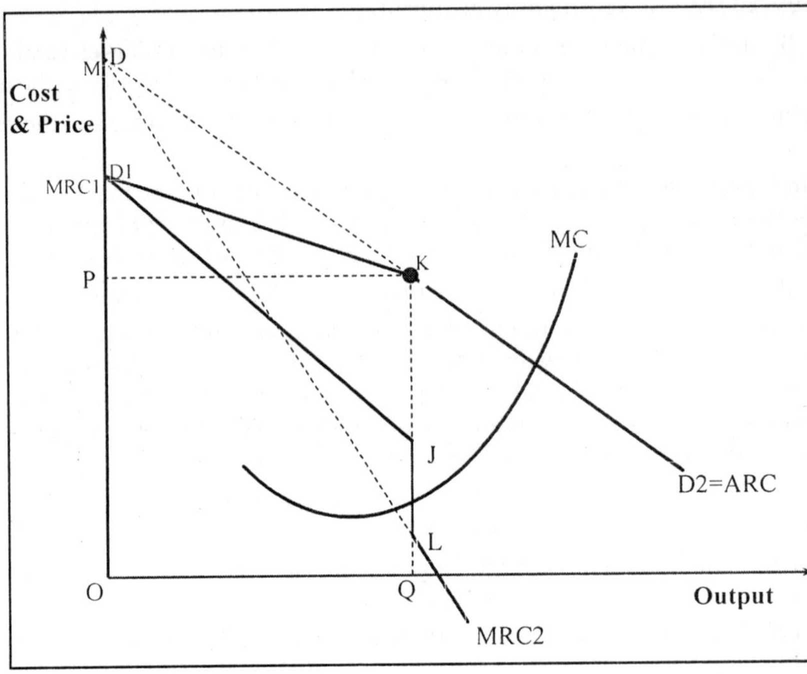

Figure 22.12: Kinked demand curve theory explaining price rigidity under oligopoly competition (during trade depression).

If the price is decreased below OP, the firm cannot increase the sales and profit as expected because, other firms also reduce the price of the product. Hence, the firm which initially reduces the price of the product will not experience the increase in sales and profits, as expected. This makes the portion of demand curve KD_2 (below OP price) less elastic or more inelastic, *i.e.*, a greater decrease in price of the product will not increase the quantity demanded of the product as expected.

So, above the price OP, the demand curve of the firm is D_1K and it is more elastic and below the price OP, the demand curve of the firm is KD_2 and it is less elastic. In other words, the demand curve consists of two segments *viz.*,

- ☆ a relatively more elastic segment for price increases above kink 'K' and
- ☆ a relatively less elastic segment for price decreases below kink 'K'

The relative elasticities of these two segments is directly based on the interdependent decision-making of oligopolistic firms.

So, when there is less scope for the significant increase in the profits either by increasing or decreasing the price of the product, the firm generally think of maintaining same price level OP and this is termed as price rigidity, indicated by a kink 'K' on the demand curve. Since, on the demand curve, there is kink at 'K', where the firm maintains price rigidity, the demand curve is called as Kinked demand curve.

The MRC (MRC_1MRC_2) of the firm is also shown in the Figure 22.12. The peculiarity is that, there is gap or discontinuity in MRC below the kink indicated by 'JL' between MRC_1 and MRC_2. This is due to the change in the E_D *i.e.*, from the more elastic nature of the demand (above kink 'K') to more inelastic nature of demand (below kink 'K'). The extent of gap between MRC_1 and MRC_2 depends upon

- ☆ the extent of price rigidity followed by the firm and it varies directly.
- ☆ the E_D for the product above and below the kink 'K'. If the E_D of product is greater above K and inelasticity of demand for the product is greater below K, then the gap between J and L will be large. That means, the length of discontinuity or gap depends upon the relative elasticities of two segments *i.e.*, D_1K and KD_2 and greater the difference in two elasticities, the greater the discontinuity. So, the gap 'JL' on the MRC implies the extent of price rigidity the firm follows, indicated by a kink on the demand curve. (To have a better understanding, please see the Appendix given at the end of this chapter).

- the other factors like number of rivals, relative size of rivals, differences among the rivals' products, extent of collusion and number of buyers.

DD_2 and M-MRC_2 are the demand curve or ARC and MRC respectively in an imperfect competition. Since, in oligopoly market situation, there is price rigidity and hence, there is a kink at price OP on the demand curve or ARC. Above OP price, the demand curve or ARC is more elastic and below OP, the demand curve or ARC is less elastic. Hence, the demand curve or ARC takes a new shape as D_1D_2 with a kink at 'K'. Hence, MRC starts below M *i.e.* at MRC_1, but have a gap or discontinuity JL below the kink 'K', thereby, MRC_1MRC_2 is the MRC of a oligopoly firm.

MC curve is also shown in the Figure 22.12. The point of optimality is MC=MR. The MC curve intersects the MRC curve *i.e.*, MRC_1MRC_2 between the gap JL, implying that, the firm produces OQ level of output and fix the price of the product at OP. It is important to note that, as long as MC curve intersects the MRC_1MRC_2 curve between the gap J and L, there is no need for the firm to change the price of the product, as the firm is already producing equilibrium output OQ. Hence, it will keep the price rigid at OP level. So, each segment of the demand curve has its own MRC segment. This actually means that, the MRC of an oligopolistic firm, labeled as MRC_1MRC_2 in the Figure 22.12 shows the following three distinct segments:

- ***Top Segment:*** The flatter top portion of the MRC corresponds to the more elastic demand curve generated by price increases.
- ***Middle Segment:*** The vertical middle segment of the MRC connecting the top and bottom segments that occurs at the equilibrium quantity OQ corresponds with the kink on the demand curve at OP price.
- ***Bottom Segment:*** The steeper bottom portion of the MRC corresponds to the less elastic demand curve generated by price decreases.

5.1. Price and Output Determination

Price and output determination with reference to kinked demand curve theory can be studied under two different types of trade situations *viz.*, during trade depression and when trade is under boom conditions

1. When Trade is Under Depression Conditions

The above explanation of kinked demand curve (Figure 22.12) holds good during trade depression conditions. It is when the trade is relatively under depression, an oligopolist will be reluctant to either increase or decrease the price of the product, as both measures will not fetch him higher profits as expected. Hence, he will prefer to maintain price rigidity for the product.

2. When Trade is Under Boom Conditions

When trade is experiencing boom conditions, the relevant kinked demand curve or ARC will be in the opposite direction (Figure 22.13). This implies that, if one firm raises the price, others will follow and if it lowers the price, others will not follow the suit. This is because, since trade is already under boom conditions, the rivalry firms feel no need to

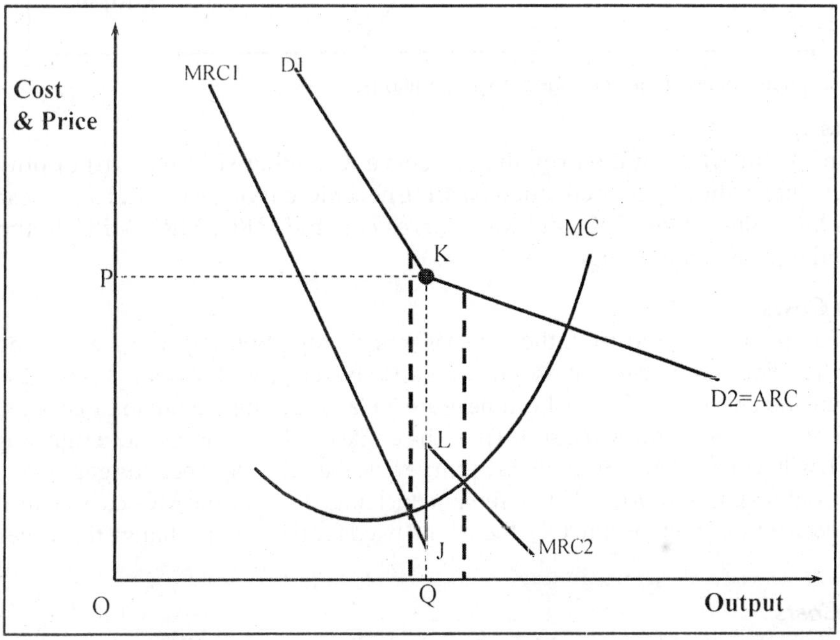

Figure 22.13: Kinked demand curve theory explaining price-output determination under boom conditions of trade.

expand sales by cutting the price. So, the tendency to keep the price rigid is of less certain under boom conditions. So, under boom conditions, the upper portion of the ARC *i.e.*, D_1K is more steep, while the lower portion of ARC *i.e.*, KD_2 is more flatter.

The MRC again comprises of two parts *i.e.*, MRC_1J and $LMRC_2$. The MRC *i.e.*, MRC_1MRC_2 has a gap or discontinuity below the kink *i.e.*, LJ, but the right hand portion of the MRC *i.e.*, $LMRC_2$ begins at L, which is located at higher level than lower part of the left hand part of MRC (MRC_1J), *i.e.*, above J, where left hand part ends.

The MC curve passes through the gap LJ of the MRC, *i.e.*, MRC_1MRC_2. But, while passing, it cuts both the parts of the MRC. So, it implies, profits will be increased even when price of the product is increased or decreased. So, an oligopoly firm can increase the price above OP and earns more profits, but selling less quantity of output (<OQ) or the firm may go alone by reducing the price below OP and still can increase profits through selling more amount of output (>OQ), as trade is under boom conditions. So, under boom conditions of trade, in oligopoly industry, prices will not remain stable or price rigidity is not seen.

Factors Influencing Price of the Product in Oligopoly Market

Since, the firms in oligopoly industry tend to keep the price level rigid at point 'K' on the demand curve, it is interesting to know the influences of the changes in cost structure and changes in demand conditions for the product on the price rigidity. This is explained through the Figures 22.14 and 22.15.

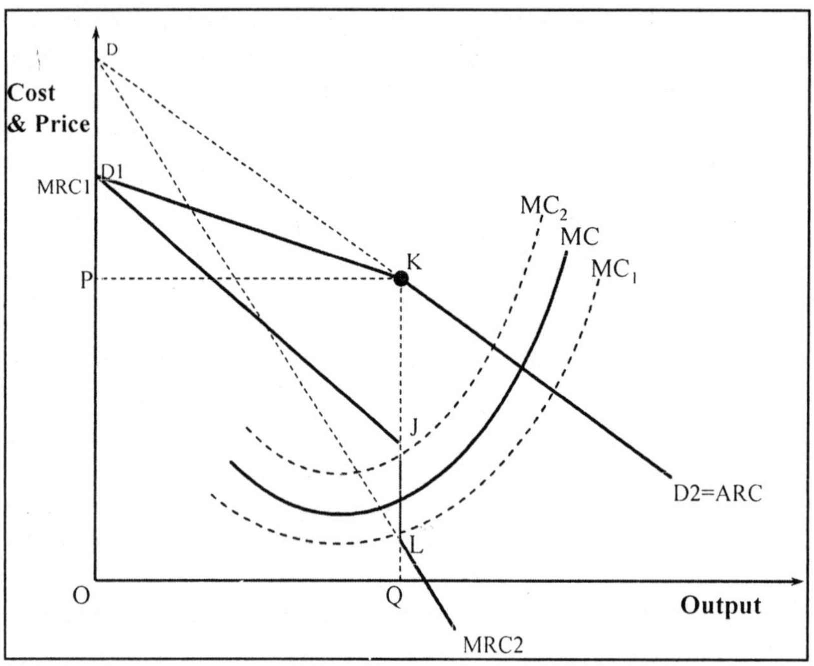

Figure 22.14: Concept of price rigidity under changed cost conditions.

(a) Changes in Costs

Due to changes in cost structure of the firm, the MC curve will either shift upward or downward. It is shown through the Figure 22.14 that, if the cost of production falls, the new MC curve is MC_1 and if the cost of production rises, the new MC curve is MC_2. Since, these MC_1 and MC_2 curves cuts the MRC, MRC_1MRC_2 in the gap 'JL' the profit maximizing output and price will not change.

Case 1-Decrease in Costs

It is interesting that, if costs are reduced, the new MC curve, MC_1 will fall down and generally cut the MRC, MRC_1MRC_2 in the gap 'JL'. This is because, as costs fall, the gap between JL will widen. This is because, as costs fall, the lower portion of the demand curve *i.e.*, KD_2 will become more inelastic because, a fall in price of one firm (due to fall in costs), will make other firms to follow the same suit. Thus, the angle D_1KD_2 becomes more right angle at 'K' and hence, the gap between J and L will widen. Hence, the new MC_1 curve cuts the MRC between the gap JL. So, the net result is, the firm will produce same output OQ and price OP will remain unchanged. Since, the MC_1 curve cuts the MRC in the same gap 'JL' at the same level of equilibrium output OQ, there is no need for the firm to change the price (keep the price level rigid).

Case 2-Increase in Costs

Similarly, if costs rises, the new MC curve, MC_2 rises upward and cut the MRC in the gap 'JL'. This is because, as costs rises, the gap between J and L will further widen. This is because, as costs rises, the upper portion of the demand curve *i.e.*, D_1K will become more elastic because, a rise in price of one firm (due to increase in costs), will make other firms not to follow the same suit. Hence, MRC becomes more flat or elastic above 'K', (*i.e.*, D_1K portion) and thereby, the gap increases between J and L. Hence, the new MC_2 curve cuts the MRC between the gap J and L. So, the net result is the firm will produce same output OQ and price OP will remain unchanged.

So, there will be price stability or rigidity under oligopoly even when costs change, so long as the MC curve cuts the MRC in its discontinuity or gap portion. However, the chances of existence of price rigidity is more, where there is reduction in costs rather than there is rise in costs. The vertical segment of MRC (*i.e.*, 'JL') is the key to analyze the short-run production activity of an oligopolistic firm. The firm can equate MR with MC and thus maximize profit, even though MC increases or decreases. If MC increases a bit, the profit-maximizing price and quantity remain unchanged at OP and OQ. Similarly, even if MC decreases a bit, the profit-maximizing price and quantity also remain at OP and OQ. So, it leads to price rigidity again.

(b) Changes in demand

So far, we studied that, even when costs changes (increase or decrease) in the firm, there will be price rigidity and there is no change in equilibrium level of output. Similarly, even due to changes in demand conditions for the product (increase or decrease), price rigidity will be maintained. This is shown through Figure 22.15.

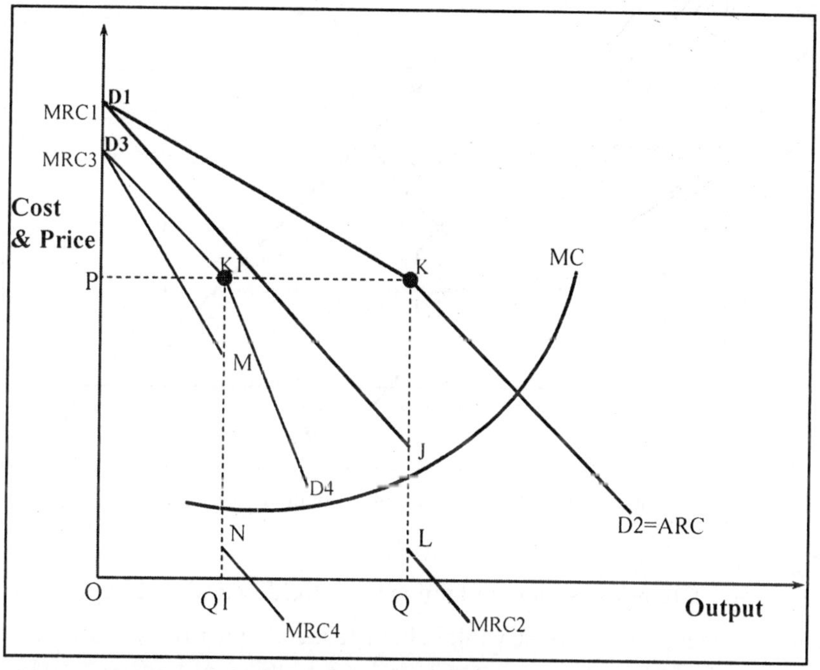

Figure 22.15: Concept of price rigidity under changed demand conditions.

Case 1-Decrease in Demand for the Product

- In the Figure 22.15, D_1D_2 is the original demand curve or ARC of the firm and MRC_1MRC_2 is its corresponding MRC.
- Suppose, there is decrease in demand for the product in the market and hence, D_3D_4 is the new demand curve and MRC_3MRC_4 is its corresponding MRC.
- When demand decreases, if the firm lowers the price, other rival firms also follow the same suit. So, the lower portion of the new demand curve *i.e.*, K_1D_4 becomes more inelastic than the lower portion of old demand curve KD_2. So, this will make the angle of D_3D_4 demand curve more a right angle at K_1. As a result, the gap or discontinuity in new MRC_3MRC_4 curve become much wider compared to MRC_1MRC_2 curve, *i.e.*, MN>JL.
- The MC curve will intersect the new MRC *i.e.*,, MRC_3MRC_4 between the gap MN, as in old MRC *i.e.*, MRC_1MRC_2 between JL.
- So, same price level or price rigidity is followed by the firm even due to decrease in demand for the product. This is so because, the level of kinks *i.e.*, K and K_1 of the two demand curves lie at the same level, *i.e.*, at OP level of price. But, the output sold in the market falls from OQ to OQ_1, as demand for the product is decreased in the market but, keeping price level same.

Case 2-Increase in Demand for the Product

The same explanation can be reversed in case, if the demand for the product is increased in the market, by assuming D_3D_4 and MRC_3MRC_4 as old demand curve and old MRC respectively and D_1D_2 and MRC_1MRC_2 as new demand curve and new MRC respectively (Figure 22.15). Even then, the price rigidity is maintained. But, output sold in the market increases from OQ_1 to OQ, due to increase in demand for the product, but keeping price level same.

Case 3-Demand for the Product is more Elastic

If the demand for the product is more elastic and if the demand for the product increases, the price will increase. This is because, when there is increase in demand for the product, the upper portion of new demand curve *i.e.*, D_1K portion of D_1D_2 demand curve becomes more elastic and hence, the angle between D_1K and KD_2 *i.e.*, at point K, the angle become more obtuse, *i.e.*, move away from right angle (Figure 22.16). So, for its corresponding MRC *i.e.*, MRC_1MRC_2, the gap between J and L will decrease. Hence the MC curve cuts the MRC_1MRC_2 curve above the gap JL, thereby, the price will increase to QP_1 and output sold in the market also increases from OQ_1 to OQ.

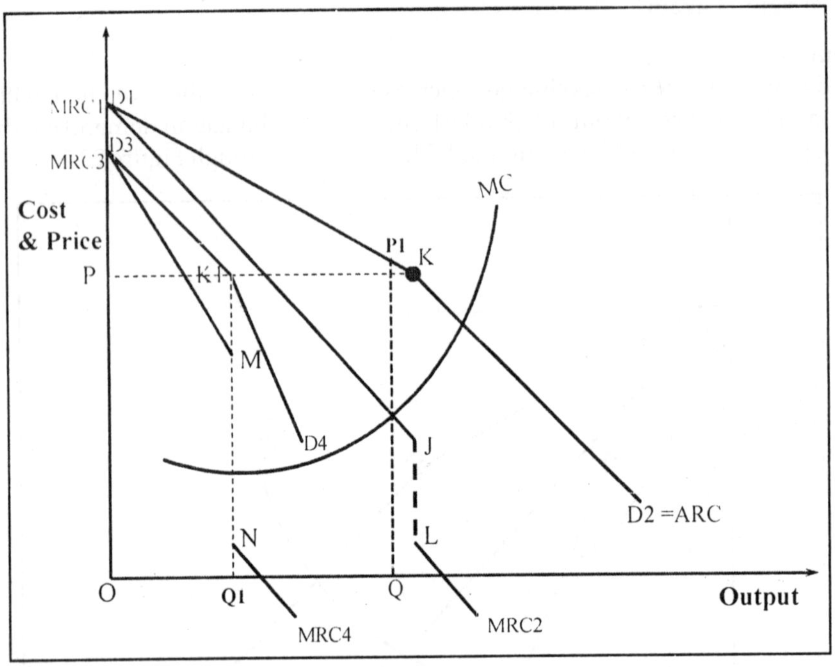

Figure 22.16: Price increase for the product - when the demand for product is more elastic.

Reasons for Price Rigidity

The following are the important reasons for maintaining price rigidity under oligopoly competition:

- Individual sellers in oligopolistic industry might have realized that, price wars adversely affect their profits.
- The firms desire to stick on to the current output-price policies and any variation leads to unnecessary insecurity and uncertainties in the business.
- Price rigidity may prevent the threat of entry of new firms into the industry.
- The firms can concentrate more on non-price competition, instead of involving in price wars across the firms.
- The dominant firm price leadership may not enable small firms to change the prices.

Shortcomings of the Kinked Demand Curve Model

The kinked-demand curve theory, however, is considered as an incomplete theory of oligopoly for the following reasons:

- It does not explain, how the oligopolist finds the kinked point on the market demand curve. That means, this analysis only indicates, why price do not change. It does not explain why or how the 'kink' price was determined or established.
- This theory does not allow for the possibility that, price increases by one oligopolist are matched by other oligopolists, a practice that has been frequently observed. That means, in the practical situation, oligopoly prices are often not rigid, especially in the upward direction.
- This theory does not consider the possibility that, oligopolists collude in setting output and price. Thus, this theory does not apply to the cases of collusion of oligopoly firms *i.e.*, cartel and leadership models.
- George Stigler criticized that, the kinked demand analysis only suggests why prices remain sticky and does not describe the mechanism that establishes the kink and how the kink can reform once prices change.

☆ At times of inflation, the firms will not keep the price rigid due to rise in costs.

☆ This theory applies only, when the firms experience depression in the trade conditions. When there are boom conditions in trade, this theory is not applicable, as firms will raise price of the product.

6. Collusion

Many real world markets have a relatively small number of competing firms. When there are large number of competing firms in an industry, each individual firm's output and pricing decisions are unlikely to have a major effect on its rivals. In such case, an individual firm can significantly increase its market share by offering price discounts on the product, while only taking a negligible share from each of its rivals. However, when there are relatively few firms, as in case of oligopoly competition, then an output-price policy of one firm will have a marked impact on the output-price policies of its rival firms and they are likely to react. So, the output-price policy of a firm are closely watched by the rivalry firms, thereby, the firm have to plan strategically to achieve the desired targets in the business. So, the degree of competition in a market is influenced not only by the number and size of the firms, but also how the firms behave. Due to its own virtue, setting prices independently is very rare or almost non-existent in the oligopolistic markets. In general, two options are before the firms in an oligopolistic industry *viz.*, all the firms act non-cooperatively and all the firms act cooperatively. When all the firms act non-cooperatively, it implies stiff competition exists between the firms leading to price wars. However, when all the firms prefer cooperative action to avoid the uncertainty arising out of interdependence of firms, they collectively plan business strategies, thereby, all the firms mutually benefit to the maximum possible extent. So, when all the firms in an oligopolistic industry cooperate with each other towards achieving a common goal, they are said to be engaged in collusion. So, under collusion, all the oligopoly firms jointly set their output and pricing policies, divide the entire market among them and plan the other business decisions on joint basis. Collusion in a oligopoly industry is easier to achieve, when there are only a less number of firms and their individual outputs can be monitored, when market demand for the product can be easily estimated and when the demand for the product is less elastic. The collusion among the firms may be explicit in the form of formal agreements (Explicit collusion or Overt collusion) or it may be tacit (Secret or Implicit). A formal agreement or explicit collusion is one, when the oligopolists agree after discussion to follow certain common rules of conduct in regard to price and output determination. But, more often, we find that, the agreement between the firms is a tacit one, as in most of the countries a formal or open or explicit collusion agreements to form monopolies are illegal. Implicit price collusion also exists when multiple firms make the same pricing decisions even though they have not consulted with one another. Though the consequences of both formal and tacit collusions are the same, they are treated differently under competition law. So, considering the nature of agreements between the firms, there are two forms of collusion in oligopoly *viz.*, Cartel and Price leadership. In a cartel type of collusive oligopoly, firms jointly fix the price and output policy through formal agreements. But, under the price leadership type of collusive oligopoly, the collusion is a tacit one, where one firm sets the price and others follow it. The one which sets the price is a price leader and the others who follow it are its followers. The follower firms adopt the price set by the leader, even though they have to depart from their profit maximizing position, as they think that it is to their advantage not to compete with their leader and between themselves. Sometimes, oligopolistic firms adopt facilitating practices that can contribute to cooperation and collusion even though the firms do not formally agree to cooperate.

A. Cartel

A cartel model of oligopoly is a model that assumes that oligopolies act as if they were a monopoly and set a price to maximize profit. Thus, 'cartel' is an organization of independent firms, producing similar products, which work together to make output and price decisions. So, a cartel is a group of firms that creates a formal written agreement specifying how much each member will produce and charge. Basically, the term 'cartel' was used for the agreement in which there exist a common agency, which alone undertook the selling operations of all the firms that were party to the agreement. But, now-a-days, all types of formal and informal agreements reached among the oligopolistic firms of an industry are known as Cartels. For example, the OPEC is the most significant international cartel. OPEC members meet regularly to decide, how much oil each member of the cartel will be allowed to produce. Thus, in a cartel, several firms form an organization in order to act, so that, they may capture the benefits that would accrue to a monopolist in that market. So, to ensure monopoly benefits, the cartel will reduce output and increase price compared to what the members would produce as individuals in order to increase total profits. If each firm in an oligopoly sells a homogenous product like oil, the demand curve that each firm faces will be horizontal at the market price. If, however, the oil-producing firms form a cartel like OPEC to determine their output and price, they will jointly face a downward-sloping market demand curve, just like a monopolist. In fact, the cartel's profit-maximizing decision is the same as that of a monopolist. The profits will then be distributed among the cartel members according to some agreed-upon system. The conditions that give rise to an oligopolistic market are also conducive to the formation of a cartel, in particular, cartels tend to arise in markets, where there are few firms and each firm has a significant share in the market. Thus, oligopolistic firms join a cartel to increase their market power and members work together to determine jointly the level of output that each member will produce and or the price that each member will charge. Cartels may be voluntary or compulsory and open or secret depending upon the agreement across the firms in the industry. Since these cartels restrain competition among the member firms, their formation have been made illegal by the Governments in some countries passing laws against them. For instance, the formation of a cartel is illegal in Pakistan and in the U.S. and prohibited under both Irish and EU competition law. However, they are still formed in these countries through secret devices and by adopting some means or by the other, thereby, shrewd businessmen are able to evade the anti monopoly laws. In a cartel, the firms mainly aim at two important aspects.

(*i*) Maximization of profits – called as Joint profit maximization cartel.

(*ii*) Market sharing of firms – called as Market sharing cartel.

(i) Joint Profit Maximization Cartel

An extreme form of collusion is found, when all the member firms agree to surrender completely their rights of price and output determination to a 'Central Agency', so as to secure maximum 'joint profits'. Thus, the Central Agency determines the output quotas of each firm, price to be charged and distribution of profits among the firms. Hence, Central Agency acts like a single monopoly for that homogenous product, whose main aim is to maximize the joint profits of the oligopolistic industry. This type of cartel is treated as the 'Perfect cartel'. Thus, under perfect cartel type of collusive oligopoly, the price and output determination of the whole industry as well as of each member firm is determined by the Central Agency, so as to achieve maximum joint profits for the member firms. Here, the Central Agency will first determine the profitable output and price of the product by estimating the market demand curve for the product (aggregate customers demand curve) and its MRC. The summation of MC curves of different firms gives the cumulative MC curve for the cartel and based on the equality of MRC of cartel and cumulative MC curve, it derives the equilibrium output and price for the product in the industry, thereby, joint profits are maximized. The same is explained through the Figure 22.17, where we assumed only two firms formed as a perfect cartel and producing a homogenous product.

- In the Figure 22.17, MC_1 and ATC_1 are the per unit cost curves of firm 1 and MC_2 and ATC_2 are the per unit cost curves of firm 2. The summation of MC_1 and MC_2 gives the market SMC curve.
- ARC and MRC are the market AR and MR curves respectively.
- For entire market, the profit maximization point is SMC=MRC i.e, at E and its corresponding point on ARC is P, indicating that, at OA=QP price, the market/industry earns super normal profits by producing OQ output.
- Now, the cartel decides to allocate this output OQ between two firms in such a way that, the market MRC equals the MC curves of two firms *i.e.*, MC_1 and MC_2.
- This is done by drawing a line from E to M parallel to X-axis and this line touches the MC_1 of firm 1 at E_1 where, it produces OQ_1 level of output and touches the MC_2 of firm 2 at E_2, where it produces OQ_2 level of output. Thus, firm 1 produces OQ_1 level of output and firm 2 produces OQ_2 level of output, such that, their sum matches total output of the industry *i.e.*, $OQ=OQ_1+OQ_2$.

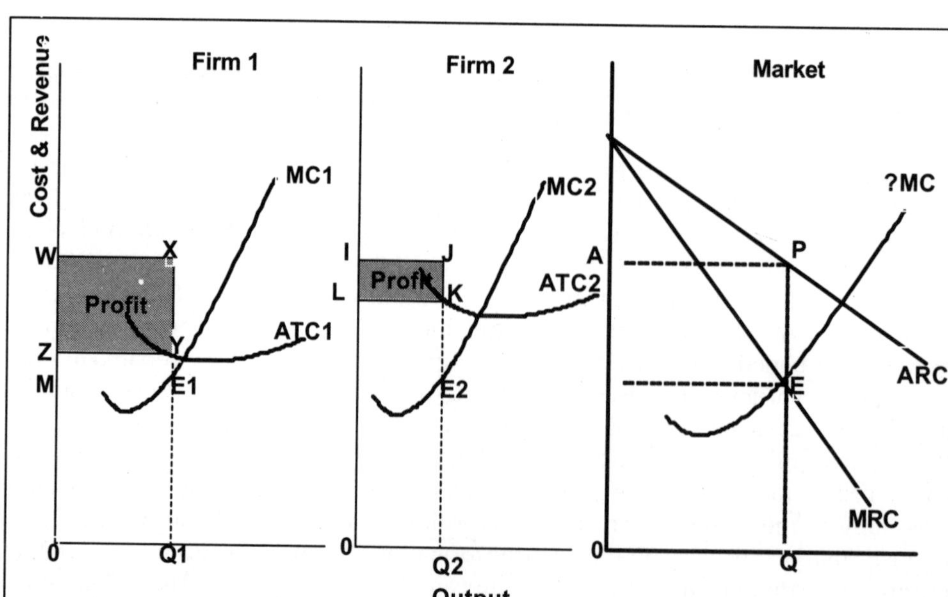

Figure 22.17: Joint profit maximization cartel.

- Both the firms sells these outputs OQ_1 and OQ_2 at OP level of price. As the cost of production is lower in firm 1, it produces more output *i.e.*, OQ_1 compared to firm 2, OQ_2 ($OQ_1 > OQ_2$). But, this does not mean that, firm 1 will be getting more profit than firm 2. The joint maximum profit earned by the two firms is WXYZ + IJKL. This joint profit will be pooled into a fund and distributed by the Central Agency to the firms according to the agreement made between the two firms at the time of formation of cartel. So, this type of cartel formation enables both the firms to maximize their profits.

The output quota to be produced by each firm is decided by the Central Agency in such a way that, the total costs of the output produced should be minimum. In fact, under perfect cartel, the Central Agency determine the separate outputs to be produced by the various firms and the price they have to charge is in the same way as a monopolist operating multiple plants would do.

(ii) Market Sharing Cartel

In reality, the existence of the so called 'perfect cartel' is quite rare. In a perfect cartel, not only the price, but also the output to be produced by each member of a cartel is decided by a Central Agency and profits made in all of them are pooled together and distributed among the members according to the terms of a prior agreement. But, when cartels are loose, instead of being perfect, and in such cases, the distribution of profits and fixation of outputs of individual firms are not determined in a manner perfect cartel does. So, the difficulties encountered by the member firms either in terms of agreeing on the price to charge, allocating output and profits among members and avoiding cheating, make a market-sharing cartel to emerge in a loose type of cartel. So, in a market-sharing cartel, the member firms agree only on how to share the market. Each firm then operates only in one area or region agreed without encroaching on the others' territories. Under certain simplifying assumptions, a market-sharing cartel can also result in the monopoly solution. This is because, each firm can set its prices knowing that its 'rivals' will not undercut them. A market-sharing cartel may be no more than an agreement among firms not to approach each other's customers or not to sell to those in a particular area. It may involve secretly allocating specific territories to one another or agreeing lists of which customers are to be allocated to which firm. Thus, in this form of cartel, the firms collude for sharing the market area, but keep their individual identity regarding their output, pricing, selling activities and other decisions. The market sharing cartel may work in two different forms *viz.*, Non–price competition cartel and Quota system of cartel.

(a) Non-price Competition Cartel or Market Sharing by Non-price Competition

Under market sharing by non price competition, only a uniform price is set and the member firms are free to produce and sell the amount of outputs, which will maximize their individual profits. Though the firms agree not to sell at a price below the fixed price, they are free to vary the style of their product and the advertising expenditure and to promote sales in other ways. That is, the price being a fixed datum, the firms compete on upon non-price basis. If the different member firms have identical costs, then the cartel price will be fixed, such that, it ensures maximization of joint profits. But, when the firms differ in their cost structure, which is more common, then the cartel price will be fixed by bargaining between the firms and it is so fixed such that, it will ensure some profits to high cost firms. However, this form of cartel is inherently unstable because, generally, a low cost firm cheats a high cost firms and thereby, charges a lower price than the common agreed price. So, the customers will get attracted to the low cost firm, thereby, the profits of other firms are affected. When other firms come to know about this, they will leave the cartel. Hence, a price war will start and ultimately a low cost firm will remain in the industry.

(b) Quota System of Cartel or Market Sharing by Output Quota

In this system, an agreement will be reached between the oligopolistic firms regarding quota of output to be produced and sold by each of them at the agreed price. The quotas are decided through the bargaining power between the firms. Thus, in this form of cartel, the firms enter into agreement for selling their product at the common agreed price, but the main agreement is with reference to quota of output, such that, each firm will get profits on its sales. The same is explained through the Figure 22.18, where we assumed only two firms formed as a cartel based on output quota and producing a homogenous product.

- Let us assume, there are two identical firms selling a homogeneous product and the market is shared equally between these two firms.
- D=ARC is the market demand curve/ARC and D/MRC is its MRC and it is assumed that the market demand curve is known to the cartel.
- D/MRC also represents the demand curve of each firm and 'mrc' is its corresponding MRC.
- 'mc' and ATC are the per unit cost curves of both the firms and it is assumed that, the cost structure of both the firms remains same.
- ΣMC is the aggregate MC curve of the market/industry, which is the summation of 'mc' curves of both the firms. This ΣMC curve intersects the market MRC *i.e.*, D/MRC at E and its corresponding point on market

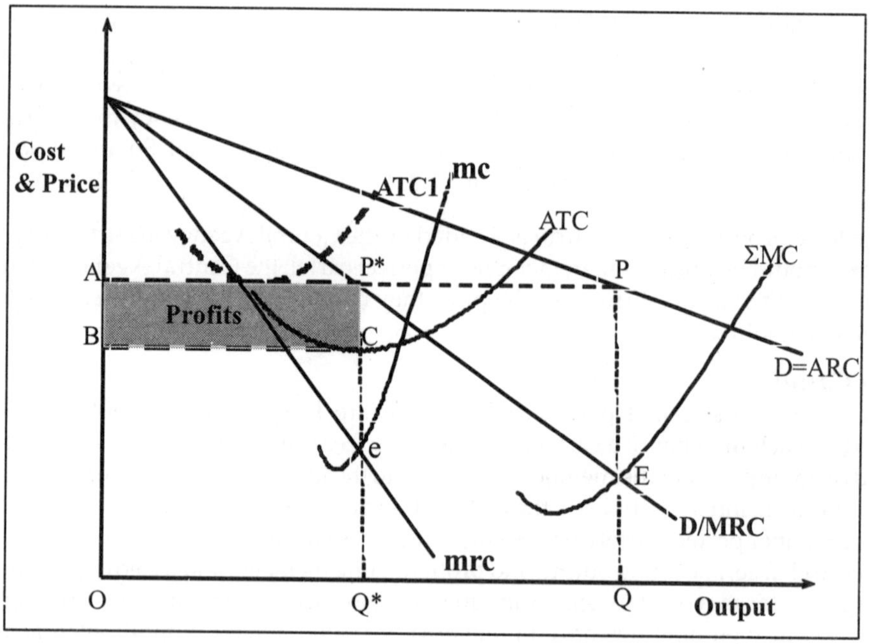

Figure 22.18: Quota system of cartel.

demand curve/ARC is P, indicating that, the industry produces OQ level of output and fix the price of the product as QP. Now the industry has to allocate or fix the quota of this total output OQ between the two firms.

- As shown in the Figure 22.18, at firm level the 'mc' curve intersects the 'mrc' at 'e' and its corresponding point on the firm's demand curve D/MRC is P*, indicating that, each firm produces OQ* level of output and sell the product at Q*P* price. So, both the firms sell the product at the same price Q*P*, which is equal to QP.
- Since, each firm will share the total output equally, the total output of both the firms is 2xOQ*=OQ. Each firm will earn a profit of ABCP*. Therefore, the total profit of the cartel=2x ABCP*.

In the real world, there may be more than two firms, each may have different cost structure, and the market may not be shared equally. Then, in such case, we cannot have the monopoly solution as discussed above. The firm with a greater capacity or operating in an inferior territory may demand a greater share of the market. The result will then depend upon their bargaining power and the possibility of incursions into each other's territory cannot be excluded and this will ultimately lead to breaking up the cartel regarding market share agreement.

In the above analysis (Figure 22.18), we studied the collusive oligopoly taking into consideration only two firms, without any threat of entry of new firms into the industry. But, with the entry of new firms into the industry, the ATC will rise to ATC_1 (due to increase in demand for factor services) thereby, each firm earns only normal profit instead of super-normal profits. Hence, each firm will sell the output less than OQ*. To check the entry of new firms, the existing firms will charge the price less than Q*P*, but above the ATC, such that, the new firms if they enter the industry cannot recover the ATC_1 from the price quoted by earlier two firms (below Q*P*). Hence, the existing two firms will share the market and earn only larger profits (but less than earlier super normal profits, ABCP*). By doing so, new firms may not enter the industry.

Cartels on an International Scale

Cartels are usually formed on an international scale and the most famous international cartel is the OPEC. It is an international organization of more than a dozen nations, primarily located in the Middle East, but also in Africa and Central America, that controls a sizeable portion of the world's petroleum reserves. This control of reserves gives it a great deal of influence over the petroleum market. This cartel attempts to set prices high enough to earn member countries significant profits, but not so high as to encourage dramatic increases in oil exploration or the pursuit of alternative energy sources. It controls prices by setting production quotas for member countries. The International cartels come in three varieties:

- ***Private firm cartel:*** This form of cartel includes two or more private firms from two or more different countries that seek to control the market for a particular product. This form of cartel is highly profit-motive and involves secret collusion among firms. Such cartels are inevitably illegal and are often prosecuted by domestic law enforcement agencies.
- ***Export cartel:*** This type of a cartel includes two or more private firms in a particular county, often an industrial trade group, that seek to control the exports of a product to other countries. These cartels are often exempted from domestic antitrust laws and might even be encouraged by domestic governments.

- *Government cartel:* This form of cartel is exemplified by OPEC and includes the Governments of two or more countries that seek to control the market for a particular product. These cartels, like most international organizations, often seek multiple economic, political and social objectives. That is, they might not control the market exclusively for profit and economic gain, but may be to gain political favour from other countries.

Conditions Conducive for Formation of Cartels

They include:

- Only a small number of firms exist in the industry, so that, the cartel can easily supervise their actions.
- Strong barriers to entry for new firms to protect the monopoly power of existing firms in the long run
- Stable market demand for the product
- High price inelastic demand for the product, which fetches significant TR to the cartel.
- The cartel should be capable of monitoring each firm's output, so that, it can regulate the total supply of the output into the market and further facilitates to identify the firms cheating on output quotas.

Merits of Formation of Cartel

The following are the advantages of the formation of cartels by the few oligopolists:

- In a cartel, several firms form an organization in order to act, as if they were a monopolist, so that, they may capture the monopoly profits in that market.
- Collusion reduces uncertainty, increases profits and may prohibit the entry of new rivals.
- A cartel may reduce the chance of a price war breaking out, particularly during a general business recession.

Possible Break-downs of Cartels

Most cartel arrangements experience difficulties and tensions and some producer cartels collapse completely. First of all, in many parts of the world, formation of cartels is illegal. Even if they are legal, once established, they are difficult to maintain. The problem is that, there is built-in incentive for each cartel member to cheat and the possible ways of break down of cartel are listed out here under.

- By producing more output than it has agreed to produce, a cartel member can increase its share of the cartel's profits.
- The cartels are unstable, especially, when the cartel members greatly differ in their cost structure. This is because, the low cost firms always have the tendency to reduce the price of the product, so as to maximize their profits, which ultimately result in the collapse of the collusive agreement.
- Some cartel members with high cost structure deliberately produce less quantity of output than the agreed output quota and thereby, fix a high price for the product to cover the costs.
- If the entry of the firms into the oligopolistic industry is free, the instability of the cartels is intensified, as the new entrants may not join the cartel and may fix a lower price of the product to sell a larger quantity.
- There may be disputes about how to share out the profits.
- Falling market demand during a slow down or recession creates excess capacity in the industry and puts pressure on individual firms to cut prices to maintain their revenue. Ex., collapse of the coffee export cartel.
- Large number of firms and differing demand and cost conditions among the firms in the industry also contribute to break down of cartel.
- In the cartel, each member is likely to have its own priorities, thereby, making the policy suiting to all the members of cartel is difficult.
- As long as the cartel price remains intact, the cheating firm may sell additional output at the cartel price and, thus, reap additional profits.
- Cartels can be destroyed by an outsider with technological superiority.

Conditions Necessary for a Cartel to be Stable (Maintainable)

They include,

- There are few firms in the industry.
- There are significant barriers to entry.
- An identical product is produced.
- There are few opportunities to keep actions secret.
- There are no legal barriers to sharing agreements.

B. Price Leadership Model

One possible answer to the question of where the 'kink' price came from and a meaningful explanation for oligopoly pricing and output is offered by the price leadership model. The key behavioural assumption of this model is that, one firm (the dominant firm) sets the market price and all other firms in the industry (the follower or fringe firms) take this price as given. The price leader is generally a very large or dominant firm or a firm with the lowest cost of production of the product. It often happens that, price leadership is established as a result of price war in which one firm emerges as the winner.

Price leadership is regarded as an imperfect collusion among the oligopolistic firms, where all firms follow the lead of one firm. The firm that takes the initiative of setting the price and announcing changes in price from time to time is called as Price leader. Other firms in the industry, which match the leader's price and variation over time are called as Price followers or Fringe firms. The leader firm fixes the price by tacit understanding rather than formal agreement. It fixes a uniform price for the homogenous product produced by all the firms. In case of differentiated product also, the leader firm can fix a uniform price, so as to avoid unnecessary price conflicts in the industry. The fringe firms are simply price takers and equate that price to their own MCs. The leader firm can maintain its leadership by pursuing a definite and a consistent price policy, *i.e.*, by using its power with restraint. The following are the price leadership tactics practiced by the leading firm:

- ☆ Leader firm changes the price only, when cost and demand conditions have been altered significantly and industry-wide.
- ☆ Impending price adjustments are often communicated through publications, meetings etc.
- ☆ The new price may be below the short-run profit-maximizing level, so as to check the entry of new firms.

Types of Price Leadership Models

There are four important types of price leadership models:

(i) The Low-Cost Price Leadership Model

In this leadership model, the firm, which produces the product at lowest cost, will act as a leader to other firms in deciding the price of the product. In this model, the firms can share the market equally or unequally. Let us examine each case, as explained below.

(a) The Low-Cost Price Leadership Model with Equal Market Shares among the Firms

As shown in the Figure 22.19,

- ☆ Let us assume there are two firms *viz.*, firm 1 and firm 2 and firm 1 enjoys low cost structure than firm 2.
- ☆ D is the market demand curve/ARC and D/MRC is its corresponding MRC and assume that, market demand curve is known to both the firms.

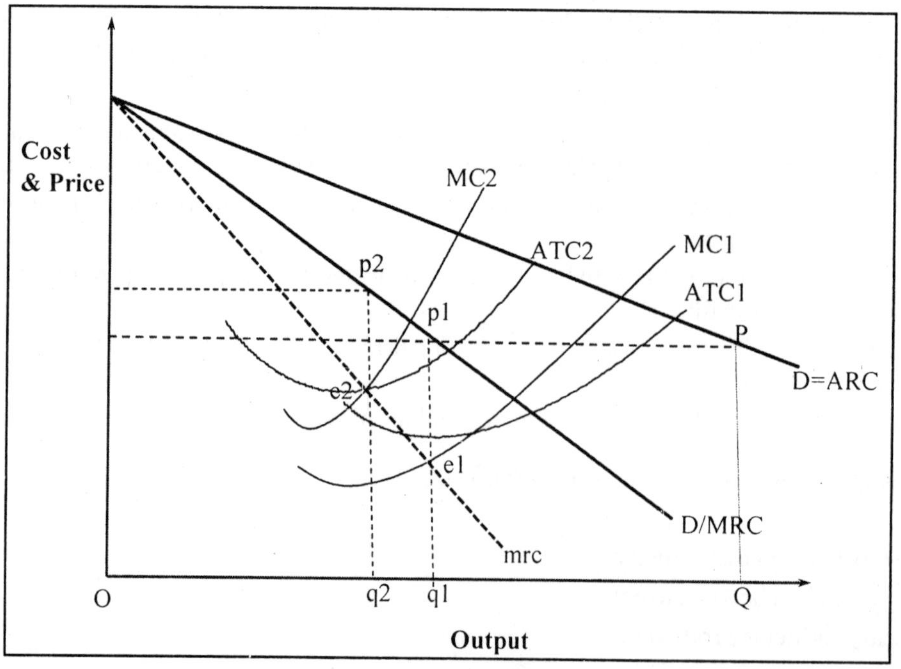

Figure 22.19: Low cost price leadership model with equal market shares.

☆ D/MRC also represents each firm's demand curve and 'mrc' is their corresponding MRC.

☆ ATC_1 and MC_1 are the per unit cost curves of firm1 and ATC_2 and MC_2 are the per unit cost curves of the firm 2. As firm 1 is the low cost firm, its per unit cost curves are below the high cost firm 2.

☆ MC_1 cuts the 'mrc' at 'e1' and it implies that, the firm 1 produces Oq_1 level of output and sell the product at q_1p_1 price.

☆ MC_2 cuts the same 'mrc' at e_2 and it implies that, firm 2 produces Oq_2 level of output and sell the product at q_2p_2 price.

☆ This indicates that, a low cost firm 1 charge low price (q_1p_1) compared to high cost firm 2 (q_2p_2). The low cost firm 1 sells more output Oq_1 compared to high cost firm 2 *i.e.*, Oq_2.

☆ As there is a tacit agreement between the two firms that, high cost firm 2 has to follow the price policy of low cost firm 1.

☆ So, firm 2 will have to sell the product at low price q_1p_1. Since, it has to sell the output at low price q_1p_1, it has to increase output to Oq_1.

☆ Firm 1 will earn more profits, as it sells all its output Oq_1 at its price q_1p_1. But, firm 2 will get less profits than firm 1 because, it will sell the output at q_1p_1 price of firm 1, but incurs high costs in producing the product. Further, firm 2 will not definitely stick to q_2p_2 price, because, since the product is homogenous, all the customers will prefer the product of firm 1, which is supplied at low price q_1p_1.

☆ Since, both the firms sell Oq_1 output at price q_1p_1, total market/industry output is 2^*Oq_1 =OQ.

☆ If two firms act independently, the firm 1 can drive out firm 2 from the market/industry by quoting a low price to the product *i.e.*, below q_1p_1, such that, the new price may lie below the ATC_2 of firm 2, but above ATC_1 of firm 1. If this is the case, the firm 2 cannot recover ATC_2 and hence, it has to leave the industry. Then, firm 1 becomes the monopoly firm. But in such case, the firm 1 has to face some legal problems from the Government, as it imposes several restrictions to prevent the exploitation of its monopoly behaviour.

☆ Hence, firm 1 fixes q_1p_1 price and tolerates the firm 2 in market competition to share the market equally and gain maximum profits.

(b) The Low-Cost Price Leadership Model with Unequal Market Shares among the Firms

As shown in the Figure 22.20,

☆ If the two firms have unequal market shares, they will have different demand curves or ARCs, different MRCs and different cost curves. The demand curve of low cost firm 1 is more elastic and demand curve of high cost firm 2 is less elastic or more inelastic.

☆ In the Figure 22.20, D_1=ARC_1 and MRC_1 are the demand and MR curves of firm 1 and D_2=ARC_2 and MRC_2 are the demand and MR curves of firm 2.

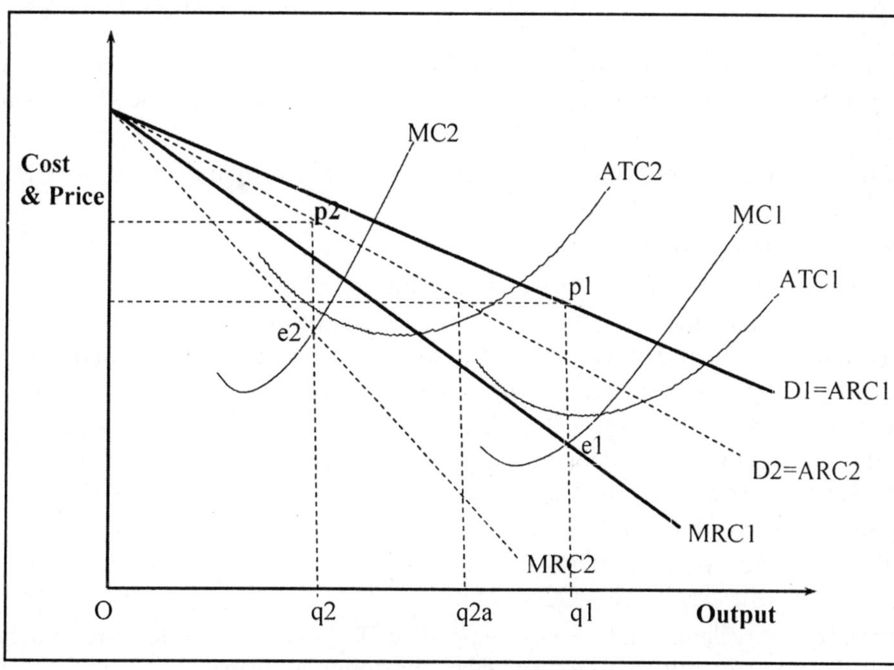

Figure 22.20: Low cost price leadership model with unequal market shares.

- For low cost firm 1, MC_1 cuts the MRC_1 at e_1 indicating that, firm 1 produces Oq_1 level of output and sell the product at q_1p_1 price.
- For high cost firm 2, MC_2 cuts the MRC_2 at e_2 indicating that, firm 2 produces Oq_2 level of output and sell the product at q_2p_2 price.
- Thus, a low cost firm 1 sells the product at low price q_1p_1 compared to high cost firm 2 at high price q_2p_2.
- As per the tacit agreement, the high cost firm 2 must follow the price policy of low cost firm 1 and hence, it has to sell its output at q_1p_1 price of firm 1.
- Since, firm 2 has to sell the product at low price (q_1p_1 price), it will increase the output to Oq_{2a} (to compensate the loss due to selling at low price), but it will increase the cost on the part of firm 2.
- So, firm 1 will sell output Oq_1 at q_1p_1 price and firm 2 will sell output Oq_{2a} at q_1p_1 price (same price).
- Firm 2 will not definitely stick to q_2p_2 price for selling Oq_2 output because, since, the product is homogenous, all the customers will prefer the product of firm 1 which is supplied at low price q_1p_1.
- If the firms act independently, the firm 1 can drive out firm 2 from the market/industry by quoting a low price to the product *i.e.*, below q_1p_1 such that, the new price may lie below the ATC_2 of firm 2, but above ATC_1 of firm 1.
- So, when the firms have unequal market shares, the high cost firm will sell the output at the low price fixed by the low cost firm, but increase the quantum of sale of output to compensate the loss due to selling at low price.

(ii) The Dominant Firm Price Leadership Model

This is the most common type of price leadership model. It is a case of implicit price collusion, where the largest or most dominant firm takes the lead in setting prices and other firms with lower market shares follow the pricing changes prompted by the dominant firm. In this model, the firm size is highly skewed, resulting in a dominant firm that exists alongside a competitive fringe of much smaller firms, typically supplying a relatively standardized product. This type of leadership model is commonly seen with the major mortgage lenders and petrol retailers, where most suppliers follow the pricing strategies of leading firms. So, in this case, the single large dominant firm will be the price maker for the entire industry, but the small firms are simply price takers. The dominant firm sets its own price on the basis of industry demand not served by the fringe suppliers, thereby, providing a price umbrella for the latter. So, the leader firm is large and powerful enough to fix a price, which all other firms will be forced to follow. Each follower firm takes this market price as given and produces the output at which MC=MR. Here, the dominant firm acts as a monopolist, who maximizes profit by taking the supply curve of the followers as given. The small firms sell as much product as they like and the remaining market is filled by the large or dominant firm. If the market share of the followers goes up, the monopoly power of the leader suffers. Therefore, in this version, the equilibrium price is lower than what would be obtained by a pure monopolist. The dominant firm can maintain its dominance in the market by innovating on 'non-price competitive areas', *i.e.*, new brands, product improvements, promotion, distribution, dealer concessions, free gifts, easier credit terms, free home delivery, after sales services, longer period of guarantee etc., or by keeping the market price low enough to deter entry of new firms into the industry. However, when this leader firm sets a very low price, it may force some of the firms to leave the industry. The concept of dominant firm price leadership model is explained through Figure 22.21.

- As small firms are simply price takers given by the dominant firm, their demand curve is perfectly elastic at that price.
- DD_1 is the market demand curve.
- Suppose, the dominant firm sets the price of the product as OP.
- This price OP is passed on to the small firms to sell as much output as they can sell in the market. At this OP price, PC becomes the demand curve=ARC=MRC for small firms (as in the case of perfect competition).
- The summation of MC curves of all small firms combined laterally is ΣMCs and it intersects the MRC curve 'PC' of small firms at C, which implies that, all the small firms will supply PC quantity of product into the market and it matches the market demand. So, at this price OP, dominant firm will not supply any product into the market. So, we can say, point P is the starting point of demand curve for dominant firm.
- So, to supply the product into the market by the dominant firm, it will reduce the price to OP_1. Now P_1C_1 becomes the demand curve for small firms. The ΣMCs intersects the demand curve P_1C_1 at B. This implies that, all small firms will supply P_1B (=OQ_S) quantity of product into the market at OP_1 price.
- But, since, total quantity demanded (market demand) at OP_1 price is P_1C_1(=OQ), the small firms could supply only P_1B. So, BC_1=Q_SQ quantity will be supplied by the dominant firm.

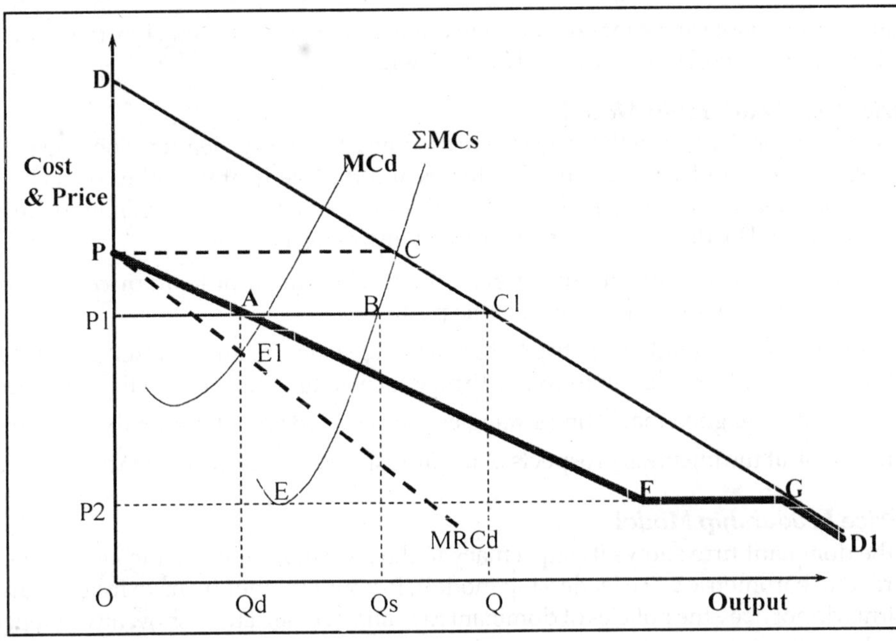

Figure 22.21: Dominant firm price leadership model.

- Let us take a point A on the demand curve P_1C_1, such that, $P_1A=BC_1$. From A, we draw a line perpendicular to X-axis, which implies that, $OQ_d=P_1A$ is the quantity to be supplied by the dominant firm and remaining quantity $Q_dQ=AC_1$ will be supplied by the small firms.
- So, PA is a part of demand curve for dominant firm.
- To derive the remaining part of the demand curve of the dominant firm, let us assume that, the dominant firm still reduce the price of the product to OP_2. This facilitates the small firms to sell the product even at this price, because MRC=ΣMCs with reference to point E. (here, MRC of small firms=P_2G). Below this price OP_2, small firms will not sell the product because, ΣMCs>MRC.
- Thus, if price OP_2 is fixed by the dominant firm, the small firms would sell P_2E quantity and dominant firm will sell remaining EG quantity of output in the market.
- Between OP_1 and OP_2 prices, the demand curve for dominant firm slopes downward from left to right indicating that, it sells more quantities of output at lower prices.
- Taking this OP_2 price into consideration, the downward falling dominant firm's demand curve coincides with P_2G demand curve of small firms at F and equates with market demand curve DD_1 through G. So, $PAFGD_1$ is the demand curve of dominant firm.

Profit Maximization

If the dominant firm fixes the price of the product at OP_1, it maximizes the profits by producing the output such that, its MC_d intersects the MRC_d at E_1, where the dominant firm produces OQ_d output and sell the product at $OP_1=Q_dA$ price. At this price OP_1, the small firms will be in equilibrium, when their ΣMCs intersects MRC of small firms *i.e.*, P_1C_1 at B, where the small firms sell OQ_S level of output. So, the total output produced in the industry at OP_1 level of price is $OQ_d+OQ_S=OQ=P_1C_1$.

If the dominant firm fixes the price of the product at OP_2, small firms will produce P_2E quantity of output, where ΣMCs = MRC of small firms *i.e.*, P_2G. The remaining output *i.e.*, EG will be produced by the dominant firm to realize maximum profits. So, the total output produced in the industry at OP_2 level of price is $P_2E + EG = P_2G$.

If the dominant firm fixes the price of the product below OP_2, small firms will not sell any output because ΣMCs>MRC and the entire output in the industry is to be supplied by the dominant firm only. This practice of a large dominant firm driving smaller firms out of the market by temporarily selling the product at an artificially low price is called Predatory pricing. Such behaviour can be considered as illegal and it can be checked effectively with the passage of antimonopoly legislation.

So, the dominant firm price leadership model will function because, the small firms behave passively as price takers. But, this does not mean that, dominant firm will act like a monopoly firm. The dominant firm will also look into output of the small firms, their cost structure etc., and taking these aspects into considerations, it will set the price of the product and make small firms viable in the market. Of course, if the small firms do not obey the leader or dominant firm regarding its price fixation strategies, this model will breakdown.

(iii) The Barometric Price Leadership Model

In this model, an old, experienced and the largest firm assumes the role of a leader. The leader firm is regarded as a wise firm, which sets the price reflecting the market forces and the needs of the other firms in the industry. Any changes in price cannot be done by a barometric firm in an arbitrary fashion, as the dominant price leader can do. The barometric price leadership will be developed due to the following reasons:

- ☆ The firm may be capable of withstanding greater fluctuations in output and prices by formulating suitable strategies, which the industry experienced in the past.
- ☆ Most firms do not have enough expertise to forecast cost structure and market demand conditions for the product. So, they leave these decisions to one firm, which is capable of doing the same on accurate lines.
- ☆ The firm may be having good marketing strategies compared to other firms in the industry.
- ☆ The firm may look at the business prospects of the industry in a better way compared to other firms.

(iv) Exploitative Price Leadership Model

In this model, the dominant firm shows its supremacy in the industry by following aggressive price leadership. Unlike dominant firm and barometric price leadership models, it compels other firms to follow it and accept the price fixed by it. If other firms do not obey the policies of dominant firm and in case, if they show any independence regarding their price-output policies, the dominant firm threatens them and coerces them to follow its leadership.

It is clear from the above price leadership models, the price leadership arises for a firm, if it enjoys lowest cost structure or having comparative advantage in terms of productive capacity. This is so because, the lowest cost firm or the firm with highest productive capacity can withstand the losses of a price war in a better way compared to other firms in the industry. Such firm can increase market share by snatching large and profitable markets from conservative rivals. Besides these, long-term profit history, sound management and product innovation by a firm may also lead to price leadership.

Difficulties of Price Leadership Model

The following are the difficulties faced under price leadership models:

- ☆ It is very difficult on the part of the leader firm to correctly assess the reactions of its followers regarding output-price policies in the industry.
- ☆ The rival firms may disobey the guidelines of the leader firm and secretly charge lower prices for the product, so as to increase their market share. They may even offer some price cutting devices like rebates, favourable credit terms, money back guarantees, after delivery free services etc., and try to attract the customers of other firms in the industry.
- ☆ The rival firms may indulge in non-price competition to boost the sales of their products.
- ☆ The high price set by the leader firm may attract new entrants into the industry and these new entrants may not accept the price leadership.
- ☆ If a cost-effective production technology is available to all the firms in the industry, the rivals may not obey the policies of the leader firm.

7. Game Theory

Strategic behaviour of firms in oligopoly competition can be analyzed by using the mathematical techniques of Game theory. This theory provides a description of oligopolistic behaviour as a series of strategic moves and countermoves. This theory was developed by John Von Neumann and Oskar Morgenstern. This theory explains that, 'cooperation' rather than 'competition' between the firms can lead to a better outcome that is beneficial to the firms in the oligopoly competition. So, this theory helps to explain mutual interdependence among the firms in the oligopolistic industry. Thus, the concept of game theory is used to analyze the behaviour of the firms, who try for optimal position in the business through the use of strategic behaviour, being fully aware of their mutual interdependence with the other players in the game and attempting to anticipate the decisions of those other players.

For example, there are two giant firms A and B in a pesticide industry and suppose that, firm A decides for advertising its own product. The decision of firm A not only influences sales and profits of firm A, but also affects the sales and profits of its competitor firm B. Now, competing firm B will take a counter action and it also goes for advertising its own product, which influences its sales and profits and also affects the sales and profits of firm A. Much like a game of chess, each firm plans moves and counter moves the strategies based on the actions or anticipated actions of competitor. Let us assume that, each firm spends Rs. 50 crores on their respective advertising campaign. The benefits realized by each firm from its advertising campaign depends upon the actions of other firm. The possible outcomes of this game (profits due to advertising or without advertising) between the two firms in terms of advertising campaign is shown through the Figure 22.22. It is evident from the Figure 22.22 that,

- ☆ *Case 1:* The lower right square (Square IV) indicates that, if neither Firm A nor Firm B decide to advertise, then each firm receives Rs. 300 crores profit. The top rectangular part in the Square IV represent the profits Rs. 300 crores of Firm A and the bottom rectangular part in the Square IV represent the profits Rs. 300 crores of Firm B. So, the total industry profit in this case is Rs. 600 crores, divided evenly between the two firms, when both firms do not go for advertising.
- ☆ *Case 2:* If both the firms A and B go for advertising the product by spending Rs. 50 crores each on advertising, then each firm gets Rs. 250 crores profit. So, the advertising by each firm effectively cancels out the profits realized by the other firm. So, neither gains greater sales or market share, but both end up reducing profit by

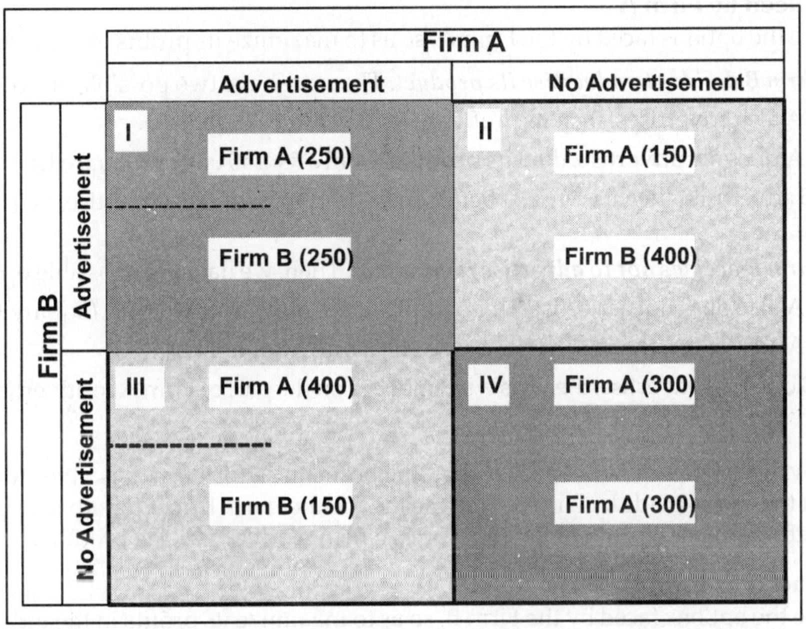

Figure 22.22: Game theory solution for duopoly competition.

Rs. 50 crores, the amount of the advertising expense. The top rectangular part in the Square I, represent the profits Rs. 250 crores of Firm A and the bottom rectangular part in the Square I represent the profits Rs. 250 crores of Firm B. So, the total industry profit in this case is Rs. 500 crores, divided evenly between the two firms. That means, there is a decline in total profits of Rs. 100 crores, when both the firms goes for advertising, which is the total advertising expense of both the firms.

- ☆ *Case 3:* Say, if only Firm A advertises, but Firm B does not, then Firm A receives Rs. 400 crores profit and Firm B receives only Rs. 150 crores profit. This outcome is shown through the Square III, where top rectangular part represent the profits Rs. 400 crores of Firm A and the bottom rectangular part represent the profits Rs. 150 crores of Firm B. So, Firm A gets higher profits and Firm B has a big drop in profits because, the advertising by Firm A attracts the customers away from Firm B. The total profit of the industry for this case is Rs. 550 crores, which reflects the significant positive impact of Rs. 50 crores advertisement cost of Firm A.
- ☆ *Case 4:* The situation is exactly reversed, if Firm B advertises, but Firm A does not. In this case, Firm B receives Rs. 400 crores profit and Firm B receives only Rs. 150 crores profit. This outcome is shown through the Square II, where top rectangular part represent the profits Rs. 150 crores of Firm A and the bottom rectangular part represent the profits Rs. 400 crores of Firm B. So, Firm B gets higher profits and Firm A has a big drop in profits because, the advertising by Firm B attracts the customers away from Firm A. The total profit of the industry for this case is Rs. 550 crores, which reflects the significant positive impact of Rs. 50 crores advertisement cost of Firm B.

Considering the above possible cases, it is interesting to know, whether Firm A goes for advertising or not. What actions firm B might take considering the incremental income due to its advertizing and possible impact, if firm A also goes for advertizing. So, to ascertain the most profitable outcome, let us examine the following options:

- If neither firm advertises, then total industry profit is Rs. 600 crores as shown in Square IV. If the two firms had a mutual understanding and cooperative or collusive arrangement, then both the firms opt not to go for advertising their products, as this outcome maximizes the industry profits (Rs. 600 crores) than any other option. So, each firm can equally share the total industry profit. So, each avoids the Rs. 50 crores advertising expenses and ends up with Rs. 300 crores profit. So, both the firms are benefitted, if neither advertises.
- If both the firms do not have a collusive arrangement, Firm A must select the advertising strategy, which is not known to the firm B and the outcome selected by the firm A must be best for firm A irrespective of the decision taken by the firm B. Similarly, Firm B must select the advertising strategy, which is not known to the firm A and the outcome selected by the firm B must be best for firm B irrespective of the decision taken by the firm A.

So, if both the firms do not have collusive arrangements, then we have to ascertain the options of Firm A and Firm B separately considering their game in the field of advertising, as shown through the Figure 22.22. So, based on the relative outcomes, we can decide, what firm A has to do or not to do and similarly, what firm B has to do or not to do.

Possible Options Faced by Firm A

The following are the options faced by the Firm A, so as to maximize its profits in the industry:

- ***Option 1-Firm B decides to advertise its product:*** Then, we have two possible options for Firm A.
  - If Firm A also advertises, then its profit are Rs. 250 crores (Square I).
  - If Firm A does not advertise, then its profits are only Rs. 150 crores (Square II).

  So, if Firm B advertises, then it is wise or better on the part of Firm A to advertise, as it can earn an extra Rs. 100 crores profit.
- ***Option 2-Firm B decides not to advertise its product:*** Then, we have two possible options for Firm A.
  - If Firm A also does not advertises, then its profit are only Rs. 300 crores (Square IV).
  - If Firm A advertises, then its profits are Rs. 400 crores (Square III).

  So, if Firm B does not advertise, then it is wise or better on the part of Firm A to advertise, as it can earn an extra Rs. 100 crores profit.

So, from the above two options, we can infer that, the best choice for Firm A is, therefore, to advertise its own product, regardless of the choice made by Firm B. That means, if either Firm B goes for advertising or does not advertise its product, Firm A definitely gets a profit of Rs. 100 crores, if firm A advertises the product.

Possible Options Faced by Firm B

The following are the options faced by the Firm B, so as to maximize its profits in the industry:

- ***Option 1-Firm A decides to advertise its product:*** Then, we have two possible options for Firm B.
  - If Firm B also advertises, then its profit are Rs. 250 crores (Square I).
  - If Firm B does not advertise, then its profits are only Rs. 150 crores (Square III).

  So, if Firm A advertises, then it is wise or better on the part of Firm B to advertise, as it can earn an extra Rs. 100 crores profit.
- ***Option 2-Firm A decides not to advertise its product:*** Then, we have two possible options for Firm B.
  - If Firm B also does not advertises, then its profit are only Rs. 300 crores (Square IV).
  - If Firm B advertises, then its profits are Rs. 400 crores (Square II).

  So, if Firm A does not advertise, then it is wise or better on the part of Firm B to advertise, as it can earn an extra Rs. 100 crores profit.

So, from the above two options, we can infer that, the best choice for Firm B is, therefore, to advertise its own product, regardless of the choice made by Firm A. That means, if either Firm A goes for advertising or does not advertise its product, Firm B definitely gets a profit of Rs. 100 crores, if firm B advertises the product.

So, the end result is that, if the two firms are competitive, then it is better for both the firms to decide to go for advertising their respective products. In so doing, they end up with less profit (Rs. 250 crores each), than if they had colluded and jointly decided not to advertise (Rs. 300 crores each). This infers that, game theory illustrates the key problem of interdependent decision-making feature in oligopoly. However, in practical situation, things are not so simple, as more number of firms may enter the competition and several other factors may also influence the decision making of a firm. However, the concept explained above illustrates the general principle of Game theory.

8. Cost Plus Pricing Model

This is a really dumb pricing strategy in which the firm always sets its product prices equal to ATC + some percentage markup. However, the firm that uses this obsolete strategy will surely fail, since it fails to take into account the pricing strategies of its competitors.

22.5. Theory of Contestable Markets

Over the past several years, the focus of the market structure theory has changed from the number of firms in the industry to the issue of the extent of entry into and exit from the industry. In this context, the Theory of Contestable markets came into existence, developed by the William Baumol, Panzor and Willing. According to this theory, the firms operating under imperfect competition say, monopoly and oligopoly tend to fix the price of the product, as if they are operating under perfect competition *i.e.*, price=ATC. This is because, in monopoly and oligopoly, the price fixed by the firms is generally higher than the price fixed under perfect market competition, thereby, they earn super-normal profits even in the long run. New firms gets attracted to these super-normal profits and enters into the industry, thereby, the firms in the industry experiences only normal profits. So, to over come the threat of entry of new firms, the existing firms even under imperfect competition are forced to charge only competitive price, thereby, reap only normal profits. So, the threat of entry of new firms into the industry will make the industry operate close to the perfect competitive price and output and in such case, the market is said to be contestable market. So, a contestable market is a market structure, where there is more freedom of entry and exit of firms in the industry. Thus, contestable market model is a model of oligopolies where barriers to entry and exit, not market structure, determine price and output decisions and a competitive price is set. So, a contestable market is characterized by free entry and free exit of firms. Even if the industry contains only one firm, it will set a competitive price, to check the entry of new firms. Thus, contestability may constrain incumbent firms from exploiting their market power. It also refers to a market structure with low sunk costs. The markets that have become more contestable in the recent period include, internet service providers, electricity supply, gas supply, parcel delivery, low cost domestic airlines etc.

Essential Conditions or Features of a Contestable Market

A market structure is said to be contestable market, if it satisfies the following conditions:

- ☆ There is extremely easy entry into the industry and cost less exit from the industry.
- ☆ The new firms entered the industry can produce the same product at the same unit costs, as the existing firms.
- ☆ The firms leaving the industry can easily dispose of their fixed assets by selling or using them else where.
- ☆ The new firms entering the industry should not be in a disadvantageous position when compared to existing firms with reference to production technology and quality of the product produced.
- ☆ Firms are able to diversify their activities, thereby, they reduce or eliminate losses.
- ☆ Free entry and cost less exit from the market encourages hit and run behaviour of some firms. That means, a firm enters the market for a short period of time, makes super-normal profits and then leaves the industry. This fear of attack from a hit and run firm brings about efficiency in the market, as firms aim to suppress the super-normal profits that attract hit and run firms to enter the industry. This threat makes the existing firms to plan and act like under perfect market competition.

From the above conditions, we can infer that, in the contestable market, the number of firms is not so important, but the threat of competition should be sufficient enough to keep the prices low and prevent monopoly power. So, the less the number of barriers to enter the industry, the more contestable the market is. If the entry barriers are completely absent, then perfect contestable market exists even under monopoly or oligopolistic market situation.

Pricing Under Contestable Market

To explain the pricing behaviour under contestable market situation, let us assume a duopoly market situation. A shown in the Figure 22.23,

- ☆ DD is the market demand curve for the product.
- ☆ The two firms together produce OQ quantity of output.
- ☆ The two firms earn OP price for the product and earns normal profits only, as price equals the ATC of the firms.
- ☆ The two firms can quote higher price for the product (above ATC) and can earn super-normal profits. But, th may attract new entrants into the industry and the price may fall drastically.
- ☆ So, the firms even under duopoly market competition behave like the firms under perfect competition earn only normal profits, so as to avoid the threat of entry of new firms into the market.

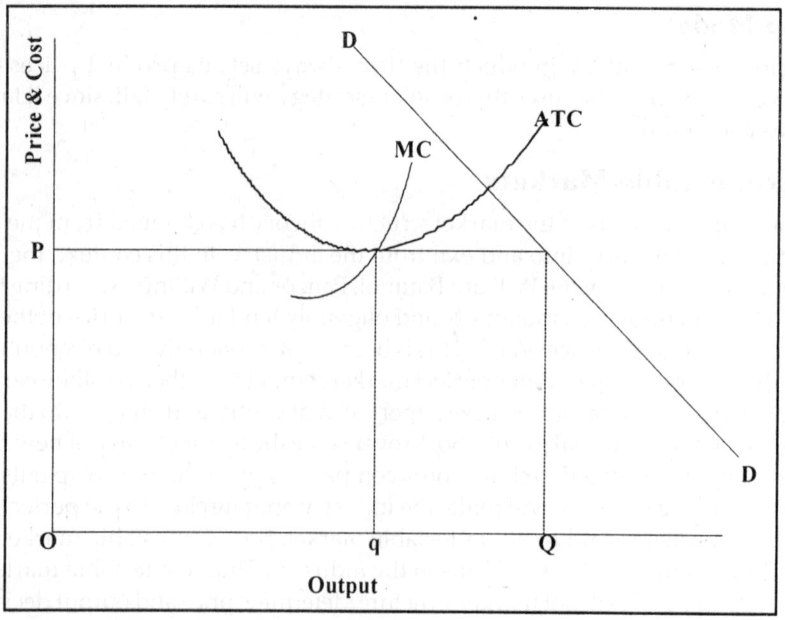

Figure 22.23: Contestable market - Fixing the competitive price to prevent the threat of new entrants under imperfect competition.

Having discussed the concept of contestable markets, we can draw the following conclusions:

- ☆ The presence of potential new entrants makes the existing firms in the industry to behave in a non-competitive way.
- ☆ The existing firms in the industry will earn only normal profits even though, they are capable of earning super-normal profits.
- ☆ The presence of potential entrants makes the existing firms to produce at the lowest possible ATC and charge the price equal to ATC, thereby, indicating that, the contestability encourages allocative efficiency of resources

Criticisms of Contestable Markets

The theory of contestable market is also subjected to criticisms on the following grounds:

- ☆ In practicality, no market is perfectly contestable, *i.e.*, there will be some barriers regarding the entry of new firms into the industry. For example, in oligopoly market competition, there will be huge sunk costs in the industries and these sunk costs acts as a barrier for the new entrants.
- ☆ Because of involvement of sunk costs, the new entrants will set a high price above the ATC to cover the sunk costs, thereby, normal profits are not derived under contestable markets. This is advantageous for the existing firms, as high price facilitates them to earn super-normal profits.

22.6. Cartel Model vis-à-vis Contestable Market Model

The cartel model is appropriate for oligopolists that collude, set a monopoly price, and prevent market entry. But, the contestable market model describes oligopolies that set a competitive price and have no barriers to entry. Both models use strategic pricing decisions, where firms set their price based on the expected reactions of other firms. So, oligopolies are concentrated industries with two extremes. At one extreme is the cartel, in essence, acting as a monopolist. At the other extreme, firms compete for small contestable markets in response to observed profits. In between are a number of alternative models, all of which stress the interdependence of oligopolistic firms and they give price results between the two extremes.

22.7. Advantages of Oligopoly

Oligopoly competition offers the following advantages:

- ☆ The giant business firms gain massive profits and they can retain them for the long haul.
- ☆ The firms enjoy economics of large scale that reduce production costs and prices. So, the large firms go for massive production with lower LRATC.
- ☆ The firms aim at promoting innovations that advance the level of technology, expand the production capabilities, promote living standards of the people and finally leads to economic growth and development.

Both motive and opportunity guide the oligopoly firms to pursue innovation. The interdependency among the few firms ensures motive and access to abundant resources ensures opportunity for the firms to pursue innovation.

- Even at times of economic hardship, the oligopoly firms can survive and manage in the market, as they enjoy so much power in setting prices for the products.
- There is an incentive to engage in R&D
- When compared to pure monopoly or perfect competition, oligopolies offer a wide range of differentiated products at the customers' disposal.

22.8. Disadvantages of Oligopoly

The following are the disadvantages of oligopoly competition in the economy:

- ***Small output and high prices:*** When compared to perfect competition, oligopolists set the prices at higher level and output at lower level *i.e.*, profit-maximizing oligopolists are likely to price above MC.
- ***Restrictions on the entry of new firms:*** Like monopoly, there is a restriction on the entry of new firms in an oligopolistic industry, thereby, the oligopolies earn super-normal profits even in the long run.
- ***Lower production efficiency:*** Some Economists argue that, there is a low level of production efficiency in oligopoly, as there is no tendency for the oligopolists to build optimum scale of plant and operate them at the optimum rates of output. However, the Schumpeterian hypothesis states that, there is high tendency of innovation and technological advancement in oligopolistic industries, thereby, the production efficiency increases. This will offset the loss of consumer's surplus from too high prices.
- ***Higher selling costs:*** In order to attract the customers towards their products, the oligopolistic firms may engage in aggressive and extensive sales promotion efforts by means of advertisement and this burden is shifted on to the customers through charging higher prices for the product.
- ***Loss of welfare:*** Under oligopoly, huge amounts of money are poured into sales promotion activities and from this point of view, oligopolies fares fairly badly in terms of economic welfare. The oligopolists push non-price competition beyond socially desirable limits.
- ***Strategic behaviour:*** It can force firms into deadlocks that waste resources.
- ***Product differentiation and advertising:*** They may pose a real danger of waste and inefficiency.
- ***Inequalities of income and wealth:*** Oligopoly firms tend to be inefficient in the allocation of resources and it promotes concentration, thereby, it leads to inequalities of income and wealth.
- ***Power in the hands of a wealthy few:*** Oligopoly competition tends to increase the concentration of wealth and income among the few people.
- ***Less creativity:*** When wealth is concentrated in the hands of few people, it is difficult for new ideas to come into play. Even if there are new ideas, it is unable to introduce them because of interdependency among the firms.
- ***Setting prices:*** The few big firms are capable of setting prices for the products, which is negative from the welfare of the society.

22.9. Desirability of Oligopoly Competition

It is hard to make concrete statements regarding price-output policy under oligopoly. However, from the point of view of the society, an oligopolistic structure provides a fair degree of competition in the market, if the oligopolists do not collude. However, (tacit) collusion is more common among the firms to set price and/or quantity. In such context, it is not socially desirable, as the few colluded firms aim at monopoly profits in the business.

APPENDIX

Kinked Demand Curve Theory – MRC with Discontinuity Lies Below ARC

As discussed earlier, the ARC and MRC do not have a smooth downward slope under oligopoly, unlike monopoly and monopolistic competitions. Both these curves acquire kinks. Because of interdependency of few firms under oligopoly competition, if one firm raises the price of the product, the other firms will not increase the price, as they lose the market share. But, when one firm reduces the price of the product, other firms too reduce the price of the product, thereby, the firm that first reduces the price will not experience the increase in demand for the product as expected.

As shown in Panel A of the Figure given below, if a firm raises the price of the product, its rival firms will not follow the suit, thereby, its demand curve (ARC) becomes more flat (elastic) after 'K'. So, its consequent MRC rises discontinuously from m1 to m2 and then persists its course (below ARC) at the new higher level.

Alternatively, as shown in Panel B of the Figure, if a firm reduces the price of the product, its rival firms also follow the suit, thereby, the firm that first reduces the price of the product is not able to enhance its sales as expected. So, its ARC becomes less elastic from 'K' onwards, thereby, its consequent MRC falls vertically from m1 to m2 and then slopes down below ARC.

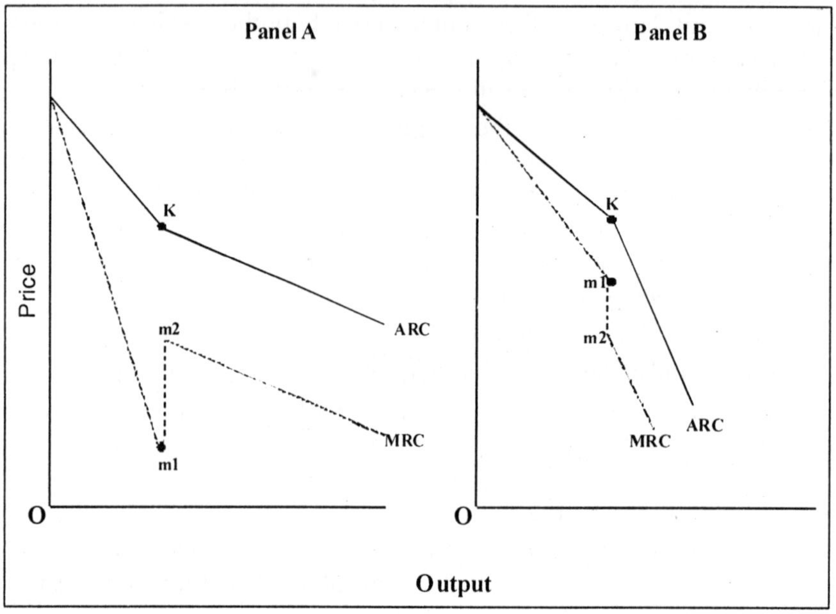

Kinked Demand Curve Theory – MRC with Discontinuity Lies Below ARC

23

Monopolistic Competition

Perfect competition and monopoly are the two extreme forms of market competition, the former being extreme form of perfection and the latter being extreme form of imperfection. These two competitions are purely theoretical and not practically seen. However, between these two extremes, there is a Monopolistic competition, which is more common and practically seen in the economy. This monopolistic competition combines the features of both perfect competition and monopoly. As in perfect competition, large numbers of firms (not as large as in perfect competition) exist in the market. Each seller has a small share in the total market. New firms are free to enter the market and existing firms are free to quit the market at any given period of time. But, the products of different sellers are differentiated on the basis of brands, thereby, price is not uniform in the market. This product differentiation gives rise to an element of monopoly to the firm over the competing product. As such, the firm of a particular brand can increase the price of its product knowing fully well that, his brand-loyal customers are not going to leave the firm. This is possible only because, the products have no perfect substitutes. Since, however, all the brands are of close substitutes to one another, the seller will lose some of his customers to his competitors. So, each firm do not enjoy monopoly status strictly because, the producer is not a single individual and there is no restriction regarding the entry of new firms. Thus, the monopolistic competition is a mix of perfect competition and monopoly. The same is presented below, how the monopolistic competition approaches perfect competition and monopoly and how it is different from them.

Views	*Perfect Competition*	*Monopoly*
Positive	Monopolistic competition approaches perfect competition because, ☆ Presence of large number of firms ☆ Each firm has a small share in the total market and ☆ Freedom for entry and exit of firms	Monopolistic competition approaches monopoly because, ☆ Each firm has some control over market, as the products are differentiated and ☆ The products are not perfect substitutes. ☆ ARC and MRC slopes downward from left to right
Negative	Monopolistic competition differs from perfect competition because, ☆ Products of different sellers are differentiated on the basis of brands and ☆ Price of the product is not uniform.	Monopolistic competition differs from monopoly because, ☆ Producer is not a single individual. ☆ There is no restriction on entry and exit of firms and ☆ Products are close substitutes.

So, monopolistic competition is a case of imperfect competition, where there are many or large number of firms selling a differentiated or heterogeneous product. It is the best approximation of perfect competition that the real world offers. Monopolistic competition has been introduced by an American Economist Prof. Edward Chamberlin, in his book *'Theory of Monopolistic Competition'* published in 1933. Before Chamberlin, perfect competition and monopoly were regarded as the only two alternatives of market competition. But, according to Chamberlin, in most of the real world, economic situations of both perfect competition and monopoly elements are present. So, according to Chamberlin, monopolistic competition is a blend of perfect competition and monopoly and is a more realistic model than either perfect competition or monopoly. Almost all retail operations including restaurants, fast food centers, clothing stores, fertilizers, pesticides and convenience stores are in this form of market. Small businesses in all sectors fall in this category. Perhaps, half of the economy's total production comes from monopolistically competitive firms. In this form of competition, starting a business is relatively easily, but staying in the business is not. This is because, the firm requires an ability to convince the customers that the product is different and better than that of competitors. The following are the popular definitions of monopolistic competition:

'Monopolistic competition is a situation in which there are many sellers of a particular product, but the product of each seller is in some way differentiated in the minds of consumers from the product of every other seller'.

– (Leftwhitch)

'Monopolistic competition is found in the industry, where there is a large number of small sellers selling differentiated but close substitute products'.

– (J.S. Bain)

23.1. Features of Monopolistic Competition

The features of monopolistic competition include,

1. Existence of Large Number of Firms

There exist a large number of firms or sellers in the market. Since their number is large, no one has the control over the total supply of the product in the market. The output-price policies of each firm are independent. That means, a change in the output-price policy of one firm will not show any influence on the output-price policy of other firm. Thus, there is no interdependence among the firms in the industry, unlike in oligopoly market situation. That means, each firm act independently in planning its own output-price policies because, there is product differentiation across the firms.

2. Product Differentiation

Product differentiation is the primary reason that each firm operating in a monopolistically competitive market is able to create a little monopoly all to itself. Each firm sells differentiated product, which are close substitutes to one another. The product of one firm is different from that of the other firm in terms of physical differences, quality differences, imaginary differences, purchase-benefit differences, features etc. So, each firm has an absolute monopoly in the production and sale of that differentiated product. The main aim of product differentiation across the firms is to capture a large number of customers by the firm towards its product. Since, the products are close substitutes across the firms, the E_C is high, unlike monopoly. This product differentiation further helps the firms to enjoy the trade protection in terms of copy rights, trademarks, industrial designs etc. This product differentiation can be brought by the firms in various ways as given below.

- ☆ By using different types and quality of raw material and technology in producing the product.
- ☆ Differences in features, attractiveness, durability etc (*Physical product differentiation*). For example, electronic goods can easily be physically differentiated.
- ☆ By offering various kinds of benefits to customers like free service, home delivery of product, acceptance of returned goods, discounts on price, distinctive packaging etc (*Marketing differentiation*). For example, breakfast cereals can easily be differentiated through packaging.
- ☆ By effective advertisement and propaganda about the product.
- ☆ Special significance to products through enjoying copy rights, trademarks, industrial designs etc.
- ☆ The firms can bring differentiation through the enhancing the skills of their employees, imparting trainings etc (*Human capital differentiation*).
- ☆ The firms can bring differentiation in the mode of distribution of their products to the customers (*Differentiation through distribution*). *For example,* distribution via mail order or through internet shopping such as Amazon.com.

In monopolistic competition, pricing of the product is not a problem, but product differentiation is the main essence to stay in the industry. This is because, with product differentiation only the firm can survive in the industry.

3. Freedom of Entry and Exit of Firms

There is freedom for the firms regarding the entry into and exit from the industry. But, these firms are not 'perfectly' mobile as compared to perfect competition. Similarly, unlike monopoly, they are largely unrestricted by Government rules and regulations, start-up costs or other substantial barriers to entry. Since, there is more scope for producing differentiated products and for gaining substantial profits, the tendency of entry of firms into the industry is more rather than exit of firms from the industry. Each new firm produces a differentiated product under its own brand name. This freedom of entry and exit of firms increases competition. Further, the ease of entry simply prevents the firms from maintaining super normal profits in the long run. If the firms enjoy super normal profits in the business in the short run, new firms will soon enter the industry, thereby, the profits will be reduced to normal level.

4. Nature of Revenue Curves

Demand curve or ARC of firms under this market competition slope downward from left to right indicating that, they can sell more output by reducing price. It is relatively elastic in monopolistic competition because, each firm faces competition from a large number of firms producing the products, which are very close substitutes. However, ARC is not perfectly elastic (as in perfect competition) because, the output of each firm is slightly different from that of other firms and monopolistically competitive goods are close substitutes, but not perfect substitutes. It is not highly inelastic (as in case of monopoly) because, there is product differentiation in monopolistic competition. So, the ARC in monopolistic competition is rather more elastic (flat slope) *i.e.*, less elastic compared to perfect competition and more elastic compared to monopoly market situation (Figure 23.1). So, when a firm raises the price, the buyers may switch over to other firms product and when the firm lowers or decreases the price, the demand for the firm's product will increase, as other firms' products may become more costly to the customers. So, the demand curve of a firm in monopolistic competition appears to be more elastic with a flat slope, mainly due to product differentiation. The extent of flatness of slope of ARC depends upon the following aspects:

- ☆ ***E_C for the product:*** If the product has more E_C, the demand curve is more flat and *vice versa*.
- ☆ ***Number of firms selling the same product and their output share in market demand*:** If this is more, then again the E_C is more, and hence, the ARC of a firm for the product is more elastic.

As shown in the Panel A of the Figure 23.1, ARC slopes downward from left to right and hence, MRC also slopes downward from left to right with a smooth and flat slope and lies below it. The ARC or demand curve is more elastic in monopolistic competition compared to monopoly market situation (Panel B) mainly due to product differentiation. Thus, monopolistic competition resembles monopoly in the aspect that, firm exercises some control over price, but there is more control in case of monopoly and less control in case of monopolistic competition. This product differentiation also leads to price differentiation across the firms.

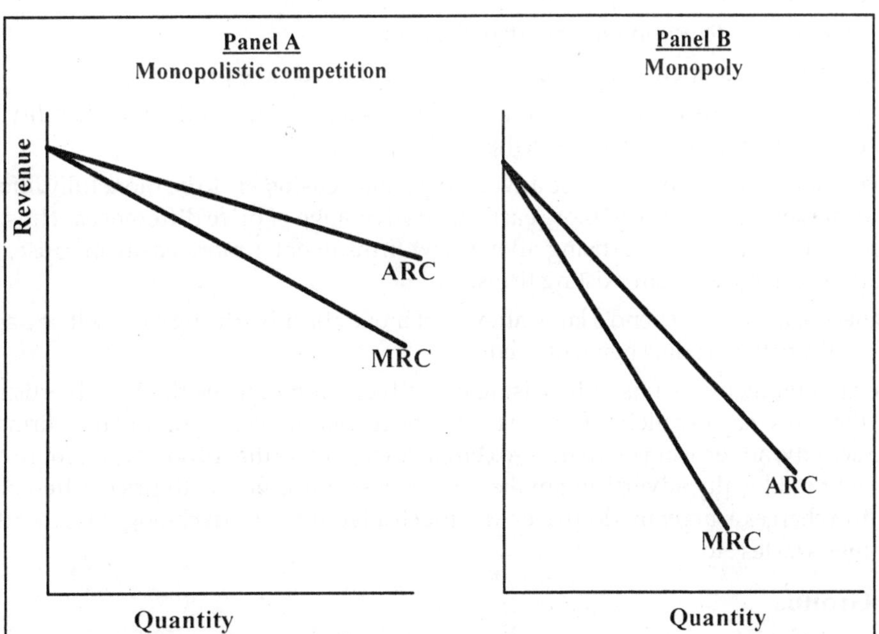

Figure 23.1: Revenue curves under Monopolistic competition and Monopoly.

5. Price Maker

A firm under monopolistic competition is a price maker with some degree of control over price. So, unlike perfect competition, a monopolistically competitive firm has the ability to raise or lower the price a little, but not at a significant level. As in monopoly, the price received by a firm under monopolistically competition is greater than its MR, as ARC is above MRC. In monopolistic competition, due to product differentiation, if one firm lowers the price, there is increase in demand for the product offered by that firm, thereby, it attracts the customers of other firms. Contrary wise, if a firm increases the price of the product, the customers will shift to other firms' close substitute product. So, the firm enjoys some control over the price of the product in monopolistic competition. unlike in perfect competition, where the individual firm has absolutely no control on the price of the product, because of homogenous nature of the product. So, in monopolistic competition, a firm has to choose a proper output–price policy that maximizes the profits. However, in monopoly, the firm enjoys more control over the price compared to monopolistic competition, as there are absolutely no substitute goods in monopoly situation.

6. Two Dimensional Competition among the Firms

Under this form of market structure, competition among the firms takes place in two different ways *viz.*, price competition and non-price competition. In price competition, firms compete with each other on the basis of price of the product. Each firm has its own output-price policy, thereby, it fixes the price of the product, so as to take advantage of higher sales. In non-price competition, to increase the sales, the firms compete by adopting product differentiation or through increasing selling costs. Selling costs are the costs incurred by the firm on advertising and other forms of sale promotion. This non-price competition has become an integral part of marketing of products and through this competition, the firm tries to make a favourable shift in demand for the product and tries to capture the market. So, any action a firm takes to increase demand for its product other than cutting its price refers to non-price competition. The important types of non-price competition include, advertising the product, better service, product guarantees, free home delivery, more attractive packaging etc. Due to the presence of more number of firms producing the same product, but with differentiation, the non–price competition is significant among the firms. Among different types of non-price competition, advertisement expenditure is the prominent one. The costs incurred towards advertising and sales promotion by the firm will influence both the costs and returns of the business. Hence, the firm will adjust its outlay towards selling costs, besides planning output–price policy. This adjustment of funds towards selling costs is one of the characteristic features of monoplastic competition, unlike perfect and monopoly competitions, where there are no selling costs. However, compared to oligopoly, non–price competition is lower in case of monopolistic competition, as there is absence of mutual interdependency of firms in monopolistic competition. Regarding non-price competition (advertising), there are both positive and negative views. The positive views supporting advertising include,

- ☆ advertising is informative to the buyers about the products and their features. This information allows customers to make better choices about what to buy and thus, enhances the ability of markets to allocate resources efficiently.
- ☆ advertising increases sales of the product
- ☆ advertising contributes to economic growth of the firms
- ☆ advertising increases competition between the firms and lowers the prices
- ☆ advertising provides consumers with the valuable information on product availability, quality and price that they need to make efficient choices in the market place.
- ☆ advertising fosters competition. Because advertising allows customers to be more fully informed about all the firms in the market, customers can more easily take advantage of price differences. Thus, each firm has less market power. In addition, advertising allows new firms to enter more easily because, it gives entrants a means to attract customers from existing firms.
- ☆ the willingness of the firm to spend a large amount of money on advertising can itself be a signal to consumers about the quality of the product being offered.

However, there are arguments against advertising *viz.*, advertising reduces the E_p, advertising costs raises the price of the product, the firms may use their influence to bias the media, affect the entry of new firms into the industry, it is not a productive activity, advertising often tries to convince customers that products are more different than they truly are, by fostering brand loyalty advertising makes buyers less concerned with price differences among similar goods thereby, each firm charges a larger markup over MC, much advertising is psychological rather than informational, it manipulates customer's tastes etc.

7. Existence of Groups

Firms under monopolistic competition together form a group. They cannot be called as an 'industry'. Since, there is product differentiation, each firm is an industry by itself. That means, since products are not homogeneous there is no industry, as in the case of perfect competition. Therefore, Chamberlin introduced the term 'group'. A group is a cluster of firms producing very closely related, but differentiated products. The collection of firms that produce same type of products with high E_C will constitute a 'group' or a 'product group'. A firm under this market competition is a price maker. So, in place of Marshallian concept of industry, Chamberlin introduced the concept of 'Group' under monopolistic competition. Thus, an 'industry' means a number of firms producing an identical product, but a 'group' means a number of firms producing differentiated products, which are closely related.

8. Large Number of Buyers

There is large number of buyers in the market. Each buyer prefers to buy a specific brand of product. So in order to attract more and more customers towards its product, the firms introduce the product with distinctive features suiting to different tastes and preferences.

9. Absence of Interdependency among the Firms

In this competition, since there is large number of firms, each firm's share is negligible in the total industry's output. The firms sell differentiated products, thereby, they enjoy monopoly power to a certain extent. Thus, the firms formulate independent policies regarding price and output. Any action of one of the firms regarding change in price or output has no significant effect on the other firms. Thus, the firms are independent and not interdependent.

10. Market Information

In this competition, buyers do not have knowledge about output-price policy of different firms, but they have relatively complete information about the prices of the products, different brands, features etc., of different firms. On the contrary, each seller has complete information about production techniques and the prices charged by other rivalry firms.

11. Profit Maximization

In short run, the optimum level of output is determined at the point of equality of SRMC=MR. This optimum level of output, in turn, determines the price charged (on the ARC) and average unit cost (on the SRATC curve). The profit is the excess of TR area over SRTC area. In long run, the firm is in equilibrium, where ARC is tangent to the LRATC, thereby, the firm enjoys normal profits only. Firms may, however, retain some profits by using more non-price action.

12. Economic Effects

The economic effects of monopolistic competition includes an overall undesirable loss of allocative and productive efficiency, as the customers pay more and are able to buy less than in perfect competition. However, the effect is not as serious as in monopoly because, the product differentiation in monopolistic competition minimizes these losses. However, some waste is present in excess capacity and in going for heavy non-price competition.

13. Less Market Control

The large number of firms under monopolistic competition limits the control of any one firm over market price. It also prevents price fixing at higher level, because of the difficulties involved in getting a large number of firms to act together.

23.2. Price-Output Determination of a Firm Under Monopolistic Competition in Short Run and Long Run

Profit maximization with respect to output and price by the individual firm under monopolistic competition is governed by similar principles as mentioned in other market conditions.

1. Price-Output Determination of a Firm Under Monopolistic Competition in the Short Run

The price-output determination of a monopolistically competitive firm in short run production programme is based on the following assumptions:

- ☆ The number of firms are large and they act independently.
- ☆ There is product differentiation across the firms.
- ☆ The ARC of the firm is more elastic.
- ☆ Entry of new firms is blocked into the industry.

Case 1: TR>SRTC or Price >SRATC

In the Figure 23.2, on X-axis, output and on Y-axis, revenue and costs are taken. MRC and ARC are downward sloping curves, the ARC is a higher level sloping down curve and MRC is a lower level sloping down curve. SRMC curve cuts the MRC at point 'E'. So, the firm is in equilibrium at 'E' and produces OQ quantity of output by fixing the price of the product at OC=QP. At this price and output, the firm realizes super-normal profits shown by the shaded rectangle PABC. This is because, at OQ level of output, QP=OC price is above SRATC curve by PA portion between ARC

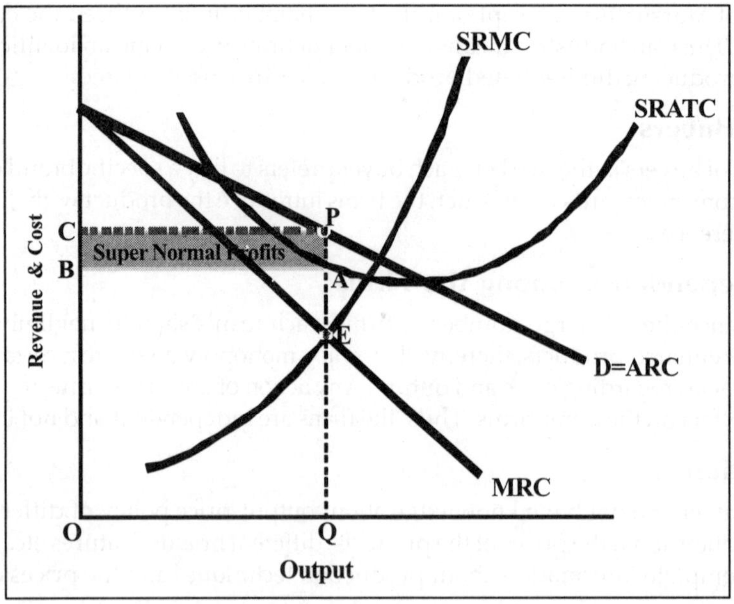

Figure 23.2: Price determination under monopolistic competition in short run (Case: TR>SRTC or Price > SRATC).

and SRATC curve. Beyond OQ level of output, MR<SRMC and below OQ level of output, MR>SRMC. So, to derive maximum profits, the firm has to produce the output OQ, where MR=SRMC.

Estimation of Profits

The output OQ is supplied in the market at price OC=QP. At this level of output, the difference between AR and SRATC is PA *i.e.*, average profit. So,

Total profit = Average profit x Output = PA x OQ = PABC

Case 2: TR=SRTC or Price=SRATC

In the Figure 23.3, on X-axis, output and on Y-axis, revenue and costs are taken. MRC and ARC are downward sloping curves, the ARC is a higher level sloping down curve and MRC is a lower level sloping down curve. SRMC curve cuts the MRC at point 'E'. So, the firm is in equilibrium at 'E' and produces OQ quantity of output by fixing the price of the product at OC=QP. At this price and output, the firm realizes normal profits or the firm becomes break even, since price=SRATC. Beyond OQ level of output, MR<SRMC and below OQ level of output, MR>SRMC and at OQ output, MR=SRMC and the firm derives only normal profits, as price=SRATC.

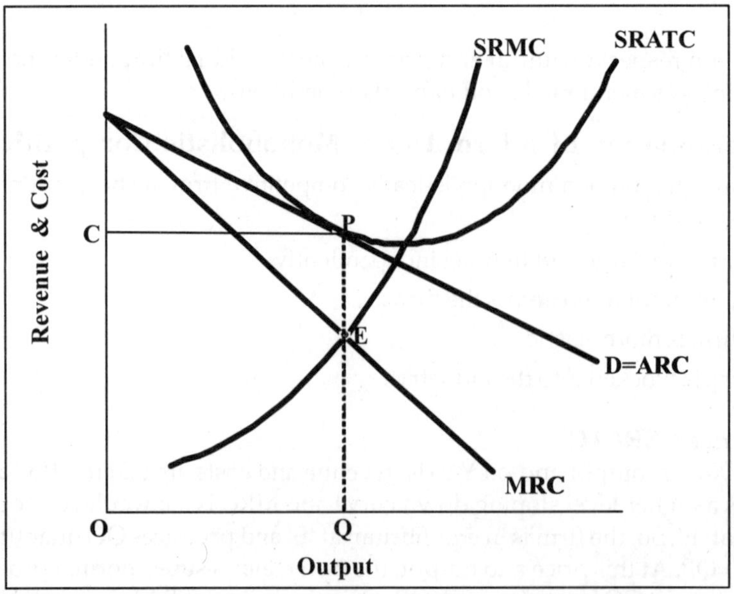

Figure 23.3: Price determination under monopolistic competition in short run (Case: TR=SRTC or Price = SRATC).

Case 3: SRTC>TR>SRTVC or SRATC>Price>SRAVC

In the Figure 23.4, on X-axis, output and on Y-axis, revenue and costs are taken. MRC and ARC are downward sloping curves, the ARC is a higher level sloping down curve and MRC is a lower level sloping down curve. SRMC curve cuts the MRC at point 'E'. So, the firm is in equilibrium at 'E' and produces OQ quantity of output by fixing the price of the product at OC=QP. At this price and output, the firm incurs losses as shown by the shaded rectangle PABC. This is because, at OQ level of output, QP price is below SRATC curve by PA portion between SRATC curve and ARC. Here, price QP covers only SRAVC, but not SRATC. In the short run, the firm can continue the business temporarily, as price per unit of output covers atleast SRAVC and not SRATC. However, if the price of the product falls below SRAVC, the firm better stop the business temporarily because, it will incur huge losses, as even explicit costs (SRAVC) are also not recovered. Beyond OQ level of output, MR<SRMC and hence, the firm should not produce output beyond OQ.

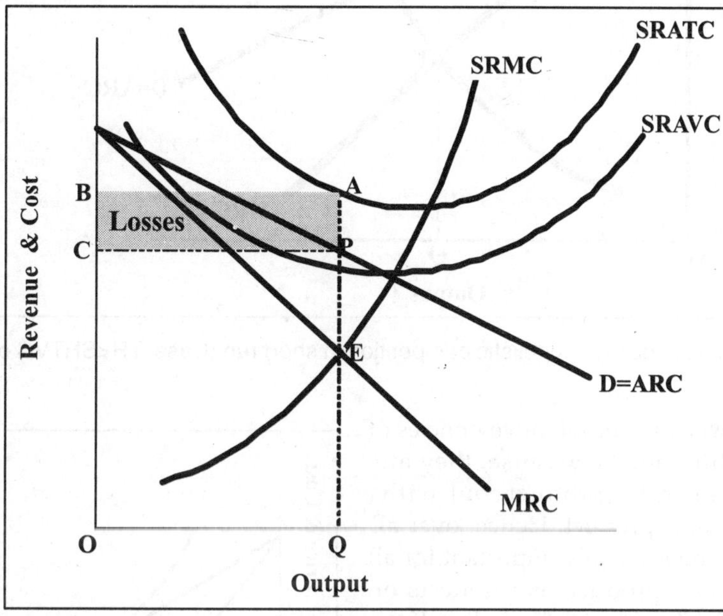

Figure 23.4: Price determination under monopolistic competition in short run (Case: SRTC>TR>SRTVC or SRATC>Price>SRAVC).

Estimation of Losses

The output OQ is supplied in the market at price OC=QP. At this level of output, the difference between ARC and SRATC is PA *i.e.*, average loss. So,

Total loss = Average loss x Output = PA x OQ = PABC

Case 4: TR<SRTVC or Price<SRAVC

In the Figure 23.5, on X-axis, output and on Y-axis, revenue and costs are taken. MRC and ARC are downward sloping curves, the ARC is a higher level sloping down curve and MRC is a lower level sloping down curve. MC curve cuts the MRC at point 'E'. So, the firm is in equilibrium at 'E' and produces OQ quantity of output by fixing the price of the product at OC=QP. At this price and output, the firm incurs huge losses, as shown by the shaded rectangle PABC. This is because, at OQ level of output, QP price is below SRAVC curve. So, the price QP could not cover even SRAVC and hence, the firm better stop the business temporarily because, he is incurring huge losses.

Estimation of Losses

The output OQ is supplied in the market at price OC=QP. At this level of output, the difference between ARC and SRATC is PA *i.e.*, average loss. So,

Total loss = Average loss x Output = PA x OQ = PABC

It is important to note that, due to product differentiation, all the firms will not charge the same price for the same product. This implies, each firm will have its own price-output policy, thereby, they act independently and try to maximize their profits.

2. Price-Output Determination of a Firm Under Monopolistic Competition in the Long Run

In the long run, the adjustment process under monopolistic competition is similar to that of perfect competition. In the group, each firm will try to maximize the profits when LRMC=MR. Even though, each firm produces a similar product, but because of product differentiation, each firm follows its own output–price policy and try to maximize the

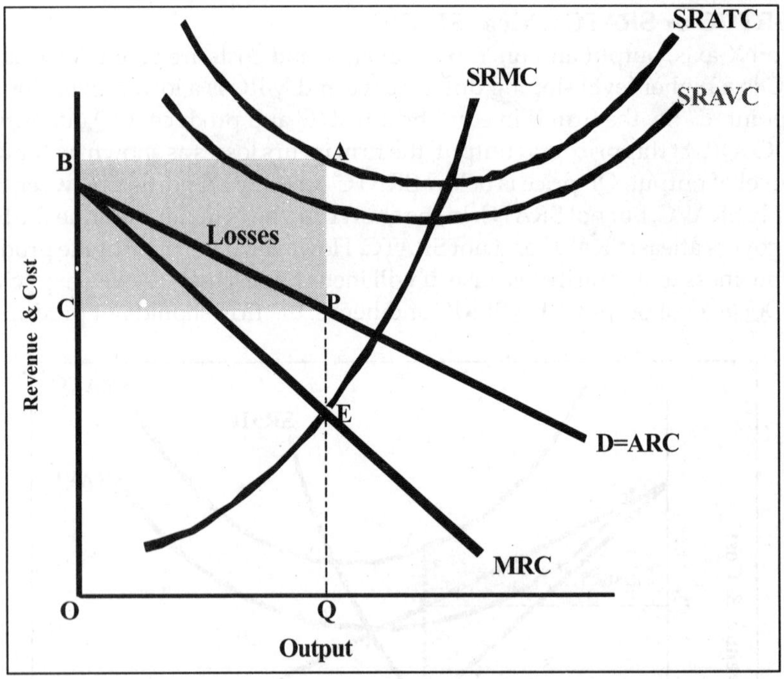

Figure 23.5: Price determination under monopolistic competition in short run (Case: TR<SRTVC or Price<SRAVC).

profits. But, at the group level, the output–price policies of different firms may not differ much, because, they are producing more or less a similar product, but with differentiation within the same product. Hence, over all output–price policy of a group is equally important for all the firms producing the same product, as it reflects or influences the individual firm's output-price policy. If the total output exceeds the total demand in the group, then price of the product will fall and hence, the demand curve or ARC will fall downwards. If ARC falls below the LRATC of the firm, that firm should leave the group because, the price could not cover LRATC of the firm in the long run. On the contrary, if the total demand exceeds total output in the group, price will increase for the product and hence, the demand curve or ARC will rise upward above LRATC and hence, the firm will earn super-normal profits. But, earning super-normal profits is not possible under monopolistic competition in the long run, as in perfect competition. This is because, if the firms earn super-normal profits in the short run, new firms gets attracted to it and they will enter the group and existing firms will expand. So, with the entry of new firms into the group, the supply of the product into the market increases, thereby, price falls. At the same time, with the entry of new firms, the demand for factor services will increase and thereby, the cost structure will increase leading to rise in cost curves. Thus, fall in revenue curves and rise in cost curves will squeeze the super-normal profits of the firm in the long run. The same is explained through Figure 23.6.

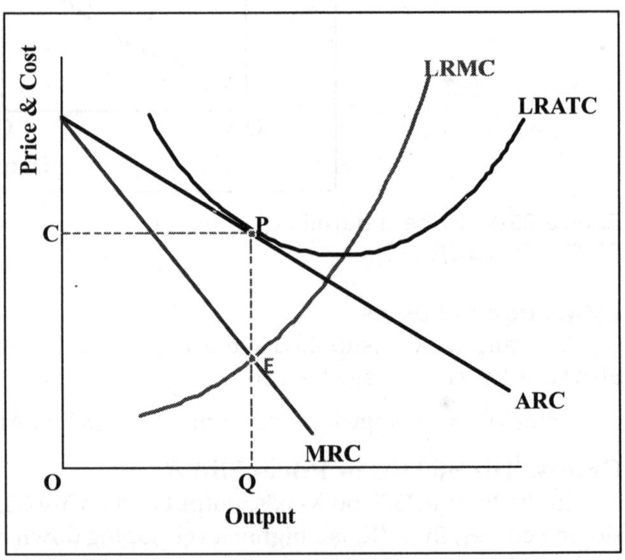

Figure 23.6: Price determination under monopolistic competition in the long run.

In the Figure 23.6, LRMC cuts the MRC at point E, which is the point of equilibrium and its corresponding point of ARC is P, implying that, the firm under monopolistic competition produces OQ quantity of output and sells the product at QP=OC price. It is important to note that, ARC in the long run will be more elastic (*i.e.*, more flatter) compared to short run, since there are more number of substitutes for the product in the long run. Since, at the point of equilibrium, the ARC lies tangent to the LRATC, thereby, the firm will earn only normal profits. So, the point of equilibrium of a firm under monopolistic competition in the long run will satisfy two essential conditions *viz.*, LRMC=MR and ARC=LRATC. This infers that, the firm will earn only normal profits in the long run. But, in short run, there is only one condition of equilibrium *i.e.*, SRMC=MR.

23.3. Chamberlin's Alternative Approach of 'Group' Equilibrium Under Monopolistic Competition

As discussed earlier, Chamberlin rightly introduced the term 'group' or 'product group' in monopolistic competition in place of the traditional concept of 'industry' in the theory of value. The term 'industry' is in perfect tune with reference to perfect market competition, as all the firms under perfect competition produces and sells homogenous product. Even in case of monopoly competition, the concept of 'industry' is applicable, as the single firm itself represents the industry and there are no substitutes for the product under consideration. Thus, an 'industry' refers to the group of firms producing a homogenous product. On the contrary, under monopolistic competition, since there is a greater degree of product differentiation, each individual firm can act like an industry (as a monopoly firm), but with the existence of close substitutes. Hence, it is ridiculous to call fertilizer industry, pesticides industry, bi-cycle industry, automobile industry in an analytical sense. So, it is appropriate to use the term 'group' in place of 'industry' under monopolistic competition.

Group Equilibrium

Chamberlin explained the concept of 'group equilibrium' by considering two different demand curves with reference to a representative firm. A representative firm is a firm, which is assumed for the simplicity of the analysis and represents each and every firm in the group regarding the price-output equilibrium. The two demand curves include dd curve and DD curve.

1. 'dd' Curve

It represents an individual firm's demand curve. It is, otherwise, called as 'Perceived demand curve' or 'ARC'. It is drawn for a firm on the basis that, the rival firms in the group will keep the prices of their products constant. That means, it guides the representative firm, how much quantity of output it can sell at different prices of the product, provided other rival firms in the group will keep the prices of their products constant. Thus, this demand curve is associated with large group (more number of firms) because, in such case only, the change in price of the product of a representative firm will not influence the prices of the products of its rival firms. This type of demand curve is also called as Subjective or Imagined or Expected demand curve, as it guides the business firm to imagine its demand curve due to changes in prices of its product, but keeping the prices of the products of its rival firms constant. This demand curve resembles the demand curve or ARC of the monopolist and it is more elastic than the group demand cure DD (Figure 23.7). The elasticity of 'dd' curve varies inversely with the brand preferences. That means, if the customers have strong brand preferences, the perceived demand curve of the firm will be less elastic and *vice versa*.

2. 'DD' Curve

It represents the group demand curve. It is, otherwise, called as 'Proportional demand curve' or 'ARC'. It is also called as *'Mutatis Mutandis'* demand curve because, it assumes all the firms in the group change the price of the product in the same direction and in the same proportion *i.e.,* they charge same price for the product, as charged by the representative firm. So, the proportional demand curve faced by a representative firm will be less elastic compared to its own perceived demand curve. (Figure 23.7). This proportional demand curve also slopes downward from left to right, as the market demand for the product varies inversely with the price. So, this proportional demand curve resembles the

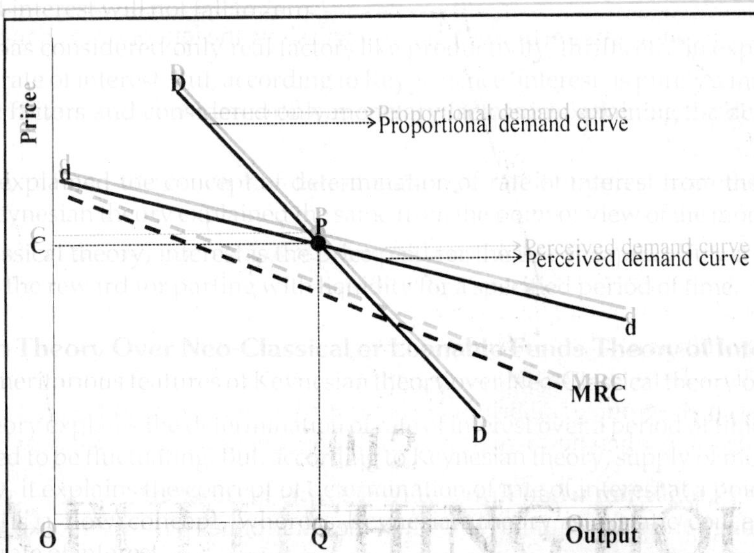

Figure 23.7: Chamberlin's approach: Two demand curves under monopolistic competition.

industry demand curve under perfect competition. The proportional demand curve faced by a firm represents the proportionate share of firm's demand in the total market demand for the product. So, if there are more number of firms in the group producing a particular product (say, a pesticide), the proportionate share of each firm in the total market demand of the product will be low, thereby, the proportional demand curve faced by an individual firm will shift towards left side and *vice versa*. Similarly, if more number of firms enter the industry, the proportional demand curve will also shift towards left side.

So, under monopolistic competition, both perceived demand curve and proportional demand curve faced by a representative firm slopes downward from left to right, but perceived demand curve is more elastic compared to proportional demand curve. In perfect competition, the perceived demand curve is perfectly elastic, whereas, proportional demand curve is a downward sloping curve from left to right. Considering these two demand curves, under monopolistic competition, the representative firm tends to lower the price of its product, till its DD curve is like the dd curve. In the Figure 23.7,

- ☆ The two demand curves dd and DD intersect each other at point 'P' with reference to OP price and OQ quantity demanded. That means, at OP price of the product, the proportional share of each of the firm of the total market demand is OQ. That means, assuming all the firms charge the same price for the product, they all produce and sell OQ quantity of output and each charge OP price. So, point P is with reference to proportional demand curve and thereby, point 'P' lies on the proportional demand curve, DD.
- ☆ So, from this point 'P' on the proportional demand curve, we can draw the perceived demand curve of the representative firm. As discussed earlier, with reference to perceived demand curve, the representative firm will find out different quantities of output it can sell at different prices of its product, provided other firms will not change the price of their products. So, when a representative firm lowers the price below QP=OC, it can increase the sales of the product because, other firms will not reduce the price and thereby, it attracts the customers of other firms. On the contrary, if the representative firm increases the price of the product above OC=QP, the customers will move to other firms, as they will not change the prices of their products, thereby, the representative firm will lose some of its customers. So, the perceived demand curve of the firm, dd will be more elastic compared to proportional demand curve, as other firms will not change the prices of their products. That means, proportional demand curve is less elastic compared to perceived demand curve, as with reference to proportional demand curve the assumption is that, all the firms will change the prices of the products in the same direction and in the same proportion.
- ☆ So, the perceived demand curve 'dd' is constructed with a flat slope, passing through point 'P' of the proportional demand curve.

Short Run Firm's Equilibrium

According to Chamberlin, the short run equilibrium under monopolistic competition involves two essential conditions:

- ☆ The price-output combination, where SRMC intersects perceived MRC
- ☆ The price-output combination, where perceived demand curve intersects proportional demand curve.

When the above two conditions are fulfilled, no firm will show the tendency to change its output and price. In the Figure 23.8,

- ☆ The above two conditions of short run firm's equilibrium are fulfilled with reference to points E and P respectively.
- ☆ So, the representative firm will sell OQ proportion of total market demand and charge OC=QP price for the product.
- ☆ The price OC=QP is above the SRATC by PA and thereby, the representative firm earns PABC super-normal profits.

Long Run Group Equilibrium

In the short run, if the firms enjoy super-normal profits, new firms will get attracted to these profits and enter the group. So, in the long run, the group equilibrium can be studied under two conditions *viz.*, when the entry of new firms is blocked into the group and when new firms enter the group. Let us discuss each case here under.

1. Long Run Group Equilibrium when Entry of New Firms is Blocked

The concept of long run group equilibrium, when the entry of new firms is blocked into the group is explained through Figure 23.9.

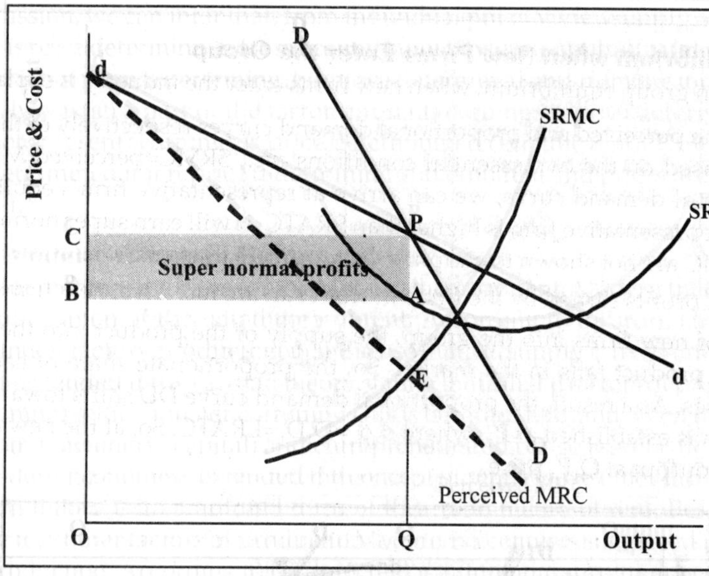

Figure 23.8: Short run firm's equilibrium under monopolistic competition – Proportional and Perceived demand curves approach.

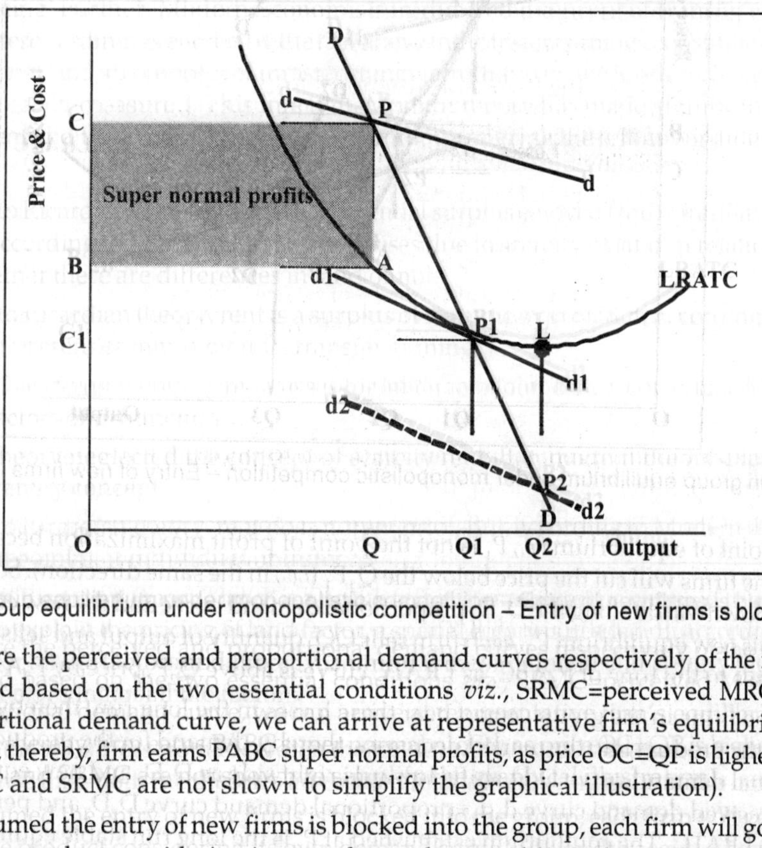

Figure 23.9: Long run group equilibrium under monopolistic competition – Entry of new firms is blocked.

- ☆ dd and DD are the perceived and proportional demand curves respectively of the representative firm in the short run, and based on the two essential conditions *viz.*, SRMC=perceived MRC and perceived demand curve=proportional demand curve, we can arrive at representative firm's equilibrium output, OQ selling at OC=QP price, thereby, firm earns PABC super normal profits, as price OC=QP is higher than SRATC. (Perceived MRC, SRATC and SRMC are not shown to simplify the graphical illustration).
- ☆ Since, we assumed the entry of new firms is blocked into the group, each firm will go for a price cut to increase the sales of the product considering its own demand curve 'dd', on the assumption that, other firms will not react to its action. Instead of increasing the quantity demanded on the 'dd' curve, each firm will slope downwards along the proportional demand curve and this continues till the new perceived demand curve d_1d_1 lies tangent to the LRATC at P_1.
- ☆ At this point P_1, the perceived demand curve d_1d_1 = proportional demand curve and it is the point of long run group equilibrium, where perceived demand curve d_1d_1 will lie tangent to LRATC. So, each firm will sell OQ_1 quantity of output at $Q_1P_1 = OC_1$ price and earns only normal profits, as price Q_1P_1 (=OC_1)=LRATC.
- ☆ If the firms still cut the price below Q_1P_1=OC_1 price, then they incur losses, as price Q_2P_2 is less than LRATC. So, no firm will cut the price below Q_1P_1=OC_1.
- ☆ So, in the long run group equilibrium, the firms will enjoy only normal profits and at the point of equilibrium, no firm show the tendency to change the equilibrium output-price policy.

2. Long Run Group Equilibrium when New Firms Enter the Group

The concept of long run group equilibrium, when new firms enter the industry is explained through Figure 23.10,

- ☆ dd and DD are the perceived and proportional demand curves respectively of the representative firm in the short run and based on the two essential conditions *viz.*, SRMC=perceived MRC and perceived demand curve=proportional demand curve, we can arrive at representative firm's equilibrium and since the price received by the representative firm is higher than SRATC, it will earn super normal profits. (Perceived MRC, SRATC and SRMC are not shown to simplify the graphical illustration).
- ☆ As super normal profits earned by the firms in short run are lucrative, new firms will enter the group.
- ☆ With the entry of new firms into the group, the supply of the product into the market increases, thereby, demand for the product falls in the market. So, the proportionate share of each firm in the total market supplies decreases. As a result, the proportional demand curve DD shifts towards left as D_1D_1 and thereby, new equilibrium is established at P_1, where $d_1d_1=D_1D_1$=LRATC. So, at the new equilibrium, each firm sells OQ_1 quantity of output at Q_1P_1 price.

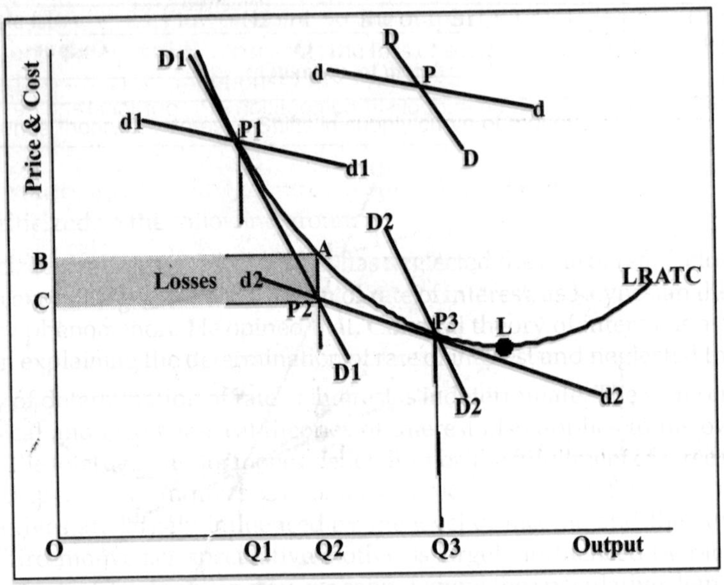

Figure 23.10: Long run group equilibrium under monopolistic competition – Entry of new firms into the industry.

- ☆ But, this point of equilibrium *i.e.*, P_1 is not the point of profit maximization because, to boost the sales of the product, the firms will cut the price below the Q_1P_1 (*i.e.*, in the same direction). So, the 'd_1d_1' curve of each firm will slide downwards along the D_1D_1 proportional demand curve and the new equilibrium will be established at P_2. At this new equilibrium P_2, each firm sells OQ_2 quantity of output and sells the product at Q_2P_2 price and incurs losses to the tune of P_2ABC, as LRATC curve is above Q_2P_2 price by P_2A.
- ☆ Some of the firms in the group cannot bear these losses in the long run, thereby, they leave the industry. So, the supply of the product into the market decreases, thereby, demand for the product gradually increases. So, the proportional demand curve D_1D_1 shifts towards right side as D_2D_2 and new equilibrium is established at P_3, where perceived demand curve d_2d_2=proportional demand curve D_2D_2 and perceived demand curve d_2d_2 is tangent to LRATC. The equilibrium established at P_3 is the long run stable equilibrium, where each firm sells OQ_3 quantity of total output at Q_3P_3 price and each firm derives only normal profits.
- ☆ From the above discussion, we can infer that, at the point of group equilibrium in the long run, dd curve must be tangent to the LRATC and at the point of tangency, DD curve must intersect both dd curve and LRATC curve.

23.4. Concept of Excess Capacity

Under perfect competition, the firm's equilibrium in the long run is at the point where, LRMC=MRC=ARC=minimum point of LRATC, thereby, the firm earns only normal profits. So, the super-normal profits enjoyed by the firms in the short run are completely squeezed out with the entry of new firms in the long run and the firm attains equilibrium on the minimum point of the LRATC curve. That means, in the long run, the ARC is tangential to the LRATC at its lowest point. (Panel A of the Figure 23.11). So, at point 'E', the firm produces OQ_1 quantity of output and sell the same at OP long run price and enjoys normal profits. The output OQ_1 so produced is considered as the most efficient or ideal or optimum output.

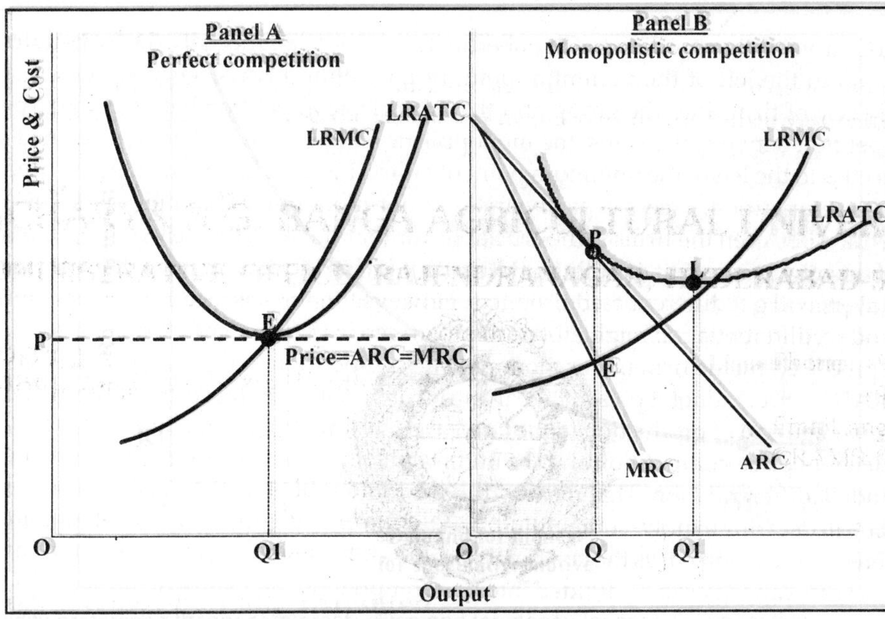

Figure 23.11: Excess capacity under monopolistic competition.

But, under monopolistically competitive market, the ARC is not perfectly elastic as in case of perfect competition, rather it slopes downward from left to right. That means, ARC is less elastic in monopolistic competition compared to perfect competition. Since, ARC is downward sloping, it cannot be tangential to LRATC at its lowest point (Panel B of the Figure 23.11). So, under monopolistic competition, the equilibrium of the firm in the long run is where, LRMC=MRC *i.e.*, at 'E' and its corresponding point on the ARC indicates the price of the product QP, which is tangential to the LRATC, while LRATC is sloping downward, (*i.e.*, at P) rather than at its minimum point L. So, the firm even under monopolistic competition will earn only normal profits in the long run, but the output 'OQ' is not an optimum or ideal or most efficient output, as the firm is not operating at the minimum point on the LRATC, as in case of perfect competition. So, no firm under monopolistic competition will produce optimum or most efficient output (OQ_1) in the long run and if the firm increases the output from OQ to OQ_1 *i.e.*, upto lowest point of LRATC 'L', it leads to LRMC>MRC. This infers that, a firm under monopolistically competitive market in the long run will operate at less than optimum size, thereby, a gap exists between optimum or ideal output (OQ_1) and actual output (OQ). This gap refers to unutilized capacity or excess capacity left over by the monopolistically competitive firm in the long run. So, in the Panel B of the Figure 23.11, OQ is the equilibrium output and it is lower than the most efficient output OQ_1, thereby, QQ_1 is the unutilized capacity or excess capacity left over by the business firm to attain ideal or optimum output corresponding to the minimum point of LRATC. But, the firm will not utilize its excess capacity to produce extra QQ_1 output, as for producing this output, LRMC exceeds MRC. So, for a monopolistically competitive firm, the output produced is lower, but price is higher and reverse in the case for the firm under perfect competition. This is due to the existence of excess capacity under monopolistic competition and this arises only in the long run production programme.

The most important reason for the excess capacity of a firm under monopolistic competition is the downward sloping ARC. When the ARC slopes downwards, it cannot be tangent to the LRATC at its minimum position and it lies tangent to the LRATC to the left of the minimum of falling LRATC. The downward falling of ARC is due to product differentiation under monopolistic competition and this excess capacity is tolerated in the context of product differentiation, as the customers gets a wide variety of products. Since, the product is homogenous under perfect competition, the ARC is a horizontal line parallel to X-axis and thereby, it lies tangent to LRATC at its minimum point and hence, there is no excess capacity under perfect competition. So, the excess capacity under monopolistic competition can be considered as a social benefit in the context of providing differentiated products to the customers and it is weighed against the cost to the society.

23.5. Chamberlin's Concept of Excess Capacity

As discussed earlier, under monopolistic competition, the long run equilibrium of the firm will be at the point, where LRMC=MRC and the ARC lies tangent to the left of the minimum point of falling LRATC curve. So, at the point of long run equilibrium, the firm will not operate at the minimum point of LRATC curve, as ARC is a downward sloping curve. But, under perfect competition, since ARC is perfectly elastic, the firm's equilibrium will be attained at the minimum point of the LRATC curve. So, in the long run, for the firm under monopolistic competition, ideal output or most efficient output or optimum output is not produced, as the output produced corresponds to the left of the minimum point of the falling LRATC curve. But, according to Chamberlin, the output produced by the firm under monopolistic competition in the long run to the left of the minimum point of the falling LRATC curve represents 'ideal output' provided, if there is free entry of firms into the group and there is price-competition between the firms. That means, when there is price competition between the firms, the monopolistic competition will not create any excess capacity, even if the output produced is to the left of the minimum point of falling LRATC. Even when there is free entry of firms, but in the absence of price competition, excess capacity arises under monopolistic competition. This infers that, the presence of price-competition between the firms is the crucial factor that decides the ideal output under monopolistic competition to the left of the minimum point of falling LRATC curve. This is because, when there is price competition between the firms, the buyers will get differentiated products. However, in the absence of price competition, but with the free entry of firms, product differentiation might not be there, as seen in perfect market competition. So, according to Chamberlin, the output produced by the firm under monopolistic competition in the long run, which corresponds to the minimum point of LRATC curve, cannot represent 'ideal output' involving product differentiation. So, the difference between actual long-run average cost of production to the left of the minimum point of falling LRATC curve due to free entry of firms and prevalence of price-competition and the minimum long-run average cost of production indicates the cost difference due to product differentiation. That means, due to product differentiation, the long run average cost of production of the product will rise (due to price-competition) and this output of the firm under monopolistic competition in the long run is considered as ideal output, as the customers prefers or desires to have the differentiated products. So, Chamberlin's analysis of excess capacity can be divided into two components *viz.*, entry of the new firms into the group with price-competition and entry of the new firms into the group with non-price competition.

1. Concept of Excess Capacity with Entry of New Firms and Price-Competition between the Firms

The concept of excess capacity with the entry of new firms and prevalence of price-competition between the firms is explained through Figure 23.12. In the Figure 23.12,

- ☆ dd and DD are the perceived and proportional demand curves respectively of the representative firm in the short run and based on the two essential conditions *viz.*, SRMC=perceived MRC and perceived demand curve=proportional demand curve, we can arrive at representative firm's equilibrium and since the price received by the representative firm is higher than SRATC, it will earn super normal profits. (Perceived MRC, SRATC and SRMC are not shown to simplify the graphical illustration).
- ☆ As super normal profits earned by the firms in short run are lucrative, new firms will enter the group.
- ☆ With the entry of new firms into the group, the supply of the product into the market increases, thereby, demand for the product falls in the market. So, the proportionate share of each firm in the total market supplies decreases. As a result, the proportional demand curve DD shifts towards left as D_1D_1 and thereby, new equilibrium is established at P_1, where d_1d_1=D_1D_1=LRATC. So, at the new equilibrium, each firm sells OQ_1 quantity of output at Q_1P_1 price.
- ☆ But, this point of equilibrium *i.e.*, P_1 is not the point of profit maximization because, to boost the sales of the product, the firms will cut the price below the Q_1P_1 (*i.e.*, in the same direction). So, the 'd_1d_1' curve of each firm

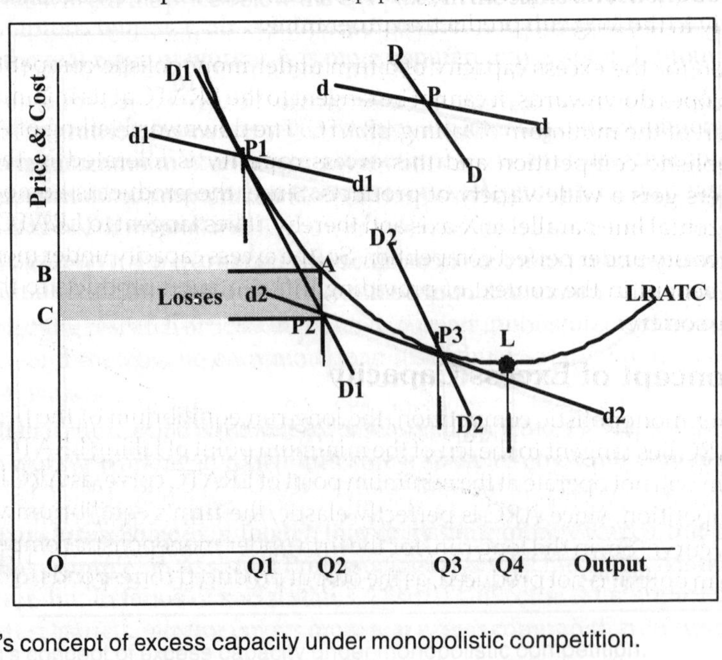

Figure 23.12: Chamberlin's concept of excess capacity under monopolistic competition.

will slide downwards along the D_1D_1 proportional demand curve and the new equilibrium will be established at P_2. At this new equilibrium P_2, each firm sells OQ_2 quantity of output and sells the product at Q_2P_2 price and incurs losses to the tune of P_2ABC, as LRATC curve is above Q_2P_2 price by P_2A.

☆ Some of the firms in the industry cannot bear these losses in the long run, thereby, they leave the industry. So, supply of the product into the market decreases, thereby, demand for the product gradually increases. So, the proportional demand curve D_1D_1 shifts towards right side as D_2D_2 and new equilibrium is established at P_3, where perceived demand curve d_2d_2=proportional demand curve D_2D_2 and perceived demand curve d_2d_2 is tangent to LRATC. The équilibrium established at P_3 is the long run stable equilibrium, where each firm sells OQ_3 quantity of total output at Q_3P_3 price and each firm derives only normal profits.

☆ So, according to Chamberlin, the output produced OQ_3 at the long run stable equilibrium is the ideal output of the firm under monopolistic competition in the long run and thereby, there is no excess capacity.

☆ But, under perfect competition, the ideal output is produced at minimum point of LRATC (at 'L') *i.e.*, OQ_4.

☆ So, the difference between monopolistic competition output and perfect competition output is Q_3Q_4 and this output difference arises due to increase in costs due to product differentiation under monopolistic competition, whereas under perfect competition, the product is homogenous. The customers are willing to pay this cost difference for the products produced under monopolistic competition, as they get differentiated products, and thereby, the output OQ_3 corresponding to the left of the minimum point of falling LRATC represents 'ideal output'. Further, this product differentiation leads to price-competition between the firms and thereby, when there is free entry of firms and price-competition between the firms, there is no excess capacity for the firms under monopolistic competition, as the output produced (OQ_3) corresponding to the left of the minimum point of the falling LRATC is regarded as 'ideal output'.

2. Concept of Excess Capacity with Entry of New Firms and with Non-price Competition

According to Chamberlin, the concept of excess capacity will arise in the absence of price-competition between the firms in the long run, despite free entry of new firms into the group. Chamberlin opined, price-competition will not arise under the monopolistic competition because of the following reasons:

☆ Firms may aim for normal profits instead of aiming for super-normal profits.

☆ The business firms instead of going for price competition may follow the policy of 'to live and let live' and thereby, follow stable prices.

☆ The business firms may have formal or tacit agreements regarding the maintenance of stable price.

☆ Since, price of the product is the indicator of the quality of the product, the firms will not go for price cut, as low price of the product may give the impression on the part of the buyers that the product is of inferior quality. Such perception among the buyers will affect the profits of the business firm, if the firms resort to price cuts.

☆ Every business firm will maintain business ethics and decide not to compete unnecessarily on the price cuts, but try to compete on the degree of product differentiation and on non-price competition basis.

The concept of excess capacity with the entry of new firms and absence of price competition between the firms is explained through the Figure 23.12. In the Figure 23.12,

☆ dd and DD are the perceived and proportional demand curves respectively of the representative firm in the short run and based on the two essential conditions *viz.*, SRMC=perceived MRC and perceived demand curve=proportional demand curve, we can arrive at representative firm's equilibrium and since the price received by the representative firm is higher than SRATC, it will earn super normal profits. (Perceived MRC, SRATC and SRMC are not shown to simplify the graphical illustration).

☆ As super normal profits earned by the firms in short run are lucrative, new firms will enter the group.

☆ With the entry of new firms into the group, the supply of the product into the market increases, thereby, demand for the product falls in the market. So, the proportionate share of each firm in the total market supplies decreases. As a result, the proportional demand curve DD shifts towards left as D_1D_1 and thereby, new equilibrium is established at P_1, where d_1d_1=D_1D_1=LRATC. So, at the new equilibrium, each firm sells OQ_1 quantity of output at Q_1P_1 price.

☆ So, with the entry of new firms and due to absence of price-competition, the firms will enjoy normal profits only with reference to point P_1, thereby, each firm will supply OQ_1 quantity of output and sell the same at Q_1P_1 price. This point P_1 is the stable equilibrium in the absence of price competition between the firms in the group.

☆ So, in the absence of price-competition, there is excess capacity for the firms under monopolistic competition in the long run.

So, the above two concepts indicate that, OQ_3 is the ideal output for each firm under monopolistic competition in the presence of price competition, but each firm will produce only OQ_1 output in the absence of price competition, thereby, there is excess capacity for the firms under monopolistic competition in the absence of price competition.

23.6. Non-Price Competition

In the above discussion, we analyzed the price-output policy of each firm under short run and long run production programmes and it represents price-competition under monopolistic competition. This is because, we have considered the quality of the product constant in the production programmes. But, under monopolistic competition, the firms not only compete on price-basis, but may be on several criteria like product features, quality, advertisement etc., and the competition between the firms on the basis of these criteria constitute non-price competition. It also refers to the efforts made by the firm to increase the demand for its product, other than cutting the price of the product. The non-price competition makes the firm's ARC more inelastic (or) less elastic, as it increases the cost of production. The examples under non-price competition includes better service, product guarantees, free home delivery, more attractive packaging, product differentiation, selling costs (advertising costs) etc.

This feature of non price competition distinguishes the monopolistic competition from perfect competition and monopoly. The concept of selling costs will not arise under perfect competition, as all the firms deal with homogenous product and a single firm has to sell any quantity of the product at the ruling market price. Even if competition arises between the firms to sell the product, it leads to reduction in price until the new equlibrium price is reached. If a firm under perfect competition goes for advertising the product, the buyers will get stimulated and purchase the product from any of the sellers, as the product is homogenous, thereby, the firm which advertises about the product may lose in the form of advertising costs. However, if all the firms in an industry commonly advertise about the product, they all may get mutually benefitted through increased sales of the product. Such advertising under perfect competition is called as Promotional advertising or Informative advertising and not Competitive advertising. For example, a Dairy cooperative, Cashew board, Coffee board, Tea board etc., advertises about their products and this advertising is called Promotional advertising or Informative advertising. Thus, under perfect competition, if there is any role for advertising, it would be to get consumers aware of the product rather than to persuade them to change their preferences. Within this context, the role which is expected to be played by advertising is an informative one, but not persuasive, and it is just for those firms which are newcomers to the market.

Under monopoly, as a single firm itself represents the industry and the product produced by that firm has no close substitutes, the concept of non-price competition will not arise. But, the firm will undertake some advertisement, so as to make the customers aware about the product and its features. Thus, the advertisement costs constitute only insignificant proportion of total costs incurred by the firm on the product. So, the advertising even under monopoly also refers to Promotional advertising or Informative advertising and not Competitive advertising. For example, Monsanto firm advertises about Bt cotton seed and this refers to Promotional or Informative advertising. We know, pure monopoly and perfect competition are opposite in the sense that, the firm in a perfectly competitive market has so many competitors, whereas the firm under pure monopoly market structure has no competitors at all. So, there is complete absence of competition for a firm under pure monopoly. It should be acknowledged that, the firm operating under the pure monopoly market structure is not completely insulated from the actions taken by other industries in the economy. It is assumed that, all commodities are rivals for the consumer's favour. For example, if a firm is a monopolist in the supply of steel in a certain market, it would still face considerable competition from other firms producing aluminum or plastics. So, the product of the monopoly firm also faces a competition from other products produced by other industries. So, to influence the consumer tastes, a monopolist often spends considerable expenditures on advertising in order to influence the demand curve and shift it to the right.

In oligopoly market structure, there exists small number of firms and a great deal of interdependence among them. So, it is assumed that, each firm designs its policies with the consideration of their effect on its competitors. Competition in the oligopolistic market structure is non-price oriented. The firm tends to use advertising and differentiation in many product characteristics rather than price, as a competitive weapon because, it seems to view price cutting as a dangerous tactic. The firm as a less risky way to keep customers away from competitors uses advertising and product differentiation. Firms use advertising to differentiate their products from those of their competitors. In this way customers may be oriented to become brand loyal and so to stick to it, even though the products of all firms in the industry are much the same. The example, which can be effectively cited here, is the cigarette case, where various brands of cigarettes are quite similar, although not identical. The cigarettes industry spends a large amount on advertising to build a particular brand image in the consumers' minds. Within this context, advertising can promote a psychological differentiation. Similarly, advertising can perceptually differentiate life insurance services delivered by different insurance companies.

So, considering the mutual interdependence among the firms regarding their investment, output and pricing policies, they rely heavily on the advertising as a main weapon (even when compared to monopolistic competition) to make the demand curve decisive or deterministic. However, the advertising expenditures of some oligopolistic industries have been expanded beyond the levels that are socially optimal. Sometimes, advertising expenditures only have the effect of increasing the costs of the entire industry, since one firm's advertising causes other firms to increase their advertising.

However, under monopolistic competition, the non-price competition or selling costs assume crucial role in boosting the sales of the product. As discussed earlier, Chamberlin opined that, firms under monopolistic competition often resort for non-price competition only and price-competition will not arise much. But, compared to differentiated oligopoly, the non-price competition is low in monopolistic competition because, the concept of mutual interdependence of firms will not arise under monopolistic competition and each firm will not closely watch the output-price policies of rival firms, as they are relatively larger in number. But, still the non-price competition plays a crucial role in determining the price-output policies of the firms operating under monopolistic competition.

So, both under monopolistic competition and differentiated oligopoly competition, the advertising about the product serves as an important competitive weapon that increases the sales of the product of one firm at the cost of the rivalry firms. Hence, the advertising under monopolistic competition and differentiated oligopoly competition refers to Competitive advertising rather than simply promotional advertising. For example, various pesticide firms, fertilizer firms etc., advertises about their products and such advertising refers to competitive advertising.

It is essential to note that, of these two types of advertising *viz.*, Promotional advertising and Competitive advertising, the former is socially advantageous, as it educates the customers about the technical aspects of the product. On the other hand, competitive advertising though leads to increased sales of the product from the firms' side, it leads to wastage of resources due to cut-throat competition among the firms under oligopoly and monopolistic competitions.

The above discussion clearly infers that, monopolistic competition is significantly influenced by non-price competition. According to Cairncross, non price competition is, otherwise, called as *'Service competition'*. In this chapter, non-price competition is discussed with reference to selling costs (advertising costs) and product differentiation.

23.6.1. Selling Costs

In the words of Chamberlin, *'selling costs are defined as costs incurred in order to alter the position or shape of the demand curve for a product.'* The selling costs include expenditures made on salesmanship, advertisement, propaganda, window displays, demonstration of products, free gifts of samples, margins granted to dealers to increase the sales of the product etc. But, according to Boulding, selling costs are not to be treated as an addition to the total costs incurred on the production of the product, but as a reduction from revenue. Similarly, Bilas opined that, selling costs are to be considered on the revenue side as 'negative revenue' because, they are incurred to increase TR and total profits by increasing the volume of the sales. Before analyzing the equilibrium of a firm under monopolistic competition with selling costs, it is essential to differentiate between selling costs and production costs (Table 23.1).

Table 23.1: Differences between selling costs and production costs under monopolistic competition.

Selling Costs	*Production Costs*
These are the costs incurred by the firms to increase the sales of the product in the market.	These are the costs incurred by the firm to produce the product.
Selling costs generally include costs or expenditure on advertising, salesmanship, expenses for window displays, margins given to dealers to promote sales etc.	Production costs generally include costs or expenditure on raw materials, human labour, machinery labour, power, fuel etc.
The costs incurred by the firm which determine or alter the demand curve for the product directly are called selling costs.	The costs incurred by the firm which do not determine the demand curve for the product are called production costs.
Selling costs determine the demand for the product in the market.	Production costs determine the supply of the product into the market.
Selling costs are meant to adapt demand to the product *i.e.*, they manipulate the demand.	Production costs are meant to adapt the product to demand *i.e.*, they manipulate the product or level of output.
Selling costs creates and manipulates the shift in demand for the product.	Production costs are meant for creation of utilities, which would satisfy the latent demand of the buyers.
With the increase in selling costs, the demand for the product increases in the market, thereby, price of the product will rise in the market. So, increase in selling costs leads to increase in price of the product.	With increase in production costs, the volume of output increases, supply of output into the market increases and thereby, in the context of given level of demand, the market price tends to fall. So, increase in production costs leads to fall in price of the product in the market.

However, the distinction between selling costs and production costs may not be clear cut always. The following are the examples:

- ☆ An attractive packing material may increase the sales of the product. So, the cost of packing can be taken as production cost and even it can be considered as selling costs because of attractive packing material, the sales are increased. But, it cannot be included in both the costs.
- ☆ Transportation cost cannot always be considered as selling costs because, transportation costs creates place utility for the product, thereby, logically, it has to be considered as production cost and not as selling cost.

Considering the above two examples, we can infer that, it is very difficult to clearly differentiate between selling costs and production costs. But, Chamberlin has given a distinct place for selling costs under monopolistic competition and he confirmed selling costs to only advertising costs. However, in determining the price of the product, it should cover both production cum selling costs of production, so as to continue the business.

23.6.1.1. Effect of Selling Costs (Advertising Costs) on the Demand for the Product

As explained earlier, selling costs (advertising costs) are meant to adapt the demand to the product. That means, selling costs manipulates the shifts in demand curve in favour of the product sales. The main objective of advertising the product is to strengthen the brand loyalty of the existing customers and to win new customers of other brands. So, advertisement alters E_C for the product and thereby, helps the business firm to expand the sales of its product. In general, a business firm will go for repeated advertising about the product because, an advertisement seen once will have a negligible impact on the customer towards the product. The effect of advertisement on the demand for the product of a firm is explained through the Figure 23.13.

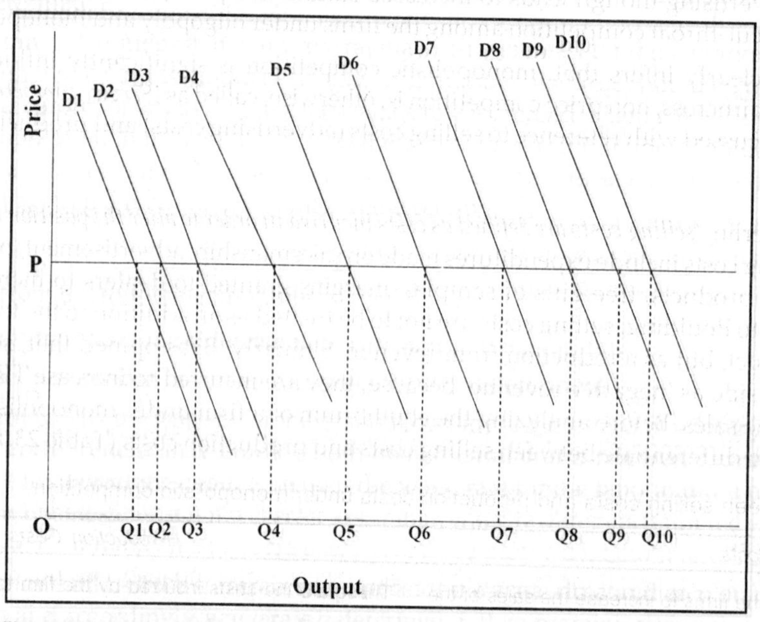

Figure 23.13: Effect of selling costs (Advertising costs) on demand for the product under monopolistic competition.

- ☆ D_1 is the demand curve for the product, when the firm do not resort for any advertisement expenditure about the product. So, when there is no advertisement expenditure, the firm sells OQ_1 quantity of output at OP price.
- ☆ When the firm goes for advertising the product, it implies, the firm wishes to increase the sales of the product at a given price. So, D_2 is the new demand curve when the firm resort for advertising the product and thereby, at the same OP price, the firm sells OQ_2 quantity of the product. So, due to advertising the product, the quantum of sales of the product has increased from OQ_1 to OQ_2.
- ☆ If the firm still increases the advertising expenditure in the same proportion, the new demand curves are D_3 to D_{10} and thereby, the quantum of sales of product has increased up to OQ_{10}.
- ☆ A close examination of the Figure 23.13 reveals that, between OQ_1 to OQ_4 levels of output sold in the market, the advertising expenditure yielded increasing returns in terms of increasing sales of output *i.e.*, $Q_1Q_2<Q_2Q_3<Q_3Q_4$. If the advertising expenditure is further increased, it leads to constant returns in terms of increasing sales of output *i.e.*, $Q_3Q_4=Q_4Q_5=Q_5Q_6$. But, beyond OQ_7 quantum of sales of the product, the increase in advertising expenditure yielded diminishing returns in terms of increasing sales of the output *i.e.*, $Q_6Q_7>Q_7Q_8>Q_8Q_9>Q_9Q_{10}$.

- ☆ Thus, increasing advertising expenditure in the same proportion will not increase the output sales in the same proportion. That means, at the beginning levels of advertising expenditure, increasing returns will be experienced from the successive additional units of advertising expenditure and after a certain level of sales of output, diminishing returns will be experienced from the successive additional units of advertising expenditure. The following are the important reasons for the increasing returns and diminishing returns with reference to quantum of sales of output from the additional advertising expenditure.

Reasons for Increasing Returns from Additional Advertising Expenditure

- ☆ The initial levels of advertising expenditure stimulates the customers and makes them aware about the product and its features, thereby, the sales of the product will increase.
- ☆ The customers, who prefer the product of other firms draw the comparison with the advertising product and if it is found superior in terms of features and quality, they shift towards this product.

Reasons for Diminishing Returns from Additional Advertising Expenditure

- ☆ If the firm goes on increasing the advertising expenditure, diminishing returns are likely to set in for the additional advertising expenditure because, by that time, the potential customers have already purchased and used the product
- ☆ The product may have only replacement sales and new sales may decline in the market. To induce the old customers to purchase the same product again, it requires huge amount of selling costs on the part of the firm.
- ☆ Other firms might have released the same product with additional features and offering the same at low cost than this product.
- ☆ The MU derived by the customer from the use of this product will decrease, as the customer goes on using the same product, but the MU of other firms' products increases.

23.6.1.2. Average Selling Cost (ASC) and Marginal Selling Cost (MSC) Curves

ASC refers to the selling cost per unit of output or product. It is synonymous to the ATC curve of the firm and it is also an 'U' shaped curve (Panel A of the Figure 23.14). As discussed above, if selling costs (advertising costs) are continuously increased, it will lead to increasing returns and diminishing returns with reference to quantum of sales of the output or product, thereby, it gives a 'U' shape to the ASC curve. To be accurate, like production costs, the selling costs are also subjected to the three sequential stages of returns *i.e.*, increasing returns, constant returns and diminishing returns. At the initial (low) levels of sales programme, the ASC is high. But, with gradual increase in sales, when the selling costs experience increasing returns in terms of quantum of sales, the ASC falls rapidly. When the selling costs experience constant returns in terms of quantum of sales, the ASC curve remains flat. When the selling costs experience diminishing returns in terms of quantum of sales, the ASC curve starts rising upward, thus giving a 'U' shape to the ASC curve. Thus, the downward sloping portion of the ASC curve implies, increase in selling costs leads to a more than proportionate increase in demand for the product. The flat portion of the ASC curve implies, demand tends to increase for the product in the same proportion to the increase in total selling costs. The upward portion of the ASC curve implies, increase in selling costs leads to a less than proportionate increase in demand for the product.

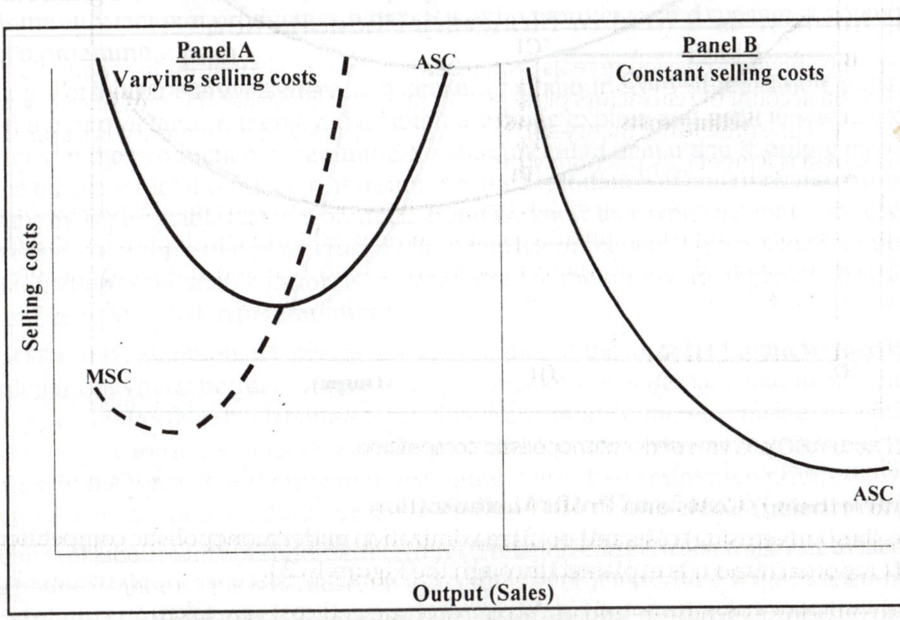

Figure 23.14: ASC and MSC curves of a firm under monopolistic competition.

In Panel A, the MSC curve is also shown. It refers to the additional quantity of output sold by the firm due to additional unit of selling cost. It is, otherwise, defined as the addition made to total selling outlay, for expanding the sale of one extra unit of output. It is also an 'U' shaped curve and its behaviour is similar to the SRMC curve discussed in cost analysis (Figure 15.5).

In the above discussion, we studied the behaviour of ASC curve when a firm goes on increasing the selling costs (advertising costs). But, in some cases, some firms will keep the selling costs fixed *i.e.*, they allocate a fixed amount of fund meant for selling costs towards boosting the sales of the product. In such a case, the ASC curve will slope continuously downward from left to right (Panel B of the Figure 23.14) and here, the behaviour of ASC curve is similar to AFC curve of cost analysis discussed earlier (Figure 15.5). So, when selling costs are considered as a fixed amount, the ASC curve will continuously fall with increase in sales of the output, thereby, it will attain the shape of a rectangular hyperbola.

23.6.1.3. Combined Cost (CC)

Chamberlin introduced the concept of CC, as a sum of production costs and selling costs and he considered this CC as a vital one to determine the price-output policy of a monopolistic firm. So, CC = Production cost + Selling cost. This CC is also called as TC.

(a) Production Costs

Production costs are the costs incurred to adjust supply to demand and these costs are not related to the sales promotion aspects of the business firm. As discussed in the earlier Chapter 18, normal profits are included in the production costs.

(b) Selling Costs

Selling costs are the costs incurred to adjust demand to supply and these costs are related to the sales promotion aspects of the business firm. That part of normal profit, which is involved in the sales promotion, is also selling cost.

The concept of CC is explained through Figure 23.15. Both Average Combined Cost (ACC) and Average Production Cost (APC) curves are 'U' shaped curves and ACC curve is above APC curve, as ACC includes the component of ASC. That means ACC=APC+ASC. So, the vertical distance between the ACC and APC curves represent ASC at each level of output. At OQ_1 level of output transacted in the market, Q_1C_1 is the ACC, Q_1P_1 is the APC and thereby, ASC = $Q_1C_1 - Q_1P_1 = C_1P_1$.

It is clear from the Figure 23.15 that, the area between the ACC and APC indicates total selling costs and the area below the APC measures the total production cost. So, with reference to OQ_1 level of output, the total selling cost is ABC_1P_1 and total production cost is OAP_1Q_1. So, the total combined cost is OBC_1Q_1.

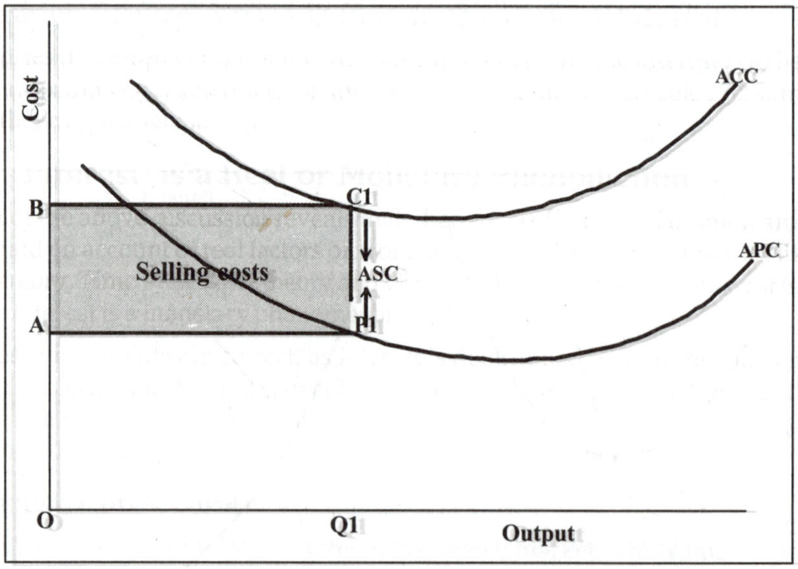

Figure 23.15: APC, ACC and ASC of a firm under monopolistic competition.

23.6.1.4. Selling (Advertising) Costs and Profit Maximization

The influence of selling (advertising) costs and profit maximization under monopolistic competition can be studied through the TR and TC approach and it is explained through the Figure 16.23.

- ☆ On X-axis, advertising expenditure and on Y-axis, revenue and costs are taken.
- ☆ 45^0 line represent the advertising costs (AC) incurred by the firm.

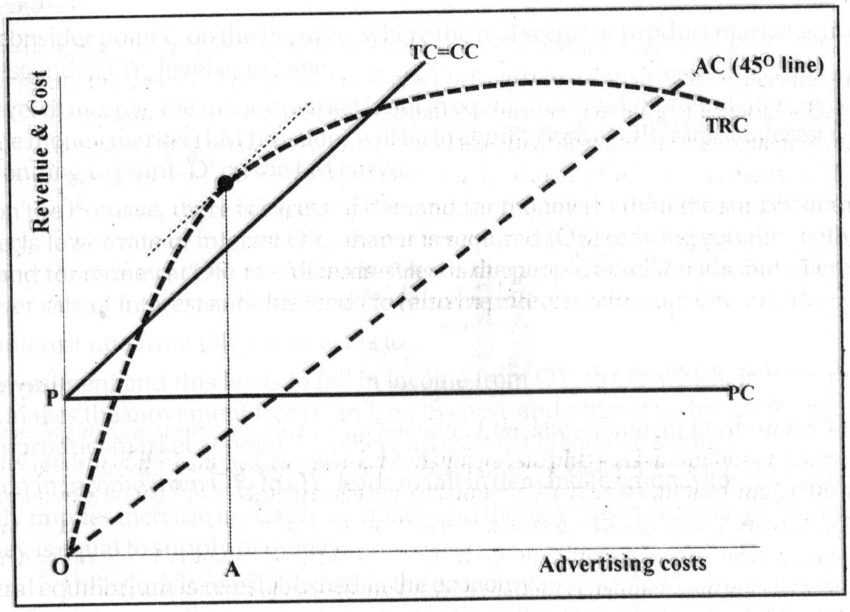

Figure 23.16: Advertising costs and Profit maximization of a firm under monopolistic competition.

☆ Production costs (PC) are shown as horizontal line parallel to X-axis, as it is independent of advertising outlay.

☆ The sum of production costs and advertising costs represent CC or TC.

☆ As per TR and TC approach, the profits are maximum, when the gap between TRC and TC (=CC) curve is maximum and this is with reference to OA level of advertising costs.

☆ This infers that, if the firm spends OA level of advertising costs, it ensures maximum profits to the firm under monopolistic competition.

23.6.1.5. Gap between ACC and APC

If the firm allocates a fixed level of selling costs, the ASC will continuously slope downward from left to right (as discussed under Panel B of the Figure 23.14) as rectangular hyperbola and thereby, the gap between ACC and APC will decline at higher quantum of sales of output (Panel A of the Figure 23.17). As discussed earlier, if the firm goes on increasing the selling costs, the selling costs will be subjected to increasing returns, constant returns and diminishing returns with reference to the output sold in the market. In such case, the gap between ACC and APC curves will gradually increase with increase in quantum of sales of the product (Panel B of the Figure 23.17). On the other hand, if

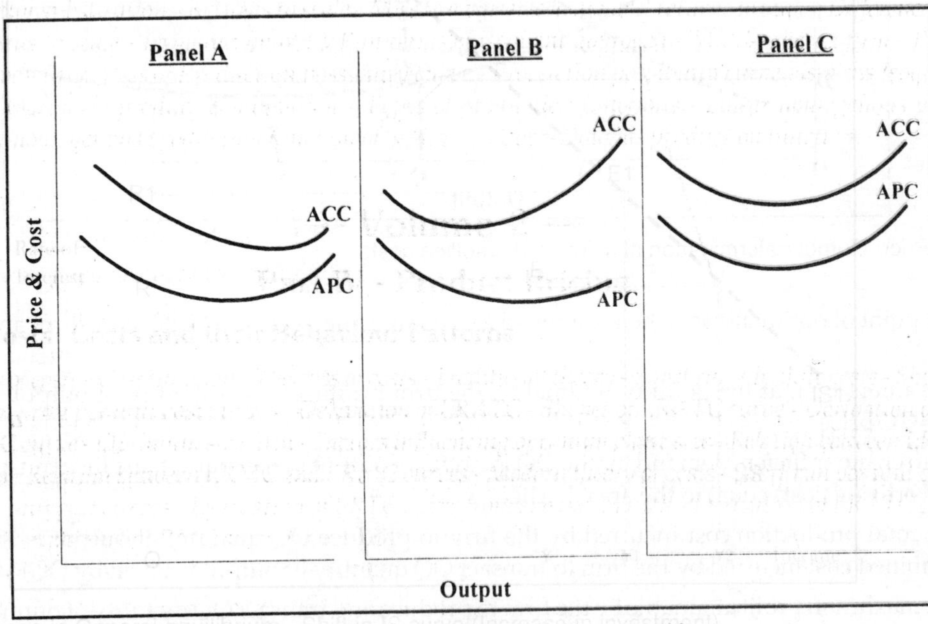

Figure 23.17: Gap between ACC and APC of a firm under monopolistic competition.

the selling costs increase in the same proportion as the increase in the quantum of sales of the output, the ASC will remain constant at different quantum of sales of output and thereby, the gap between ACC and APC will remain constant or same at different levels of sales of output. (Panel C of the Figure 23.17).

23.6.1.6. Price-Output Determination under Selling Costs

In the earlier pages, we discussed that, selling (advertising costs) increases the demand for the product. But, besides increasing demand, the selling costs also add to the total cost or combined cost of production of the product. In such a case, it is essential to discuss how much advertising expenditure a firm or a group should incur to ensure optimum level of price and output. In the following cases, let us discuss the price-output determination of a firm under selling costs and price-output determination of a group under selling costs. Here, it is important that, in both the cases, the price of the product, the amount of selling costs and design of the product are kept constant.

1. Price-Output Determination of a Firm Under Selling Costs

Considering the importance of selling costs in boosting the sales of the product and thereby, profits under monopolistic competition, the profit maximization of a firm is influenced by three important variables *viz.*, output, price and selling costs. Thus, to attain profit maximization, a firm has to allocate relevant selling costs (advertising costs), so as to create demand for the product. So, considering the selling costs incurred in the production programme, a firm can maximize the profits, measured as follows:

Net returns (profits) = (Output x Price) – (Production cost + Selling (Advertising) cost)

To maximize the profits, the firm has to follow the MR and MC approach, as explained through the Figure 23.18.

- ☆ APC curve and ACC curve indicate the APC and ACC of the firm respectively.
- ☆ The gap between ACC and APC has widened at higher quantum of sales of output, indicating that selling costs (advertising costs) are subjected to increasing, constant and diminishing returns with reference to increase in quantum of sales of output, thereby, ASC increases with increase in quantum of sales of output.
- ☆ Since, price of the product is assumed constant, the price line is drawn parallel to X-axis and thereby, price=ARC=MRC=OC.

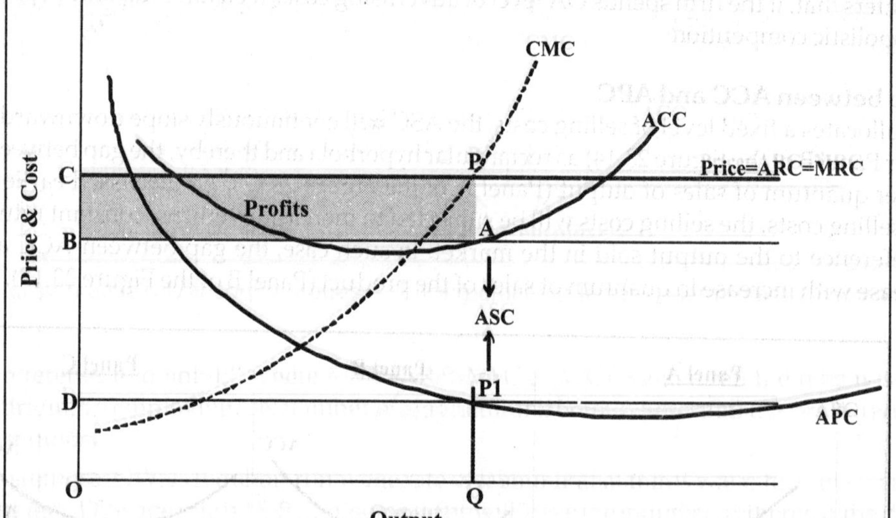

Figure 23.18: Price-Output determination of a firm under selling costs.

- ☆ The point of optimality is, when combined MC (CMC) intersects the MRC of the firm and it is with reference to 'P'.
- ☆ This infers that, at the point of optimality, the firm produces OQ quantity of output and sell the same at QP=OC price.
- ☆ So, to create sufficient demand for the OQ quantity of output in the market, the firm must incur the selling cost (advertising cost) equal to the area P_1ABD.
- ☆ The total production cost incurred by the firm to produce OQ quantity of output is given by OQP_1D and combined cost incurred by the firm to transact OQ quantity of output is given by $OQP_1D + P_1ABD = OQAB$.
- ☆ The maximum profit attainable by the firm to produce and transact OQ quantity of output is equal to the area PABC, as price of the product is above the ACC by PA portion.

- ☆ The firm should not produce and sell the output more than OQ, as beyond OQ, the CMC rises above MRC and thereby, the firm incurs losses.
- ☆ Similarly, the firm should not produce and sell the output less than OQ, as below OQ, the price=MRC is above the CMC and hence, it is better for the firm to increase the output up to OQ.
- ☆ So, OQ is the equilibrium output the firm must produce and advertise and at the price QP=OC, the firm maximizes the profits.

2. Price-Output Determination of a Group Under Selling Costs

Let us assume, all the firms in a group have identical cost-structures and sell the product at the same price. So, in the Figure 23.19, we explained the concept of price-output determination of a group under selling costs (advertising costs) considering a representative firm, as it represents all the firms in the group.

- ☆ Since, price of the product is assumed to be constant, the price line is drawn parallel to X-axis, thereby, price=ARC=MRC of a representative firm.
- ☆ Let us suppose, the representative firm sells OQ quantity of output at OP level of price, when the firm incurs no advertising costs on the product. (Therefore, the total quantity of output sold by all the firms (n) in the group amounts to (OQxn), when no firm goes for advertising the product).
- ☆ Let us assume, the representative firm alone decides to go for advertising the product and other firms will not follow the suit. So, the firm moves from APC curve to ACC curve. The point of optimality is given by when CMC curve intersects the MRC curve and it is at point E_1. So, the representative firm alone sells OQ_1 output at OP price and earns PE_1A_1X super normal profits.
- ☆ Noticing the super normal profits earned by the representative firm, other firms now compete with the representative firm in advertising the product otherwise, their customers will move towards representative firm.
- ☆ So, due to rise in advertising costs, the ACC and CMC curves will shift upward as ACC_1 and CMC_1 and the point of optimality is when CMC_1 intersects the MRC *i.e.*, at point E_2. So, now the representative firm alone will sell only OQ_2 quantity of output at OP price and earn only normal profits, as price line is tangent to ACC_1 at E_2. So, the total quantity of output sold by all the firms (n) in the group amounts to OQ_2xn, when all the firms in the group goes for advertising the product.
- ☆ It is interesting that, when all the firms goes for advertising the product, they could sell (OQ_2xn) quantity of output *i.e.*, more than (OQxn) quantity of output when no firm goes for advertising the product, but each firm earns only normal profits.
- ☆ Had, only representative firm goes for advertising the product and not the rivalry firms, it could sell more quantity of output OQ_1 than OQ quantity of each of other firms and it earns super-normal profits PE_1A_1X. Further, the equilibrium output OQ_2 is less than what the representative firm would have sold (OQ_1), had rival firms not followed initial advertising.

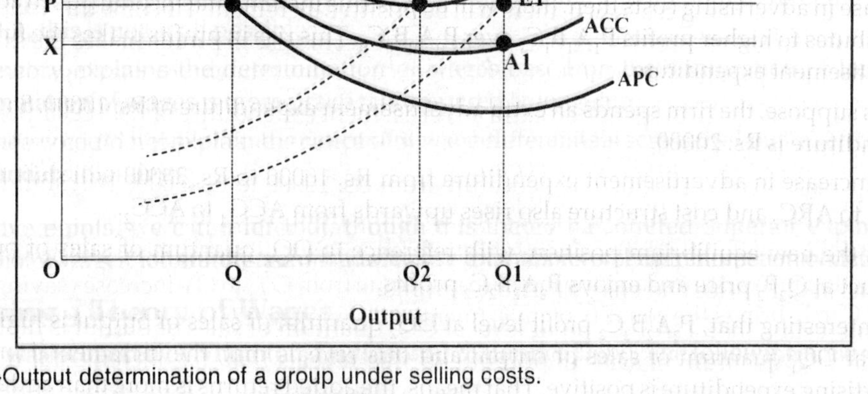

Figure 23.19: Price-Output determination of a group under selling costs.

3. Price-Output Determination of a Firm Under Variable Selling Costs and Price of the Product

In the above analysis, we assumed price of the product, amount of selling costs and design of the product as constant. We shall now explain the price-output determination of a firm considering variable selling costs and price of the product and it is shown through the Figure 23.20.

- ✰ Let us assume, the firm has allocated three levels of selling costs *viz.*, Rs. 10000, Rs. 20000 and Rs. 30000.
- ✰ The increasing levels of selling costs not only increase the ARCs (demand curves) upwards or towards right side, but also raise the cost structure. So, let us consider, the three levels of ARCs due to rise in selling costs as ARC_1, ARC_2 and ARC_3 and the rise in cost structure is indicated by three levels of ACC curves *viz.*, ACC_1, ACC_2 and ACC_3. So, these ACC curves include both APC and ASC.
- ✰ MRCs and MCs are not shown in the figure, only to simplify the graphical presentation.
- ✰ Let us assume ARC_0 is the demand curve of the firm and APC curve is the APC of the firm, when the firm do not incur any advertising cost.
- ✰ Considering MR and MC approach, it is clear that, at OQ_0 quantum of sales of output, the firm sells the product at Q_0P_0 price and enjoys $P_0A_0B_0C_0$ profits.

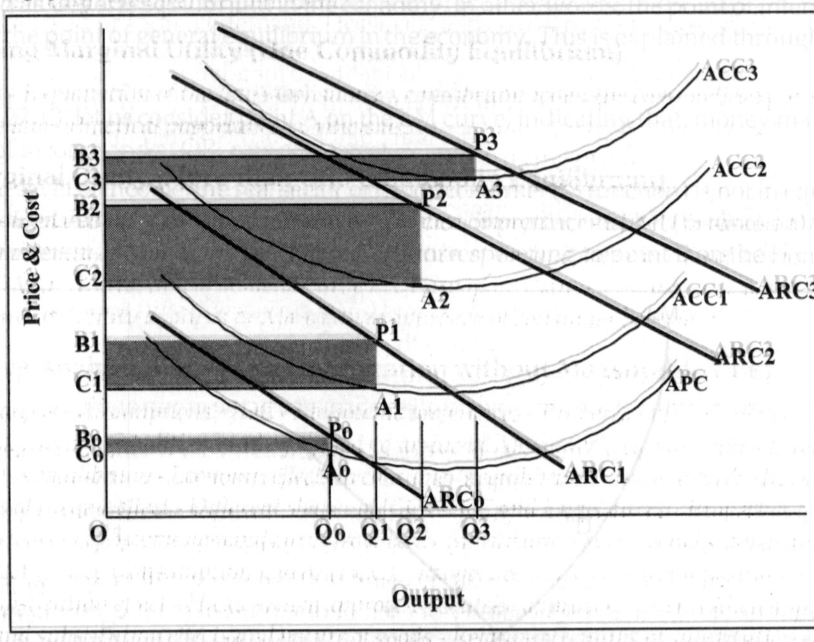

Figure 23.20: Price-Output determination of a firm under variable selling costs and price of the product.

- ✰ So, when there is no selling costs, the firm enjoys super normal profits, $P_0A_0B_0C_0$ by selling OQ_O output at Q_OP_O price. In order to increase the profits, the firm goes for advertising the product and thereby, ACC curve will rise upward (ACC_1). This advertising costs (Rs. 10000) stimulate the customers and thereby, the demand for the product increases (ARC_1) and hence, quantum of sales will be increased upto OQ_1.
- ✰ At OQ_1 quantum of sales of output, the firm sells the product at Q_1P_1 price and enjoys $P_1A_1B_1C_1$ profits.
- ✰ It infers that, due to advertising costs, the profits enjoyed by the firm $P_1A_1B_1C_1$ is more than $P_0A_0B_0C_0$.
- ✰ It is important to note that, the increase in profits from $P_0A_0B_0C_0$ to $P_1A_1B_1C_1$ will occur only when the incremental income is positive. That means, as long as the added returns is greater than added costs due to increase in advertising costs then, there will be positive incremental income due to advertising costs and this contributes to higher profits $P_1A_1B_1C_1$ over $P_0A_0B_0C_0$. This rise in profits makes the firms to undertake further advertisement expenditure.
- ✰ Let us suppose, the firm spends an extra advertisement expenditure of Rs. 10000. So, the total advertisement expenditure is Rs. 20000.
- ✰ This increase in advertisement expenditure from Rs. 10000 to Rs. 20000 will shift the ARC upwards from ARC_1 to ARC_2 and cost structure also rises upwards from ACC_1 to ACC_2.
- ✰ So, at the new equilibrium position, with reference to OQ_2 quantum of sales of output, the firm sells the product at Q_2P_2 price and enjoys $P_2A_2B_2C_2$ profits.
- ✰ It is interesting that, $P_2A_2B_2C_2$ profit level at OQ_2 quantum of sales of output is higher than $P_1A_1B_1C_1$ profit level at OQ_1 quantum of sales of output and this reveals that, the incremental income derived from the advertising expenditure is positive. That means, the added returns is more than added costs of the additional advertisement expenditure. This infers that, the additional advertisement is showing increasing MRs.

- ☆ In this way, the firm goes on increasing the advertising expenditure, so long as the added returns or MRs derived from the additional advertising expenditure is greater than additional cost of advertising expenditure. The firm will gain maximum profits from the advertising expenditure, when added returns or MRs derived from the additional advertising expenditure is equal to the additional cost or MC of advertising expenditure.
- ☆ If the firm still increases the advertising expenditure by Rs. 10000 (total advertising is raised now to Rs. 30000), the ARC will shift upward from ARC_2 to ARC_3 and cost structure also rises from ACC_2 to ACC_3.
- ☆ So, at the new equilibrium position, with reference to OQ_3 quantum of sales of output, the firm sells the product at Q_3P_3 price and enjoys $P_3A_3B_3C_3$ profits.
- ☆ It is interesting that, $P_3A_3B_3C_3$ profits at OQ_3 quantum of sales of output is lower than $P_2A_2B_2C_2$ profits at OQ_2 quantum of sales of output and this reveals that, the added returns is less than added costs of the additional advertising expenditure. This infers that, the additional advertising expenditure is showing diminishing MRs.
- ☆ So, it is better for the firm not to increase the additional advertising expenditure from Rs. 20000 to Rs. 30000, even though the firm is deriving profits from Rs. 30000 advertising expenditure.
- ☆ So, the firm will be in equilibrium by incurring Rs. 20000 on advertising expenditure at which its profits are largest *i.e.*, $P_2A_2B_2C_2$, where the firm sells OQ_2 quantity of output and sells the product at Q_2P_2 price. Thus, Rs. 20000 is considered as the optimal advertising expenditure for the firm. So, the firm selects OQ_2 quantity of output, Q_2P_2 price for the product and Rs. 20000 as optimal advertising expenditure, as these yields maximum-maximorum level of profits.

23.6.2. Product Differentiation

Product differentiation is one important feature that influences the profit maximization of a firm under monopolistic competition. A firm can bring product differentiation in different ways like changing the features of the product, offering various kinds of services to the customers etc. This product differentiation will increase the cost of production and thereby, it leads to shifts in cost and revenue curves of the firm. As shown in the Figure 23.21,

- ☆ D_1D_1 is the ARC and $LRATC_1$ is the LRATC curve of the firm when there is no product differentiation. (MRC and LRMC curves are not shown to simplify the graphical illustration).
- ☆ As per the MR=MC approach, the firm's equilibrium will be at OQ_1 level of output and the firm sells the product at Q_1P_1 or OC price, thereby enjoys P_1A_1BC super normal profits.
- ☆ Assume the firm decided to go for product differentiation, thereby, its demand curve and LRATC will shift towards right. The new demand curve D_2D_2 is less elastic compared to D_1D_1 curve.
- ☆ Given the fixed price OC, the firm due to product differentiation earns more super normal profits P_2A_2CD compared to P_1A_1BC profits in the absence of product differentiation.
- ☆ It is interesting to note that, even though the cost structure is increased due to product differentiation (LRATC curve shifted towards right side), the firm still enjoys higher super normal profits ($P_2A_2CD>P_1A_1BC$), as the firms sells more quantity of output ($OQ_2>OQ_1$) due to increase in demand for the product.

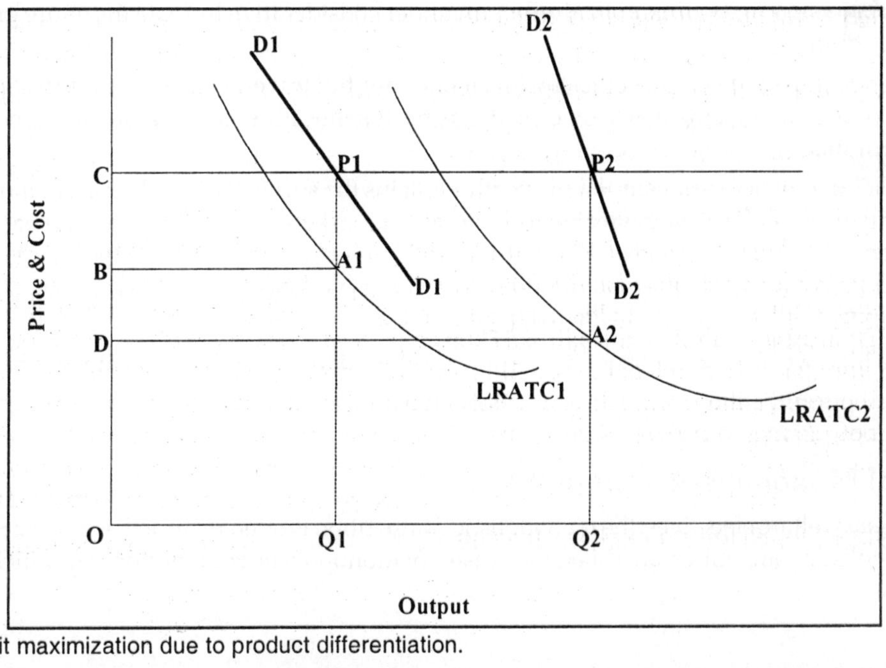

Figure 23.21: Profit maximization due to product differentiation.

So, firms under monopolistic competition attempt to juggle three factors in efforts to maximize profits (typically by trial and error):

- ☆ price based competition
- ☆ non-price competition based upon advertising (attempts to influence the demand curve for the product)
- ☆ non-price competition based upon product differentiation (developing characteristics to make product stand out).

The above discussion infers that, under monopolistic competition, the firms can enjoy more profits through price cutting or through increasing selling costs (advertising costs). But, among these two options, it is felt better by the firms to go for advertising about the product rather than price cutting because of the following reasons:

- ☆ If advertising costs have not yielded fruitful results, the firms can cut back the advertisement costs. But, once the price of the product is reduced and if it has not yielded desired results, it cannot be increased suddenly in the market.
- ☆ Frequent price variations by a firm lead to rivalry among the firms.
- ☆ There is no guarantee for the firms to increase the profits by price cutting because, other firms may go for still large price reductions. This is not the case with advertising costs because, advertising about the product by the firm tries to stimulate the customers to purchase the product and promote brand loyalty towards the product and these desirable changes are permanent on the part of the customers.

However, from the customers point of view, price cutting is beneficial rather than the advertising costs, because of the following reasons:

- ☆ The customers will get differentiated products at low price.
- ☆ Advertisement is an extra cost for the firm and the same is reflected as higher price of the product. This is more burdensome for the customers.

23.7. Merits of Monopolistic Competition

Monopolistic competition can bring the following advantages:

- ☆ As an economic model of competition, monopolistic competition is more realistic than several other models of market structure, as many familiar and common place markets have many of the characteristics of this model.
- ☆ It incorporates the facts of product differentiation and selling costs.
- ☆ It can be easily used for the analysis of duopoly and oligopoly.
- ☆ There are no significant barriers to entry therefore, markets are relatively contestable.
- ☆ Product differentiation creates diversity, choice and utility.
- ☆ Under monopolistic competition it is possible to see that even when each individual firm produces under conditions of increasing returns, not only the firm under consideration but also the entire group of firms can be in equilibrium.
- ☆ Monopolistic competition is more efficient than monopoly, but less efficient than perfect competition both in terms of allocation and production. However, it may be dynamically efficient, as it offers innovative processes and new products.
- ☆ The existence of monopolistic competition partly explains the survival of small firms in modern economies. The majority of small firms in the real world operate in markets that could be said to be monopolistically competitive.
- ☆ Monopolistic competition is able to show that, even when each individual firm is producing under increasing returns, it still earns only normal profit in the long run.
- ☆ Prices are a bit lower in monopolistic competition compared to pure monopoly due to substitution effect.
- ☆ In monopolistic competition we are able to consider the interaction between several interdependent variables on the basis of which a firm takes its decisions.

23.8. Wastes of Monopolistic Competition

In the light of social welfare objectives, the monopolistic competition is often regarded as a 'wasteful competition'. The defects of monopolistic competition are called the Wastes of monopolistic competition. They include,

1. Excess Capacity

Even in the long run production programme, the firm under monopolistic competition do not expand the output up to the minimum point of LRATC. That means, the firm will not enjoy all the economies of large scale production and thereby, do not produce optimum output. The firm will stop producing the output even when LRATC is falling *i.e.*, to the left of the minimum point of falling LRATC. So, output is less and price of the product is higher under monopolistic competition. So, the firm will have unutilized capacity or excess capacity and this is a sign of inefficiency. That means, the firms under monopolistic competition will not utilize the fullest capacity of plants. So, when the firms under monopolistic competition do not utilize the fullest capacity, there will be under utilization of productive resources by the existing plants and that obviously amounts to a significant economic waste from the nation's point of view. So, the excess capacity of the firms under monopolistic competition has the following implications:

- ☆ Under-allocation of resources
- ☆ The total production from the nation's point of view is far less than social desirable output and thereby, the country remains much poorer due to monopolistic competition than what it would have been had its economic position been under perfect market competition.
- ☆ Since the output produced by the firms under monopolistic competition is lower than the social desirable output, the prices of the product tends to be higher when compared to perfect market competition. So, due to excess capacity of firms under monopolistic competition, the consumer's welfare is adversely affected.
- ☆ When the firms under monopolistic competition do not enjoy all the economies of scale, it leads to inefficient utilization of resources, thereby, affects the quality of the output when compared to the quality of output produced under perfect competition.
- ☆ Since the output produced under monopolistic competition is lower than the social desirable output, it leads to unemployment and under employment of resources in the economy.

2. Competitive Advertising

Under monopolistic competition, the firms spend huge amounts on advertisement and this type of advertising is regarded as Competitive advertising, in the sense that, each firm tries to compete out its rival firms through advertising heavily about its product(s). The firms' investment on competitive advertisement increases the cost and this burden is shifted on to the customers through charging higher prices for the product. The following are the negative implications due to competitive advertising under monopolistic competition:

- ☆ Waste of huge amount of resources.
- ☆ Price of the product at the customer level is increased.
- ☆ Leads to psychological dissatisfaction on the part of the customers, who are not affordable to purchase the high priced advertised products.

3. Cross Transport

Because of product differentiation and brand loyalty of the customers, the firms will have to transport the products from one area to another. There will be inter-state movement of goods and even between the countries, the products are exported and imported. These huge transportation costs raise the price of the products at the customers' level.

4. Insufficient Specialization

Because of product differentiation and non-price competition, the firms under monopolistic competition do not specialize sufficiently in the production of products, thereby, the advantages of full specialization cannot be reaped. This can also be considered as a waste, as the firms do not take risks and cannot expand the production programme up to the minimum point of LRATC.

5. Higher Cost of Production

Some of the firms operating under monopolistic competition may have high cost structure and these firms will survive in the market because of the attachment of the customers to their products due to product differentiation, brand loyalty etc. These firms will charge high price for their products to cover the high costs. So, under monopolistic competition, even in the long run, the firms will survive by charging higher price for their differentiated products to cover the higher costs. But, under perfect competition, such firms will have to leave the industry in the long run because of homogenous nature of the product.

6. Unemployment

Because of excess capacity of firms under monopolistic competition, the firms will not produce optimum scale of output in the long run. So, the resources are not fully exploited under this competition. These adversely affect the employment of resources in the economy.

7. No Standardization of Products

Because of product differentiation, the products produced by the firms do not have uniform standards and this infers that, the economies of large scale are not obtained under monopolistic competition. This lack of uniform standards for the products will create confusion among the customers regarding the right selection of products.

8. Price Higher than MC

Under monopolistic competition since, ARC slopes downward from left to right, MRC also slopes downward and lies below it. So, at the point of optimal output, price>MC. But under perfect competition, price=MC. So, under monopolistic competition, customers' welfare is affected, as price>MC and this is regarded as allocative inefficiency. This price>MC under monopolistic competition leads to loss of social welfare and the same is explained through Figure 23.22. OQ is the optimal output produced under monopolistic competition, where LRMC=MRC at point E, and with reference to point E or optimal output OQ, the price OC=QP>LRMC and this price OC=QP is situated left to the minimum point of LRATC (L). Had the firm under monopolistic competition produces the output with reference to minimum point of LRATC, 'L', (as discussed under perfect competition), the output OQ_1 will be produced. But, this is not the case under monopolistic competition, as ARC is downward sloping and not parallel to X-axis as in perfect competition and hence, it could not be tangent to the LRATC at its lowest point. So, the output QQ_1 represents excess capacity of the firm under monopolistic competition, and the area PEE_1 represents the loss of social welfare due to price hike under monopolistic competition (QP) compared to perfect competition (Q_1L).

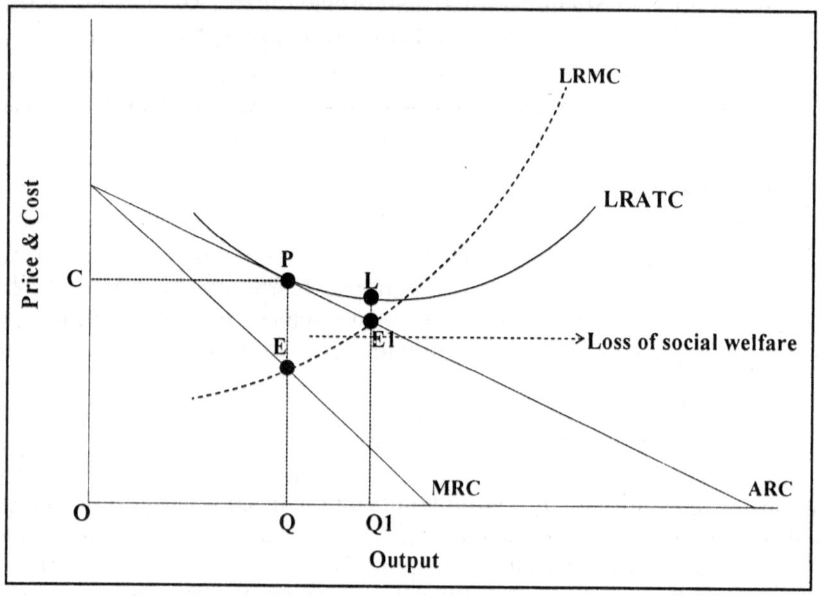

Figure 23.22: Loss of social welfare under monopolistic competition compared to perfect competition.

From the above wastes, we can infer that, the presence of monopolistic competition will affect the customers adversely and even from the nation's point of view it poses threats like less output, high unemployment, unnecessary wastage of money in terms of non-price competition between the firms etc. But, Chamberlin justified the monopolistic competition and he opined that, so long as customers want different varieties of products and product differentiation satisfies customers' wants, these wastes need not be called as wastes.

9. Number of Firms

Another way in which monopolistic competition may be socially inefficient is that the number of firms in the market may not be the 'ideal' one. That means, there may be too much or too little entry.

10. Business-Stealing Externality Compensates Product-Variety Externality

In monopolistic competition, the consumers get some consumer surplus from the introduction of a new (differentiated) product due to the entry of a new firm and this refers to positive externality on consumers (product-variety externality). However, with the entry of new firm(s), other existing firms lose customers and profits, thereby, entry of a new firm imposes a negative externality on existing firms (business-stealing externality). Thus, the business-stealing externality compensates the product-variety externality in the monopolistic competition due to the entry of new firms into the industry.

11. Influence of Retail Dealers

The theory of monopolistic competition fails to take into account the fact that, the demand by final consumers is largely influenced by the retail dealers because, the consumers themselves are not fully aware of the technical qualities of the product.

12. Demand for Raw Materials

The theory of monopolistic competition fails to fully account for the determination of equilibrium quantities and prices of commodities like raw materials and other inputs. To a large extent, their demand is governed by a combination of the technical quality, price and timely availability rather than by brand name etc. For example, given the technical quality of an input (say, fertilizer), its demand is governed more by its price and availability than its brand name.

13. Differentiation also Leads to Waste

Some differentiation does not create utility but generates unnecessary waste, such as excess packaging. Under perfect competition, each seller produces an identical (homogenous) product that too at lower ATC than the monopolistically competitive firm. This had led some Socialist Economists to complain that, there is needless product differentiation in monopolistically competitive groups, aimed more at deceiving and confusing buyers rather than serving them, which raises prices for customers and provides them with little benefit.

14. Fixed Factors Cannot be Fully Exploited

The firms can never fully exploit their fixed factors because, mass production is difficult. This means they are productively inefficient in both the long and short run. However, this is may be outweighed by the advantages of diversity and choice.

23.9. Desirability of Monopolistic Competition

The earlier discussion pertaining to the profit maximizing behaviour of a monopolistically competitive firm implies that, the price associated with the product (at the equilibrium or the profit maximizing output) is higher than MC (which equals the MR). Thus, the production under monopolistic competition does not take place to the point, where price equals MC of production. With increased production, price charged (which is higher than MR at every level of output) is successively falling (downward falling ARC), while the MC of production is rising. So, if a monopolistically competitive firm were to stop production where price is equal to MC (as in the case of perfect market competition, where price=ARC=MRC=MC), the output produced would be greater than when it stops production where MR=MC (*i.e.*, at profit maximizing level of output). The net result of the profit maximizing decisions of monopolistically competitive firms is that, the price charged under monopolistic competition is higher than under perfect competition. In addition, quantity of the commodity produced under monopolistic competition is simultaneously lower than under perfect market competition. Thus, both on the basis of price charged and output produced, monopolistic competition is less socially desirable than perfect market competition.

23.10. Market Failure

Having discussed different market structures in-detail, it is essential to discuss the concept of market failure, which arises when the market is unable to achieve an efficient and equitable allocation of resources. The following are the important sources of market failure:

- ☆ ***Provision of public goods:*** A public good is one, which is made freely available to all consumers. Examples include *street lighting, parks, roads etc*. For such free goods, no private firm is able to make a profit. So, Government should ensure that, such goods are provided at the socially desired level.
- ☆ ***Income distribution:*** The market structures will not necessarily ensure equitable distribution of incomes. So, this may motivate the Government to introduce policies to redistribute wealth through measures like income taxes, social security benefits etc.
- ☆ ***Mis-use of market power:*** The misuse of market power by a natural monopoly result in inefficient allocation of resources, which reduce community welfare. So, Government generally regulates monopoly and enforce laws preventing cartels.
- ☆ ***Information asymmetries:*** Information asymmetries between producers and consumers can lead to market failure and reduce community welfare.

23.11. Differences between Different Market Structures

So, far, we studied the different forms of market competition in terms of their features and price and output determination. It is high time now to have the comparative analysis across these market structures, so as to enable the readers for better understanding them (Table 23.2).

Table 23.2: Differences between different market structures

Items	*Perfect Competition*	*Monopoly*	*Oligopoly*	*Monopolistic Competition*
Number of firms	Infinite number of firms.	Single firm	Few firms	Large number of firms (but not as large as under perfect market competition)
Nature of the products	All firm produce homogenous products, which are perfect substitutes. So, the group of firms represents an industry.	The single firm produces a unique product, with out any close substitutes. This single firm represents the industry.	Few firms sell branded products, which are close substitutes of each other. Some oligopolistic industries produce homogenous or standardized products (steel, zinc, copper, cement etc), whereas, others produce differentiated products (automobiles, detergents, greeting cards etc). As few firms produces an homogenous product (Pure oligopoly) together constitute an industry.	Each firm sells differentiated product, which are close substitutes to one another. The product of one firm is different from that of the other firm in terms of physical differences, quality differences, imaginary differences, purchase-benefit differences, features etc. As all the firms produces differentiated products, they constitute a 'group'.
Product differentiation	The product is perfectly homogenous in terms of size, shape, taste, colour, ingredients, quality etc. So, the products are perfect substitutes, thereby, there is zero product differentiation.	The product produced by a monopoly firm is unique and with out close substitutes. So, in monopoly, there is absolute product differentiation, in the sense that, there is no available substitute for a monopolized product.	Few firms produce either homogenous product or differentiated products. In case of differentiated oligopoly, products are highly differentiated.	Compared to oligopoly, the products are medium differentiated, but still they are close substitutes.
E_c for the product	E_c is infinite, as all firms produces and sells homogeneous products, which are perfect substitutes.	There are no close substitutes for the product produced by the monopoly firm, thereby, the E_c is zero.	In case of pure oligopoly, the E_c is infinite and in case of differentiated oligopoly, E_c is significantly high.	E_c is high, as the firms produces and sells differentiated products, which are close substitutes.
Free entry and exit of firms	There are absolutely no restrictions regarding the entry and exit of firms into an industry. During short run, when the existing firms enjoy super-normal profits, new firms enter the industry. When the existing firms in the industry suffer from losses, some firms will leave the industry.	The existing firm will have firm control on the market and hence, there are strong barriers for other firms to enter the industry. Output-pricing strategies, control over the raw material, legal rights in the form of patents, copy rights, trade secrets etc., acts as strong barriers for other firms into the industry.	There are strong barriers regarding the entry of new firms into the industry. Extensive control over essential and specialized inputs, high capital requirement due to huge plant costs, exclusive patent rights on the raw materials or products produced, economies of scale enjoyed by only few large firms, heavy selling costs etc., acts as strong barriers for other firms into the industry.	There is complete freedom for the firms for entry and exit from the group. Since each firm is small in size and is producing differentiated products under their brand name, there is freedom of entry and exit of firms and this increases competition.
Perfect knowledge	All the buyers and sellers have perfect information about prices and products of competing firms. This perfect knowledge among the sellers and buyers ensures prevalence of uniform price for the product in the market.	The technology or process of production of a product is maintained secrecy, as it ensures trade protection for the monopoly firm.	Oligopoly firms have perfect knowledge about their own cost structures and demand functions, but their inter-firm information may be incomplete. Buyers have only imperfect knowledge about cost, price and product quality.	In this competition, buyers do not have knowledge about output-price policy of different firms, but they have relatively complete information about the prices of the product of different firms, brands, features etc. On the contrary, each seller has complete information about production techniques and the prices charged by other rivalry firms.

Contd...

Table 23.2–Contd...

Items	*Perfect Competition*	*Monopoly*	*Oligopoly*	*Monopolistic Competition*
Price taker or Price maker	The price of commodity is decided by the market forces of demand and supply at the industry level and the same price is passed on to the individual firms. Hence, individual firm is a price taker and not a price maker.	Firm is a price maker. The single-firm status gives the monopolist an extensive market control, thereby, he is either price maker or quantity adjuster in the market and not a price taker.	Due to mutual interdependence between the firms, oligopolies are price makers rather than price takers, as the few sellers mutually dominate the pricing decisions.	A firm under monopolistic competition is a price maker, with some degree of control over price.
ARC or demand curve	For an industry as a whole, the demand curve slopes downward from left to right. But, for an individual firm, the demand curve or ARC both in the short run and long run is perfectly elastic *i.e.*, it lies parallel to X-axis.	The industry or firm's demand curve or ARC is downward sloping from left to right indicating that, monopolist sells large quantities of a commodity or output at lower prices or price reductions.	An oligopolist faces a downward sloping demand curve, but its price elasticity depends upon the rivals' reaction to changes in its investment, output and price. So, the demand curve of a firm under oligopoly is indeterminate because, any strategy formulated by the rival firms, may affect or change the demand curve. The demand curve will have a kink at the present price and this kinked demand curve is more elastic for price increase and less elastic for price fall or decrease. It is more elastic than under simple monopoly and not perfectly elastic as under perfect competition.	The ARC in monopolistic competition is rather more elastic with a flat slope, but less elastic compared to perfect competition and more elastic compared to monopoly market situation.
MRC	Since, ARC is constant, MRC must be constant and coincides with ARC. Under perfect market competition, price=ARC=MRC=Demand curve.	MRC slopes downward from left to right and remain below ARC and the numerical slope of MRC is twice as steep as ARC.	The MRC of the firm will have a gap or discontinuity below the point of kink of ARC. This is due to the change in the elasticity of demand i.e., from the more elastic nature of the demand (above kink) to more inelastic nature of demand (below kink).	Since, ARC slopes downward from left to right, the MRC also slopes downward from left to right with a smooth and flat slope and lies below it.
Price fixation by the firm	The price determined at the industry level will be passed on to the individual firms, thereby, individual firms are simply price takers and they have no role in fixing or changing the price of the product. This ensures prevalence of uniform price for the product in the market.	Monopolist is either price maker or output decider. Considering different elasticities of demand for the product in different markets, the monopolist sometimes may adopt price discrimination. He may even fix different prices for the same product among different categories of consumers in the same market segment.	Oligopolies are price makers rather than price takers, as the few sellers mutually take the pricing decisions. Since, there exists mutual interdependence among the firms and the price-output policies of one firm influences the fortunes of other firms, there will not be any unique pattern of price behaviour for each firm in the oligopoly market situation. There exists both the co-operative and competitive trends among the business firms regarding the fixation of output-price policies.	The firm enjoys some control over the price of the product due to product differentiation, unlike in perfect competition where the individual firm has absolutely no control on the price of the product, because of homogenous nature of the product. However, in monopoly, the firm enjoys still more control over price when compared to monopolistic competition, as there are absolutely no substitute goods in monopoly situation.

Contd...

Table 23.2–Contd...

Items	*Perfect Competition*	*Monopoly*	*Oligopoly*	*Monopolistic Competition*
Market power	Perfectly competitive firms have zero market power when it comes to setting prices, as they are simply price takers for the price settled at industry level.	A monopolist has considerable, although not unlimited market power. He is a price maker, as he can set the price of the product to his maximum advantage. Likewise, the monopolist has complete control over the supply of the product. But, the monopolist is either a price maker or output decider. The two primary factors determining monopoly market power are the firm's demand curve and its cost structure.	Since, the firms generally decide the output-price policies of the industry on collusion basis, the firms enjoy significant market power. Further, if the few firms produce differentiated products, each firm should be considered as a monopoly firm, as no other firm is producing the same product. So, by producing the differentiated products by all the oligopoly firms, the firms will enjoy significant monopoly power, as in monopoly competition.	The large number of firms under monopolistic competition limits their power in fixing higher price for the products produced by them. This is because of the difficulties involved in getting a large number of firms to act together. However, the product differentiation across the firms may enable them to execute some market power in setting prices for the products to their maximum advantage.
Supply curve	In short run production programme, for the firm, the portion of the SRMC curve above SRAVC curve represents the short run supply curve. In long run, the portion of the LRMC curve above LRATC curve represents long run supply curve. However, for an industry, the supply curve slopes downward from left to right.	Since, monopolist is either price maker or output decider, we cannot draw a unique supply curve for the monopoly firm. Moreover, the monopolist's output decision depends not only on MC, but also on the demand curve. So, there is no one-to-one relationship between price and the quantity supplied.	Since, firms may collude or compete each other in an industry, the firm's output decision depends not only on MC, but also on the demand curve and firms are either price makers or output deciders, we cannot have a unique supply curve for the firm under oligopoly competition.	Since, the firm can either decide the output or price, we cannot draw a unique supply curve for the firm.
Firm's equilibrium	The firm is in equilibrium at the level of output, where MC is equal to MR. Since in perfect competition, MR is equal to AR or price, therefore, when MC is equal to MR, it is also equal to AR or price at that equilibrium position, i.e., MC=MR=AR=Price. At the point of equilibrium, MC curve is rising and it cuts the MRC from below. MC curve can neither be falling nor constant and it should be rising and then only, the firm under perfect competition is under equilibrium. That means, the firm is in equilibrium only when the MC curve cuts the MRC from below, while MC curve is rising. Since the MRC is a horizontal straight line parallel to X-axis, the second order condition of firm is equilibrium i.e., MC=MR is satisfied only when MC curve is rising.	The firm is in equilibrium at the level of output, where MC is equal to MR. So, at the point of equilibrium, MC=MR and MC curve cuts the downward falling MRC. The MC curve may be falling (if the firm experiences increasing returns or decreasing costs) or rising (if the firm experiences decreasing returns or increasing costs) or constant (if the firm experiences constant returns or constant costs), but for stability of the equilibrium, when MC curve is falling, it should fall at slow pace than MRC. If this fails, there can be no equilibrium what so ever. The second order condition of firm's equilibrium i.e., MC curve should cut MRC from below at the equilibrium point, is satisfied in all the three cases i.e., when MC curve is rising, remaining constant or falling. Since, MRC is a continuous falling curve firm left to right, the second order equilibrium condition is fulfilled even when MC curve is rising or remaining constant or falling.	An oligopoly firm maximizes its profits by producing the output, where MR equals MC.	In short run, the condition of firm's equilibrium is SRMC=MR. But, the point of equilibrium of a firm under monopolistic competition in the long run involves two essential conditions *viz.*, LRMC=MR and ARC=LRATC. This infers that, the firm will earn only normal profits in the long run.

Contd...

Table 23.2–Contd...

Items	*Perfect Competition*	*Monopoly*	*Oligopoly*	*Monopolistic Competition*
MC and Price	In equilibrium position, the price charged by the firm equals to MC, as MRC is equal to ARC.	In equilibrium position, the price charged by the firm is above MC, as ARC is above MRC.	In equilibrium position, (under Cartel and Price leadership models), the price charged by the firm is above MC, as ARC is above MRC.	In equilibrium position, the price charged by the firm is above MC, as ARC is above MRC.
Profits potential	In the long run, the firm earns only normal profit. The super normal profits enjoyed by the firms in the short run will be swept away in the long run, as new firms enter the industry.	The firm can earn supernormal profits even in the long run, as there are no competitors for the firm producing the same product and there are strong barriers for the entry of firms into the industry.	Since barriers to entry for new firms are very strong in oligopoly competition, the existing firms can earn super-normal profits even in the long run.	The firms enjoy super normal profits in the short run. In long run, the firm is in equilibrium, where ARC is tangent to the downward sloping LRATC, thereby, the firms enjoy normal profits only. Firms may, however, retain some profits by using more non-price action.
Price and output	When compared to monopoly, under perfect competition, output is higher and price is lower.	When compared to perfect competition, output is lower and price is higher.	The firms generally collude to maximize the profits either fixing higher price for the product or sharing the market or fixing the output across the firms.	Output is lower and price is higher
Firm's equilibrium in long run with reference to LRATC curve	The firm is in long-run equilibrium at the minimum point of the LRATC curve. That means, a firm under perfect competition operates at its full capacity in the long run. So, no excess capacity is left over, since the firm under perfect competition is producing up to the point where MC=LRATC. That means, the firm employs optimum plant size in the long run to achieve equilibrium.	The firm is in long-run equilibrium at the level of output, where LRATC curve is still declining and has not reached the minimum point. So, the firm operates at less than full capacity in the long run. So, some excess capacity or reserve capacity is left over, since the firm is not operating up to the minimum point of LRATC. That means, the firm employs less than full optimum plant size in the long run to achieve equilibrium.	If there is no product differentiation, the customers do not have any particular preference for any firms. So, the firms have to charge the same price for the product in the long run. The price charged by each firm in the long run may lie in between monopoly price and competitive price. If there is product differentiation, each firm acts like a monopoly and need not fear about its rivals' action. The firm which produces superior quality product earns super normal profits even in the long run and thereby, charges price more than LRATC.	The firm is in long-run equilibrium at the level of output, where LRATC curve is still declining and has not reached the minimum point. So, the firm operates at less than full capacity in the long run. So, some excess capacity or reserve capacity is left over, since the firm is not operating up to the minimum point of LRATC.
Long run profit	Zero	May be high	Varies	Zero
Consumer's surplus	Under perfect competition, equilibrium output is higher and equilibrium price is lower, thereby the consumer's surplus is more	In monopoly, equilibrium output is lower and equilibrium price is higher, thereby the consumer's surplus is less.	Since, the firms generally collude with an objective of gaining monopoly gains, the consumer's surplus is less.	Since, equilibrium output is lower and equilibrium price is higher, thereby the consumer's surplus is less.
Resource allocation	The resource allocation is optimum and hence, social welfare is maximum.	The resources are misallocated, thereby causing a loss of social welfare.	The resources are not efficiently utilized, as the firms produce less output and sell the same at higher price. In fact, oligopoly tends to be the worst efficiency offender in the real world, because perfect competition does not exist, monopolistic competition inefficiency is minor and	The resources are not efficiently utilized, as the firms will not produce optimum scale of output. That means, the firms will stop producing the output, even when LRATC is falling *i.e.*, to the left of the minimum point of LRATC. But, the resource inefficiency is less

Contd...

Table 23.2–Contd...

Items	*Perfect Competition*	*Monopoly*	*Oligopoly*	*Monopolistic Competition*
			monopoly inefficiency has the potential for being so bad that it is inevitably subject to corrective Government regulation.	compared to oligopoly and monopoly competitions.
Economically efficient	More, as price equals MC in long run equilibrium and firms does not have excess capacity.	Less compared to perfect competition, as price exceeds MC in long run equilibrium and firms have excess capacity.	Less compared to perfect competition, as price exceeds MC in long run equilibrium and firms have excess capacity.	Less compared to perfect competition, as price exceeds MC in long run equilibrium and firms have excess capacity.
Consumer's welfare	It is considered to be more under perfect market competition. This is because, a firm under perfect competition ☆ Produces homogenous product ☆ Products are good substitutes ☆ Operates up to optimum plant size ☆ Can increase the profits by increasing the output rather than increasing the price of the product.	Monopoly is equally beneficial in promoting consumer's welfare. But, there is a misconception that, monopolist exploits the consumers by charging higher prices. The major difference between monopoly and perfect market competition is that, under monopoly, the firm produces sub-optimum output, thereby, price of product under monopoly is greater than price of product under perfect competition. But, monopoly price is not always higher than the price under perfect competition, but generally it is. Even it is possible that, the monopoly price is lower than the price under perfect competition, as the firms may enjoy economies of large scale production, the firm can go for significant investments on R&D, thereby, it can employ cost effective production techniques, procurement of raw material in bulk at low cost, efficient marketing strategies etc.	As there is no other alternative for the consumer to purchase the products from the few firms, the firms try to exploit the customers. This is much severe, if the firms collude and formulate higher prices for the products.	It is safe guarded in the sense of providing products with differentiated and attractive features, but the firm charges the price>MC, inferring that, consumer gets exploited under monopolistic competition.
Product Promotion & Advertising	None or minimal	Some, not directed at rivals, but to increase sales	Heavy	Considerable
Government intervention	There is complete absence of Government regulation regarding the output, supply and pricing mechanism of the products. There is also no collusion among buyers or sellers.	Government regulation is indispensable, especially when the monopolist exploits the consumers in charging higher prices for the product.	Government regulation is essential, especially at times of illegal collusion of few firms and aiming for monopoly profits in the business.	Like perfect competition, they are largely unrestricted by Government rules and regulations.
Frequency in reality	Rare	Rare	Limited	Widespread

Part V

Factor Pricing

24

Theory of Distribution

In the earlier chapters, we dealt with the concepts of producer's behaviour and consumer's behaviour with reference to production and consumption of products. These concepts will arise only, when there is involvement of various factors and factor services in the production of products. The factors and factor services, which involve in the production of products are called Factors of production and they fall under four important categories *viz.*, Land, Labour, Capital and Management.

24.1. Features of Factors of Production

The following are the important features of different factors of production:

Land

Land includes all natural resources that human being get free from air, water and land. In short, the terms 'land' refers all that nature has created on the earth, above the earth and below the earth's surface. All aspects of economic life *i.e.*, agriculture, trade and industry are generally influenced by natural resources, which is called as 'Land' in Economics. As a factor of production, it is quite peculiar, as it possess the following important features:

- ☆ Land is free gift of nature
- ☆ Land is limited in area
- ☆ Land is permanent in existence
- ☆ Land lacks geographical mobility
- ☆ Land is of infinite variety

Labour

It can be defined as any exertion of mind or body undergone partly or wholly with a view to earning some good other than the pleasure derived directly from the work. In economic sense, labour refers to any type of work performed by a labourer with an intention to earn income. As a factor of production, it is quite peculiar, as it possess the following peculiarities:

- ☆ Labour is inseparable from the labourer
- ☆ Labourer sells his services not himself

- ☆ Labour is more perishable than other commodity
- ☆ Labourer has not the same power of bargaining as their employers
- ☆ Less efficient compared to machinery
- ☆ Less mobile
- ☆ Supply independent of its demand
- ☆ Labourers differs in efficiency

Capital

Capital is that part of person's wealth, other than land, which yields an income or which aids in the production of further wealth. Capital is produced means of production. It is man-made instrument of production. Just like land and labour, capital as factor of production is not original. Since, it is man-made, it is not freely available. Karl Marx defined capital as 'Crystallized labour' in his famous book, Das Kapital. As a factor of production, it is quite peculiar, as it possess the following important features:

- ☆ Capital is a man-made factor of production.
- ☆ It involves time element.
- ☆ Capital may be fixed (machinery usage in agriculture) or it may be working (seed, fertilizer usage in agriculture).

Organization (Management)

Organization of enterprises means to plan a business, to start it and run it. It means to bring all the other three factors of production *i.e.*, land, labour and capital together to undertake a business or production process. Organization implies not only running the business, but also shouldering the loss, if any. The person, who undertakes the responsibilities for organizing or managing the business is called as Organizer or Entrepreneur. The following are the important functions of an organizer:

- ☆ The organizer initiates a business or production considering the level of resources available with him.
- ☆ He organizes or combines the other three factors of production *i.e.*, land, labour and capital to start the business or production.
- ☆ He directs and supervises the production on a systematic line.
- ☆ He controls the business activities, so that, they are tuned towards the achievement of desired goals and objectives.
- ☆ He bears risks and uncertainities in the business.
- ☆ He will plan for new innovations in the business.

24.2. Distribution

For employing these factors of production in the production programme, remuneration has to be paid for the services rendered by them and the remuneration paid for different factors of production are differently named. Land is rewarded in the form of rent, labour gets wages, capital receives reward as interest and finally, the reward for the managerial services offered by the entrepreneur is called profit. These rewards to the different factors of production serve as income for their owners and with these incomes, they demand for the products produced in the market economy. This payment of rewards to the factors of production refers to Distribution. The process of distribution of rewards to different factors of production may be studied either from Functional basis or Personal basis.

24.3. Functional Distribution Vs Personal Distribution

The distinguishing features of functional distribution and personal distribution are discussed here under.

1.Functional Distribution

It refers to the distinct shares of the total income received by different factors of production, in accordance with the different functions performed by them. That means, in functional distribution, we study how the factors of production are rewarded for their services in the process of production, *i.e.*, how the rent of land, wages of labour, interest for capital and profit for management are determined. This Functional distribution is also called as Factor-share distribution, as it explains the share of total income derived by each factor of production in the economy. So, in functional distribution, we are no longer concerned with the individuals and their individual incomes, but we are concerned with the factors of production and how much each factor of production is earning in the industry. Thus, we can infer that, the total income derived in the economy is the function of incomes earned by all the factors of production and this is represented as,

$$Y = f(R,W,I,P)$$

where,

Y= Total income

R= Share of land factor (Rent)

W= Share of labour factor (Wages)

I= Share of capital (Interest)

P= Share of management (Profits).

Thus, functional distribution studies the remuneration received by each factor of production. The important aspect here is that, we will not study the individuals and income of each individual, but studies only in terms of activities or functions performed by them. That means, we study the remuneration paid per labour, remuneration paid per land usage, remuneration paid for using the capital and remuneration paid for managerial service contributed. This is so because, in some cases like, 'Peasant proprietor' *i.e.*, the farmer will contribute all the four factors of production *i.e.*, he will provide land, labour, allocate his own capital and his own managerial services in the production programme. Here also, we will not study what the total income earned by that individual in his production programme, but rather, we will study the share of each factor of production in the total income earned by the individual. Hence, the functional distribution is, otherwise, called as Theory of factor pricing. In the words of Professor Jan Pen, *'in functional distribution, we are no longer concerned with individuals and their individual incomes, but with factors of production, labour, capital, land and something also that may, perhaps, best be called 'entrepreneurial activity'. The theory examines how these factors of production are remunerated. It is primarily concerned with the price of a unit of labour, a unit of capital, a unit of land and being an extension of price theory, it is sometimes called the theory of factor prices'*. The concept of functional distribution can be categorized into Macro theory of distribution and Micro theory of distribution.

(a) Macro Theory of Distribution

In general, functional distribution is dealt as a 'macro' concept. This theory of macro distribution is also known as 'Theory of distributive shares'. If the concept of functional distribution is studied at macro level *i.e.*, if we analyze the share of rents to land, share of labourers' income, share of capitalists income and share of profits to the entrepreneurs of the total National Income, it refers to Macro theory of distribution. This concept helps to study the effect of economic development on the relative shares of different factors of production. For example, the theories like Classical theory of employment, Keynes theory of employment etc., falls under this category of functional distribution.

(b) Micro Theory of Distribution

If the concept of functional distribution is studied at micro level say, with reference to a firm or an industry, it refers to Micro theory of distribution. That means, we study the share of rents to land, share of labourers' income, share of capitalists income and share of profits to the entrepreneur of the total income determined at the firm or an industry level.

2. Personal Distribution

In personal distribution, we study the income earned by various individuals in the economy through performing economic activities *i.e.*, distribution of national income among various individuals in the economy. So, personal distribution of national income is otherwise called as 'Size-wise distribution of incomes'. The concept of personal distribution is dealt as a 'micro' approach. Since, the national income is not distributed equally among the individuals in the country, it leads to income inequalities among the people. So, the concept of personal distribution helps to study, how personal incomes are determined and to analyze the inequalities of income and wealth across the individuals in the country. This helps to study the causes for inequality and plan the strategies to overcome the same. Thus, the concept of equality and inequality of income distribution and social justice is basically concerned with the personal distribution.

The above explanation infers that, there exist certain differences between functional distribution and personal distribution. It is important to note that, the personal distribution in an economy depends upon the functional distribution or functional distribution is only a part of the theory of personal distribution. This is because, if one person controls several factors, his personal income will be much higher, say, the case of 'peasant proprietor'. Since, he controls all the four factor of production, his personal income is much higher. Similarly, the personal distribution of income also depends upon the prices of factors and quantity of factors possessed by him. For example, if a person possess four acres of land, his personal income depends upon the product between remuneration paid per acre of land × number of acres of land. So, we can say, functional distribution is a part of personal distribution and functional distribution influences the personal distribution. For example, if the prices of factors of production are higher (functional distribution is higher), then the personal distribution is higher. If personal distribution is higher, the income of individual will be higher. So, there will be greater demand for the products, services, leading to more investment, more employment, more production and more national income and thus, leads to high standard of living of people in the economy. However, the theory of factor distribution deals mainly with the functional distribution and not with personal distribution. It offers useful explanation regarding the determination of payments for factors of production *i.e.*, rent, wages, interest and profits.

24.4. Why a Separate Theory of Distribution

The theory of factor distribution is essentially a pricing process of factors *i.e.*, it explains the determination of prices of different factors of production. Hence, the Theory of factor distribution is also called as Theory of factor pricing. But, there are some fundamental differences between factor pricing and product pricing, as the factors of production have certain distinctive features when compared to those of ordinary products or commodities. The following points highlights and justifies the essentiality of having a separate theory of distribution to analyze the pricing process of factors of production:

- A firm demands a factor or factor service not for its own sake, but because, the consumers will demand for the product manufactured or derived from that factor or factor service. Hence, the demand for factors of production is called as Derived or Indirect demand. However, in case of product demand, the consumer purchases the product for personal consumption. Hence, product demand is a Direct demand.
- The four factors of production *viz.*, land, labour, capital and management will be paid remuneration as rent, wages, interest and profits respectively. These remunerations are paid to the factors of production, not for the factors as such, but for the services they contribute towards the production of the product. For example, we pay rent for the land not in terms of land value, but in terms of using the land for a period of time and after using the land, it has to be returned back to the owner. Similarly, we pay wages for labour for the services we extract from them. For capital, we pay interest because, the lender has deprived from the use of that money for a considerable period of time and after using the borrowed capital, we have to return the original amount of capital back to the lender. So, in factor pricing, we deal with the payments made for the services we derive from the use of the factors of production and not for owning the factors as such. This is, where, factor pricing also differs from product pricing. In product pricing, we pay the price for owning the product itself, but in factor pricing, we own the factor services and not the factors as such.
- Product pricing is influenced by demand and supply forces. Similarly, the factor prices are also determined by demand and supply forces. But, there are slight differences between product pricing and factor pricing. In product pricing, the process is purely on an objective plan, as the commodities are inanimate objects and we decide purely on demand and supply conditions. But, in factor pricing, in addition to demand and supply forces, we have to take into consideration the subjective aspect, as the factors are either human or have a human agency behind them in regulating their supply. For example, labour and management are purely human, and land and capital are controlled by human beings. But, in case of product pricing, though the product supply is regulated by human beings, the product itself is inanimate.
- The supply of the product depends upon the money cost of production, while the supply of factor depends upon opportunity cost. A factor is supplied into a particular production activity, if its opportunity cost is low in its next best alternative use, compared to the present use or present production activity.
- The pricing of the factors like labour and management are also influenced by social and human factors, besides the demand and supply forces. But, the product pricing is little influenced by social and human factors and mainly influenced by demand and supply forces.
- The demand for the factors is influenced by their productivity, while the demand for the products is influenced by MU.
- Factors of production are always jointly demanded, as they are collectively employed and coordinated in the production process, whereas, products are not jointly demanded by the consumers.
- The main objective of studying the factor pricing process is to minimize the costs in the production programme, while the objective of studying the product pricing is to maximize the profits in the production programme.
- In product pricing, we simply analyze the pricing process of products in different markets. But, in factor pricing, besides studying the above aspect, we also deal with the issue of 'equity' in remuneration of factors.
- When price of the product increases in the market, the supply of the product also increases. But, this is not always true in case of factor pricing, as we discussed earlier in the Figure 9.59, where the supply curve of labour takes a backward bending shape, when increased wage rate is offered to him. This implies that, the supply of labour increases with increase in wage rate, but this is only upto a certain level of wage rate. If still higher wage rate is offered, the labour prefers more of leisure hours or less number of working hours.

24.5. Importance of Factor Pricing

The concept of factor pricing is important considering the following grounds:

- Factor pricing determines the level of employment of factors in the production programme. For example, a fall in price of a factor, increases the demand from the entrepreneurs to employ that factor in the production programme.
- Factor pricing determines the allocation of resources in the production programme in accordance with their opportunity cost.
- Factor pricing guides the farmer in analyzing the resource use efficiency in the production programme. Without pricing, the resource use efficiency cannot be studied.
- Factor pricing forms the basis for studying the functional distribution of factors in the economy and in turn, helps to estimate the personal distribution.
- Factor pricing guides the farmer in planning for substitution of resources, so as to minimize the cost in the production programme.

24.6. Theories of Factor Pricing or Distribution

Basically, two theories explain the determination of factor payments *viz.*, Marginal productivity theory of distribution and Modern theory or Demand and Supply theory of distribution.

I. Marginal Productivity Theory of Distribution

This theory has been regarded as the General theory of distribution. The concept of marginal productivity theory was first applied to study pricing of land factor (rent) by Ricardo and West. But, this concept has not gained much popularity, until it was rediscovered by the Economists like Jevons, Menger, Walras, Wicksell, Marshall, Wicksteed and J.B. Clark. Among these Economists, J.B. Clark's name was most widely associated for the refinement and development of this theory. In the words of J.B. Clark, '*under static conditions, every factor including the entrepreneur would get a remuneration equal to its marginal product*'.

We know, the entrepreneur buys the services of various factors of production to employ them in the production programme. So, the entrepreneur is the agent through whom the factors of production will receive the rewards or remuneration. The entrepreneur generally employs the factors of production that contribute higher marginal productivity and he substitutes the factor with higher marginal productivity in place of the factor with lower marginal productivity. The concept of the theory of marginal productivity involves the following propositions:

- The marginal productivity of the factor determines the price of the factor.
- In the long run, the price of the factor tends to equate to its Average Revenue Product (ARP) and Marginal Revenue Product (MRP).
- The point of optimal allocation of a factor in the production programme is when the remuneration paid to the factor (Marginal Factor Cost (MFC)) equals the MRP.
- When all the factors receive their respective shares in accordance with their MPPs, the TPP will be exhausted.

Assumptions

This theory is based on the following assumptions:

- All the units of a factor are homogenous, thereby, all the factor units are perfectly substitutable.
- The technology is given, though the scales and proportions of factors may change.
- There is perfect mobility of factors of production between different employments.
- There is perfect competition both in the factor market and product market.
- The factor units are perfectly divisible.
- The marginal productivity of factor is measurable.
- The firm aims at profit maximization through efficient allocation of resources.
- This theory considers long run analysis, as the price of the factor tends to be equal to ARP and MRP.
- There is full employment equilibrium in the economy.
- This theory is based on the concept of LDR.
- The bargaining power of the seller and buyer of the factor service are equal.
- There is no Government intervention in the fixation of factor's remuneration like minimum wage rate etc.

Basic Concepts of Productivity

The factor productivity can be studied from two angles *viz.*, physical productivity and revenue productivity.

1. Physical Productivity

Physical productivity refers to the amount of output produced in physical terms per unit of factor employed. Physical productivity again involves two components *viz.*, APP and MPP.

(a) APP

It indicates the technical efficiency of a variable factor. It refers to the ratio between TPP and number of units or quantity of a variable factor employed in the production programme. It is given by,

$$APP = (TPP/n) = (TPP/Q_X)$$

(b) MPP

It refers to the additional output produced in a production programme due to the employment of additional unit of a variable factor. It is given by, $TPP_n - TPP_{n-1}$.

2. Revenue Productivity

Revenue productivity refers to the amount of output produced in monetary terms per unit of factor employed. From the computation of components of physical productivity, we can easily measure the revenue productivity of the factor employed in the production programme. The components of revenue productivity include, ARP, MRP and Marginal Value Product (MVP) or Value of Marginal Product (VMP).

(a) ARP

It refers to the ratio between TR and number of units or quantity of factor resource employed in the production programme. It is given by, $ARP = (TR/n) = (TR/Q_X)$.

(b) MRP

It can be computed in two ways.

Method 1

MRP is the additional revenue the firm gets by selling an additional unit of output. So,

$MRP = TR_n - TR_{n-1}$. So, MRP can be found by taking the difference between the TRs before and after selling the additional unit of output. For example, if the farmer sells 10 units of output at Rs. 20/unit, then TR = Rs 200. If the farmer sells 11 units of the same output at Rs. 19, then TR = Rs 209. So, MRP=209-200 = Rs. 9.

In perfect competition, since additional output is sold by the firm at the same price, MRP is constant and it is equal to price or ARP of the product (Table 18.2). But, under imperfect competition, ARP or price>MRP (Table 18.3).

Method 2

The MRP can also be studied as the net addition made to the TR due to the employment of an additional unit or quantity of a factor in the production programme, assuming other factors to be fixed at a given level of technology. So, MRP is obtained by multiplying the MPP with the MR. It is given by, MRP=MPP x MR.

(c) MVP or VMP

When MPP is expressed in its monetary terms, it is called MVP or VMP. It is calculated by multiplying MPP of the factor by the price of output. It is given by,

$$MVP\ (\text{or } VMP) = MPP \times P_Y.$$

For example, a farmer in a production programme employed two machines and 50 labourers to produce 1000 units of output. By employing two machines along with 51 labourers, if he produces 1005 units of output, then the MPP of additional unit of labour is five units of output. This additional five units of output produced by employing an additional unit of labour refers to MPP of labour resource. If the price of each unit of output is Rs 10, then MVP or VMP is given by, MVP or VMP = 5 x 10 = Rs. 50.

So, MVP means the monetary value of additional unit of output obtained by the employment of additional unit of factor of production.

From the above discussion, we can easily differentiate between MRP and MVP (or VMP). In MRP, we consider MPP to be multiplied by MR, but in the latter *i.e.*, in MVP or VMP, we consider MPP to be multiplied by the price of the output. So, under perfect market competition, for the product, MRP=MVP (or VMP) because, MR is equal to Price of the product (P_Y). But, under imperfect competition, MVP (or VMP)> MRP, because, Price>MR. So, we can infer that, under perfect competition, [MRP=MVP(or VMP)]= MPP x MR (or Price). But, under imperfect competition, MRP=MPP x MR and MVP (or VMP)=MPP x P_Y, thereby, MRP≠MVP(or VMP) and MVP(or VMP)> MRP, as price>MR.

The above concepts involved in Marginal productivity theory of distribution are measured assuming the following data with respect to perfect market competition (Table 24.1).

Table 24.1: Physical and Revenue productivity under perfect competition assuming the firm transacting paddy

Units of Labour	TPP (Units)	APP = (TPP/n) (Units)	MPP = (ΔTPP/Δn) (Units)	Price/Unit = AR=MR (Rs)	TRP = TR*P_Y (Rs)	ARP = APP*P_Y (Rs)	MRP = MPP*MR (Rs)	MVP or VMP = MPP*P_Y (Rs)
1	8	8.00	-	1000	8000	8000.00	–	–
2	16	8.00	8	1000	16000	8000.00	8000	8000
3	26	8.67	10	1000	26000	8666.67	10000	10000
4	36	9.00	10	1000	36000	9000.00	10000	10000
5	45	9.00	9	1000	45000	9000.00	9000	9000
6	53	8.83	8	1000	53000	8833.33	8000	8000
7	60	8.57	7	1000	60000	8571.43	7000	7000
8	64	8.00	4	1000	64000	8000.00	4000	4000
9	66	7.33	2	1000	66000	7333.33	2000	2000
10	66	6.60	0	1000	66000	6600.00	0	0

As discussed under the laws of returns, we know that, TPP first increases at increasing rate, then at constant rate and at diminishing rate. It, thus, follows the MPP pattern. The same concept is explained through Table 24.1. The marginal and average relationship with respect to the physical productivity of a labour factor (Panel A of the Figure 24.1) clearly infers the following interesting points:

- When MPP increases, APP also increases
- When MPP = APP, APP is maximum
- After the point of intersection of MPP and APP, when MPP declines, APP also declines, but lies above MPP

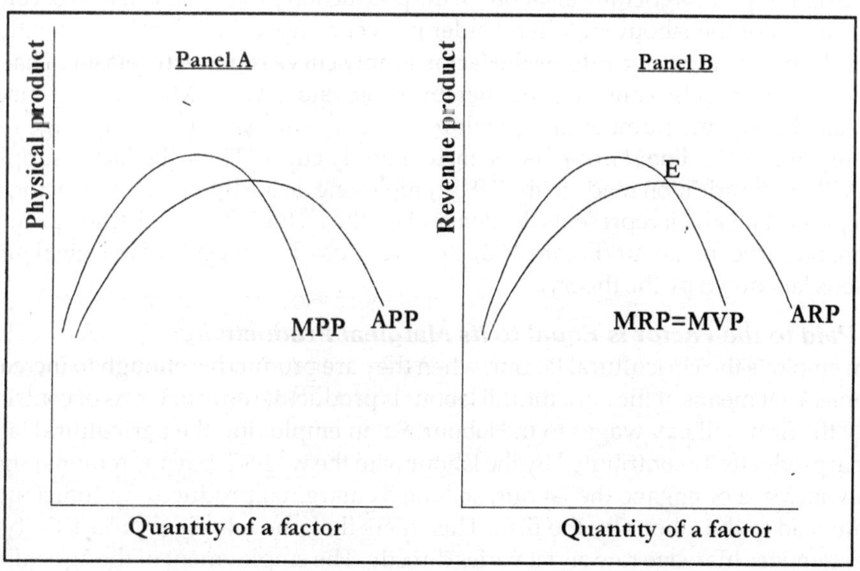

Figure 24.1: ARP and MRP=MVP curves under Perfect competition resemble APP and MPP curves of LDR concept.

The same concept is discussed in terms of revenue productivity and the findings of MRP and ARP are shown through the Table 24.1. The behaviour or relationship of MRP and ARP concepts of revenue productivity are similar to MPP and APP relationship of the concepts of physical productivity and hence, the MRP and ARP curves shows the same pattern and relationship (Panel B of the Figure 24.1) as shown by MPP and APP curves of Panel A of the Figure 24.1. We know, MVP (or VMP)=MRP under perfect market competition and this MRP represents the demand curve for a factor.

So, the definite average-marginal relationship between APP and MPP brings about a similar relationship between ARP and MRP and this infers that, the relationship between ARP and MRP is based on the concept of LDR. Thus, both ARP and MRP are inverted 'U' shaped curves and their relationship infers the following interesting points:

- ☆ When MRP (=MVP or VMP) increases, ARP also increases
- ☆ When MRP(=MVP or VMP) = ARP, ARP is maximum
- ☆ After the point of intersection of MRP and ARP, when MRP declines, ARP also declines, but lies above MRP

As shown in the Panel B, upto point 'E', both MRP (rises and falls) and ARP increases and at point E, when MRP and ARP intersect each other, ARP is maximum and later, beyond point E, MRP falls rapidly compared to ARP and lies below ARP. So, the explanation about the relationship between MRP and ARP is in accordance with the concept of LDR.

Another view is that, the relationship between MRP (=MVP or VMP) and ARP is similar to MC and ATC curves. But, the main difference is that, MRP and ARP are 'inverted U' shaped curves, but MC and ATC curves are 'U' shaped. The relationship between MRP and ARP infers that, when the farmer goes on increasing the variable factor, the efficiency of the variable factor first increases and so, both MRP and ARP increases, but beyond a certain point, both MRP and ARP falls, as the efficiency of the variable factor declines due to lack of effective supervision from the entrepreneur. When both MRP and ARP are rising, MRP is above the ARP, as MRP is related to one additional variable factor and ARP is related to all the units of a variable factor employed in the production programme. For similar reasons, when both MRP and ARP falls, MRP falls rapidly compared to ARP and lies below ARP.

Explanation of the Theory

Let us discuss the main propositions of the theory with reference to a labour factor. Before that, at this juncture, it is essential to introduce two important aspects *viz.*, Average Factor Cost (AFC) and MFC.

AFC

It is the ratio between Total Factor Cost (TFC) and number of units or quantity of factor (labour) employed in the production programme. It is given by, AFC = (TFC/n) = (TFC/Q_X)

MFC

It is the additional cost incurred by the farmer in the production programme by employing an additional unit of a factor resource (labour). So, it is the rate of change of AFC. It is given by, MFC = $\Delta TFC/\Delta n$ (or) $TFC_n - TFC_{n-1}$.

We know, if a firm employs the agricultural labour in the production programme, it is a cost on the part of the firm, but it is revenue on the part of the labour supplier. Under perfect competition, the firm can employ any number of agricultural labour at the prevailing wage rate, as the factor supply curve is perfectly elastic. That means, when wage rate is constant, AFC and MFC are also constant, and thereby, wage rate = AFC = MFC and the curve lies parallel to the X-axis. So, the firm can employ any number of agricultural labour as it likes at a given wage rate=AFC=MFC and thereby, the wage rate=AFC=MFC line represents the factor supply curve. Thus, the factor supply curve is perfectly elastic (Figure 24.2). MRP is the addition made to the TR by employing an additional unit of factor, when the quantities of other factors are kept constant and it represents the demand for the factor in the production programme. That means, MRP is the firm's demand curve for labour (Figure 24.2). Let us discuss the concepts of marginal productivity of labour in terms of propositions laid down by the theory.

(a) Remuneration Paid to the Factor is Equal to its Marginal Productivity

We know, a firm employs the agricultural labour, when they are productive enough to increase the output in the production programme. That means, if the agricultural labour is productive only in terms of contributing or increasing the output, then only the firm will pay wages to the labour. So, in employing the agricultural labour, the firm often compares the marginal productivity contributed by the labour with the wages it pays as remuneration to the labour and the firm continuously increase or engage the labour, so long as marginal productivity contributed by the labour is higher than wage rate paid to the labour by the firm. This infers that, the marginal productivity contributed by the agricultural labour determines his wage rate and it varies directly. The employment of the agricultural labour in a firm increases, so long as the marginal productivity contributed by the labour is greater than wage rate paid to the labour and the firm will come to equilibrium or stop employing additional labour, when the marginal productivity of labour is equal to his wage rate. To generalize the statement, a firm will engage the factor up to the point, where the MC of the factor *i.e.*, MFC (from the firm side) or when the reward for the factor (from the factor side) is equal to its marginal productivity. At this point of equilibrium, the firm will minimize the cost in the production programme. This is because, when marginal productivity of the agricultural labour is greater than the wage rate, it implies, the firm can still engage more number of agricultural labour in the production programme. But, if marginal productivity of labour is less than the wage rate paid by the firm, it is better for the firm to remove excess labour, as they are not contributing much to the total output of the firm in proportion of the wage rate they receive, and thereby, the firm will attain equilibrium when marginal productivity of labour is equal to his wage rate. Thus, the firm will attain equilibrium in the production programme with reference to the employment of factor, when the marginal (revenue) productivity (MRP) is equal to the cost of employing additional unit of factor (MFC).

(b) Firm's Equilibrium in the Short Run

In Figure 24.2, MRP=MVP or VMP is an inverted 'U' shaped curve and WW[1] refers to the wage rate paid to the agricultural labour=AFC (Average wage rate)=MFC (Marginal wage rate) and it is an horizontal straight line parallel to X-axis, assuming perfect competition both in the product and factor markets respectively. It is clear from the Figure 24.2 that,

- We know, MRP=VMP under perfect market competition and MRP indicates the firm's demand curve for labour.
- From the firm's view point, at a given wage rate OW, the factor's supply curve is perfectly elastic *i.e.*, WW_1.
- So, from the above two curves *i.e.*, MRP (= VMP) and WW_1, the firm can draw the information pertaining to the price of the product and wage rate of the agricultural labour.
- At point 'E', MRP=MFC or wage rate derived by the agricultural labour and this infers that, the firm is in equilibrium by employing ON number of agricultural labour.

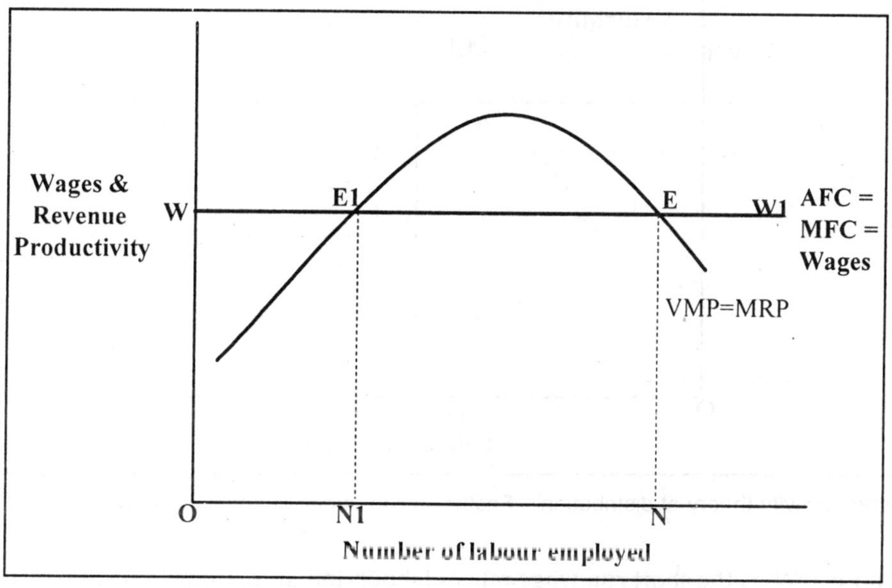

Figure 24.2: Marginal productivity theory of distribution – Factor pricing under Perfect competitive product and factor markets (Short run).

- Below ON number of agricultural labour, MRP of agricultural labour is greater than wage rate paid per labour and thereby, the firm can still employ number of labour up to ON to attain equilibrium. Even at point E_1, the firm is in short run equilibrium, where MRP=MFC, with reference to ON_1 level of employment of labour. But, beyond E_1, the MRP>MFC and thereby, the firm can still increase the employment up to ON number of labour, where MRP is again equal to MFC, and thereby, the firm is in equilibrium.
- If the firm employs more than ON number of agricultural labour, the MRP of agricultural labour is less than the wage rate paid per labour, thereby, the firm must necessarily reduce the number of labour to ON, so as to ensure equilibrium or minimization of costs in the production programme.
- This infers that, in a competitive market, the firm will be in equilibrium when the marginal productivity or MRP of a factor is equal to the MFC (or) price paid per factor with reference to employment of a factor in the short run production programme.
- So, the essential condition for firm's equilibrium in short run with respect to the employment of a factor is that, the demand curve or MRP curve must cut the supply curve of the factor or MFC curve from above, when MRP is falling.

(c) Long Run Industry Equilibrium

When a firm is in short run equilibrium, it does not mean that, the industry is also in equilibrium with respect to employment of a factor. When the firm is in short run equilibrium, it implies MRP=MFC and at this point of equilibrium, the firm maximizes the super-normal profits. These super-normal profits are lucrative and thereby, new firms enter the industry. Consequently, the demand for the factor increases, and thereby, the wage rate of agricultural labour will increase. Since, new firms enter the industry, the supply of the output into the market will increase, thereby, the price of the product will fall in the market and hence, MRP falls. When MRP falls, ARP also falls. This will continue till MRP=ARP=MFC (=AFC=wage rate) and this is the condition of industry equilibrium in the long run (Figure 24.3). That means, in the long run, the essential condition for the firm's equilibrium with reference to employment of labour factor is MRP=ARP=MFC and ARP must be tangent to the MFC (=AFC=Wage rate) at the point of equilibrium. But, in short run, the essential condition for firm's equilibrium is only MRP=MFC. As shown in the Figure 24.3,

- With reference to point 'E', where MRP=ARP=MFC (=AFC=wage rate), the firm is said to be in long run equilibrium, by employing ON number of agricultural labour, where all the firms derive only normal profits in the industry.
- Let us suppose, OW_1 is the short run wage rate of labour and at this wage rate, the firm incurs losses, as ARP is less than OW_1 wage rate by E_1A_1. So, some firms will leave the industry, thereby, the demand for agricultural labour will decrease and the wage rate will fall from OW_1 to OW and again the firm's equilibrium is attained.

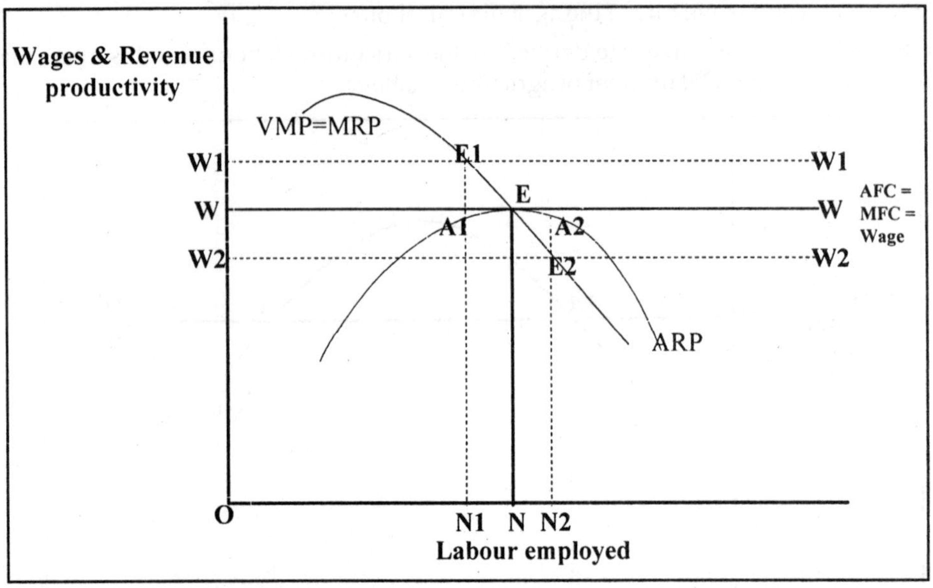

Figure 24.3: Marginal productivity theory of distribution – Factor pricing under Perfect competitive product and factor markets (Long run).

- Let us suppose, OW_2 is the short run wage rate of labour and at this wage rate, the firm will enjoy super-normal profits, as ARP is higher than OW_2 by E_2A_2. New firms get attracted to these super-normal profits and enter the industry. As a result, the demand for the agricultural labour will increase, thereby, wage rate increases from OW_2 to OW. Further, with the increase in supply of the product into the market due to the entry of new firms into the industry, the price of the product will fall and thereby, ARP will fall and gets equated to OW wage rate. So, due to increase in demand for the factor and fall in demand for the product, again the firm will attain long run equilibrium, where, MRP=ARP=MFC (=AFC=OW wage rate).
- Hence, in the long run, the industry will be in equilibrium when MRP=ARP=MFC and ARP is tangent to the MFC at the point of equilibrium, where all the firms will earn only normal profits.

(d) Payment for the Factor Across Different Firms

As discussed under more than one commodity equilibrium or Principle of proportionality, similarly in respect of a firm with reference to employment of more than one factor, we can discuss about producer's equilibrium. For example, if a labour's productivity is more in paddy production compared to maize production, he will be shifted or allocated to the paddy enterprise and any labour who contributes more towards the maize production, he will be allocated under maize production. This is assumed earlier that, factors of production are easily mobile across the enterprises or firms. In the firm, the point of equilibrium under short run is given by MRP=MFC. The same concept of equilibrium is applicable for all the factors employed in the production programme, thereby, MRP of factor A=MFC (= Price) of factor A, MRP of factor B=MFC (=Price) of factor B, MRP of factor C=MFC (= Price) of factor C and so on. It is otherwise expressed as,

[MRP of factor A/MFC (= Price) of factor A] = [MRP of factor B/MFC (= Price) of factor B] = [MRP of factor C/MFC (= Price) of factor C] and so on.

When the above equilibrium is attained with reference to all the factors employed in the production programme then, it ensures LCC of factors.

Criticisms of Marginal Productivity Theory

This theory has been criticized on the following grounds:

- ☆ ***All units of a factor are not homogeneous:*** The assumption that, all the units of a factor are homogeneous is unrealistic. This is because, we know that, the efficiency differs from agricultural labour to agricultural labour. Similarly, one piece of land differs from other in fertility status. It is, therefore, not correct to assume that, the different units of the same factor are homogeneous. In fact, heterogeneity and non-homogeneity is the rule. Hence, no two factor-units are homogeneous, thereby, they are non-substitutable for each other. For example, an expert functioning in a fruit processing industry cannot be employed in a sugar factory.
- ☆ ***Factors are not perfectly mobile:*** This theory assumes that, there is perfect mobility of factors between different employments and places. But, in reality, factors are mostly immobile. There is no automatic movement of factor units from one industry or place to other. For example, the mobility of labour across different occupations is influenced not only by the wage rate, but by several other factors like, seniority, pensions, permanency of the job, work-leisure ratio, sociological and ethical aspects etc. The greater the degree of specialization in the industry, the less is the factor mobility from one industry to other. Due to less mobility, the units of same factor are not paid equally in proportion to their marginal productivities across the occupations and across the places.
- ☆ ***There is no perfect competition:*** This theory is based on another unrealistic assumption of perfect competition both in factor and product markets. We know, imperfect competition is the rule and it leads to exploitation of factors, as they are paid much less than their marginal productivity. Prof. Robinson has applied this marginal productivity theory to imperfect competition.
- ☆ ***Factors are not fully employed:*** This theory assumes the existence of full employment equilibrium in the economy. But, unemployment and underemployment is the rule. This is because, we practically observe that, factor services are paid less than their marginal productivity. This is explained by Keynes based on the theory of effective demand.
- ☆ ***All factors are not perfectly divisible:*** The assumption that, factors are perfectly divisible and therefore, can be increased by small amounts does not hold good. This is because, it is not possible to increase the labour service by fraction amounts, such as, half time or one and a half times etc.
- ☆ ***Production is not a result of one factor alone:*** It is not possible to calculate the marginal productivity of each factor unit separately. This is because, the production of a commodity cannot be attributed to any one factor *i.e.*, land or labour or capital. Rather, it is always a result of coordinated and combined efforts of different factors.
- ☆ ***Profit motive is not the only motive:*** This theory assumes that, the firms are motivated by the maximization of profits. Hence, more units of a factor service are employed when the firm finds the MRP of the factor is higher than its price. But, as pointed out by Schumpeter, the firm's action is guided by the desire to get things done in the firm, rather than influenced by profit motive alone. So, besides profit maximization, the firm may have other priorities like social welfare, building a commercial empire, over throwing the rivals, joy of taking up business, successful handling of risks and uncertainities etc., and these influences the entrepreneur to take up the business.
- ☆ ***Not applicable in the short-run:*** This theory is applicable only in the long run, as the rewards of the factor services tend to equal their MRP. But, in reality, we are more concerned with the short run problems. As remarked by the Keynes, *'in long run we all are dead'*. So, this assumption of long run analysis makes the concept of factor pricing unrealistic.
- ☆ ***Neglects technical progress:*** This theory fails to throw light on the influence of technical changes in increasing the marginal productivity of factor service. Prof. Hicks explained that, the labour saving innovation tends to raise the marginal productivity of capital relative to that of labour. The opposite may happen in case of capital saving innovations. Thus, Keynes criticized this theory as a *'static concept'*.
- ☆ ***Supply of factors is not fixed:*** This theory assumes that, supply of factors is perfectly elastic even in the long run. But, the supply of factors is fixed in the short run and not in the long run. Moreover, this theory studied the factor pricing from demand side only and neglecting the supply side. Hence, this theory cannot be applied to the factor markets, which requires explanation from both demand and supply forces.
- ☆ ***Sum of factor payments is not equal to the total product:*** Since, each factor is paid according to its marginal product, it follows that, sum of all payments to factors will be equal to total product. But, in reality, the sum of marginal products is greater than the total product. Thus, there is a surplus, which is the result of cooperative working of all the factors. If we increase one factor in a production programme, extra output we generate is not only due to the use of extra factor, but due to working of that extra unit with the existing factors, thereby, leading to higher output. Likewise, if at any time, an important factor unit is withdrawn from the productive process, it may completely disorganize the business. The withdrawal of this factor unit will diminish the total product by more than the marginal product of that unit (capital/machinery) and sometimes totally

(land). Thus, the theory of marginal productivity does not provide the correct measure of pricing the factor services.

- ☆ ***Ignores ethical aspect:*** This theory deals with only positive aspect of analysis pertaining to process of factor pricing and it does not have ethical aspect. It is not concerned with ethical justification of payment of remuneration to the factors.
- ☆ ***Employment level depends upon effective demand:*** This theory supports the Pigou's version that, employment of labour can be increased by reducing the wage rate. But, Keynes criticized this version and opined that, when wage rate is reduced, income of the labour will decrease, thereby, effective demand declines, leading to unemployment and under employment problems in the economy.
- ☆ ***Changing factor proportions is not easy:*** According to Hobson, it is not quite easy to change the factor proportions or substitute one factor for another factor in a production programme. This is because, the factor proportion in a production programme is determined by the technological changes. Moreover, some factors will combine in a fixed proportion. For example, if the farmer employs tractor for ploughing the land, tractor hours and driver hours combine in a fixed proportion. We cannot employ two drivers for one tractor, even though drivers available in more number and their wage rate is low.
- ☆ ***Reward for the factor determines productivity:*** One of the important propositions of the marginal productivity theory is that, the marginal productivity of a factor determines its payment or remuneration. So, we can infer that, marginal productivity of the factor is the cause and remuneration paid to the factor is the effect. But, according to Prof. Sydney Webb, the proposition also applies in the reverse way. For example, when an agricultural labour is provided with higher wage rate, his income will increase, standard of living will increase, consumes more quality food and thereby, his productivity in the job work will increase. So, here, remuneration paid to the labour is the cause and marginal productivity contributed by him is the effect. Thus, the proposition is also valid in the reverse way.
- ☆ ***Measuring MRP is a difficult task:*** According to this theory, marginal productivity of the factor determines its remuneration. But, measuring MRP is not an easy task. We know, in several Governmental organizations, even if we remove one labour or worker from the organization, the TPP will not fall and this implies that, MRP of the worker is zero. So, the organization should not pay any wage or salary to that worker. But, this is absurd to say so. Similarly, in some businesses, there is only one entrepreneur and to measure the MRP of entrepreneur, we cannot remove the entrepreneur from the business and moreover, we cannot employ another entrepreneur in the business.
- ☆ ***No explanation how remuneration is fixed for the factors:*** This theory simply offers the explanation regarding the level of employment of a factor (agricultural labour) at a given wage rate. But, this does not offer explanation regarding the determination of wage rate for labour. Thus, this theory does not explain how factor's remuneration is determined.
- ☆ ***LDR not common in manufacturing industries:*** This theory assumes that, as the firm goes on employing the factors, LDR sets in to a factor. But, in case of manufacturing industries, where improved technology is adopted, the factors will contribute increasing returns and LDR is held in check.
- ☆ ***Mostly applicable for pricing of labour factor:*** This theory is mostly applicable for pricing of labour factor and loosely extended for other factors of production. The pricing of other factors like rent and capital is totally distinct compared to pricing of labour factor, as these two factors have distinct features compared to labour factor. Moreover, if the entrepreneur gets loss in the farm business, we cannot say the MRP is negative for the entrepreneurial factors.
- ☆ ***Explains only functional distribution:*** This theory explains only functional distribution of income and does not explain the personal distribution of income and income inequalities in the earnings of the people.

Considering the above criticisms, we can infer that, this theory is of little practical use. This theory assumes too much, that too relied on unrealistic assumptions. So, this concept of marginal productivity is of no use in dealing with the pricing and problems of distribution. According to Benham, '*the marginal productivity theory is a micro economic theory and needs to be supplemented by a general theory of output and employment such as that of Keynes. Moreover, some of its assumptions as we have seen, are incorrect or apply only to a part of the economy. Nevertheless, I repeat that, in my view of opinion, it is useful as a first approximation and there is no other general theory of distribution which is superior to it*'. From the words of Benham, we can infer that, this theory has an important significance (though it has not offered full and valid explanation about pricing of factors), as marginal productivity is the most important economic factor that governs the price of the factor. So, in spite of the above criticisms, this theory has offered a reasonable and understandable concept of explaining the pricing of factors.

Factor Pricing Under Imperfect Competition

The concept of marginal productivity theory explains factor pricing by assuming perfect market competition both in the factor and product markets. But, the real world is of imperfect competition. So, let us study the factor pricing assuming imperfect competition and we have three different cases as given below:

1. When the factor market is perfectly competitive and the product market is imperfectly competitive or monopolistic.
2. When there is a monopsony in the factor market and perfect competition in the product market.
3. When there is the monopsony in the factor market and monopoly in the product market.

1. When the Factor Market is Perfectly Competitive and Product Market is Under Monopolistic Competition

When the factor market is perfectly competitive, it implies, the firm can purchase any quantity of the factor at the same price. That means, in terms of costs, the factor cost/unit remains same, irrespective of the quantity of the factors engaged by the firm. Hence, supply curve of the factor or AFC or MFC lies parallel to the X-axis and it is perfectly elastic (Figure 24.4).

- ☆ Since, we assume product market is operating under monopolistic competition, MRP curve lies below the MVP or VMP curve. This is because, under monopolistic competition, price>MR and hence, VMP>MRP and these two curves slopes downward from left to right.
- ☆ As discussed earlier, with reference to factor employment, the firm will be in equilibrium, when MRP=MFC and it is at point 'E', where the firm employs OQ quantity of factor at OP=QE price in perfect competitive factor market.

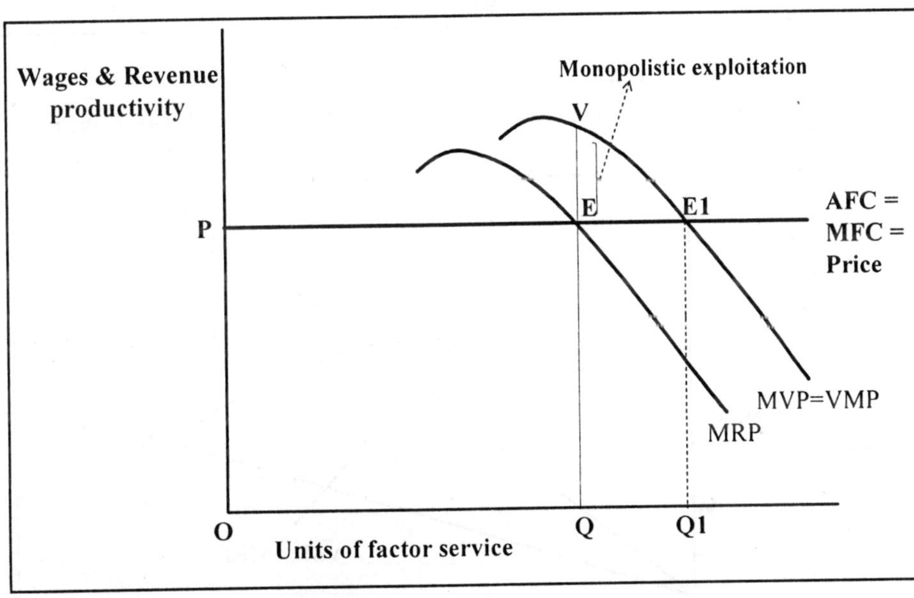

Figure 24.4: Factor pricing under imperfect competition
Case - Factor Market is perfect competitive and Product market is monopolistic (Imperfect)

- ☆ Thus, MRP=MFC is OP=QE and it is less than its MVP or VMP, QV by 'VE'. This infers that, the factor service gets less compared to the value of its marginal product. So, the factor gets exploited by VE and this is called Monopolistic exploitation. So, if the factor is paid less than MVP or VMP under conditions of perfect competitive factor market and imperfect (monopolistic) competitive product market, it is called Monopolistic exploitation.
- ☆ Since, the product market is operating under monopolistic competition, only less quantity of factor is employed in the production programme *i.e.*, OQ.
- ☆ Had the product market operates under perfect competition, the firm employs more quantity of factor *i.e.*, OQ_1, where MFC = MRP (=VMP) at E_1 *i.e.*, what the factor gets is equal to the VMP.
- ☆ This infers that, when the product market is imperfect, the demand for the factor is less by QQ_1 compared to product market under perfect competition.

2. Monopsony in the Factor Market and Perfect Competition in the Product Market

Monopsony in a factor market implies, the firm is the single buyer (employer) of the factor. Monopsony (from Ancient Greek words, '*monos*' means 'single' + '*opsônia*' means 'purchase') is a market form with only one buyer, but facing many sellers. It is an instance of imperfect competition, symmetrical to the case of a monopoly, in which there is only one seller facing many buyers. The term 'monopsony' was first introduced by Joan Robinson. Monopsonist is the sole employer of the factor and no one will compete with him in purchasing the factor service. So, he will have control on the employment of the factor. For example, if he needs more of the factor service (say, labour), he will pay more price (wages). So, both AFC and MFC are upward sloping from left to right and AFC lies below MFC, as MFC refers to one additional unit of factor, whereas AFC refers to all units of factors employed. It is clear from the Table 24.2 and Figure 24.5 that,

Table 24.2: AFC and MFC under imperfect competition in factor market

Units of Factor	*Price of Factor Service*	*TFC*	*AFC (TFC/X)*	*MFC (ΔTFC/ΔX)*
1	50	50	50	–
2	51	102	51	52
3	52	156	52	54
4	53	212	53	56
5	54	270	54	58
6	55	330	55	60
7	56	392	56	62
8	57	456	57	64
9	58	522	58	66
10	59	590	59	68

- ☆ MFC>AFC=Price of the factor service.
- ☆ Since, we assumed perfect competition in the product market, MVP(=VMP)=MRP, as Price=MR under perfect competition.

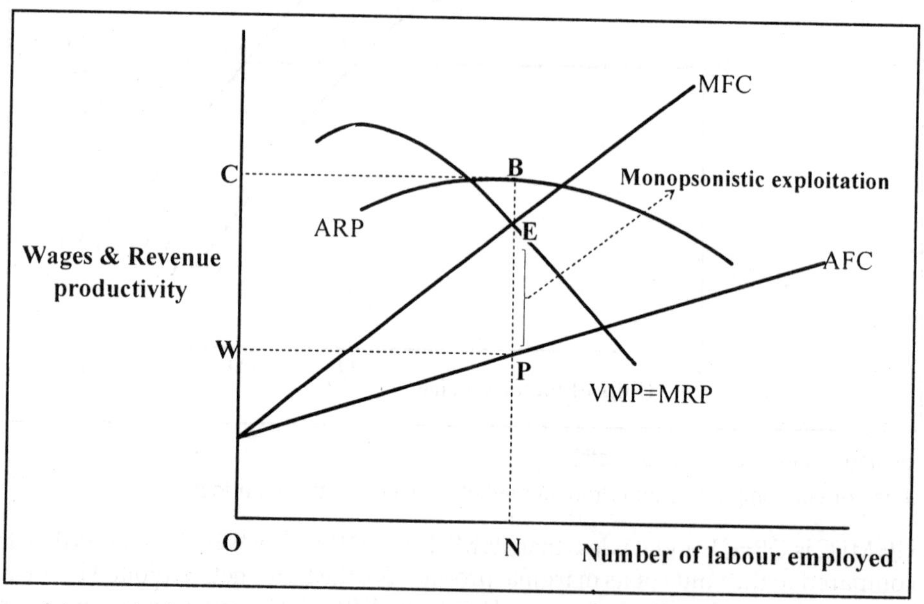

Figure 24.5: Factor pricing under imperfect competition case – Factor Market is monopsony and Product market is perfect competitive.

- ☆ The firm is in equilibrium at point 'E', where MRP=VMP curve intersects the MFC curve from above. This implies that, the firm employs ON number of labour, but the average wage received by the factor (AFC) *i.e.*, OW=NP is less than MRP=VMP, NE by PE.
- ☆ Thus, the labour contributes more to the employer and receives fewer wages from the employer (*i.e.*, NE>NP).
- ☆ This infers that, being a monopsonist, the firm exploits the labour factor and earns super-normal profits PWBC. This exploitation of labour factor by PE refers to monopsonistic exploitation of factor. So, if labour is paid less than MRP=VMP under conditions of monopsony in factor market and perfect competition in product market, it is called Monopsonistic exploitation.

Under perfect competition in factor market, the firm maximizes its profits by employing more and more quantities of the factor till the MRP of the factor is equal to the MFC=AFC or price paid for the factor (as discussed under Figure 24.4). But, under imperfect factor market (*i.e.*, monopsony in factor market), a monopsonist maximizes his profit by paying less (AFC) to the factor than the MRP=VMP derived from that factor, as the product market is under perfect competition. In other words, the factor contributes more to the employer than the wage it receives from the employer. This is called monopsonistic exploitation of the factor service by the employer.

3. Monopsony in the Factor Market and Monopoly in the Product Market

As discussed above, when there is monopsony in the factor market, it implies, MFC>AFC. Similarly, when there is monopoly in the product market, MVP(=VMP)>MRP. As shown in the Figure 24.6,

- The firm will be in equilibrium at point E, where the MRP curve intersects the MFC curve from above. This implies that, the firm employs ON number of labour by paying NP price (wage rate) to the factor service, which is less than MRP of the factor, NE, *i.e.*, by PE.
- Since, MRP *i.e.*, NE is greater than NP price (wage rate) paid to factor service by PE, the factor is paid less than the marginal productivity derived from that factor or the factor contributes more to the employer than the price it receives from the employer and this is called Monopsonistic exploitation of the factor service by the employer.

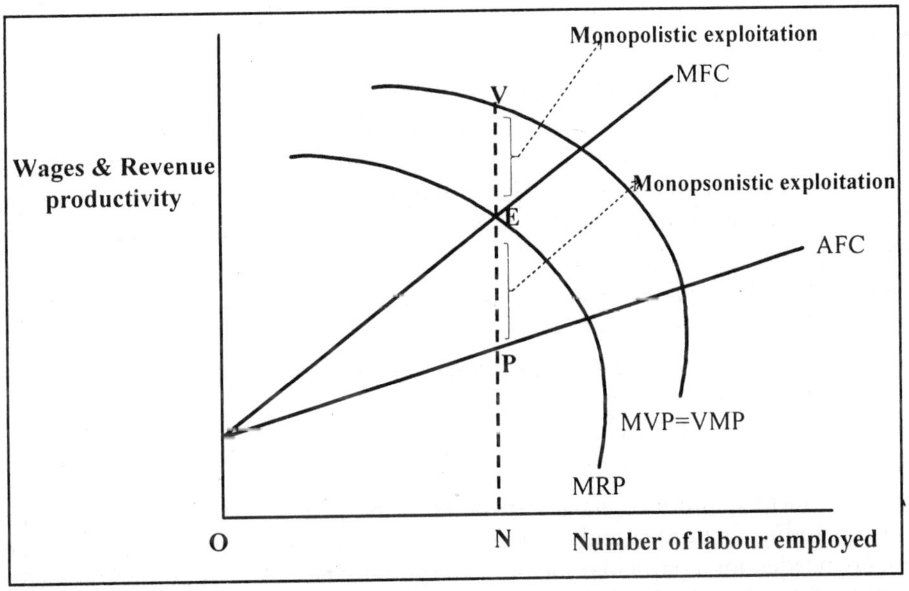

Figure 24.6: Factor pricing under imperfect competition case – Factor Market is monopsony and Product market is monopoly.

- On the other hand, due to monopoly position in the product market, the MVP(=VMP)>MRP of the factor, since price>MR under monopoly competition. Hence, the firm further exploits the factor units employed to the extent of EV amount. This exploitation of factor *i.e.*, difference between MVP(=VMP) and MRP=MFC by VE refers to Monopolistic exploitation. Thus, the total exploitation is PV.
- So, in the case of monopsony in the factor market and monopoly in the product market, the factor employed in the production process by the firm is doubly exploited:
    - First, due to the excess of MRP over the AFC=price of the factor (Monopsonistic exploitation), and
    - Second, due to the excess of the VMP over the MRP=MFC of the factor (Monopolistic exploitation).

Reasons for Monopsony in Factor Market

Monopsonistic situation may arise in the following ways:

- The firm may require a specialized factor service to produce the output. In the absence of that factor, the firm cannot produce that product. For example, if there is only one sugar factory in an area, it enjoys monopsonistic competition, as it is the sole purchaser of raw material in that area.
- Immobility of factors of production
- Ignorance of the factor supplier about the availability of alternate sources of employment.

Measures to Overcome Monopsonistic Exploitation

The following are the measures to over come monopsonistic exploitation of a business firm:

- ☆ The factor suppliers may organize themselves to demand a particular price for their service from the entrepreneur on par with the marginal productivity they contribute to the firm.
- ☆ The Government should fix minimum prices for the factors, by passing Minimum Wages Act, to over come their exploitation.

II. Modern Theory of Distribution or Demand and Supply Theory of Distribution

The marginal productivity theory explains only the factor equilibrium both in short run and long run production programmes *i.e.*, how many labourers the firm will employ at a given wage level in order to maximize the profits. But, it won't explain, how the wage rate is determined. That means, marginal productivity theory explains the reward for a factor, only from the demand side and not from the supply side. It is further based on several unrealistic assumptions like perfect competition, all units of a factor are homogenous, perfect divisibility of factors, perfect mobility of factors, perfect substitutability of factors etc. Hence, this theory could not offer satisfactory explanation to determine the pricing of factors. The theory that offered satisfactory explanation to factor pricing is Demand and Supply theory. Just as the product price is determined by demand and supply forces, this theory explains that, factor prices are also determined by demand and supply forces in the market. But, there are certain points of difference between demand and supply of factors and products.

Demand Side of a Factor

The following are the few points connected with the demand for a factor service:

- ☆ In case of products, the demand is a direct demand, but in case of factors, the demand is derived demand. This is because, the firms or producers demand factors because, there is a demand for the products which are derived from the services of these factors. Hence, the demand for factors is only a derived demand and not direct demand, as the factors do not cater to the needs of the customers directly, but indirectly through producing the products.
- ☆ The demand for a factor is a joint interdependent demand, as the combined and coordinated efforts of all the factors contribute to production of output.
- ☆ If the demand for the product increases, the demand for a factor, which involves in the production of that product also increases and *vice-versa*.
- ☆ If the demand for the product is elastic, the demand for the factor, which produce that product is also elastic and *vice-versa*.
- ☆ The demand for a factor of production will depend upon the quantity of other cooperant factors of production required in the production process and it varies directly. That means, if more number of cooperant factors of production are needed for a particular factor, then the demand will be high for that factor. For example, if demand increases for the use of machinery in the production programme, demand also increases for other factors like diesel, skilled workers etc.
- ☆ If the factor has cheaper substitutes, the E_D for that factor will be high and *vice versa*.
- ☆ The demand price for a factor depends upon the value of the finished product and it varies directly. That means, if the value of finished product is more, the demand for the factor involved in the production of that product will be more and its *vice versa*.
- ☆ The more productive is the factor, the higher will be the demand for that factor to be employed in the production process. But, if the factor usage is continuously increased in the production programme, the marginal productivity of that factor will gradually decline. So, if the marginal productivity of that factor decreases, the firm has to employ more of that factor to produce the same level of output and *vice versa*.
- ☆ If the factor employed constitutes only a small proportion of the total quantity of all the factors employed in the production programme, the demand for that factor is less elastic and *vice versa*.
- ☆ The quantity of factor employed in the production programme varies inversely with its price level.
- ☆ The price of a factor is not determined based on an individual firm's demand, but based on the industry demand. We can have a demand curve for a factor at industry level summing up the demand curves for that factor of all the firms in that industry (Figure 24.7). If one factor is used in several industries, we can have a demand curve for that factor by summing up the demand curves of all the industries.
- ☆ Panel A of the Figure 24.7 shows the employment of factor in a firm. MRP is the demand curve for that factor (labour) for a firm and it is inverted U shaped, indicating that, if the firm employs more of that factor continuously, its marginal productivity (MRP) initially increases and then decreases. The entrepreneur goes

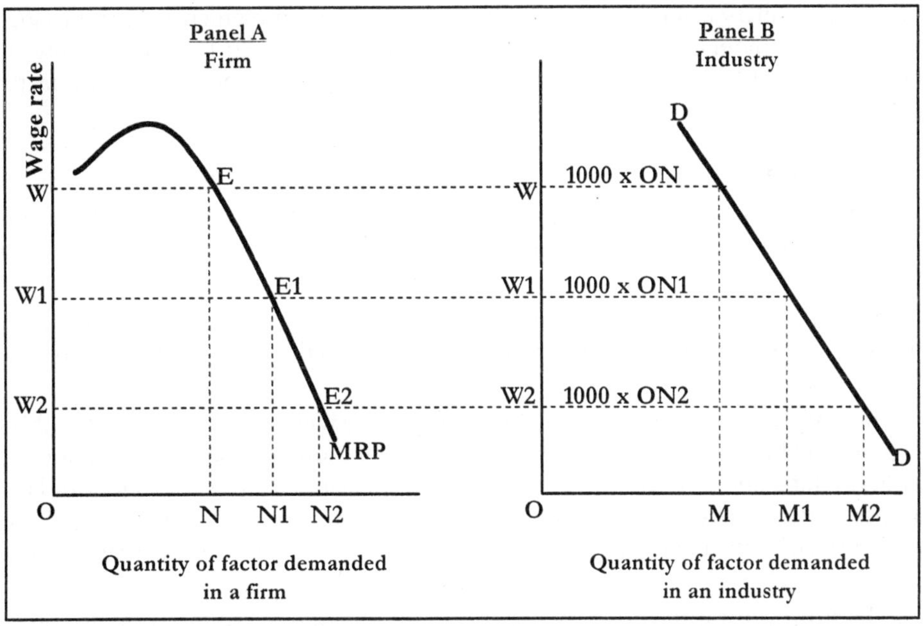

Figure 24.7: Determination of demand curve for a factor at industry level.

on employing this factor (labour) as long as the MRP of labour is greater than MFC and the firm is at equilibrium when MRP=MFC.

- ☆ When the wage rate is OW, the firm is in equilibrium at E, where MRP=MFC(=wage rate) and the firm employs or demands ON number of labour.
- ☆ Similarly, when wage rate is decreased to OW_1, the firm is in equilibrium at E_1 and firm employs or demands ON_1 number of labour.
- ☆ When the wage rate is still decreased to OW_2, the firm is in equilibrium at E_2 and firm employs or demands ON_2 number of labour.
- ☆ The line connecting the equilibrium points, E, E_1, E_2, represents the diminishing marginal productivity of labour (MRP), as the firm goes on employing more number of labour in the production programme.
- ☆ But, for determining the price of a factor, it is not the demand of an individual firm that matters, but the demand of all the firms in the industry that matters. So the industry demand curve for a factor is derived by taking into consideration the demand of all the firms employing that factor. This is shown in Panel B of the Figure 24.7, considering lateral summation of demand curves or MRP curves of all the firms in the industry.
- ☆ Let us assume that, there are 1000 firms in the industry, employing the same factor (labour). In both the Panels of the Figure 24.7, wage rate is considered on Y-axis, but in Panel A, X-axis represents the number of labour employed in a firm and in Panel B, X-axis represents the number of labour employed at industry level. This is so because, in Panel A, the demand for labour is in one firm, whereas, in Panel B, the demand for labour is in 1000 firms or industry.
- ☆ So, at wage rate OW, the total demand for labour in industry is equal to 1000 firms x ON=OM. Similarly, at wage rates OW_1 and OW_2 the total demand for labour in industry is equal to 1000 firms x ON_1=OM_1 and 1000 firms x ON_2=OM_2 respectively.
- ☆ From Panel B, it is clear that, the demand curve DD for labour in the industry also slopes downward from left to right like MRP. This is due to the downward falling of MRP curve from left to right of all the firms in the relevant portion of DD curve. This industry demand curve will be inelastic in the short run and more elastic in the long run.
- ☆ This implies that, the more a factor is employed, the marginal productivity of the factor diminishes and this is called as Law of diminishing marginal productivity.
- ☆ In Panel B of the Figure 24.7, the demand curve for the factor in the industry slopes downward from left to right. Here, we move along the same demand curve due to changes in wage structure of the labour. This change refers to the change in number of labour demanded by the industry. But, there is also another important change called 'change in demand' and this involves altogether shift in the demand curve for labour not due to changes in wage structure of labour, but due to other factors mentioned below.
  - ❖ If there is an increase in demand for the final product, the demand for labour will increase even at the same wage rate and *vice versa*. So, the demand curve for labour in the industry will shift towards right.

- When the price of a substitute product increases or price of complementary product increases, the demand for the product under consideration will increase in the market. So, to produce more of this product, more labour will be employed even at the same wage rate. So, the demand curve for the factor will shift towards right.

Supply Side of a Factor

If price or remuneration increases, the supply of the factor also increases and in this case, it is almost similar to supply of the product. But, the supply curve of the factor is not completely identical with supply curve of the product. This is explained here under.

- ***Land:*** In general, supply of the land is limited and hence, the supply curve is parallel to Y-axis and this is from an economy point of view. But, from firm or industry point of view, the supply of land can be increased, if more rent is paid by the entrepreneur.
- ***Labour:*** The supply of labour service will increase, if more wage rate is offered. But, this is not always true. This is because, at higher wage rate, the labour will prefer more number of leisure hours and contribute less number of working hours. This is because, they have inelastic cash requirements and hence, they work for less number of working hours at higher wage rate and earn same level of income and prefers to enjoy more number of leisure hours. Hence, in this case, the supply curve will have a backward bending slope (Figure 9.59).
- ***Capital:*** The supply of capital not only depends upon the rate of interest, but also on the level of income, business expectations etc.
- ***Entrepreneur:*** The supply of entrepreneur not only influenced by the profits, but also depends upon various socio-economic factors.

Besides wage rate, other factors like population, geographical conditions, cost of education, training, labour efficiency, leisure preference etc., also influence the supply of labour. Even non-economic considerations like traditions, customs, habits etc., also influence the labour supply. So, taking into consideration the above aspects, the supply of a factor (labour) in the firm is not similar to the supply of a product by the firm. However, for simplicity sake, we can assume, supply of factor increases with increase in remuneration and hence, the supply curve slopes upward from left to right.

Determination of Equilibrium Price for a Factor in an Industry

Having studied the demand curve and supply curve of the factor, we need both the curves to determine the price of the factor. The equilibrium price of the factor in the industry is at the point of equality of demand and supply curves of a factor.

- In the Figure 24.8, DD and SS are the demand and supply curves of the factor (labour). These two curves intersect at point 'E' and corresponding points on Y-axis and X-axis implies that, at OW wage rate of labour, ON number of labour are employed in the industry.

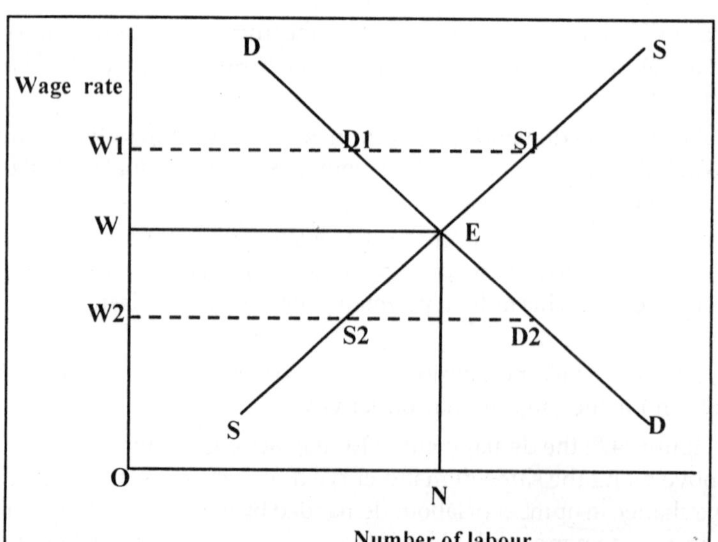

Figure 24.8: Demand and Supply theory of distribution – Determination of equilibrium price for a factor.

☆ If wage rate is increased to OW_1, then the demand for labour (W_1D_1) is less than supply of the labour (W_1S_1) by D_1S_1. So D_1S_1 labour will remain unemployed. So, some labour show willingness to work even at low wage rate, thereby, wage rate gradually decreases from OW_1 to OW and ensures equilibrium employment of labour.

☆ If wage rate is decreased to OW_2, then the demand for labour (W_2D_2) is greater than supply of labour (W_2S_2) by S_2D_2. So, for some firms, labour is not available at lower wage rate. So, some firms even try to engage labour by offering higher wage rate. Hence, slowly the wage rate will be increased to OW, as demand for labour is more than supply of labour. So at OW wage rate, both demand and supply of labour will be equal and all firms can engage labour.

So, this is how, the price of the factor is determined in the industry based on both demand and supply forces. So, the Modern Economists considered this theory has offered satisfactory explanations for pricing the factors with compared to Marginal productivity theory of distribution.

25

Rent

In the previous chapter, we studied the general theory of distribution and now, we are in a position to analyze the pricing of different factors of production *viz.*, land, labour, capital and management. The factor, 'land' is distinct among the four factors of production, in the sense that, it is natural in its source compared to other three factors of production. So, in Economics, land refers to all those gifts of nature in their original form, which can neither be reproduced by man nor destroyed by him. This infers that, land does not mean soil alone fit for cultivation of crops, but also includes other natural resources like forest resources, mineral resources, water, lower strata of the earth etc. Considering these special features, land is an important factor of production and to be more accurate, it is an indispensable factor of production. In the absence of this factor, no other factors of production can be employed in the production programme, in an economic sense.

25.1. Contract Rent Vs. Economic Rent

The term 'rent' is used to denote the remuneration paid for the use of land. The same term 'rent' is used to denote the hiring charges of a machinery, car, house etc. So, in an ordinary language, we use the term 'rent' in the sense of cost of hiring of an asset for a certain fixed period of time. But, in economic sense, the term 'rent' is used in a special sense. In the above examples *i.e.*, when a farmer pays remuneration for taking the land for lease purpose, when a person hires a taxi, when a person pays rent to the house owner etc., these payments fall under Contract rent or Gross rent, as it refers to the total payment made by the person for hiring these assets for a certain fixed period of time. Thus, contract rent refers to that rent, which is mutually agreed upon between the owner and the user. It is so called because say, if the farmer pays the (gross) rent to the land-lord, it not only includes the payment made towards the use of land alone, but also includes other payments like, interest for the use of capital invested on that piece of land, depreciation and maintenance charges of assets created on land, wages for management and other services, payment towards profits for bearing the risks of leasing and investing etc. So, the contract rent or gross rent covers all these payments. So, among all the payments made by the farmer, the one payment that is meant for using the land alone refers to 'Rent' in Economics or Economic rent. So, Economic rent is that part of total payments made by the farmer for the use of land alone, whereas, Contract rent or Gross rent refers to the total payment made by the farmer to the land lord. So, contract rent or gross rent is a much broader term compared to economic rent and economic rent is only a part of the contract rent. We can have a better understanding of the term 'rent', by drawing the distinction between contract rent and economic rent (Table 25.1).

Table 25.1: Differences between Contract rent and Economic rent

Contract Rent	*Economic Rent*
It is the total contractual payment made by the tenant to the owner for the use of owner's property for a certain fixed period of time.	It is the part of the contractual payment made by the tenant to the owner for using the land alone.
Contract rent = Economic rent + Interest + Depreciation and maintenance charges + Wages + Profits for the risk involved.	Economic rent = Contract rent – (Interest + Depreciation and maintenance charges + wages + profits for the risk involved).
Contract rent is higher than economic rent.	Economic rent is lower than contract rent.
Contract rent is a gross rent.	Economic rent is a pure rent.

25.2. Different Concepts of Rent

In layman's language, the term rent is used in the sense of contract rent. But, in economic sense, the term rent is used in the sense of 'economic rent' only. There are two different approaches *viz.*, Classical approach and Modern approach to explain the concept of economic rent.

1. Classical Approach

According to both David Ricardo and Alfred Marshall, 'rent' is a payment made for the use of land alone. In the words of David Ricardo, *'rent is that portion of the produce of the earth, which is paid to the land lord for the use of original and indestructible powers of the soil'*. According to Alfred Marshall, *'rent is the income derived by the owner of land and other free gifts of nature'*. He opined that, the supply of land is perfectly inelastic from the economy point of view, but considering a firm or an industry, the supply of land can be increased, if it is paid with higher remuneration, thereby, it is not perfectly inelastic. David Ricardo differentiated the contract rent and economic rent and he opined economic rent is the rent paid for the use of land alone, unlike contract rent, which includes other items like interest on capital, wages for the supervision of land, depreciation and maintenance charges etc. Marshall extended the concept of rent to the payments made for the use of man-made machinery, whose supply is also inelastic like that of land in the short run and he named this payment as 'Quasi-rent'.

2. Modern Approach

Modern Economists like Boulding and Joan Robinson have extended the concept of rent even to the other factors of production, whose supply is also inelastic. They opined that, rent arises as a payment to any factor of production over and above its transfer earnings or supply price. That means, economic rent is the excess of payment made to any factor of production over and above its transfer earnings. The transfer earnings refers to the minimum payment that is required for the factor to continue in its present usage. Say, for example, on a piece of land, a farmer cultivates paddy and earns Rs. 10,000 and in its next best use, it earns Rs. 8000 by producing maize. So, in its present use, the land yields Rs. 2000 more than its next best use. So, this surplus of Rs. 2000 income, which the land yields in its present use, refers to economic rent. This infers that, the farmer must get at least Rs. 8000 income to put the land into cultivation for cultivating paddy. That means, if the farmer cultivates paddy, it must yield equal to or more than Rs. 8000, otherwise, the farmer will start producing maize. If paddy crop is projected to yield Rs. 10000, the farmer shifts to paddy crop instead of maize, as it contributes Rs. 2000 income extra over Rs. 8000. So, Rs. 8000 is the transfer earnings of the land factor and the excess of income the land yields to the farmer over and above the transfer earnings is the ' economic rent' *i.e.*, Rs. 2000. So, Rs. 8000 is the minimum earnings that makes the farmer to continue the use of land factor in its present or current usage (*i.e.*, cultivating paddy). If the actual earnings fall below Rs. 8000, then the farmer starts cultivating maize rather than paddy. Thus, the concept of transfer earnings guides in explaining the economic rent earned by the factor. Thus, according to modern approach, economic rent is the surplus or excess of earnings of any factor of production over its transfer earnings or economic rent is the difference between present earnings and transfer earnings.

25.3. Definitions of Rent

Having studied the different approaches to study the concept of rent, it is revealed that, different Economists perceived the concept of 'rent' differently. In this context, it is essential to study the definitions of rent, as given by the Classical and Modern Economists.

Classical Definitions

'Rent is that portion of the produce of the earth which is paid to the landlord for the use of the original and indestructible powers of the soil'.

– *(Ricardo)*

'Rent may be defined as the income which accures from the ownership of the land and of other free gift of nature'.

– (Thomas)

'Rent is the price for the use of land'

– (Carver)

'The income derived from the ownership of the land and other free gift of nature is commonly called rent'.

– (Marshall)

Modern Definitions

Modern theory of rent is not confined to land alone, but is applicable to others factors of production as well. In modern theory, economic rent is defined as a surplus payment made to some units of a factor over and above their transfer earnings. In other words, economic rent is the excess of present actual earnings over and above the transfer earnings or supply price. Thus,

Economic rent = Present earnings – Transfer earnings

'Economic rent may be defined as any payments to a factor of which is in excess of the minimum amount necessary to keep the factor in its present occupations'.

– (Boulding)

'Rent is difference between actual payment to a factor and its supply price or transfer earnings'.

– (Hibdon)

In short, economic rent is the reward paid in excess over transfer earnings to keep a particular factor in its current use, whether that factor would be land, labour or capital. Thus, the concept of economic rent is applicable to any factor of production.

25.4. Theories of Rent

There are two popular theories of rent *viz.*, Ricardian theory and Modern theory. Let us discuss one by one here under.

I. Ricardian Theory of Rent or Classical Theory of Rent

Different Classical Economists like West, Torrents, Malthus and Ricardo independently formulated the theory of rent and among them, Ricardian theory of rent gained more popularity though the principle behind all the theories is the same. The Ricardian theory of rent proposed by David Ricardo is based on the ideas of the theories propounded by West and Malthus. According to David Ricardo, *'rent is that portion of the produce of the earth, which is paid to the land lord for the use of original and indestructible powers of the soil'*. That means, a tenant pays rent to the land lord for the use of natural productive properties of the soil, which are indestructible in nature. Since the productive properties of the soil are natural and which do not involve any cost of production, 'rent' is a unique factor payment, which is not determined by the cost of production. Hence, rent is a true surplus. According to Ricardo, 'rent is a differential surplus', *i.e.*, surplus earned by more fertile plot of the land in comparison with less fertile plot of the land. That means, the extra produce the farmer gets from a superior (more fertile) plot of the land over the farmer producing the same produce from a less fertile plot of the land constitute rent. From this theory of rent, we can learn the following reasons for earning economic rent:

Economic Rent is a True Surplus

According to Ricardo, economic rent is the true surplus left over after meeting the expenses of cultivation like payment to labour, capital and entrepreneur. This concept of economic rent arises due to niggardliness of nature.

Differences in Fertility Status of the Soils

Since, the plots of land differ in their fertility status, rent arise for superior (high fertility status) plots of land when less fertile plots of land are brought under cultivation. Thus, the fertile land starts earning rent, as soon as the land next in fertility is cultivated. That means, the land with highest fertility status earns highest rent and the land with lowest fertility status earns no rent.

Assumptions

Ricardian theory of rent is based on the certain assumptions, which are enumerated below.

- ☆ Rent is paid for the original and indestructible powers of the soil. As no cost of production is involved in the provision of natural productive properties of soil, it has no supply price. The original powers of land means the fertility or qualities that are not created by the human efforts. These original powers are indestructible and this implies, the fertility of land remains for ever, irrespective of the extent of crop cultivation.

- ✩ The emergence of land has no social cost and opportunity cost.
- ✩ The supply of land is perfectly inelastic. That means, even if high price (rent) is quoted for the use of land, its supply cannot be increased. Thus, total supply of land is unresponsive to the changes in rent. Since supply of land is fixed, demand is the sole factor that determines the rent for a land.
- ✩ There is perfect competition in the use of land and in the market for land produce.
- ✩ This theory applies only in the long run and not in the short run period.
- ✩ Alternative uses of land have been ruled out. That means, to analyze the concept of economic rent on different plots of the land *i.e.*, on different fertility levels, same crop has to be cultivated.
- ✩ The land is not uniform in terms of fertility. That means, with reference to fertility status, land is a heterogeneous factor of production.
- ✩ This theory assumes LDR applies to the agricultural production and population in the country increases in a geometrical progression. This implies, this theory is based on the Malthusian theory of population.
- ✩ Technology of production is given or constant.
- ✩ Land is used for cultivation in the order of fertility gradient and the order is diminishing. That means, most fertile land is used first and it is followed by subsequent grade and so on. Ricardo assumed that, if all the lands are of uniform fertility, then no rent arises in using for crop cultivation.
- ✩ In every country, there exists no rent land or marginal land and rent for land is calculated as a surplus over no rent land or marginal land.
- ✩ It assumed that, transfer earnings of the land is zero.

Considering the above assumptions, we can infer that, rent for land is determined by the fertility differences across different grades of land. We know, some lands are more fertile and some are less fertile and this fertility difference across the lands yields rent to superior land over inferior land. That means, the producer's surplus of superior land is considered as 'rent'. Since, rent is determined for a superior land on comparison with the inferior land, Ricardian theory is, otherwise, called as 'Differential theory of rent' or 'Theory of differential advantage'.

Concepts of Ricardian Theory of Rent

In the assumptions discussed earlier, Ricardo mentioned 'marginal land' or 'no rent land' and this represents the grade of land, below which no inferior land is used for cultivation purpose. The lands with higher fertility levels above the marginal land are called Intra-marginal lands or Super-marginal lands. The surplus between the value of outputs derived from Intra-marginal lands and marginal land determines the rent for Intra-marginal lands. Thus, rent arises for a land as a surplus over and above the marginal land. Let us discuss the concept of Ricardian theory of rent assuming two techniques of cultivation *viz.*, under extensive cultivation and under intensive cultivation.

1. Concept of Ricardian Theory of Rent Under Extensive Cultivation

Under extensive cultivation, more and more output of a crop is produced by bringing new lands into cultivation. When new lands were brought into cultivation, since they differ in their fertility status, rent arises for fertile lands (Intra-marginal lands) over and above the marginal land. This emergence of rent for intra-marginal lands is well-explained by an example (Table 25.2).

Table 25.2: Concept of differential rent under extensive crop (paddy) cultivation.

Grades of Land Based on Fertility Status	*Quantity of (Paddy) Output (Units)*	*Price of Paddy (Rs/Unit)*	*TR (Rs/-)*	*Rent = Differential Surplus (Rent Over 'D' Land)*
A	1000	1000	10,00,000	10,00,000-1,00,000 = 9,00,000
B	800	1000	8,00,000	8,00,000-1,00,000 = 7,00,000
C	500	1000	5,00,000	5,00,000-1,00,000 = 4,00,000
D	100	1000	1,00,000	1,00,000-1,00,000 = 0

Suppose, in a village, there are four grades of land A, B, C and D and assume that, they are arranged in the descending order of their fertility status, *i.e.*, 'A' land being most fertile, 'B' land less fertile than A, 'C' land less fertile than B and 'D' land is least fertile, thereby, it is a marginal land or no-rent land. Let us assume, farmers cultivate paddy on these four grades of land. As assumed earlier, the land with highest fertility status (*i.e.*, 'A' land) is brought into cultivation first and as the concept of extensive cultivation is being studied, to increase the production of paddy, more and more acreage under land 'A' category is brought under cultivation. So, the farmers will start cultivating 'A' land to produce paddy to meet their food security needs. As long as the extent of supply of 'A' land is abundant in achieving desired level of output, this 'A' land will not earn any rent for the farmers. As shown in the Table 25.2, up to 1000 units of production of paddy, no rent arises for 'A' land. But, the conditions in the economy will not remain status-quo. With the increase in population in a geometric fashion, the pressure on land increases to produce more output of paddy to ensure food security needs for the mounting population. So, next grade land *i.e.*, 'B' land is brought under cultivation to increase the output of paddy. That means, when the production of paddy is to be increased to 1800 units due to rise in population, the existing 'A' land is not sufficient and thereby, the farmer starts cultivating 'B' land to meet the food security needs of the population. As shown in the table, as long as the demand for paddy production is 1800 units, the farmers cultivate both 'A' land and 'B' land and if the demand for paddy still increases beyond 1800 units due to further increase in population, now the farmers has to depend upon 'C' land to increase the production of paddy, which is less fertile compared to 'B' land. Assuming the same cost of cultivation of paddy in 'A' land and 'B' land, but still 'A' land contributes more production of paddy (1000 units) over 'B' land (800 units) *i.e.*, by 200 units. This producer's surplus *i.e.*, 200 units of paddy in 'A' land over 'B' land is due to superior fertility status of 'A' land over 'B' land and this surplus is regarded as 'rent'. In terms of value, it amounts to Rs. 2,00,000

The farmers start cultivating 'C' land, when the demand for paddy production increases over 1800 units due to increase in population. This 'C' land yields 500 units of output and being an inferior land compared to 'A' land and 'B' land, 'A' land earns a surplus of 500 units of output and 'B' land earns a surplus of 300 units of output and these are due to differences in fertility status across the different grades of land. This producer's surplus of 500 units of output of 'A' land over 'C' land and 300 units of output of 'B' land over 'C' land refers to rents derived by 'A' land and 'B' land respectively and in terms of value, they amount to Rs. 5,00,000 and Rs. 3,00,000 respectively.

If the demand for paddy production still increases above 2300 units, the farmers bring 'D' land into cultivation to meet the food security needs of the mounting population. This 'D' land yields only 100 units of paddy output, being the most inferior land compared to other three grades of land *i.e.*, A, B and C. When 'D' land is employed to produce paddy, the other three categories of land *i.e.*, A, B and C starts earning surplus to the tune of 900, 700 and 400 units respectively and in terms of value, the surplus or rent amounts to Rs. 900000, Rs. 700000 and Rs. 400000 respectively. This infers that, the 'C' land, which does not reap any rent earlier, also earns rent now, when 'D' land is brought under cultivation.

The following are the inferences we can derive from the above concept and Table 25.2

- ✰ More fertile lands yield more and more rent to the farmers, when more and more infertile lands are brought under cultivation.
- ✰ The higher the fertility status of Intra-marginal lands, the higher the rent they yield in the production process.
- ✰ 'D' land will not yield any rent to the farmers, as it is the least fertile land brought into cultivation for paddy production and beyond that, no additional land was allocated under paddy cultivation. Hence, this 'D' land is called as 'no rent land' or 'marginal land'. It is so called because, the revenue earned from 'D' land is just equal to the cost of cultivation of the crop. So, the lands which are brought into cultivation even before the marginal land was put into cultivation are regarded as Intra-marginal land or Super-marginal lands. So, the Intra-marginal lands or Super-marginal lands earn the rent equal to the surplus of their yield (or revenue) over the yield (or revenue) of marginal land. Hence, the rent derived as per the Ricardian theory is the 'differential rent' and it is given by,

  Differential rent = Yield of Intra marginal land – Yield of marginal land.

  = TR of Intra marginal land – TR of marginal land.

The differential rents gained by the intra-marginal lands as a surplus over marginal land is shown diagrammatically through Figure 25.1. The differential rents earned by the three grades of lands *i.e.*, A, B and C are shown as shaded areas. Grade 'D' land is a marginal land or no rent land and no land below this grade was put into paddy cultivation and the returns earned from this land just equals the cost of cultivation of paddy. This infers that, under extensive cultivation, rent emerges for intra-marginal lands, as a surplus of output or TR over and above marginal land.

The same concept can also be explained through a competitive market model under different categories of land and the illustrations are shown through Figure 25.2. In different Panels of the Figure 25.2, the price line is drawn parallel to X-axis (for a farm under perfect competition) and the per-unit cost curves of paddy cultivation for different grades of land are also shown. We know, the point of optimality for a production programme is MC=MR.

- ✰ In Panel A, when only 'A' grade land is under cultivation, the price of output OP_1 is shown as a horizontal line parallel to X-axis. This price line OP_1 also represents ARC and MRC under perfect competition. Considering this price line, the farm will be under equilibrium at E_1, where the farm under 'A' grade land produces OA_1 output of paddy and earns no rent from this land, as there is no surplus revenue over the ATC_A.
- ✰ Due to increase in demand for paddy cultivation in the context of food security needs, farmers now bring 'B' grade land under cultivation (Panel B) and 'A' land is subjected to intensive cultivation. Due to increase in

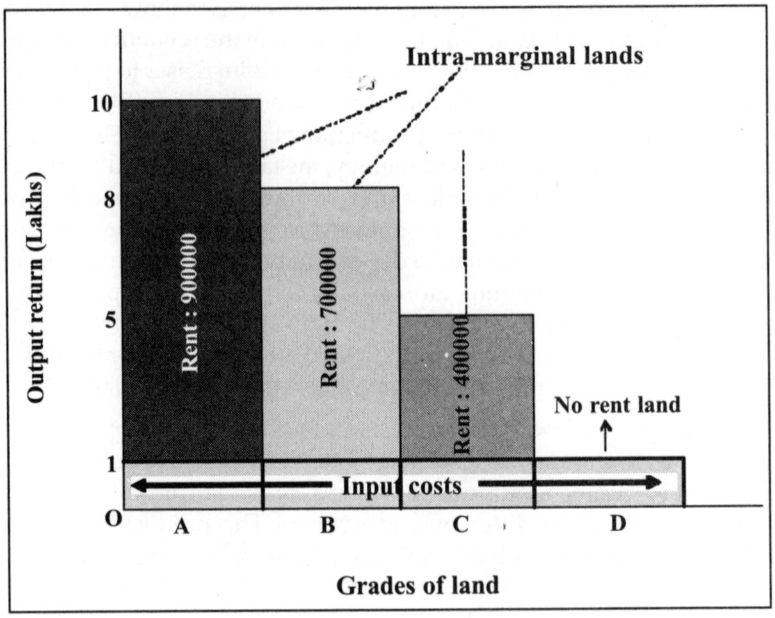

Figure 25.1: Ricardian theory – Rent as 'Differential surplus' under extensive cultivation.

demand for land produce (paddy), the price of paddy increases to OP_2 from OP_1. We discussed earlier, rent emerges as a differential rent in Ricardian theory, thereby, rent is the surplus revenue over cost. So, when 'B' land is brought under cultivation, 'A' land starts earning rent, as it enjoys surplus over cost. This also infers that, 'B' land will be brought under cultivation only when the price of paddy reaches OP_2 and if the demand for the paddy production is beyond OA_1. 'B' land is under equilibrium when MC=MR *i.e.*, at E_1, where 'B' land contributes OB_1 output at OP_2 price and it will not earn any rent, as it enjoys no surplus revenue over ATC_B. But, at this price OP_2 'A' land starts earning rent, as it enjoys surplus revenue over cost *i.e.*, $OP_2 > ATC_A$ by P_2a_1, as the equilibrium point shifts to E_2 and the A land produces OA_2 output ($OA_2 > OA_1$) due to intensive cultivation. This surplus, P_2a_1 is regarded as 'rent' claimed by the A land. Here, 'B' land is considered as no-rent land or marginal land.

- ☆ Again, when the demand increases for paddy, the farmers will bring 'C' land under cultivation (Panel C) and both 'A' land and 'B' land are subjected to intensive cultivation. So, due to increase in demand for paddy, its price increases to OP_3 from OP_2. When this 'C' land is brought under cultivation, 'A' land still enjoys more

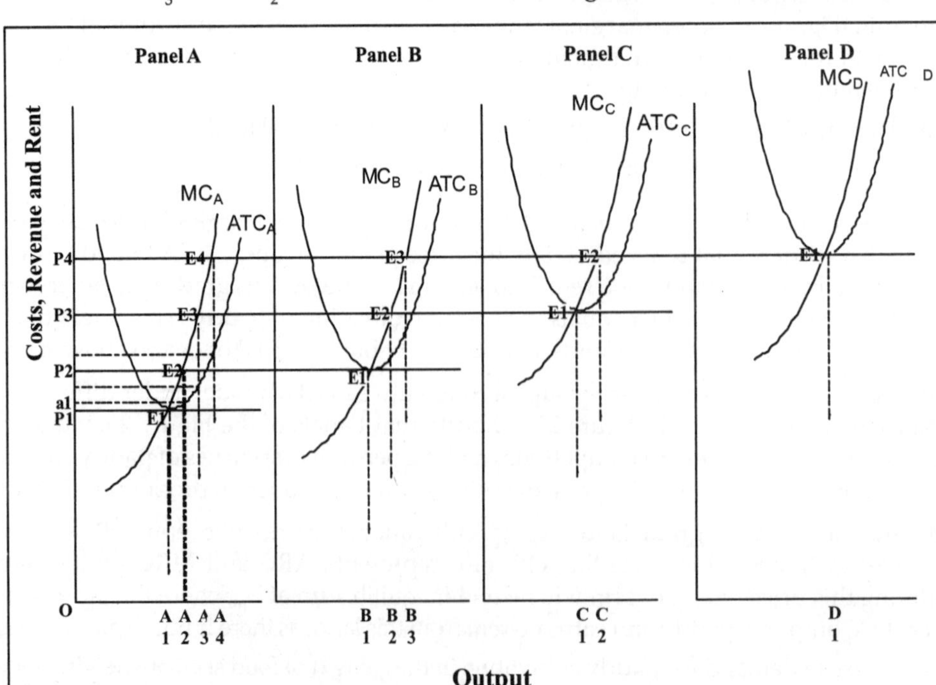

Figure 25.2: Ricardian theory – 'Rent' as a surplus of revenue over cost.

rent and 'B' land starts enjoying rent, as they enjoy surplus revenue over their respective cost structures. This also infers that, 'C' land will be brought under cultivation only, when the price of paddy reaches OP_3 and if the total output required for the market is above OA_2 of 'A' land + OB_1 of 'B' land. 'C' land will be in equilibrium at E_1 (Panel C), where MC=MR and it contributes OC_1 output at OP_3 price and it will not earn any rent, as it enjoys no surplus revenue over ATC_C. But, at this price OP_3, 'A' land earns still higher rent than earlier (when price is OP_2), and 'B' land starts earning rent, as the equilibrium points shifts to E_3 and E_2 with respect to 'A' land and 'B' land respectively, since they are subjected to intensive cultivation. Thus, output increases both from 'A' land and 'B' land from OA_2 to OA_3 and OB_1 to OB_2 respectively. Here, 'C' land is considered as no-rent land or marginal land.

☆ Let us assume that, demand for paddy still increases, thereby, the farmers will bring 'D' land into cultivation (Panel D) and 'A' land, 'B' land and 'C' land are subjected to intensive cultivation. So, due to increase in demand for paddy, its price increases from OP_3 to OP_4. When 'D' land is brought under cultivation, now 'A' land and 'B' land still enjoys higher rent, and 'C' land starts enjoying rent, as they enjoy surplus revenue over their respective cost structures. This also infers that, 'D' land will be brought under cultivation only, when the price of paddy reaches OP_4 and if the total output required for the market is above OA_3 of 'A' land + OB_2 of 'B' land + OC_1 of 'C' land. 'D' land will be in equilibrium at E_1 (Panel D), where MC=MR and it contributes OD_1 output at OP_4 price and it will not earn any rent, as it enjoys no surplus revenue over ATC_D. But, at this price OP_4, 'A' land and 'B' land earn still higher rents than earlier (when price is OP_3) and 'C' land starts earning rent, as the equilibrium points shifts to E_4, E_3 and E_2 with respect to 'A' land, 'B' land and 'C' land respectively, since they are subjected to intensive cultivation. Thus, output increases from 'A' land, 'B' land and 'C' land from OA_3 to OA_4, OB_2 to OB_3 and OC_1 to OC_2 respectively. Here, 'D' land is considered as no-rent land or marginal land.

So, from the above discussion, we can infer that,

☆ 'A' land will earn the highest rent followed by 'B' land and then followed by 'C' land.

☆ 'A' land, 'B' land and 'C' land are considered as Intra-marginal lands or Super-marginal lands, as they enjoy surplus revenue over their respective cost structures.

☆ 'D' land is no-rent land or marginal land, as it enjoys no surplus revenue over cost. That means, no further land below the grade 'D' land is brought under paddy cultivation.

☆ 'A' land to enjoy the highest rent will produce OA_4 quantity of output, 'B' land will produce OB_3 output and enjoy the rent less than 'A' land, but greater than 'C' land and 'C' land will produce OC_2 output to enjoy the rent above 'D' land.

☆ So, with the increase in demand for paddy, more and more land will be brought under cultivation. Due to increase in demand for paddy, its price will increase in the market and thereby, it will increase the rents earned by the superior lands over no rent land or marginal land. This also infers two interesting points *viz.*, demand for the product mainly influences the price of the product and this price will influence the rent. That means, price is the cause and rent is the effect or we can say, rent is a price-determined surplus. In the words of Ricardo, *'rent is paid because the price of corn is high and not the price is high because, rent is paid.* This also infers that, rent is price-determined and not price-determining.

☆ Rent arises as a differential surplus payment *i.e.*, it arises as a differential surplus yield of more fertile land over less fertile land. Thus, according to Ricardian theory, rent refers to Differential rent and it arises owing to differences in fertility of land. That means, the surplus that arises due to fertility difference between the marginal and intra-marginal land is the differential rent. It is accrued generally under extensive cultivation of land. This concept of Differential rent was first stated by David Ricardo.

☆ Rent does not form a part of the cost of production. Thus, rent of land is the earnings made over and above the cost of production. Since, rent of land does not enter the cost of production, it, therefore, does not determine the price.

2. Concept of Ricardian Theory of Rent Under Intensive Cultivation

We know, under intensive cultivation, output of the crop is increased by applying increased doses of resources and resource services. In this context, Ricardo assumed LDR operates in agriculture, which implies that, output increases at a diminishing rate, if the farmer continuously increases the input dosage in the crop production programme. This indicates that, if the farmer employs more and more of the resources (say, labour and capital, the level of other factors remains constant), the MPP goes on diminishing. As shown in the Table 25.3, if the farmer initially employs first dosage of labour and capital, it contributes 10 units of output (TPP) of paddy. If the farmer wishes to increase the output of paddy, he still employs a higher dosage of labour and capital resources and thereby, TPP increases to 18 units. That means, second dosage of labour and capital contributed only 8 units of MPP, implying diminishing MPP in the production programme. If still the farmer wishes to increase the output, he employs higher doses of labour and capital and accordingly third and fourth dosages leads to higher TPP to a tune of 24 units and 27 units respectively implying diminishing MPP of third and fourth doses of labour and capital resources *i.e.*, 6 units and 3 units respectively. If the farmer still increases the labour and capital dosage, the MPP falls below zero and hence, he stops increasing the dosage beyond fourth dosage. The Table 25.3 infers that, the first, second and third dosages of labour and capital earn rent over fourth dosage of labour and capital and hence, these three dosages are called intra-marginal units. Their rent is equal to the surplus of MPP over the MPP of last dosage (fourth dosage). Thus, first, second and third dosages of labour and capital earns rent *i.e.*, 7, 5 and 3 units of paddy over fourth dosage of labour and capital and these rents are not due to differences in soil fertility, but due to differences in MPP contributed by each additional unit dosage of labour and capital applied on the same piece of land. Assuming price per unit of paddy is Rs. 1000, then in terms of value, the rents earned by the first, second and third dosages of labour and capital amounts to Rs. 7000, Rs. 5000 and Rs. 3000 respectively over the fourth dosage. Thus, even on the same piece of land under a given status of soil fertility conditions, rent arises under intensive cultivation due to the operation of diminishing MRs of different units of labour and capital dosages. The same concept is shown through Figure 25.3, and the rent emerges for the first, second and third dosages of labour and capital are shown as shaded areas. Thus, according to Ricardo, rent arises even under intensive cultivation, as a surplus of MPP of each additional dosage of input over the marginal dosage of input.

Table 25.3: Concept of differential rent under intensive crop (paddy) cultivation.

Dosages of Labour and Capital	*MPP of Paddy Output (Units)*	*TPP of Paddy Output (Units)*	*Price of Paddy (Rs/unit)*	*TR(Rs/-)*	*Rent = Differential Surplus (Rent over 4th dosage) (Rs)*
1st	10	10	1000	10000	10000-3000 = 7000
2nd	8	18	1000	8000	8000-3000=5000
3rd	6	24	1000	6000	6000-3000=3000
4th	3	27	1000	3000	3000-3000 = 0

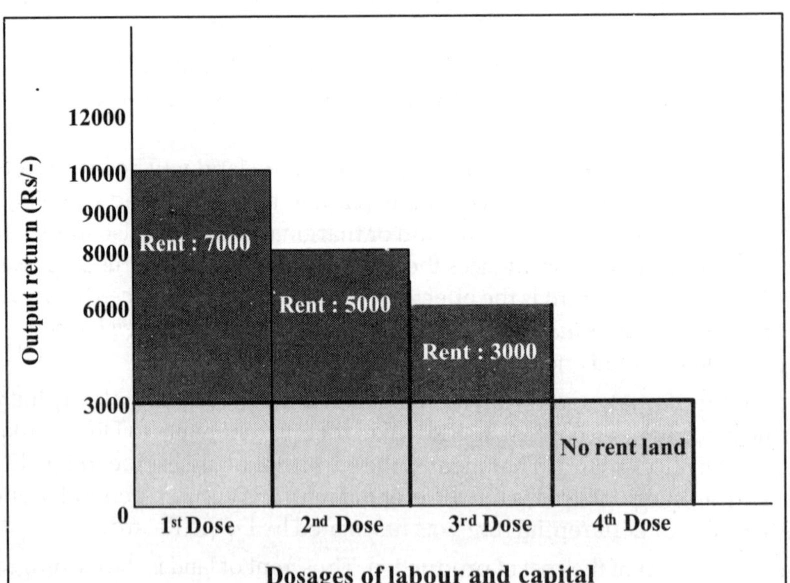

Figure 25.3: Ricardian theory – Rent as 'Differential surplus' under intensive cultivation.

Rent and Price

According to Ricardo, rent of the land is determined by the price of the output and hence, rent does not enter into price. That means, rent is price-determined and not price-determining. He opined that, rent of the land increases, when price of the output increases and not price of output is high because, rent for land is high. In the words of Ricardo, '*Corn is not high, because a rent is paid, but a rent is paid because corn is high and it has justly been observed that no reduction would take place in the price of corn although land lords should forego the whole of their rent.*' He further opined that, cost of production of crop on marginal land determines the price and this ensures intra marginal lands to enjoy surplus returns over cost as rent, thereby, marginal land receives no rent. Hence, rent does not enter price. Since surplus returns over cost are treated as rent, it implies that, rent is not treated as an element of cost. Since price is based on ATC, which does not include rent, it is obvious that, rent is not price-determining. As discussed under the Figure 25.2, if price increases, rent increases for intra-marginal lands and this infers that, rent is price determined.

Criticisms Against Ricardian Theory of Rent

Both the assumptions and the concept formulated by Ricardo are subjected to severe criticisms and they are discussed here under.

- ☆ ***No original and indestructible powers of the soil:*** Ricardo's definition for rent laid too much emphasis on the original and indestructible powers of the soil and this is subjected to severe criticism by Stonier and Hague. This is because, if the farmers continuously cultivate the crops, the natural fertility status of the lands gets exploited and exhausted. Hence, the surplus yields from the lands are due to artificial application of manures, chemical fertilizers and other modern techniques of production. So, the concept of 'original and indestructible powers of the soil' is vague or carries no meaning. However, the disciple of Ricardo, Prof. Haney has answered this criticism saying that, there are certain natural elements that go with land like climate, sunshine, rainfall etc., which cannot be increased or destroyed. He further opined that, Ricardo's expression of 'original and indestructible powers of the soil' refers to inelastic supply of land to change in prices.
- ☆ ***No historical proof:*** Ricardo assumed that, the land was put into cultivation in a historical sequence *i.e.*, from most fertile land to the least fertile land (marginal land). But, according to Carey and Roscher, there is no such historical proof that land is cultivated in that order assumed by Ricardo. In practicality, the lands which are most convenient from production point of view and in terms of marketing of produce are cultivated first and the sequence of cultivation is not based on the fertility status of the land.
- ☆ ***Neglects scarcity rent:*** According to Ricardo, rent emerges due to differences in fertility status across the lands. But, this is partially true. This is because, compared to fertility status of land, scarcity is the most important factor that determines the rent for land. Since, supply of land is fixed, it is the demand that plays the crucial role in influencing rent for land. On the other hand, if there is abundant supply of land, then it receives no rent, either the land is fertile or not. So, it is not the fertility of land, but the scarcity that primarily influences the rent for land.
- ☆ ***Absence of perfect competition:*** Ricardo assumed perfect competition for the use of land and in the product market. But, this is not realistic, as the rent charged by the land lords from the farmers is more than economic rent.
- ☆ ***Rent also arises in short run:*** According to Ricardo, the theory is applicable only in the long run. But, according to Marshall, rent also arises in the short run (on machinery and other man made appliances) and he named it as Quasi-rent.
- ☆ ***There is no marginal land:*** There is a lot of controversy and contention on the concept of no-rent land, assumed by Ricardo. But, in reality, there may not be a no-rent land because, for meeting the food security needs of the mounting population, even the worst lands will be brought under cultivation and they might also earn rent on account of increase in demand for such lands. Moreover, besides farming, the land can be put into several other uses and hence, there will not be any no-rent land. So, it absurd to assume no-rent land.
- ☆ ***LDR can be delayed:*** Ricardo assumed LDR concept in explaining the emergence of rent in the cultivation of crops. But, Ricardo could not anticipate that, the LDR can be effectively checked by the adoption of technological and managerial improvements in the production programme.
- ☆ ***Rent enters into cost of production:*** David Ricardo opined that, rent does not enter cost of production and hence, it will not influence price of the output. He opined that, rent is high when price of output is high and rent is low, when price of output is low. He said that, cost of production on marginal land determines the price and this ensures intra-marginal lands to enjoy surplus over their respective cost structures, thereby, they enjoy rent. Thus, marginal land do not enjoy rent, as price is equal to cost of production and hence, rent does not enter price. But, this relationship between rent and price was criticized by the Modern Economists and they opined that, rent enters the cost of production and thereby, it enters the price. This is because, for an individual farmer, the rent paid by him to the land lord is the cost to the farmer and thereby, rent enters the cost of production and hence, the price. Similarly, for an individual farmer, the supply of land is perfectly elastic at a given rental value. So, payment of rent by the farmer enters the cost of production and thereby, determines the price of the product. Thus, rent is price determining and not price-determined. But, considering the entire economy as a whole, the supply of land is perfectly inelastic, thereby, it has no transfer earnings and the entire earnings are considered as rent, which do not enter into the cost of production and hence, do not enter into the price. But, we are mostly concerned with individual farmer and not with entire economy in studying the relationship between rent and price.
- ☆ ***Rent arises even for other factors of production:*** According to Ricardo, rent emerges as a payment for the use of land factor only. But, according to Modern Economists, besides land, other factors also earn rent as a surplus over their respective transfer earnings.

- ☆ ***Differential surplus arises for other factors also:*** The concept of rent for land as a differential surplus also arises in case of payments for other factors of production like wages, interest and profits and thereby, 'rent' need not be attributable to land factor alone.
- ☆ ***Alternative uses of land not considered:*** Ricardo assumed that, land factor is used to cultivate only one crop and alternative uses of land has been ruled out. That means, Ricardo takes the transfer earnings of land factor as zero. But, in reality, no farmer will keep his land idle and he tries for various alternate crop enterprises to achieve profit maximization.
- ☆ ***Rent arises, even if all lands are equally fertile:*** According to Modern Economists, rent emerges, even if all lands are equally fertile and this is due to scarcity factor.
- ☆ ***Could not explain determination of rent:*** Ricardian theory simply explains how rent arises as a differential surplus between high fertile land and low fertile land. Thus, it explains the cause of origin of rent. But, it does not explain how rent is determined for the land factor. In this context, Modern theory outweighs Ricardian theory in the sense that, it explains the determination of rent for land factor based on demand and supply forces.

The severe criticisms levelled against the Ricardian theory does not mean that, the theory is futile, rather, it occupies a very pride place in the economic theory. The aspects like increasing population, increasing demand for food, bringing new lands into cultivation etc., are practically seen even today and thereby, this theory has not lost its vitality and instructiveness. According to Ricardo, a land lord earns rent without performing any efforts or economic activity. So, the abolition of idle landlordism in many countries is due to the influence of Ricardian theory of rent. Moreover, this theory also formed the basis for the Modern theory of rent.

II. Modern Theory of Rent

According to the Classical Economists, a separate theory is necessary to explain the concept of rent, as land factor is fundamentally different compared to other factors of production. But, according to Modern Economists, all factors of production are alike and there is no basic difference between them, thereby, no separate theory is needed to discuss about the concept of rent for land factor alone. The peculiarities mentioned by Classical Economists with reference to land factor are also found in other factors as well. The following are the common features across different factors of production:

- ☆ ***Inelastic supply:*** According to Ricardo, the supply of land is perfectly inelastic. That means, the supply of land is limited. This feature, we can also observe in case of skilled labour and professional entrepreneurs.
- ☆ ***Mobility of factors:*** Classical Economists assumed that, land is immobile and they considered it as a specific factor of production. A factor is said to be specific (immobile), if it is put to only one use. But, in the context of Economics, physical mobility of factors is of less significance and mobility (non-specificity) refers to the use of the factors across different uses of production and it is more important compared to physical mobility. For example, we can use a plot of land for cultivating different crops like paddy, cotton, chillies, wheat, maize etc., for constructing a factory, building etc., and in this context, we can say, land is a mobile (non-specific) factor. Similarly, all casual labour are mobile, as they can be adjusted over different farm operations across different enterprises.

In some cases, land is considered as a specific (immobile) factor. For example, if large-scale investments are made on a plot of land to take up mango orchard, then that plot of land becomes specific or immobile till the mango produce is realized completely (say, 20 years). After removing the mango orchard, again the land becomes non-specific or mobile. Similarly, skilled labour is a specific (immobile) factor in a particular production programme and they are not mobile across the enterprises. For example, a plant breeder cannot execute the operations in a sugar processing factory. Even casual labour may also become specific or immobile, if they enter into contract with the organization. Likewise, capital and entrepreneur may be specific or non-specific.

Since, the Classical Economists considered land is always a specific (immobile) factor and this is the only specific factor among different factors of production, they proposed a separate theory of rent to explain the land factor. But, according to Modern Economists, a factor cannot be completely specific and completely non-specific. A specific factor in due-course becomes a non-specific factor. That means, every factor is mobile or immobile.

This infers that, all the factors are exactly alike in terms of their features, thereby, their payments are to be determined by the same fundamental principle. Economists like Marshall, Joan Robinson and Boulding have contributed new ideas and views and suggested modifications, amplifications and refinements to the Ricardian theory of rent and proposed 'Modern theory of rent', which lies on the following five propositions:

1. Rent arises due to scarcity of land.
2. Rent arises as a surplus in all the factors.

3. Rent arises as a surplus for a factor above the transfer earnings in its present employment.
4. Rent varies based on supply elasticity of factors.
5. Rent does enter into price.

1. Rent Arises Due to Scarcity of Land

According to Ricardian theory, rent arises due to differences in fertility status across the lands and thereby, it is called as 'Differential rent'. But, rent may also arise due to scarcity of land in relation to demand and this is called as Scarcity rent, which Ricardian theory failed to explain. Scarcity rent refers to the price paid for the use of the homogeneous land when its supply is limited in relation to demand. If all units of land are homogeneous, but demand exceeds supply, the entire land will earn economic rent by virtue of its scarcity. We know, when the population in the country increases, demand for food also increases and to meet the food security needs of the mounting population, even the worst lands and no rent lands are also brought into cultivation and even such lands also starts earning rent. At one point of time, there will be scarcity of land in relation to demand and the rent the land earns due to its scarcity factor constitutes scarcity rent. This concept of scarcity rent was proposed by Malthus and he framed the following assumptions:

- All land is homogenous.
- The land is specific or immobile and scarce.
- The price for land is determined by the demand and supply forces.
- The supply of land is perfectly inelastic both in the short run and long run.

Considering the above assumptions, the phenomenon of scarcity rent is explained through the Figure 25.4. It is clear from the Figure 25.4 that,

- Since, supply of land is fixed, the supply curve SS is parallel to Y-axis, indicating that, the supply of land is fixed at OQ.
- Let us assume, the demand curve for land is D_1D_1 and it intersects the supply curve 'SS' at E_1, where the land earns OR_1 rent.
- If the demand for the use of land is increased, the demand curve will shift upward as D_2D_2 and it intersects the supply curve 'SS' at E_2, where the land earns OR_2 rent.
- If the demand for the use of land is still increased to D_3D_3 rent for the land also increased to OR_3.
- So, when the supply of land is relatively fixed, with the rise in demand for the use of land due to growth in population, it will still intensify the relative scarcity of supply of land and thereby, rent earned by the land will increase with increase in demand.
- Thus, rent increases due to relative scarcity of land in relation to increase in demand only and thereby, this rent is called as Scarcity rent. So, the rent is demand-determined.

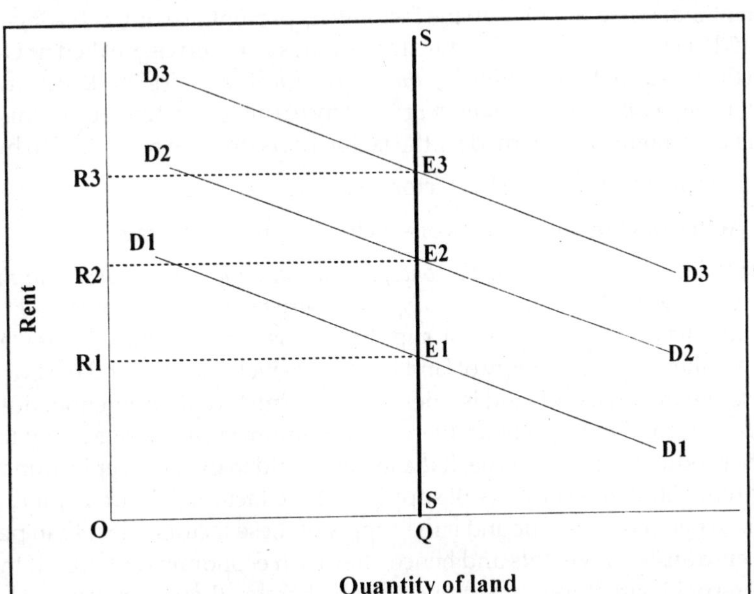

Figure 25.4: Modern theory – 'Rent' arises due to scarcity of land.

But, the interesting aspect is, Ricardo too recognized that, rent is demand-determined. But, he assumed that, when the demand for land increases, the farmers bring new lands into cultivation and thereby, superior fertile land starts earning rent over inferior lands. But, according to the Modern concept, when the demand for land increases, it will intensify its relative scarcity of supply, thereby, rent increases. Thus, according to Ricardo, rent arises as a differential surplus, but according to Modern Economists, rent arises because of scarcity of (homogenous) land in relation to demand. Further, like differential rent, scarcity rent is also a producer's surplus.

2. Rent Arises as a Surplus in all the Factors

Modern Economists agree with Ricardian concept of rent as 'surplus' returns over cost. But, they disagree that, 'land' is the only factor of production that earns rent. They opined that, besides land, even other factors of production also earn rent, as the feature of inelasticity of supply is also seen among them in the short run. Thus, inelasticity of supply is not the peculiar feature of land alone and it is also the feature of other factors of production like labour, capital and entrepreneur and this feature ensures them to earn rent. So, rent is not the peculiar earning of land alone, rather it is the 'generalized surplus', as it can be realized irrespective of the type of factor of production. For example, a skilled labour is working in a sugar factory and he is earning Rs. 3500 per month. But, due to increase in demand for the workers, his salary was increased to Rs. 5000 per month. Since, the supply of skilled workers cannot be increased in the short run, he earns Rs. 1500 more and this surplus income is called rent. Further, workers vary in their efficiency and the more efficient worker enjoys more surplus compared to less efficient worker. Similarly, an entrepreneur is operating a business with a profit margin of Rs. 1,00,000. Due to sudden rise in prices of the product (boom conditions of trade), his profit margin was increased to Rs. 1,20,000. This surplus income of Rs. 20000 is considered as 'rent' in the short period for entrepreneur. Further, due to some situational advantages of land, say, nearer to processing industries, nearer to market etc., some entrepreneurs earn more profits compared to others. So, rent element in profits is said to be 'rent of ability'. Similarly, rent element also emerges in case of capital. For example, superior machinery yield more income compared to inferior and outdated machinery. This differential surplus in capital refers to rent. Similarly, if the rate of interest decreases in the economy, it induces the investors to go for investment in the economy. But, even at such low rate of interest, some investors prefer to save the money keeping in view of the poor business expectations. So, such supra-marginal investors earn surplus or rent at low rate of interest. So, the concept of inelasticity of supply is seen among all the factors of production, but the degree of inelasticity of supply varies across the factors. For example, the supply of land is perfectly inelastic (economy point of view), less than perfectly elastic (industry point of view), perfectly elastic (firm point of view) and for other factors of production, the elasticity of supply varies between perfect elastic supply and perfect inelastic supply. So, all the factors earn rent and the proportion of rent earned varies across the factors depending upon the degree of elasticity of supply. The more is the inelasticity of supply of a factor, the more is the share of the rent among the total earnings of the factor and *vice versa*. In the words of Marshall, *'if the supply of any factor of production is limited and incapable of such increase by man's effort in any given period of time, then the income to be derived from it is to be regarded as of the nature of rent'.*

It is important that, for a factor to be employed in a production programme, certain minimum payment is essential and this should be sufficient enough to continue the factor to retain in the production programme. This minimum payment for the factor is called Minimum supply price or Transfer earnings, introduced by Joan Robinson. It is defined as, the minimum payment that is needed to keep the factor in its present use or it is the income that a particular factor could earn, if it is allocated in its next best use. So, transfer earnings is, otherwise, called as Opportunity Cost or Income foregone alternative. In the words of Joan Robinson, *'the price which is necessary to retain a given unit of a factor in a certain industry may be called its 'Transfer earnings' or 'Transfer price'*. Any earning over and above this minimum supply price or Transfer earnings is called economic rent earned by the factor. So, economic rent is given by,

Economic rent = Remuneration paid to the factor – Transfer earnings

Economic rent = Actual or Current earnings of a factor – Transfer earnings

In the words of Joan Robinson, *'the essence of the conception of rent is the conception of a surplus earned by a particular part of a factor of production over and above the minimum earnings necessary to induce it to do its work'*. It is interesting that, since land is a free gift of nature, its transfer earnings or supply price is zero. So, the total reward paid to the land factor becomes economic rent. Further, since the supply of land is perfectly inelastic, demand plays a decisive role in influencing the economic rent. Thus, rent in the case of land is a demand-determined phenomenon. But, in case of other factors of production like labour, capital and entrepreneur, there is a minimum supply price and if that payment is made only, they get employed in the production programme. If the factor is paid in excess over its transfer earnings, then it is said to be earning economic rent. Since, the elasticity of supply of these factors (labour, capital, and entrepreneur) ranges between perfectly elastic to perfectly inelastic and if the supply of these factors is less than perfectly elastic, they will be rewarded more than their transfer payments and hence, they earn economic rent. But, if these factors have perfectly elastic supply, they will be paid with minimum remuneration, thereby, their economic rents will be minimum or even wiped out. On the other hand, if the supply of the factor (land) remains perfectly inelastic all the time, it earns economic rent. So, in this context, land is a different factor compared to other factors of production, as economic rent becomes its permanent feature, as its supply is always perfectly inelastic (at economy level), where as, other factors also earn economic rent, but it varies (not as a permanent feature) depending upon their elasticity of supply. In this context, Marshall has rightly remarked that, *'rent of land is seen, not as a thing by itself, but as the leading species of a large genus'*.

3. Rent Arises as a Surplus for a Factor above the Transfer Earnings in its Present Employment

As discussed in the earlier proposition, rent is measured for a factor as a surplus of present earnings above its transfer earnings in its present employment. Thus, even Modern Economists hold that, rent is a differential surplus. But, the major difference between Modern theory and Ricardian theory of rent is that, in Ricardian theory, rent arises as a differential surplus between intra-marginal lands and marginal land, but in Modern theory, rent is differential surplus between total remuneration received by the factor and its transfer earnings. Morever, Modern Economists considered 'rent' as a generalized concept, unlike Ricardian theory, which states that, rent arises for land factor alone. The concept of transfer earnings can be better explained with the help of following examples:

- ✰ If a plot of land contributes Rs.10000 income to the farmer in paddy cultivation and it could earn Rs.8000, if it is employed for the cultivation of maize, then the plot of land must get at least Rs. 8000, so as to retain it in the cultivation of paddy. On the other hand, if the plot of land yields less than Rs.8000 in paddy cultivation, it is better employing it for the cultivation of maize. So, the minimum earning the plot of land must yield is Rs.8000, so as to continue its present usage in paddy cultivation is called Transfer earnings. Since, by employing the plot of land in paddy cultivation, it is yielding Rs.10000 income then, Rs.10000 – Rs.8000 = Rs.2000 is the surplus earned by the land over transfer earnings. This surplus is called economic rent.
- ✰ If agricultural labour earns Rs.100 per day in one occupation and if he can earn the same in the next best alternative occupation, then there is no element of economic rent because, present earnings of a factor (labour) is equal to the transfer earnings.
- ✰ If a minimum of five per cent rate of interest will make the person to lend the capital, but if he gets six percent interest rate, then the difference of one percent is the economic rent for the factor, as transfer earnings is only five per cent rate of interest.
- ✰ If a skilled worker in a sugar factory earns Rs.2000 in the present job and if he could not find any alternate source of employment, then all the present earnings *i.e.* Rs.2000 will constitute economic rent because, transfer earnings is zero. But, if the skilled worker could get Rs.5000 in an alternate job, then it would be wrong to say that, the economic rent is negative (-Rs.3000) because, the skilled worker will shift to the new job, where he derives Rs.5000 income.
- ✰ If an entrepreneur is taking up a poultry unit with a profit of Rs.50000 and in the next best alternative venture, he can derive Rs.44000, then the difference of Rs.6000 constitute the economic rent.

The above examples clearly infers that, economic rent is the surplus of present earnings over transfer earnings and it accrues to all the factors of production. According to Boulding, *'economic rent is the excess of minimum amount necessary to keep that factor in its present occupation'*. J.S. Mill defined economic rent as, *'the extra gain which a producer or dealer obtains through superior talents for business or superior arrangements are very much of a kind similar to rent'*.

4. Rent Varies Based on Supply Elasticity of Factors

The above discussion clearly infers that, the extent of economic rent derived by a factor depends upon its specificity of usage. If a factor is more specific in its usage, that means, it cannot be employed in alternative sources, thereby transfer earnings are zero. So, all the earnings of the factor in its present usage constitute economic rent. However, in case of non-specific factors, they can be employed across different sources of employment and thereby, they have transfer earnings. So the economic rent enjoyed by the specific factors is more than non-specific factors, as for the specific factors, transfer earnings are zero. This is given by,

Specific factors: Economic rent = Actual earnings in the present employment

Non-specific factors: Economic rent = (Actual earnings in the present employment – Transfer earnings)

As discussed earlier, the elasticity of supply of land varies as perfectly elastic (for a firm), less than perfectly elastic (for an industry) and perfectly inelastic (from economy point of view) whereas, in case of other factors of production, the supply elasticity ranges between perfect elastic supply and perfect inelastic supply. Thus, it is important to note that, according to Modern theory, rent is the surplus earned by the land (of course, any factor) even when its supply is less than perfectly elastic. But, according to Ricardian theory, rent arises for land factor alone, that too, when the supply of the land factor is only perfectly inelastic. So, from the point of view of elasticity of supply of land i.e., perfectly elastic, less than perfectly elastic and perfectly inelastic, let us discuss the concept of economic rent as a surplus over transfer earnings through Figures 25.5 to 25.7.

(a) When the Supply of Land is Perfectly Inelastic

The supply of land will be perfectly inelastic from the economy point of view and thereby, its supply curve SS_1 will be parallel to Y-axis (Figure 25.5). That means, the supply of land cannot be increased or decreased and it has no alternative use from the view point of economy. It is clear from the Figure 25.5 that,

- D_1D_1 is the initial demand curve for land and it intersects the supply curve SS_1 at E_1, where OS quantity of land demanded and supplied and OP_1 is the rent paid for the use of land.
- At the point E_1 the present earnings of the land factor is OSE_1P_1 and transfer earnings (area below supply curve from origin and upto equilibrium point) is zero and thereby, the entire present earnings OSE_1P_1 constitute economic rent.
- If the demand for land factor increases, D_2D_2 is the new demand curve and the point equilibrium is E_2, where again OS quantity of land is demanded and supplied, but the rent paid for the use of land is increased to OP_2. So, at this point E_2, the present earnings of the land factor is OSE_2P_2 and again transfer earnings (area below supply curve from origin and upto equilibrium point) is zero, thereby, the entire present earnings OSE_2P_2 represent economic rent (area above the supply curve from Y-axis and upto equilibrium price of the factor).
- Thus, when the supply of land (or any factor) is perfectly inelastic, then the present earnings of the factor is equal to economic rent, as transfer earnings are zero.

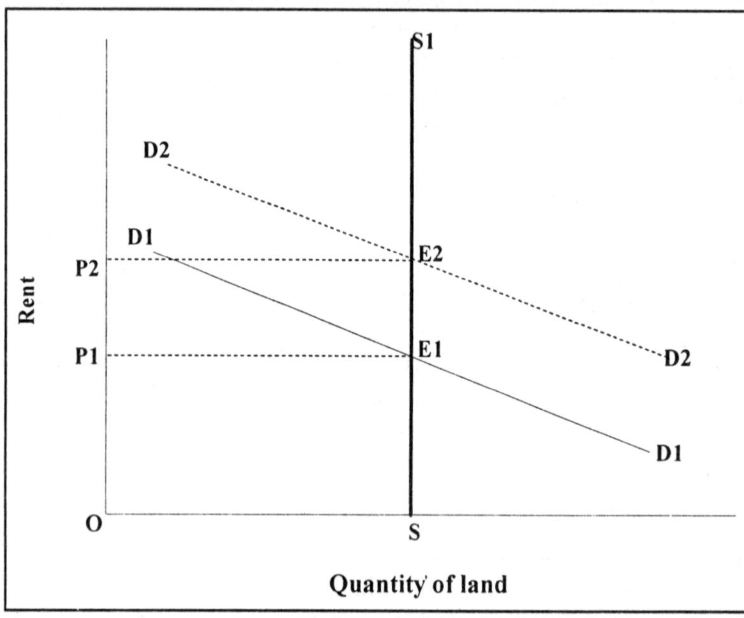

Figure 25.5: Case of Perfect inelastic supply of land (Economy view) Present earnings=Economic rent .

(b) When the Supply of Land is Perfect Elastic

The supply of land will be perfectly elastic from the individual firm's point of view and here, the supply curve SS_1 will be parallel to X-axis (Figure 25.6). D_1D_1 is initial demand curve for land and it intersects the supply curve SS_1 at E_1, where OQ_1 quantity of land is demanded and supplied and OS is the rent paid for the use of land. At this point E_1, quantity of land demanded and supplied is OQ_1, present (actual) earnings and transfer earnings of the land are same, *i.e.*, represented by the area OQ_1E_1S. Since, present (actual) earnings are equal to transfer earnings, economic rent earned by the land factor is zero.

If the demand for the land factor is increased to D_2D_2, the point of equilibrium is E_2, where OQ_2 is the quantity of land demanded and supplied, but the price remains fixed at OS. So again, actual earnings are equal to transfer earnings (area below supply curve from origin and upto equilibrium point) *i.e.*, OQ_2E_2S, thereby, economic rent (area above the supply curve from Y-axis and upto equilibrium price of the factor) is zero. This infers that, when the supply of land is perfectly elastic, it earns no economic rent, as actual earnings are equal to transfer earnings.

(c) When the Supply of Land is Less than Perfectly Elastic

The supply of land will be less than perfectly elastic from the industry point of view and in such case, the supply curve SS_1 of land will have an upward slope from left to right (Figure 25.7). This implies that, with increase in price (rent) of land, supply of land also increases. It is clear from the Figure 25.7 that,

- D_1D_1 is the initial demand curve for land and it intersects the supply curve SS_1 at E_1, where, OQ_1 quantity of land is demanded and supplied and OP_1 is the rent paid for the use of land.

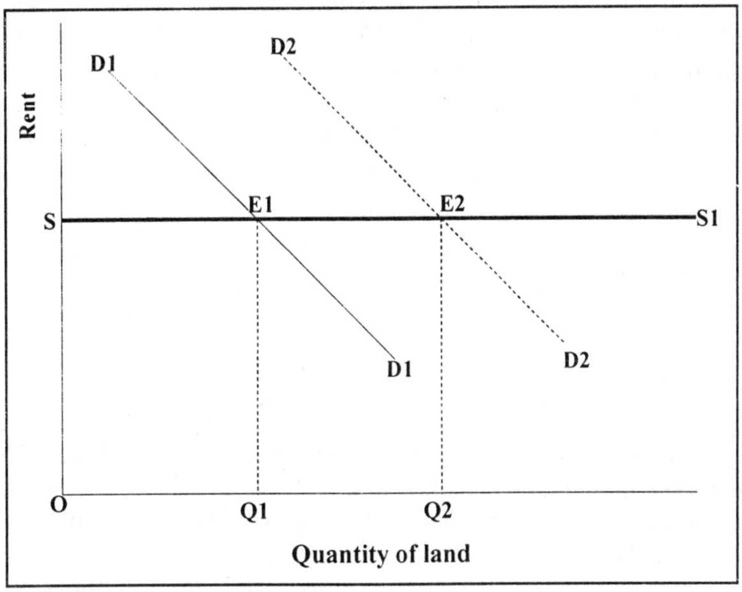

Figure 25.6: Case of Perfect elastic supply of land - (Individual firm point of view) (Actual earnings = Transfer earnings) = Zero Economic rent

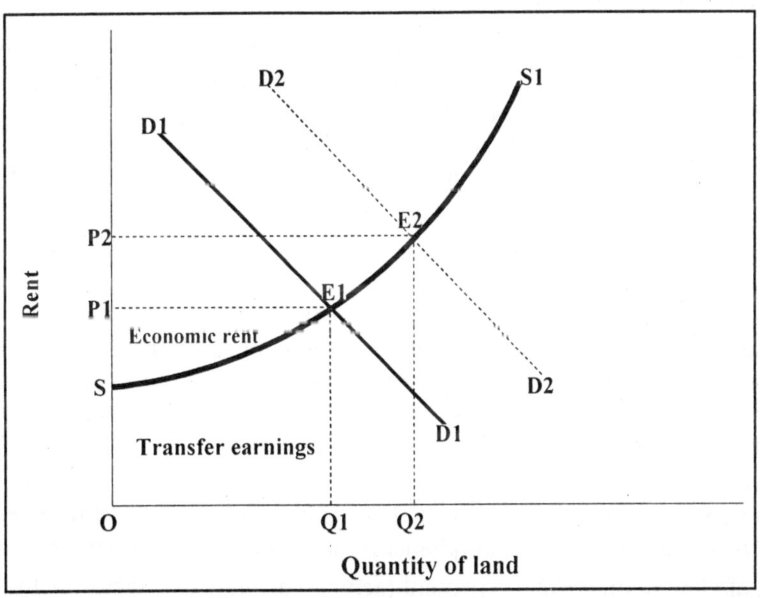

Figure 25.7: Case of less than perfect elastic supply of land - (Industry point of view) (Actual earnings – Transfer earnings = Economic rent).

- ☆ With reference to point E_1, where OQ_1 is the quantity of land demanded and supplied, present earnings of the land factor is $OQ_1E_1P_1$ and transfer earnings (area below supply curve from origin and equilibrium quantity of factor demanded and supplied) is OQ_1E_1S, thereby, the economic rent (area above the supply curve from Y-axis and upto equilibrium price of the factor) earned by the land factor is $OQ_1E_1P_1 - OQ_1E_1S = P_1E_1S$.
- ☆ If the demand for the land factor increases, D_2D_2 is the new demand curve and the point of equilibrium is E_2, where OQ_2 quantity of land is demanded and supplied and OP_2 is the rent paid for the use of land.
- ☆ With reference to point E_2, where OQ_2 is the quantity of land demanded and supplied, present earnings of the land factor is increased to $OQ_2E_2P_2$. So, due to increase in demand, economic rent of land increases to P_2E_2S *i.e.*, $P_2E_2S > P_1E_1S$.

As discussed above, like land (industry point of view), for other factors of production like labour, capital and entrepreneur, their elasticity of supply ranges between perfectly elastic to perfectly inelastic, thereby, they earn economic rent as a surplus between present earnings and transfer earnings. This infers that, even other factors also earn economic rent like land and thus rent earnings is not only the peculiar feature of land as given by the Ricardian theory. The above discussion infers the following interesting points:

- ☆ When the supply of land factor is perfectly inelastic, its entire (present) earnings are economic rent, as transfer earnings are zero.
- ☆ When the supply of the land factor is perfectly elastic, its entire earnings will be equal to transfer earnings, thereby, economic rent is zero.
- ☆ When the supply of land factor is less than perfectly elastic, its present earnings include a part of economic rent.
- ☆ Greater the degree of inelasticity of supply of a factor, the greater is the element of economic rent.
- ☆ A factor will earn economic rent only, when its supply is less than perfectly elastic and perfectly inelastic.
- ☆ The rent for a land factor is determined by the demand and supply forces.

5. Rent does Enter into Price

According to Ricardian theory,

- ☆ Rent is price determined and not price determining. That means, as price increases, rent increases and not the other way round. This is true even from commonsense point of view. For example, when the price of paddy is increased in the market, the demand for cultivating paddy from the farmer's side will increase and thereby, the farmers (tenants) must pay higher rents to the landlords. Thus, rent follows the price and not price follows the rent. So price is the 'cause' and rent is the 'effect'.
- ☆ Rent is measured as a surplus revenue over the cost. Price is determined at the cost of production level of the marginal land and thereby, intra-marginal lands earn differential rents over marginal land.
- ☆ For intra-marginal lands, price lies above their cost of production and thereby, rent does not the price.

But, the Modern Economists contradicts this view and opined that, rent does enter into price and rent is price determining. This is because, for an individual farmer, the rent paid by him to the landlord is the cost to the farmer and thereby, rent enters the cost of production and hence, the price. Further, the land can be put to several alternative uses and thereby, the concept of transfer earnings will arise for the land factor. So, the farmer has to pay at least these transfer earnings to the land factor, so as to retain it in the current production programme and these transfer payments constitute cost on the part of the farmer and thus, enters into price. Whether rent enters price or not depends upon the view point from which we look and let us discuss the following cases:

- ☆ From the point of view of an individual farmer, the prices paid by him towards all the factors of production (including use of land) constitute cost of production, and hence, rent enters into price. If the individual farmer is cultivating a leased land (as tenant), he has to pay rent to the landlord, which is the cost on the part of the farmer and thereby, rent enters into price. Even in case of land owner too, rent enters the price, but rent appears as an imputed cost on the part of the land owner. This imputed rental value of land is nothing but the payment the land owner could receive, if he lease out his land to others.
- ☆ From the point of view of economy as a whole, the land is fixed in its supply, thereby, its transfer earnings are zero. So, the present or actual earnings of the land is its economic rent. But, from the view point of the economy as a whole, land has no price and hence, rent does not enter into price.
- ☆ From the point of view of an industry, present (actual) earnings of a factor include two components *viz.*, transfer earnings and economic rent. We know for an industry, the supply of land is less than perfectly elastic and hence, in order to attract the land from other uses, the entrepreneur has to pay higher amounts such that, the amount is at least equal what the land factor gets in its next profitable use *i.e.*, the entrepreneur should pay the amount equal to transfer earnings of the land factor. If the entrepreneur could not pay at least this amount of transfer earnings. the land factor will not get leased for him. For example, if the prices for paddy increases in the market, then the demand increases from the farmer's side for paddy cultivation and thereby, the farmers, who wish to cultivate paddy will try to take the land for lease from the land lords. Say, if the area under maize has to be diverted towards paddy cultivation, the farmers, who wish to cultivate paddy has to pay at least what a piece of land can get under maize cultivation. So, this transfer price has to be paid by the farmers and it is a cost on the part of them and thereby, it enters the price. Thus, transfer price is price determining, as it is treated as a supply price of a factor into the present use. But, economic rent is also paid to the factor above the transfer price to attract and keep that factor in its present use and prevent it from being transferred to next best alternative use and it is price determined. However, from individual firm point of view, whatever is incurred or spent to employ the factor in the present use is the cost of production. Since, price is determined on the basis of cost, economic rent also enters the price.
- ☆ When the demand for land factor increases, all the lands including marginal lands will be brought under cultivation and marginal lands have to be paid rent. Such rent enters the price.

From above discussion, we can infer that, from individual point of view, (supply of land is perfectly elastic), rent enters the price and it is price determining, from economy point of view (supply of land is perfectly inelastic), rent does not enter the price and it is not price determining, but price determined and from the industry point of view (supply of land is less than perfectly elastic) a part of the factor (present) earnings is price determining (transfer earnings) and another part of the factor (present) earnings is price determined (economic rent). However according to Davenport, neither rent is price determined nor it is price determining and both rent and prices are influenced by scarcity of land in relation to its demand.

Ricardian Theory Vs Modern Theory of Rent

Ricardian theory of rent forms the basis for the Modern theory of rent. Modern theory of rent is often regarded as amplification and modification of Ricardian theory of rent. According to Ricardo, land as a factor of production is peculiar compared to other factors of production in the sense that, its supply is fixed and thereby, it earn rent. Modern Economists amplified or extended the Ricardian theory stating that, all the factors of production earn rent as a surplus of present (actual) earnings over its transfer earnings. This is because, like land, even other factors of production like supply of (skilled) labour, machinery (capital) and entrepreneurship is also less elastic in the short run, thereby, they also earn rent. Thus, Modern Economists extended the concept of rent even to other factors of production and thereby, we can consider Modern theory is an amplified form of Ricardian theory of rent. Besides extending the concept of Ricardian theory of rent to other factors of production, Modern Economists suggested useful improvements in better pricing of factors of production. According to Ricardo, rent is a differential surplus arises for intra-marginal lands over marginal land. But, this theory neglected the concept of scarcity rent, which the Modern Economists introduced to explain the factor pricing. Further, Modern Economists introduced the concept of transfer earnings and according to them, the surplus of present earnings earned by the factor above the transfer earnings constitute economic rent. Compared to the concept of marginal land, the concept of transfer earnings introduced by the Modern Economists is more meaningful and further, it can be easily measured. This infers that, Modern theory has made significant improvements over the Ricardian theory of rent and this can better understood through analyzing the following differences between these two theories:

- ✰ According to Ricardian theory, rent is a differential surplus enjoyed by the intra-marginal lands over marginal land. But, according to Modern theory, rent arises due to scarcity of land in relation to its demand. It does not matter whether there are differences in land or not.
- ✰ According to Ricardian theory, rent is a surplus of revenue over cost. But, according to Modern theory, rent is a surplus of present earnings over its transfer earnings.
- ✰ According to Ricardian theory, rent arises for land factor alone. But, according to Modern theory, rent arises for all the factors of production.
- ✰ Ricardian theory neglected the concept of scarcity rent. But, modern theory explained the pricing of factor based on scarcity concept.
- ✰ According to Ricardian theory, rent does not enter price. But, according to Modern theory, rent does enter into price from the point of individual or industry.
- ✰ According to Ricardian theory, 'land' is a peculiar factor of production compared to other factors of production and hence to explain the pricing of land factor, a special theory is needed. But, according to Modern Economists rent arises for all the factors and thereby, a uniform theory (demand and supply theory) is sufficient enough to explain the pricing of all the factors.

Though Modern theory differs from Ricardian theory in several aspects, there exists some similarities between the two theories in the sense that, both the theories advocated rent arises as a surplus and it is an unearned income for the land lords.

25.5. Quasi-Rent

The concept of quasi rent was proposed by Alfred Marshall. According to David Ricardo, land is a peculiar factor of production, in the sense that, it is fixed both in the short run and long run production programmes and thereby, it is the only factor of production, which earns rent. But, Marshall extended the concept of rent even to other factors of production like machinery and other appliances made by man, which are fixed only in the short run. Marshall has named this rent as 'quasi rent', which arises due to fixity of factors (other than land) only in the short run. So, compared to rent, quasi rent is only a temporary phenomenon that arises only in the short run and disappears in the long run. According to Marshall, *'quasi rent is the income derived from machines and other appliances made by man'*. In the words of Stonier and Hague, *'the supply of machines is fixed in the short run, whether they are paid much money or little, so they earn a kind of rent. In the long run this rent disappears for it is not true rent, but only a ephemeral reward a quasi rent'*. According to Marshall, in the short run any returns earned by the machine in excess of the TVC or prime costs of running it, it can be regarded quasi rent.

We know in a short run production programme, there are two types of resources *viz.*, fixed resources and variable resources. Fixed resources are the resources, whose level remains constant irrespective of the level of output from the production programme. But, the level of variable resources changes with the output level produced and varies directly. The fixed resource like machinery (and other man made appliances) earn quasi rent in the short run production programme, when the returns caused by the use of machinery are in excess of prime costs (variable costs) of running it. However, in the long run, the quasi rent disappears, as no resource is fixed and all the resources are variable in the long run. According to Marshall, the excess earning which a factor earns over and above its SRTVC refers to quasi rent. So, as long as the price of the product covers the SRAVC, the machinery earns the quasi rent in the production programme and if the price gets equal to the SRAVC, the factor will not earn any quasi rent in the production programme. Thus,

Quasi rent = TR in the short run - TVC in the short run or SRTVC.

Quasi rent = Price in the short run - AVC in the short run or SRAVC.

- ☆ Figure 25.8 explains the quasi rent earned by the machinery in the short run production programme.
- ☆ Assuming the perfect market competition, revenue and cost curves are drawn. At OP_1 price, the price line for a firm is drawn parallel to X-axis and it also represents the demand curve = ARC_1 = MRC_1. SRATC, SRAVC and SRMC are the short run ATC, AVC and MC curves respectively.
- ☆ The point of equilibrium is when SRMC = MRC_1 *i.e.* at E_1, where the firm produces OQ_1 output and sell the product at OP_1 price per unit. At this equilibrium level of output OQ_1, price OP_1 is greater than SRAVC by E_1A_1 and thereby, the machinery earns quasi rent equal to $P_1E_1A_1C$. But, note that, the firm is experiencing losses at this output level OQ_1, where price OP_1 is less than SRATC.
- ☆ Let us assume, if the demand for the product falls, then price of the product decreases to OP_2 in the market and the point of equilibrium is E_2, where the firm produces OQ_2 output and sell the product at OP_2 price per unit. At this equilibrium output OQ_2, price (OP_2) is equal to SRAVC thereby, no quasi rent is earned.
- ☆ If the demand for the product increases, price of the product increases to OP_3 in the market and the point of equilibrium is E_3, where the firm produces OQ_3 output and sells the product at OP_3 price per unit and it is greater than SRAVC by E_3A_3 and thereby, the firm earns quasi rent equal to $P_3E_3A_3D$. So, at this level of output OQ_3, the quasi rent earned by the factor is equal to the extent of AFC *i.e.* the gap between SRATC and SRAVC curves.
- ☆ If the demand for the product is still increased, price of the product increases to OP_4 in the market and the point of equilibrium is E_4, where the firm produces OQ_4 output and sells the product at OP_4 price per unit. At this equilibrium output OQ_4, price (OP_4) is greater than SRAVC by E_4A_4 and thereby, the firm earns quasi rent equal to $P_4E_4A_4E$. This quasi rent is abnormally high as it is in excess of the fixed investment.
- ☆ The above discussion infers that, quasi rent is price-determined. That means, as price of the product increases in the short run, the surplus of the price over the SRAVC also increases and thus, quasi rent also increases.
- ☆ In the long run, the concept of quasi rent will not arise, as the factor will not remain fixed in the production programme. In the words of Marshall, *'in the long run any quasi rent is expected to and generally does yield a normal rate of interest on the free capital, represented by a sum of money that was invested in producing it'.*

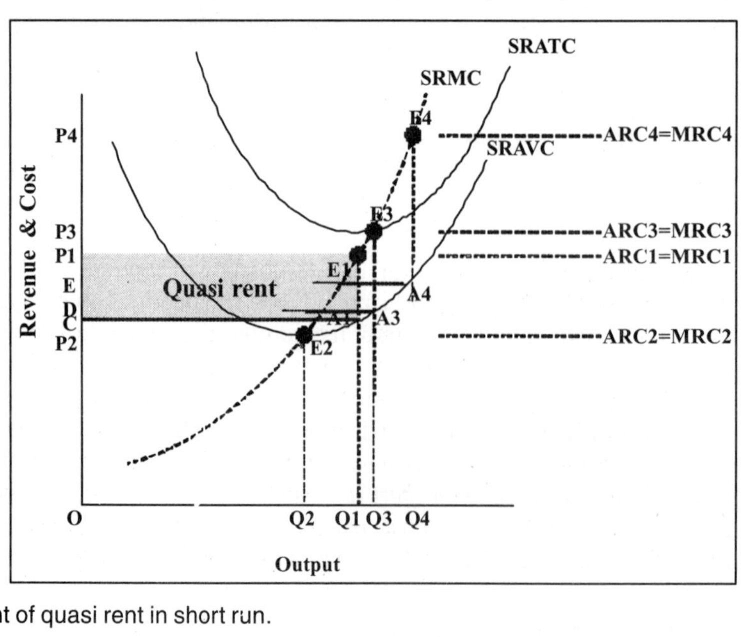

Figure 25.8: Measurement of quasi rent in short run.

The concept of quasi rent discussed above with reference to machinery is not only peculiar to machinery and man made appliances only, but equally applies to skilled labour, professional entrepreneurs, engineers, doctors, actors etc., as their supply remains inelastic in the short run. For example, if an entrepreneur gets some extra profits in the short run due to his innovation in the production programme, the extra profits constitute quasi rent. But, in the long run, the quasi rent disappears, as such innovation becomes more common in the market. Mrs. John Robinson further explained the concept of quasi rent on the basis of transfer earnings and she opined that, *'the excess or surplus earnings of a factor (man made factor) in the short run above the transfer earnings'*.

Differences between Rent and Quasi Rent

The differences between rent and quasi rent are presented here under (Table 25.4).

Table 25.4: Differences between rent and quasi rent.

Rent	*Quasi Rent*
Rent accrues to land and other factors, whose supply is fixed both in the short run and long run.	Quasi rent accrues to man-made factors, skilled workers, professional entrepreneurs, doctors, engineers, etc. whose supply is fixed only in the short run.
It is a payment to the factor owner to supply the factor for its employment in the production programme.	It is not a payment to induce the factor to get employed in the production programme.
It is a permanent phenomenon, as it arises both in the short run and long run.	It is a temporary phenomenon, as it arises only in the short run.
Rent is also earned by free gifts of nature like land.	It is earned by man-made appliances and machinery.
Rent does enter into price (Modern theory) from the point of view of individual firm and industry. Hence, rent is price determining.	It does not enter into the cost and hence, it is price determined.
Rent can never become zero.	Quasi rent can become zero, when price is equal to SRAVC.
Rent is the difference between surplus earnings of TR over TC.	Quasi rent is the difference between TR and SRTVC.
Rent is an unearned income.	It is a necessary payment by the firm to produce a given level of output.
According to Ricardian theory, rent is the result of differential fertility. According to Modern theory, rent is due to relative scarcity of land in relation to its demand.	Quasi rent is the result of scarcity or fixity of machinery and man-made appliances in relation to demand in the short run.

Similarities between Rent and Quasi Rent

The concepts of both rent and quasi rent have the following similarities:

- ☆ Both rent and quasi rent are the surplus or excess earnings.
- ☆ Both rent and quasi rent are influenced by rise in demand for land and machinery respectively.
- ☆ Both rent and quasi rent accrue due to relative fixity of land and machinery respectively in relation to their respective demands.

Differences between Quasi Rent and Interest

The differences between quasi rent and interest are presented here under (Table 25.5).

Table 25.5: Differences between quasi rent and interest.

Quasi Rent	*Interest*
It is the return of the factor with reference to sunk capital	It is the return of the factor with reference to floating capital.
It arises only in the short run.	It arises both in the short run and long run.
It does not enter the cost and hence, it is price determined.	It does enter the cost and hence, it is price determining.
If a factor is fixed only in the short run, it earns quasi rent.	If the factor is not fixed both in short run and long run, it earns interest.
Quasi rent can became zero when price = SRAVC	Interest can never become zero.

Having discussed the differences between rent, quasi rent and interest, it infers that, when a factor is fixed both in the short run and long run it earns rent, when a factor is fixed only in the short run and not in long run it earns quasi rent and when a factor is not fixed both in short run and long run it earns interest. In the words of Marshall, *'each group shades in the other gradually and thus the rent of land is seen not a thing by itself but as the leading species of a large genus'*.

Differences between Rent and Profit

The differences between rent and profit are presented here under (Table 25.6).

Table 25.6: Differences between rent and profit.

Rent	*Profit*
Anticipated.	Unanticipated.
Can be capitalized.	Cannot be capitalized.
Can persist.	Cannot persist.
Part of the cost.	Not a cost.
Stable or fixed for a period of time	Greatly fluctuates

25.6. Economic Rent and Transfer Earnings

These concepts can be well explained with the help of an example. An agricultural worker is employed in a sugar factory and he is paid Rs.10000 per month. If he is employed in a dairy cooperative, he will be paid only Rs.7500. Then, Rs.7500 is his transfer earnings. This is because, if the worker has to be employed in the sugar factory, a minimum of Rs.7500 is to be paid to him, otherwise (if < Rs.7500) he will seek employment in the dairy cooperative unit. The actual earnings or present earnings of the worker is Rs.10000 in the sugar factory. So, the worker is paid an extra of Rs.2500 in a sugar factory compared to dairy cooperative unit, so as to attract and induce him to work in the sugar factory. Then, Rs.2500 is the economic rent. So, we can infer that,

Economic rent = Present earnings – Transfer earnings

2500 = 10000 - 7500

From the above explanation, we can clearly define the transfer earnings and economic rent. Transfer earnings are the minimum payment a factor has to be paid, so as to prevent it to transfer to next best use. It is otherwise called as Opportunity cost. In the graphical representation, transfer earnings are shown as the area below supply curve from origin and equilibrium quantity of factor demanded and supplied, OAEQ (Figure 25.9). Economic rent is the surplus income earned by the factor over and above the transfer earnings and this is paid to the factor to attract and retain the factor in the present use or employment. In the graphical representation, it is shown as the area above the supply curve from Y-axis and upto equilibrium price of the factor, AWE (Figure 25.9). So, present earnings is given by OQEW. The concept of economic rent and transfer earnings of a factor can be better explained from the following cases:

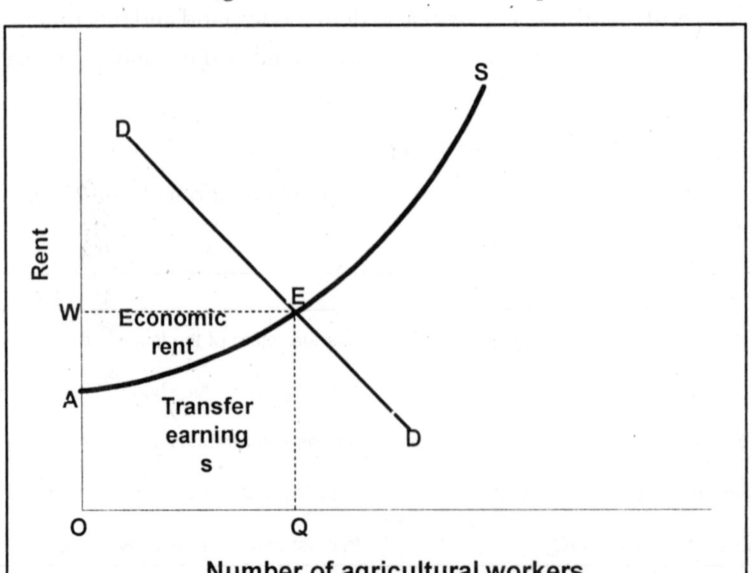

Figure 25.9: Economic rent and Transfer earnings – Case of less than perfectly elastic supply of factor.

Case 1: When the Supply of Factor is Less than Perfectly Elastic

As shown in the Figure 25.9,

- ☆ Starting at point 'A' and moving towards 'E', as the wage rate of the agricultural worker is increased, more and more number of workers join the sugar factory. At each higher wage rate, the newly joined workers are paid just enough to persuade them to join in the sugar factory *i.e.*, their wage rate is entire transfer earnings.
- ☆ But the workers, who are already in the sugar factory will earn economic rent *i.e.*, they derive more than minimum (transfer earnings) necessary to keep them employed in the factory.
- ☆ So, at the wage rate OW, the portion OAEQ represents transfer earnings of all the workers and the portion AWE represents the economic rent earned by all the workers in the sugar factory.

Case II: When the Supply of the Factor is Perfectly Inelastic

As shown in the Figure 25.10, when the supply of the factor (land) is perfectly inelastic (from the point of view of economy), there are no transfer earnings for the factor and all the earnings of the factor represents economic rent *i.e.*, OQER.

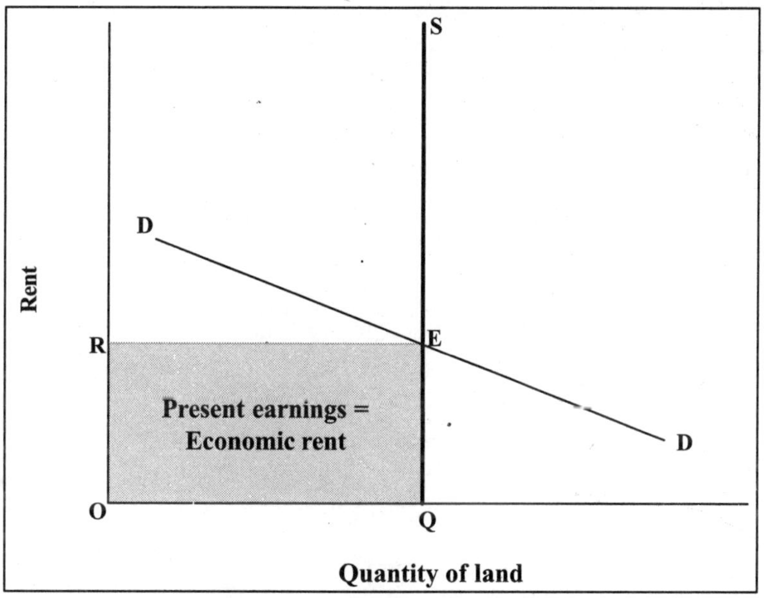

Figure 25.10: Zero Transfer earnings – Case of Perfectly inelastic supply of land.

Case III: When the Supply of the Factor is Perfectly Elastic

As shown in the Figure 25.11, when the supply of the factor (land) is perfectly elastic (from the point of view of individual firm), there is no economic rent and present earnings of the factor include only transfer earnings (OQES).

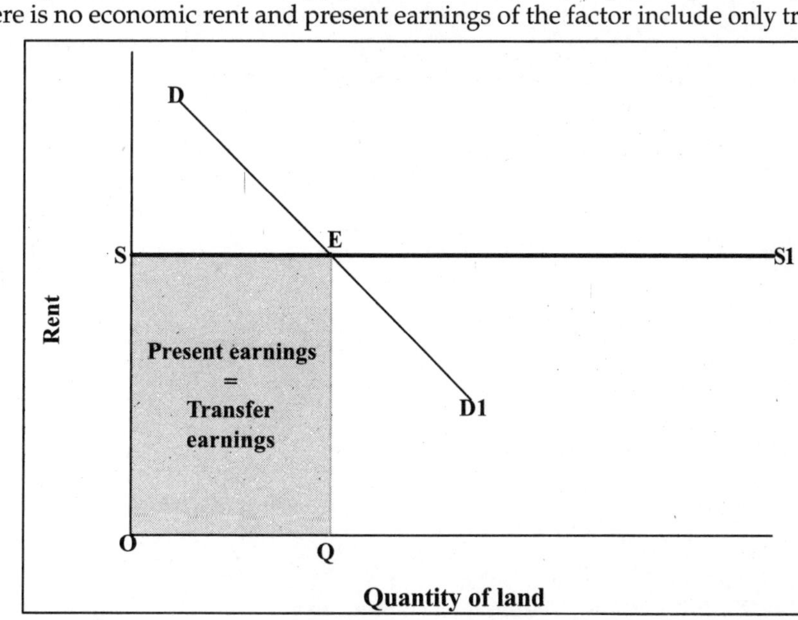

Figure 25.11: Zero Economic rent – Case of Perfectly elastic supply of land.

This infers that, the less elastic the supply curve is, the more is the proportion of economic rent. On the other hand, the more elastic the supply curve is, the more is proportion of transfer earnings.

26

Wages

Wages are the remuneration paid to the 'labour' factor employed in the production programme. In Economics, 'labour' refers to *'any exertion of mind or body undergone partly or wholly with a view to earning some good other than the pleasure derived directly from the work'*. (Marshall). Thus, all types of work, whether, it is physical or mental executed by the workers, salaried persons, employees etc., constitute 'labour' and the salaries or emoluments or fees or commission or bonus etc., derived by them refers to 'wages' in Economics. But, in ordinary language, we generally differentiate between wage and salary, where the former is a remuneration paid to hired labour and the latter represents the remuneration for white collar job holders. Similarly, other terms like bonus, emoluments, fees, commission etc., indicate only social distinction, but in the context of Economics, all the beneficiaries are said to receive 'wages'. In the words of Benham, *'a wage may be defined as a sum of money paid under contract by an employer to a worker for services rendered'*. According to Rothschild, *'wage is the price for a particular commodity viz.,* labour services'.

26.1. Different Types of Wages

Wages are paid to the labour in different modes and they are discussed here under.

1. Time Wages

These are the wages paid to the labour in accordance with the number of hours of work. For example, wages are paid to an agricultural labour for ploughing the land for five hours a day. The time wages are also paid daily, weekly, fortnightly, monthly etc.

2. Piece Wages

These are the wages paid to the labour in accordance with the quantity of output turned out in a production programme. For example, a worker is paid by a sugar factory in accordance to the number of quintals of sugar manufactured. If the price/unit of sugar is Rs. 1500, then the worker will be paid accordingly in terms of amount of sugar processed by him in the factory.

3. Task Wages

In this mode of payment, wage payment is fixed for each and every task in the production programme. For the labour, the task is allocated, such that, it has to be compelted from the start to its end within the standard time. Examples include, digging bore well in the farm, paddy transplanting in about 10 acres of land, construction of farm godowns etc.

4. Cash (Money) Wages

When wages are paid to the labour in terms of money cash, it is called Cash wages. Usually, in urban areas, the workers are paid wages in liquid cash. Money wages are also called as Nominal wages. For example, if a worker in a sugar factory is paid with Rs. 5000/month, it refers to money wages or nominal wages.

5. Kind Wages

When wages are paid to the labour in terms of kind component say, in terms of rice, wheat, pulses etc, it is called Kind wages. Even today, in rural areas, the agricultural labourers are often paid in terms of kind wages.

6. Real Wages

Nominal or Money wages does not measure the real earnings of the worker. This is because, sometimes, the worker enjoys other benefits offered by the sugar factory like free quarters, medical facilities, vehicle allowance, telephone allowance etc., and all these earnings (benefits) along with the money wage the worker enjoys constitute Real wages. So, in this case, the real wages of the worker is more than money wages. Thus, nominal wages refer only to money wages of the worker, whereas, real wages represents necessaries, comforts and luxuries enjoyed by the worker in return for his services contributed in the production programme.

The concept of real wages can also be studied from the point of view of the purchasing power the worker enjoys. The purchasing power (real wage) can be studied based on two components *viz.*, money wage and price level in the economy. This is given by, $R_W = M_W/P_L$, where, R_W = Real wage of the worker, M_W = Money wage of the worker, P_L = Price level in the economy. The following are the interesting points we can infer from the above equation:

- With the rise in price level in the economy (inflation), the real wage of the worker falls and *vice versa*.
- At the constancy of general price level, when the money-wage rate of the worker increases, his real wage also increases.
- When the price level in the economy increases more than money wage, real wage of the worker will decline and *vice versa*. If this is the case in the economy, then the Government should intervene and announce daily allowance to the worker, so as to compensate his loss in real wage.

There is another way of measurement of real wage of the worker and it is given by, $R_W = M_W$/Cost of living index. Here, cost of living index refers to the consumer's price index in the economy. Thus, the real wage determines the standard of living of the workers. Hence, compared to money wage, real wage is the true indicator of worker's prosperity. The comparison between money wages and real wages also indicates another aspect *i.e.*, it guides the worker in choosing different job alternatives. For example, if a worker finds real wages are more in one job, but money wages are less compared to another job, then the worker should choose the former job and not the latter. However, besides general price level or consumer price index, there are several factors that influence the real wages and they are discussed here under.

- ***Facilities at working place:*** If the facilities provided at the working place are good, then the worker enjoys more real wage. For example, if the workers A and B are employed in two different organizations and drawing the same salary as Rs. 10000 each, and if the worker A is provided with good facilities like infrastructure, laptop computer, air-conditioned room etc., and that of the other is not equipped with such facilities, then the real wage of worker A is more compared to worker B, though they draw same nominal (money) wages.
- ***Fringe benefits:*** These are the benefits enjoyed by the worker for the service rendered by him in an organization. For example, an organization may offer several benefits to the workers like free accommodation in the quarters, free medical aid, subsidized canteen, free transport service, telephone facility, pension etc., and these benefits add to the real wages of the workers.
- ***Scope for extra earning:*** If the worker has the scope to make for extra income, while rendering service in an organization or industry, then his real wage will be high. For example, a college professor can earn extra income by publishing research articles in reputed journals, publishing text books, evaluating answer sheets of students etc., and thereby, he earn more than his regular salary. So, the chances of extra earning will increase the real wages.
- ***Nature of job:*** If the nature of job is more risky or seasonal, then money wage will be high, but real wage is low. For example, a worker working in coal mines, forest area etc., enjoy more money wages, but real wages are low.
- ***Social prestige or status:*** Some jobs, though they carry same money wages, but in terms of social status, they greatly differ. For example, an Assistant Professor joined in the University has the same scale on par with the District Collector, but in terms of social status, District Collector is far superior compared to an Assistant Professor. So, the District Collector enjoys more real wages compared to an Assistant Professor.

- *Timely payment of wages:* If the workers are paid with the wages regularly and promptly, then his real wages are considered to be more than the workers, who are not paid with the wages regularly and promptly.
- *Possibility of promotions in the job:* If the worker has the bright chance of getting the promotion in his job duties, then his real income will be higher. If a Professor in the University has the bright chance of becoming a Vice–Chancellor of the University, then he enjoy more real wages than his money wages.
- *Period and cost of training:* When a job needs a long period and huge cost for training, then the real wage for the worker in that job will be lower.
- *Trade expenses:* When a job needs high trade expenses, then the real wages for that worker will be low. We know, a dental doctor or a surgeon have to establish and maintain their own hospitals or dispensaries, thereby, the trade expenses are very high and hence, they enjoy less real wages. In this context, an Assistant Professor in the University will enjoy high real wages compared to a dental doctor or a surgeon.
- *Climatic conditions:* For the workers, working under adverse or unhealthy or unfavourable climatic conditions, enjoy less real wages compared to the workers working under healthy environment.
- *Nature of jobs:* If the job is permanent, then the worker enjoys more real wages and for the worker, who is temporarily employed, the real wages are low. Similarly, compared to the real wage of a skilled worker working in a sugar factory, the real wage of a casual agricultural labour is low.

26.2. Why a Separate Theory of Wages

In the Chapter 24, we discussed in-detail about the general theory of factor distribution to determine the pricing of factors of production, but several Economists of different eras have propounded separate theories of wages. This is so because, compared to other factors of production, labour factor enjoys some distinctive features and they are discussed here under.

- Labour is inseparable from labourer himself. That means, when labour service is to be engaged in the production programme, the 'agent' (labour) must also go to the place of the work, as labour service is a human service.
- The engagement of labour in the work depends upon the willingness of the labour, physical fitness and mental alertness of the labour. His feelings and emotions counts much towards fulfilling the assigned tasks and in turning out quality work. Besides above, the execution of labour work also depends upon the working conditions, pleasantness and monetary compensation he derives in the production programme.
- The demand for labour for an activity depends upon the MPP, whereas, the demand for a product depends upon the MU.
- The employment of labour in a production programme is influenced by the ethical, institutional, behavioural and sociological aspects. Further, other aspects like pensions, gratuity, accommodation facilities, medical facilities, recreational facilities, vacations, job security, promotions, leisure hours etc., also influence the employment of a labour in an organization.
- Labour is the only factor of production, which acts as both producer and consumer of goods produced in the production programme.
- Since, supply of labour is always more than demand for labour in any organization, there is low bargaining power on the part of labour, thereby, the labour are being exploited and get low wages compared to their productivity in the production programme. So, the concept of demand and supply theory will not exactly determine the price for labour factor, as it supports the exploitation of labour by the entrepreneurs. Further, labour is the most perishable factor of production and we know that, when the commodity is highly perishable, demand is the major force that plays crucial role in the determination of price of the commodity. So, when the demand for labour (perishable factor) is less compared to the supply, their price will be low and this infers that, labour gets exploited in his employment.
- Pricing of labour factor involves Government intervention in the form of minimum wage level, so as to avoid their exploitation. This is because of excess supply of labour over demand and hence, the labour has weak bargaining power. Further, the labour is a perishable commodity, thereby, their exploitation is severe in the context of market imperfections and absence of trade unions. However, this concept of Government intervention in the form of minimum rent minimum interest and minimum profits will not arise in case of other factors of production like land, capital and entrepreneur (management) respectively.
- Supply curve of labour will not show perfect (direct) relationship between wage rate of labour and number of working hours of labour. This is because, up to a certain level of increasing wage rate, the number of working hours contributed by the labour will increase, but at a higher level of wage rate, the number of working hours

of labour will decrease, as they prefer more number of leisure hours and less number of working hours (Figure 9.59). Thus, at a higher level of wage rate, there exists inverse relationship between wage rate and number of working hours of labour and this is due to inelastic cash requirements on the part of the labour. Thus, there is no standard functional relationship between wage rate and number of working hours (labour supply) contributed by the labour. Moreover, the supply of labour depends not only on the wage rate, but also on various factors like population, working age-groups, sex ratio, ethical and social considerations, desire to work by the labour etc. However, in case of product market, there exists direct functional relationship between price and product supply into the market.

- ☆ The bargaining power of the labour can be increased, if they form trade union and such formation of trade unionism is not possible in case of other factors of production.
- ☆ The supply of labour is comparatively inelastic when compared to other factors of production. That means, when the demand for the labour suddenly increases, the supply of labour cannot be increased overnight or few days or few months. However, compared to labour factor, the supply of (durable) products can be increased rather quickly into the market due to increase in demand.
- ☆ The concept of bilateral monopoly *i.e.*, monopsony in the demand for labour by the entrepreneur and monopoly in the supply of labour by the labour supplier can be seen only in the labour market and not possible with reference to other factors of production. Such imperfections in the labour market are due to weak bargaining power on the part of labour, absence of trade unions, less mobility of labour, entrepreneur's may enjoy monopsonistic power, differences in skills and efficiencies of labour etc.

Considering the above peculiar features of labour factor in contrast to other factors of production, it is justifiable to have a separate theory to analyze the pricing of labour factor. However, it is essential to remember that, wages paid to the labour is a price paid for the services rendered by them, thereby, the concept of demand and supply theory also applies to wage determination of labour. But, several theories have been proposed to determine the wages of labour factor and all these theories have their own merits and demerits in explaining the concept.

26.3. Different Theories of Wages

The above discussion infers that, labour is a peculiar factor of production. The labour is highly perishable when compared to other factors of production, as even the landlord may withhold the supply of land and capitalist may withhold the supply of capital, but the labour must necessarily supply his services, whatever price or wage is available in the market. This peculiar feature has prompted the earlier Economists to propose different theories to explain the determination of wages and they are discussed here under.

1. The Subsistence Theory of Wages

This theory was proposed by Classical Economists like Ricardo and Malthus and developed by Adam Smith. This theory is also called as 'Iron law of wages' or 'Brazen law of wages' by a German Economist, Lassalle. The following are the important features of this theory:

- ☆ This theory was based on the Malthusian theory of population, which states that, population increases at a faster pace compared to food production and food production is subjected to LDR.
- ☆ In the long run, the wage rates tend to be equal to the money that is sufficient for subsistence of the workers. These wages are called Natural level of wages.
- ☆ If the wage rate exceeds the subsistence level, then the supply of labour will increase and thereby, due to competition among the labourers for employment, it will bring down the wage level back to the subsistence level.
- ☆ If the wage rate falls below the subsistence level, some labour gets unemployed, they could not be able to marry, thereby, supply of labour population will decline in comparison with demand for labour and gradually wage rate will be increased up to the subsistence level.
- ☆ According to Ricardo, the presence of trade unions will not increase the wage level. This is because, the wage level in the economy is determined by the size of population of workers *i.e.*, the number of workers. If the wage rate of workers is above the subsistence level, then labour supply will increase and due to rise in competition among the workers, the wage level will decline back to the subsistence level. On the contrary, if the wage rate is below the subsistence level, the population of workers will fall and thereby, supply decreases and hence, wage rate will gradually increase up to the subsistence level.
- ☆ This theory assumes that, the supply of labour is perfectly elastic, thereby, even with increase or decrease in demand for labour, the wage rate of labour remains same and it is just sufficient to meet the subsistence needs of the worker and his family (Figure 26.1).
- ☆ This theory assumes that, the subsistence wage rates will exist only in the long run.

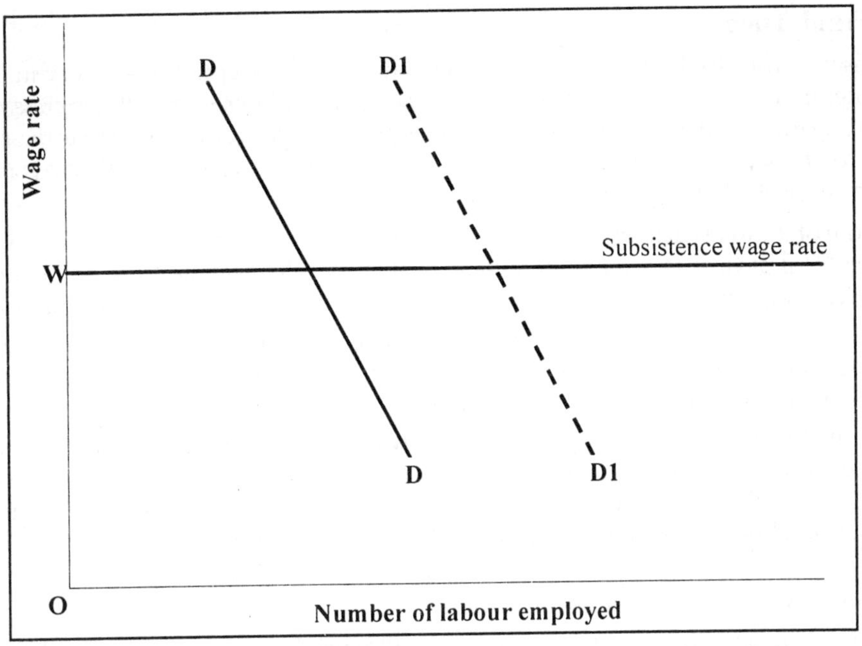

Figure 26.1: Subsistence theory of wages – Perfectly elastic supply curve and wage determination.

Merits

This theory has its own merits and they include,

- ✰ This theory still holds good in over populous countries, where wage rates for most of the workers are just sufficient to meet their subsistence requirements.
- ✰ This theory explains that, there is a certain minimum level of wage rate (subsistence wage rate), below which, the wage rate should not fall.

Demerits

This theory has been criticized on the following grounds:

- ✰ This theory does not explain differential wages in different employments.
- ✰ All the workers cannot have a common minimum wage rate across different employments and further, the family size and wants vary across the workers.
- ✰ This theory explained the concept of wage determination from supply side ignoring the demand side. That means, with reference to demand side, the entrepreneur will assign the work to the worker depending upon the nature and quantum of work load and productivity of labour rather than based on the subsistence wage rate required for the worker.
- ✰ This theory explains the adjustment of wage rates over life time of a generation (long run) and does not explain wage fluctuations over a short period of time.
- ✰ It is unjust to say that, wages will match the subsistence level of the worker and his family. Higher wages are needed for the working class to enhance their productive efficiency.
- ✰ This theory explains that, when the wage rate increases above the subsistence level, population increases and labour supply will increase. But, when workers enjoy higher wage rates, they prefer to have fewer children, so that, they can maintain higher standard of living in the society.
- ✰ This theory failed to take into consideration the role of trade unions in increasing the wage rate.
- ✰ This theory is pessimistic in its outlook because, even in the long run, the wage rates are subsistence wage rates despite workers can show better improvements in their efficiency. This also infers that, even in the long run, the working conditions cannot be improved. But, with the advent of modern technologies in the long run, the working conditions will improve and the actual wage rates will be higher than subsistence wage rates.

Considering the above criticisms, it infers that, this theory has not offered satisfactory explanation for the determination of wages. However, it points out certain minimum level of wages, below which the wages should not fall in the economy, so as to meet the subsistence requirements of the workers.

2. The Wages Fund Theory

This theory was proposed by John Stuart Mill. He explained the concept of this theory in *'Principles of Political Economy'* and he states that, *'wages depend upon the demand and supply of labour, as it is often expressed, on the proportion to population and capital. By population, is here meant the number only of the labouring class or rather of those who work for hire and by capital, circulating capital and not even the whole of that, but the part which is expanded on the direct purchase of the labour'*. The following are the important features of this theory:

- In a firm, out of the total earnings, the entrepreneur must set aside some amount of capital, after meeting the requirements for payments of rent, interest and profits, so as to meet the wage payments of the labour. This amount of capital is to be maintained every year in the firm and hence, it is called as a wages fund.
- Since, the wages fund is determinate or fixed, the number of labour engaged at a point of time is also fixed, at a given wage rate. If the labour wants to get a higher wage rate, there is no other alternative except to reduce the employment of labour.
- As a short run theory of wages, it suggests that, the wage rate is determined based on demand and supply of labour. These are not actually demand and supply schedules as used in Modern Economics. Here, demand for labour refers to the size of wage fund the entrepreneur has to be maintained and supply of labour refers to the number of labour the entrepreneur wishes to engage. So, the wage rate is determined based on the formula,

  Wage rate = Wage fund / Number of labour

  Thus, in short run, wage fund is a fixed amount and the supply of labour is also fixed, thereby, the wage rate is uniquely determined.
- The above formula also suggest that, if wage rates are to be increased for the workers, there are two possible options *viz.*, either wage fund should be increased or number of workers should be reduced. But, the former is not under the control of workers. Even, if the wage rate is increased, again it will be counteracted by the rise in workers' population in the country. So, even the presence of trade unions cannot succeed in raising the wage rate. Further, even if the wages are increased in one country, they are likely to fall in another country, as the wage fund was a fixed amount of capital in the economy.

Demerits

This theory suffers from the following limitations:

- This theory does not explain the determination of wages in the long run.
- This theory does not explain how the level of wage fund is determined.
- In general, a firm will fix the wage rate for the labour considering the level of working capital employed in the business. The entrepreneur may employ both owned capital and borrowed capital in the business and considering the total working capital, he fixes the wage rate for the labour and the number of labour to be employed in the business. So, it is not correct to say that, wage rate depends upon the wage fund. Further, the same level of wage fund may not be carried every year to employ the labour and to fix the wage rate.
- This theory assumes that, first the entrepreneur decides the wage fund and then plan the number of workers to be employed and accordingly, wage rate is determined. But, in practical situation, based on the prevailing wage rate, the entrepreneur plans either the labour-intensive technology or capital-intensive technology and thereby, accordingly the level of employment of workers and wage fund are determined.
- Since, wage fund is a fixed amount, this theory has ignored the role of trade unions in bargaining for higher wages.
- This theory has ignored the wage differentials across the workers in accordance with their working efficiencies.
- This theory wrongly assumed that, wage fund can be increased only at the expense of profits (*i.e.*, share of the entrepreneur). But, if wage fund is increased, wage rate of workers increase, income of the workers increase, standard of living increases, quality of food intake increases, productivity of the workers increase and thereby, profits in the business increases. So, the increase in wage fund leads to increase in profits in the business.

Even, J.S. Mill himself became aware of the wrong assumptions formulated in the theory and he even reformulated them. So, considering the faulty assumptions and concept of the theory, it has not offered satisfactory explanation for the determination of wages. However, it contains an element of truth, as in case of under developed countries and in high populous countries, unless national income is increased, the wage rate cannot be increased.

3. The Residual Claimant Theory of Wages

This theory was proposed by an American Economist, Prof. F.A. Walker in his book, *'Political Economy'*. According to Walker, wages are the residue left over and paid to the labour after meeting the remunerations of other factors of production. In the words of Walker, *'labouring class receive all the help to produce, subject to deductions on the three several counts. First, rent is deducted and rent must come out before the question of wages is considered. Secondly, remuneration must be deducted for the use of capital and the third and last deduction is profits, the remuneration of the entrepreneur. These three shares being out of the product of industry, the whole remaining body of wealth is the property of the labouring class'*. In the words of Jevon, *'the wages of a working man are ultimately coincident with what he produces, after the deduction of rent, taxes and the interest on capital'*. Thus, after meeting the remunerations of all other factors of production, the remaining income of the entrepreneur will goes to the labour as wages. This theory admits that, the efficiency of the labour counts much in the increase in total output of the firm and thereby, wages increase through greater efficiency of the labour. In this context, this theory is considered more optimistic in its outlook, when compared to the earlier Subsistence theory and Wages fund theory.

Demerits

This theory has been criticized on the following grcunds:

- ✰ This theory ignored the influence of supply of labour on wages.
- ✰ In practical sense, the claimant of the residual goes to the entrepreneur and not to the labour.
- ✰ This theory ignored the role of trade unions in increasing the bargaining power of labour.

Considering the above criticisms, we can infer that, this theory has not offered valid explanation for the determination of wages. However, this theory has rightly quoted that, increase in efficiency of the labourers will lead to increase in wages through increase in the total output.

4. The Standard of Living Theory of Wages

This theory is a refined form of Subsistence theory of wages and it was proposed by Torrens. The following are the important features of this theory:

- ✰ This theory states that, wage rates of labour tend to confirm to the standard of living of the workers. That means, the standard of living of the workers determines the wage rate. Here, the standard of living not only refers to the basic necessaries of the labour, but also includes education, comforts and luxuries for which the labour is habituated. For example, if a skilled labour is habituated to have an own house, bike, healthy nutritious food etc., his wage level is also determined that, he can meet all these wants. That means, the standard of living of the worker determines the wage level and to maintain this standard of living, the worker tries to maintain his wage level.
- ✰ The higher the wage rate, the more will be the efficiency and productivity of the labour and this higher productivity also justifies the need for a higher wage rate.
- ✰ This theory is considered optimistic over Subsistence theory of wages, as it considers efficiency and productivity of the labour in determining the wage rate of the labour.

Demerits

This theory has been criticized on the following grounds:

- ✰ This theory explains that, the standard of living of the labour determine his ruling wage rate. But, even wage rate also influences the standard of living of the labour. So, the concept can be true even in the reverse way.
- ✰ It is not rational to conclude that, the standard of living alone will determine the wage rate of the labour, as the wage rate of the labour is influenced by several other factors like efficiency, profits of the business, seniority, business fluctuations etc.
- ✰ This theory takes into account only the supply side of labour and neglected the demand side of labour.
- ✰ This theory suggests that, first the entrepreneur must know the living standards of the labour to fix their wage rate. But, this is not at all possible.
- ✰ This theory explains the determination of wages based on the assumption of long run. But, short-run determination of wages is more practical compared to long-run.
- ✰ This theory could not explain the concept of wage differentials across the labour, even though they maintain same standard of living.

From the above points, we can infer that, though this theory has offered superior explanation compared to Subsistence theory of wages, it too suffers from inadequate explanations of the phenomenon of determination of wages.

5. The Bargaining Theory of Wages

This theory was proposed by Thomton, Davidson and Maurice Dobb. The following are the important features of this theory:

- The level of wage rate of the labour depends upon the bargaining power of the labour and it varies directly.
- From the entrepreneur's side, there will be a maximum level of wage rate and from the labour side, there will be a minimum level of wage rate. The maximum level of wage rate implies that, it is the highest possible wage rate the entrepreneur can give to the labour. If the wage rate of the labour goes beyond that maximum level, the entrepreneur will decide to reduce the number of labour and may plan for machinery usage in the production programme. Regarding minimum wage rate from the labour side, it is the wage rate, below which, no labour is willing to accept the work in the production programme. In general, due to bargaining between the entrepreneur and the labour, a wage rate is fixed in between maximum level and minimum level wage rate.

Merits

Compared to the earlier theories, this theory has the following merits:

- This theory has recognized the role of trade unions in increasing the bargaining power of workers to get higher wages.
- This theory also explained the wage differentials across the workers in accordance with the bargaining power of the labour.

However, this theory too suffers from the drawbacks in the sense that, the entrepreneur and labour greatly differ in their economic position and thereby, the labour has comparatively weak bargaining power. Further, the supply of labour is also relatively more compared to the demand for labour and thereby, the labour is easily exploited. Thus, this theory of determination of wages is not supported both from moral and economic grounds.

6. The Marginal Productivity Theory of Wages

The General theory of factor distribution has been extended to wages paid to labour and this is called Marginal productivity theory of wages. This theory was proposed by a German Economist, Von Thunen and later promoted by Walras, J.B. Clark, Taussig and others. This theory states that, the entrepreneur generally employs the labour factor with higher marginal productivity and he substitute the labour with higher marginal productivity in place of the labour with lower marginal productivity. This theory is based on the following propositions:

- The marginal productivity of the labour determines the price (wage) of the labour.
- In the long run, the wages paid to the labour tends to equate to its ARP and MRP.
- The point of optimal allocation of labour in the production programme is when the remuneration paid to the labour (MFC) equals the MRP.

Assumptions

This theory is based on the following assumptions:

- All the labour are homogenous, thereby, they are perfectly substitutable.
- The technology is given, though the scales and proportions of labour factor may change.
- There is perfect mobility of labour between different employments.
- There is perfect competition both in the labour market and product market.
- The marginal productivity of labour is measurable.
- The firm aims at profit maximization through efficient allocation of labour.
- This theory considers long run analysis, as the price paid to the labour tends to be equal to ARP and MRP.
- There is full employment equilibrium in the economy.
- This theory is based on the concept of LDR.
- The bargaining power of the seller and buyer of the labour service are equal.
- There is no Government intervention in the fixation of labour's remuneration like Minimum wage rate etc.

Explanation of the Theory

Let us discuss the main propositions of the theory here under.

(a) Remuneration Paid to the Labour is Equal to its Marginal Productivity

We know, a firm employs the agricultural labour, when they are productive enough to increase the output in the production programme. That means, if the agricultural labour is productive only in terms of contributing or increasing the output, then only the firm will engage and pay wages to the labour. So, in employing the agricultural labour, the firm often compares the productivity contributed by the labour with the wages it pays as remuneration to the labour and the firm continuously increase or engage the labour, so long as marginal productivity contributed by the labour is higher than wage rate paid to the labour by the firm. This infers that, the marginal productivity contributed by the agricultural labour determines his wage rate and it varies directly. The employment of the agricultural labour in a firm increases, so long as the marginal productivity contributed by the labour is greater than wage rate paid to the labour and the firm will come to equilibrium or stop employing additional labour, when the marginal productivity of labour is equal to his wage rate. To generalize the statement, a firm will engage the agricultural labour up to the point, where the MFC of the agricultural labour (from the firm side) or when the reward for the labour (from the factor side) is equal to his marginal productivity. At this point of equilibrium, the firm will minimize the cost in the production programme. This is because, when marginal productivity of the agricultural labour is greater than the wage rate, it implies, the firm can still engage more number of agricultural labour in the production programme. But, if marginal productivity of labour is less than the wage rate paid by the firm, it is better for the firm to remove excess labour, as they are not contributing much to the total output of the firm in proportion to the wage rate they receive, and thereby, the firm will attain equilibrium when marginal productivity of labour is equal to his wage rate. Thus, the firm will attain equilibrium in the production programme with reference to the employment of agricultural labour, when the marginal (revenue) productivity (MRP) is equal to the cost of employing additional unit of labour (MFC).

(b) Firm's Equilibrium in the Short Run

In Figure 26.2, MRP=MVP or VMP is an inverted 'U' shaped curve and WW_1 refers to the wage rate paid to the agricultural labour=AFC (Average wage rate)=MFC (Marginal wage rate) and it is an horizontal straight line parallel to X-axis, assuming perfect competition both in the product and labour markets respectively. It is clear from the Figure 26.2 that,

- ☆ We know, MRP=VMP under perfect market competition and MRP indicates the firm's demand curve for labour.
- ☆ From the firm's view point, at a given wage rate OW, the factor's supply curve is perfectly elastic *i.e.*, WW_1.
- ☆ So, from the above two curves *i.e.*, MRP (= VMP) and WW_1, the firm can draw the information pertaining to the price of the product and wage rate of the agricultural labour.
- ☆ At point 'E', MRP=MFC or wage rate derived by the agricultural labour and this infers that, the firm is in equilibrium by employing ON number of agricultural labour.
- ☆ Below ON number of agricultural labour, MRP of agricultural labour is greater than wage rate paid per labour and thereby, the firm can still employ number of labour up to ON to attain equilibrium. Even at point E_1, the firm is in short run equilibrium, where MRP=MFC, with reference to ON_1 level of employment of labour. But, beyond E_1, the MRP>MFC and thereby, the firm can still increase the employment up to ON number of labour, where MRP is again equal to MFC, and thereby, the firm is in equilibrium.
- ☆ If the firm employs more than ON number of agricultural labour, the MRP of agricultural labour is less than the wage rate paid per labour, thereby, the firm must necessarily reduce the number of labour to ON, so as to ensure equilibrium or minimization of costs in the production programme.

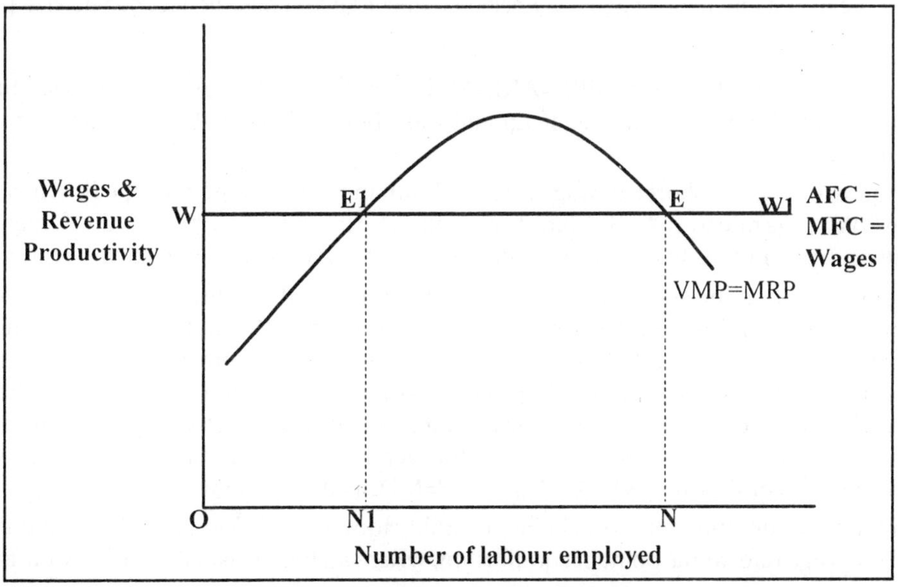

Figure 26.2: Marginal productivity theory of wages – Pricing of labour factor under perfect competitive product and labour markets (Short run).

- This infers that, in a competitive market, the firm will be in equilibrium when the marginal productivity or MRP of a factor is equal to the MFC (or) price paid per factor with reference to employment of a factor in the short run production programme.
- So, the essential condition for firm's equilibrium in short run with respect to the employment of a factor is that, the demand curve or MRP curve must cut the supply curve of the factor or MFC curve from above, when MRP is falling.

(c) Long Run Industry Equilibrium

When a firm is in short run equilibrium, it does not mean that, the industry is also in equilibrium with respect to employment of a factor. When the firm is in short run equilibrium, it implies MRP=MFC and at this point of equilibrium, the firm maximizes the super-normal profits. These super-normal profits are lucrative and thereby, new firms enter the industry. Consequently, the demand for the factor increases, and thereby, the wage rate of agricultural labour will increase. Since, new firms enter the industry, the supply of the output into the market will increase, thereby, the price of the product will fall in the market and hence, MRP falls. When MRP falls, ARP also falls. This will continue till MRP=ARP=MFC (=AFC=wage rate) and this is the condition of industry equilibrium in the long run (Figure 26.3). That means, in the long run, the essential condition for the firm's equilibrium with reference to employment of labour factor is MRP=ARP=MFC and ARP must be tangent to the MFC (=AFC=Wage rate) at the point of equilibrium. But, in short run, the essential condition for firm's equilibrium is only MRP=MFC. As shown in the Figure 26.3.

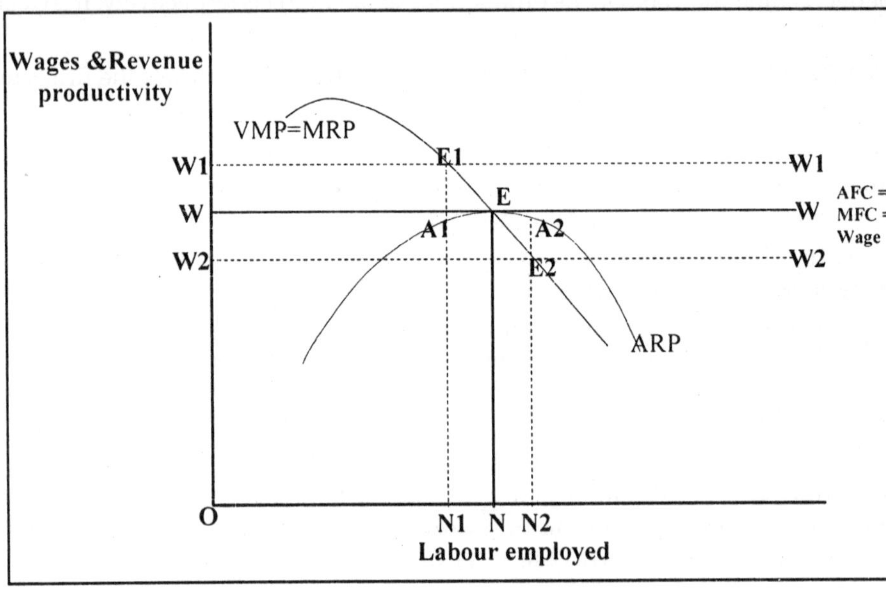

Figure 26.3: Marginal productivity theory of wages – Pricing of labour factor under perfect competitive product and labour markets (Long run).

- With reference to point 'E', where MRP=ARP=MFC (=AFC=wage rate), the firm is said to be in long run equilibrium, by employing ON number of agricultural labour, where all the firms derive only normal profits in the industry.
- Let us suppose, OW_1 is the short run wage rate of labour and at this wage rate, the firm incurs losses, as ARP is less than OW_1 wage rate by E_1A_1. So, some firms will leave the industry, thereby, the demand for agricultural labour will decrease and the wage rate will fall from OW_1 to OW and again the firm's equilibrium is attained.
- Let us suppose, OW_2 is the short run wage rate of labour and at this wage rate the firm will enjoy super-normal profits, as ARP is higher than OW_2 by E_2A_2. New firms get attracted to these super-normal profits and enter the industry. As a result, the demand for the agricultural labour will increase, thereby, wage rate increases from OW_2 to OW. Further, with the increase in supply of the product into the market due to the entry of new firms into the industry, the price of the product will fall and thereby, ARP will fall and gets equated to OW wage rate. So, due to increase in demand for the factor and fall in demand for the product, again the firm will attain long run equilibrium, where, MRP=ARP=MFC (=AFC=OW wage rate).
- Hence, in the long run, the industry will be in equilibrium when MRP=ARP=MFC and ARP is tangent to MFC(=AFC=Wage rate) at the point of equilibrium, where all the firms will earn only normal profits.

Criticisms

This theory has been criticized on the following grounds:

- ☆ ***All units of labour are not homogeneous:*** The assumption that, all the units of labour are homogeneous is unrealistic. This is because, we know that, the efficiency differs from agricultural labour to agricultural labour. It is, therefore, not correct to assume that, the different units of the labour are homogeneous. In fact, heterogeneity and non-homogeneity is the rule. Hence, no two factor-units are homogeneous, thereby, they are non-substitutable for each other. For example, an expert functioning in a fruit processing industry cannot be employed in a sugar factory.
- ☆ ***Labour are not perfectly mobile:*** This theory assumes that, there is perfect mobility of labour between different employments and places. But, in reality, labour are mostly immobile. There is no automatic movement of labour from one industry or place to other. For example, the mobility of labour across different occupations is influenced not only by the wage rate, but by several other labours like, seniority, pensions, permanency of the job, work-leisure ratio, sociological and ethical aspects etc. The greater the degree of specialization in the industry, the less is the labour mobility from one industry to other. Due to less mobility, the labour are not paid equally in proportion to their marginal productivities across the occupations and across the places.
- ☆ ***There is no perfect competition:*** This theory is based on another unrealistic assumption of perfect competition both in labour and product markets. We know, imperfect competition is the rule and it leads to exploitation of labour, as they are paid much less than their marginal productivity.
- ☆ ***Factors are not fully employed:*** This theory assumes the existence of full employment equilibrium in the economy. But, unemployment and underemployment is the rule. This is because, we practically observe that, labour are paid less than their marginal productivity.
- ☆ ***Labour are not perfectly divisible:*** The assumption that, labour are perfectly divisible and therefore, can be increased by small amounts does not hold good. This is because, it is not possible to increase the labour service by fraction amounts, such as, half time or one and a half times etc.
- ☆ ***Production is not a result of labour factor alone:*** It is not possible to calculate the marginal productivity of each labour unit separately. This is because, the production of a commodity cannot be attributed to labour factor alone. Rather, it is always a result of coordinated and combined efforts of different factors.
- ☆ ***Not applicable in the short-run:*** This theory is applicable only in long run, as the rewards of the labour services tend to equal their MRP. But, in reality, we are more concerned with the short run problems. As remarked by the Keynes, *'in long run we all are dead'*. So, this assumption of long run analysis makes the concept of pricing of labour factor unrealistic.
- ☆ ***Neglects technical progress:*** This theory fails to throw light on the influence of technical changes in increasing the marginal productivity of labour. Prof. Hicks explained that, the labour saving innovation tends to raise the marginal productivity of capital relative to that of labour. The opposite may happen in case of capital saving innovations. Thus, Keynes criticized this theory as a *'static concept'*.
- ☆ ***Supply of factors is not fixed:*** This theory assumes that, supply of labour is perfectly elastic even in the long run. But, the supply of labour is fixed in the short run and not in the long run. Moreover, this theory studied the pricing of labour factor from demand side only and neglecting the supply side. Hence, this theory cannot be applied to the labour market, which requires explanation from both demand and supply forces.
- ☆ ***Ignores ethical aspect:*** This theory deals with only positive aspect of analysis pertaining to process of pricing of labour factor and it does not have ethical aspect. It is not concerned with ethical justification of payment of remuneration to the labour factor.
- ☆ ***Employment level depends upon effective demand:*** This theory supports the Pigou's version that, employment of labour can be increased by reducing the wage rate. But, Keynes criticized this version and opined that, when wage rate is reduced, income of the labour will decrease, thereby, effective demand declines, leading to unemployment and under employment problems in the economy.
- ☆ ***Reward for the factor determines productivity:*** One of the important propositions of the marginal productivity theory is that, the marginal productivity of the labour factor determines its payment or remuneration. So, we can infer that, marginal productivity of the factor is the cause and remuneration paid to the factor is the effect But, according to Prof. Sydney Webb, the proposition also applies in the reverse way. For example, when an agricultural labour is provided with higher wage rate, his income will increase, standard of living will increase, consumes more quality food and thereby, his productivity in the job work will increase. So, here remuneration paid to the labour is the cause and marginal productivity contributed by him is the effect. Thus the proposition is also valid in the reverse way.
- ☆ ***Measuring MRP is a difficult task:*** According to this theory, marginal productivity of the labour factor determines its remuneration. But, measuring MRP is not an easy task. We know, in several Governmental organizations, even if we remove one labour or worker from the organization, the TPP will not fall and this

implies that, MRP of the labour is zero. So, the organization should not pay any wage or salary to that labour. But, this is absurd to say so.

- ***No explanation how remuneration is fixed for the labour factor:*** This theory simply offers the explanation regarding the level of employment of a factor (agricultural labour) at a given wage rate. But, this does not offer explanation regarding the determination of wage rate for labour. Thus, this theory does not explain how factor's remuneration is determined.
- ***LDR not common in manufacturing industries:*** This theory assumes that, as the firm goes on employing the labour factor, LDR sets in to a factor. But, in case of manufacturing industries, where improved technology is adopted, the labour factor will contribute increasing returns and LDR is held in check.
- ***Explains only functional distribution:*** This theory explains only functional distribution of income and does not explain the personal distribution of income and income inequalities in the earnings of the people.

Discounted Marginal Productivity Theory of Wages

This concept of discounted marginal productivity was proposed by Taussig. According to him, the labour receives the wages not equal to their marginal (revenue) productivity, but equal to the discounted marginal (revenue) productivity. This is because, the entrepreneur will pay the wages to the labour even before realizing the output from the business and thereby, he pays the wages by deducting the interest from the marginal (revenue) productivity such that, wages paid will be equal to the discounted (marginal) productivity of the labour. Thus, the entrepreneur does not pay the full amount of the expected MRP of labour. He deducts a certain percentage from the final output in order to compensate himself for the risk he takes in making advance payments to the labour employed. The extent of deduction from the final output is equal to the current market rate of interest. But, this method of determination of wages is difficult to follow in practical situation.

7. Modern Theory of Wages

The Classical theories of wages discussed earlier are either defective or inadequate in explaining the pricing of labour factor. Further, they are one sided (*i.e.*, explained the concept either from demand side or supply side) and they could not explain how the wage rate is determined, but rather simply explains, how many number of labour the firm will employ at a given wage rate. Among the Classical theories, Marginal productivity theory to some extent offered satisfactory explanation, but even this theory has offered explanation with reference to demand side only, neglecting the supply concept of labour. Further, it assumed several unrealistic assumptions. So, the Modern Economists opined that, wages of the labour factor are to be determined based on the interaction between both demand and supply forces in an industry. But, there are certain points of difference between demand and supply of labour factor and products and they are discussed here under.

Demand for a Labour Factor

The following are the few points connected with the demand for labour factor:

- In case of products, the demand is a direct demand, but in case of labour, the demand is derived demand. This is because, the firms or producers demand labour because, there is a demand for the products, which are derived from the services of these labour. Hence, the demand for labour is only a derived demand and not direct demand, as the labour factor does not cater to the needs of the customers directly, but indirectly through producing the products.
- The demand for a labour is a joint interdependent demand, as the combined and coordinated efforts of the labour with other factors of production contribute to production of output.
- If the demand for the product increases, the demand for a labour that involves in the production of that product also increases and *vice-versa*.
- If the demand for the product is elastic, the demand for the labour factor that produces that product is also elastic and *vice-versa*.
- The demand for labour will depend upon the quantity of other cooperant factors of production required in the production process and it varies directly.
- If the labour has cheaper substitutes, the E_D for labour will be high and *vice versa*.
- The demand price for labour depends upon the value of the finished product and it varies directly. That means, if the value of finished product is more, the demand for the labour involved in the production of that product will be more and *vice versa*.
- The more productive is the labour, the higher will be the demand for the labour to be employed in the production process. But, if the labour employment is continuously increased in the production programme, the marginal productivity of the labour will gradually decline. So, if the marginal productivity of the labour decreases, the firm has to employ more of the labour to produce the same level of output and *vice versa*.

- If the number of labour employed constitutes only a small proportion of the total quantity of all factors employed in the production programme, the demand for the labour factor is less elastic and *vice versa.*
- The number of labour employed in the production programme varies inversely with the wage level.
- The wage rate of labour is not determined based on an individual firm's demand, but based on the industry demand. We can have a demand curve for labour at industry level by summing up the demand curves for labour of all the firms in that industry (Figure 26.4).

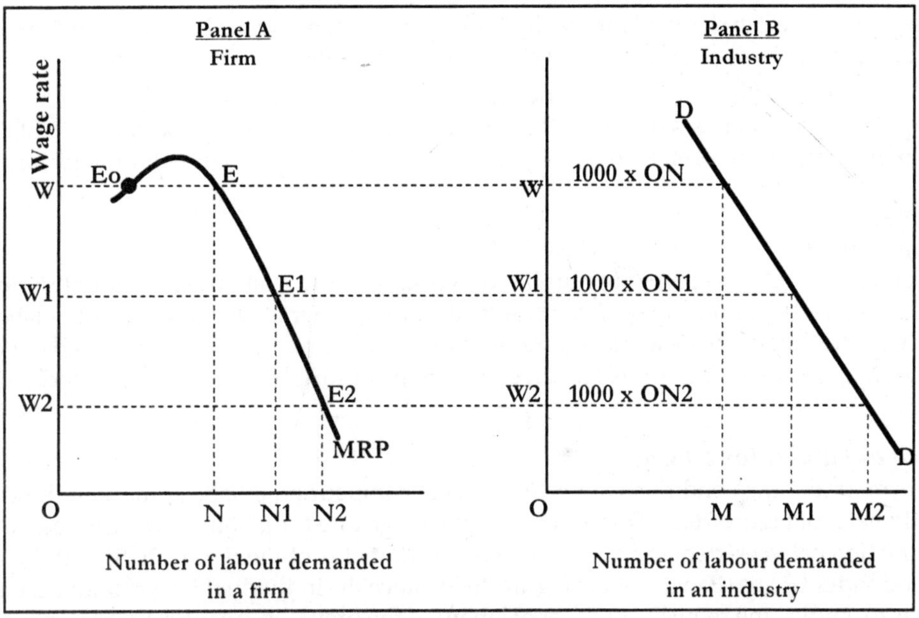

Figure 26.4: Determination of demand curve for labour at industry level.

- Panel A of the Figure 26.4 shows the employment of labour in a firm. MRP is the demand curve for labour for a firm and it is inverted U shaped, indicating that, if the firm employs more number of labour continuously, its marginal productivity (MRP) initially increases and then decreases. The entrepreneur goes on employing this factor (labour) as long as the MRP of labour is greater than MFC and the firm is at equilibrium when MRP=MFC.
- When the wage rate is OW, the firm is in equilibrium at E, where MRP=MFC(=wage rate) and the firm employs or demands ON number of labour.
- Similarly, when wage rate is decreased to OW_1, the firm is in equilibrium at E_1 and it employs or demands ON_1 number of labour.
- When the wage rate is still decreased to OW_2, the firm is in equilibrium at E_2 and it employs or demands ON_2 number of labour.
- The line connecting the equilibrium points, E, E_1, E_2, represents the diminishing marginal productivity of labour (MRP), as the firm goes on employing more number of labour in the production programme.
- But, for determining the wage rate of labour, it is not the demand of an individual firm that matters, but the demand of all the firms in the industry that matters. So the industry demand curve for labour is derived by taking into consideration the demand of all the firms employing that labour. This is shown in Panel B of the Figure 26.4, considering lateral summation of demand curves or MRP curves of all the firms in the industry.
- Let us assume that, there are 1000 firms in the industry, employing the same factor (labour). In both the Panels of the Figure 26.4, wage rate is considered on Y-axis, but in Panel A, X-axis represents the number of labour employed in a firm and in Panel B, X-axis represents the number of labour employed at industry level. This is so because, in Panel A, the demand for labour is in one firm, whereas, in Panel B, the demand for labour is in 1000 firms or industry.
- So, at wage rate OW, the total demand for labour in industry is equal to 1000 firms x ON=OM. Similarly, at wage rates OW_1 and OW_2 the total demand for labour in industry is equal to 1000 firms x ON_1=OM_1 and 1000 firms x ON_2=OM_2 respectively.
- From Panel B, it is clear that, the demand curve DD for labour in the industry also slopes downward from left to right like MRP. This is due to the downward falling of MRP curve from left to right of all the firms in the relevant portion of DD curve. This industry demand curve will be inelastic in the short run and more elastic in the long run.

- ✩ This implies that, the more of a labour factor is employed, the marginal productivity of the labour diminishes and this is called as Law of diminishing marginal productivity.
- ✩ In Panel B, the demand curve for the labour in the industry slopes downward from left to right. Here, we move along the same demand curve due to changes in wage structure of the labour. This change refers to the change in number of labour demanded by the industry. But, there is also another important change called 'change in demand' and this involves altogether shift in the demand curve for labour not due to changes in wage structure of labour, but due to other factors mentioned below,
- ✩ If there is an increase in demand for the final product, the demand for labour will increase even at the same wage rate and *vice versa*. So, the demand curve for labour in the industry will shift towards right.
- ✩ When the price of a substitute product increases or price of complementary product increases, the demand for the product under consideration will increase in the market. So, to produce more of this product, more labour will be employed even at the same wage rate. So, the demand curve for the factors will shift towards right.

Supply of Labour Factor

The supply of labour factor refers to the number of working hours a labour can offer at various wage rates. In general, there exists a direct relationship between wage rate and number of working hours the labour supplies. That means, at higher wage rate, the labour works for more number of hours and *vice versa*, thereby, the shape of supply curve for labour is an upward sloping curve. However, the shape of supply curve of labour varies from the following view points:

(a) Supply Curve of Labour for a Firm

For a given firm, under the conditions of perfect market competition in labour market, the supply of labour is perfectly elastic (Panel B of Figure 26.5). This is because, the wage rate for labour is determined at the industry level based on the interaction between demand and supply forces (Panel A of the Figure 26.5) and the same wage rate is passed on to the individual firms (Panel B of the Figure 26.5). Since the individual firm cannot influence the wage rate of the labour, it has to employ the requisite number of labour at the prevailing wage rate. That means, the demand for labour by an individual firm cannot influence the wage rate of the labour. For example, a paddy farmer alone cannot influence the wage rate of the agricultural labour. As shown in the Panel B of the Figure 26.5, 'SS' is the supply curve of labour parallel to X-axis and this shows that, the paddy farmer can employ as many agricultural labourer as he needs at a given wage rate OW.

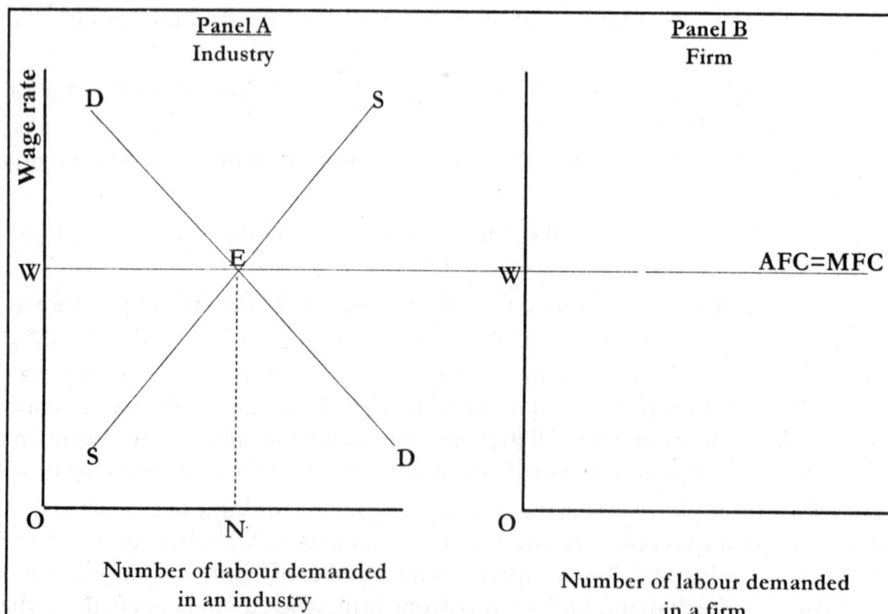

Figure 26.5: Supply curve of labour for a firm and an industry.

As shown in the Panel B of the Figure 26.5, 'SS' is the supply curve of labour parallel to X-axis and this shows that, the paddy farmer can employ as many agricultural labourer as he needs at a given wage rate OW.

(b) Supply Curve of Labour for an Industry

As shown in the Panel A of the Figure 26.5, the supply curve of labour (SS) is an upward sloping curve from left to right, indicating that, with increase in wage rate, the supply of labour also increases and *vice versa*. That means, if an industry wants to attract more labour, it has to offer higher wage-rate, so that, it can attract labour of other industries. So, the supply curve of labour for an industry is not perfectly elastic, but rather, it is more elastic, as higher wages will attract labour of other industries to a particular industry. However, even at higher wage rate, the supply of labour to a particular industry will not increase so rapidly, as several factors besides wage rate also influence the shift of labour across the industries and they are discussed here under.

- ☆ ***Job security:*** If an industry provides more job security to the workers in the context of permanency, gratuity, pension etc., the occupational mobility of the labour from other industries to this industry will be high.
- ☆ ***Skill and efficiency of labour:*** If the labour are less skilled, their occupational mobility is quite high and relatively easy compared to the skilled labour. On the other hand, for the skilled labour, their occupational mobility depends upon their expertise, seniority of service, attractive pay norms of other industries, trainings to be offered by other industries, their challenging attitude towards work in new areas or assignments etc.
- ☆ ***Cost in shifting:*** If the shift of labour from one industry to another industry involves huge cost in the form of loss of seniority, loss in free accommodation facilities, loss in free health service facility etc, then the occupational mobility will be low.

Derivation of Supply Curve for Labour in an Industry

The IDC technique can be employed to derive supply curve for labour for an industry. We know, a labour will supply his service in an industry based on various factors like nature of the work, wage rate offered by the industry, number of working hours offered by the industry, leisure hours the labour prefers etc.

In the Figure 26.6, hours of leisure and work are taken on X-axis and income is taken on Y-axis. For a given working hours (or leisure hours enjoyed) and the income earned by the worker, I_1L is the Income-Leisure line. The slope of this I_1L line indicates wage rate/hour. This I_1L line touches the IDC I_1 at Q_1, where the labour earns ON_1 or Q_1L_1 income for performing LL_1 working hours or enjoying OL_1 leisure hours.

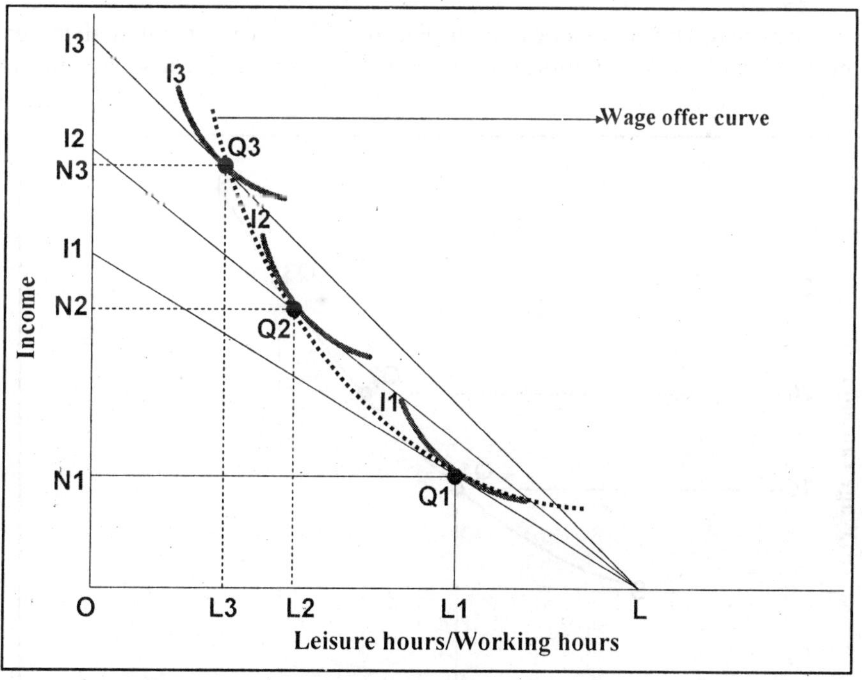

Figure 26.6: Application of IDC technique to derive wage-offer curve.

Let us suppose, the industry increased the wage rate and the labour shows the tendency to work for more number of working hours to earn more income. So, the new Income-Leisure line is I_2L. This new Income-Leisure line I_2L touches the point L on X-axis because, the total number of leisure hours and working hours is constant for a day. So, with the increase in income due to increase in wage rate, the labour derives more satisfaction and hence, he moves to higher IDC I_2, where I_2L line touches IDC I_2 at Q_2. So, it is clear that, when more wage rate is offered, the labour works for more number of hours *i.e.*, $LL_2>LL_1$ and thereby, his leisure hours are decreased *i.e.*, $OL_2<OL_1$ and thereby, he derived higher income $ON_2>ON_1$ or $L_2Q_2>L_1Q_1$.

If still the wage rate is increased, the labour contributes still more number of working hours *i.e.*, $LL_3>LL_2$ and thereby leisure hours are decreased *i.e.*, $OL_3<OL_2$ and the labour derives a higher income, $ON_3>ON_2$ or $L_3Q_3>L_2Q_2$. Note that, if the wage rate is increased, the income-leisure line becomes more steeper and touches the higher IDC.

If we connect all the equilibrium points, Q_1, Q_2 and Q_3 of different incomes at different wage rates and working hours, it gives rise to a Wage–Offer curve. It is important that, this Wage-Offer curve is not synonymous to supply curve and we have to derive the supply curve from this Wage–Offer curve. For the Figure 26.6, the data is assumed and depicted in Table 26.1.

Table 26.1: Supply schedule of labour for an industry.

Equilibrium Point	*Wage Rate/hour (Rs)*	*Working Hours*	*Leisure Hours*
Q_1	$OI_1/OL=15$	$L_1L=6$	$OL_1=18$
Q_2	$OI_2/OL=20$	$L_2L=8$	$OL_2=16$
Q_3	$OI_3/OL=25$	$L_3L=10$	$OL_3=14$

From the Figure 26.7, it is clear that, if wage rate is increased in an industry, the labour contributes more number of working hours and hence, the supply curve of labour will have a positive slope. This is mainly due to the following two reasons:

- ☆ ***Substitution effect:*** When wage rate is increased by the industry, the labour prefers to work for more number of hours and this indicates that, the labour has substituted working hours for leisure hours. So, due to substitution effect, working hours preference is more than leisure hours preference. Thus, due to substitution effect, the wage hours to leisure hours ratio will increase.
- ☆ ***Income effect:*** When more wage rate is offered by the industry, the labour contributes more number of working hours and thereby, he earns more income and with this earned income he becomes potentially better-off and thereby, he prefers more number of leisure hours compared to working hours. This is due to income effect in which, leisure hours preference is more than working hours preference. Thus, due to income effect, the wage hours to leisure hours ratio will decrease.

But, the supply curve will have a positive slope as in the Figure 26.7, indicating that, with increase in wage rate, the labour contributes more number of working hours and substitute leisure hours because, substitution effect is stronger than income effect.

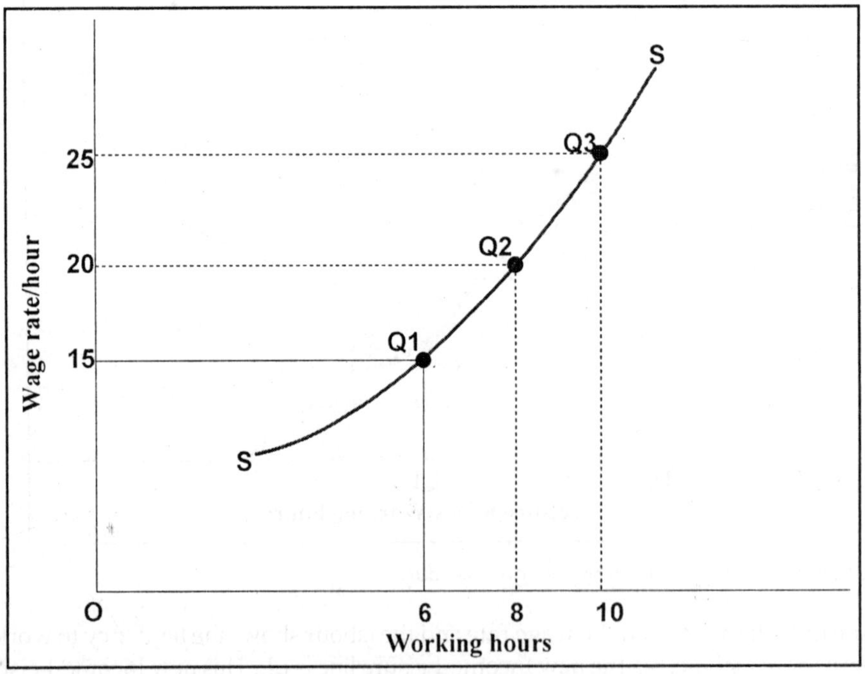

Figure 26.7: Application of IDC technique to derive supply curve for labour for an industry from wage offer curve.

(c) Supply Curve of Labour for an Economy as a Whole

Compared to the industry level, at economy level, the supply of labour depends upon still more factors *viz.*, economic and political factors, population growth, sex ratio, traditions, customs, society's attitude regarding women employment, geographical environment, age-structure, work preference, leisure preference, attractive pay norms, expertise, trainings received, presence of strong trade unions etc. At economy level, it is generally noted that, if the wage rate increases, the labour contributes more number of working hours to earn more income. But, it is interesting to say that, if wage rate is still increased, the number of working hours contributed by the labour will not increase proportionately, rather they decrease because, at one point of time, the labour prefers leisure hours to working hours.

In the Figure 26.8, when the wage rate is Rs.15/hr, the labour contributed six working hours, when the wage rate increased to Rs.20/hr, then the number of working hours are increased to eight, and when the wage rate is still increased to Rs.30/hr, the labour contributed 12 working hours. But, when the wage rate is further increased to Rs.32/hr, now the labour contributed less number of working hours *i.e.*, only 11.25 working hours. This infers that, at higher wage rates, the labour prefers more of leisure hours compared to working hours. This is because, the labour will have inelastic cash requirements at higher wage rates. That means, when the wage rate is Rs.30/hr, the labour contributed 12 working hours and thereby, he derived Rs. 360/- income. But, when the wage rate increased to Rs.32/hr, he contributed only 11.25 working hours and still he derives same amount of income *i.e.*, 360/-. So, the labour prefers more number of leisure hours and less number of working hours at very high wage rate, because of inelastic cash requirements.

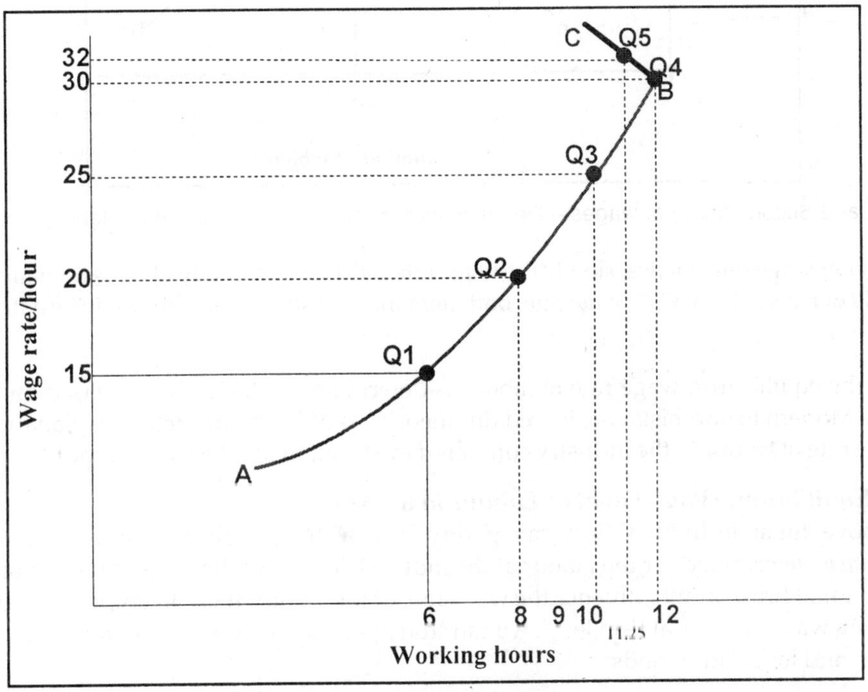

Figure 26.8: Supply curve of labour for an economy as a whole (Backward bending supply curve).

From the Figure 26.8, it is clear that, up to AB portion of supply curve, substitution effect is greater than income effect and hence, the supply curve has positive slope. But, between B and C portion of supply curve, income effect is greater than substitution effect and hence, the supply curve has negative slope and thereby, at very high wage rate, the labour prefers leisure hours to working hours.

However, among different shapes of supply curves of labour discussed above, the upward sloping supply curve is generally treated as a normal shape in the long run and the backward bending shape is only an exceptional or abnormal shape. So, for all analytical purposes, we assume upward sloping supply curve of labour both in the short run and long run and it is to be noted that, in the short run, the supply curve of labour is less elastic and in the long run it is relatively elastic.

Determination of Equilibrium Wage Level of Labour in an Industry

Having studied the demand curve and supply curve of the labour, we need both the curves to determine the equilibrium wage rate of the labour. The equilibrium wage rate of labour in the industry is at the point of equality of demand and supply curves of labour.

- ✩ In the Figure 26.9, DD and SS are the demand and supply curves of the labour. These two curves intersect at point 'E' and corresponding points on Y-axis and X-axis implies that, at OW wage rate of labour, ON number of labourers are employed in the industry.
- ✩ If wage rate is increased to OW_1, then the demand for labour (W_1D_1) is less than supply of the labour (W_1S_1) by D_1S_1. So D_1S_1 labour will remain unemployed. So, some labour show willingness to work even at low wage rate, thereby, wage rate gradually decreases from OW_1 to OW and ensures full employment of labour.
- ✩ If wage rate is decreased to OW_2, then the demand for labour (W_2D_2) is greater than supply of labour (W_2S_2) by S_2D_2. So, for some firms, labour is not available at lower wage rate. So, some firms even try to engage labour by

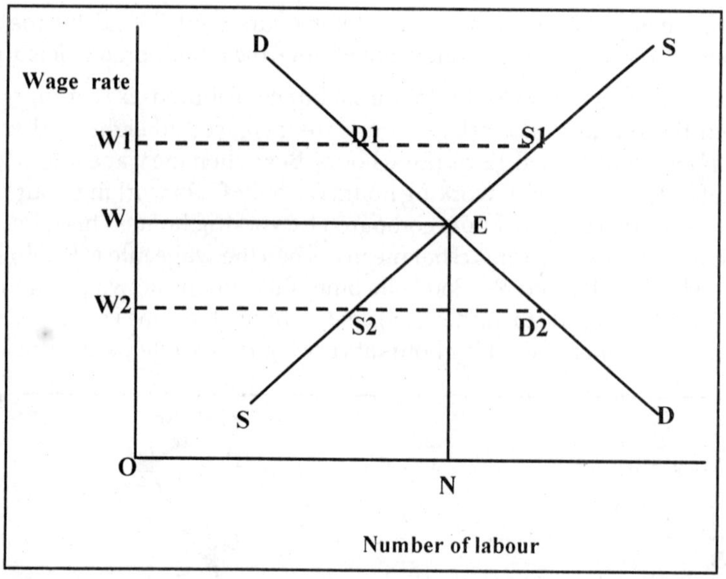

Figure 26.9: Demand and Supply theory of wages – Determination of equilibrium wage rate of labour.

offering higher wage rate. Hence, slowly the wage rate will be increased to OW, as demand for labour is more than supply of labour. So at OW wage rate, both demand and supply of labour will be equal and all firms can engage labour.

So, this is how, the equilibrium wage rate of labour is determined in the industry based on both demand and supply forces. So, the Modern Economists considered this theory has offered satisfactory explanation for determining the equilibrium wage rate of labour in the industry compared to Marginal productivity theory.

Determination of Equilibrium Wage Level of Labour in a Firm

As discussed above, for an individual firm (say, paddy farmer), the supply of labour is perfectly elastic. This is because, the wage rate is determined for the labour at the industry level and the same is passed on to the individual firms. Since, the individual firm cannot influence the wage rate of the labour, it has to employ the requisite number of labour at the prevailing wage rate. So, at firm level, we can study the equilibrium level of number of labour employed considering short run and long run periods.

(a) Equilibrium of Firm in the Short Run Regarding the Employment of Labour

We know, in a short run production programme, the supply of labour cannot be fully adjusted to the firm's demand, thereby, the firm can attain super-normal profits or normal profits or even attain losses due to employment of labour at a given wage rate. This depends upon the wage structure of the labour derived from the industry and the MRP and ARP of the labour factor from the firm's point of view. These cases are shown in Figure 26.10 through Panels A, B and C.

In Panel A, for firm A, at OW wage rate (passed on to the firm from the industry), the point of equilibrium is at 'E', where MRP intersects the wage line or MFC=AFC from above, indicating that, the firm employs ON number of labour. At this ON level of labour employment, the firm earns super-normal profits PEBC, as the ARP contributed by the labour factor is more than the wage rate they receive *i.e.*, by PE.

In Panel B, for firm B, at the same wage rate OW, the point of equilibrium is at 'E', where MRP intersects the wage line=MFC=AFC from above, indicating that, the firm employs ON number of labour. At this ON level of employment of labour, the firm earns only normal profits, as the ARP contributed by the labour is equal to the wage rate they receive.

In Panel C, for firm C, at the same wage rate OW, the point of equilibrium is at 'E', where MRP intersects the wage line=MFC=AFC from above, indicating that, the firm employs ON number of labour. At this ON level of employment of labour, the firm incurs losses PEBC, as the ARP contributed by the labour factor is lower than the wage rate they receive from the entrepreneur of the firm *i.e.*, by PE.

(b) Equilibrium of the Firm in the Long Run Regarding the Employment of Labour

In the short run, if the firms enjoy super-normal profits with respect to labour factor, new firms get attracted to these profits and enter the industry thereby, the demand for the labour factor increases. So, the wage rate of the labour increases and hence, the firms in the long run will earn only normal profits. The same is explained through Figure 26.11. At OW wage rate, we know MFC=AFC for a firm (under perfect competition). So, the point of equilibrium is 'E' where MRP=MFC (=AFC=Wage rate), where the firm employs ON number of labour. At this point of equilibrium, 'E',

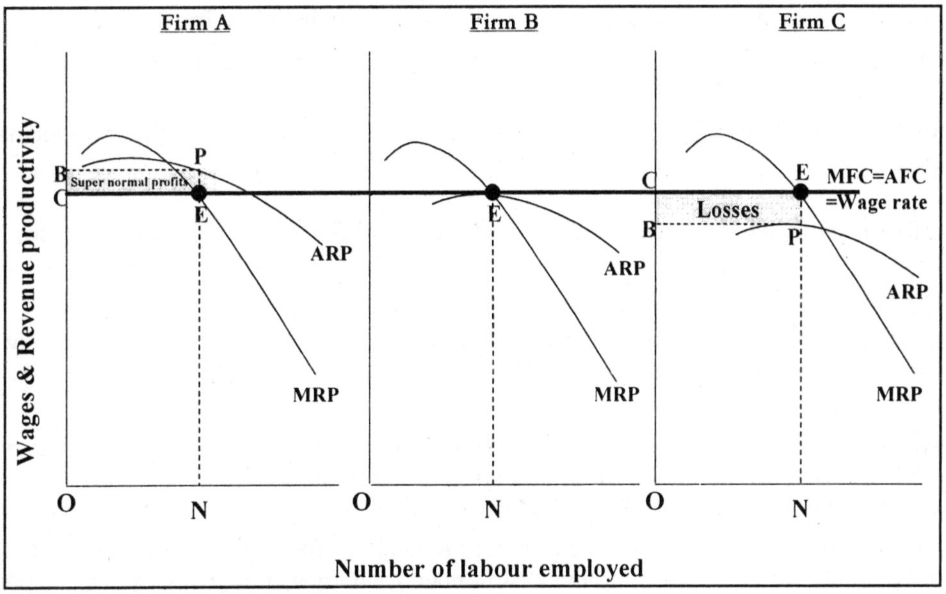

Figure 26.10: Equilibrium of firm in the short run regarding the employment of labour.

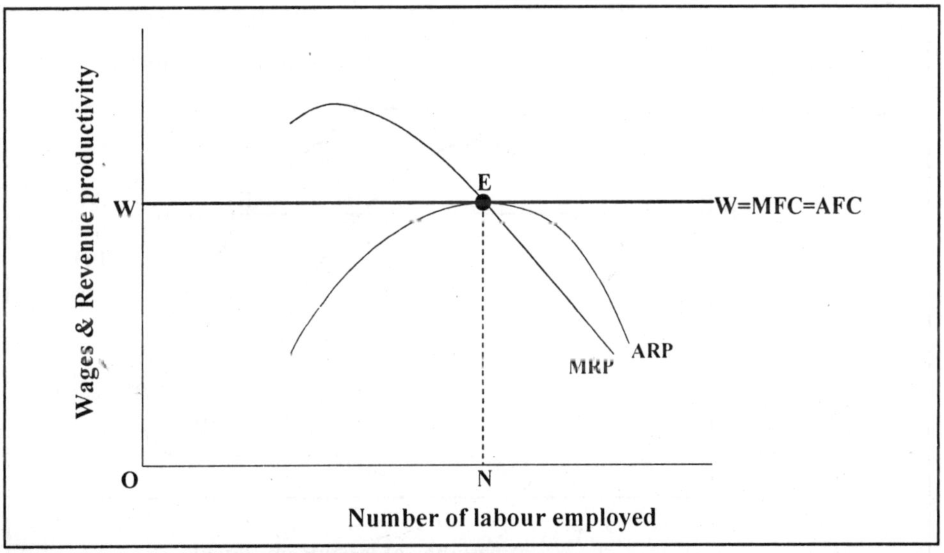

Figure 26.11: Equilibrium of firm in the long run regarding the employment of labour.

there is another essential condition *i.e.*, ARP = AFC, where the firm earns only normal profits in the long run. So, there are two essential conditions that are to be fulfilled to ensure firm's equilibrium in the long run, so as to enjoy normal profits with respect to labour employment *i.e.*,

- ☆ MRP = MFC
- ☆ ARP = AFC (*i.e.*, ARP should lie tangent to the equilibrium position of MRP=MFC=AFC)

Since, MFC = AFC under conditions of perfect competition in labour market, we can represent the above two essential conditions as MRP=MFC=ARP=AFC.

26.4. Exploitation of Labour

Before discussing the concept of exploitation of labour, it is essential to note the following two important aspects:

- ☆ When there is perfect competition both in labour and product market, the wages of the labour (MFC=AFC) equals to the MRP=VMP, thereby, there is no exploitation of labour factor. But, under practical conditions, imperfect competition is common both in labour and product markets and thereby, the wage rate of labour (AFC) tends to be less than the MRP or MVP=VMP contributed by the labour and this refers to exploitation of labour. In the words of Chamberlin, *'exploitation of labour means the payment of labour of a wage less than its MRP'.*
- ☆ The concept of exploitation of labour can be studied by applying the theory of marginal productivity of wages.

The concept of labour exploitation can be studied with reference to the following three cases:

1. When the labour market is perfectly competitive and the product market is imperfectly competitive or monopolistic,
2. When there is a monopsony in the labour market and perfect competition in the product market, and
3. When there is the monopsony in the labour market and monopoly in the product market.

1. When the Labour Market is Perfectly Competitive and Product Market is Under Monopolistic Competition

When the labour market is perfectly competitive, it implies, the firm can engage any number of labour at the same wage rate. That means, in terms of costs, the labour wage rate remains same, irrespective of the number of labour engaged by the firm. Hence, supply curve of the labour or AFC or MFC lies parallel to the X-axis and it is perfectly elastic (Figure 26.12).

- Since, we assume product market is operating under monopolistic competition, MRP curve lies below the MVP or VMP curve. This is because, under monopolistic competition, price>MR and hence, VMP>MRP and these two curves slopes downward from left to right.
- As discussed earlier, with reference to labour employment, the firm will be in equilibrium, when MRP=MFC and it is at point 'E', where the firm employs ON number of labour at OP=QE price (wage) in perfect competitive labour market.

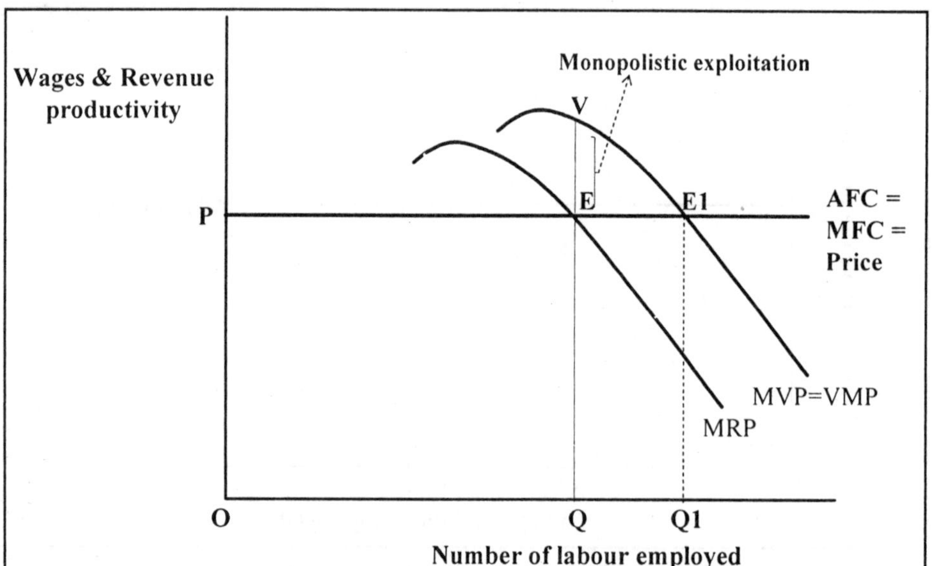

Figure 26.12: Exploitation of labour Case – Labour market is perfect competitive and Product market is monopolistic (Imperfect).

- Thus, MRP=MFC is OP=QE and it is less than its MVP or VMP, QV by 'VE'. This infers that, the labour gets less compared to the value of its marginal product. So, the labour gets exploited by VE and this is called Monopolistic exploitation. So, if the labour is paid less than MVP or VMP under conditions of perfect competitive labour market and imperfect (monopolistic) competitive product market, it is called Monopolistic exploitation.
- Since, the product market is operating under monopolistic competition, only less number of labour are employed in the production programme *i.e.*, OQ.
- Had the product market operates under perfect competition, the firm employs more number of labour *i.e.*, OQ_1, where MFC = MRP (=VMP) at E_1 *i.e.*, what the labour gets is equal to the VMP.
- This infers that, when the product market is imperfect, the demand for the labour is less by QQ_1 compared to product market under perfect competition.

2. Monopsony in the Labour Market and Perfect Competition in the Product Market

Monopsony in a labour market implies, the firm is the single buyer (employer) of the labour. That means, no one will compete with monopsonist in engaging the labour services. So, monopsonist will have control on the employment of the labour. For example, if he needs more of the labour, he will pay higher wages. So, both AFC and MFC are upward sloping from left to right and AFC lies below MFC, as MFC refers to one additional unit of labour, whereas AFC refers to all units of labour employed. It is clear from the Table 26.2 and Figure 26.13 that,

Table 26.2: AFC and MFC under imperfect competition in labour market.

Units of Labour (X)	Wage Rate of Labour	Total Labour Cost (TFC)	AFC (TFC/X)	MFC ($\Delta TFC/\Delta X$)
1	50	50	50	-
2	51	102	51	52
3	52	156	52	54
4	53	212	53	56
5	54	270	54	58
6	55	330	55	60
7	56	392	56	62
8	57	456	57	64
9	58	522	58	66
10	59	590	59	68

- ☆ MFC>AFC=Price of the labour service.
- ☆ Since, we assumed perfect competition in the product market, MVP(=VMP)=MRP, as Price=MR under perfect competition.
- ☆ The firm is in equilibrium at point 'E', where MRP=VMP curve intersects the MFC curve from above. This implies that, the firm employs ON number of labour, but the average wage received by the labour (AFC) *i.e.*, OW=NP is less than MRP=VMP, NE by PE.

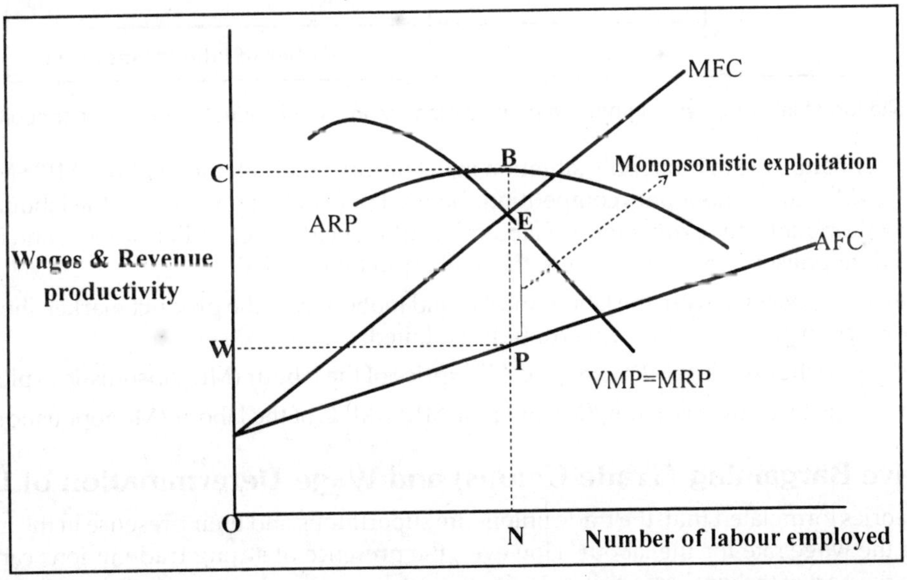

Figure 26.13: Exploitation of labour Case – Labour market is monopsony and Product market is perfect competitive.

- ☆ Thus, the labour contributes more to the employer and receives fewer wages from the employer (*i.e.*, NE>NP).
- ☆ This infers that, being a monopsonist, the firm exploits the labour labour and earns super-normal profits PWBC. This exploitation of labour labour by PE refers to monopsonistic exploitation of labour. So, if labour is paid less than MRP=VMP under conditions of monopsony in labour market and perfect competition in product market, it is called Monopsonistic exploitation.

Under perfect competition in labour market, the firm maximizes its profits by employing more and more number of labour till the MRP of the labour is equal to the MFC=AFC or wage rate paid to the labour (as discussed under Figure 26.12). But, under imperfect labour market (*i.e.*, monopsony in labour market), a monopsonist maximizes his profit by paying less to the labour (AFC) than the MRP=VMP derived from the labour, as the product market is under perfect competition. In other words, the labour contributes more to the employer than the wage it receives from the employer. This is called Monopsonistic exploitation of the labour service by the employer.

3. Monopsony in the Labour Market and Monopoly in the Product Market

As discussed above, when there is monopsony in the labour market, it implies, MFC>AFC. Similarly, when there is monopoly in the product market, MVP(=VMP)>MRP. As shown in the Figure 26.14,

☆ The firm will be in equilibrium at point E, where the MRP curve intersects the MFC curve from above. This implies that, the firm employs ON number of labour by paying NP price (wage rate) to the labour service, which is less than MRP of the labour, NE, *i.e.*, by PE.

☆ Since, MRP *i.e.*, NE is greater than NP price (wage rate) paid to labour by PE, the labour is paid less than the marginal productivity derived from that labour or the labour contributes more to the employer than the price it receives from the employer and this is called Monopsonistic exploitation of the labour service by the employer.

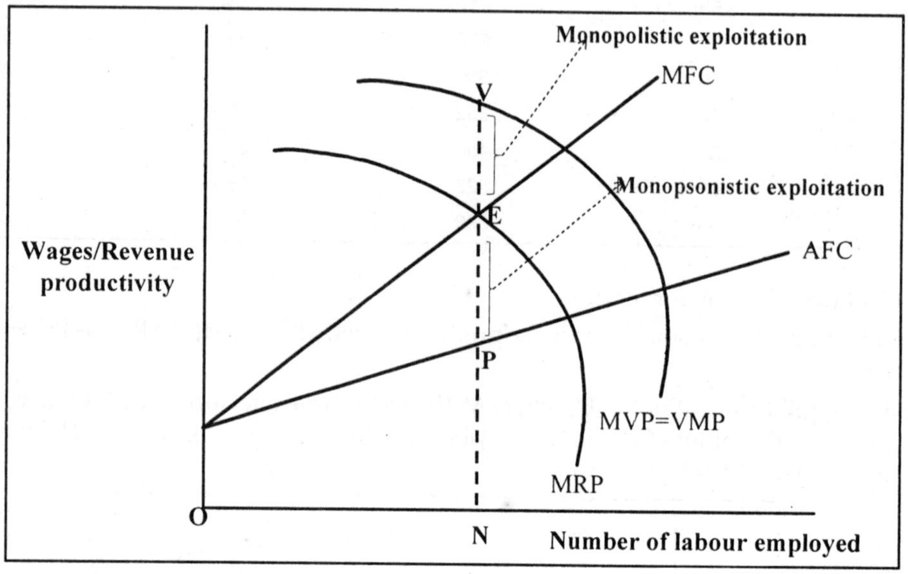

Figure 26.14: Exploitation of labour Case – Labour market is monopsony and Product market is monopoly.

☆ On the other hand, due to monopoly situation in the product market, the MVP(=VMP)>MRP of the labour, since price>MR under monopoly competition. Hence, the firm further exploits the labour employed to the extent of EV amount. This exploitation of labour *i.e.*, difference between MVP(=VMP) and MRP=MFC by VE refers to Monopolistic exploitation. Thus, the total exploitation is PV.

☆ So, in the case of monopsony in the labour market and monopoly in the product market, the labour employed in the production process by the firm is doubly exploited:

- ❖ First, due to the excess of MRP over the AFC=price of the labour (Monopsonistic exploitation), and
- ❖ Second, due to the excess of the VMP over the MRP=MFC of the labour (Monopolistic exploitation).

26.5. Collective Bargaining (Trade Unions) and Wage Determination of Labour

The earlier theories formulated that, the trade unions are superfluous and their presence in the industry could not succeed in raising the wage rate for the labour. However, the presence of strong trade unions can definitely show positive impact on the part of the workers *i.e.*, it can raise the volume of employment and can raise the wage-rate. The following are the grounds (Table 26.3) on which the earlier theories considered collective bargaining of trade unions was futile in raising the labour employment and wage rate of the labour.

Table 26.3: Criteria on which earlier theories rejected the influence of collective bargaining of trade unions in raising the wage rate and employment of labour.

Sl.No.	*Theory*	*Criteria/Concept*
1.	Subsistence theory of wages	The presence of trade unions will not increase the wage level because, wage level in the economy is influenced by the size of the population of workers *i.e.*, the number of workers. If wage rate is above the subsistence level of the workers, then labour supply will increases and due to rise in competition among the workers, the wage level will decline back to the subsistence level. On the other hand, if wage rate is below the subsistence level of workers, the population of workers fall and thereby, supply decreases and hence, wage rate will gradually increase up to the subsistence level.
2.	Wages fund theory	Since, wage fund is a fixed amount, trade unions cannot increase the wage rate. This is because, if they succeed in increasing the wage rate in one industry, it is possible only at the expense of labour employment in another industry, as wage fund is a fixed amount.

Contd...

Table 26.3–Contd...

Sl.No.	Theory	Criteria/Concept
3.	Residual claimant theory of wages	As the surplus left over after meeting the remunerations of rent, interest and profits, is paid as wages to labour, this theory too considered the collective bargaining of trade unions is futile in increasing the wages of the labour.
4.	Standard of living theory of wages	This is a refined version of the concept of Subsistence theory of wages and this theory too considered collective bargaining by trade unions is futile in increasing the wages of labour. This is because, it considers the wage rate matches the standard of living of the workers in the long run.
5.	Marginal productivity theory of wages	According to this theory, (assuming perfect competition both in factor and product markets) the MRP=MVP (=VMP) and it slopes downward from left to right and this infers that, increase in wage rate would result in unemployment of workers. So, the collective bargaining of trade unions will not succeed in raising the wage rate without causing unemployment.

The above discussion clearly infers that, all the theories proposed in the 19th century considered that, collective bargaining by trade unions has no room to raise the wage rate of the workers, thereby, in all these theories, the role of trade unions is neglected. However, Modern Economists recognized the importance of these trade unions and hence, they assumed 'absence of trade unions' while explaining the concept of wage determination through demand and supply forces concept. These trade unions, today, are becoming an integral part of the working life in the modern industrial system. The trade union can be defined as the group of workers formed as an association, so as to gain the monopolistic hold regarding their supply of labour services in the industry. The following points highlight the need for collective bargaining by the trade unions:

- To enable the employer to secure cooperation expected from workers and for keeping cordial relations between employers and employees.
- To protect the workers from exploitations by the employer.
- Mutual negotiations become essential for the solutions of the problems in the firm/industry.
- Collective bargaining puts a check on the one sided decisions of the employers.
- It is needed for maintaining a peaceful industrial atmosphere.

So, collective bargaining is a method of settling of wage disputes and other related problems of the workers by the trade union in negotiating with the entrepreneur. Since, individual labour has relatively weak bargaining power, the formation of trade unions enhances their collective strength in negotiating with the entrepreneur to address their problems, raising the wage rate, increasing the employment etc. However, when both labour market and product market are perfectly competitive, the trade unions faces a dilemma *viz.*,

- If the wage rate has to be increased through collective bargaining, it leads to labour unemployment.
- If the trade union wishes to collectively bargain for increasing employment, it is possible only at a lower wage rate.

  As shown in the Figure 26.15,
- WW is the wage line or labour supply curve=MFC=AFC and it is perfectly elastic assuming perfect competition in the labour market. MRP=MVP (=VMP) is the downward sloping curve from left to right assuming perfect competition in the product market.
- The point of equilibrium is at 'E', where MRP intersects the MFC from above and it infers that, at OW wage rate, ON are the number of labour employed.
- If the collective bargaining of trade union is for raising the wage rate of labour from OW to OW_1, the new point of equilibrium is at E_1 where, the firm employs only ON_1 number of labour. That means, by increasing the wage rate from OW to OW_1, it leads to NN_1 labour unemployment. On the other hand, if the trade union bargains for raising labour employment from ON_1 to ON, it is possible only by lowering the wage rate from OW_1 to OW.
- Thus, under perfect competitive market conditions both in labour market and product market, collective bargaining by trade union cannot raise wage rate without causing unemployment.
- It is interesting to say that, as long as the labour market is under perfect competitive conditions, the trade union faces the above dilemma. For example, even if the product market is monopoly, the trade union faces the same dilemma.

However, collective bargaining of trade union has favourable impact in the form of raising the wage rate and simultaneous increase in the employment of workers, when labour market is under imperfect competition. Let us examine the following two cases:

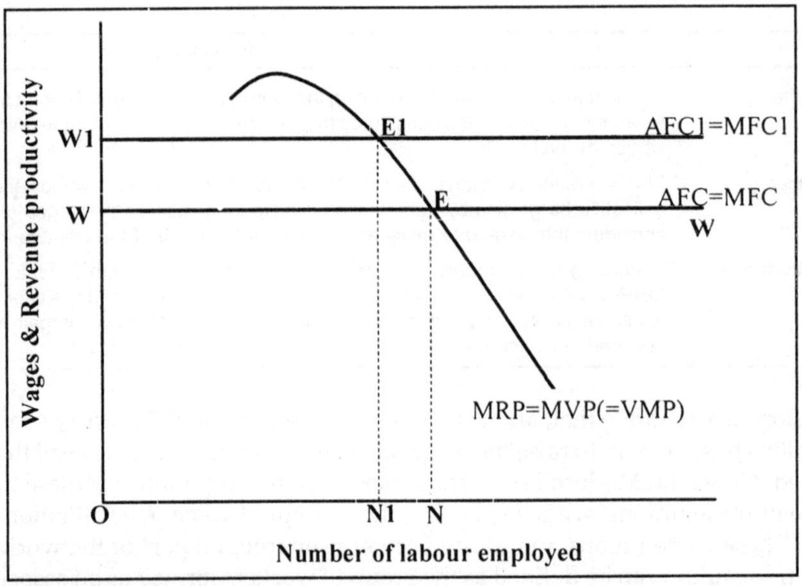

Figure 26.15: Collective bargaining by trade union to raise wage rate and consequent unemployment.
Case – Both labour market and product market are perfect competitive.

1. When Labour Market is Monopsony and Product Market is Perfect Competitive

As shown in the Figure 26.16,

- ☆ When the labour market is monopsony, MFC > AFC = price (wage) of the labour service.
- ☆ Since, we assumed perfect competition in the product market, MRP=MVP (=VMP), as price=MR under perfect competition.
- ☆ The firm will be in equilibrium at 'E', where MRP=VMP curve cuts the MFC curve from above. This implies that, the firm employs ON number of labour, but the wage rate of labour is fixed at NP = OW, *i.e.*, less than MRP (NE) contributed by the labour. So, the labour is exploited by PE and it is called Monopsonistic exploitation.
- ☆ Noticing the exploitation, the trade union bargains for raising the wage rate from OW to $OW_1 = N_1P_1$, thereby, wage rate OW_1 equals to MRP=VMP. This situation makes the trade union to succeed both in raising the wage rate (OW to OW_1) and as well as the employment of the labour from ON to ON_1.

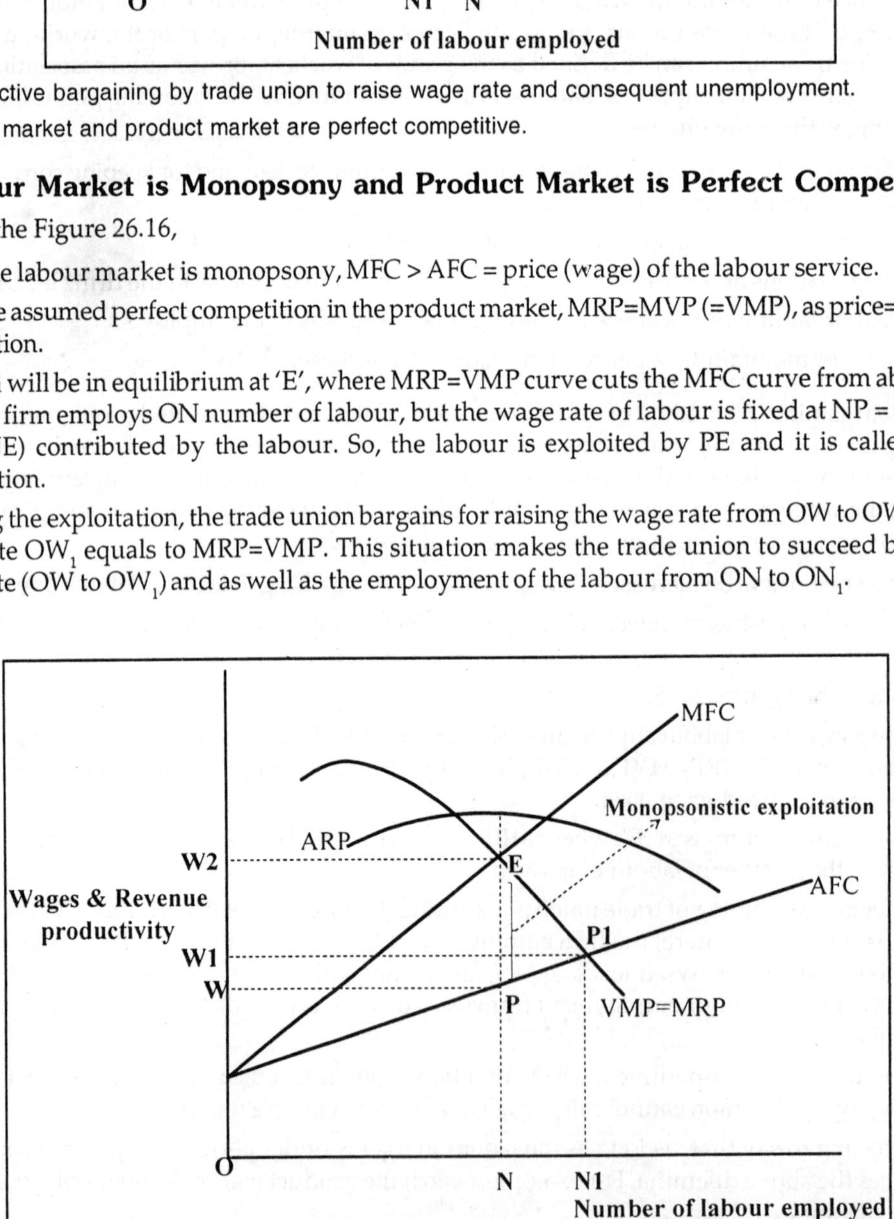

Figure 26.16: Collective bargaining by trade union to raise wage rate and to increase labour employment.
Case – Labour market is monopsony and Product market is perfect competitive.

- ☆ There is another possibility for the trade union to bargain for raising the wage rate up to OW_2=NE, where the wage rate=MRP=VMP, thereby, at the same level of employment ON, the labour can derive higher wage rate OW_2 and hence, there is no monopsonistic exploitation.

2. When Factor Market is Monopsony and Product Market is Monopoly

As shown in the Figure 26.17,

- ☆ When the factor market is monopsony, MFC>AFC (=Price (wage) of the labour service).
- ☆ Since, product market is monopoly, MVP (=VMP)>MRP, as price>MR under monopoly.
- ☆ The firm will be in equilibrium at 'E', where MRP curve cuts the MFC curve from above. This implies that, the firm employs ON number of labour, but the wage rate of labour is fixed at NP=OW *i.e.*, less than MRP (NE) contributed by the labour. So, the labour is exploited by PE and it is called as Monopsonistic exploitation.

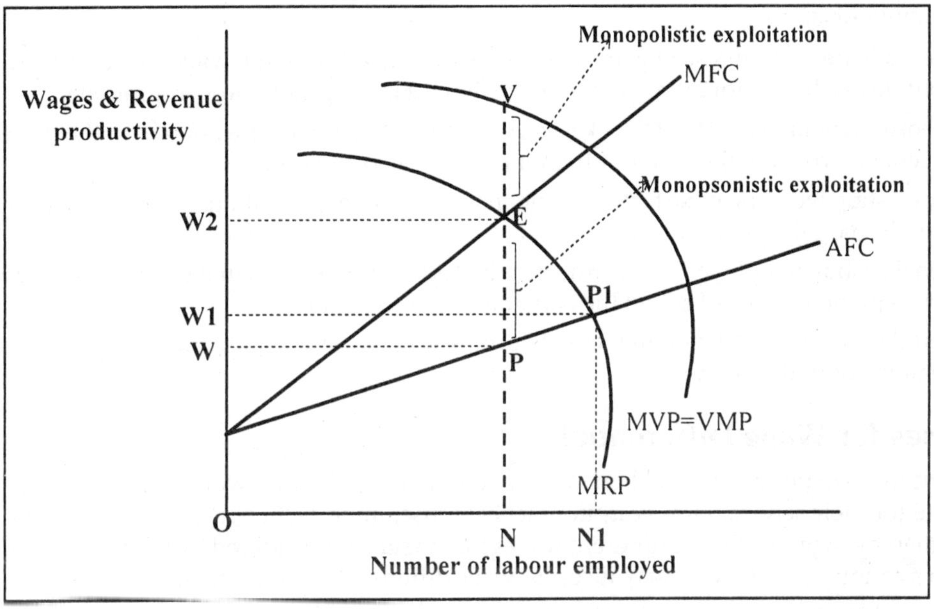

Figure 26.17: Collective bargaining by trade union to raise wage rate and to increase labour employment.
Case – Labour market is monopsony and Product market is monopoly.

- ☆ Noticing the exploitation, the trade union bargains for raising the wage rate from OW to OW_1=NP_1, thereby, wage rate OW_1 equals to MRP. This situation makes the trade union to succeed both in raising the wage rate from OW to OW_1 and as well as increasing the employment of workers from ON to ON_1.
- ☆ There is another possibility for the trade union to bargain for raising the wage rate up to OW_2 =NE, where the wage rate = MRP, thereby, at the same level of employment ON, the labour can derive higher wage rate OW_2 and hence, there is no Monopsonistic exploitation.

Thus, the above discussion infers that, the collective bargaining of trade union will succeed in both ways *i.e.*, by increasing wage rate and simultaneous increase in the level of employment of workers or increasing the wage rate still a higher level without affecting the level of employment of workers and these are possible only when the labour market is under imperfect competition. But, if the labour market is under perfect competition, trade union cannot increase the wage rate without causing unemployment. Further, trade union can bargain for increasing the wage rate of the workers through enhancing the workers' productivity and thereby, it pressurizes the entrepreneur to install modern machines, rendering training to the workers etc. It may even ask the entrepreneur to raise the price of the product, so as to increase the wage rate of the workers. The following are the circumstances under which the trade union can bargain for higher wages:

- ☆ We know, when labour market is imperfect, the entrepreneur exploit the labour by paying less wages than the MRP. So, the trade union bargains to raise the wage rate of the workers up to the point, where wage rate equals the MRP.
- ☆ The trade unions may bargain the entrepreneur to raise the productivity of the workers by providing updated machines or technology, implements, infrastructural facilities, rendering trainings to the workers, thereby, they can pressurize the entrepreneur to raise the wage rate on par with the MRP contributed by the workers.

- When the demand for the product produced is inelastic in the market, the demand for the labour involved in the production of the product is also inelastic. In such case, the trade union can bargain for higher wage rate for the workers.
- According to Prof. Marshall, if wages constitute a low per cent share of the total cost of production of the product, the trade union can bargain for higher wage rate, as it won't increase the total cost of production of product at significant level. Further, when rewards paid to other factors *i.e.*, land, capital and management are squeezable, then the trade union can bargain for higher wages to the workers.
- If trade union purposively limit the labour supply to the factory by way of restricting the new entrants both from within the country and from outside the country, the existing labour can demand for higher wages.

Limitations of Collective Bargaining of Trade Unions

The collective bargaining of trade unions will not always succeed in raising the wage rate and the following are the important limitations:

- If the trade union is less strong, the entrepreneur will not raise the wage rate and moreover, he will try to substitute machinery for labour to minimize the cost in the production programme.
- The entrepreneur may attract 'black legs', that is, the labour who are not part of the trade union, but willing to accept the work offered by the entrepreneur even at low wage rate.
- If the existing workers in the factory do not have adequate occupational mobility, the entrepreneur will not accept the bargaining of trade union.
- When the labour employed in the firm are with less experience and that too with low marginal productivity, the entrepreneur may not accept the bargaining of trade union.
- When the firm suffers from high cost of production of the product, the entrepreneur will not accept the bargaining of trade union.

26.6. Causes for Wage Differentials

Under conditions of perfect competition in the labour market, identical workers performing similar jobs will get same wage rate for their service. However, in practical sense, imperfect competition in the labour market is more common and thereby, workers derive different wages for the services rendered by them across different regions and occupations in a country and the various causes for wage differentials are discussed here under.

- ***Differences in demand conditions:*** Since, the demand for labour varies from one firm to another firm producing different goods and even among the firms that produce a homogenous good, the output levels will differ, thereby, demand level varies for the labour and accordingly, wage rates will fluctuate. The demand for labour also varies from place to place and in places, where there is abundant supply of labour, the wage rate of the labour will be low and *vice versa.*
- ***Differences in labour efficiency***: The demand for labour also depends upon the efficiency of the labour and if the labour is more efficient and skilled, the wage rate will be more and *vice versa.*
- ***Availability of substitute***: If a substitute (machinery) is available for labour, then the entrepreneur will substitute machinery usage for labour and thereby, demand for labour will decline and hence, wage rate will be low.
- ***Differences in labour supply***: In some areas, the supply of labour will be more than their demand, thereby, the wage rate will be low.
- ***Degree of mobility may be low***: Due to several factors, as discussed earlier, such as, social, cultural, ethical, economical, language barriers, family attachments etc., the degree of mobility of labour may be low from one occupation to another occupation and this causes wage differentials among the workers working at different places.
- ***Differences in education, experience and efficiency***: Workers differ in their education and experience status, thereby, they differ in working efficiency. The more is the working efficiency on the part of the workers, the more is the wage payment and *vice versa.*
- ***Permanency of the job***: In general, the workers will be prepared to work at lower wages, provided the job is permanent. When the job assignment is purely temporary, the worker bargains for higher wages.
- ***Degree of bargaining power***: If the trade union in a factory is strong enough for collective bargaining of high wage rates, its efforts will be fulfilled and the workers will get higher wages. On the other hand, if the trade union is weak, its efforts will be futile in increasing wage rate for the workers. So, the greater the bargaining power on the part of the workers (trade union), the higher the wage rates and *vice versa.*

☆ ***Risks involved:*** If the workers have to face more risk in executing the job assignment, they demand for higher wages and *vice versa*. For example, the workers functioning in forest area, local mines etc., generally demand higher wages from the organization, as they have to face more degree of risk in executing their respective job duties.

☆ ***Future prospects:*** The future prospects of the worker vary from one occupation to another occupation. If a worker enjoys more future prospects in one job, he will accept that job even at low wage rate. On the contrary, if the worker believes less job prospects for him, he will continue in the same, if he derives a higher wage rate.

☆ ***Sex difference:*** This is one of the important factors that influences wage difference across the workers. In general, men workers were paid more wages compared to women workers and this is due to several reasons as mentioned below.

- The productivity contributed by women workers is low compared to men workers, as they are physically less strong compared to men.
- Women are more burdened with other duties like family care, children's care, house care etc., thereby, they could not concentrate on job duties as efficiently as men.
- The areas of work for women labour are few compared to men labour.
- Women have to face many social and religious restrictions to come out of the family and to execute the jobs.
- Compared to men, the organizational power of women workers is weak, as they are burdened with other responsibilities, thereby, their bargaining power is low.

☆ ***Non-monetary benefits:*** The workers prefer for jobs even at low wage rates, provided they enjoy non-monetary benefits like job satisfaction, pleasantness at work, facilities provided by the organization, freedom in selecting the work schedule etc.

☆ ***Social prestige:*** Some workers desire to enjoy social prestige in rendering jobs and thereby, they even accept such jobs even at low wage rates. For example, a business man may earn more than an IAS officer, but the IAS officer enjoys more social prestige than the business man.

☆ ***In-born qualities:*** Some people like artists, singers, players etc., have inborn qualities and thereby, they earn more remuneration compared to others.

The above discussion infers that, there are several factors that influence the wage differentials across the workers. But, the major influential factor is the difference between supply and demand forces. That means, in some occupations, demand for labour is more than their supply and in some other occupations, reverse is the case. In occupations, where demand is more than supply, the wages are more and in other occupations, where supply of workers is more than demand, wages are less. Further, differences in supply of workers across the occupations arise due to imperfect mobility of the workers. That means, workers cannot move freely across the occupations, as their occupational mobility is restricted by several factors (as discussed earlier) and this is also the major causal factor for wage differentials.

26.7. Fixation of Minimum Wages

In the context of exploitation of workers and wage differentials across the workers, the concept of fixation of minimum wages for the workers came into existence. In some industries, labour are not organized, as they are relatively small in number or scattered or due to illiteracy, thereby, entrepreneur exploits such labour by paying low wage rate than their MRP. Such industries are called 'Sweating industries'. So, the demand for fixation of minimum wages to the labour calls the Government to react by passing the enactments, such that, no entrepreneur should pay the wages below the prescribed minimum wages. For example, Government of India passed the 'Minimum Wages Act' in 1948. As shown in the Figure 26.18,

☆ OW is the equilibrium wage rate at which ON number of labour are employed.

☆ If the Government fixes the minimum wage rate above OW as OW_1, then at that OW_1 wage level, W_1D_1 is the demand for labour and W_1S_1 is the supply for labour, thereby, supply of labour (W_1S_1) > Demand for labour (W_1D_1) by D_1S_1.

☆ That means, due to fixation of minimum wage level (OW_1) above equilibrium wage level (OW), D_1S_1 is the total number of labour unemployed (observed unemployment) and out of this D_1S_1 number, D_1A_1 is the decline in employment compared to equilibrium level of employment and A_1S_1 is the unemployment due to increase in number of labour supplied.

☆ Thus, observed unemployment D_1S_1 is higher than the reduction in employment D_1A_1 due to rise in minimum wage by the Government.

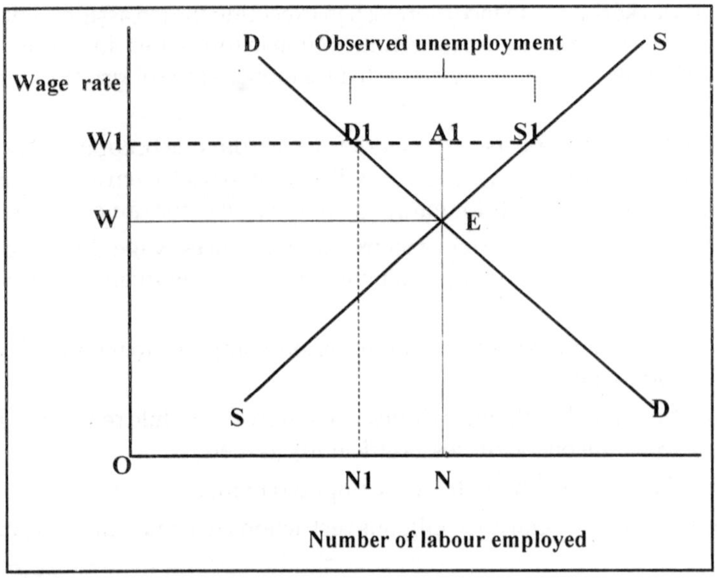

Figure 26.18: Effect of minimum wages on labour employment.

However, the extent of observed unemployment and decline in employment depends upon the relative elasticities of demand for and supply of labour force. The following are the two interesting cases:

- When the demand for labour is more inelastic than supply for labour (Panel A of the Figure 26.19), then the observed unemployment D_1S_1 is considerably higher than decline in employment N_1N due to fixation of minimum wage (OW_1) by the Government above the equilibrium wage (OW). That means, the decline in employment (N_1N) do not account much in observed unemployment (D_1S_1). Thus, at the minimum wage level, the decline in employment is low, but the supply of labour force is comparatively very high. Since, the decline in employment is at low scale, the fixation of minimum wage higher than the equilibrium wage can be justified.
- When the demand for labour is more elastic than supply of labour (Panel B of the Figure 26.19), then the decline in employment N_1N accounts much in the observed unemployment D_1S_1 due to fixation of minimum wage (OW_1) by the Government above the equilibrium wage (OW). Thus, at the minimum wage level (OW_1) the decline in employment N_1N is high and it occupies a significant share in the observed unemployment D_1S_1. Since, the decline in employment is significantly high, the fixation of minimum wage at such a high level cannot be justified.

It is important that, the minimum wage level fixed by the Government is not a fixed wage rate, but it will be fluctuating considering the inflationary pressures in the economy. So, whenever there is a change in the minimum

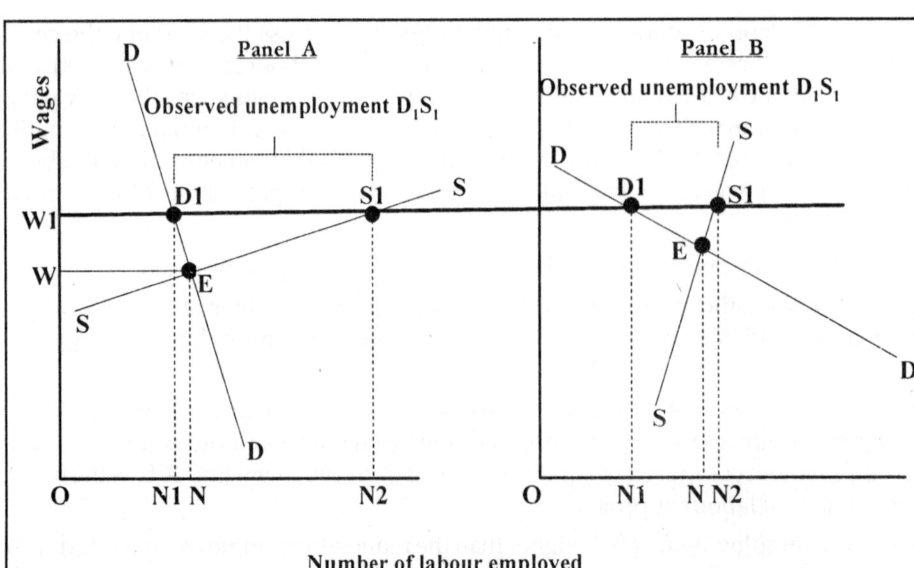

Figure 26.19: Effect of minimum wages on labour employment depending upon the relative elasticities of demand and supply forces.

wage level, the Government has to pass the legislation prescribing that wage level is minimum for the labour force till the new legislation is passed.

Merits of Fixing Minimum Wage Level

The following are the merits of fixing minimum wage level to the labour force by the Government.

- It helps the labour force to remove exploitation from the entrepreneur.
- When minimum wage is fixed above the equilibrium wage, it increases income on the part of labour, their standard of living will improve and their working efficiency will increase. When the productive capacity increases on the part of the workers, it guides the entrepreneur to update machinery and technology usage in the production programme.
- It ensures industrial peace, as trade unions has no much role to play regarding bargaining for higher wages.
- It makes the labour market perfect *i.e.*, supply curve of labour for the industry becomes perfectly elastic and it is parallel to X-axis, thereby, the industry can employ any number of labour at the minimum wage level and hence, the labour are not exploited by the entrepreneur.
- When minimum wage is fixed above the equilibrium wage, it increases income on the part of the labour, thereby, the demand for the products increases and it leads to more employment.
- It ensures equitable distribution of income among the labour force.

Demerits of Fixing Minimum Wage Level

Though fixation of minimum wage level ensures benefits both to the work force and to the entrepreneurs, it too suffers from the following limitations:

- Sometimes, if the Government fixes the minimum wage level below the equilibrium wage level, the labour are greatly exploited by the entrepreneur.
- If minimum wage is fixed such that, it is same both for men and women, then it leads to total reshuffling of labour force in the production programme, as women labour are replaced by men labour, old labour may be replaced by youth etc.
- If minimum wages are fixed at higher level compared to equilibrium wage level, then it leads to decline in profits on the part of the entrepreneur. Further, the entrepreneur will substitute machinery for labour, thereby, it leads to decline in employment of labour force
- If minimum wages for labour are announced in one sector and not in other sector, it leads to wage-disparity or income-inequality among the wage earners.

27

Interest

Interest is the remuneration paid for the capital borrowed and invested in the production programme. The concept of determination of interest is altogether different compared to the determination of remunerations paid to other factors of production. Even today, this topic is the most controversial one compared to other concepts of theory of distribution. The following are the important definitions of interest:

Payment made by the borrower of capital, by virtue of its productivity, as a reward for his abstinence.

–(Wicksell)

Price paid for the hire of loan

–(Cairncross)

Price paid for a loan

–(Benham)

Price paid for the use of capital in any market

–(Marshall)

The market rate of interest is that percentage return per year which has to be paid on any safe loan of money, which has to be yielded by any safe bond or other type of security and which has to be earned on the value of any capital asset (such as a machine, a hotel building, a patent right) in any competitive market where there are no risks or where all risk factors have already been taken care of by special premium payments to protect against risk.

–(Samuelson)

Interest is the income, which goes to the owner of capital

–(Carver)

Interest is the remuneration for mere abstinence

–(Mill)

Interest is the return from the fund of capital

–(Seligman)

Price for the use of loanable funds

–(Meyers)

Interest is a reward for parting with liquidity for a specified period

– (J.M. Keynes)

From the above definitions given by different Economists, the definition about the interest is purely individualistic, as perceived about the concept of determination of interest. Hence, several theories have been proposed to explain about the origin and determination of the equilibrium rate of interest and they are discussed in detail in the ensuing pages. However, before analyzing the different theories of interest, it is essential to differentiate between gross interest and net interest.

27.1. Gross Interest and Net Interest

Gross interest is the total payment made by the borrower to the lender for the capital borrowed by him. Net interest is the payment made by the borrower to the lender exclusively for the use of capital. Net interest is also called as Pure interest. Thus, gross interest is always higher than net interest. Gross interest includes other components like compensation for risk, compensation for inconvenience, payment for management services, compensation for the change in value of money besides net interest.

- ✰ ***Compensation for risk :*** When a lender gives the loan to the borrower, the lender will be at risk, if the borrower do not repay the loan. So, to cover the risk element borne by the lender, some compensation will be charged by the lender from the borrower. If the borrower takes the loan from the lender without providing adequate security, the risk will be more on the part of the lender and thereby, he asks for higher compensation for the risk.
- ✰ ***Compensation for inconvenience :*** Before lending loan to the borrower, the lender has to save the money. He even sacrifices some part of his consumption needs for saving the money. Similarly, the lender also feels some inconvenience after lending the money, as he cannot get back and use that money as and when he wishes. Such inconveniences borne by the lender must get compensated, thereby, he charges more than the pure or net interest.
- ✰ ***Payment for management service :*** In lending business, the lender has to fulfill some formalities like getting license, stamp duties, maintenance of accounts, maintenance of staff and in case of willful defaulters, he has to undergo certain legal formalities, frequent visits to the house of the borrower etc., and for such managerial services, he will claim compensation from the borrower.
- ✰ ***Compensation for the change in value of money :*** By the time, when borrower returns back the money to the lender, in general price level increases in the economy, thereby, the lender will lose. So, to compensate this, the lender will claim higher compensation from the borrower above the net interest.

So, the above four compensations will be charged by the lender from the borrower, besides charging net interest. So, Gross interest is given by

Gross interest = Net interest + Compensation for risk + Compensation for inconvenience +
Payment for management services + Compensation for change in value of money.

Thus, gross interest comprises of several items charged above the net interest. In practice, gross interest is charged, but in Economics, we deal with net interest only and the equilibrium between demand and supply of capital alone determines the net rate of interest. From the above discussion, we can infer that, net interest is one of the constituents of gross interest, whereas, gross interest includes other elements like compensation for risk, compensation for inconvenience, payment for management service and compensation for change in value of money. In case of Governmental loans, the interest charged by the financial institutions refers to net interest only. It is essential to note that, gross interest varies from place to place and from person to person, but net interest tends to be the same everywhere, provided there is perfect mobility of capital. Net interest is determined by the equilibrium between demand and supply of capital alone or it is the price paid for the services of capital employed or it is that part of the gross interest which is paid for the use of capital but, besides capital, other factors (compensation for risk, compensation for inconvenience, payment for management service and compensation for change in value of money) also exert influence in the determination of gross interest. In this chapter, we deal with the determination of net rate of interest. The following are the popular definitions about net interest:

Net interest is the payment from that involved saving and no work simply on the lender.

– (Mill)

Net interest is that portion of the gross interest which is paid simply for the use of capital.

– (Chapman)

Net interest is that portion of the gross interest which is paid simply for the use of capital.

– (Moreland)

27.2. Theories of Interest

Different theories have been proposed to explain the concept of interest and some theories explain, why the interest has to be paid and some theories explain how the net rate of interest is determined in the economy. They are discussed in-detail in the ensuing pages.

1. Marginal Productivity Theory of Interest

This theory was proposed by Malthus, J.B. Say and Carey. The following are the main features of this theory:

- ☆ Interest is paid for the use of capital because, capital is productive. That means, output is more in the presence of capital than without it.
- ☆ Even in the absence of capital, production or output is possible, but in the presence of capital, output further increases. For example, a farm labour can assist in production and can produce output. But, the farm labour along with machinery usage can further increase the output.
- ☆ The rate of interest is determined by the marginal productivity of capital. The concept of LDR also applies to capital and the entrepreneur goes on employing capital, as long as marginal productivity of capital is more than the interest paid for the use of capital and he stops borrowing the capital, when marginal productivity of capital equals the interest paid for the use of capital.

Demerits

This theory has the following drawbacks:

- ☆ This theory assumed perfect competition, perfect substitutability and perfect mobility of capital, which are not realistic in practical sense.
- ☆ This theory explained the concept only from the demand side and ignored the supply side. But, considering the supply side, when there is abundant supply of capital, the rate of interest will be nominal, what ever may be the productivity of capital. But, this theory suggests that, interest is paid only when the productivity of the capital is high. On the contrary, when the supply of capital is limited, the rate of interest will be high and this high interest rate is due to scarcity rather than productivity of capital.
- ☆ This theory simply explains, why the rate of interest is high or low and does not explain how the rate of interest is determined.
- ☆ This theory explains that, interest is paid since, capital is productive. But, it is difficult to measure the productivity realized due to capital factor alone because, capital in association with other factors contributes to productivity.
- ☆ This theory explains that, interest rate is influenced by productivity. But, besides productivity, other factors like economical, institutional, political and social factors also influence the rate of interest.
- ☆ The financial institutions also give consumption loans to the beneficiaries and for such loans, they charge meager rates of interest. But, this theory could not offer explanation, why interest is paid for such loans, as they are not directly productive.
- ☆ The productivity of capital varies with the nature of investment and thereby, varies across the industries. But, we know the net interest rate remains same throughout the market and the borrower can avail the loan for any productive use.

Considering the above demerits, we can infer that, this theory has not offered satisfactory explanation for the determination of net interest. Further, it offered one sided (demand side) explanation and that too could explain only, why the interest is paid for the use of capital and cannot explain the mechanism of determination of rate of interest.

2. Abstinence Theory of Interest

This theory was proposed by Nasau Senior and J.S. Mill. The following are the important features of this theory:

- ☆ Interest is the reward for saving of money. We know, a person earns income and this can be bifurcated into consumption expenditure and savings. When the person is saving money, it implies he is postponing his present consumption needs. This is nothing but a sacrifice made by the person. So, by doing sacrifice, the person is saving the money and he is postponing his present consumption needs, thereby, interest has to be paid for that person for abstinence (sacrifice).
- ☆ Thus, interest has to be paid for the capitalist to compensate the sacrifice he makes by abstaining the present consumption needs, in order to provide loan for the borrower's use.

Demerits

This theory suffers from the following limitations:

- Rich people have much income thereby, there is no question of abstinence in them, while saving and lending the money. Hence, Marshall used the term *'waiting'* in place of 'abstinence' and thereby, interest is the payment made for waiting on the part of the capitalist after lending the money till its repayment and not for abstinence. That means, to make the capitalist to wait, an inducement is essential to encourage his postponement of consumption to the future and it is in the form of interest.
- This theory ignored the concept of productivity of capital.
- The extent of sacrifice or abstinence vary from person to person and more over, it is difficult to measure thereby, it makes the concept mere subjective and not analytical.
- This theory explains the concept only from the supply side, neglecting the demand side.
- As in Marginal productivity theory, this theory too explains, why the interest is to be paid and does not explain the mechanism for the determination of rate of interest.

Considering the above demerits, we can infer that, this theory could not offer satisfactory explanation for the concept of interest, as it deals only from the supply side of capital. Further, it explains only, why interest has to be paid and does not explain how it is determined. However, this theory gave satisfactory explanation regarding the payment of interest, as a reward for temporary postponement of present consumption needs and thereby, creating savings for capital formation in the economy.

3. Agio or Austrian Theory of Interest

This theory was proposed by Bohm Bawerk of the Austrian School of Economists and he explained the concept based on time preference. He opined that, people prefer present consumption over future consumption, thereby, there is an 'agio' or premium on present consumptions goods. For example, if a borrower borrows Rs.10,000 from a lender in the current year and if he returns the same amount after one year, then the lender feels he is the loser. So, the borrower must repay more than Rs.10,000 to the lender and this infers that, the future satisfaction when viewed from the present undergoes a discount. Interest is this discount, which must be paid to induce the capitalist to lend money to the borrower and thereby, postpone his present consumption needs to a future date *i.e.*, for changing the time preference from present consumption to future consumption. This, otherwise, infers that, there is an agio or premium on the present consumption compared to future consumption. Bohm-Bawerk gave the following three important reasons, why the people prefer present consumption to future consumption:

- Compared to present, future is uncertain thereby, present wants are intensely felt by the people.
- In the present consumption, the MU of money will be high. But, in the future, with increase in income, the MU of money may fall and thereby, the consumer may derive less satisfaction.
- The technical superiority of present goods will be more than future goods.

This theory is considered superior compared to earlier theories, as it offers explanation, why interest is to be paid both for productive and unproductive loans. However, Irving Fisher refined this theory, but he still considered 'time preference' as the central element in his theory.

4. Time Preference Theory

This theory was proposed by an American Economist, Irving Fisher and he opined that, even when the future is certain, people show the tendency for present consumption over future consumption. If people prefer to save *i.e.*, if they postpone their present consumption needs, they are rewarded in the form of interest, to encourage this postponement of consumption. The following are the important features of this theory:

- Interest arises because, people prefer present consumption to the future consumption. So, the people tend to be impatient or willing to spend their income on the present consumption. The greater the degree of impatience, the more is the money spent on the present consumption rather than on future consumption and *vice versa*. So, when there is greater degree of impatience, the higher should be the rate of interest, so that, it induces the people to save and lend the money.
- The degree of impatience or willingness of the people to spend depends upon the factors like size of the income, distribution of income over time, certainty of enjoyment of income in the future, personal attributes like foresight, temperament, life expectancy of the individual etc.
- ***Size of the income*** : If size of the income is large (rich people), the individuals will tend to satisfy present wants more than future wants, thereby, they discount the future at a lower rate. So, compared to poor people, rich people will not bother much about the future.

- ☆ *Distribution of income over time*: There are three different ways of distribution of income to the individuals *viz.*, the flow of income may be uniform throughout the future, the flow of income increases in the future and the flow of income decreases in the future. When there is uniform flow of income throughout the life of the individual, the degree of impatience depends upon the level of income, temperament and foresight of the individual. If the flow of income increases in the future with the increase in age of the individual, it implies two important aspects *viz.*, the future is safe guarded and present income of the individual is at low level. So, the degree of impatience on the part of the individual to spend the money on present consumption will be high or the individual spends more on the present consumption. In such a situation, to make the individual to save the money, the rate of interest should be sufficiently high. So, the individual will lend the money at a high rate of interest. Thus, he will discount the future at higher rate of interest. Reverse is the case, when the flow of income decreases in the future with the increase in age of the individual. When the flow of income decreases in the future, it implies, the future is uncertain and present income of the individual is at high level. So, the degree of impatience will be low on the part of the individual with reference to present consumption. This indicates that, with the fall in income, in the future, the individual bothers much about the future and thereby, the individual starts saving even at low rate of interest. They lend the money even at low rate of interest and thus, they will discount the future at lower rate.
- ☆ *Certainty of enjoyment of income in the future*: If the individual is sure of enjoyment of future income, the degree of impatience will be less with reference to present consumption and he starts saving the income even at low rate of interest, thereby, he discounts the future at lower rate. The smaller the rate of interest, the smaller the time preference.
- ☆ *Foresight of the individual*: If the individual has more foresight about the future, he will be less impatient to spend the money in the present period and thereby, start saving even at lower rate of interest. So, he discounts the future at low rate and *vice versa*.
- ☆ *Life expectation of the individual*: If the longevity of the individual is more, he prefers to save the money even at low rate of interest to meet the living expenses at his old age (medical, insurance etc) thereby, he discounts the future at lower rate.

Considering the above factors, when the degree or rate of impatience or willingness or time preference is higher than market rate of interest, then the individual will borrow and spend the money to meet urgent present consumption needs. On the other hand, when the degree or rate of impatience or willingness or time preference is lower than market rate of interest, he will lend the money and gain the interest. So, the individual will borrow or lend the money till his rate of impatience or willingness equals the market rate of interest.

The other important factor that determines the rate of interest is the Marginal Efficiency of Capital (MEC). '*It is equal to the rate of discount, which would make the present value of the series of annuities given by the returns expected from the capital assets during its life just equal to its supply price*' (Keynes). According to Kurihara, '*MEC is the ratio between prospective yield of additional capital goods and their supply price*'. Let us suppose, a farmer faces two investment opportunities A and B and from these two, he has to opt for the better one. Then, considering the MEC concept, we estimate the discount rates at which the net present worth values of both the investments are equal. These discount rates measure the rate of return over cost. Then ranking among the investments is done in relation to the market rate of interest. If the discount rate (rate of return over cost) is higher than the market rate of interest for investment A and if the discount rate is higher compared to other investment B, then investment A will be considered as a project by the farmer.

Thus, this theory infers that, degree of impatience or willingness or time preference and MEC together determine the peoples' preference for present consumption over future consumption, which, in turn, determine the rate of interest because, it is paid for this preference. The same concept can be explained considering the IDCs and PPF (Figure 27.1). As shown in the Figure 27.1,

- ☆ I_1, I_2 and I_3 are the IDCs or willingness curves of different levels of consumption preferences by the individual and they are convex to origin. The steepness of IDCs indicates the community's time preference. If the IDC has absolute slope greater than unity at the 45^0 line, it implies positive time preference. If the IDC has absolute slope less than unity at the 45^0 line, it implies Negative time preference. If the IDC is symmetrical around the 45^0 line, it implies neutral time preference. In the Figure 27.1, I_1 indicates negative time preference, I_2 indicates positive time preference and I_3 indicates neutral time preference.
- ☆ PP is the PPF curve, which is concave to the origin and this indicates the net productivity of the capital.
- ☆ 45° line is shown and this indicates the constancy of income stream line.
- ☆ The equilibrium rate of interest is determined by the point of tangency between IDC and PPF curves and the point of equilibrium implies the maximum value of an income stream.

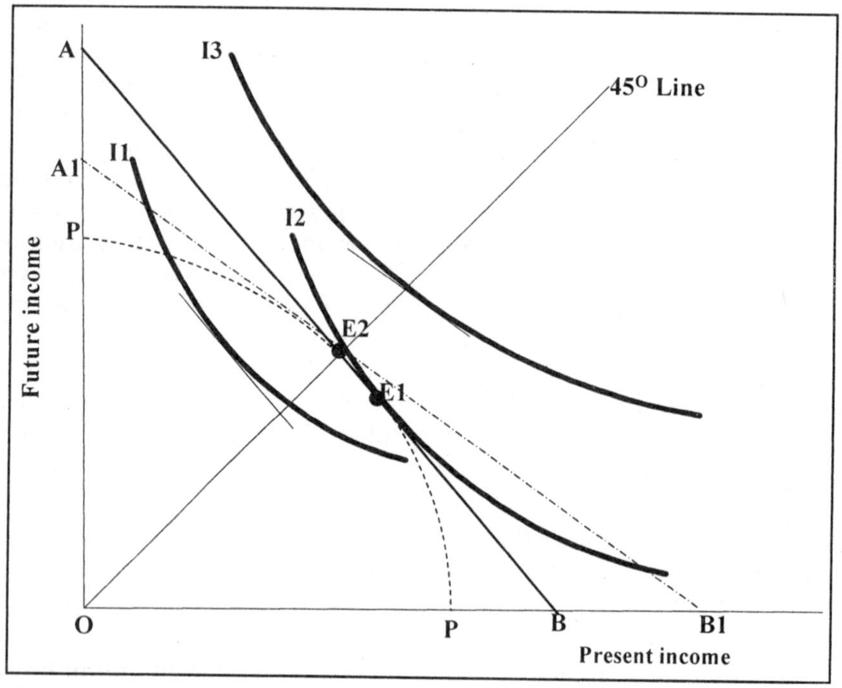

Figure 27.1: Fisher's Time preference theory – Determination of rate of interest.

- With reference to point E_1 on IDC I_2, there is positive time preference, as the slope drawn with reference to IDC I_2 *i.e.*, AB has slope greater than unity, indicating that rate of interest is positive. However, at point E_2, the rate of interest is zero, as the absolute slope of the tangent drawn A_1B_1 is minus unity. Thus, with reference to point E_2, the time preference is neutral and net productivity of capital is zero, as in case of IDC I_3.

Criticisms

Though Fisher refined the Bohm-Bawerk theory, it is too not free from the limitations and they are discussed here under.

- This theory explained the concept considering only the supply side of the capital and ignoring the demand side.
- It is not always true that the persons prefers present consumption to future consumption, but in some occasions like planning for purchase a house, performing daughter's marriage, payment of premium towards medical insurance etc., persons prefer future consumption to present consumption.
- This theory considers that, supply of capital arises due to savings only. But, there are several other ways from the point of view of the banking system that, supply of capital can be increased in the economy.
- This theory assumed constancy of discount rate, but in practical sense, discount rate fluctuates.

Considering the above demerits, we can infer that, this theory has not offered satisfactory explanation for the determination of interest.

5. Classical Theory of Interest

This theory was proposed by Marshall and Fisher and later it was refined by Pigou, Knight and Taussig. According to this theory, the rate of interest is determined by the interaction between demand for and supply of capital. Hence, this theory is also called as Demand and Supply theory of capital or Saving-Investment theory of interest. Since, this theory emphasizes physical factors like productivity, thrift etc., in the determination of interest, it is also called Real theory of interest. According to this theory, demand for capital arises for investment purposes and supply of capital arises from savings. Assuming the conditions of perfect competition, the interaction between demand and supply forces determine the rate of interest. It should be noted that, this theory further assumes savings as 'real savings' and investment as 'real investment'. Real savings means, the resources which are left over after meeting the consumption requirements and they are exclusively meant for investment purposes only. Real investment means, the investment meant for creating new capital assets like machinery, tools, equipments etc. It considers money does not play any direct role in the determination of interest. Let us examine, the demand and supply forces that determine the rate of interest.

Demand for Capital

As discussed earlier, the demand for a factor of production is derived demand and the demand for capital arises from the entrepreneurs' side and they invest the same in the creation of new productive assets. The entrepreneur will demand for capital because, it is productive in the sense that, it yields more income even after covering the interest component. But, it is essential to note that, the productivity of capital is also subjected to LDR. That means, when the entrepreneur applies more and more capital investment in the production programme, the marginal productivity of capital goes on diminishing (Panel A of the Figure 27.2), thereby, the marginal productivity curve of capital slopes downward from left to right. This marginal productivity curve refers to the demand curve for capital, inferring that, at low rates of interest, the demand for capital is more thereby, investment is more and *vice versa*. As shown in the Panel A, when the rate of interest is OR_1, the demand for capital or investment is OI_1. But, when the rate of interest decreases to OR_2, the demand for capital or investment increases to OI_2. However, besides rate of interest, other factors like growth of population, technical progress etc., also influence the demand for capital or investment.

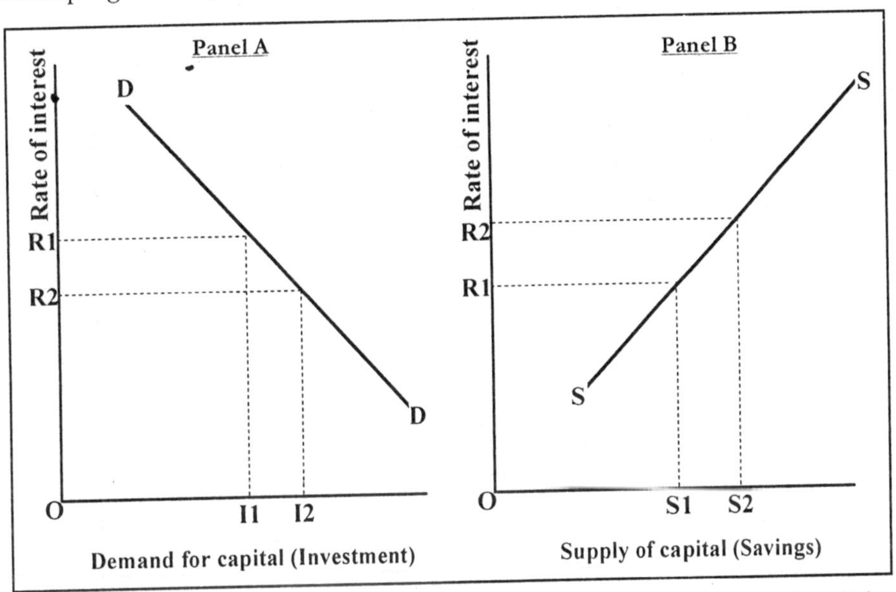

Figure 27.2: Classical theory of interest – Demand (Investment) and Supply (Savings) curves of capital.

Supply of Capital

Supply of capital into the economy is due to saving of capital, which in turn depends upon the thrift habits of the community. Thus, savings leads to capital formation. Savings in the economy is influenced by two major factors *viz.*, willingness to save and ability to save. Besides these two, other factors like socio-economic conditions, political climate, growth of population etc., also influences the level of savings in the economy. According to Classical Economists, savings is the direct function of rate of interest *i.e.*, when rate of interest increases, savings increases and *vice versa*. When savings increases, supply of capital also increases, thereby, supply of capital will also have direct function with the rate of interest. As shown in the Panel B of the Figure 27.2, the supply function of capital will have a positive slope with respect to rate of interest *i.e.*, slopes upward from left to right. That means, with increase in rate of interest, supply of capital also increases. When the rate of interest is OR_1, the savings or supply of capital is OS_1 and when the rate of interest increases to OR_2, the savings or supply of capital also increases to OS_2.

It is essential to note that, the concepts of investment and savings of capital discussed above refers to real investment and real savings respectively.

Determination of Equilibrium Rate of Interest

Assuming the given level of income, the rate of interest in the economy is determined by the interaction between demand and supply forces of capital. The equilibrium or natural rate of interest is at that point, where demand for and supply of capital are equal (Table 27.1). From the table, we can infer that,

☆ Demand for capital or investment is an inverse function of rate of interest.

Table 27.1: Determination of equilibrium or natural rate of interest

Rate of Interest (per cent)	*Demand for Capital or Investment (Rs. Crores)*	*Supply of Capital or Savings (Rs. Crores)*
12	45	110
11	55	92
9	62	81
8	**70**	**70**
7	80	61
6	94	54
5	104	42

- ☆ Supply of capital or saving is a direct function of rate of interest.
- ☆ When the rate of interest is at eight per cent, both demand for capital or investment (Rs 70 crores) and supply of capital or saving (Rs. 70 crores) and equal thereby, eight per cent interest rate is called equilibrium or natural rate of interest.
- ☆ When the interest rate in the economy is above equilibrium rate of interest (say 9 per cent), the supply of capital is more than demand for capital, thereby, the rate of interest will fall and this continues till when demand for capital is equal to supply of capital at eight per cent rate of interest.
- ☆ On the contrary, when the interest rate in the economy is below the equilibrium rate of interest (say 7 per cent), the demand for capital is more than supply of capital and thereby, gradually the interest rate will increase till demand for capital is equal to supply of capital at eight per cent rate of interest.
- ☆ The same concept is explained through Figure 27.3.

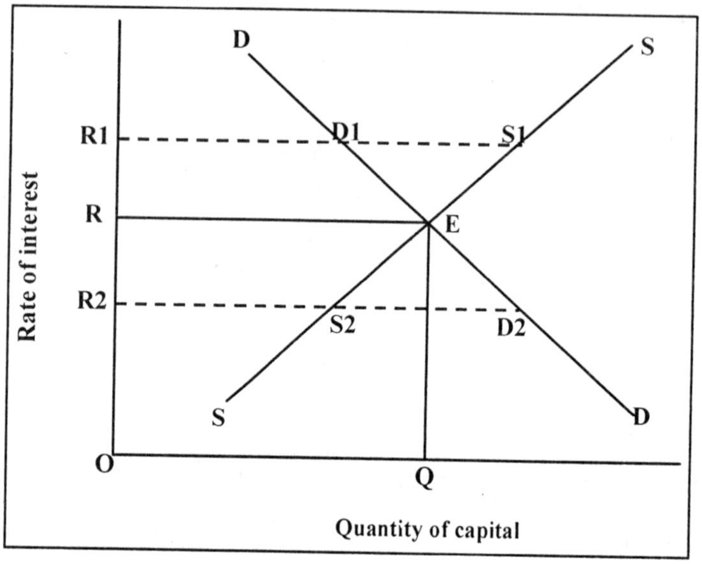

Figure 27.3: Classical theory of interest – Determination of equilibrium (natural) rate of interest.

- ☆ In the Figure 27.3, DD and SS are the demand and supply curves of the capital. These two curves intersect at point 'E' and corresponding points on Y-axis and X-axis implies that, at OR rate of interest, OQ is the amount of capital demanded and supplied in the economy.
- ☆ If the rate of interest is increased to OR_1, then the demand for capital (R_1D_1) is less than supply of the capital (R_1S_1) by D_1S_1. So, more supply of capital than its demand will lead to fall in interest rate from OR_1 to OR.
- ☆ If the rate of interest is decreased to OR_2, then the demand for capital (R_2D_2) is more than supply of the capital (R_2S_2) by D_2S_2. So, excess demand for capital over its supply will raise the interest rate from OR_2 to OR.

So, this is how, the price of the capital factor is determined in the market based on both demand and supply forces.

Criticisms

J.M. Keynes strongly criticized this Classical theory of interest. The major criticisms leveled against this theory are as under.

- ☆ This theory wrongly assumed full employment equilibrium conditions in the economy. But according to Keynes, unemployment and under employment situations in the economy are more common.
- ☆ According to Classical theory, the mechanism of equality between savings and investment is brought by the rate of interest. But according to Keynes, it is not the rate of interest, but the level of income that brings about the equality between savings and investment. This is because, when the rate of interest rises in the economy, the investment falls and thereby, income falls. When income falls, the savings on the part of the individuals will fall and thereby, the equality between savings and investment is brought due to the changes in level of income and not the rate of interest.
- ☆ This theory considered only real factors like productivity, thrift etc., and ignored monetary factors in the determination of rate of interest. According to Keynes, interest is a monetary phenomena, thereby, it should take into consideration the role of monetary factors.

- According to Classical theory, more investment in the economy is due to more savings or reduced consumption. But, we know, demand for capital goods is only a derived demand and it is derived due to the demand for consumer goods. So, if the consumption is reduced, it implies, reduced demand for consumer goods, and thereby, reduced demand for capital goods as well. So, the investment will fall when there is reduced consumption. Keynes, further explained that, investment will not increase at the expense of consumption. This is because, when investment increases, it leads to increase in income in the economy due to multiplier effect and thereby, it leads to increase in demand both for consumer and capital goods. This infers that, Keynes analysis is more realistic compared to the Classical Economists.
- This theory is narrow in its scope, as it ignores the concept of consumption loans and considered only investment loans for analysis purpose.
- According to Classical theory, the equilibrium rate of interest determined by the demand for and supply of capital prevails only in the long run and this theory is not applicable in the short run,. But, according to Keynes, the analysis should be applicable in the short run, as he considered *'we are all dead'* in the long run.
- This theory was strongly criticized by Keynes on the ground of indetermination. This theory states that, equilibrium rate of interest is determined by the interaction between demand for and supply of capital. The supply of capital depends upon the level of income and it varies directly. Since, supply of capital depends upon the level of income, it is not possible to compute equilibrium rate of interest unless we know the level of income in the economy. But, this level of income in the economy also depends upon the rate of interest. So, this circular reasoning will lead to a confusing situation and thereby, this theory has created an indeterminate solution for the determination of interest.
- This theory stated that, supply of capital arises from the current savings only. But, supply of capital also arise from past savings, bank credit etc., and Classical theory has neglected these factors in the supply schedule of capital.
- The Classical theory considered that, investment schedule is independent of savings schedule. That means, investment schedule can change without causing a change in the savings schedule. Suppose, when the investment falls in the economy due to poor business expectations, the investment demand curve will shift downward towards the origin and cuts the savings curve at the lower level. But, this change alone is not possible because, when the investment declines in the economy, income declines and thereby, savings also decline and hence, savings curve will also shift towards left. The Classical theory does not take this into account.
- This theory considered level of income is fixed in the economy. But, when a small investment is made in the economy, the income greatly increases due to multiplier effect.
- According to Classical theory, the equilibrium (natural) rate of interest and market rate of interest are equal and if any discrepancy exists between the two, it is only a temporary phenomenon. But, according to Keynes, the discrepancy between the two is not temporary and accidental. This is due to expansion or contraction of bank credit. When there is expansion of bank credit, it leads to fall in the market rate of interest below the equilibrium (natural) rate and *vice versa*. Thus, the disequaltiy between the equilibrium (natural) and market rates of interest is not accidental and thereby, there is no automatic mechanism to bring about the equality between the two.

Considering the above demerits, Keynes criticized the Classical theory of interest as indeterminate and incomplete.

6. Loanable Funds Theory of Interest

This theory is also called as Neo-Classical theory of interest expounded by Knut Wicksell and developed by Robertson, Pigou, Myrdal, Ohlin etc. The Classical theory considered real factors like productivity, thrift etc., in the determination of rate of interest and ignored the monetary factors. But, in this Neo-Classical theory, both real factors and monetary factors are considered for the determination of rate of interest and thereby, this theory is more realistic and broader compared to the Classical theory. The monetary factors considered in this theory include hoarding and dishoarding of money, credit creation through banks, consumption loans etc. So, this theory states that, rate of interest is price paid for the use of loanable funds and it is determined by the equilibrium between demand for and supply of loanable funds. Loanable funds are the amounts of funds that are meant for loaning in the money market or it is the sums of money supplied and demanded at any time in the money market or it is the fund that is available for borrowing. Thus, loanable funds is wider in scope and includes not only savings out of current income, but also bank credit, dishoarding and disinvestment. The above discussion infers that, Neo-Classical theory of interest is a monetary approach to the theory of interest, as distinguished from that of Classical theory, which altogether neglected monetary factors in the determination of rate of interest. Thus, Neo-Classical theory synthesizes both the monetary and non-monetary factors in the determination of rate of interest. So, let us discuss the demand and supply forces of loanable funds.

Demand for Loanable Funds

The demand for loanable funds will arise from three fields *viz.*, investment, consumption and hoarding.

- ☆ ***Investment*:** The bulk of demand for loanable funds will arise from business firms to invest them for purchasing new capital assets, buildings etc. The demand for loanable funds for investment purposes is interest elastic *i.e.*, more amount of loanable funds will be demanded at low rates of interest. Thus, there exists an inverse relationship between demand for loanable funds for investment and rate of interest, thereby the demand curve for investment (I) slopes downward from left to right (Figure 27.4). The loanable funds are demanded upto the point at which the expected rate of return from capital goods is equal to the rate of interest. This demand for loanable funds for investment purposes is called Investment demand.
- ☆ ***Consumption*:** The individuals also demand loanable funds for consumption purposes. In general, if the individuals wants to purchase consumer goods that are beyond the reach of their earning incomes, they demand for loanable funds. For example, if a person earns Rs.10000 per month and if he wishes to purchase a car, he demands for loanable funds. Like investment demand, this demand for loanable funds for consumption purposes is also interest elastic *i.e.*, more amount of loanable funds will be demanded at low rates of interest. Thus, there exists an inverse relationship between demand for loanable funds for consumption purposes and rate of interest, thereby the demand curve for consumption (C) slopes downward from left to right (Figure 27.4). This demand for loanable funds for consumption purpose is called Consumption demand or Dissavings.
- ☆ ***Hoarding*:** It refers to keeping the money in liquid form with the individuals. Thus, hoarded money refers to idle cash balances. We know, the income of the individuals can be bifurcated into consumption expenditure and savings. That means, when the individuals do not spend all their money income on consumption needs, they save some amount of money. With this remaining (savings) money, the individuals can purchase securities or lend the same outside to earn some interest or invest the same in purchasing capital assets etc. Another way of using the saved money is to hoard with them as idle cash balances to meet any unforeseen contingencies and to take the advantages of changes in the rate of interest in the future. This demand for money for hoarding it, is also interest elastic *i.e.*, more amount of funds will be hoarded at low rates of interest. Thus, there exists an inverse relationship between demand for money for hoarding and rate of interest. At low rates of interest, the people will keep the liquid cash with them rather than giving it for lending purpose. But, at high rates of interest, they lend the money outside and earn interest. Thus, the demand curve for boarding (H) also slopes downward from left to right (Figure 27.4).

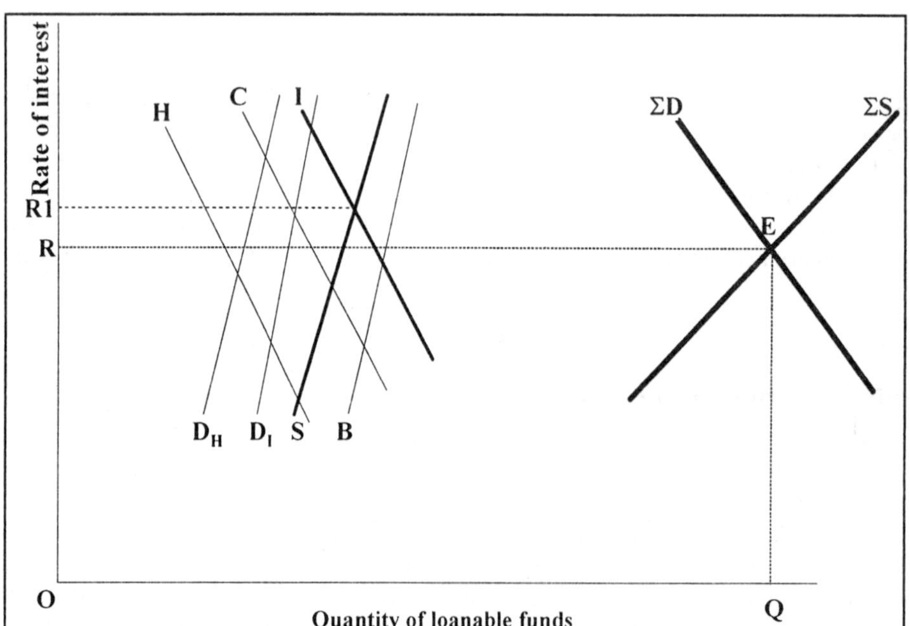

Figure 27.4: Neo-Classical theory of interest – Determination of equilibrium (market) rate of interest.

Thus, the total demand for loanable funds (ΣD) is the sum of demands for loanable funds for investment purposes (I), consumption purposes (C) and hoarding purposes (H) given by, $\Sigma D = I+C+H$. Hence, the total demand curve for loanable funds is the sum of demand curves for loanable funds for investment, consumption and hoarding purposes and this is represented as ΣD in the Figure 27.4. As the three demand curves for loanable funds for investment, consumption and hoarding slopes downward form left to right, the total demand curve for loanable funds (ΣD) also slopes downward from left to right.

Supply of Loanable Funds

According to Classical theory, supply of capital arises due to savings. But, according to Neo-Classical theory, supply of loanable funds arises from four sources *viz.*, savings, dishoarding, bank money and disinvestment. These four sources of loanable funds are discussed here under.

- ☆ ***Savings***: Among different sources of supply of loanable funds, savings done by the individuals is the most important source. The savings are of two types *viz.*, ex-ante savings and ex-post savings. Ex-ante savings are also called as Planned savings. That means, even before making the expenditure, the individuals plan their savings. Ex-post savings are also called as Unplanned savings and it is the difference between income of the preceding period (Y_{T-1}) and consumption expenditure in the present period (C_T). It is in this sense, Robertson has used the word '*savings*'. These two savings *i.e.*, planned and unplanned savings varies directly with the rate of interest. The level of savings also depends upon the level of income of the people. But, this theory assumed the level of income is constant in the economy and thereby, level of savings is influenced directly by the rate of interest. Like individuals, business firms also save a part of their earnings and these savings also vary directly with the rate of interest. So, the supply curve of savings for loanable funds (S) will have a positive slope sloping upwards from left to right. (Figure 27.4).
- ☆ ***Dishoarding*** : It is opposite of hoarding of idle cash balances. That means, it refers to the supply of idle cash balances by the individuals back to the loanable funds. This occurs at a very high rate of interest, thereby, the individuals will go for savings of idle cash balances and makes them available for demand for loanable funds for investment and consumption purposes. If the rate of interest is lower, dishoarding is very less and the individuals prefer to maintain idle cash balances. It infers that, when the individuals dishoard the idle cash balances, they are now added to the supply of loanable funds. Thus, dishoarding is interest-elastic and it varies directly and it is shown as an upward sloping curve (D_H) form left to right (Figure 27.4).
- ☆ ***Bank money*** : The credit creation is an important function of the commercial banking system. The supply of credit by the commercial banks will be more at higher rates of interest and *vice versa*. So, the supply curve of bank money (B) will slope upwards from left to right (Figure 27.4).
- ☆ ***Disinvestment*** : Disinvestment is the opposite of investment. Some firms in the economy will experience losses in the business and for these firms, the marginal productivity of capital is less than the market rate of interest. So, the firms will withdraw the investment, as they are incurring losses and lend the capital of the business outside for earning interest. This disinvestment will be high at higher rates of interest and thereby, at higher rates of interest, the supply of loanable funds will be high due to higher disinvestments. So, the supply curve of loanable funds due to disinvestment (D_I) will also slope upwards from left to right (Figure 27.4).

Thus, the total supply of loanable funds (ΣS) is the sum of supplies of loanable funds through savings (S), dishoarding (D_H), bank money (B) and disinvestment (D_I) and it is given by $\Sigma S = S + D_H + B + D_I$. Hence, the total supply curve of loanable funds is the sum of supply curves of loanable funds through savings, dishoarding, bank money and disinvestment and this is represented as ΣS in the Figure 27.4. As the four supply curves of savings, dishoarding, bank money and disinvestment slopes upward from left to right, the total supply curve of loanable funds (ΣS) also slopes upward from left to right.

Determination of Equilibrium Rate of Interest

According to Loanable funds theory, the equilibrium rate of interest is determined by the interaction between ΣD and ΣS curves and the rate of interest at which ΣD and ΣS are equal is called market rate of interest or equilibrium rate of interest. In the Figure 27.4,

- ☆ ΣD curve represents the horizontal summation of demand curves of loanable funds for investment, consumption and hoarding.
- ☆ ΣS curve represents the horizontal summation of supply curves of loanable funds due to savings, dishoarding, bank money and disinvestment.
- ☆ These two curves *i.e.*, ΣD and ΣS intersect at point E and it infers that, at OR rate of interest, OQ is the amount of loanable funds demanded and supplied n the market economy.
- ☆ This figure also helps to differentiate between Classical theory of interest and Neo-Classical theory of interest or Loanable funds theory of interest. According to Classical theory, the equilibrium or natural rate of interest (OR_1) is determined by the interaction between demand (Investment) and supply curves of capital. But, according to Neo-Classical theory, the equilibrium or market rate of interest (OR) is determined by the interaction between ΣD and ΣS curves of loanable funds. Further, according to Classical theory, OR_1 is called

as equilibrium or natural rate of interest due to the interaction between demand for (Investment) and supply of capital. But, according tc Neo-Classical theory, OR is called as the equilibrium or market rate of interest due to the interaction between demand for and supply of loanable funds. Thus, the concept of Neo-Classical theory of interest helps to differentiate between natural rate of interest and market rate of interest.

Improvement of Neo-Classical Theory Over Classical Theory of Interest

The following are the improvements of Neo-Classical theory over Classical theory in the determination of equilibrium rate of interest.

- Neo-Classical theory considers both monetary and non-monetary factors in the determination of rate of interest, whereas Classical theory considers only non-monetary factors. Since, determination of rate of interest is a monetary phenomenon, Neo-Classical theory is considered to be more realistic in its approach.
- Classical theory assumes passive role for money, whereas, Neo-Classical theory recognized active role for money in the determination of rate of interest.
- Classical theory is narrow in its scope, as it ignores the concept of consumption loans, but the same is considered in Neo-Classical theory with reference to demand for loanable funds for consumption purposes.
- According to Classical theory, savings (thrift) is the only source of supply of capital. But, according to Neo-Classical theory, besides savings, dishoarding, bank money and disinvestment also constitutes sources of supply of loanable funds.
- According to Classical theory, demand for capital is influenced by a non-monetary factor *i.e.*, productivity of capital. But, according to Neo-Classical theory, demand for loanable funds is influenced by investment, consumption and hoarding.
- Neo-Classical theory of interest helps to distinguish between natural rate of interest and market rate of interest, but the Classical theory could not offer any explanation about market rate of interest.

Criticisms

The following are the short comings of Neo-Classical theory of interest:

- The concept of Neo-Classical theory is merely an expansion of the concept of Classical theory and there is no material advancement except in considering different sources of supply of loanable funds and different demands for loanable funds.
- Hansen criticized this theory in not providing a determinate solution for the problem of determination of rate of interest. According to this theory, the market rate of interest is determined by the interaction between demand for and supply of loanable funds. However, different forms of demand for loanable funds and different sources of supply of loanable funds are themselves determined by the rate of interest. Further, the level of savings cannot be known unless the level of income is known. So, like Classical theory, this theory is also indeterminate in explaining the mechanism for the determination of rate of interest.
- This theory considers that, the supply of loanable funds can be increased by dishoarding of idle cash balances and the supply of loanable funds can be decreased by hoarding of cash balances and this infers that the loanable funds in the economy can be either increased or decreased. But, this is not actually the case, as the amount of loanable funds in an economy at a point of time are constant or fixed and equal to the total amount of money supply.
- This theory has been criticized on the ground that, it combined the monetary factors and non-monetary factors in the determination of rate of interest.
- This theory wrongly assumes the constancy of level of income. But, in reality, the level of income in the economy changes due to changes in investment.
- This theory, like the Classical theory, assumes full employment equilibriums in the economy. But, according to Keyens, unemployment and under employment are common in practical sense and existence of full employment equilibrium is only an abnormal case.
- According to this theory, demand for loanable funds for investment is interest elastic. But, besides rate of interest, it also depends upon the MEC, business expectations etc.
- Critics argued that, people in the economy save the money not only for the sake of earning interest, but for the sake of precautionary motive and in this context it is interest-inelastic.

Considering the above criticisms, this theory too has not offered satisfactory explanation for the determination of rate of interest. However, this theory is more explanatory when compared to the Classical theory.

7. Liquidity Preference Theory of Interest

This theory was developed by John Maynard Keynes, a British Economist in his book *'The General Theory of Employment, Interest and Money'* to explain the determination of rate of interest based on the interaction between demand for and supply of money. Keynes criticized the earlier theories *viz.*, Classical theory and Non-Classical theory, as both the theories considered non-monetary factors in explaining the determination of rate of interest. Though Neo-Classical theory or Loanable fund theory explained the concept of determination of rate of interest considering monetary factors also, Keynes criticized the mingling of monetary factors and non-monetary factors in explaining the concept. But, according to Keynes, interest is purely a monetary phenomenon and therefore, it should be determined based on the forces of demand for and supply of money. That means, interest is an equilibrium force in the money market. According to Keynes, *'interest is neither a reward for saving nor waiting, but it is the reward for parting with liquidity for specified period'*. That means, a borrower will pay interest to the lender because, the lender has to part with his liquid money when he lends his money to the borrower. According to Keynes, the greater the desire for liquidity, the greater is the rate of interest demanded for parting with liquidity. In the words of Keynes, *'the rate of interest is the premium which is to be offered to induce people to hold their wealth in a form other than hoarded money'*. We know, an individual can maintain money in different forms *viz.*, bonds, securities, debentures, fixed deposits, gold and silver ornaments etc. But, these forms of money cannot be easily converted to another form and some time element is indispensable to convert from one form to another form. On the other hand, if the individual maintains liquid cash, he can easily employ the same to purchase any asset *i.e.*, it can be easily converted to any other form. This maintenance of money in cash form is called 'liquidity' and the liking of the people to maintain money (liquid) cash is called liquidity preference. Let us discuss the two forces *i.e.*, demand for money and supply of money here under.

Demand for Money

According to Keynes, people demand for money (liquid cash) to fulfill three motives *viz.*, Transactions motive, Precautionary motive and Speculative motive.

(a) Transactions Motive

Individual households and firms have to make different transactions in rendering various economic activities. For example, an individual household has to purchase several things for his family till he gets his next month's salary. This is called Income motive. Similarly, a business firm has to incur expenditure on several items like payment of wages to labour, rent for the land, salaries for the employees, expenditure on raw materials etc., till the firm gets the returns from the sale of the final output. This is called the Business motive. So, this type of demand for money for meeting day to day transactions of households and business firms constitute demand for money for transactions motive. This demand for money for transactions motive depends upon the following aspects.

- ☆ ***Size or level of income*** : The demand for money for transactions motive varies directly with the level of income of the individual or a firm. That means, if the individual or firm has more level of income, the number of transactions will be more, thereby, demand for money for transactions motive will be more.
- ☆ ***Time gap between investment and expected returns from the business*** : The more is the time gap between the investment made and expected returns from the business, the more is the demand for money for transactions motive.
- ☆ ***Turn over of the business*** :The larger the turn over of the business, the larger the demand for money for transactions motive.
- ☆ ***Methods of payment of wages*** : If the wages are paid to the individuals on daily basis, the demand for money for transactions motive by the individuals will be less. So, the greater the delay in payment of wages or salaries to the individuals, the greater is the demand for the money for transactions motive.

According to Keynes, the demand for money for transactions motive is relatively interest inelastic (Panel A of the Figure 27.5), but highly income elastic. This is given by $M_1 = f(Y)$, where M_1 = Demand for money for transactions motive, Y = Level of income.

(b) Precautionary Motive

The demand for money for precautionary motive refers to the desire to hold liquid cash to meet any unforeseen contingencies. Individuals hold liquid cash to meet age old sickness, accidents, unemployment etc. Similarly, business men also hold liquid cash to meet unfavourable conditions in the business. The extent demand for money for precautionary motive depends upon the level of income of the individual or firm, nature of the enterprise, nature of the person etc.

- ☆ ***Level of income*** : Higher is the level of income of the individual, the more is the capacity of the individual to keep the money aside to meet unforeseen contingencies.

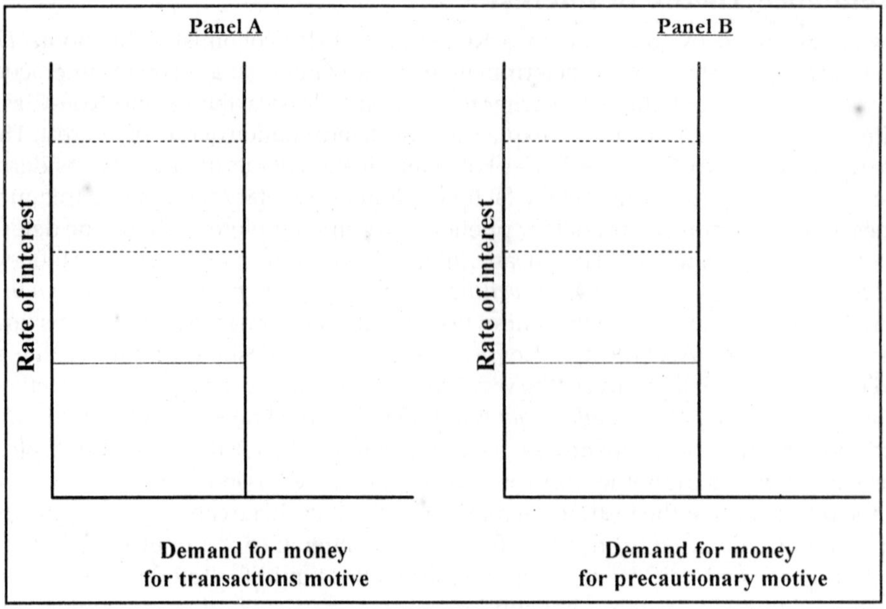

Figure 27.5: Liquidity preference theory of interest – Demand for money for transactions and precautionary motives.

- ☆ ***Nature of the enterprise*** : If the enterprise under consideration is more risky in getting the expected returns, the demand for money for precautionary motive will be high.
- ☆ ***Nature of the person*** : If the person has more future outlook and if he can predict the future and emergencies for his business, his demand for money for precautionary motive will be higher. However, this also depends upon risk bearing ability of the individual. If the individual has more risk bearing ability for the emergencies he may face in the near future, his demand for money for precautionary motive will be less and *vice versa.*

Keynes hold that, like transactions motive, the demand for money for precautionary motive is also interest inelastic (Panel B of the Figure 27.5), but highly income elastic. This is given by $M_2 = f(Y)$, where M_2 = Demand for money for precautionary motive, Y = level of income.

Since, both the demands for money for transactions and precautionary motives are income elastic, Keynes put them together and expressed them jointly as $M_1+M_2 = L_1 = f(Y)$. where M_1 = Demand for money for transactions motive, M_2= Demand for money for precaution motive, L_1 = Total demand for money for both transactions and precautionary motives.

(c) Speculative Motive

The demand for money for speculative motive refers to the desire of the individual to keep the liquid cash with him, so as to take the advantage of market movements regarding the future changes in the rate of interest and bond prices. Bond prices and rate of interest are inversely related to each other. That means, if bond prices are expected to fall (the rate of interest is expected to rise), people will sell away bonds now to avoid losses and thereby, people will like to keep more cash with them, so as to save it, when rate of interest rises in the future. On the contrary, if bond prices are expected to rise (the rate of interest is expected to fall), people will buy bonds now, to sell them when their price later actually rises and thereby, people will not like to keep the liquid cash with them. So, the speculator takes advantage of market movements regarding the future changes in the rate of interest and bond prices. A security which yields a fixed rate of interest over a period of time is called bond. For example, if a bond of value Rs.1000 carries four per cent rate of interest and if the market rate of interest is doubled (*i.e.*, 8 per cent), then the value of this bond falls to Rs.500 only. On the other hand, if the market rate of interest falls to two per cent, then the value of bond will rise to Rs.2000. So, the speculators will purchase the bonds at Rs.500 when the market rate of interest is eight per cent and sell the same for profit at Rs.2000, when market rate of interest falls to two per cent. So, liquidity preference will be more at lower rates of interest because, at lower rates of interest, people will keep liquid cash with them rather than in the form of bonds, so as to avoid capital loss.

This notion of demand for money for speculative motive is a new idea from the Keynes, which the Classical Economists could not conceptualize. According to Keynes, the speculative demand for money is a function of rate of interest and it varies inversely (Figure 27.6). As shown in the Figure 27.6, at very low rate of interest *i.e.*, OR_1, the speculative demand curve for money will become perfectly elastic i.e, the portion of the curve lies parallel to X-axis. This portion of the curve 'AB' is known as Liquidity trap and this implies that, at such a low rate of interest, people prefer to

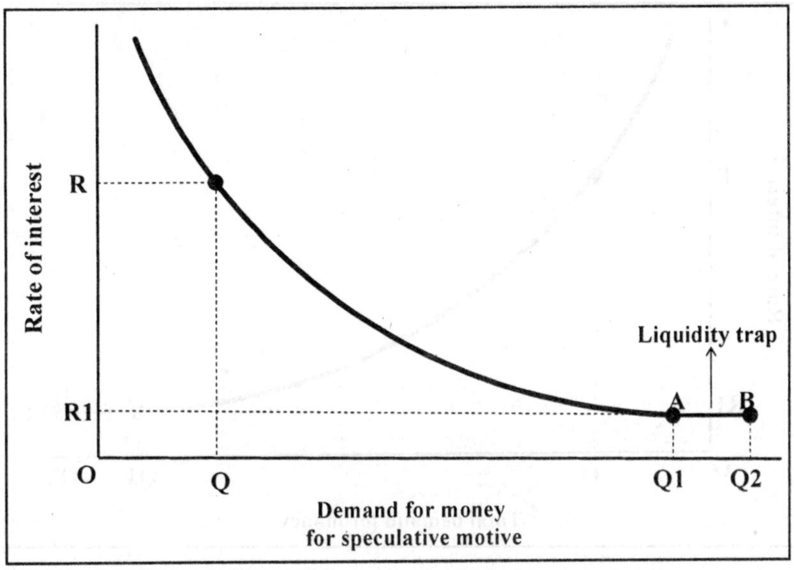

Figure 27.6: Liquidity preference theory of interest – Demand for money for speculative motive.

keep the money in liquid cash rather than investing on bonds, as purchasing bonds will yield at a definite loss. Thus, according to Keynes, there is a definite minimum limit for the rate of interest, but it will not fall to zero. But, according to Classical Economists, the rate of interest can fall upto any limit and even to zero. The Figure 27.6 also shows that, at OR rate of interest, the demand for money for speculative motive is OQ. When the demand for money for speculative motive increases from OQ to OQ_1, the rate of interest falls from OR to OR_1. Even if the demand for money for speculative motive increases from OQ_1 to OQ_2, the rate of interest will remain same at OR_1 indicating that, OR_1 is minimum limit of rate of interest. This AB portion of the curve for the demand for money for speculative motive refers to Liquidity trap. Algebraically, Keynes expressed the speculative demand for money as $L_2 = f(r)$, where L_2 = Speculative demand for money, r = rate of interest.

Total Demand for Money

The total demand for money refers to the sum of demands for money for transaction, precautionary and speculative motives. So, by adding L_1 and L_2 we get the total demand for money. Since, both the demands for money for transactions and precautionary motives are interest inelastic, Keynes ignored these two demands for money in his theory for the determination of rate of interest. Since, the demand for money for speculative purposes is interest elastic (varies inversely), Keynes emphasized the demand for liquidity preference with reference to speculative motive only. The total demand for money (M) is given by,

$$M = L_1 + L_2 = f(Y) + f(r) = M = f(Y,r)$$

Thus, total demand for money for transactions, precautionary and speculative motives depends upon the level of income and rate of interest. If the level of income remains same, the total demand for money (M) is less influenced by the first two motives *i.e.*, transactions and precautionary motives, as they are interest inelastic. But, if there is change in rate of interest, there is significant change in the total demand for money (M), as it influences the third motive (speculative motive) significantly. So, speculative demand for money is the major determinant of rate of interest compared to demand for money for transactions and precautionary motives. This total demand curve for money (M) is called 'Liquidity Preference Curve (LPC)' and it is a downward sloping curve from left to right, indicating the total demand for money to meet the three motives (Figure 27.7). The LPC (Figure 27.7) assumes the same shape like demand curve for speculative motive (Figure 27.6), as it is for only the speculative motive, the demand for money is interest elastic and for the remaining two motives *viz.*, transactions and precautionary motives, the demand for money is interest inelastic.

Supply of Money

In the short run, the supply of money is interest-inelastic. That means, no private individual can change the supply of the money into the economy, as it is purely regulated by the Government or its agency. Hence, the supply curve of money SS is parallel to Y-axis (Figure 27.8) indicating that, in the short run, the changes in the rate of interest does not effect the supply of money into the economy. It is important to note that, the total supply of money into the economy is a stock, but its supply to the community is a flow and it is influenced by rate of interest. So, considering the total supply of money into the economy as a whole, it is interest-inelastic.

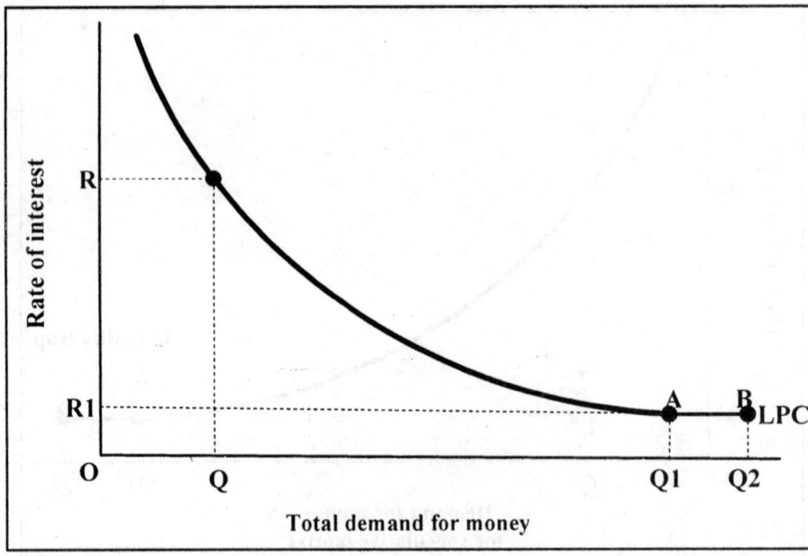

Figure 27.7: Liquidity preference theory of interest – Total demand curve (LPC) for money for three motives.

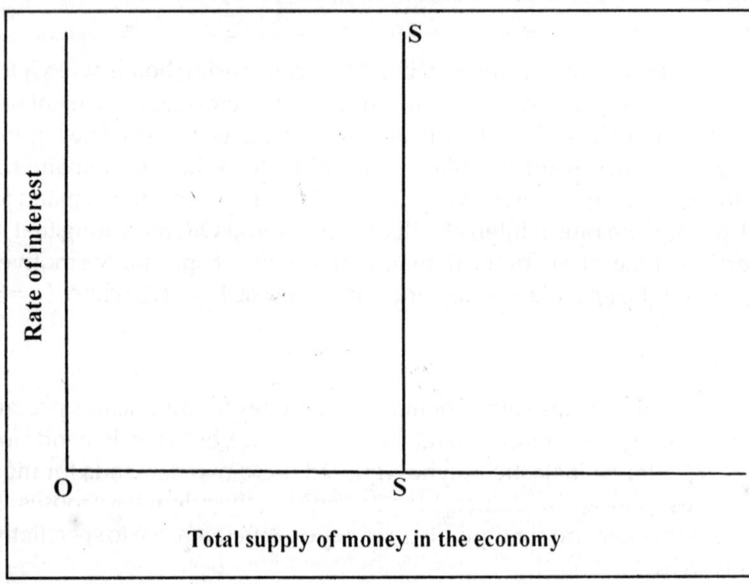

Figure 27.8: Liquidity preference theory of interest – Total supply of money in the economy.

Determination of Rate of Interest

According to Keynes, the rate of interest is determined by the interaction between demand and supply forces of money. With reference to the demand side, there is total amount of demand (LPC) from the people to fulfill the transactions, precautionary and speculative motives and with reference to the supply side, there is total amount of money supplied (SS) in the economy by the Government or the Regulating Authority, irrespective of the level of rate of interest. So, when LPC=SS, we can have equilibrium rate of interest and the same is explained through Panel A of the Figure 27.9.

- As explained earlier, LPC is a downward sloping curve from left to right, indicating that, at lower rates of interest, the total demand for money increases and *vice versa*.
- The total supply curve (SS) for money is interest-inelastic in the short run, thereby, it is a straight line parallel to Y-axis.
- These two curves *i.e.*, LPC and SS curves intersect at point E, indicating OR is the equilibrium rate of interest.
- If the interest rate in the economy is higher (OR_1) than OR, then the total demand for money (R_1L_1) is less than total supply (OS) *i.e.*, by L_1F. That means, there is excess supply of money in the economy than the total demand for money and thereby, interest rate gradually decreases from OR_1 and equals to original equilibrium rate of interest OR.
- On the contrary, if the rate of interest in the economy is lower (OR_2) than OR, then the total demand for money (R_2L_2) is more than total supply (OS) *i.e.*, by L_2G. That means, there is a shortage of supply of money in the

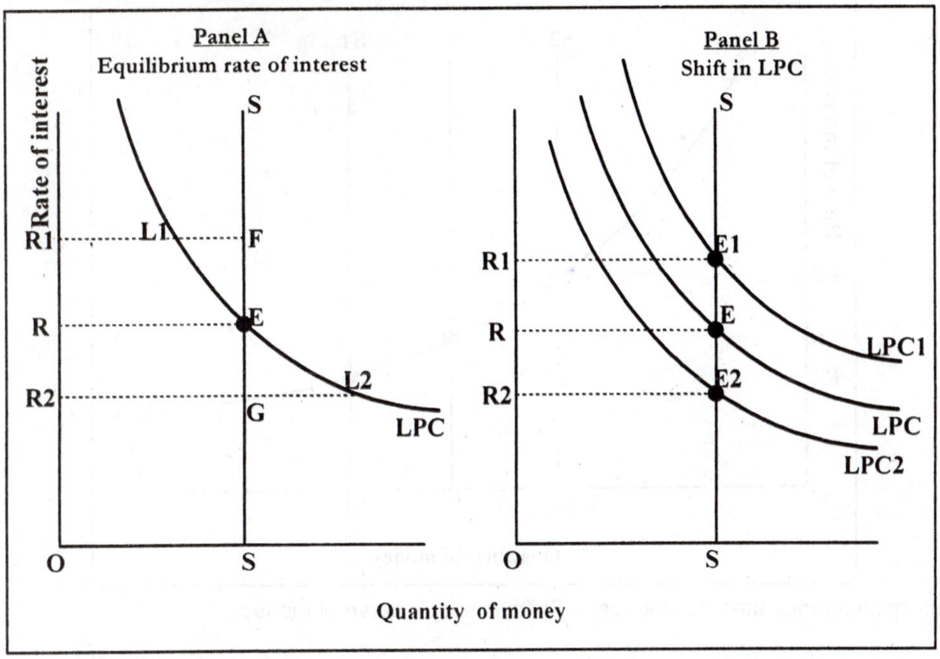

Figure 27.9: Liquidity preference theory of interest – Determination of equilibrium rate of interest.

economy compared to the total demand for money and thereby, interest rate gradually increases from OR_2 and equals to original equilibrium rate of interest OR.

- Thus, any interest rate other than the equilibrium rate of interest (OR) will lead to unstable position in the economy and again the interest rate gradually moves towards equilibrium rate of interest.

However, equilibrium interest rate will fluctuate due to shifts in total demand for money (LPC) or total supply curve for money (SS). These two cases are discussed here under.

Case 1 - Shifts in the LPC Curve

When the total demand for money (LPC) increases in the economy, it leads to an upward shift in the LPC curve and *vice versa* and the same is shown through Panel B of the Figure 27.9. As shown in the Panel B,

- When the total demand for money is LPC and supply of money is SS, the equilibrium rate of interest is OR.
- When the total demand for money increases in the economy, it leads to an upward shift in the LPC curve (LPC_1), from its original position (LPC). Accordingly LPC_1 intersects the total supply cure for money SS at E_1 thereby, equilibrium rate of interest increases from OR to OR_1.
- When the total demand for money decreases in the economy, it leads to a downward shift in the LPC curve (LPC_2) from its original position (LPC). Accordingly, LPC_2 intersects the total supply curve for money SS at E_2, thereby, equilibrium rate of interest decreases from OR to OR_2.
- Thus, when the total supply of money in the economy is constant, an increase in the total demand for money, will increase the equilibrium rate of interest and a decrease in the total demand for money will decrease the equilibrium rate of interest.

Case 2 - Changes in the Supply of Money

When the Government or Money Regulating Authority in the country increases the total supply of money it the economy, but total demand for money remains constant, the equilibrium rate of interest will fluctuate and the same is explained through Figure 27.10. As shown in the Figure 27.10,

- When the total demand for money is LPC and total supply of money is SS, then the equilibrium rate of interest is OR.
- When there is an increase in the total supply of money in the economy (S_1S_1), but at the constancy of total demand for money, then the equilibrium rate of interest will decline in the economy from OR to OR_1.
- When there is decline in the total supply of money in the economy (S_2S_2), but at the constancy of total demand for money, then the equilibrium rate of interest will increase in the economy from OR to OR_2.
- Thus, when the total demand for money remains constant in the economy, an increase in the total money supply will reduce the equilibrium rate of interest (OR_1) and a decrease in the total money supply will increase the equilibrium rate of interest (OR_2).

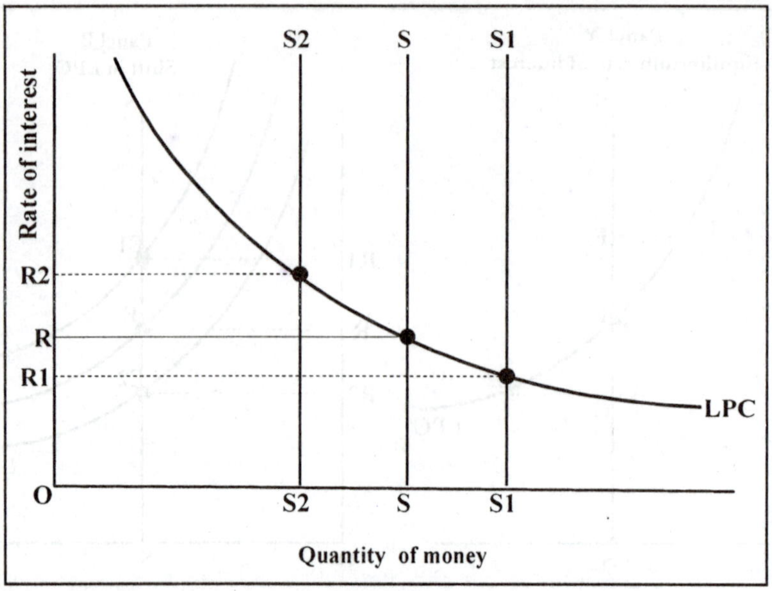

Figure 27.10: Liquidity preference theory of interest – Shifts in supply curve of money.

Criticisms

This theory has been criticized on the following grounds:

- ☆ Prof. Hazlitt criticized the Keynes theory that, it has neglected the role of real factors like productivity, time preference etc., in explaining the determination of rate of interest, as Keynesian theory considers interest is purely a monetary phenomenon. He opined that, Classical theory of interest is also wrong, as it considers only real factors in explaining the determination of rate of interest and neglected the monetary factors.
- ☆ Keynesian theory of determination of rate of interest is indeterminate. The same criticism made by Keynes against the Classical and Neo-Classical theories of interest also applies to his own theory. According to Keynesian theory, the total demand for money depends upon the fulfillment of three motives *viz.*, transactions, precautionary and speculative motives. Out of these three motives, first two motives *i.e.*, transactions and precautionary motives are largely influenced by the level of income and they are interest inelastic. But, demand for the third motive *i.e.*, speculative motive, is largely influenced by rate of interest. So, without knowing the rate of interest in the economy, we cannot know the speculative demand for money. So, in the process of determination of rate of interest, we need the rate of interest itself to estimate the speculative demand for money. Thus, the theory is indeterminate is explaining the determination of rate of interest.
- ☆ This theory explains the determination of rate of interest only in the short run and it gives no clue to the determination of rate of interest in the long run. But, for capital investment, it is the long run rather than short run rate of interest that is really significant.
- ☆ According to Keynes, interest is a reward for parting with liquidity and it is no way an inducement for saving. But, according to Jacob Viner, Keynes has overlooked the fact that, without saving, the liquidity of money will not arise and thereby, interest will not arise. In the words of Jacob Viner, *'without savings there can be no liquidity to surrender'*. Thus, the concept of rate of interest is vitally related to savings, which is neglected by Keynes
- ☆ As per the concept of Keynesian theory, the rate of interest should be the highest at times of depression in the economy because, there will be more liquidity preference due to declining prices and increase in the value of money. But, in practicality, the rate of interest will be low at times of depression in the economy. Similarly, as per the concept of Keynesian theory, the rate of interest should be lowest during the boom period in the economy, but this is not the case.
- ☆ In this theory, the concept of liquidity has vague meaning. This is because, if a person is having a short term deposit or a treasury bill, he enjoys interest and at the same time, he gets liquidity too from them. So, the concept of liquidity is confusing, as it could not correctly explain interest, as the reward for parting with liquidity.
- ☆ According to some critics, interest is not the reward for parting with liquidity, but it is the reward paid to the lender, since capital is productive.
- ☆ This theory does not explain the existence of different rates of interest prevailing in the market at the same time.

- ✰ This theory explain the determination of rate of interest, when both creditor and debtor are different persons and it does not offer explanation for the determination of rate of interest, when both debtor and creditor is the same person.
- ✰ Some critics opined that, demand for money arises not only from the three motives (transactions, precautionary and speculative motives) as suggested by Keynes, but also other factors like business expansion motive, deflationary motive etc., also influence it.

Considering the above criticisms, we can infer that, Keynesian theory is also indeterminate like earlier Classical and Neo-Classical theories and further, it offered inadequate explanation about the determination of the rate of interest, considering monetary factors alone and neglecting the real factors.

Superiority of Keynesian Theory Over Classical Theory of Interest

The following are the meritorious features of Keynesian theory over Classical theory of interest.

- ✰ Classical theory assumes full employment equilibrium in the economy, which is not realistic. But, Keynesian theory assumes less than full employment equilibrium in the economy, which is more realistic.
- ✰ According to Classical theory, money will be demanded only for transactions and precautionary motives and speculative demand for money was not considered. But, Keynes has assigned importance for speculative demand for money along with transactions and precautionary motives.
- ✰ According to Classical theory, the mechanism of equality between saving and investment is brought by the rate of interest. But, according to Keynesian theory, it is the level of income and not the rate of interest that brings the equality between saving and investment in the economy.
- ✰ According to Classical theory, savings automatically leads to investment. But, according to Keynes, investment leads to savings due to the increase in income through multiplier effect.
- ✰ Classical theory explains the concept of determination of rate of interest only in the long run and not in short run, but Keynesian theory explains the same only in the short run and not in the long run.
- ✰ According to Classical theory, saving is a virtue, as it serves as a source of capital. But, Keynes considered savings as a vice, as if one person saves, the other person's income will get reduced.
- ✰ According to Classical theory, the available capital in the economy is due to thrift habits of the community and it refers to savings. But, according to Keynes, the available capital in the economy infers liquidity preference on the part of the individuals to fulfill the three motives *viz.*, transactions, precautionary and speculative motives and thereby, they represent total demand side for money.
- ✰ Classical Economists simply considered money as a medium of exchange. But, according to Keynesian theory, money serves as a medium of exchange and further recognized hoarding function of money (for precautionary motive).
- ✰ According to Classical theory, when the investment greatly exceeds the savings, the rate of interest tends to be zero. But, according to Keynes, there is a certain minimum level of rate of interest, below which, the interest rate will not fall and at this minimum interest rate, the LPC will be perfectly elastic. So, according to Keynesian theory, the rate of interest will not fall to zero.
- ✰ Classical theory has considered only real factors like productivity, thrifts etc., in explaining the concept of determination of rate of interest. But, according to Keynes since 'interest' is purely a monetary phenomenon, he neglected real factors and considered only monetary factors in explaining the determination of rate of interest.
- ✰ Classical theory explained the concept of determination of rate of interest from the point of view of the borrower. But, Keynesian theory explained the same from the point of view of the money lender.
- ✰ According to Classical theory, interest is the price paid for the use of capital. But, according to Keynesian theory, interest is the reward for parting with liquidity for a specified period of time.

Superiority of Keynesian Theory Over Neo-Classical or Loanable Funds Theory of Interest

The following are the meritorious features of Keynesian theory over Neo-Classical theory of interest:

- ✰ Neo-Classical theory explains the determination of rate of interest over a period of time, where the supply of money is supposed to be fluctuating. But, according to Keynesian theory, supply of money in the economy is fixed and thereby, it explains the concept of determination of rate of interest at a point of time. Thus, Neo-Classical theory is a flow concept, whereas Keynesian theory is a static concept in explaining the determination of rate of interest.
- ✰ According to Neo-Classical theory, rate of interest is a reward for the use of loanable funds. But, according to Keynesian theory, interest is the reward for parting with liquidity for a specified period of time.

- ☆ Neo-Classical theory has not considered a minimum limit for the rate of interest, whereas Keynesian theory considers a certain minimum level of the rate of interest below which it will not fall.
- ☆ In Neo-Classical theory, the demand for loanable fund is studied in terms of investment, consumption and hoarding of liquid cash and it fails to recognize the speculative demand for money. But, Keynesian theory gave prime importance for speculative demand for money along with transactions and precautionary motives and emphasizes the role of expectations of the people about the rate of interest in the near future.
- ☆ Neo-Classical theory combined real factors and monetary factors in explaining the concept of determination of rate of interest. But, liquidity preference theory considered only monetary factors and ignored real factors in explaining the concept of determination of rate of interest.
- ☆ Loanable funds theory assumes full employment equilibrium in the determination of rate of interest and it is unrealistic. But, Keynesian theory explained the concept assuming less than full employment equilibrium, which is realistic.
- ☆ According to Keynesian theory, the supply of money in the economy is given and it is interest-inelastic. But, according to Loanable funds theory, savings, dishoarding, bank money and disinvestment influence the supply of loanable funds in the economy and it varies directly with the rate of interest.

8. Modern Theory of Interest

Modern theory is also called as Neo-Keynesian theory and it was proposed by Hicks and Hansen. They combined the Classical theory and Liquidity preference theory by taking into consideration the real factors of the former and monetary factors of the latter and explained the concept of determination of rate of interest. According to this theory, there are four determinants of the rate of interest *viz.*, savings and investment (from Classical theory), liquidity preference function or demand for money function and supply of money function (from Liquidity preference theory). These four variables of the earlier two theories are combined to construct two new curves called Investment-Saving (IS) curve and Demand for money and Supply of money (LM) curve. The IS curve shows the equilibrium in the real sector (product market) and LM curve shows the equilibrium in the monetary sector (money market) and thereby, the equilibrium between IS curve and LM curve will provide a determinate solution for the equilibrium rate of interest in the economy. Thus, at the equilibrium rate of interest, savings will be equal to investment and total demand for money will be equal to total supply of money.

Derivation of IS Curve

IS curve represents all possible combinations of rates of interest and levels of income at which the saving and investment are equal in the economy, such that, the real sector (product market) is in equilibrium. The derivation of IS curve is shown in the Figure 27.11.

- ☆ In Panel A, the savings curve 'S' is an increasing function of income. This curve is drawn on the Keynesian assumption that, rate of interest has little effect on saving.

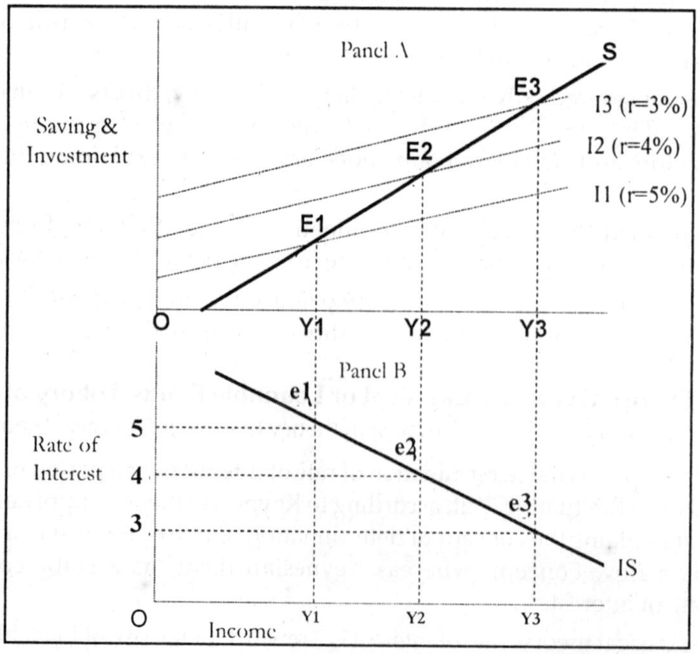

Figure 27.11: Modern theory of interest -Derivation of IS curve.

☆ Given a rate of interest, the investment increases with increase in income and we move on the same investment curve.

☆ If there is change in rate of interest, investment also changes and hence, there is a shift in the investment curve. For example, in Panel A, at four per cent rate of interest, I_2 is the investment curve. At this rate of interest, investment increases with the increase in income. If the rate of interest is decreased to three per cent, the investment curve will shift upward (*i.e.*, I_3). Similarly, when the rate of interest is raised to five per cent, the investment curve will fall to I_1.

☆ In Panel B, we draw the IS curve, by considering the level of income at different interest rates and any point on the IS curve represent a level of income at a particular rate of interest, at which saving and investment are equal.

☆ For example, in Panel A, if the rate of interest is 5 per cent, the 'S' curve intersect the I_1 curve at E_1, which determines OY_1 level of income in the economy. Similarly, at 4 per cent rate of interest, the 'S' curve intersects the I_2 curve at E_2, which determines the OY_2 level of income. At 3 per cent rate of interest, the 'S' curve intersects the I_3 curve at E_3, which determines the OY_3 level of income.

☆ If the three equilibrium points E_1, E_2 and E_3 are extended downwards into Panel B, they touch the points of e_1, e_2 and e_3 respectively representing 5, 4 and 3 percents of rate of interest at which the income levels in the economy are Y_1, Y_2 and Y_3 respectively. These three points e_1, e_2 and e_3 when connected represent IS curve.

☆ This IS curve will slope downward from left to right indicating that, as the rate of interest falls, investment increase and hence income also increases. So, there is a negative relationship between income and rate of interest in the real sector or product market of the economy.

Derivation of LM Curve

LM curve shows all possible combinations of interest rates and levels of income at which the demand and supply of money are equal, such that, the money market is in equilibrium. The LM curve is derived from the Keynes LPC (implying demand for money) and supply curve of money. The supply curve of money is perfectly interest inelastic and hence, it is parallel to Y-axis (Panel A of the Figure 27.12). The LPC (demand curve for money), is convex to origin, implying that, at low rates of interest, the liquidity preference will be more in the economy and *vice versa*. When the total demand for money increases in the economy, it leads to an upward shift in the LPC curve from its original position. When the total demand for money decreases in the economy, it leads to a downward shift in the LPC curve from its original position. Thus, when the total supply of money in the economy is constant, an increase in the total demand for money, will increase the equilibrium rate of interest and a decrease in the total demand for money will decrease the equilibrium rate of interest. The same is explained through the Panel A of the Figure 27.12.

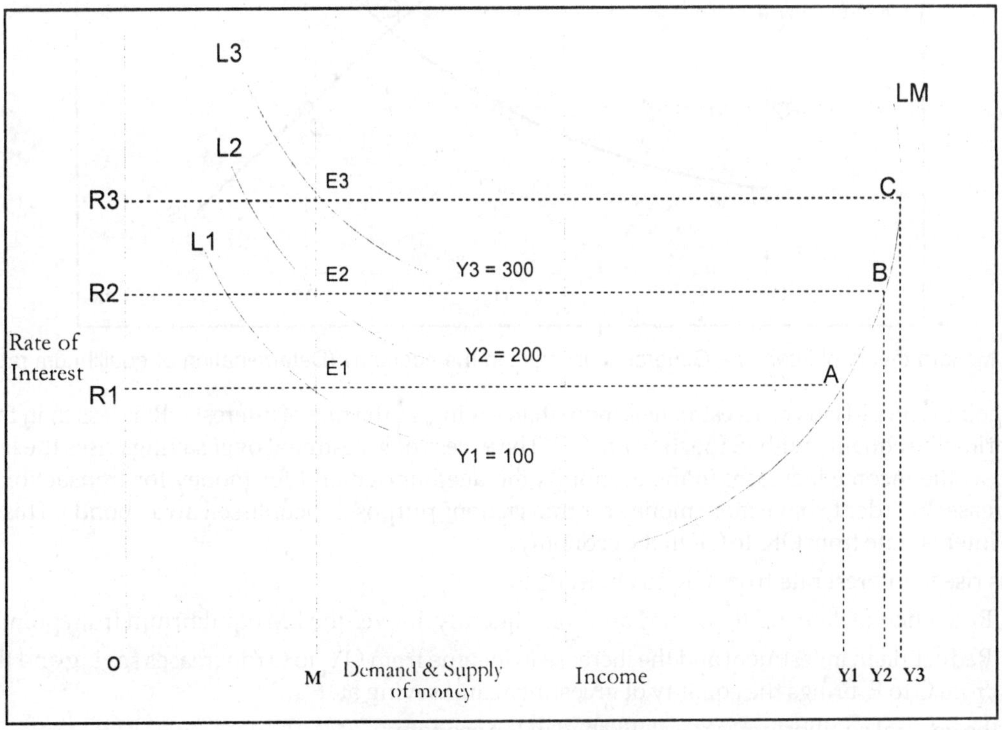

Figure 27.12: Modern theory of interest -Derivation of LM curve.

☆ In the Panel A, SS is the money supply curve and LPC_1, LPC_2 and LPC_3 are the LPCs.

☆ If the LPC is LPC_1, this demand for money equals the supply curve at E_1, *i.e.*, at OR_1 rate of interest. That means, at OR_1 rate of interest, the demand for money is equal to supply of money in the economy. If the LPC is LPC_2, it touches the supply curve at E_2 *i.e.*, at OR_2 rate of interest. Similarly, LPC_3 touches the money supply curve at E_3 *i.e.*, at OR_3 rate of interest.

☆ This information from supply curve of money, LPCs, interest rates and equilibrium points provide the data for drawing LM curve (Panel B). From the point E_1 of Panel A, where LPC_1 intersects the SS curve, we extend a dashed line horizontally to the right, and that point is A. Similarly, B, C points are determined from E_2 and E_3 equilibrium points respectively. By connecting these points A, B and C we get the LM curve. So, this LM curve shows different levels of income in the economy at different interest rates and at any point on the LM curve, demand for money is equal to supply of money.

Determination of Equilibrium Rate of Interest or General Equilibrium of Real Sector and Monetary Sector

So far, we studied the equilibrium conditions of real sector (product market) and monetary sector (money market) separately in terms of IS and LM functions respectively. When these two sectors are brought into simultaneous equilibrium, it is termed as General equilibrium in the economy. In other words, the point of intersection of both IS and LM functions, indicate the point of general equilibrium in the economy. This is explained through the following cases:

Case 1

☆ In the Figure 27.13, let us consider point A on the LM curve, indicating that, money market is in equilibrium at OY_1 level of income and at OR_2 rate of interest.

☆ But at this OY_1 level of income, the real sector or product market (IS function) is not in equilibrium, as OR_2 rate of interest is lower. At OY_1 level of income, the real sector or product market (IS function) will be in equilibrium at OR_1 rate of interest, *i.e.*, at higher rate of interest corresponding to point B on the IS curve.

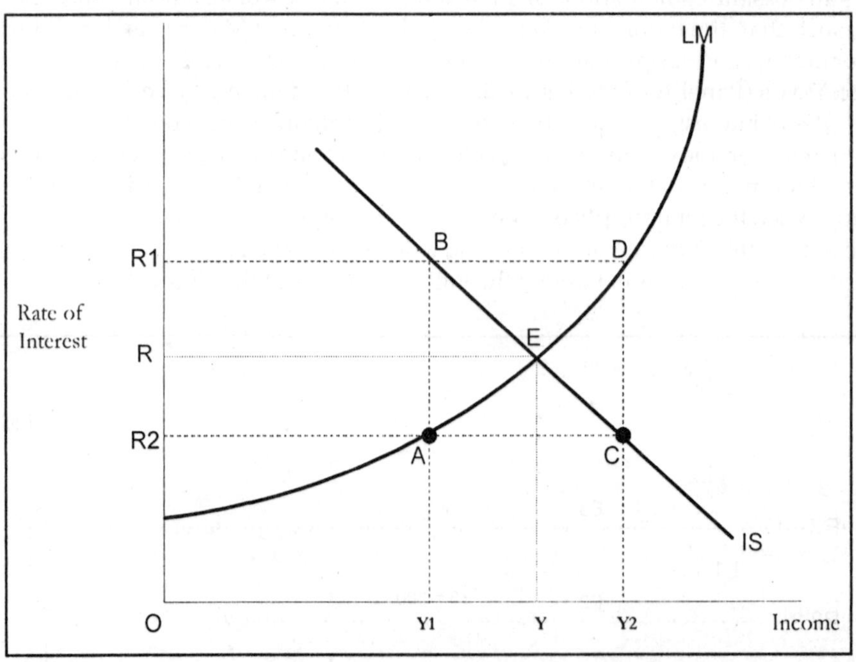

Figure 27.13: Modern theory of interest – General equilibrium in the economy (Determination of equilibrium rate of interest).

☆ At point A on LM curve, investment is more than saving, as the rate of interest OR_2 is less than that is required to bring the equality with IS function *i.e.*, OR. The excess of investment over savings rises the level of income and as the income increases in the economy, the need or demand for money for transactions motive will increase. In order to have more money for transactions purposes, people sell away bonds. This tends to raise the interest rate from OR_2 to OR in the economy.

☆ This rise in interest rate from OR_2 to OR leads to

- ❖ Reduction in demand for money and consequently, moves the LM equilibrium from point A to E
- ❖ Reduction in investment and the increase in income from OY_1 to OY increases saving and the movement from C to E, brings the equality of investment and saving at 'E'.

☆ So, the general equilibrium is re-established in the economy.

Case 2

- Now, let us consider point C on the IS curve, where the real sector or product market is in equilibrium at OR_2 rate of interest and at OY_2 level of income.
- But at OY_2 level of income, the money market is not in equilibrium, as OR_2 rate of interest is lower. At OY_2 level of income, the money market (LM function) will be in equilibrium at OR_1 rate of interest *i.e.*, at higher interest rate corresponding to point 'D' on the LM curve.
- At point C on the IS curve, there is excess of demand for money (L) than the supply of money (M) because, point C reflects lower rate of interest OR_1, than it is required (OR) to bring equality with LM function. This excess demand for money at OR_2 rate of interest leads the people to sell bonds. But, there is less demand for bonds at lower rate of interest and this tends to raise the interest rate from OR_2 to OR.
- This rise in interest rate from OR_2 to OR leads to
  - Fall in investment and this leads to fall in income from OY_2 to OY which, in turn, reduces saving. This process makes the movement from C to E on IS curve and ultimately brings the equilibrium of the real sector or product market at point 'E', where investment is equal to savings.
  - Reduction in income from OY_2 to OY_1 leads to fall in demand for money for transactions motive and this indirectly implies increase in supply of money and thereby, we move from point A to E, where demand for money is equal to supply of money.
- So, the general equilibrium is re-established in the economy.

Changes in General Equilibrium

The general equilibrium of both product and money markets discussed above is based on the static equilibrium analysis. But, the general equilibrium of income and interest rate may change either due to shift in IS or LM functions or change in both functions simultaneously. Let us discuss one by one here under.

Case 1-Shifts in IS Curve

Let us consider the disequilibrium position, where there is shift in IS curve, but LM function is constant.

- ***Increase in investment:*** With the increase in (autonomous) investment (or reduction in savings), the IS curve shift to right as IS_1 curve and the new equilibrium E_1 is established, which indicates a higher level of income OY_1 at higher rate of interest OR_1 (Figure 27.14).
- If the increase in investment is at the same rate interest OR, the increase in income is from OY to OY_2. But, this much increase in income is not possible because, with the rise in income, the rate of interest in the economy will also increase to OR_1 and hence, the equilibrium level of income is only at OY_1. Thus, the expansionary effect of increase in investment is limited by the rise in interest rate and hence, the income rises by less than the full multiplier.

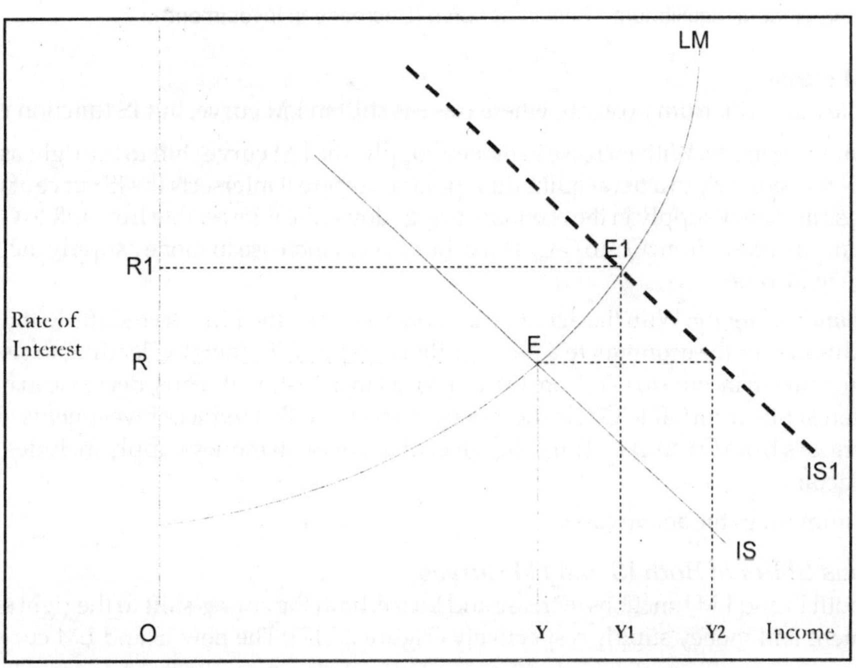

Figure 27.14: Changes in General equilibrium - Shift in IS curve (Increase in investment).

- ☆ But, if full expansionary effect of rise in income (OY_2) is experienced at lower rate of interest (OR), it will lead to inflationary rise in prices in the economy. To check this, interest rate rises in the economy, thereby, savings in the economy increases leading to leakage of multiplier process and hence, income rises by less than full multiplier effect.
- ☆ *Decrease in investment:* With the decrease in (autonomous) investment, (increase in saving), the IS curve shift towards left as IS_2 curve (Figure 27.15), and the new equilibrium E_2 is established, which indicates a lower level of income OY_3 at lower rate of interest OR_2.
- ☆ If the decrease in investment is at the same rate of interest OR, the fall in income is upto OY_4. But, this much fall in income from OY to OY_4 is not possible because, with the fall in income, gradually interest rate is decreased from OR to OR_2, to encourage investment in the economy. So, due to slight increase in investment, the income rises from OY_4 to OY_3, and thereby, the drastic fall in income can be avoided to some extent. So, due to fall in interest rate, the fullest operation of backward multiplier process is checked and hence, the new equilibrium level of income is OY_3. Thus, the contraction effect of fall in investment to some extent is protected by gradual fall in interest rate, thereby, income falls slightly from OY to OY_3 rather than from OY to OY_4.

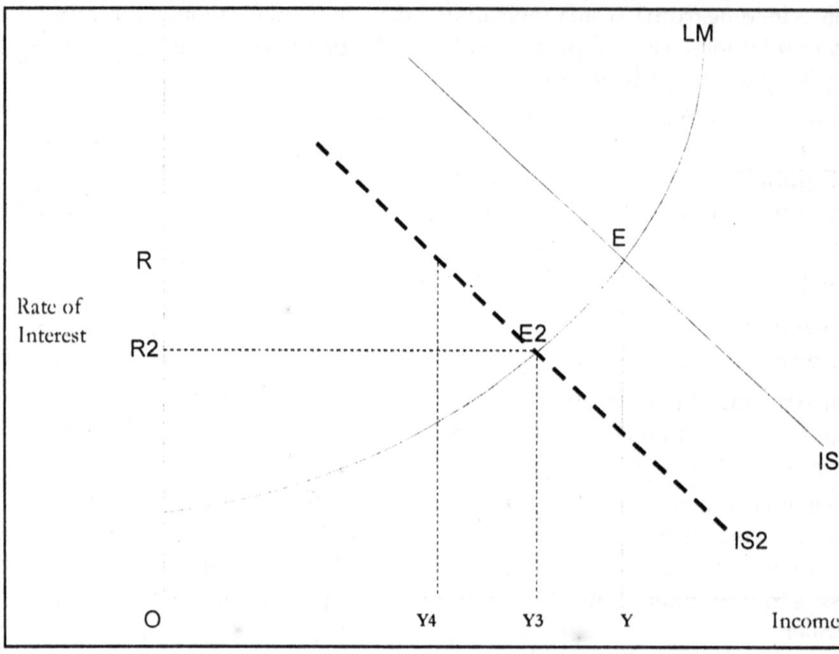

Figure 27.15: Changes in General equilibrium - Shift in IS curve (Decrease in investment).

Case 2-Shifts in LM curve

Let us consider the disequilibrium position, where there is shift in LM curve, but IS function is constant.

- ☆ *Increase in money supply:* With increase in money supply, the LM curve shift to the right as LM_1 (Figure 27.16) and it moves the economy to a new equilibrium point E_1, where it intersects the IS curve at OR_1 rate of interest. Thus, increase in money supply in the economy brings down the interest rate from OR to OR_1 and thereby, the level of income increases from OY to OY_1. Thus, the effect of increase in money supply includes fall in interest rate and rise in income.
- ☆ *Decrease in money supply:* With the decrease in money supply, the LM curve shifts to the left as LM_2 (Figure 27.16) and this moves the economy to a new equilibrium point E_2, thereby, leading to rise in interest rate to OR_2 and hence, fall in income to OY_2 from OY due to fall in investment. Thus, decrease in money supply leads to rise in interest rate from OR to OR_2 in the money market and this reduces investments, thereby, the level of income decreases from OY to OY_2. Thus, the effect of decrease in money supply includes rise in interest rate and fall in income.

The Table 27.1 summarizes the above cases.

Case 3-Simultaneous Shifts of Both IS and LM Curves

Let us suppose, both IS and LM functions increase and hence, both the curves shift to the right simultaneously due to increase in investment and money supply respectively (Figure 27.17). The new IS and LM curves are IS_1 and LM_1 curves.

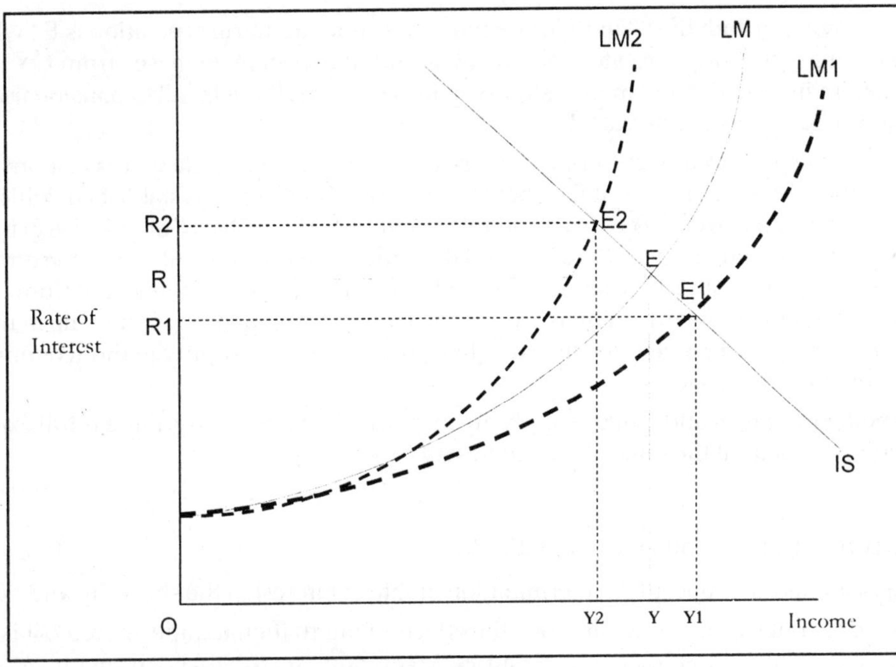

Figure 27.16: Changes in General equilibrium – Shift in LM curve.

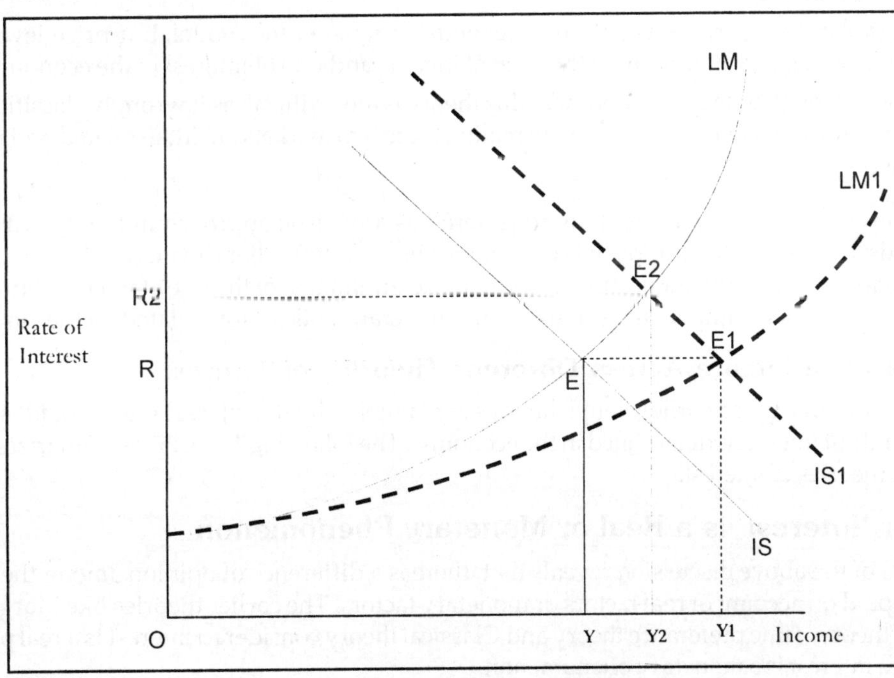

Figure 27.17: Changes in General equilibrium – Simultaneous shifts in IS and LM curve.

Table 27.1: Effects of changes in investment and money supply on the general equilibrium in the economy.

S.No.	*Cause*	*Shifts in Curve*	*Direction*	*Impact*
1.	Increase in investment	IS	Shift to right	Income rises Interest rises
2.	Decrease in investment		Shift to left	Income falls Interest falls
3.	Increase in money supply	LM	Shift to right	Income rises Interest falls
4.	Decrease in money supply		Shift to left	Income falls Interest rises

- ☆ Due to the increase in both the IS and LM functions, the new equilibrium position is E_1, where, both IS_1 and LM_1 curves intersect at original rate of interest OR. But, the income increases from OY to OY_1. So, due to increase in both investment and money supply in the economy, there is full expansionary rise in income in the economy at same rate of interest OR.
- ☆ Had there is no change in money supply (*i.e.*, assume old LM curve), the increase in investment (new IS_1 curve) will touch the old LM curve at E_2, where the general equilibrium is established, which implies, there is increase in both income (from OY to OY_2) and rate of interest (OR to OR_2). But, this rise in income from OY to OY_2 is slowed down compared to OY to OY_1 and this is due to rise in interest rate in the economy from OR to OR_2. However, if money supply is increased, so as to prevent the rise in interest rate from OR to OR_2, it leads to shift in the LM curve towards right as LM_1 and it leads to full expansion of income in the economy. That means, Y_2Y_1 increase in income is possible due to increase in money supply in the economy, which induces more investment in the economy.
- ☆ So, when both investment and money supply are increased in the economy, there is full expansionary rise in income in the economy at the same rate of interest OR.

Criticisms

This theory too suffers from the following limitations:

- ☆ This theory explains the concept of determination of rate of interest in the short run and not in the long run.
- ☆ This theory does not take into consideration the effects of international trade, thereby, it is a closed model.
- ☆ This theory assumed interest rates are too flexible, but if the interest rate is not so flexible, automatic adjustments will not take place.
- ☆ This theory considers price level as an exogenous variable in the model, but price level is an important endogenous variable in determining the level of income and rate of interest in the economy.
- ☆ According to Don Patinkin and Friedman, this theory is too artificial, as it wrongly classified the market into product market and money market. But, in reality, these two markets are inter-related and interdependent of each other.

Despite the above limitations, this theory is considered as a realistic approach in the determination of rate of interest since, it finds a determinate solution. It is rated superior over the earlier Keynesian theory, in the sense that, it takes into consideration both real and monetary factors in the determination of the rate of interest. Further, it could offer useful explanation regarding the implications of fiscal and monetary policies formulated by the Government.

27.2.1. Comparative Picture Across Different Theories of Interest

Among the theories discussed earlier, some theories explained 'why the interest is to be paid' and some theories explained 'how the rate of interest is determined in the economy'. The following Table 27.2 summarizes and differentiate across the different theories of interest:

27.3. Whether 'Interest' is a Real or Monetary Phenomenon

A close perusal of the above discussion reveals that, there is a difference of opinion among the Economists that, whether interest is paid on account of real factors or monetary factors. The earlier theories like Marginal productivity theory, Abstinence theory, Time preference theory and Classical theory considered interest is a real phenomenon. But, according to Keynes interest is a monetary phenomenon.

In fact, both these view points are correct, as interest is determined by both real factors (savings, productivity, waiting etc) and also by monetary factors (investment, consumption expenditure etc). Modern theory of interest explained the phenomenon of determination of rate of interest considering both real and monetary factors and provided a determinate solution. So, interest is both a real as well as monetary phenomenon.

27.4. Why Interest Rates Differ

In practical sense, we see different interest rates in the money market at the same time and the following are the major reasons:

- ☆ ***Period of loan***: Compared to short term loans, long term loans enjoy higher interest rate, as longer is the time period, more is the risk borne by the lender.
- ☆ ***Type of security*** : Loans obtained based on easily realizable securities like gold, government bonds etc., are charged with lower interest rates compared to the loans obtained based on the securities like land, buildings etc., as they cannot be easily realized.

Table 27.2: Comparative picture across different theories of interest.

S.No.	Theory	Proposed by	Concept	Explanations offered for
1.	Marginal productivity theory	Malthus, J.B. Say and Carey	Interest is paid for the use of capital because, capital is productive.	It offers explanation for 'why interest is to be paid'.
2.	Abstinence theory	Nasau Senior and J.S.Mill	Interest has to be paid for the capitalist to compensate the sacrifice he makes for abstaining the present consumption needs to the future.	It offers explanation for 'why interest is to be paid'.
3.	Agio or Austrain theory	Bohm Bawerk	Since people prefer present consumption over future consumption, an 'agio' or premium arises on the present consumption goods.	It offers explanation for 'why interest is to be paid'.
4.	Time preference theory	Irving fisher	The people are rewarded with interest to encourage the postponement of present consumption to future consumption.	It offers explanation for 'why interest is to be paid'.
5.	Classical theory	Pigou, Knight and Taussig	Equilibrium or natural rate of interest is determined based on the interaction between demand for (investment) and supply of (saving) capital.	It offers explanation 'how the rate of interest is determined'.
6.	Neo-Classical or Loanable funds theory	Knut Wicksell, Robertson, Pigou, Myrdal, Ohlin etc.	Equilibrium or market rate of interest is determined based on the interaction between demand for and supply of loanable funds.	It offers explanation 'how the rate of interest is determined'.
7.	Liquidity preference theory	J.M. Keynes	Equilibrium rate of interest is determined based on the interaction between demand for and supply of money.	It offers explanation 'how the rate of interest is determined'.
8.	Modern theory	Hicks and Hansen	Equilibrium rate of interest is determined based on the interaction between IS and LM curves.	It offers explanation 'how the rate of interest is determined'.

- *Imperfect market*: The presence of imperfect market conditions in the loan market leads to greater variability in the interest rates. For example, in India, private money lenders charge higher interest rates over co-operative and commercial banks for the same purposes.
- *Demand and supply conditions for capital*: The demand and supply conditions for capital are different across the places, thereby, interest rates differs.
- *Different purposes*: Depending upon the purposes for which the borrower avails the loans, interest rates will fluctuate. For self-liquidating loans, the interest rates are lower compared to partial liquidating loans.
- *Immobility of capital*: Due to immobility of capital across the places, different interest rates will prevail in the money market at the same time.
- *Repayment capacity of the borrower*: If the borrower has sufficient repayment capacity, the lender charges less rate of interest from him.
- *Social status of the borrower*: If the borrower has good social status, honesty etc., he will be charged with less rate of interest.
- *Productive use of investment*: If the lender feels that, the borrower will use the loan amount for the said productive purpose and will not divert the same towards unproductive purposes, he will charge less rate of interest.

27.5. Whether Interest Rate can Fall to Zero

Interest rates in the money market will never fall to zero. The following are the two possible circumstances under which the interest rate will fall to zero:

- When consumer spends all his income on the consumption of goods and services, savings will be zero. When savings are nil, there is no investment in the economy and when there is no investment, there is no interest.
- When the capital supply is plenty to the investors in the economy, they are least bothered about the marginal productivity of capital. In the long run, the investment opportunities will decline, thereby, the rate of interest will fall to zero.

But, the above two circumstances cannot be seen in the economy. Considering the futures needs, the individuals will definitely save the money and future investment opportunities will never decline, that too at the cost of zero marginal productivity of capital. So, the interest rates in the economy will never fall to zero.

28

Profits

Having discussed about the remunerations paid to the land, labour and capital factors, let us analyze the remuneration paid to the entrepreneur. The entrepreneur will receive remuneration in the form of profits, as he is one among the factors of production employed in the business. According to Prof. Taussig, this concept of profits has become a *'mixed and vexed one'*. We know that, the entrepreneur performs several functions *viz.*, planning, organization, direction, control, monitoring, coordination etc., in the business. Further, he will own the capital in the business. But, we cannot decisively say, for which function he is receiving the payment, thereby, profits have become a mixed income for the entrepreneur. Further, there is no unanimity among the Economists in defining 'profits', thereby, it has become a vexed one. So, Prof. Knight pointed out that *'no term or concept in economic discussion is used with a more bewildering variety of well established meaning than profit'*. Several Economists have defined 'profits' in different ways and the following are the popular definitions:

'Profits may be considered a reward for making innovations, a reward for accepting risks and uncertainties and the result of imperfections in the market structure'

– (Henry Grayson)

'The residual payment, what is left to the producer's income after all other payments have been met'.

– (Hansen)

'The surplus of current income over past cost is profit'.

– (Drucker)

'The average remuneration necessary to bring into existence and to keep in existence, a sufficient supply of entrepreneurs'.

– (Marshall)

'Profit is a determinate return for a production function performed by the entrepreneur with a superior ability than others'.

– (Walker)

From the above definitions, we can have the following meanings for the term 'profit':

- Profit is the compensation received by the entrepreneur for performing managerial functions.
- Profit is the reward earned by the entrepreneur for bearing the risk in the business.
- Profit is the residual income for entrepreneur after paying off to other factors.

☆ Profit is the difference between total sales proceeds obtained and total expenses incurred in the production programme.

Thus, there are different ways of interpreting the term 'profit' and all the interpretations explain the profits in relation to the functions performed by the entrepreneur.

28.1. Different Types of Profits

The term 'profit' is used in different senses and they are discussed here under.

1. Economic Profit and Accounting Profit

Economic profit refers to the difference between TR and total expenditure. In economic sense, the total expenditure includes both explicit costs and implicit costs. Thus,

Economic profit = TR – Total expenditure

= TR – (Total explicit cost + Total implicit costs)

But, in accounting sense, profits refer to the difference between TR and total explicit costs. So, while computing accounting profits, implicit costs are not considered. Thus, Accounting profit = TR – Total explicit costs.

2. Gross Profit and Net Profit

Gross Profit

The term 'profit' generally refer to 'gross profit'. That means the excess of income earned by the entrepreneur over the total expenses incurred in the production programme refers to gross profit. But, this 'gross profit' will include several items, which are not profits in the strict sense and they are discussed here under.

(a) Remuneration for the Factors of Production Contributed by the Entrepreneur Himself

During the course of production programme, the entrepreneur himself contribute his owned land, owned labour and owned capital. So, the implicit returns like implicit rent, implicit wages and implicit interest earned by the entrepreneur will get included in the gross profit. These implicit earnings should be deducted from the gross profits to compute net profits.

(b) Depreciation and Maintenance Charges

During the process of production, the machinery will get depreciated and thereby, the entrepreneur has to meet repairs and maintenance charges on the depreciated machinery. So, these costs should be deducted from gross profits to arrive at net profits.

(c) Extra-personal Profits

The entrepreneur may get some extra profits in the business in the form of monopoly profits due to inflationary rise in prices, due to external causes like wars etc., and all these chance profits should be deducted from the gross profits to arrive at net profits because, these profit elements are not related to the entrepreneurial ability in the strict sense.

The above three elements included in gross profit are relatively stable even in the long run and hence, any significant variation in the gross profit is due to significant variation in the net profit.

Net Profit

Net profit is the exclusive reward paid to the entrepreneur for the functions performed by him in the production programme. The entrepreneur has to perform several functions like planning, organization, coordination of the factors of production, risk and uncertainty bearing, making innovations in the production programme etc., with a view to minimize the cost or maximize the profits in the production programme. So, for executing all the above functions, the entrepreneur is rewarded with 'net profits'. It is essential to note that, in the modern days of business, the entrepreneur will insure the business with an insurance company against various risk factors thereby, risk taking is not the function of the entrepreneur, but by the insurance company. So, uncertainty bearing is one of the important functions borne by the entrepreneur and not risk taking.

From the formulae given below, we can easily differentiate gross profit and net profit accrued to the entrepreneur in conducting the business:

Gross profit = Net profit + implicit rent + implicit wages + implicit interest + depreciation and maintenance charges + non-entrepreneurial profits.

Net profit = Gross profit – (Implicit rent + implicit wages + implicit interest + depreciation and maintenance charges + non-entrepreneurial profits).

So, net profit refers to economic profit or pure business profit, as it is the reward exclusively paid for executing the entrepreneurial functions.

From the above discussion, we can infer the following interesting points:

- ☆ Gross profit includes net profit.
- ☆ Gross profit includes implicit costs whereas, net profit does not includes such costs.
- ☆ Net profit may be positive or negative. Being a residual income, if TC exceeds TR, gross profits will be negative, thereby, net profits will also be negative *i.e.*, loss. Thus, among the four factors of production, only the entrepreneur has the chance of getting the negative reward *i.e.*, losses.
- ☆ Net profit cannot be pre-determined, as it is influenced by the level of output, costs incurred and price of the product in the future. Further, the output in agriculture is influenced by climatic factors, which are not predictable accurately.
- ☆ Net profit is not a fixed remuneration, as it varies with the output, cost structure and price of the output. Thus, this remuneration fluctuates widely.
- ☆ Net profit is a residual surplus *i.e.*, the entrepreneur is rewarded after meeting the remunerations of all the other three factors of production.
- ☆ Net profit is an end payment *i.e.*, it is paid to the entrepreneur after selling the output and after realizing the TR from the business.
- ☆ Net profit is not earned in the long run under perfect competition and the profits earned in the long run include only gross profits (implicit costs).
- ☆ Loss in the business firm is computed with reference to net profits and not in terms of gross profits.

28.2. Different Theories of Profits

Several theories have been propounded to explain the concept of profits and the important theories are discussed in the ensuing pages.

1. Rent Theory of Profits

This theory was initially propounded by J.S.Mill and Senior and later it was developed by F.A.Walker. The following are the important features of this theory:

- ☆ According to Walker, as rent arises for lands due to differences in fertility status, similarly, profits arise for the entrepreneurs due to differences in their abilities. So, profit is the price paid for the ability of the entrepreneur In his words, *'profits are of the same genus as rent'*. That means, like no rent land or marginal land, there exists marginal entrepreneur and for him, the profits are zero, as TR equals to the TC in the production programme. Similarly, like intra-marginal lands, there are intra-marginal entrepreneurs and they enjoy profits compared to marginal entrepreneur, as they differ in their abilities in executing the production programmes.
- ☆ Walker, like Ricardo, opined that, profits are not a part of cost of production, as rent is not a part of cost of production. So, profits, like rent, also does not enter into price and it is a differential surplus like rent.
- ☆ The market price of the product is determined by the cost of production of the product of the marginal entrepreneur.
- ☆ Marginal entrepreneur just receive wages of management called normal profit and they are included in the cost. But, the surplus earned by intra marginal entrepreneurs does not enter the cost.

Criticisms

The following are the criticisms leveled against this theory:

- ☆ The comparison between rent and profit cannot be justified, as rent is a contractual income, but profit is a residual income. Moreover, profits can be negative, but rent cannot be negative. So, when profits are negative, it does not mean that the ability of entrepreneurs is declining.
- ☆ The profits gained in the production programme cannot be attributed solely for the ability of the entrepreneur. The profits may also be due to inflationary rise in prices of the product, situations like war, political disturbances etc.
- ☆ As explained by the Ricardian theory, there may be no rent land, but in the context of profits, the existence of marginal entrepreneur is highly questionable atleast in the long run. That means, we cannot imagine a no-profit entrepreneur in the long run.
- ☆ Rent is a contractual income fixed over a period of time, but profits is a residual income and it fluctuates. So, profits are result of uncertainty in a dynamic state, but rent arises both in static as well as dynamic states due to inelastic supply of land factor.

- We cannot justify marginal entrepreneur is the least efficient entrepreneur.
- It is wrong to say that, profit does not enter into price of the product. This may be true in the short run, but in the long run, profit does enter into the price of the product.
- According to this theory, profit is the price paid for ability. But, this theory could not offer any explanation, how the share holders in a joint stock company are getting profits without showing any business ability.
- We know, scarcity of land exists due to natural limitations, but why the scarcity exists even in case of entrepreneurs and this cannot be explained by this theory.

From the above criticisms, we can infer that, this theory could not offer satisfactory explanation regarding the origin and nature of profits.

2. Wages Theory of Profits

This theory was proposed by Taussig and Devenport. The following are the important features of this theory:

- This theory compared the entrepreneurial service with the labour service in the production programme, as the entrepreneurial service is a specialized form of labour.
- They further opined that, profits are rewarded for the entrepreneurs on account of their special abilities compared to labour.

Criticisms

This theory suffered from the following limitations:

- The entrepreneurial service cannot be compared with labour service, as entrepreneurs will take more risks in the business, unlike labour, as they get fixed payments.
- The profits of the entrepreneurs will fluctuate greatly, but the labour will get only fixed wages even when the business prospers.
- Prices of the products shows positive influence on profits, but they do not influence the wage structure of the labour.
- If the TC of the firm exceeds the TR, the profits will be negative, but the wages cannot be negative.
- According to this theory, profits are rewarded for the entrepreneurs for their special abilities. But, this theory does not offer explanation, how the shareholders in a joint stock company are getting profits without exhibiting special abilities.
- The entrepreneurs will work more mentally and less physically whereas, labour will execute the work more physically and less mentally.

3. Risk Theory of Profits

This theory was proposed by Prof.H.B.Hawley. The following are the important features of this theory:

- Profit is the reward for 'risk taking' in the business.
- Risk taking is an important function of the entrepreneur, for which he is rewarded with profits. Since, the risk is rewarded, the entrepreneur will take the risks in the production programme and the higher the risk, the greater must be the possibility of getting the profit. According to Hawley, '*the profit of an undertaking, of the residue of the product after the claims of land, capital and labour (furnished by others or by the undertaker himself) are satisfied, is not the reward of management or coordination, but of risks and responsibilities that the undertaker subjects himself to*'.
- No entrepreneur will be willing to take risks, if he gets a normal return from the business. That means, the reward for risk taking must be higher than the value of risk borne by the entrepreneur.
- According to Hawley, the entrepreneur faces four types of risks:
  - ***Replacement:*** It is, otherwise, called as depreciation and this will be included in the fixed costs of the production programme.
  - ***Risk proper*** : It is the risk of marketability of the product.
  - ***Uncertainty***: It is the non-quantifiable risk, which cannot be predicted or measured.
  - ***Obsolescence***: It is not calculable because, it is difficult to anticipate the technical progress.

All the above four types of risks will be borne by the entrepreneur and it is for undertaking these risks, the entrepreneur is rewarded with profits. So, profits are also called as 'Costs of staying in the business'.

Criticisms

The following are the criticisms leveled against the theory:

- ☆ Profits arise not only by taking risks in the business, but it is a mixed income due to ability, managerial powers, proper-decision making etc.
- ☆ There is no exact functional relationship between amount of risk undertaken and extent of profits derived. That means, risk taking always will not lead to profits.
- ☆ According to Prof. Knight, all the profits earned by the entrepreneur cannot be attributed for risk taking because, some risks can be transferred by insuring them.
- ☆ According to Carver, profits are not because, entrepreneurs efficiently tackle the risks, but because, they avoid or reduce the risks.

Though, the above criticisms leveled against this theory are appropriate, still it cannot be denied the fact that, profits is the reward for risk taking.

4. Uncertainty Bearing Theory of Profits

This theory was proposed by Frank H. Knight in his work, '*Risk, Uncertainty and Profit*' and he criticized the risk theory of profits that, profits earned by the entrepreneur cannot be attributed for risk taking because, some risks can be transferred. According to Knight, there are two types of risks *viz.*, insurable risks and non insurable risks. In the former type, the entrepreneur will insure the risks with an insurance company and thereby, he transfer the risk and hence, in this case, the risk is actually borne by the insurance company and not by the entrepreneur. For example, the risks in the business like fire accident, theft etc., are insurable. However, there are certain types of risks, which cannot be quantified or measured and thereby, they cannot be insured. Hence, they are called non-insurable risks or uncertainties. Since, (insurable) risks are transferred to the insurance company by insuring them, the entrepreneur has to face only uncertainties in the business and thereby, profit is the reward for bearing these uncertainties or non-insurable risks. Some of the examples of non-insurable risks are price uncertainty about the product, technological uncertainty, uncertainty behaviour of the competitors, uncertainty regarding the State interventions in the production and marketing aspects of the products, out break of war, earth quakes, change in consumer's tasks and preferences etc. Prof. Knight explained undesirable risks thus '*the modern economic order is built around the concept of enterprise the correlate of which life is necessary involves much uncertainty referred to as profit system*'. Thus, this theory assumes 'uncertainty' bearing is the unique factor giving rise to profits.

Criticisms

The following are the criticisms leveled against this theory:

- ☆ Besides uncertainty bearing, even other factors like ability, planning, coordination, bargaining power etc., fetches profits to the entrepreneur. Even for computing profits due to handling of uncertainty factors, they cannot be quantified.
- ☆ A monopolist can maximize the profits when he faces least uncertainty in the business.
- ☆ It is difficult to know that, whether Directors or Managers or the Shareholders, who bears the uncertainty in a joint stock company.

Thus, the Uncertainty bearing theory like the earlier theories does not furnish a comprehensive explanation for profits and thereby, it is inadequate. However, still the concept of uncertainty bearing by the entrepreneur and its influence on the profits cannot be ignored.

5. Dynamic Theory of Profits

J.B.Clark proposed the Dynamic theory of profits in the year 1900. The following are the important features of this theory:

- ☆ According to Clark, profit is the difference between selling price and cost of production of the product *i.e.*, it is the surplus over the costs.
- ☆ He opined that, under static conditions in the economy *i.e.*, no changes in capital investment, technology, managerial aspects of the business, demand and supply conditions in the economy etc., the entrepreneur will not realize real or pure or economic profits in the business. That means, dynamic changes in the above parameters will be the fundamental cause of the emergence of economic profits in the business.
- ☆ In the long run competitive equilibrium, price of the product tends to equal to its ATC, thereby, the entrepreneur gets only wages for his managerial functions and this does not constitute real profit. So, due to dynamic changes in the economy, the price will exceed the ATC and thereby, the entrepreneur will realize real or pure profits.

- Though some frictional profits emerge out due to frictions in the system, but they does not constitute real or pure or economic profits, thereby, profit is only a dynamic surplus in the economy.
- According to Clark, the following five dynamic changes may emerge in the economy that ensure real or pure profits to the entrepreneur:
  - Changes in tastes and preferences of the population and multiplicity of human wants
  - Changes in population
  - Changes in the amount of capital
  - Changes in the techniques of production
  - Changes in the forms of business organization

The above dynamic changes in the economy brings about changes in the demand and supply conditions of the products and thereby, ensures surplus revenue over the costs leading real or pure profits. Further, some entrepreneurs take advantage of these changes and goes for demand driven production and they realize more profits compared to other entrepreneurs, who fail to capture or exploit these changes. Thus, those entrepreneurs, who accept dynamic changes reap real or pure profits and inefficient or careless producers, who do not account for dynamic changes suffer from deriving real profits and sometimes, they even incur losses. Thus, profits will not arise in the static state, but arise due to dynamic changes in the economy.

Criticisms

This theory has been criticized on the following grounds:

- According to this theory, all dynamic changes in the economy will bring real or economic profits. Critics, however, opined that, not all dynamic changes, but only unpredictable changes can bring profits.
- The concept of frictional profits is vague. However, Marshall addressed this by introducing the concept of 'normal profits' that can be derived even in static state.
- Due to dynamic changes in the economy, new risk factors emerge and Clark has not offered any explanation with regard to the possible influences of new risk factors on the real or pure profits.
- Clark has mentioned only five dynamic changes in the economy that ensure real or pure profits to the entrepreneur. But, in practicality, there are many other changes that influence the pure profits in the dynamic economy and they are ignored by Clark.
- According to Taussig, this theory makes a superfluous or unnecessary or artificial distinction between profits and wages of management.
- This theory cannot determine the volume of profits.
- According to Prof. Knight, the society may be dynamic, yet the entrepreneurs may get only normal profits.

Inspite of the above criticisms, the explanation offered by this theory has the validity, as it considered dynamic aspect of the profit.

6. Innovation Theory of Profits

This theory was proposed by Prof. Joseph A. Schumpter. This theory is more or less similar to Clark's Dynamic theory of profits, but instead of five dynamic changes mentioned by the Clark, Schumpter suggested that, the changes caused by the innovations in the production process are responsible for earning real or pure profits in the business. The following are the important features of this theory:

- Profit is the reward for innovations.
- The concept of 'innovations' is used in the broadest sense compared to the five dynamic changes mentioned by the Clark. The innovations include all possible desirable changes in the production process like changes in technology, establishment of new plant, new managerial changes, better quality of raw material, better salesmanship etc., that ensures cost minimization in the production programme and thereby, the entrepreneur derives surplus revenue over the cost.
- Schumpter further explained the difference between invention and innovation. *'Innovation is much more than invention. Invention is not innovation if it is still born, that is if it is not used. An invention becomes an innovation only when it is applied to industrial progress'*. Thus, innovation means a commercial application of scientific inventions. The inventor must be a scientist, whereas, innovator need not be a scientist and the entrepreneur with his extra-ordinary ability can innovate the scientific inventions and thereby, create opportunities for himself to reap real or pure profits in the business.

- ✰ Schumpter assigned 'innovation' is the key function of the entrepreneur and profits are the reward for the innovations the entrepreneur execute in the business programme. According to Schumpter, the following are the five important innovations the entrepreneur can take up to reap real profits in the business.
    - ❖ Production of new products
    - ❖ Improving the techniques of production
    - ❖ Identifying new sources of procurement of raw material
    - ❖ Discovery of new markets
    - ❖ Re-organization of the business

The above five innovations will widen the gap between price and ATC, thereby, the entrepreneur gains real or pure profits in the business. However, such profits exists only in the short run, as in the long run, new firms gets attracted to these profits and enter the industry. So, to enjoy the real or pure profits, even in the long run, the entrepreneur has to search for new innovations on a continuous basis. So, profits are the result of innovations. So, the entrepreneurs, who are more innovative with extra-ordinary abilities will continue to reap real or pure profits in the business even in the long run. Thus, innovational profits appears, disappears by imitation and again re-appears.

- ✰ According to this theory, profits are both the cause and effect of innovations. The lure of profits makes the entrepreneur to innovate in the business. Thus, profits are the cause for innovations. If the entrepreneurs innovate, it will lead to profits. So, profits are the effect of innovations.
- ✰ According to Schumpter, 'innovation' is the function of an entrepreneur and not of the capitalist. This is because, to innovate in the business, the entrepreneur must have sound technical base of managing different activities in the business and for this he gets the capital from the capitalist. The role of the entrepreneur is distinct from the role of the capitalist, as the former simply innovates and does not undertake any risk in the business. But, the capitalist allocates the capital and thereby, risk taking is the function of the capitalist. Even if the entrepreneur performs these two functions *i.e.*, as an innovator and as a capitalist (risk taker), the profits accrued to him will be as a reward for making innovations in the business and not for allocating the capital (risk taking) in the business.

Criticisms

This theory is subjected to criticisms on the following grounds:

- ✰ 'Innovation' is only one of the functions of the entrepreneur and it is not the only function.
- ✰ While adopting new innovations in the business, many number of risk and uncertainty factors arise in the business. But, this theory neglected the risk taking function of the entrepreneur.
- ✰ Monopoly profits do not result from the introduction of innovations.
- ✰ This theory ignores uncertainty bearing as a source of profit.

Considering the above criticisms, we can infer that, the defects of this theory are also similar to Clark's theory, thereby, it is not a dependable and complete theory of profits. Inspite of the above limitations, it indicates a valid point that, innovation functions leads to profits on the part of the entrepreneur.

7. Marginal Productivity Theory of Profits

This theory was proposed by Prof. Chapmann. According to this theory, the MRP of the entrepreneur determines the profits of the business. Higher the MRP of the entrepreneur, higher is the profits. But, this theory was considered highly controversial and offered least explanation to the profits because, it is very difficult to compute the MRP of the entrepreneur. This is because, every business firm will have only one entrepreneur and for the sake of measuring MRP of the entrepreneur, we cannot employ additional entrepreneur or remove the existing the entrepreneur in the business. Further, the MRP of other factors of production also depends upon the managerial ability of the entrepreneur. Even if we still increase the number of entrepreneurs in the business to measure the MRP of each entrepreneur, it will not give the right picture, as entrepreneurs differ in their ability, innovativeness etc. Thus, this theory is criticized on the ground that, it considered only the demand side and did not consider the supply side of the entrepreneurs.

8. Modern Theory of Profits or Determination of Profits Under Perfect Competition

According to this theory, 'profits' as the reward for entrepreneurs is determined by the interaction between demand for and supply of entrepreneurs under perfect competition.

Demand for Entrepreneurs

It is interesting that, in case of land, labour and capital, we can compute the MRP of the factor for a firm. But, in case of entrepreneur as a factor of production, for each firm, there will be only one entrepreneur, thereby, it is difficult to

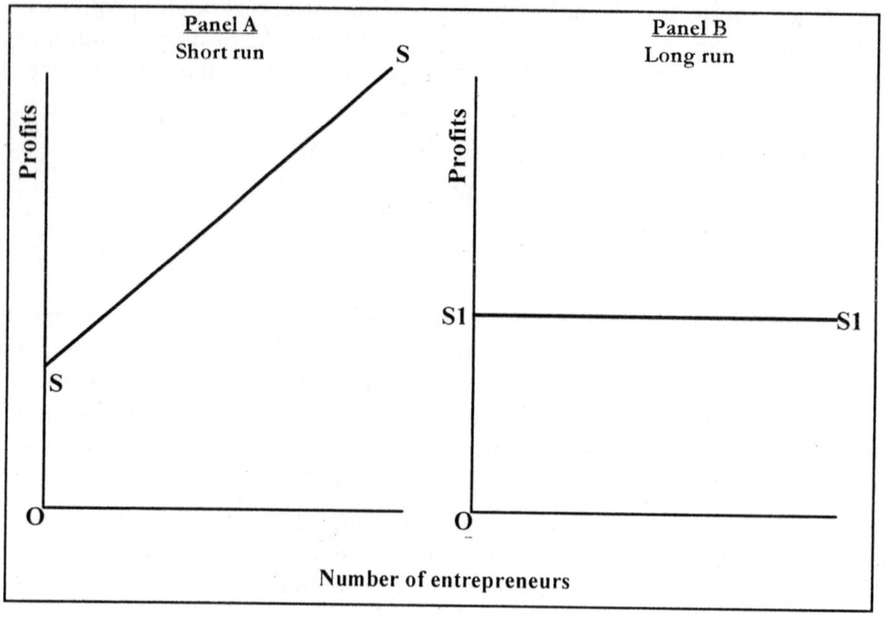

Figure 28.1: Supply of entrepreneurs in the industry.

compute the MRP of the entrepreneur at firm level. However, at industry level, we can compute the MRP of the entrepreneurs and as usual, MRP curve of the entrepreneurs will slope downwards from left to right (Figure 28.2), indicating that, with the increase in number of firms in the industry, the number of entrepreneurs will increase, output increases, supply of the product into the market increases, price per unit of the output will fall, thereby, profits per each entrepreneur will fall. The demand for entrepreneurs in the industry depends upon the following factors:

- ☆ Extent of development of the industry
- ☆ Element of uncertainty in the business
- ☆ Scale of production
- ☆ MRP of the entrepreneur

All the above four factors directly influence the demand for entrepreneurs in the industry and among these four, MRP of the entrepreneur is the important factor. As explained above, MRP curve (demand curve) of the entrepreneurs at the industry level will slope downward from left to right (Figure 28.2).

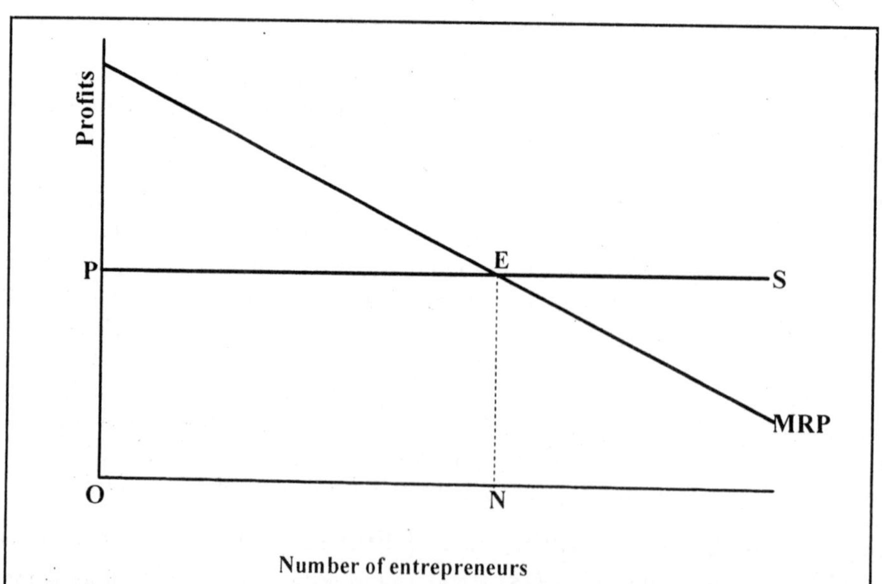

Figure 28.2: Determination of normal profits by the entrepreneurs in the long run in the industry assuming perfect competition.

Supply of Entrepreneurs

The supply of entrepreneurs to the industry depends upon the following factors:

- ☆ Availability of capital
- ☆ Existence of managerial and technical personnel
- ☆ Size of population
- ☆ Number of entrepreneurs
- ☆ Element of uncertainty in the industry
- ☆ Distribution of income
- ☆ Industrial experience
- ☆ Conditions in the society like political stability etc.

In general, profits of the industry directly influence the supply of entrepreneurs to the industry in the short run, and thereby, supply curve of entrepreneurs (SS) will rise upwards (Panel A of the Figure 28.1). But, in the long run, under the assumption of perfect competition, the supply curve of entrepreneurs (S_1S_1) will be horizontal and parallel to X-axis, as all the entrepreneurs are homogenous and of equal efficiency (Panel B of the Figure 28.1). This infers that, in the long run, all the entrepreneurs will earn only normal profits equal to that transfer earnings. So, if the entrepreneur in one industry could not earn even the minimum transfer earnings in the long run, he will quit that industry and move to some other industry to earn normal profits.

Determination of Entrepreneurial Profits in the Long Run in the Industry Assuming Perfect Competetion

Profits are determined at the point, where the demand curve and the supply curve of the entrepreneurs will interest and this is explained through Figure 28.2.

- ☆ The MRP curve (demand curve) of the entrepreneurs intersect the supply curve of labour (S) at E and it implies that, the industry employs ON number of entrepreneurs and each entrepreneur will derive OP level of profits in the long run.
- ☆ This OP level of profits represent the normal profits earned by the entrepreneurs in the long run equal to their transfer earnings.
- ☆ All the entrepreneurs earn same level of normal profits OP because, under perfect competition all the entrepreneurs are considered homogenous in terms of their efficiency.
- ☆ But, in the short run, the entrepreneurs may earn super normal profits. However, these super normal profits are competed away in the long run due to the entry of new firms into the industry (assuming perfect competition).

Determination of Entrepreneurial Profits in the Long Run in the Industry Assuming Imperfect Competetion

In practicality, imperfect competition is more common in the market and thereby, the entrepreneurs can earn super normal profits even in the long run. The same is illustrated through Figure 28.3.

- ☆ In the Figure 28.3, DD is the demand curve and SS is the supply curve of entrepreneurs.
- ☆ These two curves intersect at point E and this indicates that, ON number of entrepreneurs are employed in the industry and each entrepreneur will earn OP super normal profits.

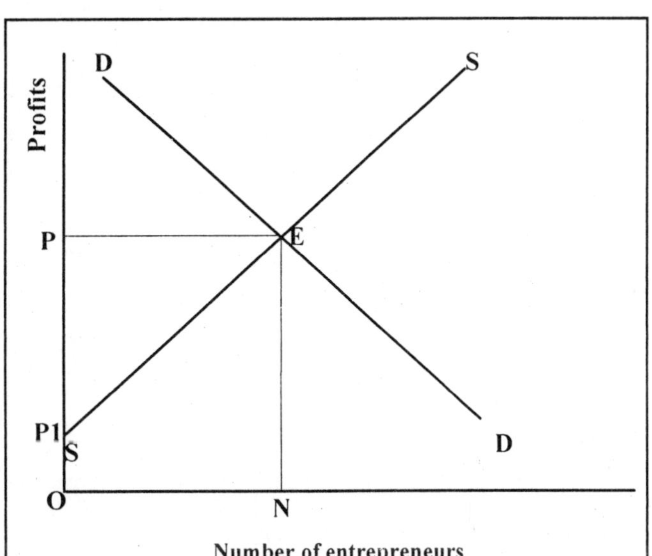

Figure 28.3: Determination of super normal profits by the entrepreneurs in the long run in the industry under imperfect competition.

- In the Figure 28.3, OP_1 is the minimum profit the entrepreneur should get to stay in the industry and if he gets the profits less than OP_1, he will leave the industry. The entrepreneur who gets this OP_1 profits in the industry is called 'Marginal entrepreneur'.
- So, the triangle PEP_1 represents the rent element in the profits gained by intra-marginal entrepreneurs.
- So, the gain of OP profits represents the super normal profits earned by the entrepreneurs even in the long run and this is due to less elastic supply of entrepreneurs. Had the supply of entrepreneurs is perfectly elastic (perfect competition) as discussed under Figure 28.2, they earn only normal profits equal to their transfer earnings. But, perfect competition does not really exist, thereby, the entrepreneurs can earn super normal profits even in the long run.

From the above discussion regarding the different theories of profits, we could infer that, none of these deals with all the aspects of the problem. Each one explains only a particular aspect of profits and ignoring the other important elements that influence the profits. More over, all the theories explained the concept of profits considering only one entrepreneur in the business firm. But, now a days, with the advent of joint stock companies, the share holders simply invest the money and realize the profits, as the activities of the companies will be looked after by the Directors and Workers. So, such profits are not explained by all the theories discussed above. However, irrespective of the form of the business firm, uncertainty bearing is the principal cause for the reward of profits considering the dynamic nature of the modern world. So, Knight's uncertainty bearing theory of profits has more relevance for the modern world, inspite of criticisms leveled against this theory.

References

Ahuja H.L., 'Modern Micro Economics- Theory and Applications', S. Chand & Company Limited, New Delhi, 2008.

Chopra P.N. 'Principles of Economics', Kalyani Publications, Ludhiana, 2003.

Jhingan M.L. 'Advanced Economic theory' (Micro and Macro Economics), Vrinda Publications (P) Limited. Delhi, 2006.

Subject Index

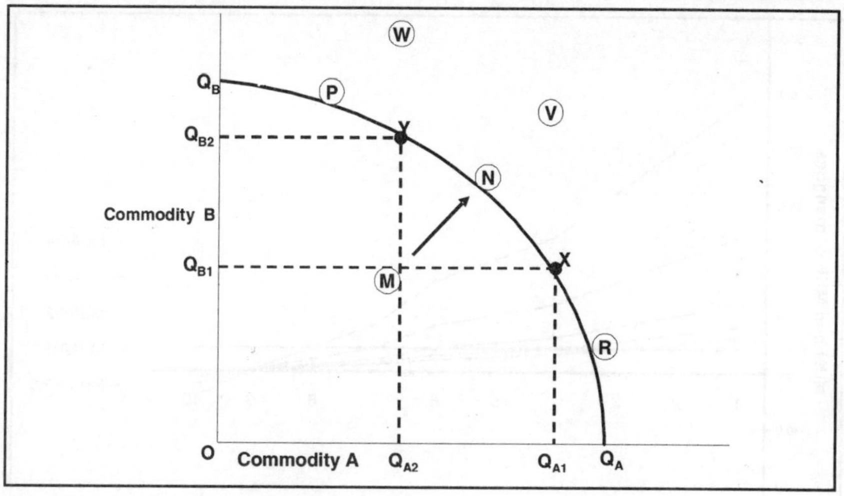

Figure 4.2: Production Possibility Frontier. **(Page 96)**

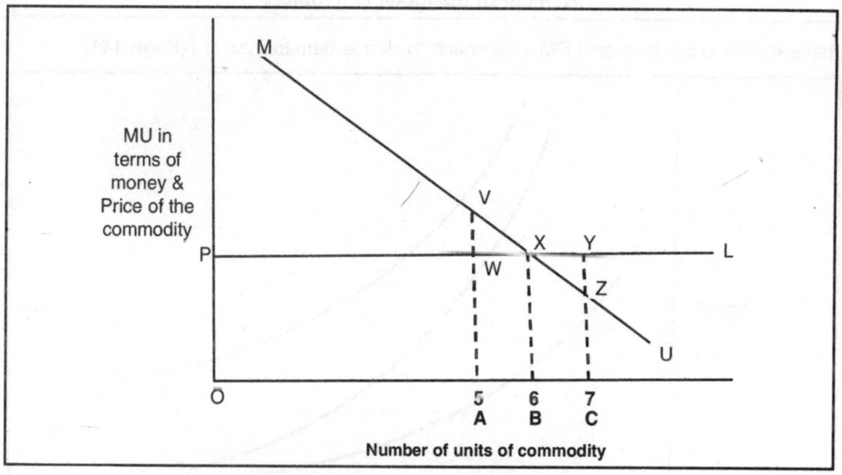

Figure 6.1: Point of consumer's equilibrium. **(Page 115)**

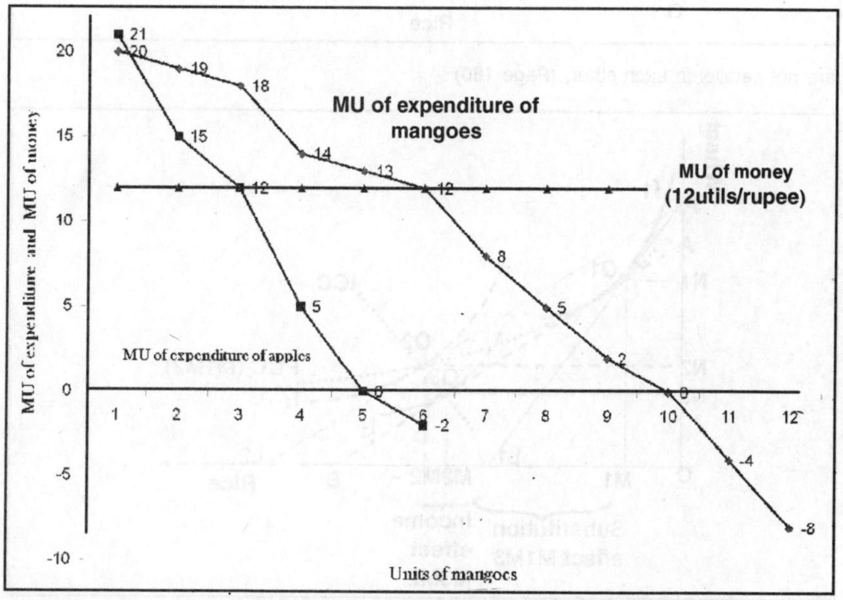

Figure 8.2: Consumer's equilibrium with respect to mangoes and apples (Method 1). **(Page 136)**

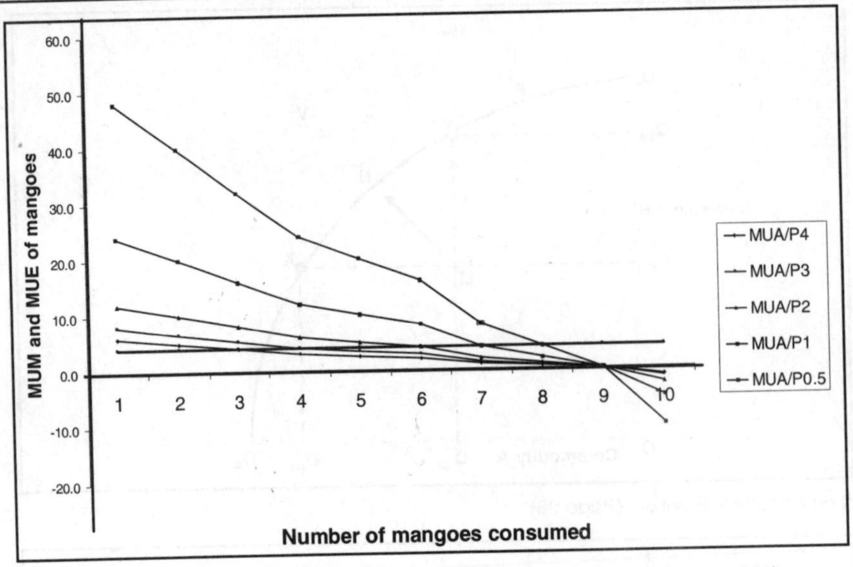

Figure 8.4: Consumer's equilibrium based on LDMU approach to derive demand curve. **(Page 141)**

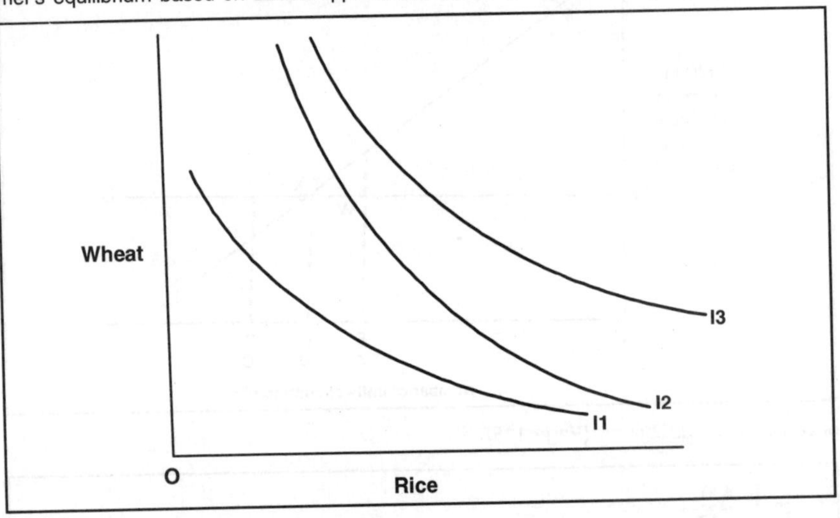

Figure 9.12: IDCs are not parallel to each other. **(Page 160)**

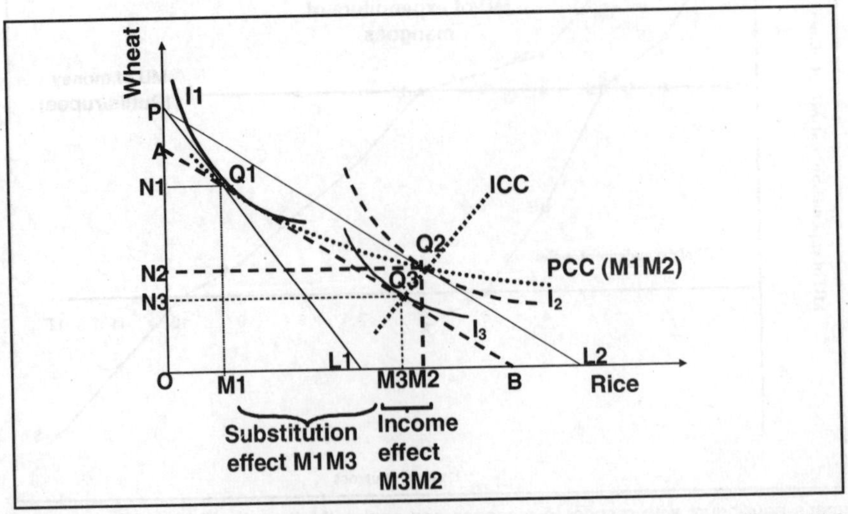

Figure 9.36: Slutsky approach : Price effect = Substitution effect + Income effect (Normal good). **(Page 187)**

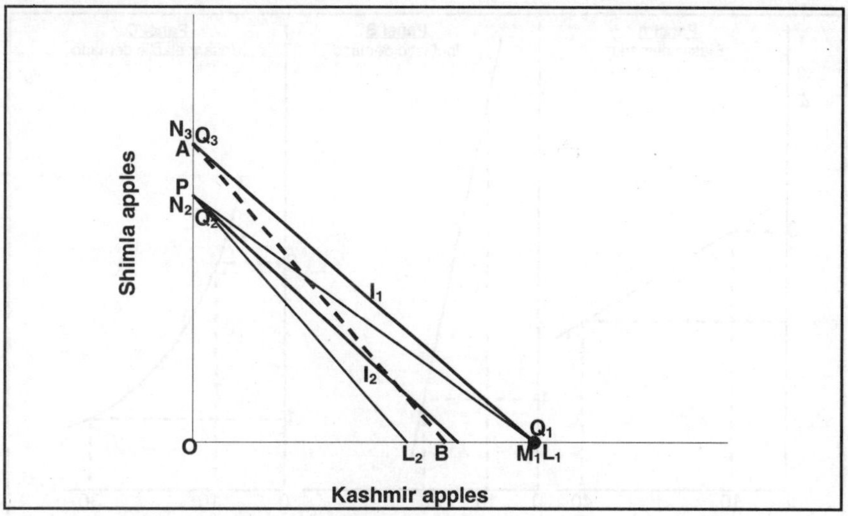

Figure 9.41: Price effect in perfect substitutes. **(Page 199)**

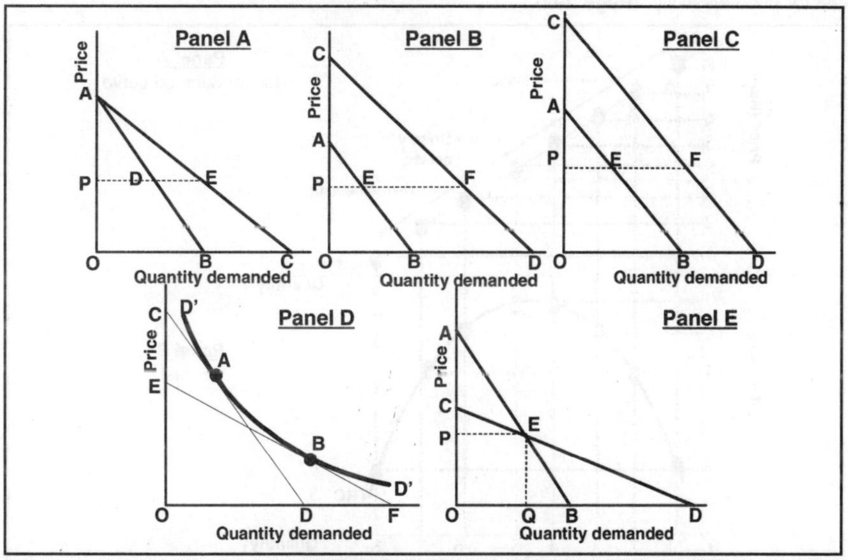

Figure 12.3: Slope of the demand curve and E_D. **(Page 270)**

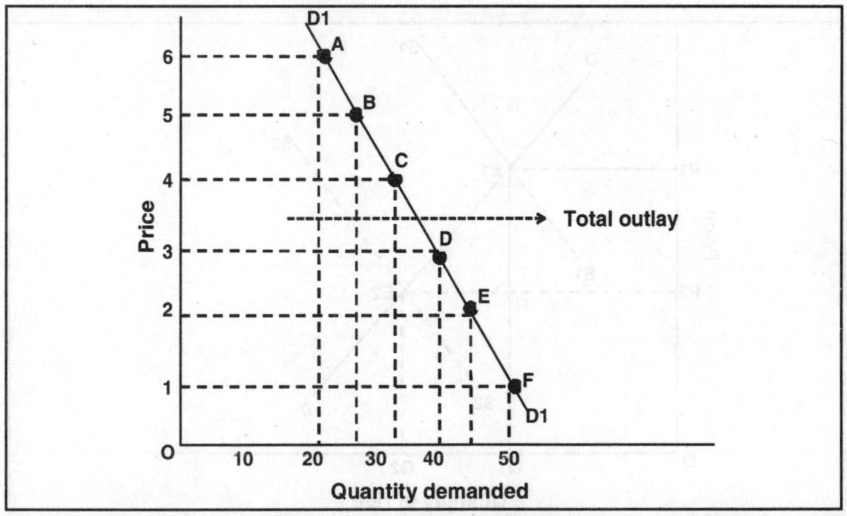

Figure 12.15: Demand curve showing total outlay of the consumer. **(Page 291)**

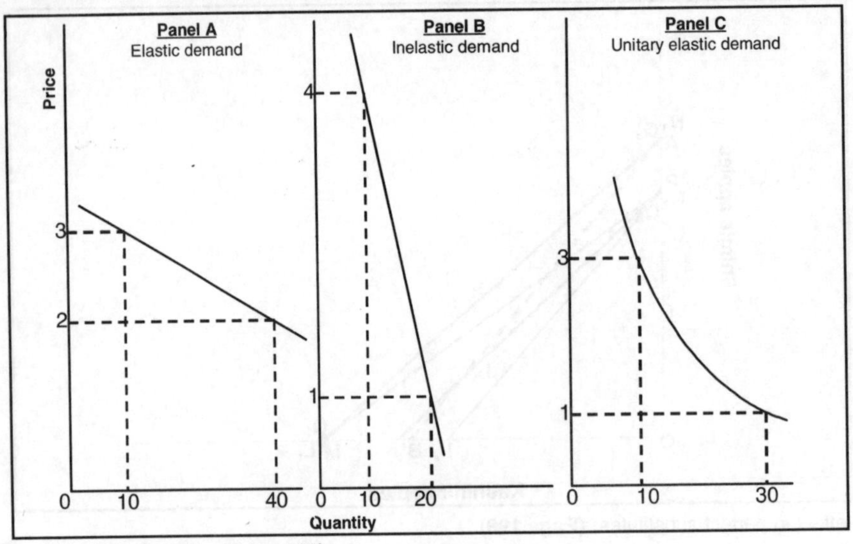

Figure 12.16: TR test for price elasticity. **(Page 292)**

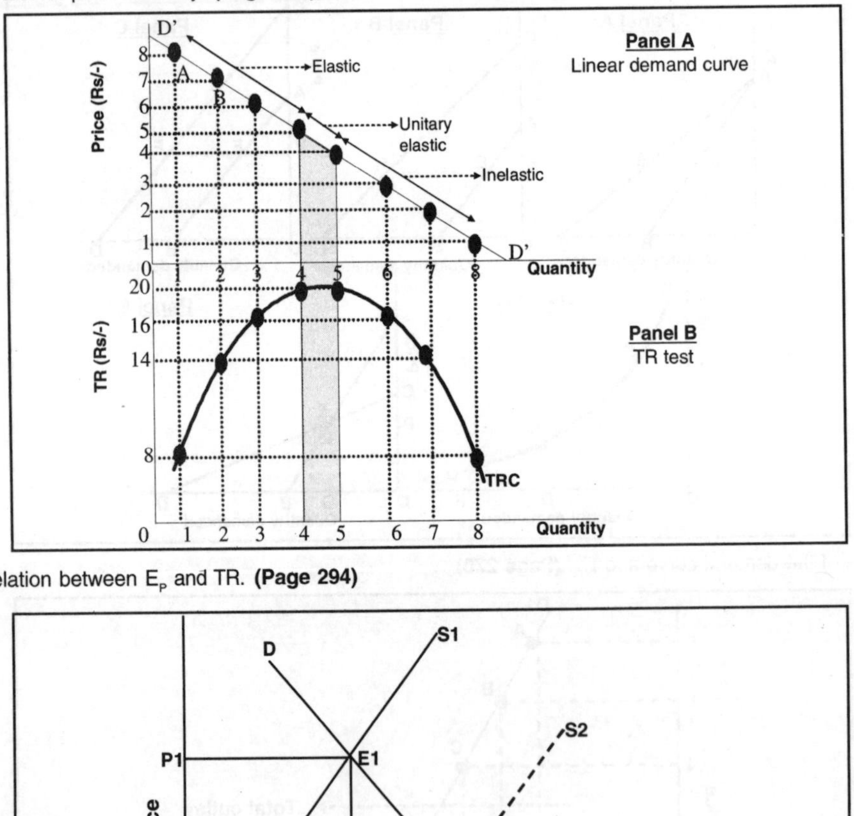

Figure 12.17: Relation between E_p and TR. **(Page 294)**

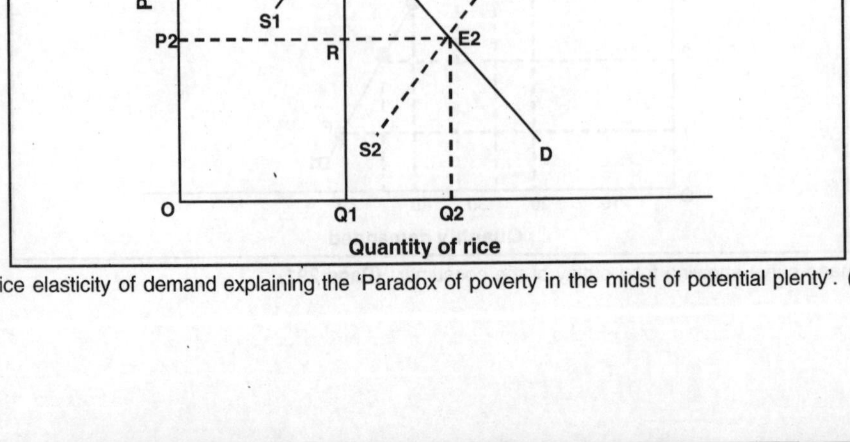

Figure 12.18: Price elasticity of demand explaining the 'Paradox of poverty in the midst of potential plenty'. **(Page 296)**

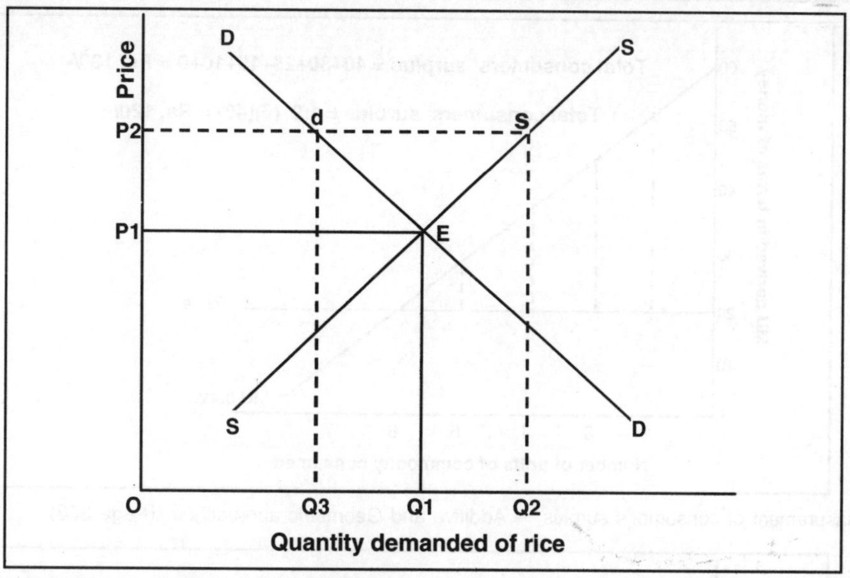

Figure 12.19: E_P guiding the procurement of rice. **(Page 297)**

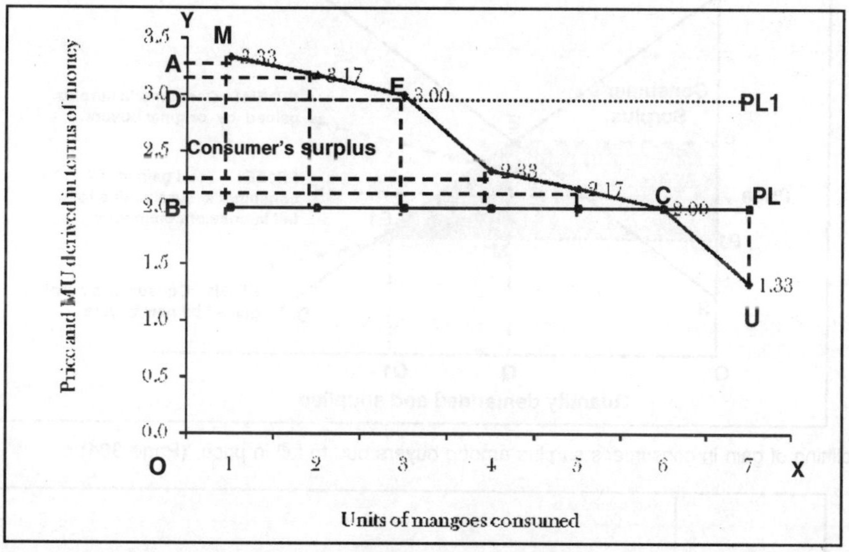

Figure 13.1: Marshallian approach of measurement of consumer's surplus. **(Page 301)**

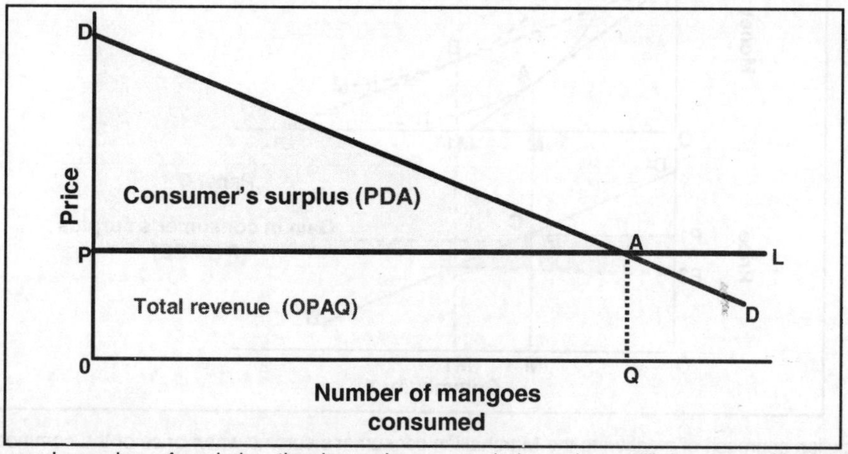

Figure 13.2: Consumer's surplus - Area below the demand curve and above the equilibrium price. **(Page 303)**

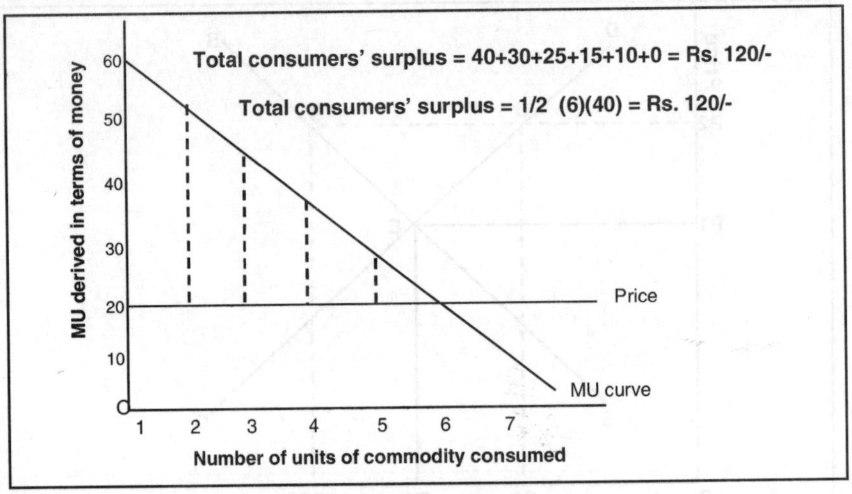

Figure 13.3: Measurement of consumer's surplus – Additive and Geometric approaches. **(Page 303)**

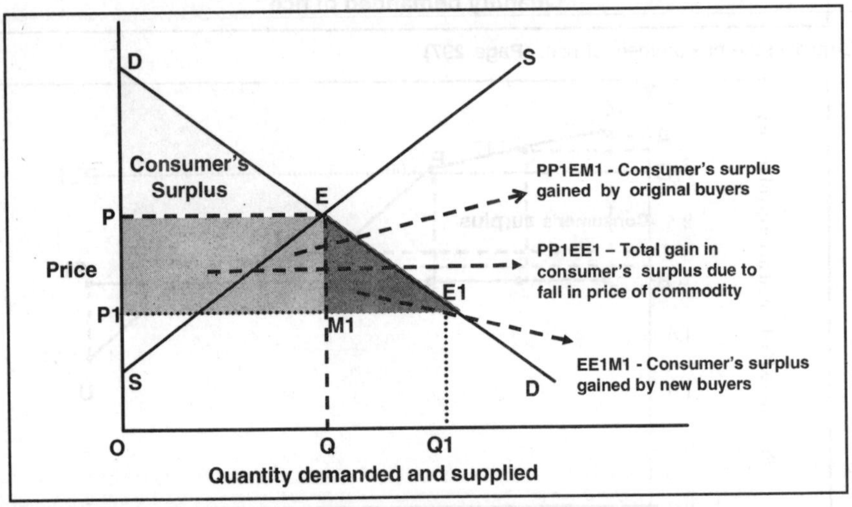

Figure 13.4: Splitting of gain in consumer's surplus among buyers due to fall in price. **(Page 304)**

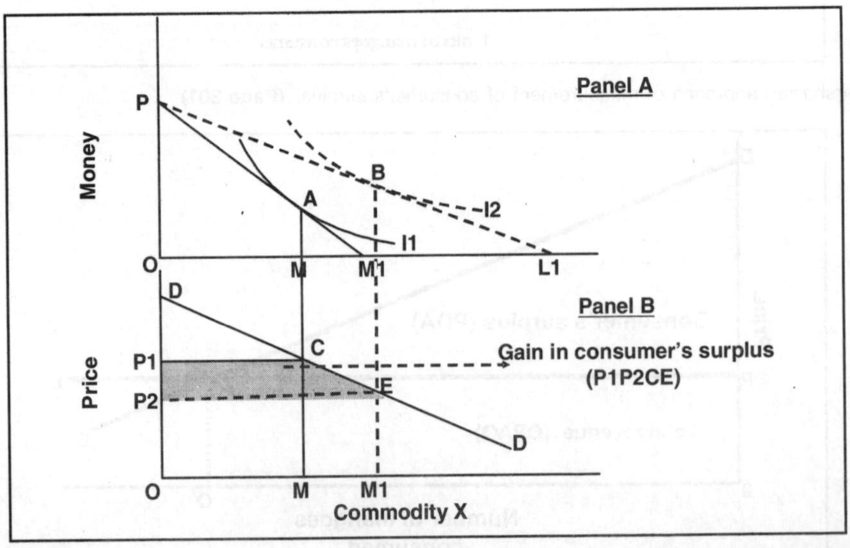

Figure 13.9: Hicksian approach of measuring the Marshallian consumer's surplus when price of the commodity changes. **(Page 308)**

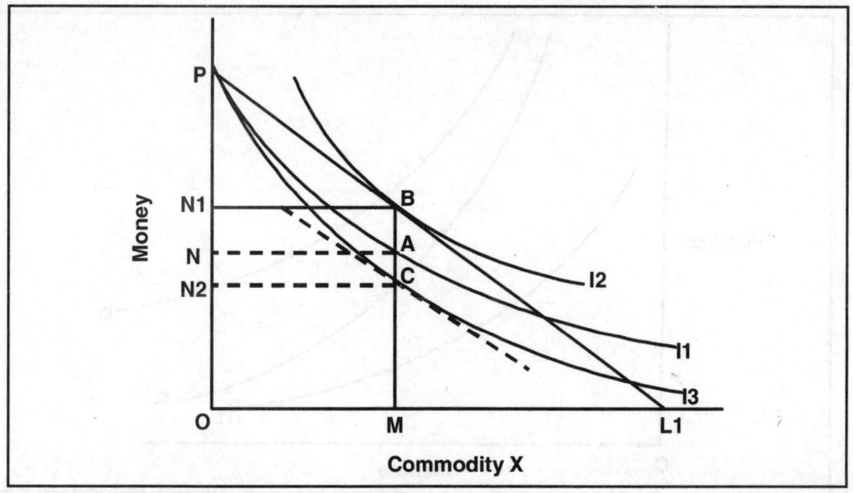

Figure 13.10: Hicksian approach of measuring the Marshallian consumer's surplus (Assuming diminishing MU of money). **(Page 308)**

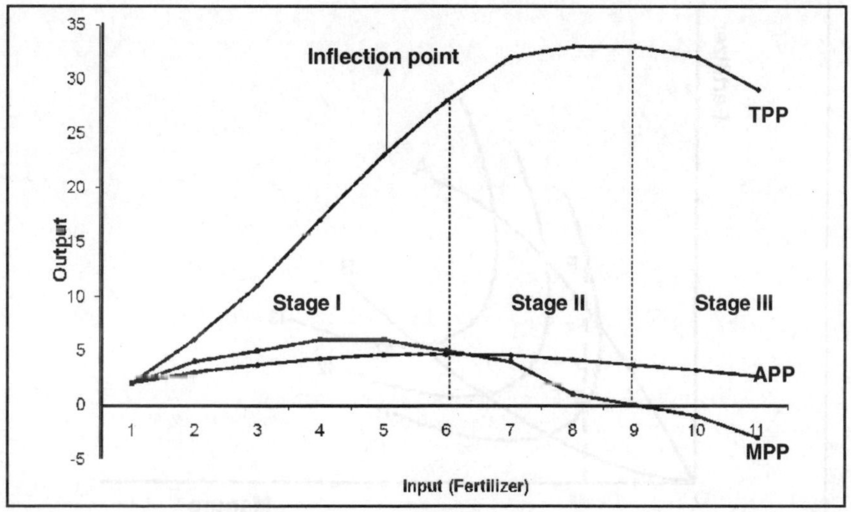

Figure 14.4: Three stages of LVP. **(Page 325)**

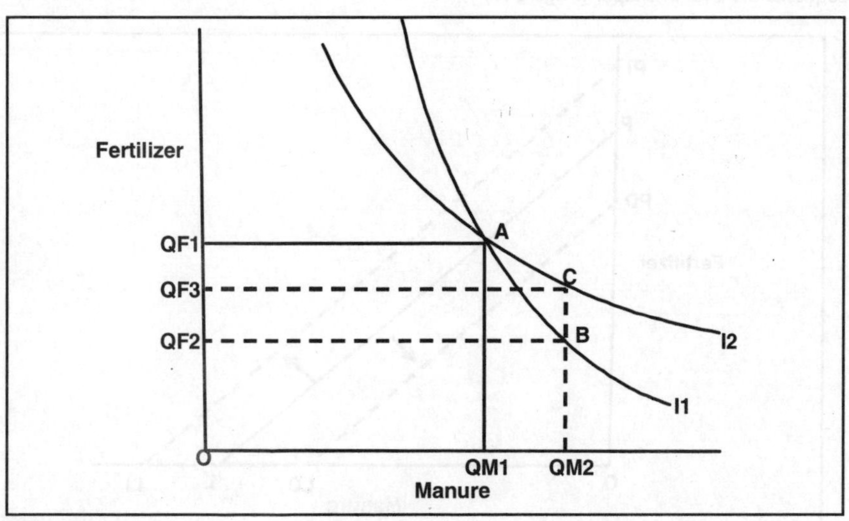

Figure 14.9: Isoquants do not intersect each other. **(Page 338)**

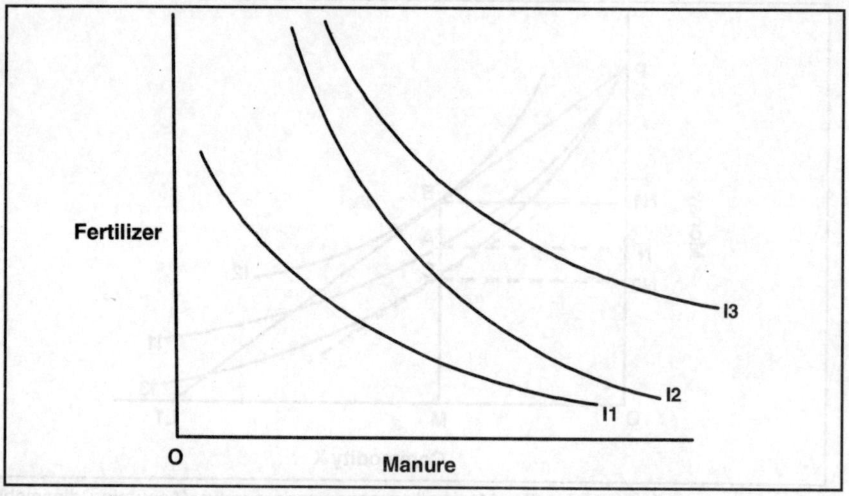

Figure 14.11: Isoquants are not parallel to each other. **(Page 339)**

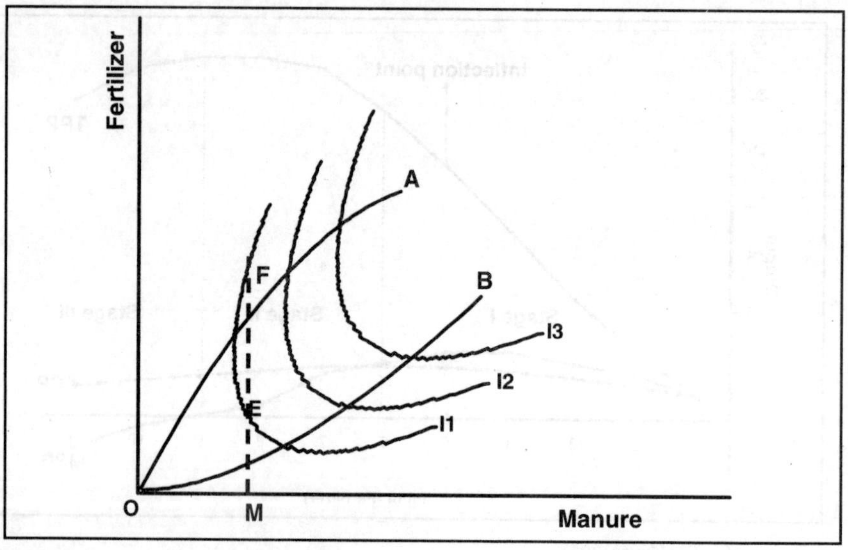

Figure 14.12: Isoquants are oval in shape. **(Page 340)**

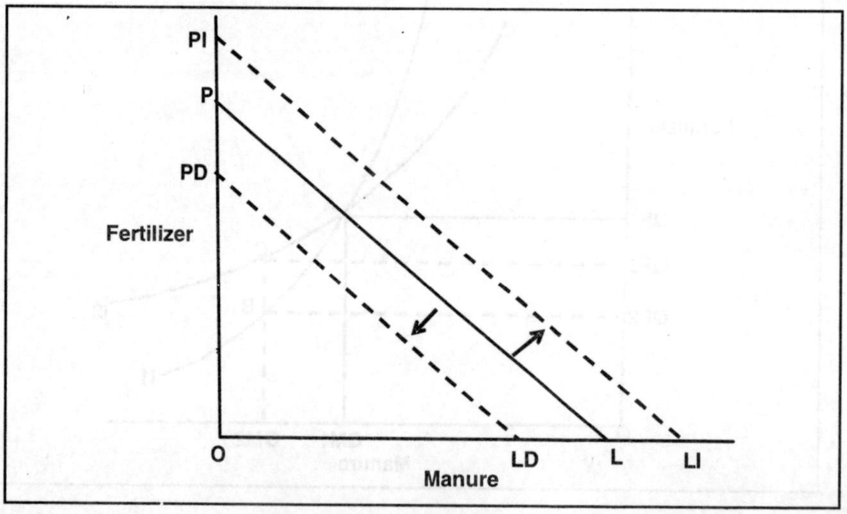

Figure 14.15: Shifts in iso-cost line due to changes in money income with the farmer. **(Page 345)**

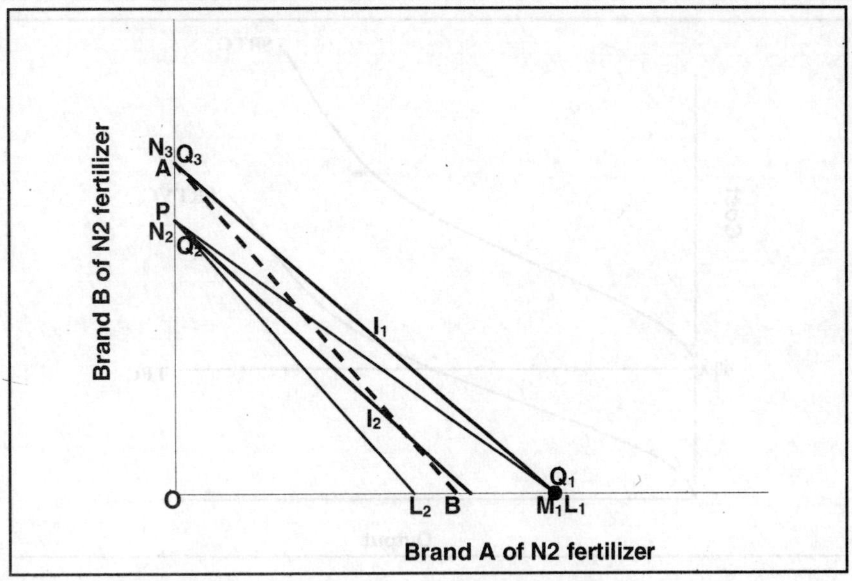

Figure 14.28: Perfect substitutes (Price effect = Substitution effect). **(Page 356)**

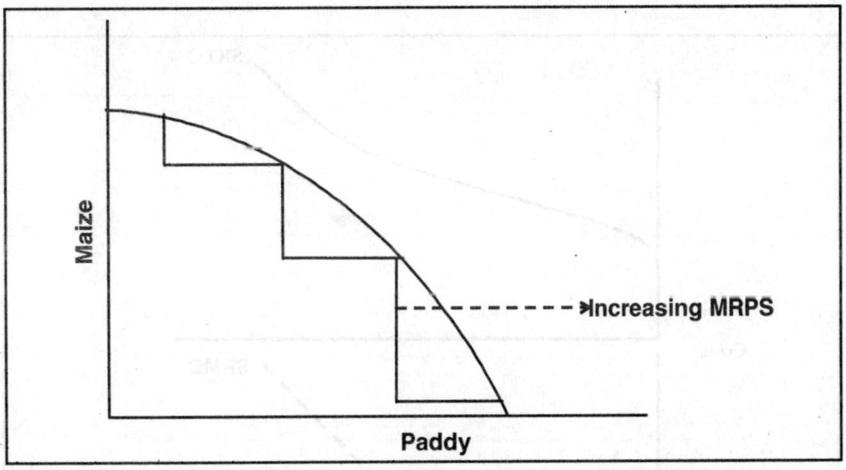

Figure 14.41: Positively sloped Production possibility curve showing increasing MRPS. **(Page 370)**

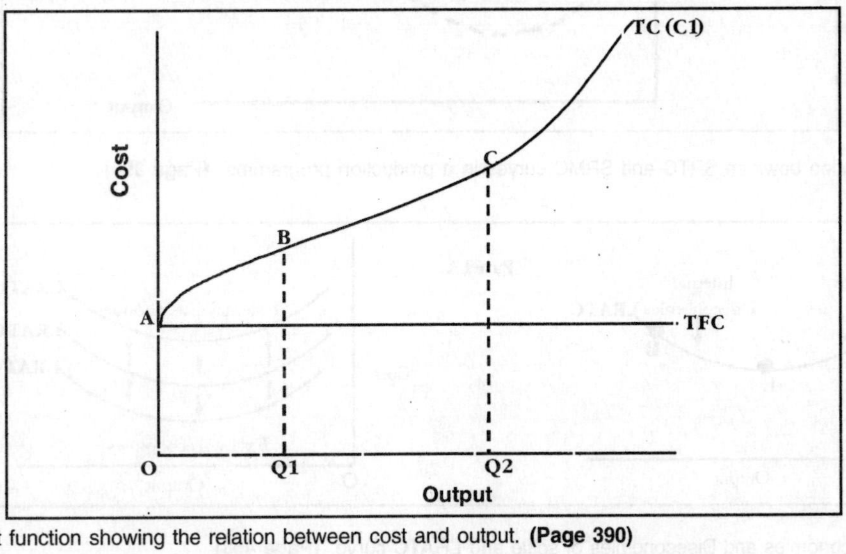

Figure 15.2: Cost function showing the relation between cost and output. **(Page 390)**

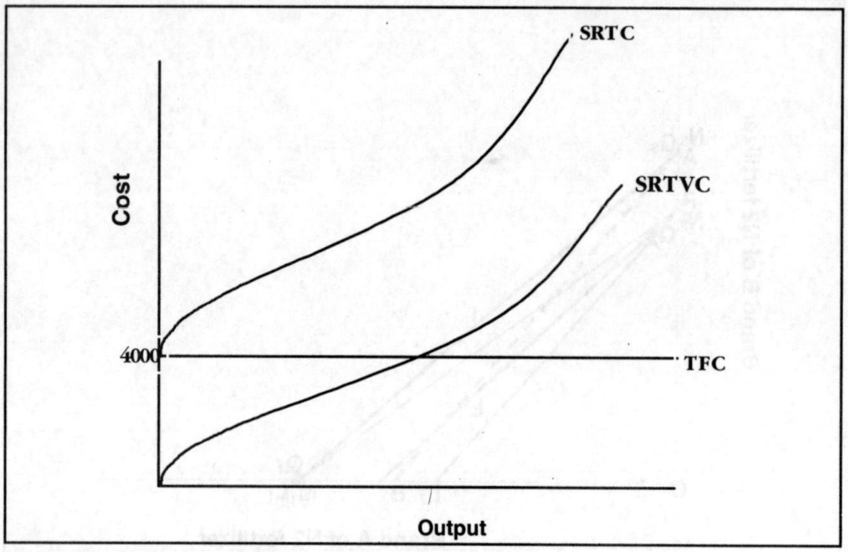

Figure 15.3: Traditional theory – Short run absolute cost curves in production of paddy. **(Page 391)**

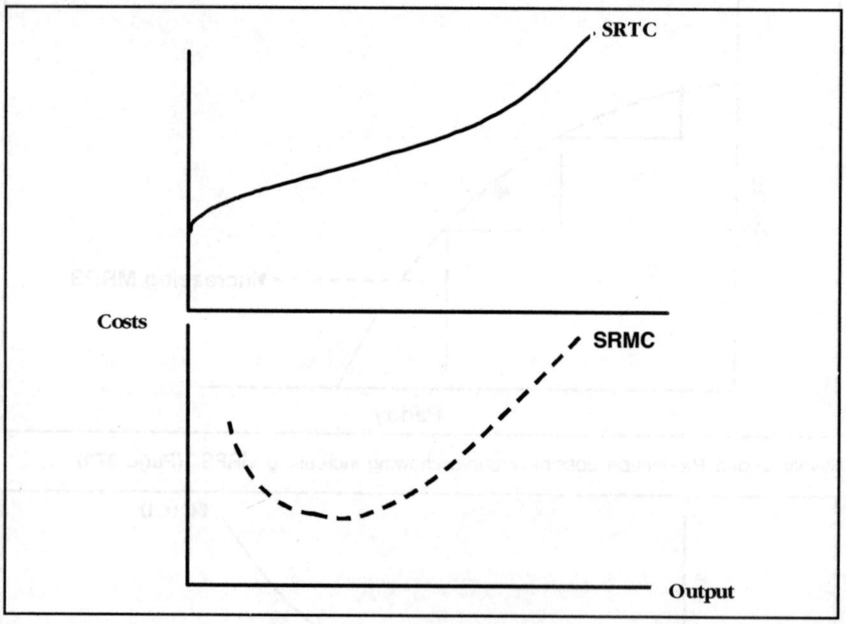

Figure 15.8: Relation between SRTC and SRMC curves in a production programme. **(Page 399)**

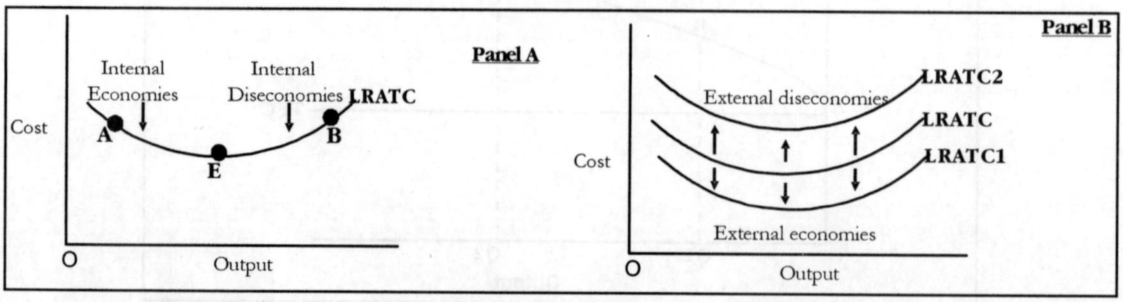

Figure 15.12: Economies and Diseconomies of scale and LRATC curve. **(Page 408)**

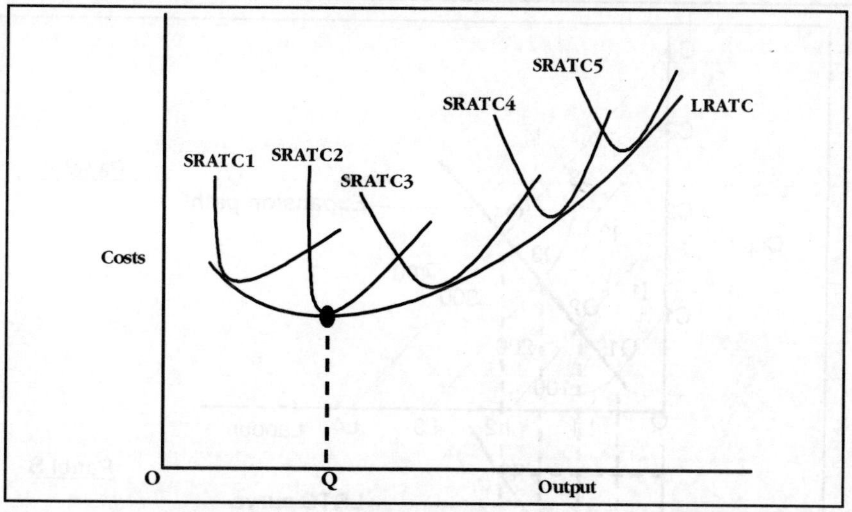

Figure 15.13: Firm operating under 'Agricultural sector' showing minimum LRATC at small scale output. **(Page 408)**

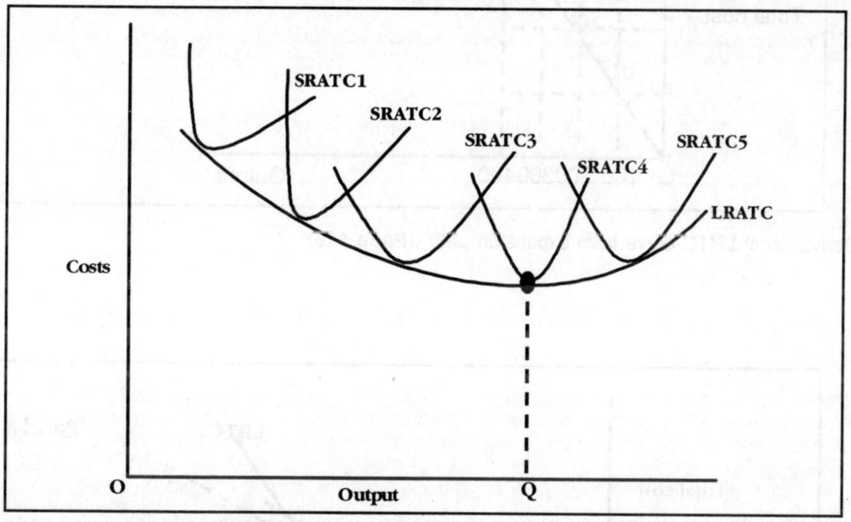

Figure 15.14: Firm operating under 'Manufacturing sector' showing minimum LRATC at large scale output. **(Page 409)**

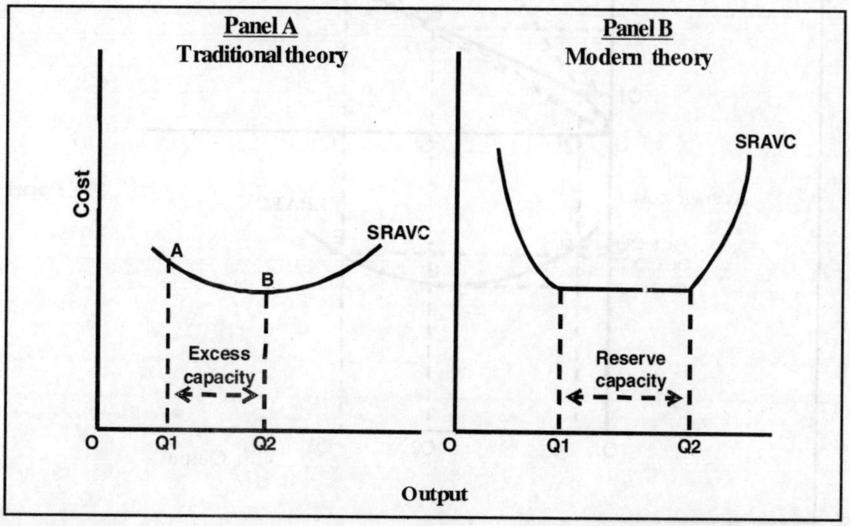

Figure 15.17: Reserve capacity (Modern theory) Vs Excess capacity (Traditional theory) of SRAVC curve. **(Page 412)**

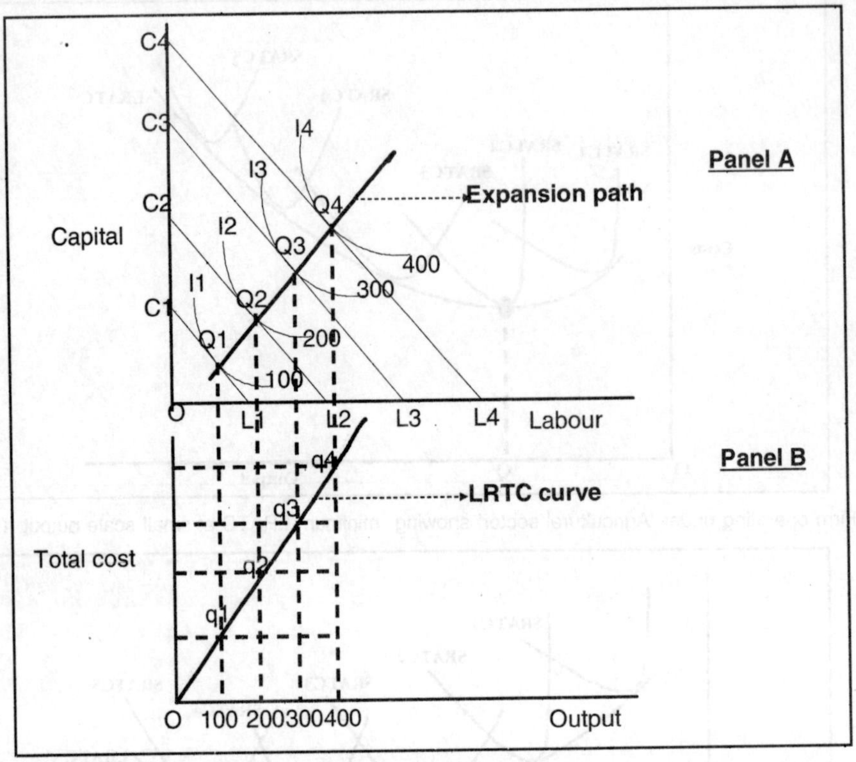

Figure 15.23: Derivation of LRTC curve from expansion path. **(Page 416)**

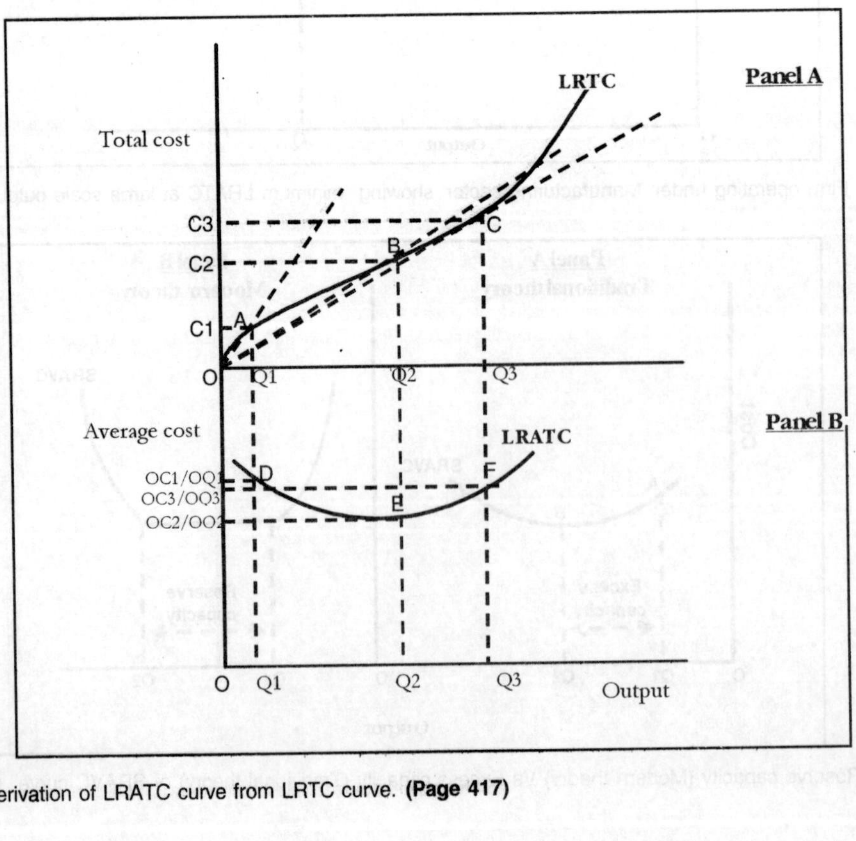

Figure 15.24: Derivation of LRATC curve from LRTC curve. **(Page 417)**

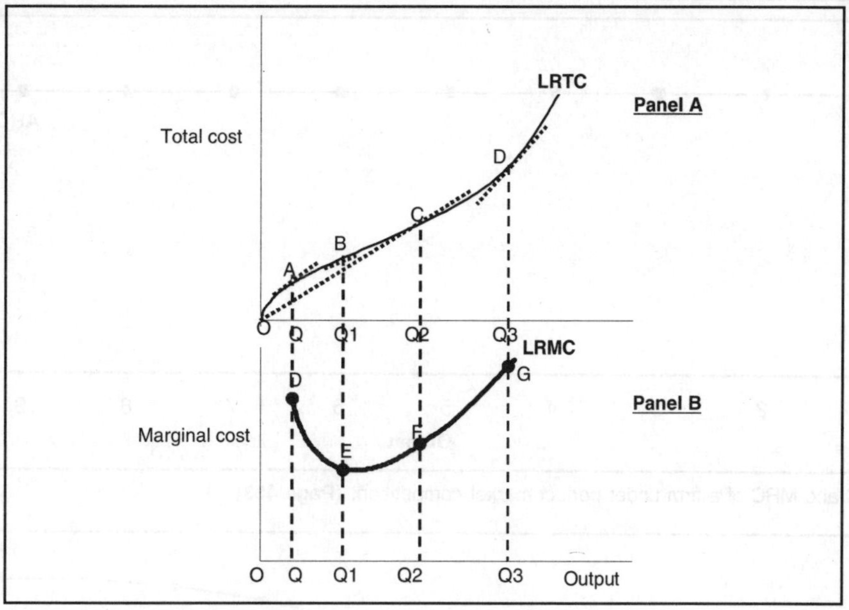

Figure 15.25: Derivation of LRMC curve from LRTC curve. **(Page 418)**

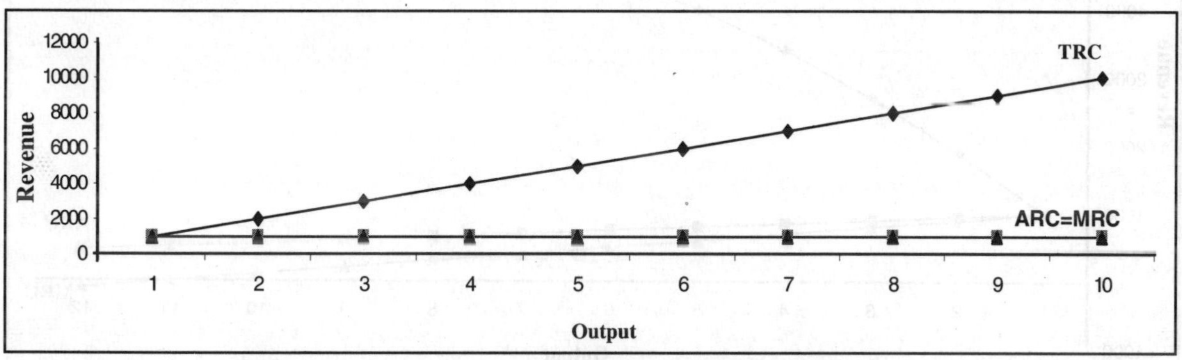

Figure 18.2: TR, AR and MR curves of a firm under perfect market competition. **(Page 483)**

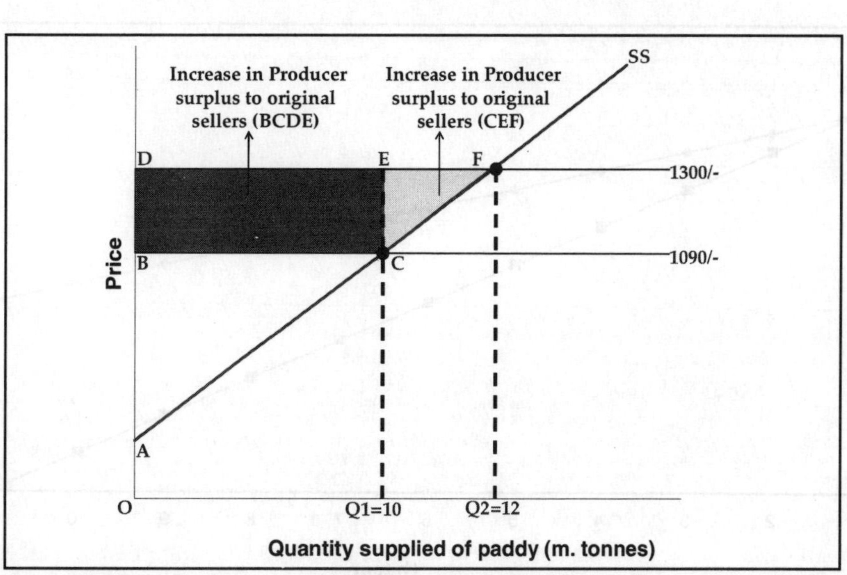

Figure 17.15: Changes in producer surplus due to change in price of paddy. **(Page 456)**

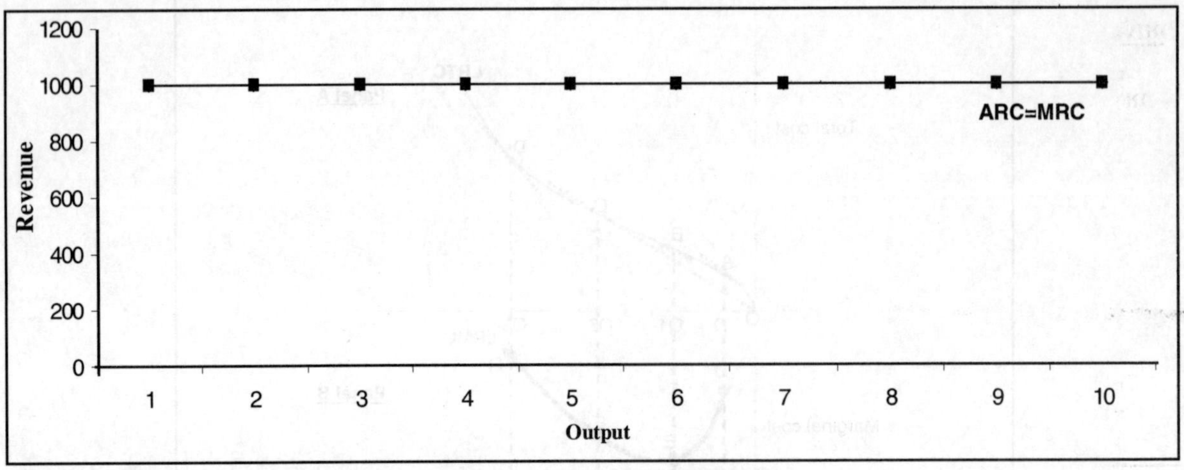

Figure 18.3: ARC and MRC of a firm under perfect market competition. **(Page 483)**

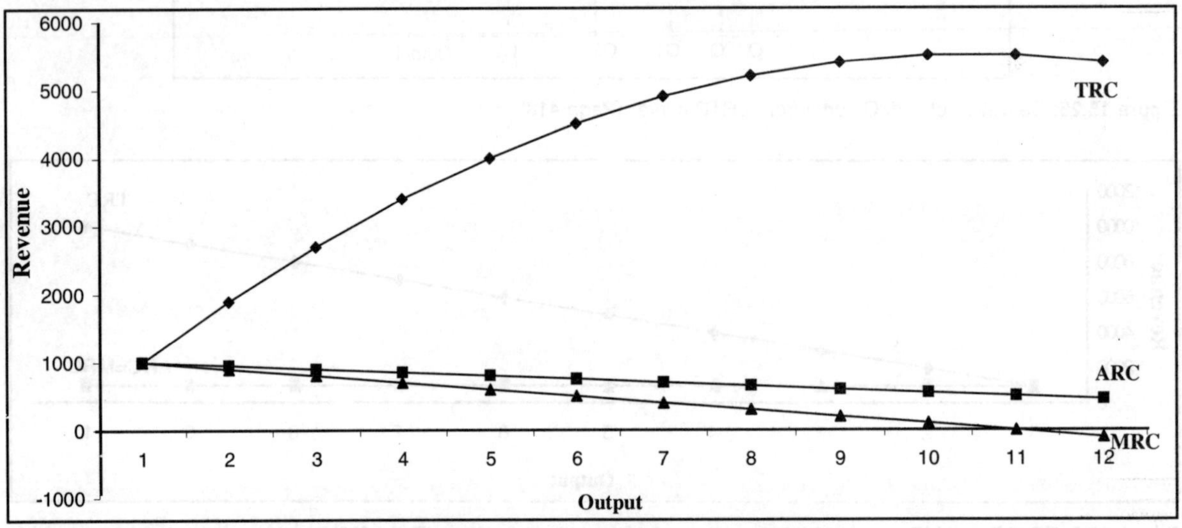

Figure 18.4: TR, AR and MR curves of a monopoly firm selling *Bt* cotton seed . **(Page 485)**

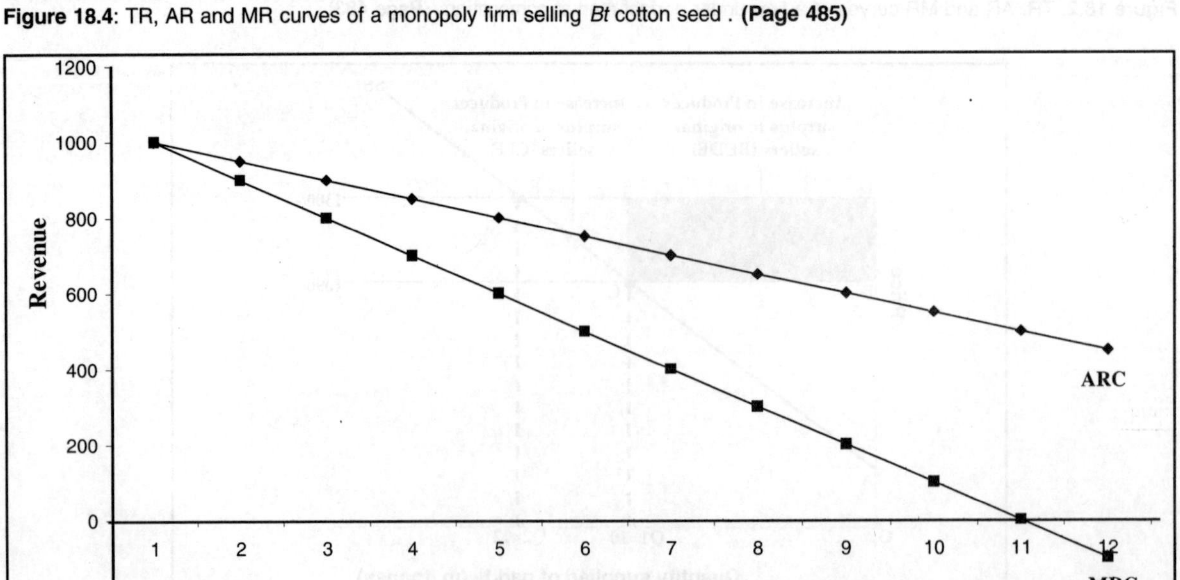

Figure 18.5: ARC and MRC of a monopoly firm selling *Bt* cotton seed. **(Page 486)**

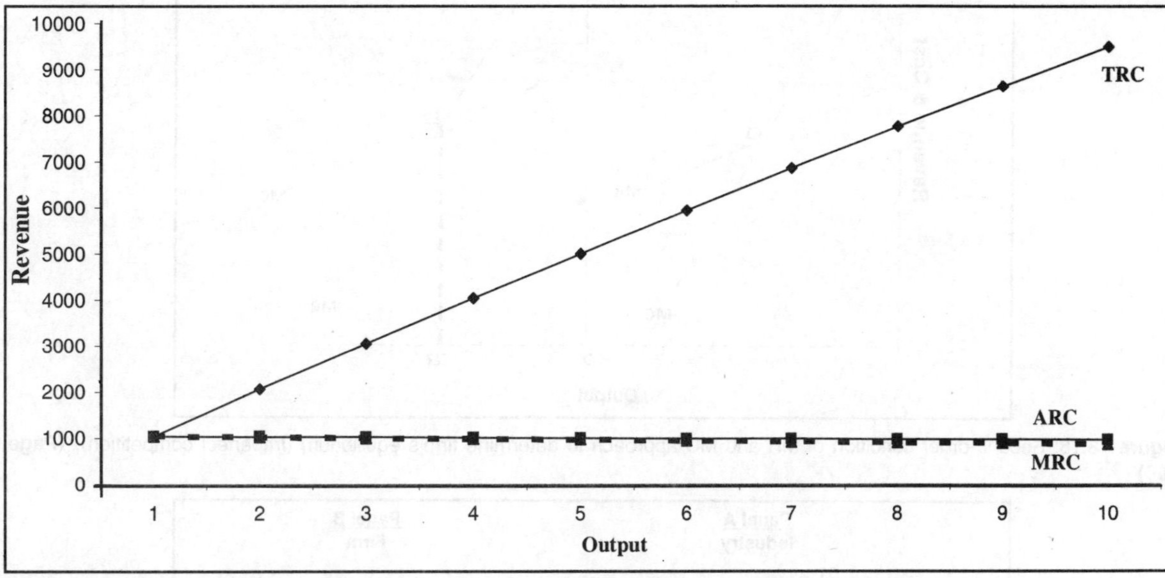

Figure 18.7: TR, AR and MR of a firm selling fertilizer under monopolistic competition. **(Page 488)**

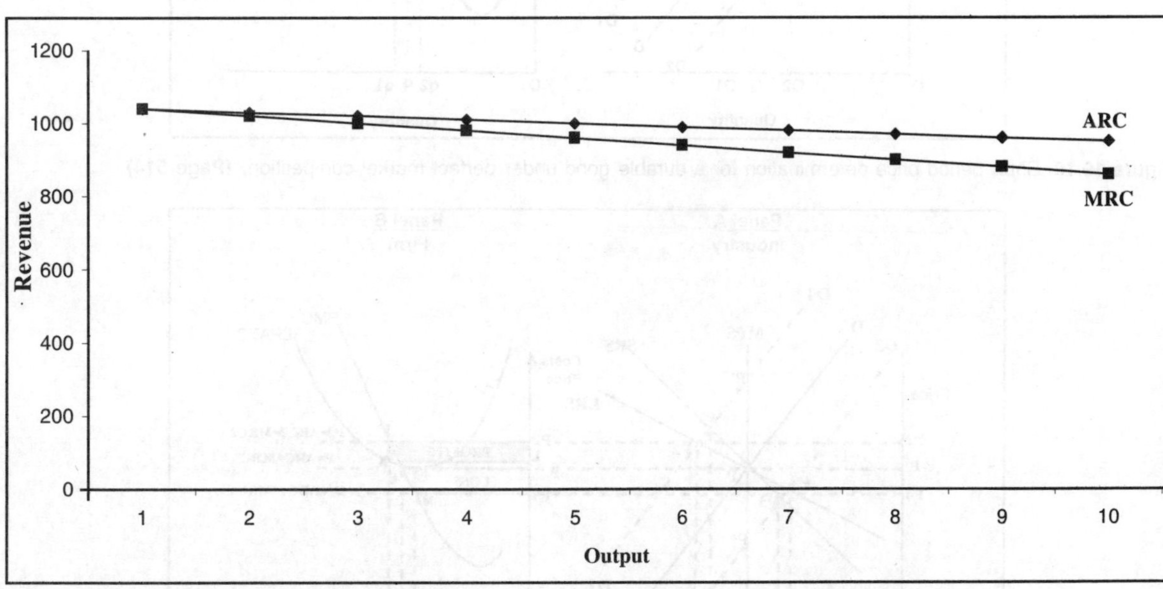

Figure 18.8: ARC and MRC of a firm selling fertilizer under monopolistic competition. **(Page 488)**

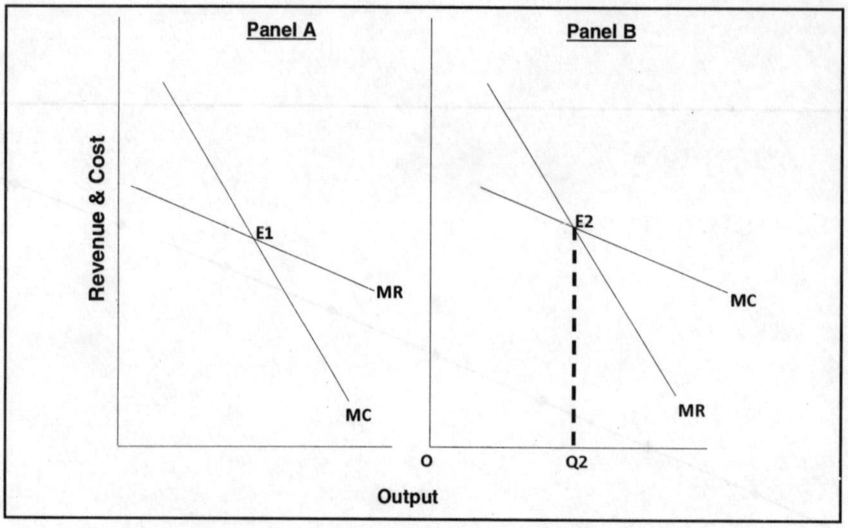

Figure 18.18: Second-order condition of MR and MC approach to determine firm's equilibrium (Imperfect competition). **(Page 500)**

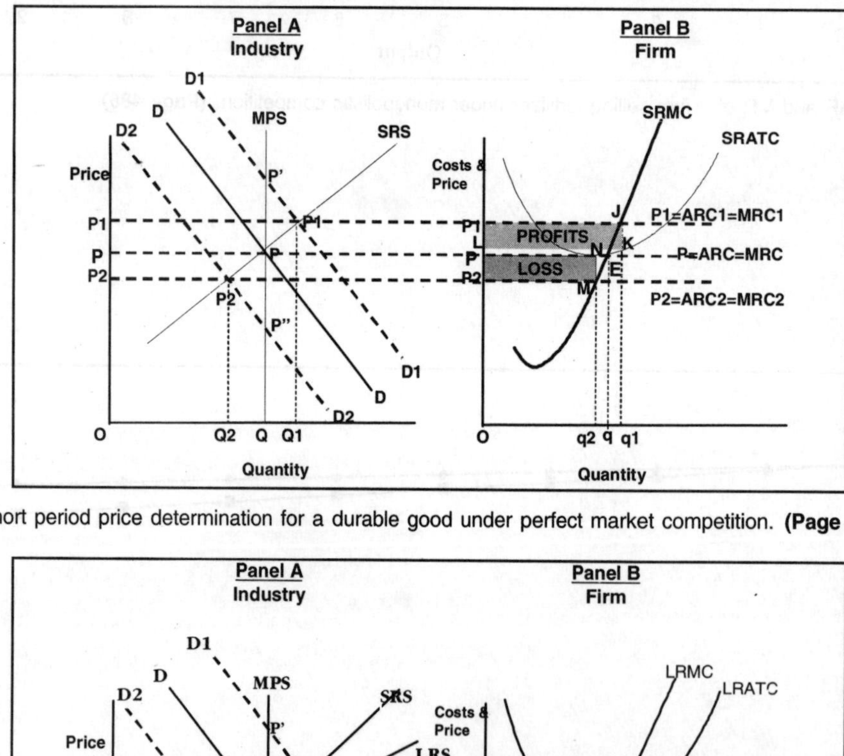

Figure 19.10: Short period price determination for a durable good under perfect market competition. **(Page 514)**

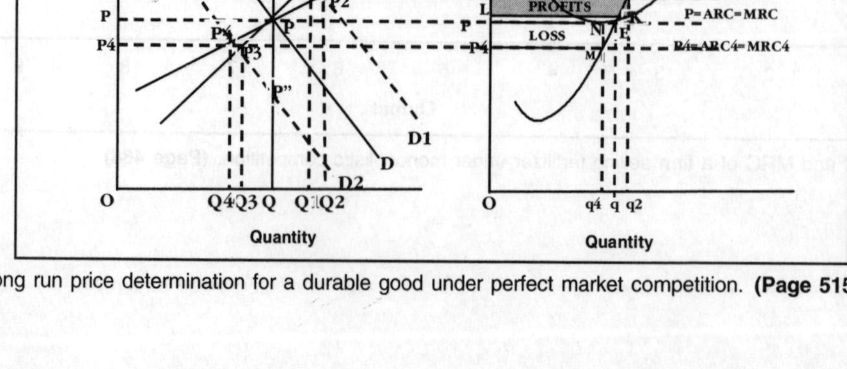

Figure 19.11: Long run price determination for a durable good under perfect market competition. **(Page 515)**

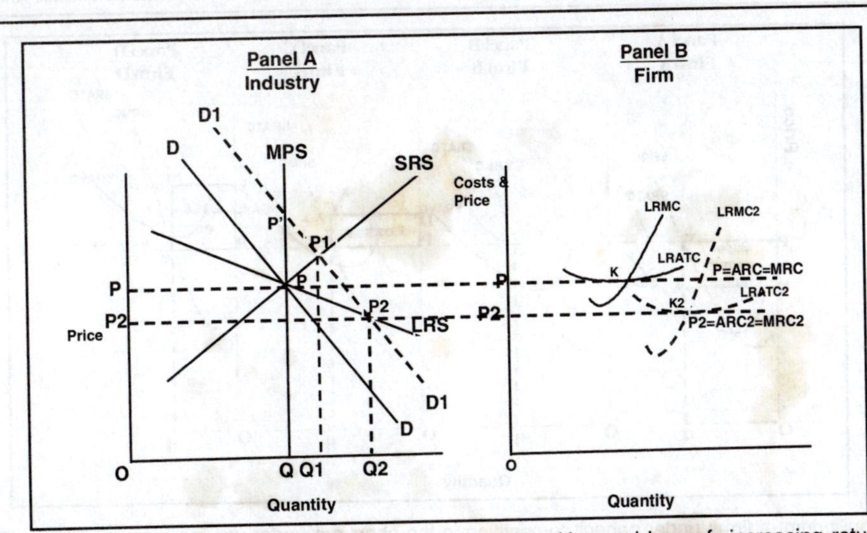

Figure 19.12: Long run price determination under perfect market competition and Law of increasing returns or decreasing costs. **(Page 517)**

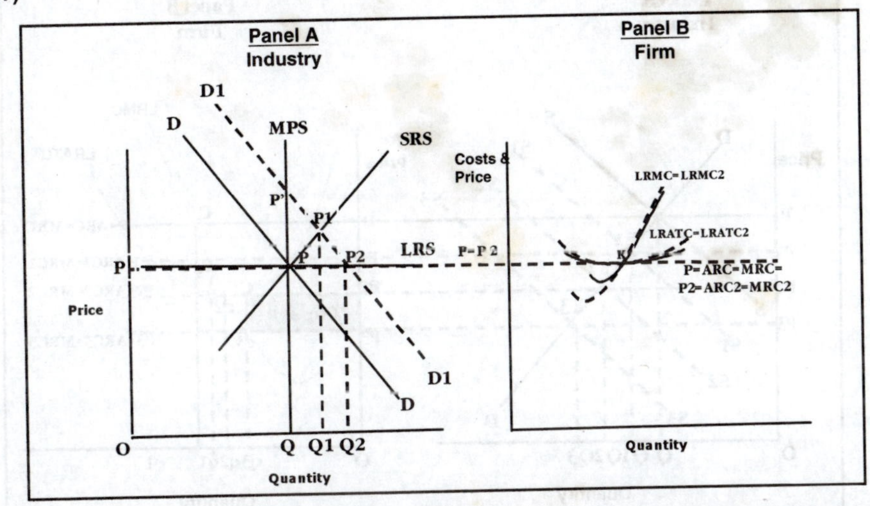

Figure 19.13: Long run price determination under perfect market competition and Law of constant returns or constant costs. **(Page 518)**

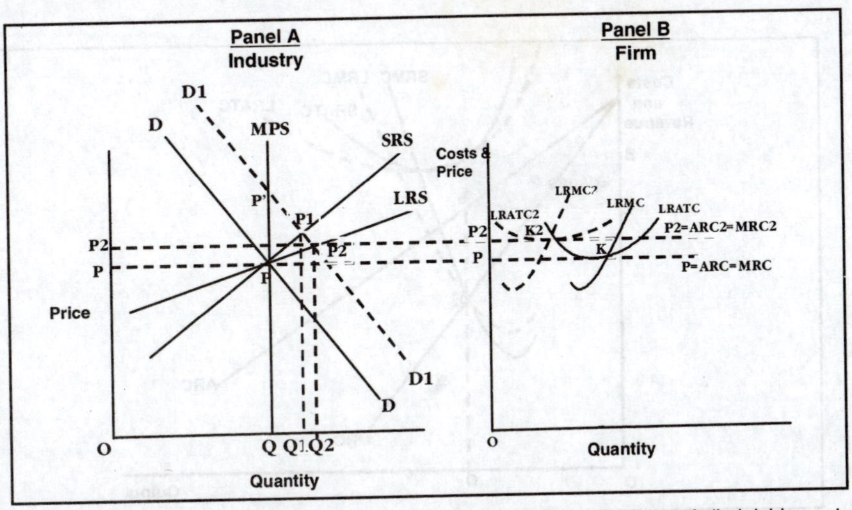

Figure 19.14: Long run price determination under perfect market competition and Law of diminishing returns or increasing costs. **(Page 519)**

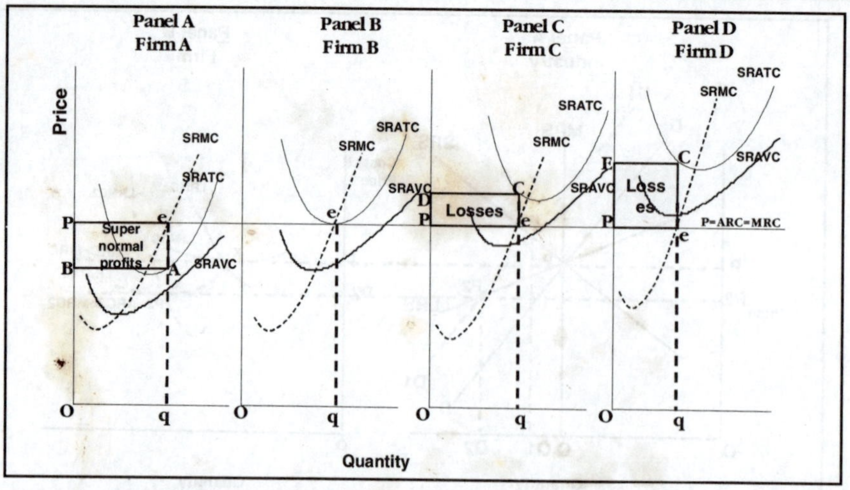

Figure 19.16: Equilibrium of firms under perfect competition in the short run under varying cost conditions. **(Page 523)**

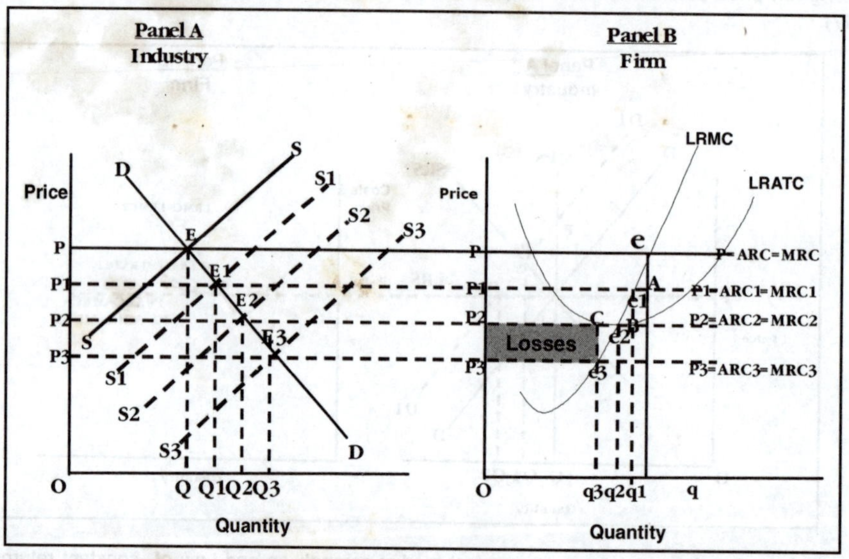

Figure 19.17: Equilibrium of firm and industry under perfect competition in the long run under identical cost conditions. **(Page 525)**

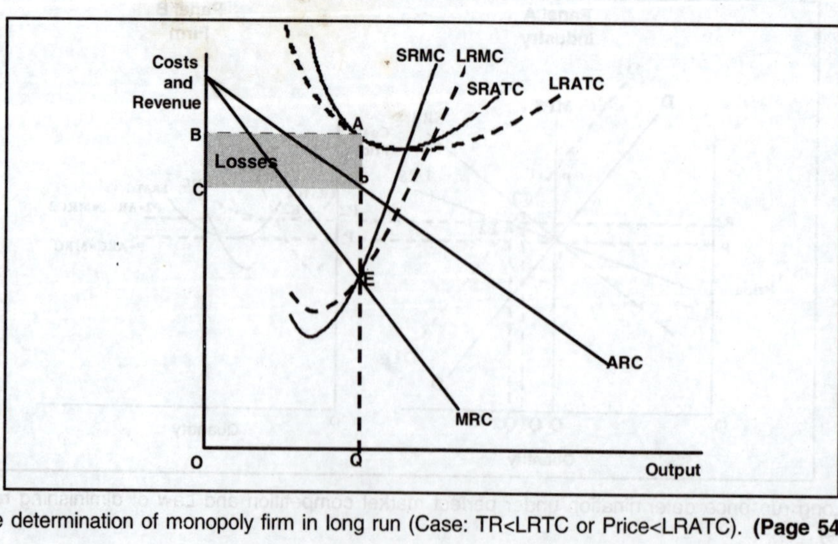

Figure 21.9: Price determination of monopoly firm in long run (Case: TR<LRTC or Price<LRATC). **(Page 543)**

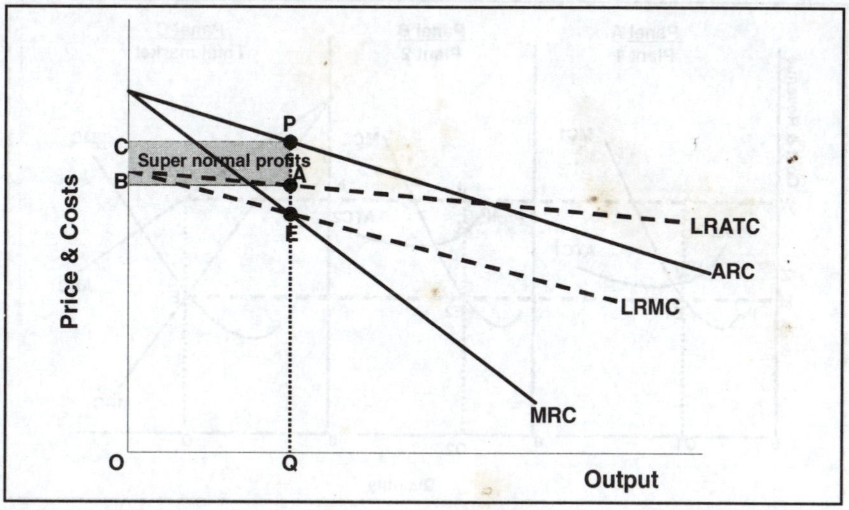

Figure 21.13: Monopoly price and Law of increasing returns. **(Page 546)**

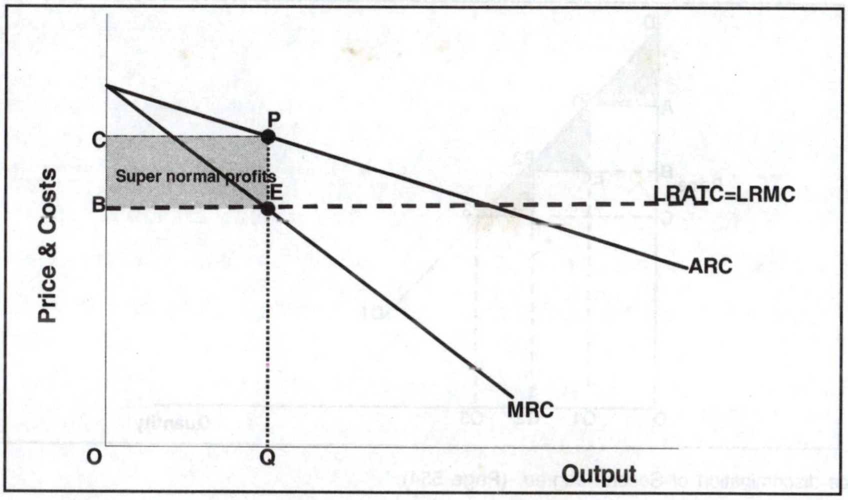

Figure 21.14: Monopoly price and Law of Constant returns. **(Page 546)**

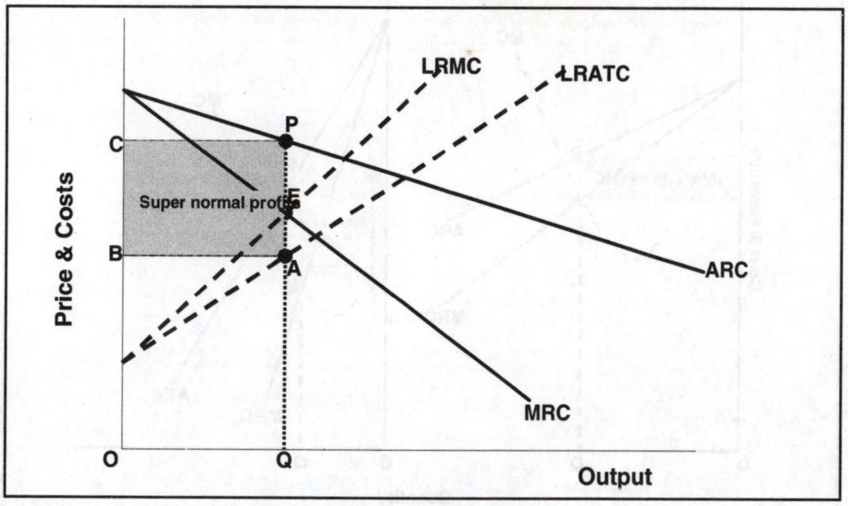

Figure 21.15: Monopoly price and Law of Diminishing returns. **(Page 546)**

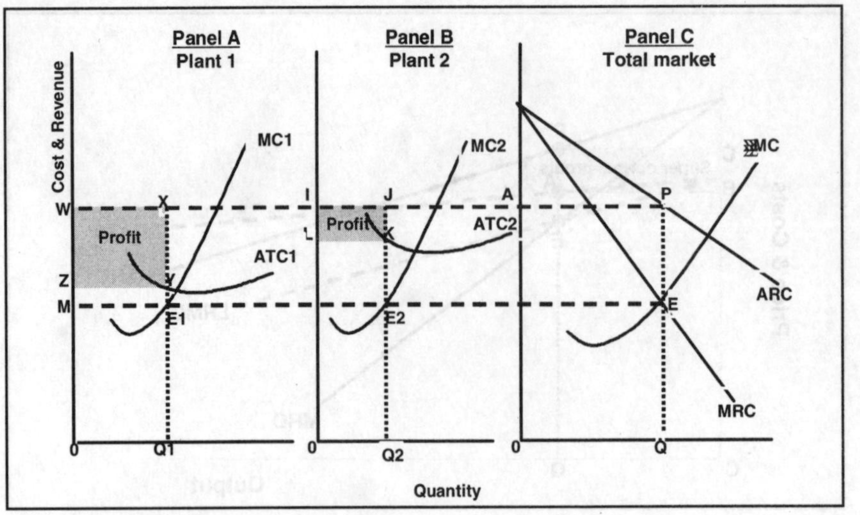

Figure 21.18: Output-Price determination in Multi-Plant Monopoly firm. **(Page 550)**

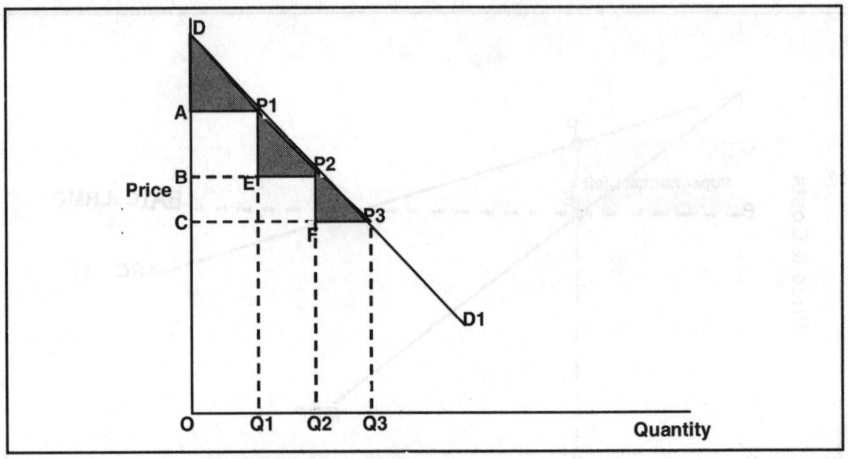

Figure 21.20: Price discrimination of Second degree. **(Page 554)**

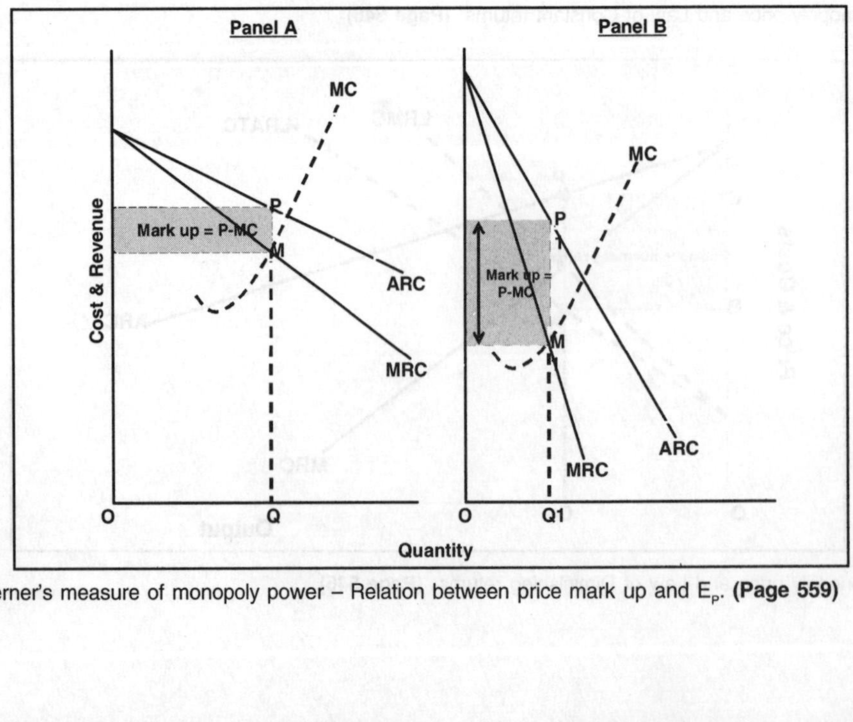

Figure 21.23: Lerner's measure of monopoly power – Relation between price mark up and E_p. **(Page 559)**

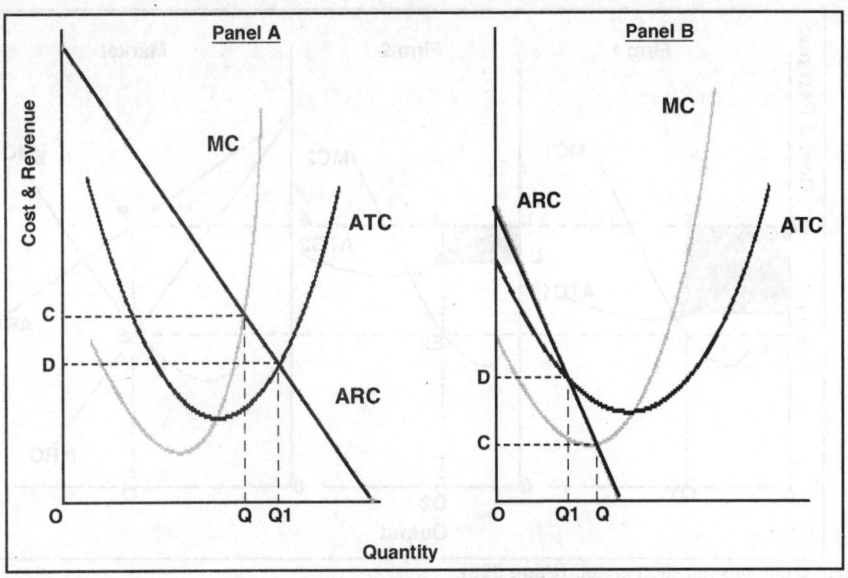

Figure 21.26: Regulation of monopoly power – Average cost pricing. **(Page 565)**

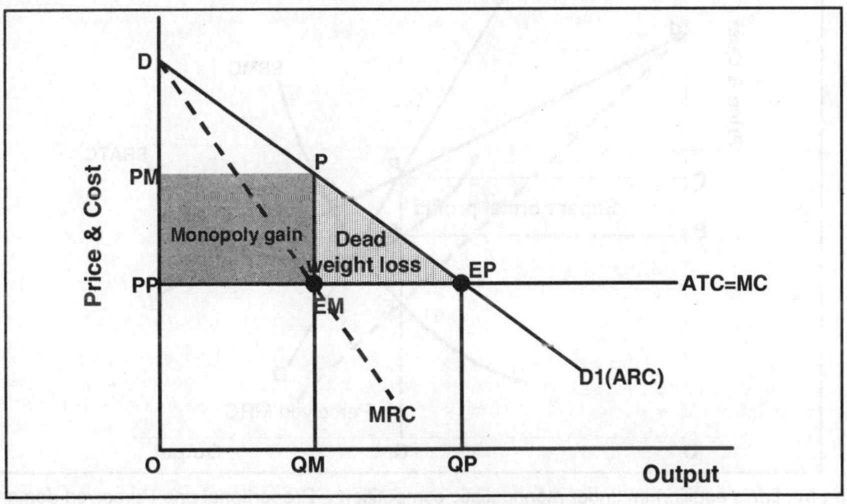

Figure 21.29: Social cost under monopoly assuming constant MC. **(Page 567)**

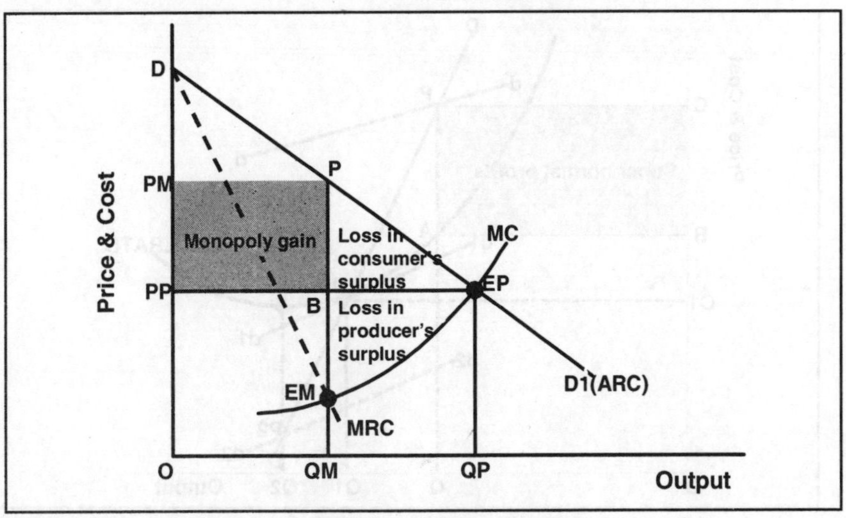

Figure 21.30: Social cost under monopoly assuming increasing MC. **(Page 568)**

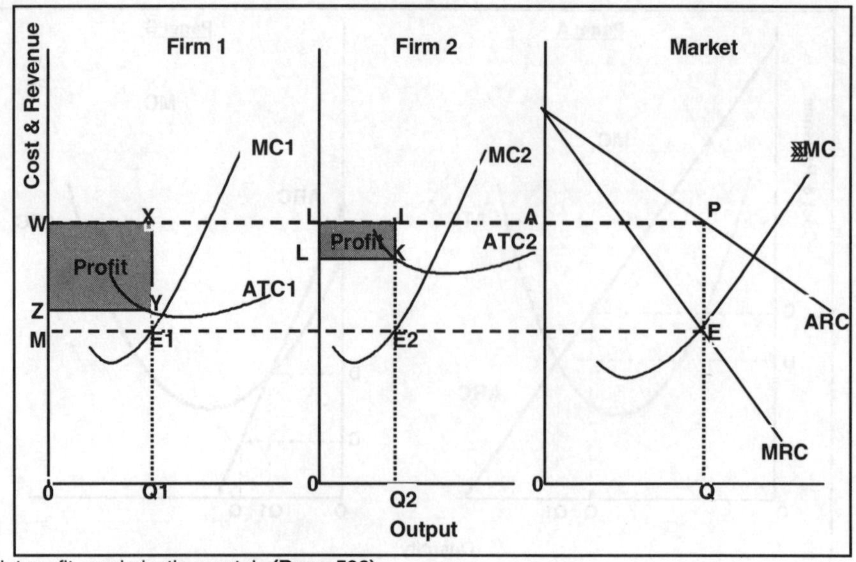

Figure 22.17: Joint profit maximization cartel. **(Page 596)**

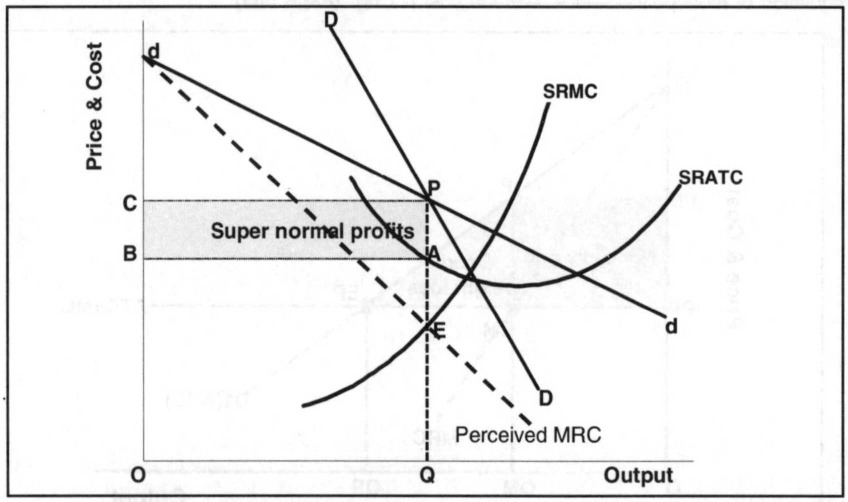

Figure 23.8: Short run firm's equilibrium under monopolistic competition – Proportional and Perceived demand curves approach. **(Page 621)**

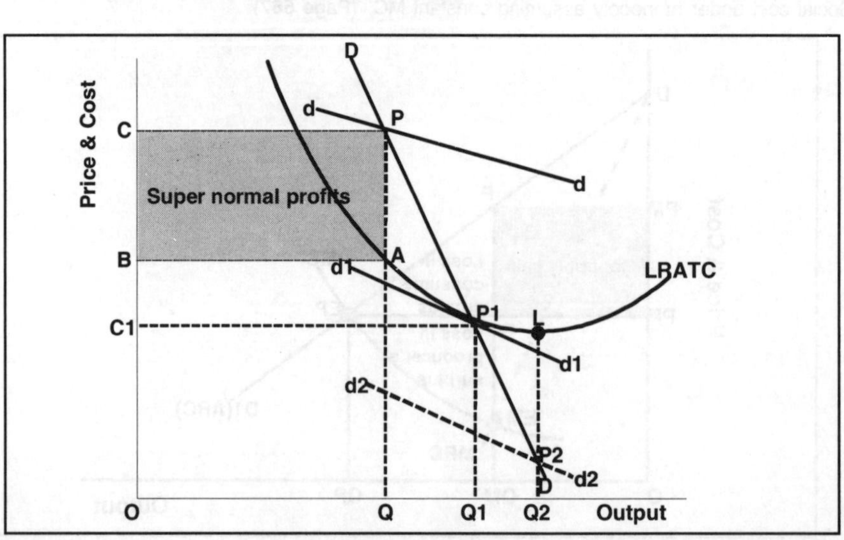

Figure 23.9: Long run group equilibrium under monopolistic competition – Entry of new firms is blocked. **(Page 621)**

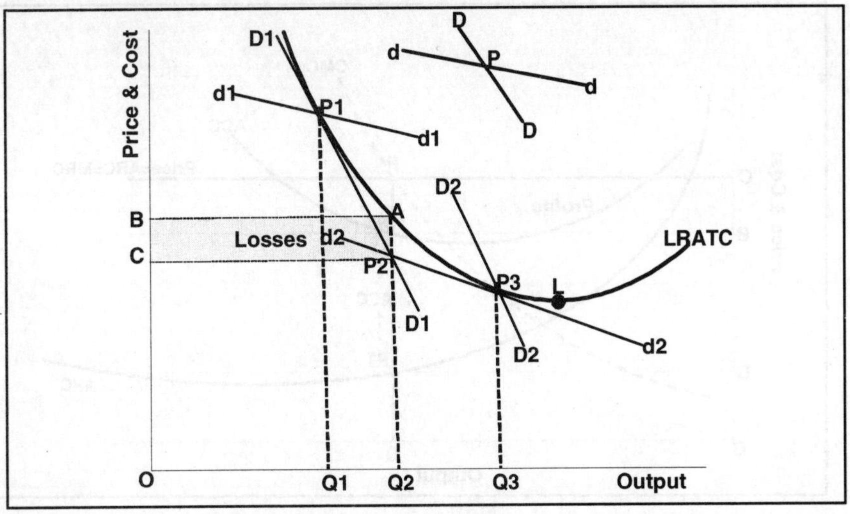

Figure 23.10: Long run group equilibrium under monopolistic competition – Entry of new firms into the industry. **(Page 622)**

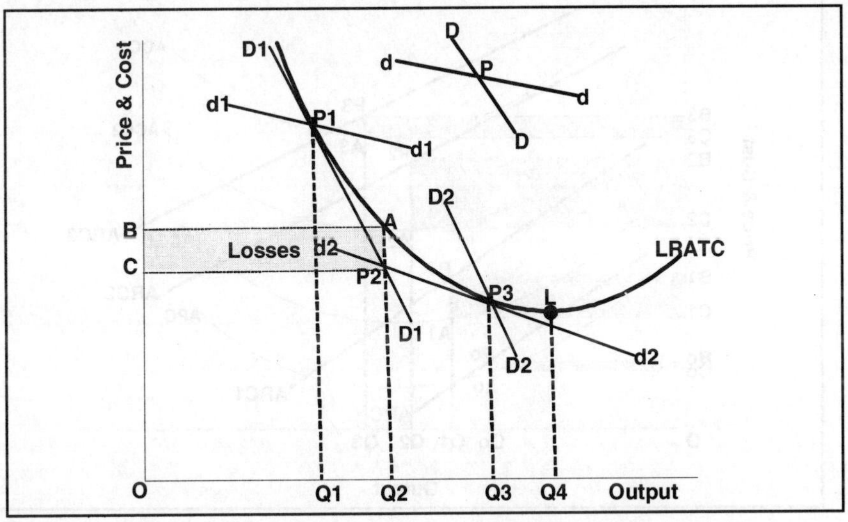

Figure 23.12: Chamberlin's concept of excess capacity under monopolistic competition. **(Page 624)**

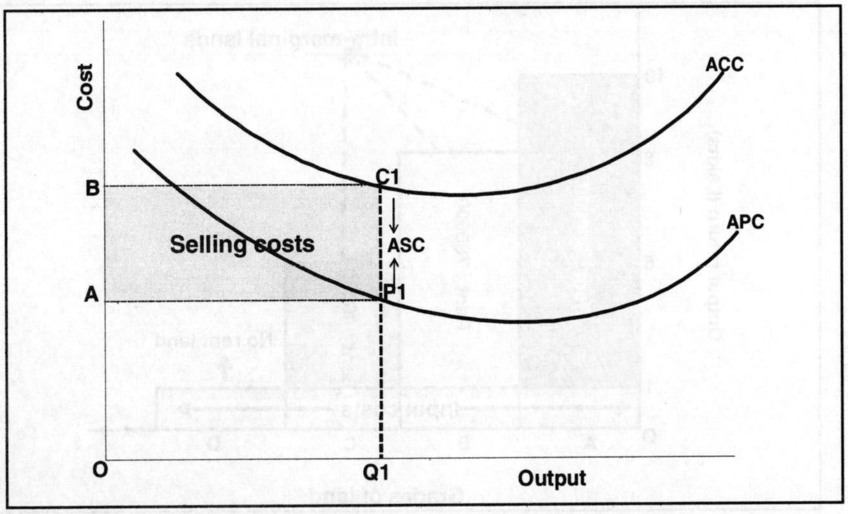

Figure 23.15: APC, ACC and ASC of a firm under monopolistic competition. **(Page 630)**

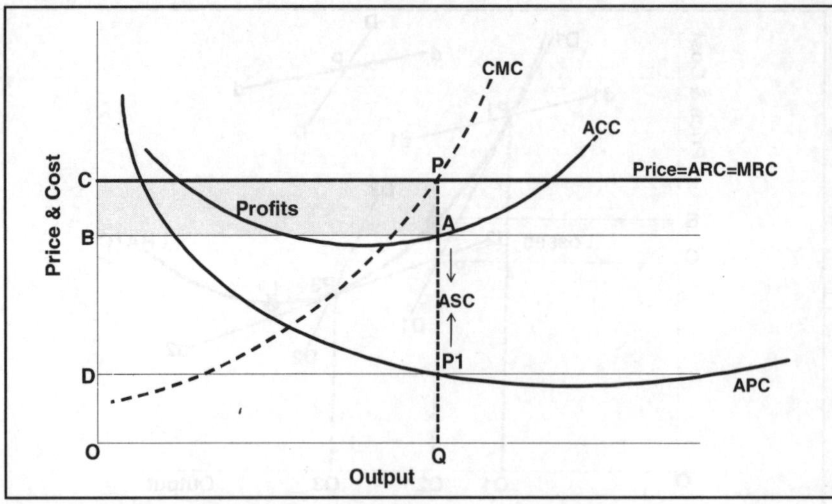

Figure 23.18: Price-Output determination of a firm under selling costs. **(Page 632)**

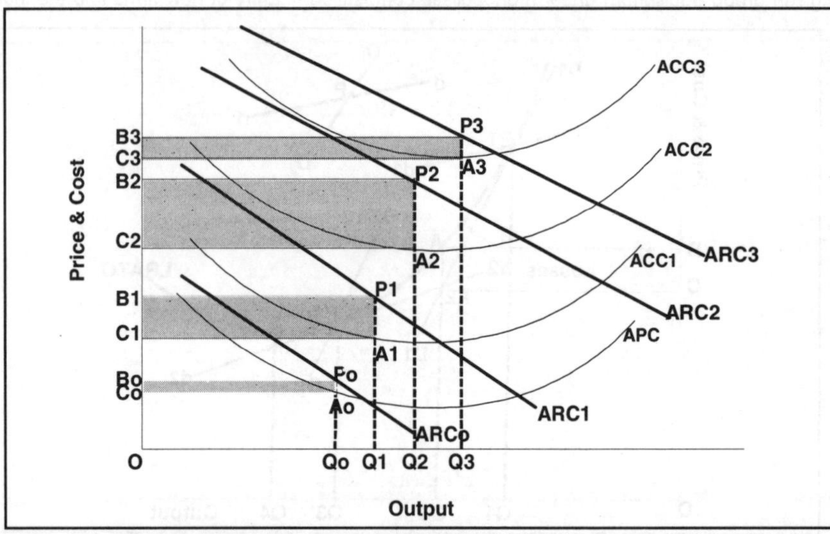

Figure 23.20: Price-Output determination of a firm under variable selling costs and price of the product. **(Page 634)**

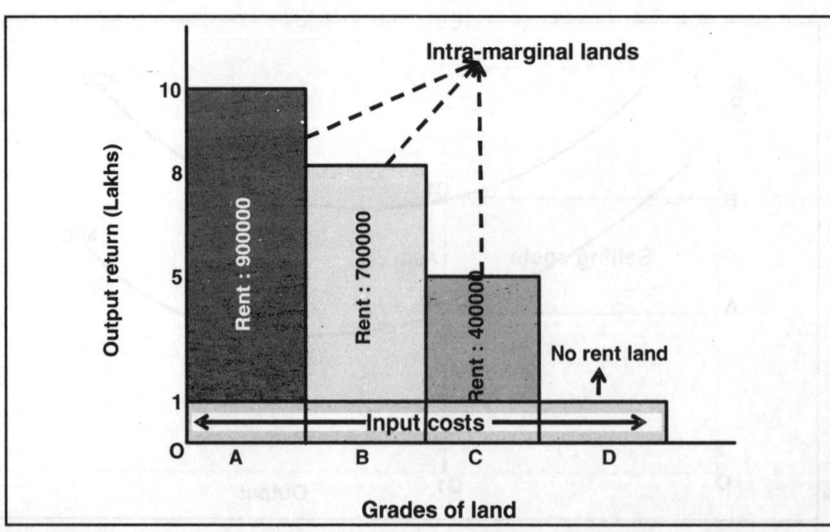

Figure 25.1: Ricardian theory – Rent as 'Differential surplus' under extensive cultivation. **(Page 672)**

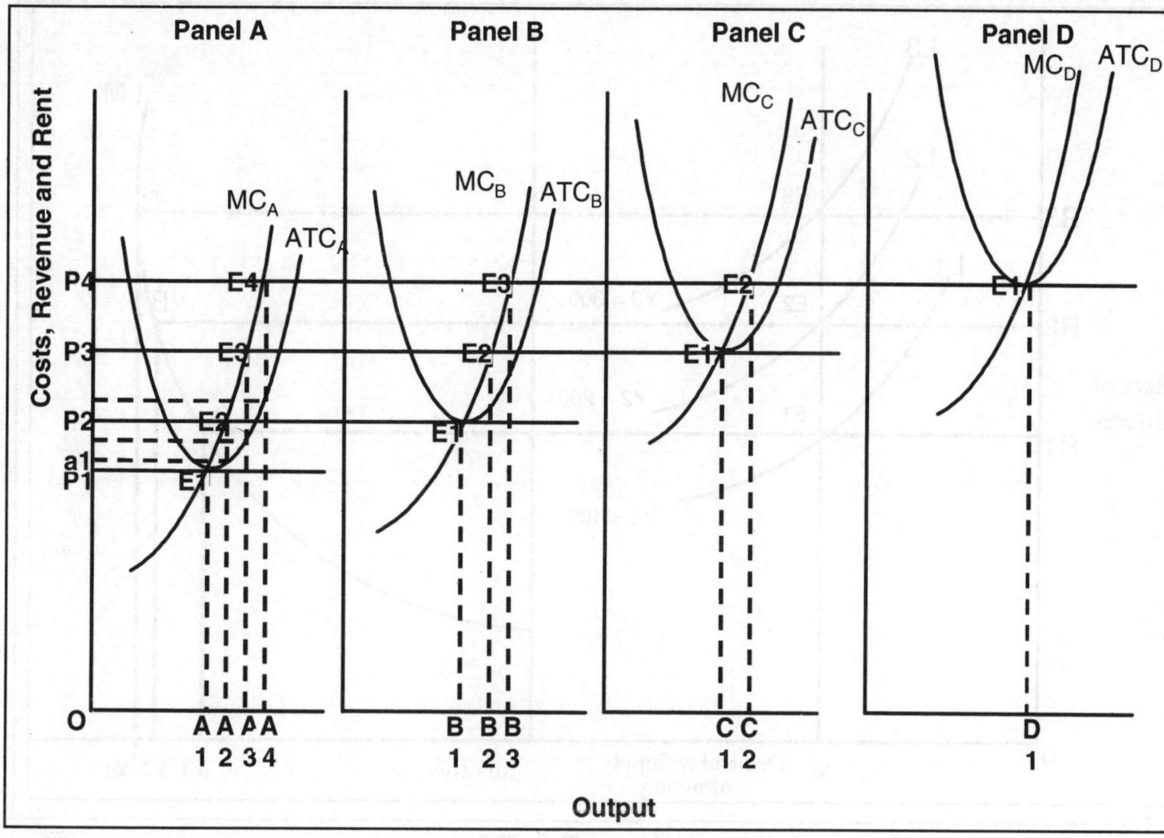

Figure 25.2: Ricardian theory – 'Rent' as a surplus of revenue over cost. **(Page 672)**

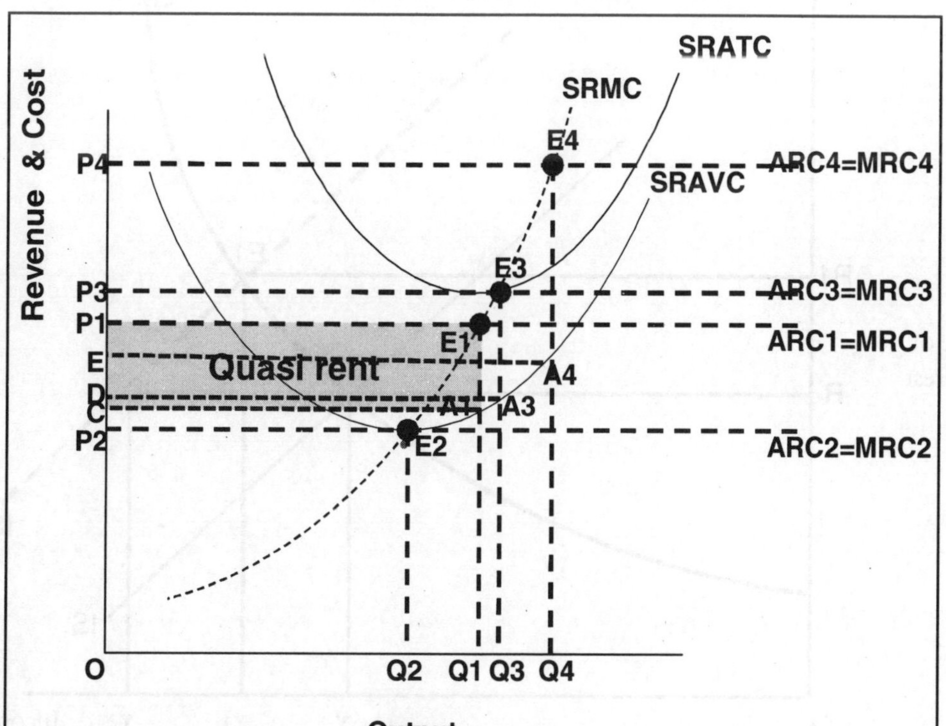

Figure 25.8: Measurement of quasi rent in short run. **(Page 684)**

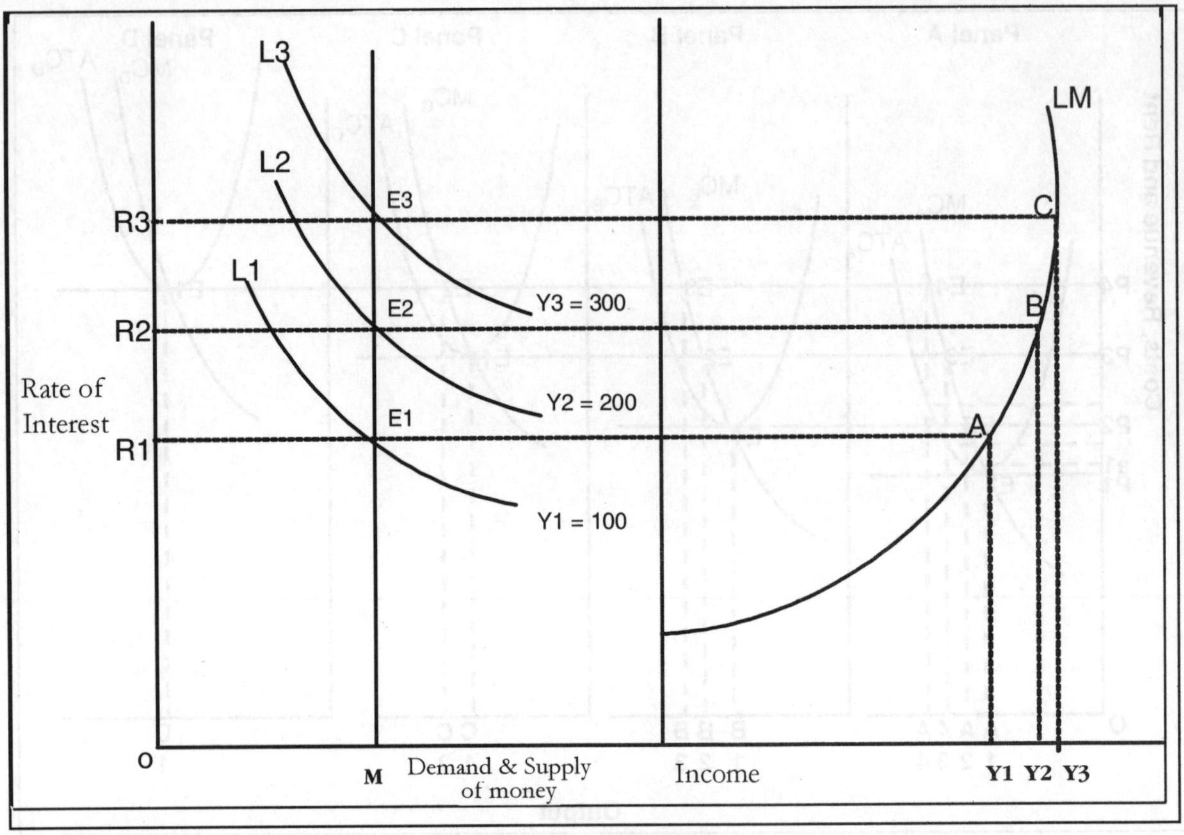

Figure 27.12: Modern theory of interest -Derivation of LM curve. **(Page 739)**

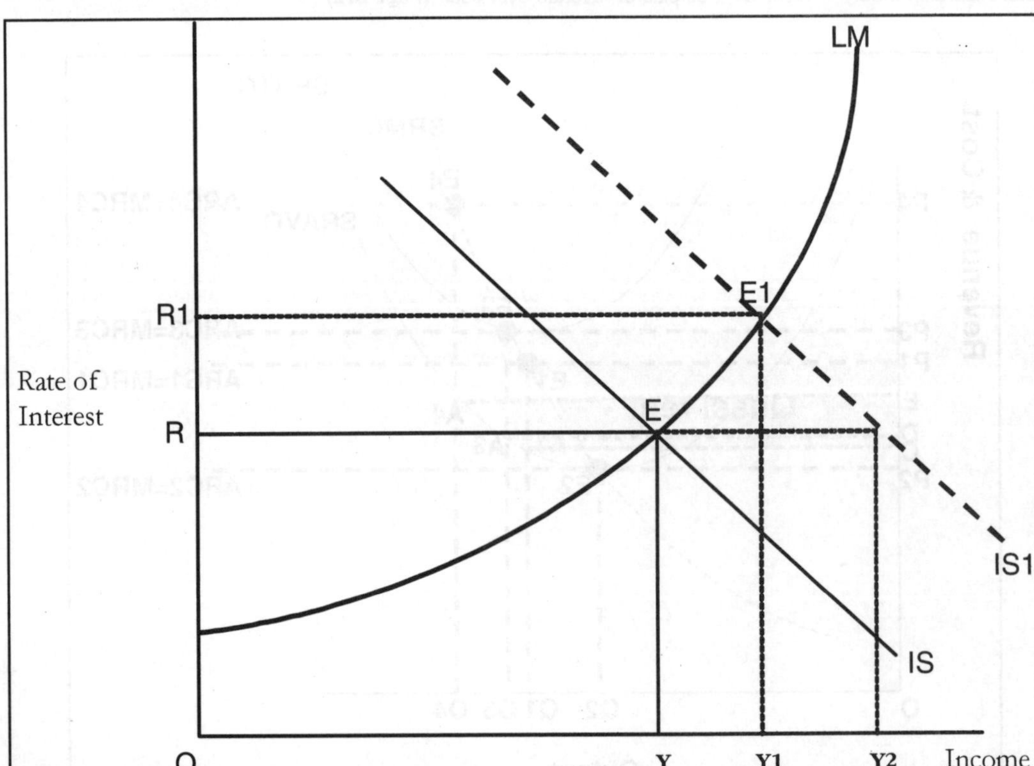

Figure 27.14: Changes in General equilibrium - Shift in IS curve (Increase in investment). **(Page 741)**

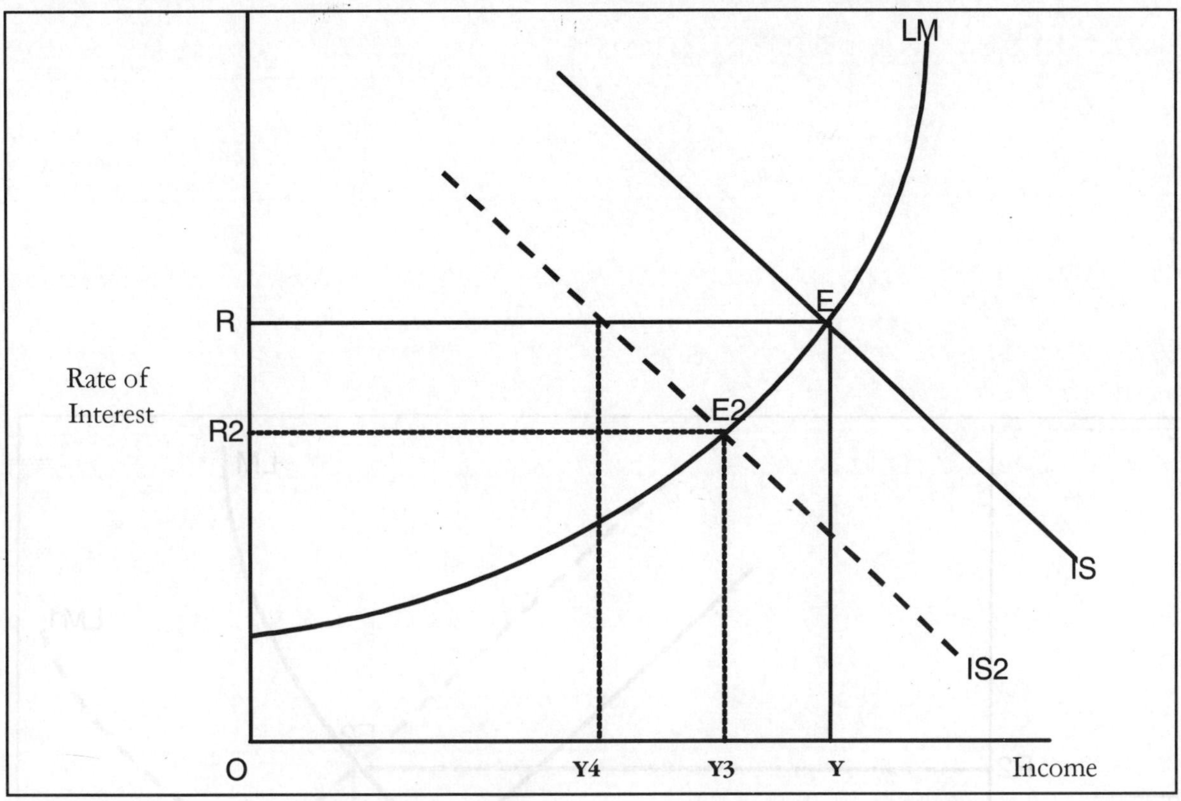

Figure 27.15: Changes in General equilibrium - Shift in IS curve (Decrease in investment). **(Page 742)**

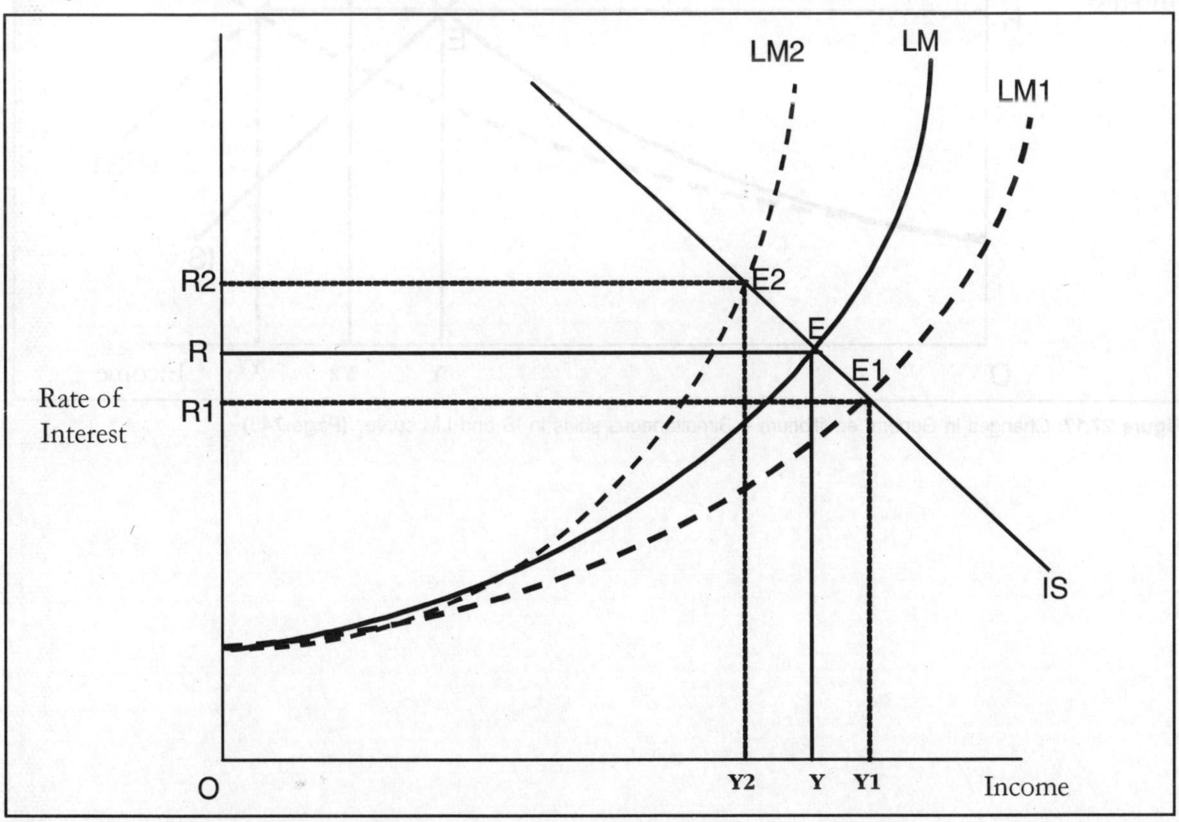

Figure 27.16: Changes in General equilibrium – Shift in LM curve. **(Page 743)**

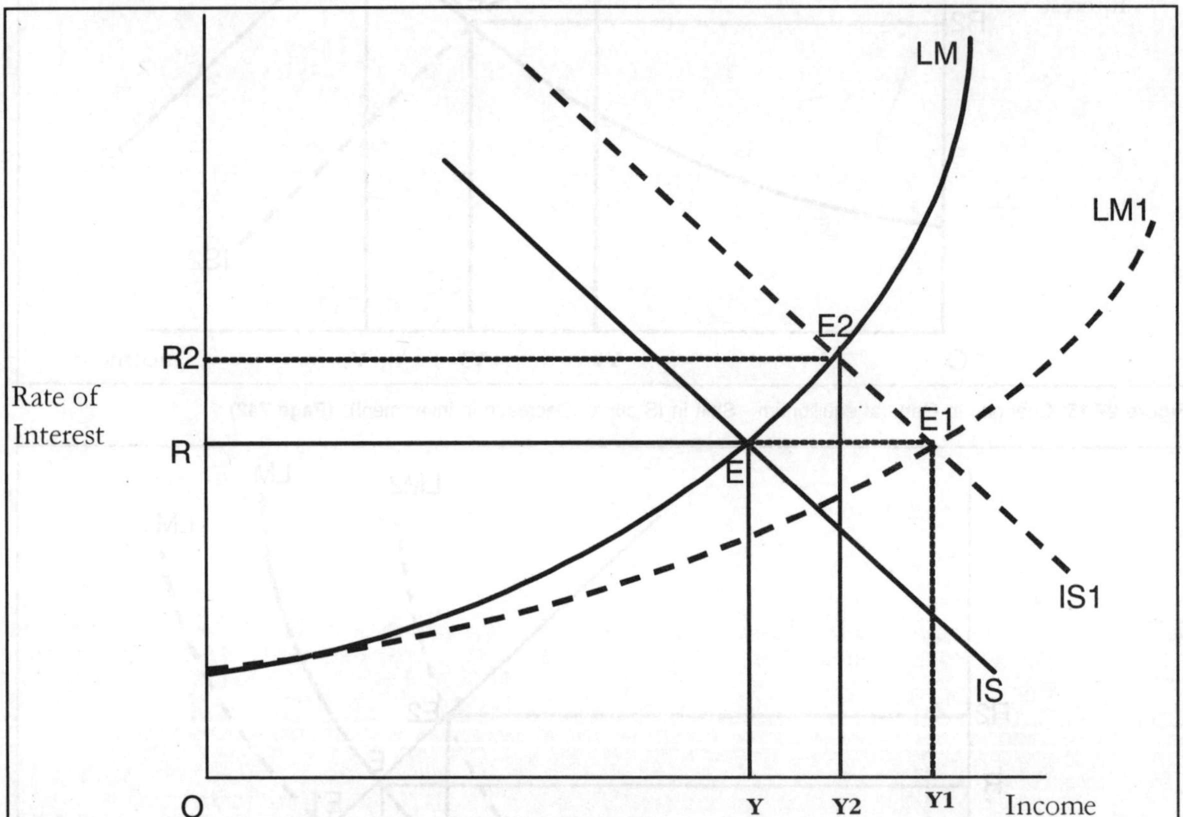

Figure 27.17: Changes in General equilibrium – Simultaneous shifts in IS and LM curve. **(Page 743)**